Dedicated to

My parents, Gopinath and Ichhamani, for their gift of life, love and living examples

My wife, Jharana, for her life-long friendship, love and support

My children, Megha, Sudhir and Subir, for their love and care

Saura C. Sahu

Contents

Preface   xi
List of Contributors   xiii
Acknowledgments   xix

1   Characterization of Nanomaterials for Toxicological Evaluation   1
      Kevin W. Powers, Maria Palazuelos, Scott C. Brown and
      Stephen M. Roberts

2   Criteria and Implementation of Physical and Chemical
      Characteristics of Nanomaterials for Human Health Effects and
      Ecological Toxicity Studies   29
      Christie M. Sayes and David B. Warheit

3   Considerations for the Design of Toxicity Studies of
      Inhaled Nanomedicines   41
      Lea Ann Dailey

4   High Aspect Ratio Nanoparticles and the Fibre
      Pathogenicity Paradigm   61
      Craig A. Poland, Rodger Duffin and Ken Donaldson

5   Application of Zinc Oxide Quantum Dots in Food Safety   81
      Tony Jin, Dazhi Sun, Howard Zhang and Hung-Jue Sue

6   Evaluation of Nanotoxicity of Foods and Drugs: Biological Properties
      of Red Elemental Selenium at Nano Size (Nano-Se) *In Vitro* and
      In Vivo   97
      Jinsong Zhang

7   **Evaluation of Toxicity of Nanostructures in Biological Systems**   115
Adam J. Gormley and Hamidreza Ghandehari

8   **Developing Bioassay Methods for Evaluating Pulmonary Hazards from Nanoscale or Fine Quartz/Titanium Dioxide Particulate Materials**   161
David B. Warheit, Kenneth L. Reed and Christie M. Sayes

9   **Nanoparticles: Is Neurotoxicity a Concern?**   171
Jianyong Wang, Wenjun Sun and Syed F. Ali

10   **Hepatotoxic Potential of Nanomaterials**   183
Saura C. Sahu

11   **Nanotoxicity in Blood: Effects of Engineered Nanomaterials on Platelets**   191
Jan Simak

12   **Sources, Fate and Effects of Engineered Nanomaterials in the Aquatic Environment**   227
David S. Barber, Nancy D. Denslow, R. Joseph Griffitt and Christopher J. Martyniuk

13   **Nanotoxicity of Metal Oxide Nanoparticles *in Vivo***   247
Weiyue Feng, Bing Wang and Yuliang Zhao

14   ***In Vivo* Hypersensitive Pulmonary Disease Models for Nanotoxicity**   271
Ken-ichiro Inoue and Hirohisa Takano

15   ***In Vivo* and *In Vitro* Models for Nanotoxicology Testing**   279
Rosalba Gornati, Elena Papis, Mario Di Gioacchino, Enrico Sabbioni, Isabella Dalle Donne, Aldo Milzani and Giovanni Bernardini

16   ***In Vitro* and *In Vivo* Toxicity Study of Nanoparticles**   303
Jayoung Jeong, Wan-Seob Cho, Seung Hee Kim and Myung-Haing Cho

17   ***In Vitro* and *In Vivo* Models for Nanotoxicity Testing**   335
Kyung O. Yu, Laura K. Braydich-Stolle, David M. Mattie, John J. Schlager and Saber M. Hussain

18   ***In Vitro* Models for Nanotoxicity Testing**   349
Yinfa Ma

19   ***In Vitro* Human Lung Cell Culture Models to Study the Toxic Potential of Nanoparticles**   379
Fabian Blank, Peter Gehr and Barbara Rothen-Rutishauser

20 **Iron Oxide Magnetic Nanoparticle Nanotoxicity: Incidence and Mechanisms** 397
Thomas R. Pisanic, Sungho Jin and Veronica I. Shubayev

21 **Toxicity Testing and Evaluation of Nanoparticles: Challenges in Risk Assessment** 427
David Y. Lai and Philip G. Sayre

22 **Evaluating Strategies For Risk Assessment of Nanomaterials** 459
Nastassja Lewinski, Huiguang Zhu and Rebekah Drezek

23 **Strategies for Risk Assessment of Nanomaterials** 499
Hae-Seong Yoon, Hyun-Kyung Kim, Dong Deuk Jang and Myung-Haing Cho

24 **Metal Nanoparticle Health Risk Assessment** 519
Mario Di Gioacchino, Nicola Verna, Rosalba Gornati, Enrico Sabbioni and Giovanni Bernardini

25 **Application of Toxicology Studies in Assessing the Health Risks of Nanomaterials in Consumer Products** 543
Joyce S. Tsuji, Fionna S. Mowat, Suresh Donthu and Maureen Reitman

26 **Safety Assessment of Engineered Nanomaterials in Direct Food Additives and Food Contact Materials** 581
Penelope A. Rice, Kimberly S. Cassidy, Jeremy Mihalov and T. Scott Thurmond

Index 597

Preface

Nanotechnology is a rapidly developing, emerging branch of modern technology. This new technology deals with materials of extremely small size, generally in the range of nanometres. The nanomaterials, with their extremely small size and high surface area associated with greater strength, stability, chemical and biological activity, find their wide range of applications in a variety of products in modern society. They are used in rapidly increasing nanoproducts, nanodevices, electronics, diagnostics and drug delivery systems. They are present in a variety of consumer products such as foods, drugs, cosmetics, food colour additives, food containers, paints and surface coatings. This trend is expected to result in an ever-increasing presence of nanoparticles in the human environment. Because of their extremely small size they are capable of entering the human body by inhalation, ingestion, skin penetration, intravenous injections and medical devices, and have the potential to interact with intracellular macromolecules. Because of their greater stability they are anticipated to remain in the body and in the environment for long periods of time. However, information on their potential adverse health effects is very limited at the present time. It is not known at what concentration or size they can exhibit toxicity. Therefore, there are obvious public safety concerns. This has led to the initiation of a new research discipline commonly known as nanotoxicology.

The main purpose of this book is to assemble up-to-date, state-of-the-art toxicological information on nanomaterials presented by recognized experts in a single edition. Therefore, it is an authoritative source of current knowledge in this area of research. The book is designed primarily for research scientists currently engaged in this field. However, it should be of interest to a variety of scientific disciplines including toxicology, genetics, medicine and pharmacology, as well as drug and food and material sciences. Also, it should be of interest to federal regulators and risk assessors of drug, food, environment and consumer products.

Nanotoxicology is an emerging new multidisciplinary field of science, and therefore there is a risk of change in its rapid development in the near future. However, its fundamental concepts and ideas as well as the experimental data are not going to change. For years to come this book will be a very valuable reference source to students and investigators in this research field to guide them in their future work.

Saura C. Sahu and Daniel A. Casciano

List of Contributors

Syed F. Ali   Division of Neurotoxicology, National Center for Toxicological Research, Food and Drug Administration, HFT-132, 3900 NCTR Rd, Jefferson, Arkansas 72079, USA

David S. Barber   Center for Environmental and Human Toxicology, Department of Physiological Sciences, University of Florida, Gainesville, FL 32611, USA

Giovanni Bernardini   Dipartimento di Biotecnologie e Scienze Molecolari, Università dell'Insubria, Dipartimento di Biologia, Università di Milano, CiSE, Università di Chieti, Italy

Fabian Blank   Telethon Institute for Child Health Research, 100 Roberts Road, Subiaco, WA 6008 Australia

Laura K. Braydich-Stolle   Applied Biotechnology Branch, 711 Human Performance Wing, Air Force Research Laboratory, Wright-Patterson Air Force Base, OH 45433-5707, USA

Scott C. Brown   Particle Engineering Research Center, University of Florida, Gainesville, FL 32611, USA

Kimberly S. Cassidy   US Food and Drug Administration, 5100 Paint Branch Parkway, HFS-275, College Park, MD 20740, USA

Myung- Haing Cho   Division of Toxicological Research, National Institute of Toxicological Research, Korea Food and Drug Administration, Seoul, Korea

Wan-Seob Cho   Division of Toxicological Research, National Institute of Toxicological Research, Korea Food and Drug Administration, Seoul, Korea

Lea Ann Dailey   King's College London, Pharmaceutical Science Division, Franklin-Wilkins Building, 150 Stamford Street, London SE1 9NH, United Kingdom

Isabella Dalle Donne   Dipartimento di Biotecnologie e Scienze Molecolari, Università dell'Insubria, Dipartimento di Biologia, Università di Milano, CiSE, Università di Chieti, Italy

Nancy D. Denslow   Center for Environmental and Human Toxicology, Department of Physiological Sciences, University of Florida, Gainesville, FL 32611, USA

Mario Di Gioacchino   G. d'Annunzio University Foundation, Ageing Research Center (CeSI), Chieti, Italy; Dipartimento di Biotecnologie e Scienze Molecolari, Università dell'Insubria, Varese, Italy

Ken Donaldson   University of Edinburgh, ELEGI Colt Laboratory, Queens Medical Research Institute, 47 Little France Crescent, Edinburgh, EH16 4TJ, United Kingdom

Suresh Donthu   Exponent, Bowie, MD, 20715, USA

Rebekah Drezek   Rice University, Department of Bioengineering MS-142, Houston, TX 77005, USA

Rodger Duffin   University of Edinburgh, ELEGI Colt Laboratory, Queens Medical Research Institute, 47 Little France Crescent, Edinburgh, EH16 4TJ, United Kingdom

Weiyue Feng   CAS Key Laboratory for Biomedical Effects of Nanomaterials and Nanosafety, and CAS Key Laboratory of Nuclear Analytical Techniques, Institute of High Energy Physics, Chinese Academy of Sciences, Beijing, 100049, China

Peter Gehr   Institute of Anatomy, Division of Histology, University of Bern, Baltzerstrasse 2, Bern, Switzerland

Hamidreza Ghandehari   Departments of Pharmaceutics & Pharmaceutical Chemistry and Bioengineering, Utah Center for Nanomedicine, Nano Institute of Utah, University of Utah, 383 Colorow Road, Room 343, Salt Lake City, UT 84108, USA

Mario Di Gioacchino   Dipartimento di Biotecnologie e Scienze Molecolari, Università dell'Insubria, Dipartimento di Biologia, Università di Milano, CiSE, Università di Chieti, Italy

Adam J. Gormley   Departments of Pharmaceutics and Pharmaceutical Chemistry and Bioengineering, Utah Center for Nanomedicine, Nano Institute of Utah, University of Utah, 383 Colorow Road, Room 343, Salt Lake City, UT 84108, USA

Rosalba Gornati Dipartimento di Biotecnologie e Scienze Molecolari, Università dell'Insubria, Dipartimento di Biologia, Università di Milano, CiSE, Università di Chieti, Italy

R. Joseph Griffitt Gulf Coast Research Center, Department of Coastal Sciences, University of Southern Mississippi, Ocean Springs, MS 39564, USA

Saber M. Hussain Applied Biotechnology Branch, 711 Human Performance Wing, Air Force Research Laboratory, Wright-Patterson Air Force Base, OH 45433-5707, USA

Ken-ichiro Inoue Environmental Health Sciences Division, National Institute for Environmental Studies, Ibaraki16-2 Onogawa, Tsukuba 305-8506, Japan

Dong Deuk Jang Risk Assessment Department, National Institute of Toxicological Research, Korea Food & Drug Administration, Seoul 122-704, Korea

Jayoung Jeong Division of Toxicological Research, National Institute of Toxicological Research, Korea Food & Drug Administration, Seoul 122–704, Korea

Sungho Jin Department of Mechanical and Aerospace Engineering, University of California, San Diego, La Jolla, CA 92093-0411, USA

Tony Jin Food Safety Intervention Technologies Research Unit, USDA, ARS Department of Agriculture, Agricultural Research Service, NAA, ERRC, 600 East Mermaid Lane, Wyndmoor PA 19038, USA

Hyun-Kyung Kim Risk Assessment Department, National Institute of Toxicological Research, Korea Food & Drug Administration, Seoul 122-704, Korea

Seung Hee Kim Division of Toxicological Research, National Institute of Toxicological Research, Korea Food & Drug Administration, Seoul 122–704, Korea

David Y. Lai US Environmental Protection Agency, 1200 Pennsylvania Avenue, NW, Washington, DC 20460, USA

Nastassja Lewinski Rice University, Department of Bioengineering MS-142, Houston, TX 77005, USA

Yinfa Ma Department of Chemistry, Missouri University of Science and Technology, Rolla, MO 65409, USA

Christopher J. Martyniuk Center for Environmental and Human Toxicology, Department of Physiological Sciences, University of Florida, Gainesville, FL 32611, USA

David M. Mattie Applied Biotechnology Branch, 711 Human Performance Wing, Air Force Research Laboratory, Wright-Patterson Air Force Base, OH 45433-5707, USA

Jeremy Mihalov   US Food and Drug Administration, 5100 Paint Branch Parkway, HFS-255, College Park, MD 20740, USA

Aldo Milzani Dipartimento di Biotecnologie e Scienze Molecolari, Università dell'Insubria, Dipartimento di Biologia, Università di Milano, CiSE, Università di Chieti, Italy

Fionna Mowat   Exponent, Menlo Park, CA 94025, USA

Maria Palazuelos Particle Engineering Research Center, University of Florida, Gainesville, FL 32611, USA

Elena Papis Dipartimento di Biotecnologie e Scienze Molecolari, Università dell'Insubria, Dipartimento di Biologia, Università di Milano, CiSE, Università di Chieti, Italy

Thomas R. Pisanic II   Magnesensors, Inc., San Diego, CA, USA

Craig Poland   University of Edinburgh, ELEGI Colt Laboratory, Queens Medical Research Institute, 47 Little France Crescent, Edinburgh, EH16 4TJ, United Kingdom

Kevin W. Powers Particle Engineering Research Center, University of Florida, Gainesville, FL 32611, USA

Kenneth L. Reed   DuPont Haskell Global Centers for Health and Environmental Sciences, Newark, DE, USA

Maureen Reitman   Exponent, Bowie, MD, 20715, USA

Penelope A. Rice   US Food and Drug Administration, 5100 Paint Branch Parkway, HFS-275, College Park, MD 20740, USA

Stephen M. Roberts   CEHT, University of Florida, Box 110885, Gainesville, FL 32611, USA

Barbara Rothen-Rutishauser   Institute of Anatomy, Division of Histology, University of Bern, Baltzerstrasse 2, Bern, Switzerland

Enrico Sabbioni Dipartimento di Biotecnologie e Scienze Molecolari, Università dell'Insubria, Dipartimento di Biologia, Università di Milano, CiSE, Università di Chieti, Italy

Saura C. Sahu   Division of Toxicology, Center for Food Safety and Applied Nutrition, Food and Drug Administration, 8301 Muirkirk Road, Laurel, MD 20708, USA

Christie M. Sayes   Department of Veterinary Physiology & Pharmacology, College of Veterinary Medicine and Biomedical Sciences, Texas A & M University, College Station, TX 77843-4466, USA

Philip G. Sayre   US Environmental Protection Agency, 1200 Pennsylvania Avenue, NW, Washington, DC 20460, USA

John J. Schlager   Applied Biotechnology Branch, 711 Human Performance Wing, Air Force Research Laboratory, Wright-Patterson Air Force Base, OH 45433-5707, USA

Veronica I. Shubayev   Department of Anesthesiology and Mechanical and Aerospace Engineering, University of California, San Diego, La Jolla, CA 92093-0629, USA

Jan Simak   US Food and Drug Administration, CBER, 1401 Rockville Pike, Rockville, MD, 20852, USA

Hung-Jue Sue   Polymer Technology Center, Mechanical Engineering Department, Texas A & M University, College Station, TX 77843, USA

Dazhi Sun   Polymer Technology Center, Mechanical Engineering Department, Texas A & M University, College Station, TX 77843, USA

Wenjun Sun   Department of Neurology, Xuanwu Hospital of the Capital Medical University, Beijing, China 100053

Hirohisa Takano   Environmental Health Sciences Division, National Institute for Environmental Studies, Ibaraki, Japan

T. Scott Thurmond   US Food and Drug Administration, 5100 Paint Branch Parkway, HFS-265, College Park, MD 20740, USA

Joyce S. Tsuji   Exponent, 15375 SE 30th Place, Suite 250, Bellevue, WA 98007, USA

Nicola Verna   G. d'Annunzio University Foundation, Ageing Research Center (CeSI), Chieti, Italy; Dipartimento di Biotecnologie e Scienze Molecolari, Università dell'Insubria, Varese, Italy

Bing Wang   CAS Key Laboratory for Biomedical Effects of Nanomaterials and Nanosafety, and CAS Key Laboratory of Nuclear Analytical Techniques, Institute of High Energy Physics, Chinese Academy of Sciences, Beijing, 100049, China

Jianyong Wang   Office of New Drugs, Center for Drug Evaluation and Research, Food and Drug Administration, HFD-540, 10903 New Hampshire Ave, Silver Spring, Maryland 20993, USA

David B. Warheit DuPont Haskell Global Centers for Health and Environmental Sciences, Newark, Delaware, USA

Hae-Seong Yoon Risk Assessment Department, National Institute of Toxicological Research, Korea Food & Drug Administration, Seoul 122-704, Korea

Kyung O. Yu Applied Biotechnology Branch, 711 Human Performance Wing, Air Force Research Laboratory, Wright-Patterson Air Force Base, OH 45433-5707, USA

Howard Zhang Food Safety Intervention Technologies Research Unit, US Department of Agriculture, Agricultural Research Service, NAA, Eastern Regional Research Center, 600 East Mermaid Lane, Wyndmoor PA 19038, USA

Jinsong Zhang School of Tea and Food Science, Anhui Agricultural University, Hefe 230036, Anhui, China

Yuliang Zhao CAS Key Laboratory for Biomedical Effects of Nanomaterials and Nanosafety, and CAS Key Laboratory of Nuclear Analytical Techniques, Institute of High Energy Physics, Chinese Academy of Sciences, Beijing, 100049, China

Huiguang Zhu Rice University, Department of Chemistry MS-60, Houston, TX 77005, USA

Acknowledgements

Saura C. Sahu expresses his sincere gratitude to the following scientists who have helped him directly or indirectly for this book:

All the participating scientists for their excellent contributions in their own areas of expertise

Dr Daniel A. Casciano, the co-editor of this book, for his advice, enthusiasm, inspiration and leadership

Dr Joseph E. LeClerc and Dr Philip W. Harvey for their encouragement and support

And finally Martin Rothlisberger, Paul Deards, Rebecca Ralf and Richard Davies of the publishing company, John Wiley & Sons, Ltd., for their cooperation and support.

1

Characterization of Nanomaterials for Toxicological Evaluation

Kevin W. Powers, Maria Palazuelos, Scott C. Brown and Stephen M. Roberts

1.1 Introduction

This chapter addresses issues related to the proper characterization of nanostructured materials (NSMs) for toxicological evaluation. Although toxicologists are familiar with biological and chemical aspects of toxicity testing, the study of NSMs poses unique challenges. Nanomaterial size, shape, dispersion, composition, surface properties, and other attributes can all potentially influence toxicity in ways that are poorly understood at present. In order to produce results that are both interpretable and reproducible, test materials must be adequately described with respect to all relevant properties, yet what constitutes a list of relevant properties is currently ambiguous, and probably varies among different NSMs.

In the absence of insight regarding key parameters affecting biological activity, general guidance is given that characterization of test materials should be as complete as practicable (Thomas and Sayre, 2005; Oberdorster *et al.*, 2005; Powers *et al.*, 2006). Full characterization of NSMs, which can include measurements of size and size distribution, shape and other morphological features (e.g. crystallinity, porosity, surface roughness), bulk chemistry of the material, solubility, surface area, state of dispersion, surface chemistry and other physico-chemical properties, would be ideal, but is time-consuming, expensive and complex. It is reasonable to expect that characterization will be limited for most studies, but this raises the question of what constitutes a minimum level of characterization necessary for the study to be scientifically valuable, or in practical terms, worthy of publication or funding.

Nanotoxicity: From In Vivo *and* In Vitro *Models to Health Risks*   Edited by Saura Sahu and Daniel Casciano
© 2009 John Wiley & Sons, Ltd

There are a number of other, practical questions related to NSM characterization that must be answered for the field of nanotoxicology to advance. For example, the properties of NSMs can change, sometimes dramatically, from the form in which they are received to the form to which test subjects or biological test systems are exposed. This raises the question of when, and under what conditions, should NSM characterization take place? Is measurement of 'as-received' material sufficient, or should the material be characterized 'as-administered' (e.g. in the vehicle in which it is administered)? Is it important to attempt to characterize the material within the biological system, and what types of measurements can be made with current technology? Can we adequately reconstruct NSM properties from *post mortem* or histological studies? Do we need to address each NSM individually, or can properties or behaviour be deduced by knowledge derived from studies of other materials? No single discipline has the expertise to answer these questions. Resolution of these issues will require the collaboration of toxicologists and other biological scientists with chemical and physical scientists, as well as engineers trained in particle technology.

One of the first attempts to address practical issues in conducting studies of the potential toxicity of NSMs was an international workshop held November 2004 at the University of Florida (Bucher *et al.*, 2005). This workshop concluded that the physical and chemical characterization of NSMs needed to be much more complete than what was typically published in the literature at the time, and urged that scientific journal editors require proper physical and chemical characterization of NSMs in nanotoxicology studies. It was also recommended that a minimum set of relevant nanomaterial characteristics be developed for toxicological studies. Subsequently, several national and international meetings/workshops/working groups have assembled to address issues related to nanotoxicity and the characterization of NSMs (Oberdorster *et al.*, 2005; Powers *et al.*, 2006; Balbus *et al.*, 2007; OECD, 2008), and recently a 'grassroots' effort to promote meaningful characterization of NSMs in toxicity studies has taken form (see http://characterization-matters.org). Although no all-inclusive characterization protocol has yet been agreed upon, there is a growing consensus regarding which NSM properties should be evaluated when conducting toxicological studies. Table 1.1 illustrates a list of properties that have been cited as important for the characterization of NSMs. Although some measurements (such as size) are always critical, the rest are not necessarily equally important, and the need for certain measurements is often dependent on the situation. Some 'as-received' properties are provided by the manufacturer or supplier of the NSMs; however, properties can change as a function of transport, handling, and environmental conditions, most notably particle size distribution and surface properties due to agglomeration, ageing and environmental exposure. The most conservative approach is to measure these properties at the point of use before initiating testing.

There are many standard analytical techniques for measuring bulk and surface physical and chemical properties. However, quantifying NSM size, shape and agglomerate structure is complicated due to the fact that these are usually distribution functions and have a propensity to change with time and environmental conditions through agglomeration. Surface properties also tend to change with the environmental surroundings and potential surface-adsorbed species. The following sections will focus on the issues and methods associated with measuring size, shape, state of dispersion, surface properties and other properties of interest in toxicological studies. Although the order in which these properties

Table 1.1   *Important properties in material characterization for toxicity studies.*

Property	Importance for toxicity testing	Comments
Particle size distribution	Essential	
Degree/state of agglomeration	Important	
Particle shape/shape distribution	Important	
Chemical composition/purity	Essential	
Solubility	Essential (where applicable)	
Surface properties		
Specific surface area/porosity	Essential	Surface roughness may be important
Surface chemistry/reactivity	Essential	
Surface adsorbed species	Important	In some cases may be the mechanism of toxicity (e.g. complement)
Surface charge/Zeta potential	Important (essential under aqueous conditions)	Especially in aqueous biological environments, may change according to the environment
Physical properties	Important	
Density	If applicable	
Crystallinity	If applicable	
Microstructure	If applicable	
Optical and electronic properties	If applicable	
Bulk powder properties	If applicable	May be important for dosimetry/exposure
Concentration	Essential	Can be measured as mass, surface area, or number concentrations

are measured may or may not be significant, the following steps are offered as a logical approach to characterizing NSMs for toxicological screening:

- gathering existing data;
- sampling;
- particle size;
- particle shape;
- dispersion;
- surface properties.

1.2   Gathering Existing Data

For most toxicological studies, the NSMs of interest are known entities or have been manufactured for some specific application or purpose. Thus, there is usually some information regarding the material's composition, method of manufacture, and intended properties. In order to design a characterization protocol, information from the manufacturer or other sources should be collected and evaluated. The list below outlines NSM properties that may be available and should be considered.

- What are the chemical composition/properties of the material?
- What is the source and/or the method of manufacture?
- What is already known about the physical state of the NSM (e.g. size, shape, porosity)? What techniques were used to derive this information?
- What is the intended purpose or application for the NSM? What properties make it valuable?
- What information or specifications has the manufacturer supplied?
- What safety or toxicity data are available for the material? Is a Material Safety Data Sheet (MSDS) available?
- Have there been previous studies published that provide analysis or characterization of the material?

Once all of the available data are collected, an assessment can be made regarding what measurements are the most important for the study under consideration. There are a number of databases that have been established to provide information on a growing number of commercial NSMs. The US Environmental Protection Agency (EPA) has initiated a Nanoscale Materials Stewardship Program (EPA, 2009) documenting the toxicity profiles and materials characterization for both commercial and research materials. The National Institute for Occupational Safety and Health (NIOSH) has established the Nanoparticle Information Library (NIL) with much the same goal. The Organisation for Economic Co-operation and Development (OECD) has established a working group with a prioritized list of 14 NSMs that will be tested as a reference base for evaluating the safety of manufactured NSMs (OECD, 2008).

1.3 Sampling

Valuable guidance on sampling of NSMs for characterization can be taken from the field of powder technology. Reliable powder sampling constitutes the first step of most powder characterization and processing procedures. Sampling particulate matter entails collecting a small amount of powder from the bulk, such that this smaller fraction best represents the physical and chemical characteristics of the entire bulk (Jillavenkatesa et al., 2001; Holdich, 2002).

It is recommended that samples of the materials 'as received' be prepared taking into account the issues listed by the National Institute of Standards and Technology (Jillavenkatesa et al., 2001):

- quantity of powder from which samples are being obtained;
- amount of sample required;
- powder characteristics, including but not limited to flow characteristics of the powder, shape and size of the particles, tendency to segregate, surface chemistry that may cause the powder to be hygroscopic, and so on;

Table 1.2 *Common errors associated with powder sampling.*

Error in powder sampling	Protocols for improvement
Non-representative sampling/improper dispersion during sample preparation	Use of specially designed instruments like spinning riffling when possible If scooping, mix sample before and take several subsamples across the bulk
Agglomeration of primary particles during sample preparation	If possible sample from a liquid suspension Use of dispersant aids when measuring the 'primary' particle size*
Contamination or introduction of artefacts during sampling/storage	Use of clean instrumentation and containers Store in properly sealed containers
Specimen degradation during storage	Store in inert gas and avoid extreme temperatures, pressures and light exposure Check stability of properties over time until sure that powder is stable

* Only used to determine the primary particle distribution. It should not be used for toxicity testing unless it is a biocompatible surfactant present along the exposure route.

- mechanical strength of the powder, i.e. are the particles friable and thus likely to fracture during transport or during sampling;
- mode by which powders are transported;
- possibility of powder contamination and acceptable limits of contamination;
- duration of time needed to conduct the sampling procedure.

It is important to note that sampling variability is greatly increased by heterogeneity in the bulk (i.e. size and shape polydispersity, flow properties, etc.), and thus special care should be practiced when dealing with such materials. NSMs subject to toxicological testing can be very diverse in nature, and while some will be very homogeneous and monodispersed (lower incidence of sampling errors), others will show opposite distributions and therefore require more thorough sampling protocols.

Despite the precautions mentioned above, there are several sources of error in the process of sampling, both systematic and random. While these errors cannot be eliminated, implementing standard protocols to handle and analyse the materials will minimize their influence on the final results. Some of the common errors associated with sample preparation, as well as recommended protocols to minimize those errors, are summarized in Table 1.2.

1.4 Particle Size Distribution

1.4.1 Definition of Size

All particles have size and shape. The size of the NSM must first be defined to classify it as a NSM. The definition of a NSM has been generally agreed upon to include any material with at least one dimension smaller than 100 nanometers (Roco, 2001; Royal Society and Royal Society of Engineering, 2004):

'The size range that holds so much interest is typically from 100 nm down to the atomic level (approximately 0.2 nm), because it is in this range (particularly at the lower end) that materials can have different or enhanced properties compared with the same materials at a larger size. The two main reasons for this change in behaviour are an increased relative surface area, and the emergence of quantum effects.' (Royal Society and Royal Society of Engineering, 2004)

Particulate systems are only occasionally perfect spheres, and even less often are they uniform in size (monodispersed). Despite this fact, in the particle technology community, particle size is most often defined as the diameter of a sphere that is equivalent in the selected property (e.g. mass), to the particle measured. This 'equivalent sphere' model enables one to conveniently plot size distributions of irregularly shaped systems using a single value (diameter) along a single axis. The property most often described is mass/volume diameter: that is, the diameter of a sphere of equal mass/volume to the particle(s) in question (Allen, 2004). In the aerosol community, the aerodynamic equivalent sphere is most often used. The aerodynamic equivalent diameter, D, is defined as the *effective spherical* diameter of a particle having the same falling velocity in air as a perfectly spherical particle of unit density (Ruzer and Harley, 2005). Other equivalent spheres such as sedimentation diameter and mobility diameter are also defined and used in aerosol characterization. Frequently the measured diameter will not precisely coincide when two different methods of measurement are used. This is why it is important to understand the physical methods used for size determination.

Size distributions are normally depicted as a log-normal histogram with particle diameter on the abscissa and the quantity or relative frequency in a particular size class depicted on the ordinate. The selection of size classes is somewhat arbitrary, but most instrument manufacturers use geometrically increasing bins. The quantity axis can depict any property, but most often is selected to represent the relative volume of particles in each size class, the relative number of particles in each size class, or the calculated surface area of the particles in that size class.

1.4.2   Why Size is Important to Toxicity

Particle size can clearly have a dramatic effect on the way in which an organism responds upon exposure to foreign materials. There are several aspects to this issue:

(a) Size can govern where in the body the particles are deposited (exposure). For example, well-established models have been developed for predicting where in the lung particulates of a given aerodynamic diameter are deposited (Bailey *et al.*, 2003; Leggett and Eckerman, 2003; Heyder, 2004). Figure 1.1 depicts the predicted percentage of particles deposited in the deep lungs as a function of three selected sizes. Note that in this case, the sizes only extend down to 200 nm.

(b) Size is a factor in the ability of the body to clear foreign particles (clearance mechanisms) (Renwick *et al.*, 2001). The ciliatory system in the lung is designed to clear particulate matter from the upper airways (National Research Council, 1989). Phagocytosis by macrophages and giant cells is a common mechanism by which the body's innate immune system attempts to clear particles of the order of a few microns or less (Lucarelli *et al.*, 2004).

(c) Size can be a factor in the ultimate fate (and location) of particles that are not cleared (translocation, fibrosis) (Oberdorster *et al.*, 1994; Nemmar *et al.*, 2001).

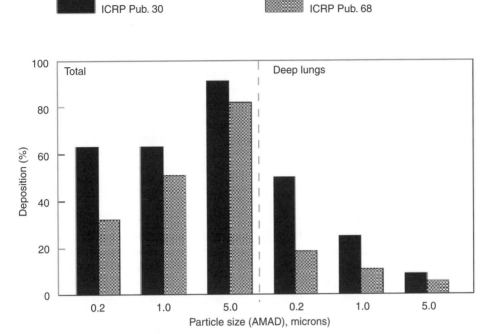

Figure 1.1 *Deposition fractions for deep lungs and total respiratory tract as a function of three particle sizes (0.20, 1.0 and 5.0 microns) as predicted by the TGLM (1RCP, 1966 and ICRB 1994b models) (Legget and Eckerman, 2003). Reprinted with permission from Oak Ridge National Laboratory*

(d) Particle size can potentially influence direct mechanisms and extent of toxicity within cells and tissues (cytotoxicity, necrosis and mutagenicity).

For particulates that strictly satisfy the criteria of nanosize, little research has been conducted regarding differential size effects in the 1–100 nm range.

1.4.3   How to Measure Size

There are a wide variety of methods for determining particle size distributions, including light scattering, centrifugal sedimentation, dynamic mobility analysis, time-of-flight, microscopy, and surface area measurements among many others. ISO committee TC24SC4 ('Particle Characterization') develops and writes standard practices addressing over ten different physical principles used for the measurement of particle size. At least six of these are applicable for size measurement in the sub-100 nm range (e.g. ISO-13321, ISO-13320, ISO-15901). To determine the size distribution of a sample, one ideally requires a representative sample in a well-dispersed system, and enough particles must be measured to achieve the desired statistical reliability. For a monodisperse system, the latter requirement is relatively easy. However, as polydispersity increases it becomes necessary to measure larger numbers of particles to portray the size distribution accurately (Masuda and Inoya, 1971). For ensemble techniques such as light scattering and centrifugal sedimentation, this requirement is easily achieved. For counting techniques, and particularly for microscopy, it is a major consideration.

1.4.3.1   Dry Bulk Powders

Many manufactured NSMs are supplied as dry bulk powders (e.g. carbon nanotubes, Nano TiO_2). Because they are nanosized and in the dry state, they are by nature highly aggregated and are difficult to redisperse to their primary particle distribution even under the best of conditions. Consequently, most manufacturers estimate the particle size of these 'as-received' powders by measuring the BET (Brunauer, Emmett and Teller) specific surface area and calculating an average particle size according to Equation (1.1) below:

$$d_p = \frac{6000}{\rho \times SSA} \tag{1.1}$$

where d_p is the average diameter of the particles in nanometers, ρ is the density in g/cm^3, and SSA is the specific surface area in m^2/g. For dry powders, this measurement is convenient, simple and useful, requiring about a gram or less of material for measurement. However, several assumptions are made in this calculation. It is assumed that the particles are spherical, uniform in size and non-porous (clearly not valid for some systems, for example carbon nanotubes). Usually additional methods (e.g. electron microscopy) are recommended to better understand the size distribution and nature of the powder. Bulk dry powders are typically aerosolized or dispersed in a liquid medium for most toxicity testing, and the dispersed particles can then be measured by additional methods.

1.4.3.2   Aerosols

Aerosols are of particular interest to the toxicologist, as inhalation is a primary route of exposure for many NSMs. Measurement technologies in aerosols have been spearheaded by research in the areas of air pollution, climate and occupational health. Consequently there is a well-developed set of tools used by researchers for the measurement of airborne particles that extends into the nanometer regime. Ensemble techniques include methods such as filtration and/or impactors. One of the simplest methods of aerosol analysis is filtration, weighing the collected particles followed by microscopic inspection. This method is particularly useful for quantifying the mass concentration of aerosols and can be used *in situ* to monitor air quality, or with personal filters to monitor occupational exposure. However, important information may be lost regarding the suspended properties (such as aggregate size) that govern the behaviour of the particles in air and when inhaled. Low-pressure impactors, where particles are separated by inertial impaction, can separate and quantify nanoparticles from larger particles, but there are difficulties in providing size resolution down to the low nanometer range, and the number of stages is limited. An advantage of impactor systems is that aerosols with nanometer sizes are collected and hence available for further analysis. Also, standard impactor stages are often selected to correlate to the inhalation properties of the aerosol.

Because nanodispersions in air are typically more dilute than liquid-borne or dry bulk powders, *in situ* aerosol measurements often rely on counting techniques such as condensation particle counters (CPCs), in which small droplets of a supersaturated atmosphere of alcohol are condensed on each airborne particle until they are large enough to be detected and counted optically. This approach can cover a relatively large dynamic range.

These particle counters do not by themselves size particles; however, they are often coupled to size-selecting instruments, such as a differential mobility analyser (DMA), that separate charged particles as a function of size by measuring their drift velocity under the action of an electric field. The combination of a DMA and a CPC is often referred to as a scanning mobility particle sizer (SMPS). Currently-available instruments can detect particles as small as 3 nm, while new developments may reach the 1 nm limit (Kim *et al.*, 2003). Aerosol mass spectrometry is a relatively new method of analysing nanometer-sized particles in which particles are vaporized and the resulting ions analysed in a mass spectrometer. Sometimes coupled with time-of-flight measurements, both size and composition of particles can be analysed simultaneously. For a more extensive review of aerosol techniques, there are several excellent references (e.g. Hinds, 1999; Baron and Willeke, 2001; Ruzer and Harley, 2005). Table 1.3 summarizes some of the common instrumentation used to provide measurement information on nanometer-scale aerosols (Aitken *et al.*, 2004; Scientific Committee on Emerging and Newly Identified Health Risks, 2005).

For dry bulk powders, BET surface area can provide an estimated average size (based on a nonporous spherical model) and has the added advantage of providing a direct measurement of specific surface area (SSA) and micro- or meso-porosity, key parameters of interest (Allen, 2004; ISO-15901-2, 2006).

Table 1.3   *Aerosol methods of measurement.*

Analysis method	Size/concentration range	Comments
Filtration	Variable dependent on filter size and airflow.	Collects and retains particles for quantification and further analysis
Condensation particle counter (CPC)	Dynamic range dependent on instrument and airflow	Counting method: number concentration of particles in size range ∼2.5–3000 nm
Differential mobility analyser (DMA)	∼2.5–900 nm	
Scanning mobility particle analyser	3–1000 nm 20–1 Mparticles/cc	Combination of CPC and DMA
Electrical low pressure impactor	30 nm–10 μm	Particles are charged and quantified by electrometer for each plate
Epiphaniometer	Surface areas/size in the range of 100 nm and below	Radioactive tagging with ^{211}Pb; signal is proportional to exposed fuchs surface area
Diffusion charger	10–1000 nm	Used for ambient air monitoring by surface area. Can underestimate size/surface area for larger particles
BET gas adsorption	∼0.01 m^2/g and higher	Size is based on calculation assuming spherical particles
Electron microscopy (SEM/TEM) EDS, EELS, SAD	SEM <10 nm TEM <1 nm	Aerosols must be collected and mounted for analysis

1.4.3.3   Liquids

For the measurement of NSMs dispersed in liquids, an ensemble method of measurement (such as centrifugal sedimentation, laser or dynamic light scattering) is normally preferred. In liquids, ensemble techniques analyse a sample by passing a stream or volume of dispersed particles through a measurement zone. Typically many millions of particles are sampled, and an algorithm or statistical analysis is used to deconvolute the measured information into a particle size distribution. As with aerosol measurements, most methods of size analysis in liquids assume that particles are spherical. Good dispersion is a key element in obtaining an accurate primary particle size distribution. In aqueous solution, there are several methods used to preserve a stable dispersion and/or to redisperse agglomerates (e.g. mixing, sonication, surface chemistry). One of the most common phenomena used to measure particle size is the interaction between light and matter (Jillavenkatesa *et al.*, 2001). Light scattering, sedimentation, acoustic measurements and other techniques are commonly used to analyse nano-sized liquid-borne particles. Laser diffraction (~40 nm–3 mm) and dynamic light scattering (~0.70 nm–6 μm) can provide particle sizes over a large dynamic range for relatively dilute dispersions. However, as particle size decreases below 100 nm, these techniques become more sensitive to errors associated with poor dispersion, shape factors and contamination. This is especially true for dynamic light scattering techniques, which are also limited in assessing broad/aggregated or multimodal particle distributions. Sedimentation techniques such as centrifugal particle sizing are useful for particles down to ~10–20 nm. The density of the particles must be known accurately, and like most other techniques, it assumes spherical geometry and a good dispersion. There are a variety of additional methods for liquid-borne particle size analysis, but most are oriented towards particles and agglomerates of relatively larger sizes (Knapp *et al.*, 1996).

1.4.3.4   Microscopy

Microscopy is one of the most powerful techniques, and is often relied upon to provide valuable information regarding size, shape and morphology. However, as mentioned above, the microscopist should ensure that enough particles are examined to provide a statistically valid representation of the full size distribution. This can be very difficult and time-consuming. It should also be noted that microscopy and image analysis normally provide only two-dimensional images, so care must be taken to avoid bias due to orientation effects. For NSMs (<100 nm), scanning electron microscopy (SEM), transmission electron microscopy (TEM), and sometimes atomic force microscopy (AFM) are normally used to capture images with the necessary resolution. Each of these methods is subject to artefacts caused by sample preparation and/or special analysis conditions. For example, high-resolution TEM requires thin sample sections or particles of limited diameter in order for the electron beam to penetrate through the sample. SEM and TEM are also usually conducted under a high vacuum, which by definition excludes volatile solvents or solvent systems. Salts or other dissolved species present in the original dispersed sample may crystallize into fine particles and obscure details.

Scanning electron microscopy (SEM) is often the method of choice to investigate particle size, shape and structure. When equipped with an electron dispersive X-ray spectrometer (EDX, EDS), elemental composition can also be determined for larger particles or aggregates. The resolution of SEM has increased to below 10 nm in recent years due to the advent of field emission sources and improved electron optics. TEM provides resolution down to a few angstroms, although sample preparation and analysis is somewhat more complex than for SEM. TEM can also be used for elemental analysis, crystal structure, and some chemical information through the use of EDS, selective area diffraction (SAD), and electron energy loss spectroscopy (EELS). AFM has the advantage of providing surface topology or force measurements in a variety of environments, such as in liquids for *in vitro* measurements.

1.4.4   Where to Measure the Size

It seems a bit odd to ask the question 'Where shall we measure the size of the particles?' until one considers that with most ensemble techniques it is actually the agglomerate size that is measured, not the primary particle size. The state of dispersion can and does change as a function of environment. Additionally, particles may be separated or 'size classified' by various means between 'the bottle and the cell'. For example, a flame pyrolysis method may be used to synthesize NSMs for various uses. Upon creation, particles begin to coagulate according to their size, concentration, temperature, residence time, pressure and a variety of other conditions. Hard aggregates may be formed through viscous or chemical sintering; fine particles may adhere to surfaces due to static charges or thermal diffusion; and large particles may settle out of the gas. The size distribution and dispersion state continues to change through transport, handling and delivery to the final application. For toxicity testing, the size distribution 'as-dosed' might be quite different than the same system 'as-generated' or 'as-received'. In the animal, physiological clearance mechanisms may reject larger particles, and membranes or vasculature may bind or exclude particles based on their physical or chemical properties, once again altering the size distribution.

Consequently, there are four points at which we believe particle characterization should be addressed:

- The first location, *perhaps the most obvious*, is the 'as-received' powder. This is also the most straightforward, but as any particle scientist will attest, is often not easy.
- The second, *perhaps the most critical*, is the 'as-dosed' or 'as exposed'. Any experimental data will be of limited use if details regarding exposure parameters are undefined.
- The third location, *perhaps the most difficult*, is at the point(s) of interaction with the organism. What is the size, shape and state of agglomeration of the particles or nanostructures as they interact with the cells, tissues or organs? We consider this the most difficult because invasive techniques usually cannot be used *in situ* without compromising the integrity of the organism and invalidating the test. Noninvasive techniques such as radio or fluorescent labelling can be useful as long as they do not alter the surface chemistry and hence the mechanisms of interaction with the organism.
- The fourth and final location, *perhaps the most useful*, is *post mortem*, or histological examination of the cells, tissues and organs exposed to the test materials. Usually this is done by microscopy. Although there are issues with artefacts, statistical reliability and interpretation, much of what we can deduce about the distribution and effects of NSMs comes from examining the cell and tissue structures.

It has been our experience that all methods of size analysis are subject to certain pitfalls or artefacts. There is no 'absolute' method of determining NSM size distributions, and multiple techniques should be used wherever possible in order to develop a more complete understanding of the system. The analyst should be knowledgeable and skilled in the techniques employed. Table 1.4 shows a number of techniques used to measure particle size in the nanometer range, along with advantages, disadvantages, and an estimate of their usefulness in the areas discussed above.

1.5   Particle Shape

1.5.1   Definition of Particle Shape

The concept of shape seems totally intuitive until one attempts to quantify any but the most geometrically regular shapes. Occasionally, particulate systems have readily definable shapes such as spheres, rods, or defined crystal morphologies. For most real systems, however, shapes are much more variable, and many 'shape factors' have been devised to attempt to quantify them. Several of the most useful include: sphericity or circularity, aspect ratio, convexity, and fractal dimension. Unfortunately, the definitions of many shape factors are not well standardized, which can potentially lead to confusion (Hentschel and Page, 2003). For example, circularity is most often defined by the perimeter to area ratio of a two-dimensional image. However, perimeter is often measured in several different ways, such as perimeter of a rectangular box drawn around the particle, perimeter of the ellipse with the same major and minor axes, or length of the outside edge of the threshold pixels. ISO has recently published a new standard on image analysis that provides standardized shape factors that can be used to evaluate particle shape (ISO-9276-6, 2008). This and other national standards should be consulted when attempting to quantify particle shape.

1.5.2   Why Shape is Important to Toxicity

Particle shape has been implicated in several forms of toxic effects, mostly relating to the inhalation toxicity of certain inorganic fibers such as asbestos (Berry *et al.*, 2005). The toxicity of quartz has been definitively connected to its crystal structure, other forms of silica having been shown to be much less toxic. Although there are a few examples of high aspect ratio fibers showing increased toxicity, it is not generally proven that all such shapes are dangerous, particularly at the nano-scale (Hart and Hesterberg, 1998). Shape, along with size and state of agglomeration, may also affect the disposition and translocation of particles in the organism. Recent research into the potential toxicity of carbon nanotubes has suggested that there may be some asbestos-like toxicity associated with carbon nanotubes longer than 20 microns (Poland *et al,* 2008; Kostarelos, 2008; Takagi *et al,* 2008).

1.5.3   How to Measure Shape

Microscopy is one of the most powerful techniques and is often relied upon exclusively to provide valuable information regarding size, shape and morphology. For NSMs (<100 nm), electron microscopy is normally required to capture images with the necessary resolution. These are currently the only techniques that provide reliable information regarding shape

Table 1.4 Common particle sizing techniques applicable to sub-100 nm particle systems.

Size measurement technique	Nominal size range	Advantages	Disadvantages	'As-received'	As-dosed	In situ	Post mortem
Dynamic light scattering[a,c]	0.7 nm–5 μm	Ensemble method also used for zeta potential	Less reliable as distribution broadens	Yes	Feasible	Maybe	No
Centrifugal sedimentation[b,c]	10 nm–10 μm	Good for broad size distributions	Need accurate density; affected by aggregation	Yes	No	No	No
Laser diffraction[b]	40 nm–3 mm	Broad dynamic range – wet or dry measurements	Assumes spherical particles – shape effects unknown	Yes	Feasible	No	No
Scanning/dynamic mobility analysis[c,i]	3 nm–1 μm	Good for size distributions	Dry low pressure technique – small sample sizes	Yes	Yes	No	No
Atomic force microscopy[c]	0.5 nm – several μm	Good resolution and 3-D imaging	Can only image surface topography	Yes	No	Possible in vitro	Yes
Electron microscopy[b,g]	0.5 nm – several μm	Good resolution and imaging	Sample preparation and vacuum	Yes	No	No	Yes
Field flow fractionation[c]	2 nm–2 μm	Good resolution of size distributions	Must be used in conjunction with other techniques (e.g. light scattering)	Yes	No	No	No
Size exclusion chromatography[d,e]	5 nm–2 μm	Good resolution	Slow; requires good calibration	Yes	No	No	Maybe
Small angle X-ray (or neutron) scattering[h,j]	1 nm–1 μm	Good for solid state embedded systems	Requires high concentrations of particles – cumbersome	Yes	No	No	No
Time of flight mass spectrometry[i]	100–60 kDaltons	Good for very molecular size particles and fragments	Expensive – representative sampling difficult	Yes	Feasible	No	No
Acoustic techniques[f]	20 nm–10 μm	Good for concentrated systems	Need at least 1 wt % of particles	Yes	No	No	No

[a] Berne and Pecora (2000); [b] Bootz et al. (2004); [c] Jillavenkatesa and Kelly (2002); [d] Fritz et al. (1997); [e] Bootz et al. (2005); [f] Dukhin et al. (1999); [g] Sjostrom et al. (1995); [h] Borchert et al. (2005); [i] Burtscher (2005); [j] Glatter and Kratky (1982).

at these length scales. Again, the microscopist should ensure that enough particles are examined to provide a statistically valid representation of the full size or shape distribution. To estimate the number of particles to be analysed for a specific powder, the National Institute of Standards and Technology refers to the mathematical theory developed by Matsuda and Iinoya in 1971 (Jillavenkatesa *et al.*, 2001). According to the equations developed in that theory, the total number of particles to analyse depends upon the standard deviation of the particle sizes, the shape of the particle size distribution, the type of distribution (i.e. number, area and volume) and the desired range for error. In order to get a mass median diameter within 5 % error with 95 % probability for a powder with a typical standard deviation of 1.60, about 61 000 particles are required (Masuda and Gotoh, 1999). While this number would be dramatically reduced to about 15 000 particles for a 10 % error, the large number of particles to be analysed for a reliable particle size distribution measurement demotes the use of EM imaging for that purpose. Nonetheless, EM images are required for shape assessment as well as to detect the presence of large particles, which are indistinguishable from agglomerates when using ensemble techniques for particle sizing.

It should also be noted that EM normally provides only two-dimensional images, so care must be taken to avoid bias due to orientation effects. High resolution microscopy is subject to artefacts caused by sample preparation and/or special analysis conditions. Quantification of shape data can be complex and there are numerous ways to express shape information, and despite attempts by international standards bodies, there is still little standardization of definitions (ISO-9276-6, 2008).

1.6   State of Dispersion

1.6.1   Definition of Dispersion

The state of dispersion of a particulate system describes the extent to which particles are agglomerated – that is, held together in groups or clusters by attractive interparticle forces. These forces have a variety of sources, the most fundamental of which are the attractive Van der Waals forces. The magnitude of these forces is a function of the fundamental atomic properties of the surface atoms, the geometry (particle shape), and the proximity of these surfaces to each other.

For most ensemble size measurement techniques, what is really measured is the size distribution of the agglomerates. Thus, the measured size distribution is highly dependent upon the state of dispersion of the system. Any ensemble particle size measurement must be interpreted in this context. Due to attractive forces (Van der Waals and others), particles will tend to agglomerate in suspension unless stabilized by surface charge or steric effects. Most aqueous suspensions of hydrophilic polymers or metal oxides will specifically adsorb or desorb hydrogen ions to generate a surface charge. Homogeneous powders that develop a surface charge high enough to overcome interparticle attraction will form more stable dispersions.

Unfortunately, the physiological or biological environments commonly found in toxicity studies can, and most likely will, affect the state of dispersion of a NSM system, and thus it is important to measure the particle size distribution in both the 'as-received' and the

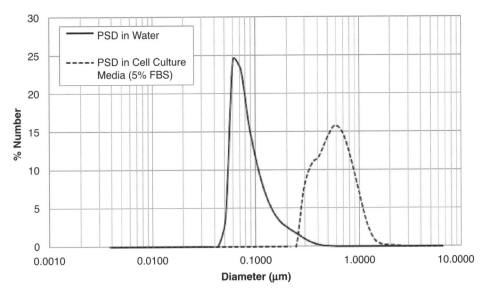

Figure 1.2 *The properties and characteristics of media surrounding nanopowders affect their state of dispersion. Graph depicts the particle size distributions (PSDs) of the same aluminum nanopowders in two different liquids, water and cell culture media. Better nanopowder dispersion was achieved in water.*

'as-dosed' stages. Figure 1.2 illustrates an example of different states of dispersion for the same aluminum nanopowder in two different media. Samples were first dispersed in water and then introduced into cell culture media following the same protocols. Particle size distribution was measured by dynamic light scattering and plotted as a relative number distribution. One can clearly see the result of agglomeration in the cell culture media when compared with the dispersion in water. Although the NSMs still retain their primary size and surface area, the nature of the particles/agglomerates as presented to the cells is clearly different than in pure water.

Particle size affects dispersion in two ways. For large particles or agglomerates, sedimentation becomes an issue. Stable, well-dispersed ultrafine and nano systems generally do not suffer from this effect. However, the smaller the particle size, the greater the relative Van der Waals forces per unit mass. This means that, once agglomerated, it becomes progressively more difficult to redisperse NSMs as size decreases. Therefore fine particle dispersions have a greater tendency towards irreversible agglomeration unless stabilized by the mechanisms discussed above.

Particle aggregates can often be redispersed to some extent by applying shear (mixing, sonication, grinding, turbulence), but unless the conditions are present for a stable dispersion (by native surface charge at a given pH, surfactants, or steric stabilization) the system will be prone to re-agglomerate rapidly. As illustrated in Figure 1.2, particles that are not stabilized by pH or surfactant may agglomerate rapidly when introduced into a different environment such as a highly buffered (high ionic strength) physiological/ biological fluid.

1.6.2   Dry Powder Dispersion

Although most physiological environments comprised wet systems, dry aerosol dispersions are associated with exposure by inhalation or deposition on the skin. If the dispersion of fine powders is difficult in solution, it is doubly so as a dry powder. In air, powders cannot be easily stabilized by charge. Although dielectric powders may support a static charge, this serves only to attract the particles to the nearest surface with an opposite (or neutral) polarity. (This is why fine dust can collect or stick on vertical surfaces.) Additives such as surfactants are generally ineffective or counterproductive in the dry state, although there are some flow aids or glidants used for dry dispersion of larger particle sizes. Fine and ultrafine powders (here defined as powders <2.5 microns and <0.1 micron, respectively) that have once been dried generally cannot be fully redispersed to their original size distributions without extreme effort.

1.6.3   How to Measure the State of Dispersion

Particle and colloid scientists have been struggling with this issue for decades. The only way to absolutely define the 'state of dispersion' is to quantify both the primary particle size (unagglomerated) distribution and the agglomerated size distribution of the system. Most often, an assessment of the state of agglomeration is made by taking two particle size measurements: one of the particles in the agglomerated state and one with the particles dispersed as close to the 'primary' particle size distribution as possible. Many have used these measurements to attempt to create a dispersion 'index' that represents with one number the state of dispersion (Jillavenkatesa and Kelly, 2002). The difficulty in assigning such numbers is that they often do not adequately take into account the effect of the polydispersity of the primary particles and the fact that one is never sure how 'well dispersed' the primary particle size distribution really is. Agglomeration is often an active (time and environment dependent) process and is also affected by mixing and other forms of shear used in the measurement process. Agglomeration rate and strength (or floc strength) studies should be pursued when evidence of agglomerate-induced effects appear relevant. Experimental protocols for these experiments have been reviewed elsewhere (Jarvis *et al.,* 2005).

Despite these limitations, some assessment can usually be made (e.g. a statistical comparison) of dispersed versus agglomerated size distributions. Measures used to disperse solutions should be carefully recorded and reproducible. Qualitative observations can also be derived from imaging. The use of multiple size analysis techniques is particularly valuable for this. Figure 1.3 illustrates an 'as-received' nano-size aluminum powder (nominal size 83 nm) analysed by laser diffraction, BET surface area and EM. What is the true size and state of agglomeration of this powder? By light scattering it has a volume median size of 14.5 microns, certainly not the nano-size attributed to it by the manufacturer. By number, the median size is 151 nm, and by surface area the size is calculated to be 80 nm.

In Figure 1.4 a series of electron micrographs (TEM and SEM) of the same powder give additional insight into the size and state of agglomeration. Note that large 'particles' can be attributed to both polydispersity and agglomeration. By using multiple techniques, it is clear that there is both significant agglomeration and polydispersity in this 'nanoparticulate' system. Finally, Figure 1.5 shows these same particles in an endosome inside a cell after

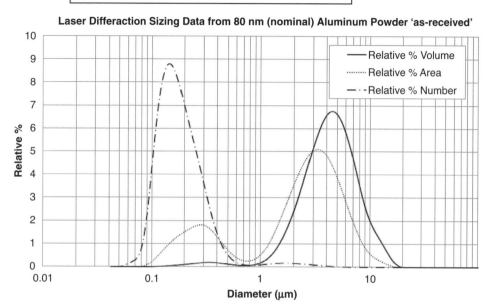

Median diameter by volume	14.4 µm
Median diameter by number	151 nm
BET surface area 25.7 m²/g	80 nm (equivalent)

Figure 1.3 *Laser diffraction size data for 'nanoscale' aluminum powder used for* in vitro *toxicity experiments. Note the apparent difference in size when depicted as a number distribution versus an area or volume distribution. Each curve, if presented by itself, would give an incomplete understanding of the particle size distribution/state of agglomeration of the sample. The three curves will overlay only for an ideal spherical, monodispersed, non-agglomerated system.*

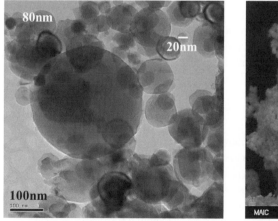

Figure 1.4 *Electron microscopy of an 'as-received' '83 nm' aluminum powder.*

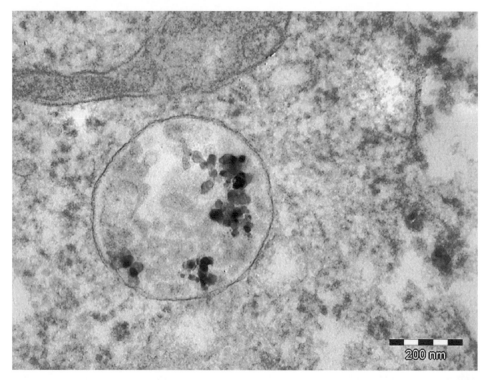

Figure 1.5 *Aluminum nano-particles inside an endosome in an A549 cell from an* in vitro *toxicity experiment.*

an *in vitro* toxicity test. Although individual particles can be resolved, it is unclear whether they were transported there as individuals or as agglomerates.

1.6.4   Dispersion and Toxicity

The tendency to agglomerate may at first seem irrelevant to the toxicologist introducing NSMs to assess cytotoxicity, immune or inflammatory response. However, if the biological response is size-dependent, the state of agglomeration may be a significant factor. For example, particle agglomerate size has been shown to play a role in the ability of macrophages to phagocytize particles, and in the tendency for particle penetration of cell membranes and translocation through the tissues, lymph or circulatory system (Nemmar *et al.,* 2001; Renwick *et al.,* 2001). There is some evidence that very fine particles may escape an immune or inflammatory response observed for larger particles of the same material (Gutwein and Webster, 2002). It is difficult to draw firm conclusions regarding size effects from existing literature due to the fact that studies are performed with different cell lines, tissues, animal models, and different types of ultrafine particles and methods of administration. The dispersion of particles during dosing and *in situ* is seldom adequately quantified and fully addressed, apart from qualitative observations made from post-exposure microscopic analysis. Clearly more detailed research is required on the specific effects of size and state of agglomeration.

1.7 Surface Properties

The properties embodied by the surface of NSMs are expected to contribute substantially to the mode and extent of their biological interaction. Surface composition, energy, charge, reactivity and opsonization – among others – clearly affect NSM interactions with biomolecules and biological systems. It is often impractical to characterize the full spectrum of surface properties for each NSM system. Because of this, it is recommended that an interactive approach to surface characterization be undertaken. It is further suggested that a sample of the NSMs ('as-received' and 'as-dosed') is stored under inert conditions (e.g. under argon gas for dry powders, and cryogenic preservation of particle suspensions) to enable future analysis if necessary. At the very minimum, the surface composition and structure of NSMs should be measured on the 'as-received' powders. Where possible, measurements should also be performed on the NSMs 'post-exposure'.

1.7.1 Surface Area

There has been a good deal of discussion regarding the role of surface area in NSM toxicology. Historically, particles have typically been dosed by mass concentration and occasionally by number concentration. Recent research (Oberdörster, 2001) has argued that surface area plays an important role in the toxicity of NSMs and is the measurement metric that best correlates with particle-induced adverse health effects. There is a growing consensus that the potential for adverse health effects is most directly proportional to particle surface area (Driscoll *et al.,* 1996; Oberdörster, 2001). Since surface area almost always scales with size (at least for nonporous materials), one could equally make the case that size is the best correlation. In fact, if one measures the mass, density and primary particle size distribution of a material, the approximate surface area can be readily calculated by assuming spherical geometry. Thus, the question is not so much what surface area of material was used to dose an animal or cell culture, but rather whether the dose should be 'normalized' by total surface area concentration rather than mass or number concentrations. For example, a 100 nm spherical aerosol of density 2.0 has a specific surface area (SSA) of 30 m^2/g, whereas a 1 micron spherical aerosol of same composition has a SSA of 3 m^2/g. Thus, ten times the mass of the larger (1 micron) particles is required to achieve the same surface area concentration.

Particle surface area is clearly an important characteristic of a NSM and should always be measured if the means are available. As described previously, the surface area of 'as-received' dry powders can be accurately measured using gas adsorption and the BET method. For particles dispersed in water or air, measuring surface area is somewhat more difficult. Aqueous dispersions can be carefully dried and the resulting powders measured by BET. Care must be exercised to ensure that the surface area is not perturbed by drying and that there are no other salts or components of the solution that influence the measurement. For liquid dispersions, titration can sometimes be used for *in situ* measurements of surface area, for example the Sears Test, a titration method for measuring surface area of colloidal silica (Sears, 1956).

Although measuring the surface area of bulk dry powders is relatively straightforward using the BET method, the measurement of surface area on very small quantities of dilute airborne particulates is more problematic. A relatively simple technique involves charging particles by ion attachment and subsequent trapping of particles in a filter within a Faraday cup, which is connected to a sensitive electrometer. Although this method provides a signal that needs to be calibrated, it gives a sensitive proxy of aerosol surface area. This is similar to the epiphaniometer, which relies on attachment and detection of a radioactive lead isotope rather than an ion, and is thus sensitive to rather low concentrations of particles (Scientific Committee on Emerging and Newly Identified Health Risks, 2007).

1.7.2   Surface Composition

The molecular composition and structure of the surface of NSMs will ultimately define their energy, charge and reactivity; however, the prediction of these properties – especially for NSMs in biological systems – is generally not feasible with current scientific approaches. Directly measuring the atomic composition of NSMs, 'as-received' and 'as-administered', is critical because many of these systems are subject to trace surface contaminants/heterogeneities that are often not detectable via bulk composition analysis. Post-exposure examination of changes in surface composition and structure will undoubtedly provide priceless clues with respect to their behaviour and the fate of these particles in biological systems. A list of techniques applicable for identifying the surface composition and atomic arrangement of NSMs is given in Table 1.5. The experimentalist should use caution when selecting a technique for NSM analysis. Many of the methods used for surface characterization require ultra-high vacuum environments ($<10^{-5}$ Pa). Under such conditions the surface properties and bonding structure of some materials have been shown to change. Because of this, techniques that operate at higher pressures are becoming more attractive for surface characterization, especially since those capable of measurements in liquids or at ambient pressures are often subject to limitations in the theories used for analysis. ESCA, XPS and SIMS, in particular, have been extensively used for characterizing NSMs, as well as correlating biomaterial surface properties to physiological endpoints (Ratner, 1996). These techniques are applicable to NSM surface examination post-exposure; however, washing and removal of biomolecules from the surface of the particles is likely to be necessary step. Care must be taken to prevent/identify artefacts from this process.

1.7.3   Surface Charge

The surface charge of NSM systems will influence the dispersion stability in aqueous solutions and can have a dramatic effect on the response of biological systems. The surface charge may reflect the native NSM surface or the adsorption of ions and biomolecules at their interface. Classically, the surface charges of particulate systems are approximated through zeta potential measurements (Adamson and Reynolds, 1997). Zeta potential refers to the sign and magnitude of charge at the shear plane, which divides the fluid envelope that associates itself with the particle and the bulk solution phase. It is usually calculated by measuring the particle mobility when subjected to an electric or acoustic field.

For 'as-received' NSMs, zeta potential measurements are normally performed in pure water with a small amount (1–10 mM) of monovalent background electrolyte. A titration as a function of pH is used to find the isoelectric point (IEP), or the pH where the zeta potential of the material is zero. It should be noted that this pH also corresponds to the point

Table 1.5 *Surface analysis techniques applicable to sub-100 nm particle systems.*

Techniques	Sample volume required	Pen. depth*	Applications	Limitations
AES	µg–mg	1–5 nm	Surface composition	Insulators cannot be used due to significant charging, surface damage, HV–UHV
EELS, ELS, HREELS	< µg	Few nm	Surface composition and elemental mapping	Done with TEM and thus requires thin sections, HV–UHV
EXAFS	Few mg	Few Å	Surface atom packing, surface–substrate interactions, colloid samples can be used	Not applicable to amorphous systems
LEED, HEED	< µg	Few Å	Surface crystal structure and phase identification	Applicable only for single crystals, HV–UHV
SEM	mg–g	Microns	Surface morphology	Sample charging, HV–UHV
SIMS	µg–mg	Few Å	Surface elemental analysis with depth profiling	Not quantitative, surface damage, low mass resolution, HV–UHV
TEM, HR-TEM, STEM	< µg	< 100 µm	Local structure and morphology	Sample preparation difficult, e-beam can damage organic materials, HV–UHV
XPS, ESCA	µg–mg	1–5 nm	Surface chemical analysis	Poor spatial resolution, not suitable for trace analysis, HV–UHV

*Penetration depth of radiation source; AES, Auger electron spectroscopy; EELS, electron energy loss spectroscopy; ELS, energy loss spectroscopy; HREELS, high resolution electron energy loss spectroscopy; EXAFS, extended X-ray absorption fine structure spectroscopy; LEED, low energy electron diffraction; HEED, high energy electron diffraction; SEM, scanning electron microscopy; SIMS, secondary ion mass spectroscopy; TEM, transmission electron spectroscopy; HR-TEM, high resolution TEM; STEM, scanning TEM; XPS, X-ray photoelectron spectroscopy; ESCA, electron spectroscopy for chemical analysis; HV, high vacuum; UHV, ultra-high vacuum.

of zero charge at the surface. Typically the IEP of a material under controlled conditions is reported along with zeta potential value (sign and magnitude) at the pH of interest (e.g. under physiological conditions). The IEP is material dependent, but typically values lie in the range from pH 3 (e.g. silica) to pH 9 (e.g. aluminum oxide). Generally, the surface charge becomes progressively more positive as pH decreases (acidic solutions), and more negative as pH increases. Figure 1.6 illustrates a typical zeta potential titration for an 'as-received' metal oxide nanopowder (DeGussa P-25 TiO_2) showing the isoelectric point at pH 6.6. Sometimes surfactants are used to change the surface charge or sterically stabilize suspensions through the use of long hydrocarbon chains attached to a polar head. However, not all surfactants are compatible with the physiological environment and may cause cell lysis and/or other adverse effects.

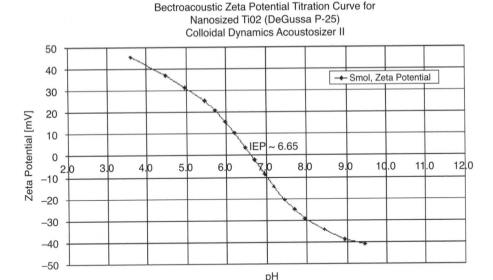

Figure 1.6   *Zeta potential titration curve for nanosize TiO₂ powder (DeGussa P-25).*

Zeta potential measurements can be performed on NSMs in biological fluids; however, care must be taken to ensure that appropriate measures are taken to avoid artefacts from the high ionic strength and biomolecules encountered in these fluids. Changes in zeta potential are influenced by the adsorption, exchange and ionization of molecules within the shear plane and at the particle surface. Table 1.6 shows a practical example of how zeta potential changes as a function of environment for several NSMs used in a recent toxicity study (Wasdo *et al.*, 2008). Note that in all cases the zeta potential decreases in absolute value, reflecting the effect of high ionic strength of the media. Also, the positive zeta potential of the aluminum powders are reversed when introduced into media, most likely reflecting the adsorption of proteins onto the NSM surface.

Table 1.6   *Isoelectric points and zeta potentials (ζ) in different environments.*

Samples	IEP*	ζ (mV) in water at pH = 7.4	ζ (mV) in media at pH = 7.4
NanoTek TiO₂	6.3	−20.2	−0.8/−1.3
P25 TiO₂	6.6	−25.1	−0.2/−1.1
Quartz	1.3	−28.5	−8.4/−10.7
Aluminum 1		+15.7/+23.6	−12.1/−16.7
Aluminum 2	9.3–9.5	+18.2/+20.3	−1.1/−4.3
Aluminum 3		+10.2/+12.7	−7.9/−10.3
Aluminum 4		+21.3/+23.5	−2.1/−3.4
Aluminum 5		+34.2/+38.5	−0.5/−3.7

*Isoelectric point: pH at which ζ = 0.

It should be noted that zeta potentials exist as distributions and are particle-size dependent. Unfortunately, most instrumentation provides only a single Smoluchowski zeta potential calculated in a size-independent manner. Potentiometric titrations can also be used to acquire particle charge information. In particular, the pKa values of particle surface functionalization groups can be determined along with information on surface charge density (functional group density).

1.7.4   Surface Reactivity

The reactive potential of NSMs can have consequences for interacting cells and biological species. Moreover, NSMs that are able to participate actively in oxidation/reduction reactions can largely skew many biological assays.

The surface reactivity of NSMs can be measured through comparative microcalorimetry, via the use of probe molecules that are monitored for either degradation or changes in oxidative state, or through a number of electrochemical methods. All of these techniques can potentially be used to monitor particle reactivity in biological fluids – although losses in sensitivity and artefacts are likely. The choice of method will depend on the types of molecular transformations that occur at the particle surface. Comparisons between the surface compositions of the 'as-received' particles and those exposed to biological systems can provide further insight for selection.

The surface energy and wettability of NSM systems can be important for understanding NSM aggregation, dissolution, opsonization and bioaccumulation behaviour. In the biomaterials community, the role of surface energy in implant biocompatibility has been recognized for several decades (Ratner, 1996). The surface energy of NSM systems can be measured through heat of immersion microcalorimetry studies or through contact angle measurements with various liquids. Dynamic and static contact angle measurements can be performed to determine directly the particle wettability within biological fluids. Phospholipids, proteins and other biomolecules are known to adsorb to surfaces in physiological fluids and change their wettability and sometimes biodistribution characteristics. Phase partitioning experiments can also be designed to determine relative surface wettability in biologically relevant fluids (Neumann and Spelt, 1996).

1.8   Recommended Practices

Basic particle characterization by the measurement of particle size and shape is not a new analysis problem. Particle scientists have dealt with the issue for decades and have developed recommended practices that are recorded in literature, texts, and national and international standards (ISO, ASTM, USP and ANSI). Many of these practices have been adapted for difficult measurement environments similar to those facing the toxicology community. For basic particle characterization, several fundamental principles include:

- *Sampling.* Due to the large number of particles in a given mass of a NSM system, it is imperative that the particles measured are representative of the bulk material. This is the first and foremost principle of particle characterization. The broader the size distribution, the more significant the errors will be if the sample is not representative.

- *Dispersion.* Primary particle characteristics should be measured in the most dispersed state achievable. Some estimate of the state of dispersion should be made at the point of interaction with the organism, and the state of agglomeration should be assessed whenever the environment changes throughout toxicological testing.
- *Statistical reliability.* Enough particles must be measured to ensure that the desired limits of precision and accuracy are achieved. For broad particle size distributions this may amount to tens of thousands of particles.
- The *physical principle of measurement* should be selected to be most appropriate to the application. For example, if measuring airborne deposition of a dry powder into the lungs of an animal subject, the principle of measurement might be a time of flight, cascade impaction or some other technique that measures aerodynamic diameters of the powder.
- The *particle size* should be measured under conditions as close as possible to the environment at the point of application. For example, if *in vitro* cell studies are being conducted, the particle size should be measured in cell culture media or at least in the same pH and ionic strength conditions.
- *Surface characteristics* and reactivity of NSMs are critical elements in their toxicity. At a minimum, measurements should be attempted to quantify surface charge, wettability, protein adsorption and the potential for the formation of reactive oxygen species.

1.9   Conclusions

A firm understanding of the properties of NSMs is essential for their characterization and for determining toxicological effects in the physiological environment. This is particularly true when trying to determine the mechanism of toxicity. The issues involved become quite complex. Outside of the body, the environmental conditions can be manipulated to promote the maximum dispersion for measuring size and surface chemistry. In the biological environment, however, one is restricted to the conditions under which the organism lives. Often, properties can only be measured after NSMs are removed or tissues fixed, potentially introducing artefacts into the measurements. Until better techniques for characterization in these environments are developed, researchers must use the tools available to reconstruct the particle properties *in situ* and how they interact with cells, tissues and organs. All tests should include basic characterization of size, shape, surface properties and state of dispersion. The acceptable level of characterization will depend on the objective of the study, and the nature of the NSM will determine the specific parameters to be measured.

References

Adamson NJ, Reynolds EC (1997) Rules relating electrophoretic mobility, charge and molecular size of peptides and proteins. *J Chromatogr B* **699**, 133–147.

Aitken R, Creely K, Tran C (2004) *Nanoparticles: An Occupational Hygiene Review*. Research Report 274, Health and Safety Executive (ed.), Institute of Occupational Medicine. HSE Books: London.

Allen T (2004) *Particle Size Measurement, Vol I: Powders Sampling and Particle Size Measurement* (5th edn). Chapman and Hall: London.

Bailey MR, Ansoborlo E, Guilmette RA, Paquet F (2003) Practical application of the ICRP Human Respiratory Tract Model. *Radiat Prot Dosimetry* **105**, 71–76 (2003).

Balbus JM, Maynard AD, Colvin VL, Castranova V, Daston GP, Denison RA, Dreher KL, Goering PL, Goldberg AM, Kulinowski KM, Monteiro-Riviere NA, Oberdorster G, Omenn GS, Pinkerton KE, Ramos KS, Rest KM, Sass JB, Silbergeld EK, Wong BA (2007) Meeting report: Hazard assessment for nanoparticles – Report from an interdisciplinary workshop. *Environ Health Persp* **115**, 1654–1659.

Baron PA, Willeke K (2001) *Aerosol Measurement Principles, Techniques, and Applications.* John Wiley & Sons, Ltd: Chichester.

Berne BJ, Pecora R (2000) *Dynamic Light Scattering: With Applications to Chemistry, Biology, and Physics.* Dover Publications: Minoela, NY.

Berry CC, Dalby MJ, McCloy D, Affrossman S (2005) The fibroblast response to tubes exhibiting internal nanotopography. *Biomaterials* **26**, 4985–4992.

Bootz A, Vogel V, Schubert D, Kreuter J (2004) Comparison of scanning electron microscopy, dynamic light scattering and analytical ultracentrifugation for the sizing of poly(butyl cyanoacrylate) nanoparticles. *Eur J Pharm Biopharm* **57**, 369–375.

Bootz A, Russ T, Gores F, Karas M, Kreuter J (2005) Molecular weights of poly(butyl cyanoacrylate) nanoparticles determined by mass spectrometry and size exclusion chromatography. *Eur J Pharm Biopharm* **60**, 391–399.

Borchert H, Shevehenko EV, Robert A, Mekis I, Kornowski A, Grubel G, Weller H (2005) Determination of nanocrystal sizes: A comparison of TEM, SAXS, and XRD studies of highly monodisperse COPt3 particles. *Langmuir* **21**, 1931–1936.

Bucher J, Masten S, Moudgil B, Powers K, Roberts S, Walker N (2005) *Developing Experimental Approaches for the Evaluation of Toxicological Interactions of Nanoscale Materials.* Workshop Report, University of Florida.

Burtscher H (2005) Physical characterization of particulate emissions from diesel engines: a review. *J Aerosol Sci* **36**, 896–932.

Driscoll KE, Howard BW, Carter JM, Asquith T, Johnston C, Detilleux P, Kunkel SL, Isfort RJ (1996) Alpha-quartz-induced chemokine expression by rat lung epithelial cells – effects of in vivo and in vitro particle exposure. *Am J Pathol* **149**, 1627–1637.

Dukhin AS, Shilov VN, Ohshima H, Goetz PJ (1999) Electroacoustic phenomena in concentrated dispersions: New theory and CVI experiment. *Langmuir* **15**, 6692–6706.

Fritz H, Maier M, Bayer E (1997) Cationic polystyrene nanoparticles: preparation and characterization of a model drug carrier system for antisense oligonucleotides. *J Colloid Interface Sci* **195**, 272–288.

Glatter O, Kratky O (1982) *Small Angle X-Ray Scattering.* Academic Press: London.

Gutwein LG, Webster TJ (2002) Osteoblast and chrondrocyte proliferation in the presence of alumina and titania nanoparticles. *J Nanopart Res* **4**, 231–238.

Hart GA, Hesterberg TW (1998) In vitro toxicity of respirable-size particles of diatomaceous earth and crystalline silica compared with asbestos and titanium dioxide. *J Occup Environ Med* **40**, 29–42.

Hentschel ML, Page NW (2003) Selection of descriptors for particle shape characterization. *Part and Part Syst Char* **20**, 25–38.

Heyder J (2004) Deposition of inhaled particles in the human respiratory tract and consequences for regional targeting in respiratory drug delivery. *Proc Am Thorac Soc* **1**, 315–320.

Hinds WC (1999) *Aerosol Technology: Properties, Behavior, and Measurement of Airborne Particles* (2nd edn). Wiley-Interscience: Chichester.

Holdich R (2002) *Fundamentals of Particle Technology.* Midland Information Technology & Publishing: Nottingham.

ISO-9276-6 (2008) *Representation of results of particle size analysis - Part 6: Descriptive and quantitative representation of particle shape and morphology.* International Organization for Standardization: Geneva.

ISO-15901-2 (2006) *Pore size distribution and porosity of solid materials by mercury porosimetry and gas adsorption - Part 2: Analysis of mesopores and macropores by gas adsorption.* International Organization for Standardization: Geneva.

Jarvis P, Jefferson B, Gregory J, Parsons SA (2005) A review of floc strength and breakage. *Water Res* **39**, 3121–3137.

Jillavenkatesa A, Dapkunas S, Lum LSH, NIST (2001) *Particle Size Characterization.* US Dept. of Commerce, Technology Administration, National Institute of Standards and Technology: Gaithersburg, MD.

Jillavenkatesa A, Kelly JF (2002) Nanopowder characterization: challenges and future directions. *J Nanopart Res* **4**, 463–468.

Kim CS, Okuyama K, de la Mora JF (2003) Performance evaluation of an improved particle sice magnifier (PSM) for single nanoparticle detection. *Aerosol Sci and Tech* **37**, 791–803.

Knapp JZ, Barber TA, Lieberman A (1996) *Liquid and Surface-Borne Particle Measurement Handbook.* Marcel Dekker: New York.

Kostarelos K (2008) The long and short of carbon nanotube toxicity. *Nat Biotechnol* **26**, 774–776.

Leggett RW, Eckerman KF (2003) *Dosimetric Significance of the ICRP's Updated Guidance and Models, 1989-2003, and Implications for U.S. Federal Guidance.* ORNL/TM-2003/207.

Lucarelli M, Gatti AM, Savarino G, Quattroni P, Martinelli L, Monari E, Boraschi D (2004) Innate defence functions of macrophages can be biased by nano-sized ceramic and metallic particles. *Eur Cytokine Netw* **15**, 339–346.

Masuda H, Inoya K (1971) Theoretical study of the scatter of experimental data due to particle-size distribution. *J of Chem Eng of Japan* **4**, 60–66.

Masuda H, Gotoh K (1999) Study on the sample size required for the estimation of mean particle diameter. *Adv Powder Technol* **10**, 159–173.

Nemmar A, Vanbilloen H, Hoylaerts MF, Hoet PH, Verbruggen A, Nemery B (2001) Passage of intratracheally instilled ultrafine particles from the lung into the systemic circulation in hamster. *Am J Respir Crit Care Med* **164**, 1665–1668.

Neumann AW, Spelt JK (1996) *Applied Surface Thermodynamics.* Marcel Dekker: New York.

National Research Council (1989) *Biological Markers in Pulmonary Toxicology.* National Academies Press: Washington, DC.

Oberdorster G (2001) Pulmonary effects of inhaled ultrafine particles. *Int Arch Occ Env Hea* **74**, 1–8.

Oberdorster G, Ferin J, Lehnert BE (1994) Correlation between particle size, in vivo particle persistence, and lung injury. *Environ Health Perspect* **102** (Suppl. 5), 173–179.

Oberdorster G, Maynard A, Donaldson K, Castranova V, Fitzpatrick J, Ausman K, Carter J, Karn B, Kreyling W, Lai D, Olin S, Monteiro-Riviere N, Warheit D, Yang H (2005) Principles for characterizing the potential human health effects from exposure to nanomaterials: elements of a screening strategy. *Particle and Fibre Toxicology* **2**, 8.

Organisation for Economic Co-operation and Development (OECD) (2008) *List of Manufactured Nanomaterials and List of Endpoints for Phase One of the OECD Testing Programme.* Series on the Safety of Manufactured Nanomaterials, Number 6. OECD: Paris

Poland CA, Duffin R, Kinloch I, Maynard A, Wallace WA, Seaton A, Stone V, Brown S, Macnee W, Donaldson K (2008) Carbon nanotubes introduced into the abdominal cavity of mice show asbestos-like pathogenicity in a pilot study. *Nat Nanotechnol* **3**, 423–428.

Powers KW, Brown SC, Krishna VB, Wasdo SC, Moudgil BM, Roberts SM (2006) Research strategies for safety evaluation of nanomaterials. Part VI. Characterization of nanoscale particles for toxicological evaluation. *Toxicol Sci* **90**, 296–303.

Ratner BD (1996) The engineering of biomaterials exhibiting recognition and specificity. *J Mol Recognit* **9**, 617–625.

Renwick LC, Donaldson K, Clouter A (2001) Impairment of alveolar macrophage phagocytosis by ultrafine particles. *Toxicol Appl Pharmacol* **172**, 119–127.

Roco MC (2001) International strategy for nanotechnology research and development. *J Nanopart Res* **3**, 353–360.

Royal Society and Royal Academy of Engineering (2004) *Nanoscience and Nanotechnologies: Opportunities and Uncertainties.* RS Policy Document 19/04. Royal Society: London.

Ruzer LS, Harley NH (2005) *Aerosols Handbook : Measurement, Dosimetry, and Health Effects.* CRC Press: London.

Scientific Committee on Emerging and Newly Identified Health Risks (SCENIHR) (2005) *The appropriateness of existing methodologies to assess the potential risks associated with engineered and adventitious products of nanotechnologies.* Health & Consumer Protection Directorate-General, European Commission: Brussels.

Scientific Committee on Emerging and Newly Identified Health Risks (SCENIHR) (2007) *The appropriateness of the risk assessment methodology in accordance with the Technical Guidance Documents for new and existing substances for assessing the risks of nanomaterials.* Health & Consumer Protection Directorate-General, European Commission: Brussels.

Sears GW (1956) Determination of specific surface area of colloidal silica by titration with sodium hydroxide. *Anal Chem* **28**, 1981–1983.

Sjostrom B, Kaplun A, Talmon Y, Cabane B (1995) Structures of nanoparticles prepared from oil-in-water emulsions. *Pharm Res* **12**, 39–48.

Takagi A, Hirose A, Nishimura T, Fukumori N, Ogata A, Ohashi N, Kitajima S, Kanno J (2008) Induction of mesothelioma in p53+/- mouse by intraperitoneal application of multi-wall carbon nanotube. *J Toxicol Sci* **33**, 105–116.

Thomas K, Sayre P (2005) Research strategies for safety evaluation of nanomaterials, part I: Evaluating the human health implications of exposure to nanoscale materials. *Toxicol Sci* **87**, 316–321.

US Environmental Protection Agency (EPA), Office of Pollution Prevention and Toxics (2009) *Nanoscale Materials Stewardship Program.* Interim Report.

Wasdo SC, Barber DS, Denslow ND, Powers KW, Palazuelos M, Stevens SM, Moudgil BM, Roberts SM (2008) Differential binding of serum proteins to nanoparticles. *Int J Nanotechnol* **5**, 92–115.

2

Criteria and Implementation of Physical and Chemical Characteristics of Nanomaterials for Human Health Effects and Ecological Toxicity Studies

Christie M. Sayes and David B. Warheit

2.1 Introduction

Nanomaterials are materials with structural features having at least one dimension in the size range of 1 to 100 nm, and that have a novel property relative to bulk counterparts. However, it is important to realize that some materials slightly larger than 100 nm also exhibit properties different than their microscale or macroscale counterparts. The characteristics described here can serve as a relevant set of characterization data for these materials, as well.

Nanomaterials include nanoparticles, nanofilms and nanocomposites. Most are synthesized or produced using a 'bottoms up' (particle growth, in either the liquid or aerosol phase, from a molecular precursor) approach in order to develop specific structural and functional features. The result is normally a highly ordered, monodispersed particle sample with a large surface area. Some nanomaterial systems are produced in the liquid phase; other systems are produced as aerosols; still other systems are produced and used as dry powders.

These novel materials are characterized using a variety of techniques, including but not limited to microscopy, spectroscopy and chromatography. These methods provide

Nanotoxicity: From In Vivo *and* In Vitro *Models to Health Risks*   Edited by Saura Sahu and Daniel Casciano
© 2009 John Wiley & Sons, Ltd

important physico-chemical information on the physical properties (i.e. size and shape, aggregation status, crystallinity) and chemical properties (i.e. surface coatings and solubility) which aid in the interpretation of toxicological and ecotoxicological responses and associated mechanisms of actions following nanomaterial exposure. When bridging the fields of nanoscience and toxicology to assess exposure-related environmental risks, it is important to recognize that the most instructive studies are derived from integrated approaches. One key property essential to understanding the effects of nanomaterials on living organisms is the interaction between cellular membranes and the surface of the nanomaterial.

Measurements of surface area, mass and particle number become increasingly important when investigating the potential toxic effects of nano-sized particles (Oberdorster *et al.*, 1992; Donaldson and MacNee, 1998; Tran *et al.*, 2000; Brown *et al.*, 2001; Donaldson *et al.*, 2002). However, more recent studies suggest that particle surface reactivity and surface coatings (or the lack thereof) may play a more important role in influencing cellular responses in *in vivo* studies (Warheit *et al.*, 2007). Most studies reported in the current literature cite inconsistencies in sizing data using different techniques. Sizing data from electron microscopy is different from size calculated from surface area analyses. Data measured in the particle's dry state is different than the particle's size in solution (i.e. wet phase). Further, it is difficult to predict a particle's variation in size because each particle has different chemical compositions and surface modifications. Because of this, it is important to recognize that all nanoparticles act differently when placed in different environments. For example, in some carbon-based nanoparticle samples, there is only a 5 % variation in sizing data; however, other metal oxide particles exhibit over 90 % variance in sizing data. Characterization data are dependent upon measurement technique, characterization phase and sample preparation.

We have divided the characterization of nanomaterials into four categories or phases. Primary characterization is performed on nanomaterials in the dry state. This phase is not relevant for nanoparticles produced and applied as particle suspensions. Secondary characterization is performed on nanomaterials in the wet phase as a solution or suspension in aqueous media, such as ultra-pure water, water from the environment (groundwater, surface water, freshwater, or saltwater), or cell culture media. Tertiary characterization is performed on particles following interactions with cells under *in vivo* or *in vitro* conditions and involves more sophisticated and non-trivial techniques than discussed here. Lastly, quaternary characterization focuses on the physical, chemical and biological transformations of the nanomaterials after exposure to a biological system. Although tertiary and quaternary phase characterizations are complex, they are the most relevant to toxicological considerations. While some material properties are similar on the nano-scale and bulk scale (e.g. chemical composition), other properties differ.

The model nanoparticle sample has three distinct features: a defined structure, monodispersity, and a large surface area. However, the feature most often mentioned is the particle size. Particle size, usually reported as diameter, is not always a useful explanatory variable for toxicology studies. The size of a nanoparticle in the dry state changes in a solvent, and may vary among solvents. There are also at least three different types of diameters that can be measured: primary particle size, hydrodynamic diameter, and aerodynamic diameter (NIST 960-1, 2001; Allen, 2004). Although these measurements provide valuable information regarding the nanoparticle's physical behaviour, additional particle parameters of relevance to toxicology studies are discussed below.

2.2   Chemical Composition

In the primary phase, the composition is one property of the nanomaterial that defines its function. Some properties are shared among nanomaterials of different chemical composition, but there is always an optimal structure for the desired purpose. Spectroscopy provides valuable information about the chemical composition of a nanomaterial. Many different types of spectroscopic measurements can be applied to each nanoparticle sample. Raman spectroscopy is used to determine type and degree of functionalization on the side-wall of a carbon nanotube. Absorption spectroscopy can provide data on the size of gold nanoshells; absorbance is red-shifted (decreased wavelength) as the thickness of the shell increases. Fluorescence spectroscopy can help to elucidate the functionality of quantum dots: the longer the quantum dot fluoresces, the greater the semiconducting effect.

2.2.1   Relation with Toxicity

Depending on the material's chemical composition, nanomaterials released into the environment could be a source of water pollution or soil contamination. For example, a particle composed of heavy metals could eventually leach metal cations into water after prolonged exposure to sunlight and humidity. This scenario is especially relevant for particles with modified surfaces. When 'capping agents' (covalently bound water-solubilizing groups) are chemically added to a nanoparticle surface, the resultant nanoparticle system is made increasingly more bioavailable. A bioavailable particle system then has the potential to be degraded in the environment or biotransformed in organisms, assuming there is biological uptake. Environmental degradation could include surface degradation and eventual particle degradation, potentially leaching heavy metal cations or hazardous inorganic/organic anions into the environment. However, it is important to realize that the environmental effects of the chemical composition (on a mass/volume basis) of nanoparticles should be similar to that of the analogous fine particle, assuming that chemical composition is the sole factor determining effects.

2.3   Surface Chemistry

In the primary phase, surface chemistry is a function of the nanomaterial's surface chemical composition and structure. Even though the concentration of surface groups is greater on nanoparticles than on fine-sized particles, it is still difficult to identify and quantify nanoparticle atomic composition due to typically limited sample size. Methods such as X-ray photoelectron spectroscopy (XPS), electron spectroscopy for chemical analysis (ESCA), and energy dispersive spectroscopy (EDX) are used for determining the chemical composition of a nanoparticle's surface layer (Ratner, 1996).

Much like how the functionality of a nanomaterial is governed by its chemical composition, the surface chemistry governs a material's affinity, that is, the initial interactions with biotic and abiotic matrices. By changing the surface chemistry, a variety of physical and chemical properties are also changed. These properties include solubility, catalysis, charge and adsorption/desorption. In particular, biomolecules in *in vivo* and *in vitro* systems can adhere to the surface of a nanoparticle and potentially dramatically change the particle's function.

2.4 Size and Size Distribution

The distribution of size within a nanoparticle sample is another engineered property. A model nanoparticle sample would be a monodisperse population of particles that corresponds to an optimized property which is size-dependent. The potential control available over particle size and shape during production facilitates small particle size distributions (e.g. coefficient of variation <5 %). As production techniques and sizing methods become more sophisticated, this size distribution will eventually be eliminated to provide particles of a single diameter. Primary versus secondary characterization of particle size requires different techniques. Electron microscopy is one method used to conduct primary characterization of particle size. Other methods include differential mobility analysis (DMA), time of flight mass spectroscopy (TOF-MS), and specific surface area (SSA) measurements. These measurements require the sample to be in the dry state. DMA is used on aerosolized nanoparticle samples in the dry phase under a nitrogen purge or dry air. SSA measurement using BET (Braunauer, Emmett and Teller) methodologies is used to attain an area measurement that can then be converted (through stoichiometry) to a primary particle size (Brunauer *et al.*, 1938). The accuracy of BET measurements increases with increased sample size (>10 mg). Since this method is non-destructive, most researchers measure their entire sample. Nitrogen, argon, carbon dioxide or krypton are absorbed onto the surface of a few aerosolized particles from the sample vial where the instrument then applies the Langmuir theory and BET equations to the monodispersed layer. There is an inverse-squared relationship between surface area and radius of the nanoparticle. As the radius of the nanoparticle decreases, the surface area increases exponentially. The SSA for engineered nanomaterials is usually hundreds of square meters per gram (reported as m^2/g). Some have proposed that properties such as the generation of reactive species and rate of dissolution can be enhanced by particles with large surface areas (Powers *et al.*, 2006).

In the secondary phase, measuring the size and size distribution of a nanoparticle sample in the wet phase is analogous to measuring its state of dispersion. The dispersion of a nanoparticle system can be reported as the number or mass of monodispersed single particles in a solvent compared with olydispersed agglomerates (Powers *et al.*, 2006). The formation of agglomerates is partly driven by particle–particle interactions known as Van der Waal's forces. Agglomerations can also be formed due to increasing the particle concentration in the solvent or changing the pH or ionic strength of the solvent (see Guzman *et al.*, 2006, for an example of the effects of particle surface potential and solution pH on agglomeration).

Lastly, the solutes present (including biomolecules) can increase particle–particle interactions. Because the potential for agglomerate formation increases once nanoparticles are introduced to a solvent, much effort may frequently be put into dispersing the agglomerated particles back into a monodispersed system. Sonication and the addition of salts or surfactants have been used in attempts to achieve this phenomenon (Jarvis *et al.*, 2005; Powers *et al.*, 2006). However, the opposite phenomenon has also been reported with natural organic matter (NOM) causing disaggregation of C_{60} crystals and aggregates, leading to changes in particle morphology and particle size under solution conditions similar to natural water (Xie *et al.*, 2008).

One of the most relevant techniques for toxicological evaluations is manually measuring particles from micrographs taken from either the transmission electron microscope (dried or cryogenic) or the scanning electron microscope. A variety of software programs (e.g. ImagePro, Media Cybernetics, Inc., Bethesda, Maryland, USA, or NanoSight, NanoSight Ltd., Salisbury, Wiltshire, UK) exist to aid in this time-consuming task. An alternative method is dynamic light scattering (DLS). Dynamic light scattering is a technique that can be used to determine the size distribution profile of small particles in solution. When light hits particles in solution or suspension that are less than 250 nm in size, the light scatters in all directions, known as Rayleigh scattering. The resultant scattered light interferes with the surrounding particles in a constructive or destructive manner. The light intensity fluctuation gives information about the aggregation and disaggregation of the particles over time (Koppel, 1972).

In the tertiary phase, just as particle characteristics differ between dry versus wet states, particle size and size distributions can differ once the nanomaterial is in a biological system. The difficulty in measuring changes in particle size and size distribution *in vivo* is partly due to the non-trivial techniques. To circumvent these difficulties, it is possible to use simulated biological fluids *in vitro* to evaluate particle size and size distributions *in vivo* (Sayes *et al.*, 2007b).

2.4.1   Relation with Toxicity

As size decreases (and surface area increases), the particle's settling velocity decreases, thus increasing its mobility, potential transport, and bioavailability in the environment. This mobility could translate to passive transport across cellular membranes. Further, there is speculation that some particular nanoparticle sizes, size ranges, or aspect ratios may transport across membranes more readily or disrupt normal macrophage phagocytosis. The nanoparticles that cross membranes will then be able to interact with DNA, RNA, the nucleus, and other organelles. If these interactions take place, initiation and promotion (the two preliminary steps in carcinoma) may be likely.

2.5   Morphology

One technique that characterizes both size and crystallinity is X-ray diffraction (XRD). Nanocrystals diffract X-rays in unique ways and their unique structure influences chemical properties. An X-ray diffraction pattern cannot exist for an amorphous sample, but in a crystalline sample both phase and grain size can be determined.

An ideal nanocrystalline sample is composed of nanoparticles that are highly uniform in size and shape. Typically, the size and shape of the nanoparticle sample is determined via microscopy. By taking multiple micrographs of the nanoparticle sample, an average size and size distribution can be acquired; however, thousands of nanoparticles must be measured to find the average for a standard distribution. Microscopy methods used in sizing nanoparticles in solutions include optical microscopy, scanning electron microscopy (SEM), transmission electron microscopy of the dried sample (dried-TEM), and transmission electron microscopy of a cryogenically flash-frozen sample (cryo-TEM, should be used as a secondary phase characterization). Dried-TEM is also one of the only methods in which the molecular structure of the sample can be identified and offers the best sizing resolution (0.4 nm). However, this method, as the name implies, is a micrograph of the sample in its dried state. This is not an accurate representation of the nanoparticle in solution/suspension, as experienced by *in vivo* or *in vitro* biological systems. Cryo-TEM produces a more accurate micrograph of what the biological system encounters when exposed to a nanoparticle sample. The sample's molecular structure cannot be determined, nor is the sizing resolution as precise as dried-TEM, but the potential for aggregation may be revealed and samples can be analysed in the solvent of interest.

2.5.1   Relation to Toxicity

A nanomaterial's shape, crystallinity and anatomical structure can influence its toxicological response. For rod, tube, needle and ribbon nanostructures, the large aspect ratio could puncture membranes, thus causing toxicity. Further, macrophage phagocytosis is compromised due to the difficulty in the cell's ability to clear the entire particle from airways, the gastro-intestinal tract or liver. A crystalline material, when compared with an amorphous material, is a highly ordered structure that may possess reactivity reminiscent of asbestos or silica – both highly crystalline materials, as well as highly reactive compounds. A nanomaterial with a core–shell anatomy will have unique properties that are different from a particle with a core and no shell. Depending on how the shell is attached to the core (micellular, covalent bonds, ionic strength), the shell may protect the environment from its nanoparticle core or may dissociate when the medium changes. Changes in medium that may dissociate a nanoparticle shell from its core include altering pH, ionic strength, heat and pressure.

2.6   Concentration and Purity

In the secondary phase, exposures during toxicological studies typically use nanomaterials as either a solid powder or a solution/suspension. For solutions/suspensions, concentration is determined by dispersing a known mass of sample into a known volume of solvent. Cryo-transmission electron microscopy is the method of choice for measurement of particles in solution. Traditional methods, such as inductively coupled plasma atomic emission spectroscopy (ICP-AES) can be used analytically to detect trace metals in the nanoparticle solution or suspension. Simply, it is a type of emission spectroscopy that uses an inductively coupled plasma to excite atoms and ions that inevitably emit electromagnetic radiation at wavelengths particular to its parent element (Mermet, 2005; Stefánsson *et al.*, 2007). The intensity of this radiation corresponds to the concentration of the element within the sample.

Toxicological studies with nanoparticles should be conducted at relevant environmental concentrations (including within water solubility) unless data are needed strictly for classification and labelling, and nanoparticle toxicity should be benchmarked against the toxicity of analogous fine-size particles (Apte *et al.*, 2007). Much of the concern with nanomaterials is due to the hypothesized 'novel toxicity or effects' based on 'nano' characteristics relative to fine-size particles. However, 'nanoparticles' have always been a part of the natural environment, and differences in behaviour, including toxicity, are likely to be substance-specific since a number of studies have reported little or no difference in effects between certain nanoparticles and analogous fine-size particles (Heinlaan *et al.*, 2008; Petersen *et al.*, 2008).

2.7 Surface Activity

In the secondary phase, the activity (or reactivity) of the particle's surface can be measured using a variety of techniques, although no one technique can be used for all nanomaterials. One metric of surface activity of a material is photocatalytic degradation (Wahi *et al.*, 2005). Another metric is delta b* using the vitamin C assay (Warheit *et al.*, 2007). Surface charge is also a relevant metric of a material's surface chemistry and can be quantified using zeta potential measurements (Sayes *et al.*, 2007a).

Perhaps the most important parameter in nanoparticle surface chemistry is the fraction of surface coating. On average, a 100 mg nanoparticle sample could be coated up to 30 % by weight with a capping agent. A secondary parameter of interest is the stability of the surface coatings. The purposes of a coating are diverse. In the medicinal and pharmaceutical field, surface coatings can potentially direct a drug to a particular cell or location within the body. In environmental conditions, the coating of a nanomaterial's surface could prevent ions from dissociating from the material's core. In all fields, the addition of a water-soluble surface-capping agent could transform a hydrophobic nanocrystal into a hydrophilic material that will not aggregate or precipitate out of solution.

Surface charge can aid in the determination of a material's dispersion characteristics. Zeta potential measurements can estimate the charge on a particle (Adamson and Gast, 1997). Zeta potential, in a colloidal system, is the difference in potential between the surface of a material dispersed in a suspension and the dispersion medium. It is related to particle stability in that, in some cases, if particle stability decreases, then surface charge may change. Further, variation in surface charge could alter the dissolution properties of a nanocolloid.

Therefore it is a useful parameter in coagulation operations. The point where the zeta potential of a particle solution/suspension is zero is the isoelectric point (IEP). Light scattering electrophoresis is commonly used to measure the zeta potential of a nanoparticle system.

One measurement used to determine the solubility of the material in the aqueous phase is conductance. This measurement, reported in Siemens (S), is a rapid method of estimating the dissolved solids content of a solution. Conductance is an indirect measure of ionic strength, which is a contributor to measurement of surface charge and zeta potential and influences the double layer. As ionic strength increases, there is an increase in conductance and a concomitant reduction in the zeta potential.

In the tertiary phase, a limited number of techniques utilizing chemical analyses can be used to determine particle reactivity *in vivo*. By monitoring the interactions between nanoparticles and proteins, changes in the surface, structure and reactivity of the particle-adsorbed protein can give valuable information at the nano–bio interface (Vertegel *et al.*, 2004; Jiang *et al.*, 2005). Some of these techniques include electron spin resonance, electron microscopy and mass spectroscopy.

Electron spin resonance (ESR), also referred to as electron paramagnetic resonance (EPR), has been shown to provide information about a nanoparticle sample's reactive species concentration. Because ESR spectroscopy is commonly used for studying chemical species that have unpaired electrons, the presence and concentration of free radicals in carbon-based nanostructures or in complexes with transition metals can be observed, even at low parts-per-billion concentrations. Electron microscopy (EM) can be used to visualize the interactions between a nanoparticle's surface and its surrounding environment. In some cases, pristine nanocrystals in an ultrapure aqueous environment tend to aggregate and 'clump' together. However, once the same particles are mixed with biological or environmental emulsifying agents, such as proteins or dissolved organic matter, nanocrystals may exist as individual particles in suspension. EM is a powerful tool used to determine these phenomena.

2.7.1   Relation to Toxicity

If a nanomaterial's surface is highly reactive in an aqueous environment, then its potential to generate reactive species (RS) is great. RS generation is one of the most anticipated side-effects of nanomaterial release into the environment. Some speculate that the toxic concern should not be placed on the nanomaterial itself, but on its potential reactions with its environmental surroundings yielding other reactive species. The reactive species could include hydroxyl radicals, the super oxide anions, and leached heavy metals and ions. RS production increases the likelihood of oxidative damage (membrane, DNA) to a cell. It should also be realized, however, that generation of RS may also be a mechanism of action for the toxicity of non-nanoparticles and so is not a phenomenon unique to nanoparticles.

2.8   Particle Translocation

Various microscopy techniques can be used to image nano-bio interfaces. Liquid atomic force microscopy (liquid-AFM) can examine the interaction of the aggregated or individually solubilized nanoparticles with natural or artificial membranes. Imaging in the aqueous phase allows a more accurate determination of the interaction between the nanoparticle and a membrane.

Information about the nanoparticle's behaviour in ecological systems can be obtained in cell-free environments. Characterization data and functionality information for nanomaterials suspended in ultra-pure water, environmentally relevant water (groundwater, surface water, freshwater and saltwater) and buffers may all differ from each other. It is possible for biomolecules (peptides or proteins) and natural organic matter to absorb onto the surface of the particle, thus increasing the hydrodynamic diameter of the nanoparticle and making it more bioavailable. Biomolecules also may provide an environment for more favorable generation of reactive species. Other dissolved components of various matrices (e.g. salts present in water) will differ among aqueous solutions and will alter the pH and ionic strength of the nanoparticle solution/suspension.

2.9   Conclusions

The toxicology of nanomaterials is a complex function of their structure and chemistry. Properties of nano-sized materials are different from properties of particles in their bulk form. The difference in physico-chemical properties of particles with different sizes but similar chemical composition has implications for risk assessments, handling and disposal guidelines, and applications development in both material science and engineering, and in biology. More data are necessary to identify relations between chemistry and toxicology.

The structure of the nanomaterial is as important as its application. In many examples, the structure defines the function and, more importantly, the material's activity. This activity can be probed in biology, electronics, or in industrial processes. No matter the chemical composition, manipulating its structure manipulates the function of the nanomaterial.

Certain rules must be engaged when determining the toxicity of nanomaterials. Firstly, nanomaterials are likely to have different chemical and physical properties from bulk materials of identical composition. It is then reasonable to expect that the biological properties of nanomaterials may be different from those of bulk materials. Secondly, manufactured nanomaterials may be unique in environmental or biological systems. The properties of a nanomaterial in non-polar solvents change when extracted into the aqueous phase; further, nanomaterials in biological fluids (stream water, buffered solutions, or cell culture media) may also behave differently. The nanoparticle surface is the part of the nanomaterial system that will have direct interactions with the biological or environmental system; therefore, the surface of the nanoparticle will influence any biological response. Lastly, a full characterization profile must be developed for any nanomaterial system used for toxicological testing. The 'model' nanoparticle rarely exists. The full character of the nanomaterial sample must be known in order to relate characteristics to biological effects within an individual study, and over time for multiple experiments.

Each technique has limitations, non-trivial sample preparation requirements, and appropriate controls; investigators should be mindful of the many parameters that change as solvent conditions, concentrations and surface chemistries change. The successful development of safe nanomaterials requires a strong collaborative effort between toxicologists, physical scientists and engineers. One important message is that no single technique may accurately describe a specific property of a material, and that characterization data should be validated using multiple techniques to facilitate evaluation of variability among techniques and studies.

Acknowledgements

CMS thanks the Department of Veterinary Physiology and Pharmacology at Texas A&M University for support of this work. CMS and DBW would like to thank the DuPont Haskell Global Centers for continuing support of nanotoxicology research.

References

Adamson AW, Gast AP (1997) *Physical Chemistry of Surfaces* (6th edn). John Wiley & Sons, Inc.: Chichester.
Allen T (2004) *Particle Size Measurement, Vol. I: Powder Sampling and Particle Size Measurement* (5th edn). Chapman & Hall: London.

Apte SC, Batley GE, Gadd GE, Casey PS (2007) Comparative toxicity of nanoparticulate ZnO, bulk ZnO, and ZnCl2 to a freshwater microalga (*Pseudokirchneriella subcapitata*): the importance of particle solubility. *Environ Sci Technol* **41**: 8484–8490.

Brown DM, Wilson MR, MacNee W, Stone V, Donaldson K (2001) Size-dependent proinflammatory effects of ultrafine polystyrene particles: a role for surface area and oxidative stress in the enhanced activity of ultrafines. *Toxicol Appl Pharmacol* **175**: 191–199.

Brunauer S, Emmett PH, Teller E (1938) Adsorption of gases in multimolecular layers. *J Am Chem Soc* **60**: 309–319.

Donaldson K, Li XY, MacNee W (1998) Ultrafine (nanometer) particle mediated lung injury. *J Aerosol Sci* **29**: 553–560.

Donaldson K, Brown D, Clouter A, Duffin R, MacNee W, Renwick L, Tran L, Stone V (2002) The pulmonary toxicology of ultrafine particles. *J Aerosol Med* **15**: 213–220.

Guzman KAD, Finnegan MP, Banfield JF (2006) Influence of surface potential on aggregation and transport of titania nanoparticles. *Environ Sci Technol* **40**: 7688–7693.

Heinlaan M, Ivask A, Blinova I, Dubourguier HC, Kahru A (2008) Toxicity of nanosized and bulk ZnO, CuO and TiO2 to bacteria *Vibrio fischeri* and crustaceans *Daphnia magna* and *Thamnocephalus platyurus*. *Chemosphere* **71**: 1308–1316.

Jarvis P, Jefferson B, Gregory J, Parsons SA (2005) A review of floc strength and breakage. *Water Res* **39**: 3121–3137.

Jiang X, Jiang J, Jin Y, Wang E, Dong S (2005) Effect of colloidal gold size on the conformational changes of adsorbed cytochrome c: Probing by circular dichroism, UV-visible, and infrared spectroscopy. *Biomacromolecules* **6**: 46–53.

Koppel DE (1972) Analysis of macromolecular polydispersity in intensity relation spectroscopy: the method of cumulants. *J. Chem. Phys.* **57**: 4814–4820.

Mermet JM (2005) Is it still possible, necessary and beneficial to perform research in ICP-atomic emission spectrometry? *J Anal At Spectrom* **20**: 11–16.

NIST 960-1 (2001) *NIST Recommended Practice Guide, Particle Size Characterization*. National Institute of Standards: Gaithersburg, MD.

Oberdorster G, Ferin J, Gelein R, Soderholm SC, Finkelstein J (1992) Role of the alveolar macrophage in lung injury-studies with ultrafine particles. *Environ Health Perspect* **97**: 193–199.

Petersen EJ, Huang Q, Weber WJ (2008) Bioaccumulation of radio-labeled carbon nanotubes by *Eisenia foetida*. *Environ Sci Technol* **42**: 3090–3095.

Powers KW, Brown SC, Krishna VB, Wasdo SC, Moudgil BM, Roberts SM (2006) Research strategies for safety evaluation of nanomaterials. Part VI. Characterization of nanoscale particles for toxicological evaluation. *Toxicological Sciences* **90**: 296–303.

Ratner BD (1996) *Biomaterials Science: An Introduction to Materials in Medicine*. Academic Press: San Diego, CA.

Sayes CM, Marchione AA, Reed KL, Warheit DB (2007a) Comparative pulmonary toxicity assessments of C60 water suspensions in rats: few differences in fullerene toxicity in vivo in contrast to in vitro profiles. *Nano Letters* **7**: 2399–2406.

Sayes CM, Reed KL, Warheit DB (2007b) Assessing toxicity of fine and nanoparticles: Comparing in vitro measurements to in vivo pulmonary toxicity profiles. *Toxicological Sciences* **97**: 163–180.

Stefánsson A, Gunnarsson I, Giroud N (2007) New methods for the direct determination of dissolved inorganic, organic and total carbon in natural waters by Reagent-Free Ion Chromatography and inductively coupled plasma atomic emission spectrometry. *Anal Chim Acta* **582**: 69–74.

Tran CL, Buchanan D, Cullen RT, Searl A, Jones AD, Donaldson K (2000) Inhalation of poorly soluble particles. II. Influence of particle surface area on inflammation and clearance. *Inhalat Toxicol* **12**: 1113–1126.

Vertegel AA, Siegel RW, Dordick JS (2004) Silica nanoparticle size influences the structure andenzymatic activity of adsorbed lysozyme. *Langmuir* **20**: 6800–6807.

Wahi RK, Yu WW, Liu Y, Mejia ML, Falkner JC, Nolte W, Colvin VL (2005) Photodegradation of Congo Red catalyzed by nanosized TiO_2. *Journal of Molecular Catalysis A: Chemical* **242**: 48–56.

Warheit DB, Webb TR, Reed KL, Frerichs S, Sayes CM (2007) Pulmonary toxicity study in rats with three forms of ultrafine-TiO_2 particles: differential responses related to surface properties. *Toxicology* **230**: 90–104.

Xie B, Xu Z, Guo W, Li Q (2008) Impact of natural organic matter on the physicochemical properties of aqueous C_{60} nanoparticles. *Environ Sci Technol* **42**: 2853–2859.

3

Considerations for the Design of Toxicity Studies of Inhaled Nanomedicines

Lea Ann Dailey

3.1 Introduction

Mirroring the current trends in particle engineering in other sectors, nanotechnology is rapidly finding its way into a wide range of healthcare applications. Wound dressings, bone cements, catheter tubes, skin creams and even athletic socks have been functionalised to exhibit antimicrobial properties through incorporation of silver nanoparticles. Quantum dots and ultra-small iron oxide nanoparticles are finding use as new imaging contrast agents in diagnostic applications. The biocompatibility of implants and the performance of surgical tools have been improved by nanostructured surface modifications (Costigan, 2006). Yet perhaps even more revolutionary and exciting is the potential of nanotechnology to improve upon the way we currently treat disease.

Abraxane®, a product that has taken a relatively old and problematic chemotherapeutic, paclitaxel, and repackaged it within albumin nanoparticles is an excellent example of the great potential of 'nanomedicines'. Due to its extreme insolubility, paclitaxel is notoriously difficult to formulate into a medicine that can be administered to patients easily. Early formulations required the use of Cremophor EL®, a polyethoxylated castor oil, and ethanol to produce an injectable emulsion of the drug. The high toxicity of Cremophor EL® limited the applicable dose a patient could receive in one treatment. Abraxane® is vastly superior to these earlier formulations because of the natural affinity of paclitaxel to the human serum protein, albumin, which allows the incorporation of the drug within the nanoparticles

Nanotoxicity: From In Vivo *and* In Vitro *Models to Health Risks*   Edited by Saura Sahu and Daniel Casciano
© 2009 John Wiley & Sons, Ltd

and eliminates the need for toxic excipients. This in turn makes it possible to administer higher, more effective doses of the chemotherapeutic in one injection. Additionally, the albumin nanoparticles are small enough ($\sim$130 nm) to enter into tumour tissue through the leaky, fenestrated tumour vasculature, whilst remaining too large to enter healthy tissue with an intact endothelial barrier (Hawkins et al., 2008). This effectively leads to passive tumour targeting with an accompanying reduction in adverse side-effects at a higher chemotherapeutic dose.

With the success of Abraxane®, the hopes of developing new successful nanomedicines run high. In recent years, the number of research publications describing new materials, delivery systems, application routes and therapeutic indications for nanomedicines has exploded, and many preliminary studies have shown promising early-stage results. Perhaps rightly so, the sole aim of the majority of these studies has been to establish proof-of-principle on a case-by-case basis for the particular nanosystem of interest. Only through success in this first objective does a potential nanomedicine have any chance of attracting the necessary interest and funding power of the pharmaceutical industry required for development on the long road towards becoming a fully approved medicine.

Unlike many other nanotechnology applications, nanomedicines by virtue of their status as medicinal products are already subject to extremely rigorous regulation with respect to their safety and efficacy. Within this regulatory framework, it is the sole responsibility of the pharmaceutical company to produce sufficient toxicological information on a new medicine to convince the regulatory bodies that the medicine is safe for human use. This unique relationship between the pharmaceutical industry and drug regulatory agencies, which is usually highly effective at safeguarding public health when it comes to healthcare products, has the consequence that public research money is rarely ever directed towards independent medicines toxicology studies, in contrast to the vast amounts of publicly funded research carried out in the environmental, occupational and consumer health sectors. In part, this may explain why the number of nanotoxicology publications dealing with the associated health risks of environmental nanoparticle exposure is overwhelming, whilst the number of research publications studying the toxicological impact of nanomedicines is pitifully small. The relative gap is even more extreme in the specific area of respiratory nanotoxicology, where the health risks of inhaled ambient particulate matter, especially the class of ultrafine particles ($<$100 nm), has been extremely well documented over the past two decades, while the number of publications investigating the possible impact of inhaled nanomedicines on respiratory health can virtually be counted on one hand.

One encouraging result of the steadily increasing public awareness of nanotoxicology issues is the interdisciplinary cross-talk of the past few years. This has been demonstrated in the number of excellent reviews published since 2006, which seek to bridge the gap between the pharmaceutical sciences and various disciplines of nanotoxicology (Hoet et al., 2004; Costigan, 2006; Scientific Committee on Emerging and Newly Identified Health Risks, 2006; Buzea et al., 2007; Sung et al., 2007; Bailey and Berkland, 2008; Card et al., 2008; De Jong and Borm, 2008; Yang et al., 2008). As many experts in nanotoxicology point out, the standardised toxicology assays used routinely in pre-clinical screening of medicines *may* identify potential negative effects of nanomedicines. However, experience has also shown that the toxicological profile of most nanoparticulate systems investigated to date is quite different to that of the bulk material (or even that of the micron-sized particulates of the same material) and requires special considerations. The main considerations include:

- *Surface reactivity*: Although most nanomedicines are designed specifically to be 'inert' and 'biocompatible', and therefore are not likely to exhibit inherently reactive surface properties themselves (in contrast to combustion-derived particulates), their high surface-area-to-mass ratios still make them far more vulnerable than bulk materials to an increased surface adsorption of reactive impurities, which if not carefully controlled could impart a surrogate surface reactivity to the nanomedicine.
- *Altered tissue and cellular distribution*: As demonstrated by Abraxane®, nanomedicines may intentionally seek to access tissues that larger particulates cannot. In some cases, the aim of a nanoparticulate drug carrier system may be to gain access to individual cells or even cell organelles, for example in non-viral gene delivery applications. Whilst much effort has been expended into increasing the uptake efficiency of such nanocarriers into their target cells, virtually nothing is known about the effects of the carrier system on the cell's internal organisation or of the effects of the biodegradation products on the cell. Secondly, few studies have been able to assess the biodistribution of drug delivery nanoparticles *in vivo* after administration by different routes. Unintended consequences of the administration of nanomedicines designed for efficient cellular uptake may also include their ability to accumulate in organs not intended for therapy, to cross the blood–brain barrier, or to influence blood coagulation pathways.

3.1.1 Chapter Aim

The aim of this chapter is to discuss respiratory nanotoxicology as it applies to current research on inhaled nanomedicines. The intention is not to provide a review of the current literature, but rather to expand upon important concepts that may shape future work in this area. These concepts can be addressed by taking a closer look at the following three questions:

1. What are inhaled nanomedicines designed to accomplish and how does their structure contribute to this aim?
2. What are the primary physico-chemical characteristics of inhaled nanomedicines that distinguish them from particles commonly studied in environmental and occupational respiratory toxicology?
3. What are the likely implications of differences in exposure of inhaled nanomedicines vs. that of ambient or manufactured nanoparticles on respiratory health?

3.2 What are Inhaled Nanomedicines Designed to Accomplish and How Does their Structure Contribute to this Aim?

3.2.1 The Quest for Lung Retention

The rationale for the design of inhalable nanomedicines is primarily the attainment of a suitable retention time for a specific drug in the lung. With the ability to control the residence time of a compound in the lung, a number of beneficial outcomes may be attained. These are summarised in Table 3.1.

Table 3.1   *Beneficial outcomes of controlling the lung residence time of a drug.*

Beneficial outcome	Example
A reduction in adverse systemic side effects may be achieved	Inhaled corticosteroids
A more effective dose of a locally acting compound may be presented to the diseased area of the lung	Inhaled antibiotics
The controlled release of a systemically acting compound may be accomplished	Inhaled pain killers
The targeting of specific cell types or areas of the lung may be possible	Inhaled chemotherapeutics or imaging agents
A reduction in dose may be achieved resulting in overall reductions in the costs of therapy	All

Sources: Hochhaus *et al*. (1998); Kelly (2003); Sung *et al*. (2007).

Lung retention is surprisingly a major issue for most inhaled medicines, regardless of whether the drug is destined for a local or systemic application. The reason for this is the extremely high permeability of the lung to a majority of compounds as a result of its immense absorption surface area and comparatively benign metabolic defence system. This high lung permeability was recently exploited by the makers of Exubera®, the first inhaled insulin. With Exubera® they showed that the rapid systemic delivery of insulin, a polypeptide of 5.7 kDa, was possible, providing a non-invasive delivery system for meal-time or 'bolus' insulin. What could not be achieved with current inhaler and formulation technology was the pulmonary delivery of a slow-releasing form of insulin, known as 'basal' insulin, which is essential for the regulation of glucose production in the liver and must be continually present in the body throughout the day at very low levels. This continued need to inject basal insulin on a daily basis is one of the several causes for Exubera's® poor economic performance and ultimate withdrawal from the market (Black *et al*., 2007).

The quest for non-invasive basal insulin delivery is not the only example that drives research in controlling the lung retention times of drugs. This issue lies at the core of nearly every drug already delivered to the lungs. From this unmet clinical need, the idea of using inhaled nanomedicines emerged as a possibly unique solution to the lung retention problem. Importantly, nanocarriers also convey the added advantage of protecting sensitive compounds from enzymatic degradation once in the lung, especially if lung residence time is increased significantly. But why focus on nanoscale drug carriers in particular?

Through efforts to develop long-circulating particulate systems for parenteral applications, data emerged showing that the greater the decrease in particle diameter <1 μm, the more successful the system was at evading clearance by the mononuclear phagocytic system (MPS); in general, the longest circulation times were achieved using particle diameters <100 nm. Particle surface chemistry also seemed to play a major role in slowing clearance by the MPS. Particles coated with hydrophilic polymers such as polyethylene glycol (known as 'stealth' systems) could be shown to reduce the surface adsorption of serum opsonin molecules (immunoglobulins, complement proteins, apolipoproteins, von Willebrand factor, thrombospondin, fibronectin, and mannose-binding protein), thus reducing opsonic activity and MPS clearance rate (Moghimi *et al*., 2001).

Following this train of thought, it was postulated that inhaled nanoparticulate drug carriers may retard alveolar macrophage clearance according to the same principle. Studies from the environmental sciences had already demonstrated the capability of various non-soluble, non-biodegradable nanoparticles to be retained in the lung for significantly long periods of time, in some cases up to several months.

3.2.2   What Do Nanomedicines Look Like?

Nanomedicines designed for inhalation therapy may comprise the drug alone – in the form of a nanocrystal or amorphous nanoparticle – or comprise a drug/excipient mixture, which can take many different forms (Table 3.2). Inhaled nanomedicines typically range from 50–500 nm in diameter. In contrast to most nanotechnology-related disciplines, which define nanoparticles as particles with at least one dimension <100 nm, nanosystems in the pharmaceutical sciences have traditionally been defined as structures with a diameter between 1–1000 nm. Of the various nanoparticle systems described in Table 3.2, those that bear the closest resemblance to the 'inert' reference particles investigated in many of the respiratory nanotoxicology studies published in recent years (e.g. polystyrene and titanium dioxide) are the polymeric and solid lipid nanoparticles. As a natural consequence, much of the discussion in the following sections will focus on these systems. Table 3.2 also provides an overview of biomaterials commonly used as excipients to manufacture the various classes of nanomedicines currently under investigation for pulmonary drug delivery purposes.

Alone, these nanoparticle systems would not have the requisite aerodynamic properties to achieve the high lung deposition needed for therapeutic use. As a result, they must be administered to the lung either as a nebulised suspension or as dry powder. Recently, several groups have engineered various sophisticated nanoparticle-containing microparticle systems, which combine the excellent aerodynamic properties of microparticles with the controlled release properties of nanoparticles. With imaginative names such as 'Trojan particles' or tiny 'cluster bombs', these microparticle systems are designed to release their drug-laden nanoparticle payload once they come into contact with the respiratory tract lining fluid (RTLF). The means of nanoparticle release from the microparticles range from simple dissolution of the water-soluble carrier material through to effervescent disintegration (Sung *et al.*, 2007; Smola *et al.*, 2008).

3.2.3   Do Nanocarriers Improve Drug Delivery to the Lung?

Within the past ten years, progress made in the field of nanoparticulate pulmonary drug delivery has been steady. Indicative of this upwards trend is the increasing number of studies reporting *in vivo* proof-of-principle in animal models. In a few cases involving liposomes, encouraging results in early-phase human trials have been reported as well. The following section will highlight selected examples of successful designs in inhaled nanomedicines according to the various drug carrier classes.

(1) *Liposomes*: Liposomal formulations designed to reduce drug side-effects, control drug release or target specific cells in the lung are by far the most advanced nanoscale drug delivery systems currently under investigation. They are the only inhaled nano-vehicles to have been studied more widely in early-phase human trials to date. These studies

Table 3.2   *Structural representations of common classes of inhaled nanomedicines and materials used in their manufacture.*

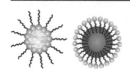

Micelles
   Materials: polymers/copolymers (poloxamers, polyesters, poly(L-amino acid)s, polyproplylene oxides), PEG-conjugated phospholipids and dendrimers, polyoxyethylated castor oil

Liposomes
   Materials: phospholipids (natural and synthetic), cholesterol, stabililisers optional (polyethylene glycol; PEG)

Solid lipid nanoparticles
   Materials: high melting triglycerides (trilaurin, tribehenate), paraffins, oils, waxes, emulsifiers (poloxamers, lecithins, polysorbates, polyethoxylated monoglycerides)

Polyplexes (and lipoplexes)
   Materials: cations – polyethylenimine (PEI), polylysine, chitosan, cationic peptides, cationic lipids (DOTAP)

Polymeric nanocapsules and nanoparticles
   Materials: poly(lactide-co-glycolide) (PLGA), polybutylcyanoacrylate, chitosan, albumin, gelatin, poly-ε-caprolactone, alginate, and stabilisers (polyvinyl alcohols, PVA; poloxamers, PEG)

Nanocrystals
   Materials: budesonide

Sources: Costigan (2006); Mozafari (2006); Kumar (2007); Leroueil *et al.* (2007); Radomska-Soukharev (2007); Smola *et al.* (2008); Warheit (2008).

have generally shown that inhaled liposomes are well-tolerated in humans and animals with no signs of overt toxicity, and some have shown promising results in terms of efficacy (Skubitz and Anderson, 2000; Cooke and Drury, 2005; Wijagkanalan *et al.*, 2008). Common to nearly all of the studies administering liposomal formulations via inhalation is the use of vesicles <500 nm in diameter. This is due to the fact that a decrease in liposome diameter corresponds directly to a substantial increase in vesicle structural stability during nebulisation, a process which subjects the formulation to either destructively high levels of shear stress or ultrasonic energy. In this regard, the use of nanovesicles does provide a significant advantage over larger liposomal structures.

(2) *Solid lipid nanoparticles*: Cousins to the liposome family of drug delivery vehicles, solid lipid nanoparticles are a more recent technological advance and consequently fewer *in vivo* studies have been carried out to date. Notably, it has been shown in a mouse model that inhaled solid lipid nanocarriers (~100 nm) can be used to prolong the release of insulin into the blood stream and reduce glucose levels significantly compared with inhaled and subcutaneous free insulin (Liu *et al.*, 2008). Another study of inhaled, radiolabelled solid lipid nanoparticles (200 nm) raised interesting questions about the fate of these drug carriers after inhalation. The particles were shown to undergo rapid lymphatic uptake within minutes, resulting in a significant distribution in the periaortic, axillar and inguinal lymph nodes. It was implied in the discussion that the therapeutic use of inhaled solid lipid nanoparticles may be restricted to applications that benefit from high levels of lymphatic uptake, such as certain forms of chemotherapy (Videira *et al.*, 2002). Further studies are certainly warranted to test the boundaries of this vehicle technology within pulmonary drug delivery.

(3) *Lipo- and polyelectrolyte complexes*: Also known simply as lipo- and polyplexes, these two related classes of nanocarrier are most widely used in non-viral gene therapy research. Whilst conventional plasmid gene delivery to the lung has shown disappointing *in vivo* results in terms of transfection efficiency, the exciting possibilities of gene knock-down therapy for lung disease using small interfering RNA (siRNA) has revitalised the sector. Currently, the most developed systems target respiratory viral infection and have been shown in murine, rat and primate models to be effective against respiratory syncytial virus (RSV) and severe acute respiratory syndrome (SARS). Indications such as inflammatory or immune conditions, cystic fibrosis and cancer are also high on the agenda. It is important to note, however, that similar to the plasmid-based gene delivery studies, inhaled lipo- or polyplexes with siRNA have not shown improved efficacy compared with intranasal delivery routes. Interestingly, some studies have even shown that intranasal delivery of naked siRNA performs just as well as siRNA combined with a vector, perhaps suggesting that this particular form of therapy may circumvent the need for a nanocarrier (Thomas *et al.*, 2007; Durcan *et al.*, 2008).

(4) *Polymeric nanoparticles*: In contrast to siRNA delivery to the lung, the effect of inhaled DNA vaccines is greatly enhance by association with a nanoparticulate carrier. Chitosan, a cationic biopolymer derived from the shells of crustaceans and composed of biodegradable β-(1-4)-linked D-glucosamine and N-acetyl-D-glucosamine units, can be formulated into DNA-binding nanoparticles which have been shown to effectively confer local immunoprotection against *M. tuberculosis* infection after intratracheal

administration to mice. Polymeric nanoparticles encapsulating the first-line antitubular agents, rifampicin, isoniazid and pyrazinamide, have also been found to be effective in eradicating tuberculosis in a guinea-pig model. In this case, three fortnightly doses of the inhaled antitubucular agents encapsulated in 240 nm sized poly(lactic-co-glycolic acid) (PLGA) nanoparticles were therapeutically equivalent to 46 oral doses of the neat drugs (Gelperina *et al.*, 2005).

As the examples above illustrate, a growing body of evidence is being accumulated to show that nanoparticle-based delivery systems for inhalation may confer therapeutic advantages in specific indications. However, in the cautious post-Exubera® regulatory environment, few of these promising technologies may make it past the proof-of-concept stage without a closer look into the potential toxicological implications of the delivery vehicles. To many in the field, it is becoming increasingly clear that an interdisciplinary cooperation between respiratory toxicologists and drug-delivery scientists may be the most effective strategy for tackling this issue. The following sections highlight areas of particular consideration when assessing the potential toxicological profile of inhaled drug delivery nanparticles.

3.3   What are the Primary Physico-chemical Characteristics of Inhaled Nanomedicines that Distinguish them from Particles Commonly Studied in Environmental and Occupational Respiratory Toxicology?

3.3.1   Biodegradation

Although all colloidal drug carrier systems are by definition non-soluble structures (i.e. solid in liquid), and thereby exhibit a surface that may be more or less solid and reactive, they are in nearly all cases designed to dissolve, erode, disintegrate, and ultimately be eliminated from the body. This property is the first to set them distinctly apart from all non-biodegradable particulates investigated in previous respiratory nanotoxicology studies. Yet, does the feature of biodegradability make nanomedicines safer to inhale than their non-soluble, non-biodegradable nanoparticulate counterparts?

With regard to long-term persistence in the lung, then the answer is yes – the biodegradable nanomedicine will eventually be reduced to its components and removed from the lung. The degradation products themselves may be removed from the lung via the normal clearance mechanisms or incorporation into metabolic pathways (for example lactic acid produced during PLGA degradation). The crucial issue of biodegradation with respect to a nanomedicine's potential toxicity profile is the speed at which degradation occurs.

The nanomedicines introduced in Table 3.2 span a nearly continuous spectrum of biodegradation rates as a function of their structure and composition. On the side of nearly instantaneous disintegration are the micellar and liposomal formulations, which are fluid crystalline structures and as such highly vulnerable to structural disruption caused by interactions with lung surfactants or other lung lining fluid components. Although liposomal integrity is in some cases maintained long enough to exert a measureable pharmacokinetic effect, it is widely accepted that the liposomal structure *in vivo* is too transient to elicit an inflammatory response.

Solid lipid nanoparticles, nanocrystals and polymeric nanoparticles (including polymeric complexes, dendrimers and nanocapsules) are structurally more stable than liposomes and bear a greater resemblance to non-soluble, non-biodegradable manufactured/ambient nanoparticles. Due to their biodegradability, it is not expected that these carriers will persist in the respiratory tract long enough to cause chronic inflammation; however, it is valid to question whether or not they are able to elicit an acute inflammatory response upon inhalation (as has been shown for non-biodegradable nanoparticles) and whether this acute response is exacerbated upon repeated administration.

If an acute response were to be defined as the measurement of respiratory inflammation within 24 hours post-administration (i.e. neutrophilic influx, increases in pro-inflammatory chemo- and cytokines, etc.), it is likely that particle degradation will not be a major factor influencing the inflammatory response. In contrast, it is probable that the surface properties of the nanocarrier will determine its acute effects. The speed of vehicle biodegradation will, on the other hand, be critical for determining whether multiple dosing is feasible, and if so, how long the dosing intervals must be to prevent accumulation and chronic inflammation with accompanying structural damage to the tissue.

Surprisingly, there is very limited information published on the degradation kinetics of biodegradable *nanoparticles*, despite a strong rationale for an accelerated degradation rate as particle size decreases. Most investigations examining biomaterial degradation kinetics are carried out with microparticles or the bulk material. An encouraging exception is provided by a study assessing the biodegradation profile of PLGA nanoparticles (size not reported) in phosphate-buffered saline at 37° C, using gel permeation chromatography to quantify the hydrolysis products. Not surprisingly, the nanoparticle degradation rate constant determined in the study was approximately twice that reported for a similar microparticle formulation, although a direct comparison was not made. Perhaps even more interesting was the fact that increasing amounts of encapsulated drug resulted in a corresponding increase in polymer degradation rate, which could be a beneficial effect in lung delivery (Birnbaum and Brannon-Peppas, 2003).

Methods to characterize nanoparticle material degradation include, for example, gel permeation chromatography coupled with refractive index detection and liquid chromatography/mass spectrometry. Atomic force microscopy (AFM) may also be used to evaluate successive morphological changes to the nanomedicine during degradation, as well as to provide quantitative measurements of the particle's surface solidity throughout the process. This is of particular interest in particles that demonstrate hydration and swelling in an aqueous environment, a factor that is certain to influence biodegradation rates in particles of such small size. AFM may also be used to characterise particle degradation (as well as *aggregation*) qualitatively in physiological fluids such as RTLF.

What happens to the degradation products after particle erosion or distintegration? If large molecules such as polymers or proteins have been used to formulate the vehicle, it will be necessary to understand the lung clearance mechanisms of these macromolecules. It has been shown that pulmonary absorbance rates of *polymers* decrease dramatically as the molecular weight increases, with macomolecules >150 kDa remaining in the lung for periods greater than 24 hours (Takada *et al.*, 1978). In contrast, some larger *proteins* have been shown to pass the barrier with relative ease (Schanker and Burton, 1976). Depending on the macromolecule in question, the degradation products may accumulate in the lung lining fluid until they have been sufficiently broken down via hydrolysis or enzymatic metabolism and can pass into the bloodstream. Certain macromolecules may even be internalised by alveolar macrophages, as was shown recently for albumin (Lombry *et al.*, 2004).

It should be noted that many 'non-polymeric' drug delivery vehicles such as solid lipid nanoparticles, drug nanocrystals or stealth liposomes also contain macromolecular polymer stabilisers to prevent aggregation. Although these stabilisers are usually present in minute amounts, the assessment of their individual residence time and potential contribution to toxicity should not be neglected.

3.3.2   Surface Properties

As stated previously, it is likely that acute lung inflammation will be driven less by biodegradability rather than by particle surface characteristics. These surface properties include chemical composition, particle hydration (swellability), charge and hydrophobicity. The combined effects of these properties will contribute to a unique 'adhesion/cohesion profile' for each individual nanoparticle formulation. The 'adhesion/cohesion profile' of a nanocarrier formulation can be thought of as the particulate equivalent of the curriculum vitae. Ideally, it will contain:

(1) *Basic surface information:*

 • chemical composition, including excipients and identification of reactive groups;
 • particle size and polydispersity;
 • surface area;
 • extent of hydration;
 • charge;
 • hydrophobicity.

(2) *Adhesion events:*

 • Identity and quantity of molecules adsorbing to the particle surface during the manufacturing process: transition metals or other reactive species may especially contribute to the surface reactivity of the particle and elicit an inflammatory response.
 • Identity and species of biomolecules adsorbing to the particle surface upon contact with the physiological environment (e.g. RTLF). These may be phospholipids, surfactant proteins, or complement components. The type and extent of biomolecule adsorption to the nanoparticle is likely to be influenced by the individual surface properties and will determine the fate of the particle in the lung by either preventing or contributing to lung clearance via cellular uptake, influencing immunogenicity of the particle, or increasing/decreasing the nanoparticle degradation rate.

(3) *Cohesion events:*

 • Aggregation during manufacturing and administration (e.g. during nebulisation). The consequences of aggregation during manufacturing and administration include material loss, lower administered doses, changes in deposition patterns in the lung, changes in drug release kinetics and thus drug bioavailability.

- Aggregation upon administration to the lung. Aggregation may occur due to changes in surface properties caused by adsorption of biomolecules to the particle surface. The consequences of particle aggregation in the lungs could be a lower particle surface area, increased clearance via phagocytosis, differences in pro-inflammatory and immunogenic potential, and changes in drug release profiles and bioavailability.

Table 3.3 provides an example of what an adhesion/cohesion profile might look like and how it might be used to design more comprehensive studies examining the relationship between specific nanoparticle surface properties and inflammation. The data have been pooled primarily from a study investigating the effects of selected drug delivery nanoparticle vehicles on inflammatory potential in the murine lung. The results of the study suggest that, in contrast to the 'ultrafine' (<100 nm) non-biodegradable polystyrene nanoparticles, the biodegradable particles were not pro-inflammatory (Dailey *et al.*, 2006). The exact mechanism of this difference remains unknown, yet a closer look at the incomplete adhesion/cohesion profile of the particles shows how much potentially important information was not defined in the original study. In future, a full characterisation of surface properties as well as adhesion and cohesion events may help to provide more conclusive evidence of surface-related mechanisms of toxicity.

3.3.3   Size and Surface Area

Particle size and surface area may be the only two properties that do *not* distinguish biodegradable nanomedicines from their manufactured or environmental counterparts. Data collected from several studies with non-biodegradable, 'inert' particles suggest that particle sizes <100 nm and their correspondingly high surface area/mass ratios may alone be enough to trigger inflammation in the lung. This cut-off size is important because it marks the boundary at which nanostructures begin to exhibit radically different characteristics when compared with their bulk materials (Borm and Kreyling, 2004).

The majority of nanomedicines reported in the literature are indeed significantly larger than 100 nm, and as a result may not be included in the 'high risk' category with regard to size and surface area. However, material engineering and manufacturing techniques for drug delivery vehicles are constantly being refined, leading to an increasing number of reports describing the production of 'ultrafine' biodegradable particles. For example, a new technique using surface acoustic wave energy to atomise polymeric solutions was able to produce monodisperse nanoparticles (~100 nm) from the biodegradable polymer, poly-ε-caprolactone (Friend *et al.*, 2008). Even smaller solid lipid nanoparticles with diameters <30 nm have been prepared by a method known as supercritical fluid extraction of emulsions. These particles are in development by Ferro and Aradigm Corporations especially for pulmonary drug delivery purposes (Chattopadhyay *et al.*, 2007).

3.3.4   Purity

One final important discrepancy between nanomedicines and manufactured/ambient nanoparticles is the degree of purity required in nanotoxicology studies. A common requirement of all nanotoxicity studies is that the particles tested should be generated under conditions that accurately reflect their state upon human exposure. For manufactured/ambient nanoparticles this should *include* all relevant impurities that could impact

Table 3.3 Example of a post hoc 'adhesion/cohesion profile' prepared using data from a three studies by Dailey et al. (2003a, 2003b, 2006) in which two biodegradable nanoparticles were tested for an acute pro-inflammatory response in the lung compared to two non-biodegradable standard nanoparticle systems.

Part 1: General Nanoparticle Surface Information

	Biodegradable nanoparticles		Non-biodegradable nanoparticles	
Nanoparticle material and size class	PLGA (ultrafine)	Modified PLGA (fine)	Polystyrene (ultrafine)	Polystyrene (fine)
Excipients	0.005% w/w Alveofact® (Bovine lung surfactant extraction)	0.05% w/w carboxymethyl celluose (CMC)	Undisclosed surfactant stabilizer	Undisclosed surfactant stabilizer
Surface reactive groups	End chain carboxylic groups, no redox reactivity	Hydroxyl groups; carboxylic groups from CMC; no redox reactivity	None	None
Polymer Mw	35 kDa	400 kDa	Mixture	Mixture
Hydrodynamic diameter (nm) in buffer:	83	216	76	220
Hydrodynamic diameter (nm) in RTLF:	~180	~300	~180	~400
Relative surface hydrophobicity in buffer:	0.16	NA	1.0	1.0
Relative surface hydrophobicity in RTLF:		Not determined		
Zeta potential (mV) in buffer:	~ −40	~ −28	~ −25	~ −25
Zeta potential (mV) in RTLF:		Not determined		
Particle hydration in buffer:	Low-Med.	Med. – High	None	None
Redox potential: Buffer or RTLF		Not determined	None	None
Radical formation: Buffer or RTLF		Not determined		

Part 2: Adhesion Events

	Biodegradable nanoparticles		Non-biodegradable nanoparticles	
	PLGA (ultrafine)	Modified PLGA (fine)	Polystyrene (ultrafine)	Polystyrene (fine)
Nanoparticle material and size class				
Adsorption of contaminants during manufacturing			Not determined	
Adsorption of RTLF components			Not determined	

Part 3: Cohesion Events

	PLGA (ultrafine)	Modified PLGA (fine)	Polystyrene (ultrafine)	Polystyrene (fine)
Aggregation during manufacturing	None	Not observed at concentrations used in study.		
Aggregation in RTLF	None		None	Significant

Part 4: Observations and comments

1. *Surface properties:*
   - The surface of the PLGA particles might have been covered to a significant extent with Alveofact®, a natural lung extract, which may have rendered the particles less visible to the lung's clearance mechanisms.
   - The relative hydrophilicity and the greater size (ergo lower surface area) of the modified PLGA particles may have also contributed to their low inflammatory potential.
   - The extremely high hydrophobicity and small size of the ultrafine polystyrene nanoparticles likely contributed to their pro-inflammatory effect.

2. *Adhesion:* As the adhesion data set is incomplete, it is impossible to evaluate whether there is a relationship between surface properties and the adhesion of molecules to the particle surface both *in vitro* and *in vivo*.

3. *Cohesion:* Particle aggregation did occur for the larger polystyrene nanoparticles in RTLF. This might have promoted increased macrophage clearance of these particles and influenced their pro-inflammatory profile.

Sources: Dailey et al. (2003a, 2003b, 2006).

respiratory health. In contrast, nanomedicines investigated for their pro-inflammatory effects must be prepared in a manner that carefully *excludes* all possibilities of contamination or impurities, as this accurately reflects the quality standards required by drug regulatory bodies for medicines manufacturing.

Since most research labs do not have the luxury of a GMP-compliant pharmaceutical manufacturing suite, it is important that measures taken to prevent contaminations during nanoparticle manufacturing and storage are included in the study protocol, are tested and reported. Examples of such measures are the use of pure, contaminant-free raw materials and sterile, endotoxin-free aqueous solutions with a low defined content of transition metals for nanoparticle preparation, in addition to aseptic nanoparticle preparation techniques. Since most drug delivery nanoparticle preparation methods include large amounts of aqueous vehicles during their manufacture and delivery, the opportunities for microbial contamination are very high. Any contamination with gram-negative bacterial-derived lipopolysaccharides will also induce an inflammatory response in the lung and could generate false positive results. Additionally, the use of solutions or buffers containing unknown contents of transition metals could result in significant adsorption of reactive species to the inert nanomedicine surface, imparting a surrogate surface reactivity to the particle. The consequence of reporting artefacts such as these in the literature may be a negative appraisal of such technologies, which could slow or ultimately block development in an area that might lead to better therapeutics of the future.

3.4   What are the Likely Implications of Differences in Exposure Patterns to Inhaled Nanomedicines vs. Manufactured or Ambient Nanoparticles on Respiratory Health?

3.4.1   Chronic vs. Acute Exposure

In terms of dose, human inhalation of ambient particulate matter from the environment is characterised by a chronic exposure to low levels of particles. Exposure to manufactured nanoparticles is often occupationally-related and can vary depending upon the type of manufacturing process, the nature of implemented safety measures, the type of nanoparticles produced and the exposure time of the employee. Inhaled doses are likely to be higher than those of ambient environmental particles, although overall exposure time is likely to be shorter and occur in bursts rather than continuously.

Depending upon the indication to be treated, the exposure to inhaled medicines may be acute *or* chronic, but the nature of exposure will always be characterised by a short duration of relatively high particle concentrations (bursts). A single administration of an inhaled vaccine may be sufficient to generate the desired immune response, with boosters required only after intervals of months or years. At the opposite end of the spectrum are systems designed for chronic, daily use such as inhaled basal insulin for diabetes treatment, and encapsulated glucocorticosteroids for asthma. In between these extremes are therapeutic strategies, such as inhaled anti-infectives, which are designed for short-term use and can have dosing intervals in the range of hours to weeks, depending upon the drug release profile of the carrier system.

The choice of nanocarrier material suitable for a specific indication should take into account the dosing frequency desired. For example, it would not make sense to design a drug delivery system that incorporates a slow-degrading polymer as the vehicle for an indication requiring chronic, daily administrations, since polymer accumulation would pose significant toxicity issues. The same polymeric system, however, might be perfectly suited to a short-term application with greater dosing intervals. Box 3.1 illustrates an example of how a preliminary approximation of the dosing interval may be determined based on the half-life of the vehicle degradation, the primary degradation mechanism, and molecular weight of the vehicle components. This example demonstrates just how vital it is to characterise carefully the particle degradation kinetics in the relevant system (i.e. in nanoparticles), as inaccuracies in the dosing interval will have a large impact on possible toxicity.

Box 3.1 Example of considerations for approximating a dosing interval using PLGA nanoparticles.

PLGA nanoparticles described in Birnbaum and Brannon-Peppas (2003) were reported to have the following characteristics:

- PLGA composition: 53 mol % D,L-lactide/47 mol % glycolide with carboxylic acid end caps (Medisorb®)
- Molecular weight: ~11 kDa
- Particle size: not provided
- Degradation mechanism: hydrolytic cleavage of the polyester bonds
- Experimentally determined degradation rate constant: 0.121 days^{-1}
- Theoretical degradation half-life: 4.13 days (in a PBS buffer solution at 37° C)

Considerations:

1. PLGA nanoparticles should undergo sufficient polymer cleavage in the aqueous environment of the RTLF (in contrast to vehicles that degrade via enzymatic cleavage, such as solid lipid nanoparticles, where the enzyme concentration may be rate-limiting).
2. The initial molecular weight is fairly low resulting in a rapid passage of any degradation products into the bloodstream.
3. Imagine that three consecutive doses of 100 mg PLGA are administered to the patient. The degradation rate constant would predict that the following amounts of polymer would be found in the lung immediately after each dose:

   - 100 mg once daily: Dose (1) 100 mg; (2) 188 mg; (3) 265 mg
   - 100 mg once weekly: Dose (1) 100 mg; (2) 115 mg; (3) 118 mg
   - 100 mg once fortnightly: Dose (1) 100 mg; (2) 100 mg; (3) 100 mg

Conclusions: If hydrolytic cleavage proceeds according to first-order kinetics (i.e. the model used to determine the rate constant) and is not affected by RTLF components, the minimum dosing interval required to avoid PLGA accumulation in the lung would be once every nine days. Please note that this approximation does not consider any other clearance mechanisms, which could significantly increase the rate of vehicle removal from the lung.

3.4.2   Lung Burden, Macrophage Overload and Maximal Tolerated Doses in a Single Administration

The design of chronic inhalation toxicity studies using particulate matter has evolved in a manner that reflects the long-term, low-level aerosol exposure patterns of ambient or occupational particulates with the so-called steady-state lung burden featuring a prominent role. The steady-state lung burden is the maximum concentration of particulate matter that does not cause impairment of alveolar macrophage clearance functions and thus does not result in lung accumulation. Lung burden values for low-toxicity particulates have been reported typically ranging between 0.3–20 mg particulates/g lung, depending on the type of particle used, although some authors have claimed that concentrations >1 mg/g lung impair macrophage function regardless of particle type (Morrow, 1986; Lewis *et al.*, 1989; Monteiro-Riviere and Tran, 2007). Such values or statements imply that there exists an 'overload threshold' for low-toxicity, non-soluble particles (at least in experimental animals), below which macrophages can function normally and above which lung defences are impaired.

The experimental determination of such an 'overload threshold' for biodegradable nanoparticles (if it exists) would be a very useful tool in the design of inhalable nanomedicines. For example, if it were shown that biodegradable nanoparticles also impair normal macrophage clearance at a specific concentration, then this value could provide formulators with realistic limits with regard to maximum tolerated doses in a single administration, maximum tolerated administration frequencies, minimum percentages of drug loading required, and minimum encapsulation efficiencies.

To illustrate the concept, let us imagine that polymeric nanoparticles of 200 nm impair macrophage clearance mechanisms at concentrations exceeding 5 mg/g lung in rats (lung ∼2 g) and the clearance half-life of the particles was 12 h (this includes clearance and biodegradation). If an antibiotic for daily administration were to be encapsulated within the nanoparticles, the maximum tolerated mass of the nanosystem depositing in the lung (i.e. after losses due to the administration technique) should be no more than 4 mg to prevent accumulation. If the formulator were equipped with this knowledge from the outset, they would be in a better position to design a suitable drug delivery system. For example, the following questions could be addressed prior to formulation:

1. Is the required dose to be delivered less than 4 mg? If yes, then...
2. What is the theoretical percentage loading required to achieve a 24 h antibiotic release?
3. Do the drug's physico-chemical properties and the formulation method allow for a sufficient encapsulation efficiency to achieve this target loading?

The current bottom-up approach to drug delivery nanoparticle design may produce a system that has no realistic chance of development with regards to potential (and as of yet unknown) limitations in dosing and toxicity. This can be illustrated by taking the thought experiment further. Let us now imagine that the antibiotic described above could be successfully encapsulated in the said nanoparticles at nearly 100 % encapsulation efficiency, but only for a loading capacity of 10 %. This would mean that only 400 µg drug are delivered for every 4 mg of nanoparticles. Would this drug dose be sufficient for a therapeutic effect? Perhaps a calculated target loading capacity of 50 % (with ∼100 % encapsulation efficiency) would be required instead. If the maximum tolerated carrier dose is known from the beginning, the formulation process may be duly influenced from the onset and the go/no-go decisions regarding formulation may be taken more rapidly.

Unfortunately, there is very little data on the clearance rate of biodegradable nanoparticle systems from the lungs of experimental animals, and no data on the potential of these systems to cause macrophage overload. Notably, one study (described previously) has undertaken to track [99M]Tc-labelled 200 nm solid lipid nanoparticles in rats after pulmonary administration via nebulisation (Videira *et al.*, 2002). The particles were shown to be removed from the lung with a rapid initial clearance half-life of ∼13.4 min, followed by a slower equilibrium phase in which the remaining ∼25 % of the particles were retained within the lung for the remainder of the study. The mechanism of this initial clearance behaviour is particularly intriguing, as a significant portion of the particles were detected in the peripheral lymph nodes within minutes of administration. The authors postulated that either the particles were able to penetrate into the interstitial space individually and were taken up by lymphatic drainage, or were phagocytised by macrophages, which then migrated into the lymphatic system. This second hypothesis contradicts data reporting that most particle-laden macrophages will leave the lung via the mucociliary elevator and will migrate only to a very small extent through the epithelial barrier into the lymph. Nonetheless, the study results imply that lung clearance, as opposed to biodegradation, is the major mechanism for removal of ∼75 % of these particles. The determination of whether biodegradation is rate-limiting in the removal of the remaining 25 % of particles will require a methodology that can track particles for a longer period of time without signal decay or label dissociation from the particle structure. However, such biphasic elimination kinetics, if they are accurate, would need to be accounted for in all dosing interval calculations.

To assess lung burden and macrophage overload in a setting that accurately reflects a drug delivery application of nanoparticles, a modified strategy of inhalation exposure must be developed. The following aspects will be of relevance to the study design:

- Administration as a bolus of high concentration, either via nebulisation of a nanoparticle suspension or aerosolisation of microparticle formulations containing nanoparticles.
- Duration of administration will be short (∼10 seconds to 30 minutes) depending on the type of aerosolisation technique used.
- Administration frequency should reflect the potential therapeutic use, taking into account the theoretical degradation half-life of the carrier and clearance kinetics.
- Blank carrier systems should be used initially to provide an idea of the maximum amount of carrier tolerated in a dosing regime.
- The actual dose per single administration will require experimental determination, as currently no knowledge on maximum tolerated doses of biodegradable nanoparticles is available. It may be noted, however, that the physico-chemical properties of the nanoparticles or the administration method may provide an upper limit to the concentrations tested (e.g. due to aggregation of the system at high concentrations; Bihari *et al.*, 2008).
- A reliable detection system for particle accumulation must be implemented, since biomaterials, unlike most ambient particulate matter, are extremely difficult to detect and quantify in a biological matrix such as the lung.

It should also be considered that some nanoparticles, especially those <100 nm, will elicit an inflammatory response at doses well below those that impair macrophage function

(Scientific Committee on Emerging and Newly Identified Health Risks, 2006). Thus, when designing inhalation exposure studies for drug delivery nanoparticle systems, it would be advisable to determine the maximum dose of blank carriers that elicit an inflammatory response in addition to macrophage impairment. The unit of dose (i.e. mass, number or specific surface area concentration) that best correlates with pro-inflammatory potential is still controversial, as was discussed in detail recently by Wittmaack (2007). Regardless of the outcome of this debate, drug delivery scientists accustomed to thinking in nominal mass doses (as are used in most traditional drug studies) may find that a number-based or surface area-based dosing range could expose dose-dependent pro-inflammatory effects otherwise not detected in the perhaps more limited mass-based dose studies.

3.4.3   Toxicity of Inhaled Nanomedicines: From *In Vivo* and *In Vitro* Models to Health Risks

The exploitation of nanotechnology to improve current inhalation therapies is making rapid and surprising progress. Only ten years ago an *in vivo* therapeutic proof-of-principle study of nanomedicines in the lung was a rarity; today they are commonplace. Surprisingly, however, little attention has been given to the potential toxicological effects of inhaled nanoparticulate drug delivery vehicles, especially those of a more substantial nature, such as the solid lipid and polymeric nanoparticles. Less than a handful of studies published to date have actually specifically addressed the pulmonary toxicity of biodegradable nanoparticles (e.g. Dailey *et al.*, 2006; Grenha *et al.*, 2007; Gul *et al.*, 2009). Obviously, it is impossible to draw any reliable conclusions from such a tiny data-set.

Luckily, the tools for evaluating the potential toxicity of drug delivery nanoparticles have been well established by the fields of occupational and environmental respiratory toxicology. Thus, it would be highly advisable for researchers wishing to examine the toxicological potential of inhaled nanomedicines to follow harmonised protocols such as those to be collected and published by the recently launched International Alliance for NanoEHS Harmonization (see www.nanoehsalliance.org/sections/Protocols). This resource has been developed to increase the reproducibility and comparability of commonly used assays in nanotoxicological experimentation, ranging from basic physico-chemical characterisation through to *in vitro* and *in vivo* testing. However, it is vital to remember that biodegradable nanocarriers will differ significantly from manufactured and ambient particles in a number of important ways. The aim of this chapter was to highlight these differences and attempt to address some of the implications they might have for test design and analysis. It is hoped that the discussion presented here will contribute to the development of an interdisciplinary, public-sector led effort to fill the knowledge gap in this particular area of respiratory nanotoxicology, which although specialised, does have important implications for the future success or failure of many exciting new therapeutic strategies.

References

Bailey MM, Berkland CJ (2008) Nanoparticle formulations in pulmonary drug delivery. *Med Res Rev* **29**, 196–212.

Bihari P, Vippola M, Schultes S, Praetner M, Khandoga AG, Reichel CA, Coester C, Tuomi T, Rehberg M, Krombach F (2008) Optimized dispersion of nanoparticles for biological *in vitro* and *in vivo* studies. *Part Fibre Toxicol* **5**: 14.

Birnbaum DT, Brannon-Peppas L (2003) Molecular weight distribution changes during degradation and release of PLGA nanoparticles containing epirubicin HCl. *J Biomater Sci Polymer Edn* **14**, 87–102.

Black C, Cummins E, Royle P, Philip S, Waugh N (2007) The clinical effectiveness and cost-effectiveness of inhaled insulin in diabetes mellitus: a systematic review and economic evaluation. *Health Technol Assess* **11**(33), 1–126.

Borm JA, Kreyling W (2004) Toxicological hazards of inhaled nanoparticles – potential implications for drug delivery. *J Nanosci Nanotech* **4**(6), 1–11.

Buzea C, Pacheco Blandino II, Robbie K (2007) Nanomaterials and nanoparticles: sources and toxicity. *Biointerphases* **2**(4), MR17–MR172.

Card JW, Zeldin DC, Bonner JC, Nestmann ER (2008) Pulmonary applications and toxicity of engineered nanoparticles. *Am J Physiol Lung Cell Mol Physiol* **295**, L400–L411.

Chattopadhyay P, Shekunov BY, Yimb D, Cipolla D, Boyd B, Farr S (2007) Production of solid lipid nanoparticle suspensions using supercritical fluid extraction of emulsions (SFEE) for pulmonary delivery using the AERx system®. *Adv Drug Del Rev* **59**, 444–453.

Cooke RW, Drury JA (2005) Reduction of oxidative stress marker in lung fluid of preterm infants after administration of intra-tracheal liposomal glutathione. *Biol Neonate* **87**, 178–180.

Costigan S (2006) *The toxicology of nanoparticles used in health care products.* Medicines and Healthcare Products Regulatory Agency, Department of Health, UK. Available at http:www.mhra.gov.uk/home/idcplg?IdcService=SS_GET_PAGE&nodeId=996, [accessed 21 October 2008].

Dailey LA, Kleemann E, Wittmar M, Gessler T, Schmehl T, Roberts C, Seeger W, Kissel T (2003a) Surfactant-free, biodegradable nanoparticles for aerosol therapy based upon the branched polyesters, DEAPA-PVAL-g-PLGA. *Pharm Res* **20**, 2011–2020.

Dailey LA, Schmehl T, Gessler T, Grimminger W, Seeger W, Kissel T (2003b) Nebulization of biodegradable nanoparticles: impact of nebulizer technology and nanoparticle characteristics on aerosol features. *J Controlled Release* **86**, 131–144.

Dailey LA, Jekel N, Fink L, Gessler T, Schmehl T, Wittmar M, Kissel T, Seeger W (2006) Investigation of the proinflammatory potential of biodegradable nanoparticle drug delivery systems in the lung. *Toxicol Appl Pharmacol* **215**, 100–108.

De Jong WH, Borm PJA (2008) Drug delivery and nanoparticles: applications and hazards. *Int J Nanomed* **3**, 133–149.

Durcan N, Murphy C, Cryan SA (2008) Inhalable siRNA: potential as a therapeutic agent in the lungs. *Mol Pharm* **5**, 559–566.

Friend JR, Yeo LY, Arifin DR, Mechler A (2008) Evaporative self-assembly assisted synthesis of polymeric nanoparticles by surface acoustic wave atomization. *Nanotechnology* **19**, 145301–145307.

Gelperina S, Kisich K, Iseman MD, Heifets L (2005) The potential advantages of nanoparticle drug delivery systems in chemotherapy of tuberculosis. *Am J Respir Crit Care Med* **172**, 1487–1490.

Grenha A, Grainger C, Dailey LA, Seijo B, Martin GP, Remuñán-López C, Forbes B (2007) Compatibility of microencapsulated chitosan nanoparticles with respiratory epithelial cell layers. *Eur J Pharm Sci* **31**, 73–84.

Gul MO, Jones SA, Dailey LA, Nacer H, Ma Y, Sadouli F, Hider RC, Araman A, Forbes B (2009) A poly(vinyl alcohol) nanoparticle platform for kinetic studies of inhaled nanoparticles. *Inhal Toxicol* **23 March**, 1–10.

Hawkins MJ, Soon-Shiong P, Desai N (2008) Protein nanoparticles as drug carriers in clinical medicine. *Adv Drug Deliv Rev* **60**, 876–885.

Hochhaus G, Suarez S, Gonzalez-Rothi RJ, Schreier H (1998) Pulmonary targeting of inhaled glucocorticosteroids: how is it influenced by formulation? *Resp Drug Del* **VI**, 45–52.

Hoet PHM, Brüske-Hohlfeld I, Salata OV (2004) Nanoparticles – known and unknown health risks. *J Nanobiotech* **2**. Available at http://www.jnanobiotechnology.com/content/2/1/12.

Kelly HW (2003) Potential adverse effects of inhaled corticosteroids. *J Allergy Clin Immunol* **112**, 469–478.

Kumar CSSR (ed.) (2007) *Nanomaterials for Medical Diagnosis and Therapy, Nanotechnologies for the Life Sciences (Band 10)*. Wiley-VCH: Weinheim.

Leroueil PR, Hong S, Mecke A, Baker JR Jr, Orr BG, Banazak Holl MM (2007) Nanoparticle interaction with biological membranes: does nanotechnology present a Janus face? *Acc Chem Res* **40**, 335–342.

Lewis TR, Morrow PE, McClellan RO, Raabe OG, Kennedy GL, Schwetz BA, Goehl TJ, Roycroft JH, Chhabra RS (1989) Establishing aerosol exposure concentrations for inhalation toxicity studies. *Tox Appl Pharmacol* **99**, 377–383.

Liu J, Gong T, Fu H, Wang C, Wang X, Chen Q, Zhang Q, He Q, Zhang Z (2008) Solid lipid nanoparticles for pulmonary delivery of insulin. *Int J Pharm* **356**, 333–344.

Lombry C, Edwards DA, Préat V, Vanbever R (2004) Alveolar macrophages are a primary barrier to pulmonary absorption of macromolecules. *Am J Physiol Lung Cell Mol Physiol* **286**, L1002–L1008.

Moghimi SM, Hunter AC, Murray JC (2001) Long-circulating and target-specific nanoparticles: theory to practice. *Pharmacol Rev* **53**, 283–318.

Monteiro-Riviere NA, Tran CL (eds) (2007) *Nanotoxicology: Characterization, Dosing and Health Effects*. Informa Healthcare: New York.

Morrow PE (1986) The setting of particulate exposure levels for chronic inhalation toxicity studies. *Int J Toxicol* **5**, 533–544.

Mozafari MR (ed.) (2006) *Nanocarrier Technologies: Frontiers of Nanotherapy*. Springer: Dordrecht.

Radomska-Soukharev A (2007) Stability of lipid excipients in solid lipid nanoparticles. *Adv Drug Deliv Rev* **59**, 411–418.

Schanker LS, Burton JA (1976) Absorption of heparin and cyanocobalamin from the rat lung. *Proc Soc Exp Biol Med* **152**, 377–380.

Scientific Committee on Emerging and Newly Identified Health Risks (SCENIHR) (2006) *The appropriateness of existing methodologies to assess the potential health risks associated with engineered and adventitious products of nanotechnologies*. European Commission Health & Consumer Protection Directorate-General, SCENIHR/002/005. Available at http://ec.europa.eu/health/ph_risk/committees/04_scenihr/docs/scenihr_o_003.pdf [accessed 12 October 2008].

Skubitz KM, Anderson PM (2000) Inhalational interleukin-2 liposomes for pulmonary metastases: a phase I clinical trial. *Anticancer Drugs* **11**, 555–563.

Smola M, Vandamme T, Sokolowski A (2008) Nanocarriers as pulmonary drug delivery systems to treat and diagnose respiratory and non respiratory diseases. *Int J Nanomed* **3**, 1–19.

Sung JC, Pulliam BL, Edwards DA (2007) Nanoparticles for drug delivery to the lungs. *TRENDS Biotech* **25**, 563–570.

Takada K, Yamamoto M, Asada S (1978) Evidence for the pulmonary absorption of fluorescent labelled macromolecular compounds. *J Pharm Dyn* **1**, 281–287.

Thomas M, Lu JJ, Chen J, Klibanov AM (2007) Non-viral siRNA delivery to the lung. *Adv Drug Del Rev* **59**, 124–133.

Videira MA, Botelho MF, Santos AC, Gouveia LF, Pedroso de Lima JJ, António AJ (2002) Lymphatic uptake of pulmonary delivered radiolabelled solid lipid nanoparticles. *J Drug Targeting* **10**, 607–613.

Warheit DB (2008) How meaningful are the results of nanotoxicity studies in the absence of adequate material characterization? *Toxicol Sci* **101**, 183–185.

Wijagkanalan W, Higuchi Y, Kawakami S, Teshima M, Sasaki H, Hashida M (2008) Enhanced anti-inflammation of inhaled dexamethasone palmitate using mannosylated liposomes in an endotoxin-induced lung inflammation model. *Mol Pharmacol* **74**, 1183–1192.

Wittmaack K (2007) In search of the most relevant parameter for quantifying lung inflammatory response to nanoparticle exposure: particle number, surface area, or what? *Environ Health Perspect* **115**, 187–194.

Yang W, Peters JI, Williams RO (2008) Inhaled nanoparticles – a current review. *Int J Pharm* **356**, 239–247.

4

High Aspect Ratio Nanoparticles and the Fibre Pathogenicity Paradigm

Craig A. Poland, Rodger Duffin and Ken Donaldson

4.1 High Aspect Ratio Nanoparticles (HARNs)

The term 'high aspect ratio' refers to a structure that has a length many times greater than its width. A high aspect ratio nanoparticle(s), or HARN, is therefore a nanoparticle with a length many times that of its width. It is within the nano range regarding width/diameter (less than 100 nm) extending to several hundreds of microns to even millimetres in length. The extremely small diameter of a nanoparticle means that it can have a high aspect ratio whilst still having both aspects (length/width) in the nanometre range. It is useful toxicologically to try to classify HARNs based upon criteria which may define how they may behave in a biological system. If HARNs are to be considered fibres and compared to asbestos, then we must look to the World Health Organization (WHO) standard description of a fibre in workplace air (World Health Organization, 1997) as having a diameter less than 3 μm, length greater than 5 μm and an aspect ratio greater than 3:1. Below this length a HARN may be considered a nanoparticle rather than a nano-fibre.

Carbon nanotubes (CNTs) are one form of HARN attracting great interest from industrial and toxicological scientists alike. This is due to their unique structure, an hexagonal lattice of elemental carbon atoms forming a graphene cylinder, nanometres in diameter and microns to millimetres in length (Donaldson *et al.*, 2006). This physical arrangement bestows exceptional structural, thermal and electrical properties on the nanotube (Donaldson *et al.*, 2006; Maynard *et al.*, 2006), rendering them desirable components for both industrial products and processes of many types. As such, CNTs represent the HARN with highest production volume and hence most occupationally relevant. Nanotubes are of considerable interest to toxicologists for both their nano-size and fibre-like dimensions, which

Nanotoxicity: From In Vivo *and* In Vitro *Models to Health Risks*   Edited by Saura Sahu and Daniel Casciano
© 2009 John Wiley & Sons, Ltd

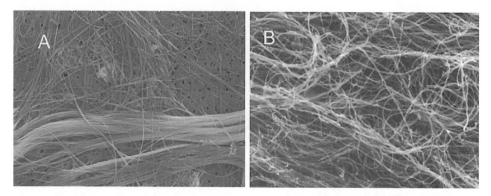

Figure 4.1   *Scanning electron microscope images of chrysotile asbestos (A) and multiwalled carbon nanotubes (B) at × 4000 and × 6000 magnification respectively, showing the similarity in appearance.*

may represent a unique inhalation hazard (Maynard *et al.*, 2006) similar to that posed by other pathogenic fibres, including asbestos (Muller *et al.*, 2005; Donaldson *et al.*, 2006; Nel *et al.*, 2006) (Figure 4.1).

4.2   Asbestos and Disease

Asbestos exposure has been identified as a causal agent in the global pandemic of neoplastic (lung cancer, mesothelioma) and non-neoplastic (asbestosis, pleural effusions, and pleural plaques) diseases of the lung and the pleura (Mossman and Churg, 1998; Mossman *et al.*, 1990; Ramos-Nino *et al.*, 2006; O'Reilly *et al.*, 2007) that occurred largely in the twentieth century. To understand the scale of the impact asbestos had and continues to have on public health, we must first understand the usefulness and therefore sheer scale of use that asbestos enjoyed. Whilst perhaps the most widely recognised fire-resistant properties of asbestos have been utilised since the time of the ancient Egyptians, it was mainly over the twentieth century that a dramatic increase in the use of asbestos (Selikoff and Lee, 1978) was seen. The reason for this widespread use of asbestos was the numerous uses asbestos could be put to and its cheapness, being essentially a mineral that was mined. The heat-resistant, robust yet flexible nature of asbestos made it an ideal material for brake-pad lining, whilst the fireproof and insulating capacity of the fibres made them extremely advantageous in a great number of buildings, including the World Trade Centres in New York. The large expansion in heavy industry during the Second World War saw a dramatic increase in the use of asbestos in ship-building, and in the economic boom of the post-war years, asbestos was used in a wide variety of products in the building trade. The legacy now is that a large proportion of buildings constructed during the mid-twentieth century contain some form of asbestos.

Shortly after the initiation of large-scale production, asbestos began to be recognised as harmful. In 1906, Montagu Murray reported pulmonary fibrosis in an autopsy of an asbestos worker (Selikoff *et al.*, 1978), and by the 1930s there was a review of over 30 cases of asbestosis (Wood and Gloyne, 1930). Lynch and Smith (1935) reported lung cancer in asbestos-silicosis, and numerous reports confirmed the link by the 1940s (Tweedale, 2002). Wagner *et al.* (1960) described a high incidence of pleural mesothelioma in the Cape crocidolite asbestos fields of South Africa, and towards the end of the twentieth century numerous epidemiological studies had confirmed the presence of asbestosis, bronchogenic carcinoma, mesothelioma and pleural plaques in populations exposed to asbestos (Tweedale, 2002). J.C. Wagner, in the MRC Pneumoconiosis Research Unit in Penarth in Wales, and J.M.G. Davis in Edinburgh, undertook similar pathological investigations aimed at dissecting the characteristic features of fibres involved in pathogenicity; to an extent, similar studies were undertaken in laboratories throughout the world. A full understanding of the asbestos hazards did not become apparent until studies on synthetic vitreous fibres (SVF) brought them together with asbestos within a single 'fibre paradigm'. The emotive and headline-grabbing nature of the asbestos issue means that any hypothetical link between the toxicology of asbestos and HARNs must be rigorously evaluated due to the potentially adverse impact on nanotechnology development (Maynard *et al.*, 2006).

4.3 Fibre Pathogenicity Paradigm: Dimension, Durability and Dose (the 3 Ds)

The relationship between fibre physicochemistry and pathogenicity has been studied in depth (Donaldson and Tran, 2004; Maynard *et al.*, 2006), and three fundamental attributes have emerged as paramount to the pathogenicity of a fibre. These attributes are referred to as the '3 Ds' – dimension, durability and dose. The importance of dimension lies in the ability of long fibres to interfere with the important macrophage clearance mechanisms in the lung. Durability (biopersistence) dictates the ability of a fibre to persist in the lungs, and dose, which is related to exposure, if sufficient, may lead to disease.

(1) **Dimension**
The dimensions of a particle are critical in determining its entrance into the body, its deposition, and its handling by cells, especially the macrophage. Long fibres (longer than about 15 μm) are too long for macrophages to enclose completely or phagocytose, which in turn leads to the cell entering an irritated or frustrated state (see below). Nanotubes, by the presence of defects in the hexagonal lattice of carbon atoms during production, are often curled, which together with Van der Waal's forces (see below) can cause tangling of nanotubes. A highly tangled CNT may have a nominal 'tangled' length of many tens of microns but its actual length across a single axis may only be a few microns; in the same way, a ball of string may be tens of metres long when wound up but still fits into the palm of your hand. A sample of CNTs which are highly curled and highly tangled is unlikely to present as a fibre, but rather as a roughly spherical particulate. As such, if the spherical agglomerate of CNTs had a small enough aerodynamic diameter to penetrate the distal airways, it is likely that it could be dealt with by the normal macrophage-mediated phagocytic clearance mechanism (Poland *et al.*, 2008). To understand which paradigm a sample of CNT may follow (particulate, nano-particulate or fibre), and as such which testing strategy and controls to use, it is essential that appropriate physico-chemical characterisation is undertaken.

(2) **Durability**

Durability or biopersistence is important in modifying the pathogenicity of long fibres. Only long fibres that are biopersistent and do not dissolve and break contribute to the long fibre dose over time (Miller *et al.*, 1999). It is thought that CNTs, due to their essentially graphene structure, may be highly biopersistent in the lung environment. The effect of dimension and durability means that biopersistent long fibres are selectively retained. A long fibre that cannot be cleared by normal lung mechanisms and that is not biosoluble will persist. Continued exposure to such fibres, even at low exposure levels, may lead to the build-up of such fibres within the lung, thereby leading to the formation of a critical dose.

(3) **Dose**

As for all toxicological processes, dose drives the response and the dose is related to exposure. Sufficient biopersistent fibres must be present to reach a threshold in order to obtain a pathological change.

4.4 Challenges of Carrying Out Toxicological Testing with HARNs

The transition from micron-sized to nano-sized particles brings with it interesting changes in the characteristics and behaviour of a particle. These alterations may bring with them beneficial attributes that may be exploited in various situations; however, they may also produce challenges for accurate toxicological testing. Such challenges may be of minor significance to testing regimes, but others, such as dispersal or interference with toxicological tests, are of critical importance to overcome if accurate results and meaningful conclusions are to be established.

4.4.1 Dispersal

The aggregation state of a particle can have significant impacts on both the activity of a particle and its translocation kinetics. As such, proper dispersal and characterisation of the actual dispersants used are an important part of both *in vivo* and *in vitro* toxicological testing.

4.4.1.1 Size

The ability of nanoparticles to translocate from their site of deposition is one of their hallmarks (Kreyling *et al.*, 2002; Nemmar *et al.*, 2004; Oberdorster *et al.*, 2004). The ability to translocate probably relies on size, and aggregates may be less likely to translocate than singlet particulates, or a small aggregate may be more likely to translocate than a larger aggregate. If aggregates are likely to 'break' into smaller aggregates or individual particles *in vivo*, then this could be important in modifying their ability to translocate.

4.4.1.2 Surface Area

Surface area is a key metric for driving inflammation in experimental studies with low-toxicity particles (Tran *et al.*, 2000), and nanoparticles by definition have a high surface area per unit mass, and so can be especially inflammogenic (Duffin *et al.*, 2002). Aggregation of nanoparticles could affect the total surface area available to interact with biological systems.

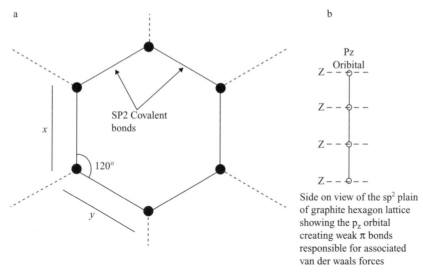

a

b

Pz
Oribital

Side on view of the sp² plain
of graphite hexagon lattice
showing the p$_z$ orbital
creating weak π bonds
responsible for associated
van der waals forces

SP2 Covalent
bonds

x

120°

y

Z -- ⊖-- --

Z -- ⊖-- --

Z -- ⊖-- --

Z -- ⊖-- --

Figure 4.2 *A schematic diagram of the graphitic structure of carbon nanotubes. (a) shows the SP2 orientation of the carbon atoms giving the distinctive 'honeycomb' hexagonal appearance of graphite. (b) shows the free electron field of the P$_z$ orbital causing weak π bonds causing associated Van der Waal's forces.*

Carbon nanotubes have a well-documented ability to stick together, potentially forming dense, tangled aggregates especially after acid purification (Li *et al.*, 2005). The source of these attractive forces lies in the structure of the nanotube, in particular the orientation of the carbon atoms. In CNTs and indeed all forms of graphite, each carbon atom is connected evenly to three other carbon atoms in the XY plane (SP2 form; see Figure 4.2a). It is this SP2 form that gives graphite the distinctive honeycomb lattice, but the presence of the weak π bond in the Z-axis, termed the P$_z$ orbital (Figure 2b), causes the associated Van der Waal's forces (Vaisman *et al.*, 2006). It is the diffuse electron field in the P$_z$ orbital which allows graphite to conduct electricity. Diamond, in contrast, has a SP3 plane which means that all the electrons are localised in the bonds, forming an insulator (Vaisman *et al.*, 2006).

These attractive forces occur over the entire surface of a nanotube, providing difficulties in dispersion. The observation that CNTs adhere together is not an adequate argument that nanotubes will not become airborne and as such are unlikely to pose a risk. Firstly, as with most industrial fibres, it is not advantageous to have poorly dispersed fibres for manufacturing processes, and so there is a tendency to try to disperse the nanotubes (Baughman *et al.*, 2002). The result of aggregation can be the suboptimal performance of products such as high-strength composites, where aggregated material provides areas of weakness and crack initiator points (Vaisman *et al.*, 2006). The even distribution of CNTs in a product is especially pertinent for their use in electrical and thermal applications, whereby the rapid and efficient dissemination of energy is essential and therefore the main advantage for using CNTs. The drive from industry, therefore, is likely to be for the formation of more easily dispersible material through the use of better production methods, dispersants and/or modification of the surface of CNTs. Furthermore, the scale and strength of aggregation, due to its inherent association with the structure of a CNT, is not seen with all forms of HARN. Indeed, nano-fibres formed of other materials such aluminium oxide or nickel nanowires do not have the same degree of Van der Waal's forces caused by their structure, and can be more readily dispersed.

The drive from industry to produce uniform dispersions of CNTs has led to numerous methods of dispersion using various organic solvents such as chloroform (Li *et al.*, 2005) and surfactants such as sodium dodecyl sulfate (Jiang *et al.*, 2003) and Tween-80 (Warheit *et al.*, 2004). However, whilst these dispersants may be very efficient, they are generally unsuitable for use in toxicological studies due to their intrinsic toxicity. When preparing a sample for use in an *in vitro* or *in vivo* system, the relevance of the dispersant is of high priority. It is also of importance to consider the route of entry as well as what entities the CNTs are likely to encounter on the way, and hence what this may do to the surface of the CNTs.

The penetration of fibres into humans is most likely to occur via the lungs, and if, like other thin fibres, CNTs are able to reach the distal parts of the lung, they will be deposited in and around the alveolar region. Upon reaching the proximal alveoli, the nanotubes would encounter and be bathed in the lung lining fluid (Hills, 2002; Meyer and Zimmerman, 2002). Lung lining fluid is composed of a colloidal suspension of various phospholipids, proteins and sugars. The lung surfactant is secreted by type II alveolar cells, and its main function is to reduce the surface tension in the alveoli, preventing collapse upon expiration (Veldhuisen *et al.*, 2000). However, once deposited, the action of the lung surfactant may also aid the dispersion of aggregated CNTs, and so size may therefore become an important factor in the lung's ability to clear such foreign particles (Renwick *et al.*, 2001). By weight, lung surfactant is 90% phospholipids (Kingma and Whitsett, 2006), of which dipalmitoylphosphatidylcholine (DPPC), a chemically inert, biocompatible phospholipid (Nii and Ishii, 2004), accounts for 40–50% of the phospholipids present. As such DPPC has been a natural candidate for use as a model surfactant in dispersing a range of particles (Gu *et al.*, 2005; Foucaud *et al.*, 2007). Indeed, in our own work analysing DPPC suspensions as a CNT surfactant, we found that whilst we could fluidise DPPC (lung surfactant requires a proper balance of fluid and rigid phospholipids) by heating past the gel–crystal interface of DPPC (Tc temperature: Kim and Franses, 2005) to form bi-layer liposomes, the suspensions were rather unstable and ineffective. The poor dispersant/surfactant properties of DPPC we found may be due, in part, to the structural difference between DPPC in lung lining fluid (as a predominantly tubular myelin structure: Kingma and Whitsett, 2006) and a synthetic preparation of DPPC (bi-layer liposomes). Other components of lung lining fluid such as other phospholipids, Ca^{2+} and specific surfactant proteins such as SP-A and SP-B (Veldhuisen *et al.*, 2000) may also contribute to the efficacy of lung lining fluid as a surfactant *in vivo*. This is because lung surfactant is not simply the action of one phospholipid, but rather the interaction of a range of components whose importance is not necessarily represented by their relative concentrations.

The use of a simple globular protein such as albumin is rapidly gaining popularity as a simple yet efficient way to disperse nanotubes, and has been successfully used in several studies (Poland *et al.*, 2008). The low inherent toxicity of bovine serum albumin (BSA) is unlikely to mask the effect of any dispersed particle, and may prove useful for other nanoparticles requiring dispersion for toxicological studies. BSA was found to possess substantial anti-oxidant activity (see Table 4.1 for comparison with the dispersants DPPC and Tween-80), very likely as a result of its thiol groups; however, albumin and other

Table 4.1   Trolox® Equivalence Antioxidant Capacity (TEAC) of dispersant preparations.

Dispersant	TEAC (mean ± SEM)
BSA	95 (11)
DPPC	7 (2)
Tween-80	45 (17)

The antioxidant capacity of BSA, DPPC and Tween-80 was evaluated using the ORAC assay. The antioxidant capacity of the dispersant is expressed in Trolox® (water soluble analogue of vitamin E) equivalents (nM) per μg/ml. BSA has a significantly higher antioxidant capacity than DPPC ($P < 0.01$) but not significantly higher than Tween-80 ($n = 5$).

thiol-rich globular proteins are ubiquitous in biological fluids, and so this does not argue against its use. Any particle depositing in the lung is unlikely to remain as a naked surface for very long, and the proteins adhering to the surface may indeed change rapidly based on a series of adsorption-displacement steps (Noh and Vogler, 2007) due to the relative concentrations of proteins in solution.

4.4.2   Interference With Other Assays

Nanoparticles, due to their individual properties and idiosyncrasies, can present a range of challenges to nanotoxicologists which extend beyond simple dispersion. There have been various studies showing that nanoparticles can indeed interfere with various assays, producing both false-positive and false-negative results (Worle-Knirsch et al., 2006).

4.4.2.1   Adsorption Effects

One of the ways in which CNTs can interfere with assays occurs through the adsorption of various compounds onto the surface, possibly modulating any subsequent detection methods. Carbon is well known for this property; indeed, its ability to bind and remove substances has been utilised in a range of processes, but in a toxicological assay this can be problematic. The scale of this problem is demonstrated in Figure 4.3 which shows the adsorptive effect of three graphitic materials, namely multi-walled carbon nanotubes (MWCNT), nano-particulate carbon black (NPCB; printex 90) and Huber carbon black (HCB) on a particle mass and surface area basis. As nanoparticles, both MWCNTs and NPCB have huge surface areas which accounts for their ability to adsorb a significant proportion of the protein in solution in a linear dose response relationship (R^2 value $= 0.9673$ and 0.9638 respectively), whilst HCB adsorbs only a small amount. In Figure 4.3b we see the same experimental procedure conducted as a function of surface area, and note that the different-sized carbon particles, when corrected for surface area, have the same adsorptive capacity.

This absorptive effect can interfere with assays by removing compounds being measured such as cytokines or the commonly-used marker of cellular toxicity, lactate dehydrogenase (LDH). This absorptive capacity is further demonstrated in Figure 4.4 where increasing doses of nano-particulate carbon (MWCNTs and NPCB) led to an apparent decrease in the level of interleukin-8 (IL-8), an important pro-inflammatory cytokine released from the alveolar type II cell line A549 below that of the media control. The increasing dose

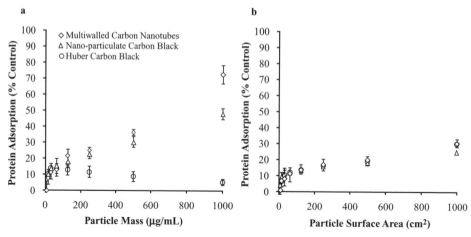

Figure 4.3 *Absorptive effect of various graphitic materials on bovine serum albumin. The absorption capacity of three graphitic materials was investigated based on both particle mass and surface area. The particles were incubated in a solution of bovine serum albumin (BSA; 1 mg/ml in phosphate buffered saline) gently mixing at 37° C for 24 hrs, after which the particles were separated and the protein concentration established. Graph (a) shows the absorptive capacity of multi-walled carbon nanotubes, nanoparticulate carbon black and Huber carbon black on a mass basis, whist graph (b) shows the particles equalised for surface area (different mass) (n = 3).*

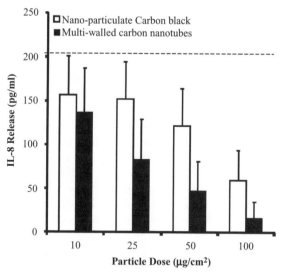

Figure 4.4 *Deleterious effect of high surface area carbon particles on assay reliability. Two forms of carbon nanoparticle, namely nanoparticulate carbon black (printex 90) and multiwalled carbon nanotubes, were incubated at varying concentrations with A549 cells to establish whether they elicit the secretion of the inflammatory cytokine. The dotted line shows the mean control levels of IL-8 (204.26 ± 36.16 pg/ml). Addition of the carbon material leads to an apparent dose-dependent decrease in the level of IL-8 protein within the media (n = 4).*

therefore appears to have a false-negative result which may not be apparent in other assays such as IL-8 gene expression via PCR or a reporter cell line.

The potential for CNTs to cause interference within an assay means that toxicity studies, as a matter of course, should be conducted using several different assays to corroborate results. Ideally these should involve different methods of detection; for example, it is known that carbon nano-particulates such as carbon black and nanotubes may interfere with toxicity assays such as LDH and MTT (a tetrazolium salt which is converted to formazan by active mitochondrial dehydrogenases in living cells and commonly used as a toxicity assay) (Worle-Knirsch *et al.*, 2006). Therefore such toxicity data could be corroborated using flow cytometry for apoptotic (Annexin V staining) and necrotic cells (Propidium Iodide staining) after the particle treatments have been removed.

4.4.3 Importance of Particle and Sample Characterisation

The wide variety of physico-chemical characteristics that exist between nanotubes/nanoparticles, produced both in industry and academia, make it essential that samples analysed for toxicological endpoints be evaluated as far as is practicable. The increased movement for the establishment of predictive structure activity relationships (SARs), whereby the pathogenic potential of a particle can be modelled and predicted based on secure paradigms, is very attractive. This is because of both increased pressure away from *in vivo* testing, both financial and ethical, and because with the vast number of manufactured nanoparticles, it is simply not feasible to test them all within *in vivo* models. In order to identify these SARs within and compare between the increasing body of nanotoxicology literature, certain information is needed to present with any associated activity.

As a minimum the particle size, state of dispersion and images of the raw particles and in suspension should always be presented for publication. Other measurements do indeed require the use of specialist equipment, such as BET (a method of measuring specific surface area of a material via the calculation of the physical adsorption of gas molecules onto a surface, developed by Brunauer, Emmett and Teller, 1938) for surface area, zeta potential for surface charge, inductively-coupled plasma mass spectrometry (ICP-MS) for metal contamination, and so on. Wherever possible these measurements should be performed, as the importance of these parameters has been borne out by a great deal of research into particle toxicity. Also of critical importance is the state of dispersal of a sample, as this relates to the mode of exposure. For example, if a particle is being introduced to alveolar cells *in vitro,* then the particle should be present in a size fraction that is respirable (<3 μm), otherwise the cell would never encounter such a particle.

Circumstances where this is not taken into account can shed doubt upon the results of an experiment and the possible creation of artefactual false-positive or negative results (Takagi *et al.*, 2008). Below is a guide to certain physicochemical characteristics that should be identified when characterising sample structures for toxicological analysis.

- **Particle size**
  Mean
  Size distribution (relevant for fibre toxicity)
- **Aspect ratio** (relationship between length and width)
- **Surface area**

- **Surface charge**
- **Contaminating metals**
- **Crystallinity**
- **State of dispersion**
  Dispersion method (e.g. surfactant used, sonication time and power)
  Stability
- **Microscopy of particles** – both primary particles and suspension
  Light
  Scanning Electron Microscopy (SEM)
  Transmission Electron Microscopy (TEM)

4.5   Testing for HARN/Fibre Effects on the Mesothelium

4.5.1   Rationale for Studying the Mesothelial Response

The almost unique susceptibility of the pleural mesothelium to fibrous dusts is well documented. Amongst pathogenic particle exposures, only asbestos and the zeolite mineral erionite is known to cause mesothelioma and other types of pleural injury, very likely as a result of the ability of the fibres to reach the pleura and have pathogenic effects there or to be retained in the region of the mesothelium. Therefore the mesothelium has been used to study the potential toxicity of fibrous dusts and is relevant in studying MWCNTs and other HARNs in relation to the fibre paradigm. There are a number of strategies for studying mesothelial responses, including using mesothelial cells *in vitro*, injecting onto the peritoneal mesothelial surface of mice and rats, and injecting into the pleural space to expose the pleural mesothelium.

4.5.2   Mesothelial Cells *In Vitro*

We confirmed the exquisite sensitivity of mesothelial cells to long, but not shortened, asbestos fibres using mesothelial cells in culture. These data (Figure 4.5) showed that the mesothelial cells were over 100 times more susceptible to long fibre-induced toxicity than to the same mass of short fibres. This highlights the primacy of fibre length over fibre chemistry in causing mesothelial cell damage, since both long and short (derived from the long by ball-milling) amosite have essentially the same bulk chemistry (Graham *et al.*, 1999).

4.5.3   Mouse Peritoneal Model: *In vivo* Testing for Effects on the Mesothelium

The peritoneal cavity has long been used as a model for assessing the toxicity and carcinogenicity of fibrous materials (Bolton *et al.*, 1982; Donaldson *et al.*, 1989). Clearance operates effectively in the peritoneal cavity, with well-documented rapid clearance of deposited particles via a number of lymphatic routes, cranially, through the diaphragm (Abu-Hijleh *et al.*, 1995). These cranial pathways include retrosternal, intercostovertebral and peritendinous drainage to the mediastinal and parathymic nodes. Such drainage will deliver intraperitoneally injected particles and fibres to the diaphragm, where long fibres are likely to accumulate at the lacunar areas as they block the stomata because they are too long to navigate them. The fibres are retained on the peritoneal aspect of the diaphragmatic

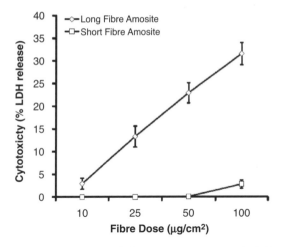

Figure 4.5 *Cytotoxic effects of long and short amosite to mesothelial cells in culture. The human mesothelial cell line MET-5A was used to test the cytotoxicity of long and short amosite asbestos samples (LFA and SFA respectively). The test particles were incubated at a range of concentrations with the mesothelial cells for a period of 24 hrs after which the percentage cytotoxicity was established using the Lactate Dehydrogenase Assay with triton-X used as a 100 % lysis control (n = 4).*

mesothelium, and no doubt further irritate by mechanical movement of the diaphragm during breathing; inflammation develops and eventually granuloma formation occurs.

4.5.3.1   Long but not Short CNTs Cause Inflammation and Diaphragmatic Granulomas after Intraperitoneal Injection

The instillation of long and short nanotubes into the peritoneal cavity showed a very dramatic inflammatory (Figure 4.6a) and granulomatous (Figure 4.6b) response with the long (longer than about 20 µm) CNTs but not with short (shorter than 5 µm) CNTs (Figure 4.6). This mimicked the length-related effects seen with long and short amosite asbestos, which were used as controls. Graphene in the form of NPCB was without activity in the assay, as was the BSA vehicle. The similarity in length-dependent toxicity seen with CNTs and asbestos suggest that CNTs might behave like asbestos in other respects regarding adverse effects on the mesothelial inflammatory response.

4.5.3.2   Scanning Electron Microscopy Morphology of the Diaphragm Surface

The effect of fibres on the mesothelium was also investigated using scanning electron microscopy by examining the peritoneal aspect of the diaphragm surface for indication of leukocytes, granulomas, or alteration in the normal serosal surface architecture. Figure 4.7 shows the normal mesothelial surface, and the mesothelial surface covered in a granuloma on the peritoneal aspect of the diaphragm following injection of long nanotubes. Large areas of leukocytes and fibroblasts are visible across large sections of the diaphragm. Closer inspection showed a dense fibrin-like meshwork surrounding the inflammatory cells.

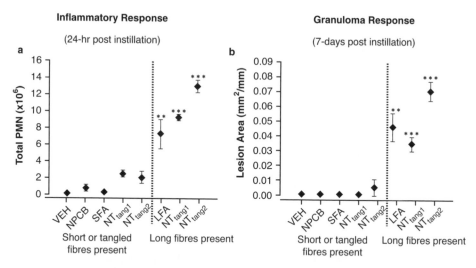

Figure 4.6 *Inflammatory cell recruitment to the peritoneal cavity and granulomatous response at the diaphragm after introduction of fibres. Female C57Bl/6 mice were intraperitoneally instilled with fibres or particle controls (50 µg). These were vehicle control (VEH, 0.5 % BSA/saline; 0.5 ml), nanoparticulate carbon black (NPCB), short fibre amosite (SFA), long fibre amosite (LFA), two curled/tangled nanotube (NT) samples of differing lengths, NT_{tang1} and NT_{tang2}, and two samples containing long NT (NT_{long1} and NT_{long2}). After 24 hrs and 7 days post-exposure the mice were killed and the peritoneal cavity lavaged. At 24 hrs the inflammatory response was evaluated using differential cell counts to establish total PMN population (a, mean ± sem). At 7 days post-exposure, the extent of granuloma formation at the peritoneal surface of the diaphragm was quantified in histological sections of excised diaphragms (b, mean ± sem). $^{**}P < 0.01$; $^{***}P < 0.001$. Reprinted from Poland et al., Nat Nanotechnol 2008, **3**: 423–428 by permission of Macmillan Publishers Ltd.*

Histological sections were taken and stained with hematoxylin and eosin for gross pathology, and Sirius red stain to show collagenous material. Figure 4.8 shows histological sections through the diaphragm with granuloma, showing a dense layer of leukocytes and fibroblasts covering much of the diaphragm surface, interspersed with giant cells. Staining of these granulomas for collagen (Figure 4.8b) shows deposition throughout the plaque forming a fibrous layer across the diaphragm. The granulomas were not uniform across the diaphragm but varied in thickness, being absent in some areas.

4.5.4   The importance of the amosite controls

The history of the long fibre amosite (LFA) and short fibre amosite (SFA) samples used here provides a basis for validation of the mouse peritoneal cavity as an assay of long CNT effects. In inhalation experiments with rats, the same LFA sample as that used here caused lung cancer, fibrosis and mesothelioma, whilst the same SFA as used here was virtually inactive in causing any of these changes at the same airborne mass concentration exposure (Davis *et al.*, 1986). Previous studies also investigated the long and short amosite

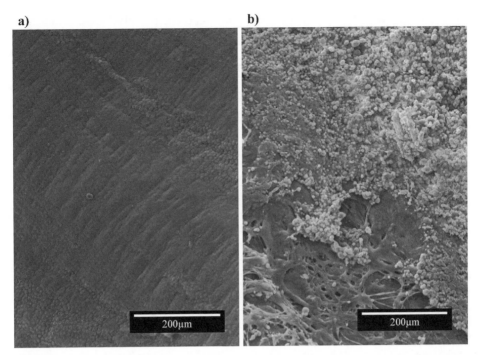

Figure 4.7 *SEM images of diaphragm surface. (a) shows the normal smooth mesothelium with lacunar area at top right and bottom left; (b) shows granuloma covering most of the diaphragm surface after injection of long nanotubes.*

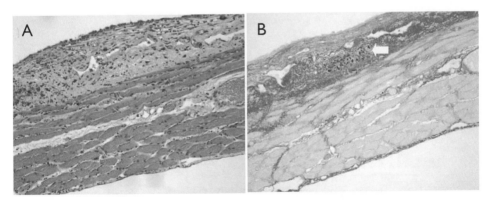

Figure 4.8 *Section through diaphragm from a mouse treated with long CNT 7 days previously. (A) Shows an H&E stained section with normal muscle layers of the diaphragm evident at the lower aspect, and a granuloma present on the upper surface. Image (B) shows the same section stained for collagen by sirius red (dark staining in the image) and shows the collagenous nature of the granulomas; a collection of CNTs can be seen in the centre of the granuloma (arrow). ×100 magnification.*

samples in the peritoneal cavity, measuring production of mesotheliomas. As might be expected, the LFA caused almost 100 % peritoneal mesotheliomas in a large group of rats which received an intraperitoneal dose; intraperitoneal SFA injection of the same mass dose resulted in only one rat developing peritoneal mesothelioma (Davis *et al.*, 1986). If both were delivered at equal mass dose, as used here into the mouse peritoneal cavity, as per the above study, we would predict that long CNTs would be more carcinogenic than LFA, since the long CNTs were even more inflammogenic and fibrogenic than LFA in the mouse peritoneal cavity.

4.6   Mechanism of the Long Fibre/CNT Effect

Reactive oxygen species are considered to be an important factor in the pathogenicity of particles, including asbestos. The ability of iron to redox cycle and generate oxidative stress and inflammation has been suggested to be a potential mechanism for adverse effects of MWCNTs (Kagan *et al.*, 2006). However, we demonstrate here that in this model, the total or water-soluble iron released by the six particle samples is not related to inflammogenic or granuloma-forming ability (Poland *et al.*, 2008). Whilst iron may be necessary in mediating oxidative damage by HARNs, it may not be sufficient for a pathogenic outcome. Retention (or failed clearance), caused by presentation in the shape of long fibres, appears necessary to allow the dose to build up to a critical level. Long fibres cause frustrated or incomplete phagocytosis (Figure 4.9) by virtue of the inability of the macrophage to close around the fibre, resulting in pro-inflammatory state in the macrophage; shorter or tangled fibres are successfully phagocytosed and cleared.

Our findings on CNT are in line with the current fibre paradigm, which has been shown to be true across very different materials: crystalline asbestos, amorphous glasses and

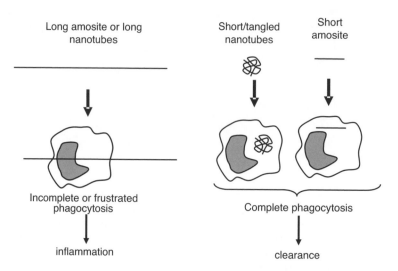

Figure 4.9   *Diagram of frustrated phagocytosis of long fibres, and the impact of fibre geometry on the macrophage response.*

organic polymers. The paradigm requires a fibre to be long and rigid to fulfil the criteria for pathogenicity. MWCNTs comprise rolled tubes of graphene (Donaldson *et al.*, 2006), and in the present study graphene presented as particles, in the form of nanoparticle carbon black, had no effect in the peritoneal cavity. Likewise, an amphibole asbestos sample (SFA), reduced by milling to short fibres, had no effect. Even tangled CNTs, although they were long (20 μm nominal length), were not inflammogenic and only minimally fibrogenic due to their tangled state. Therefore, the fact that MWCNTs can be produced in a morphology that implies an asbestos-like hazard, whilst unwelcome, was predicted (Donaldson *et al.*, 2006). The only study that has examined the release of CNT into air (Maynard *et al.*, 2004) shows images of airborne particles collected on filters that resemble the tangled MWCNTs shown to have low potency here, and that is a reassuring finding. We note that the bought samples of MWCNTs used here also had nonasbestiform morphology and little potency in the assays, and so would not be predicted to behave like asbestos in terms of pathogenicity.

We have shown that the basic tenets of the fibre toxicology paradigm are applicable to MWCNTs using a peritoneal mesothelial exposure model that is highly sensitive to long fibres. The model encompasses the key mechanisms that are likely to occur in the lungs during fibre exposure, namely impaired clearance, inflammation and mesothelial injury. However, the model is not a pulmonary one and therefore has no utility in predicting pulmonary pathological effects. On the other hand, empirically the model clearly responds to long fibres, and in view of the strong evidence uniquely linking mesothelial pathologies to pathogenic fibres, the fact that it involves direct mesothelial exposure gives the model authority. Long and thin CNTs are pathogenic in the peritoneal environment analogously to long amosite asbestos, whilst moderately long or short CNTs that are tangled are very low in potency. The study demonstrates that not all CNTs are created equal in terms of potential toxicity, and that consideration needs to be given to the possibility that long rigid forms of nanotube might, perhaps unsurprisingly, act like harmful asbestos, which has its effect through its length and its thinness. There are clearly tangled and short forms of CNTs that present very low or no asbestos-type of hazard.

4.7 Summary and Conclusions

Asbestos has left an indelible imprint on the social and medical history of occupational and environmental health and toxicology, as well as the public imagination (Tweedale, 2002). The resultant intense research scrutiny that focused on asbestos and other respirable industrial fibres produced a paradigm that explains what makes a respirable fibre pathogenic or not (Donaldson and Tran, 2004). The rise of the nanotechnology industry has been well documented and the associated concerns with its potential risk have been laid out (Royal Society and Royal Academy of Engineering, 2004), along with potential ways to ameliorate these risks (Maynard *et al.*, 2006). One identifiable risk is that high aspect ratio nanoparticles (HARNs) that have the appearance of asbestos may behave like asbestos. This concern was raised in the pages of the journal *Nature* in the form of one of the five grand challenges for safe nanotechnology (Maynard *et al.*, 2006): namely 'Assess whether fibre-shaped nanoparticles present a unique health risk'. Existing studies have pointed to a general hazard from nanotubes (Lam *et al.*, 2004; Warheit *et al.*, 2004; Muller *et al.*, 2005; Shvedova *et al.*, 2005; Li *et al.*, 2007) but this work was not directed specifically to test the hypothesis that CNTs were similar in hazard to asbestos, and so did not use suitable controls and did not focus on the mesothelioma risk, as addressed here. These studies have suggested, however, that CNTs might have considerable fibrogenic activity (Shvedova *et al.*, 2005). The fibre paradigm has been shown to be applicable to asbestos, synthetic vitreous fibres (Bernstein and Hoskins, 2006) and one organic fibre, para-aramid (Warheit *et al.*, 2000). To be pathogenic in this paradigm a fibre sample must:

(1) have an appreciable number of fibres longer than about 20 μm;
(2) be biopersistent in the lungs and not dissolve/break into shorter fibres.

A corollary of these properties is that long biopersistent fibres will have a spectrum of effects that culminate in typical asbestos pathology. These include diminished clearance of the long fibres leading to an accumulation of dose with ongoing exposure in an inhalation situation; epithelial and mesothelial injury and genotoxicity; inflammation; secondary genotoxicity; fibrosis and cancer in the lungs, pleural and peritoneal cavity. The increasing number of studies surrounding CNT toxicity and possible health effects suggest that more hazard-based research is needed, as well as a meaningful conversation between material scientists and toxicologists in considering which forms of CNT to take to the commercial marketplace. Our own research suggests that any CNT that generates long rigid wires/ropes of CNT, through its mode of preparation, should be treated as having a hazard at least equal to amphibole asbestos. Our data also show that CNTs that are short or tangled do not present an asbestos-type hazard. This study is hazard-based and there is a pressing need for exposure data for those who work with MWCNTs in order to assess and manage their risk (Donaldson *et al.*, 2006). The challenges relating to understanding the toxicology, hazards and potential risks of HARNs are numerous. However, the wealth of literature relating to the toxicology of particles (fine, ultrafine/nano) and fibres provides, if used, a strong basis with which to tackle such challenges within the 'window of opportunity' that still remains to develop safe nanotechnologies.

References

Abu-Hijleh MF, Habbal OA, Moqattash ST (1995) The role of the diaphragm in lymphatic absorption from the peritoneal cavity. *J Anat* **186**, 453–467.

Baughman RH, Zakhidov AA, de Heer WA (2002) Carbon nanotubes – the route toward applications. *Science* **297**, 787–792.

Bernstein DM, Hoskins JA (2006) The health effects of chrysotile: current perspective based upon recent data. *Regul Toxicol Pharmacol* **45**, 252–264.

Bolton RE, Davis JM, Donaldson K, Wright A (1982) Variations in the carcinogenicity of mineral fibres. *Ann Occup Hyg* **26**, 569–582.

Brunauer S, Emmett PH, Teller E (1938) Adsorption of gases in multi-molecular layers. *J Amer Chem Soc* **60**, 309–319.

Davis JM, Addison J, Bolton RE, Donaldson K, Jones AD, Smith T (1986) The pathogenicity of long versus short fibre samples of amosite asbestos administered to rats by inhalation and intraperitoneal injection. *Br J Exp Pathol* **67**, 415–430.

Donaldson K, Brown GM, Brown DM, Bolton RE, Davis JG (1989) Inflammation generating potential of long and short fiber amosite asbestos samples. *Br J Industrial Med* **46**, 271–276.

Donaldson K, Tran CL (2004) An introduction to the short-term toxicology of respirable industrial fibres. *Mutat Res* **553**, 5–9.

Donaldson K, Aitken R, Tran L, Stone V, Duffin R, Forrest G, *et al.* (2006) Carbon nanotubes: a review of their properties in relation to pulmonary toxicology and workplace safety. *Toxicol Sci* **92**, 5–22.

Duffin R, Clouter A, Brown DM, Tran CL, MacNee W, Stone V *et al.* (2002) The importance of surface area and specific reactivity in the acute pulmonary inflammatory response to particles. *Ann Occup Hyg* **46**(Suppl. 1), 242–245.

Foucaud L, Wilson MR, Brown DM, Stone V (2007) Measurement of reactive species production by nanoparticles prepared in biologically relevant media. *Toxicol Lett* **174**, 1–9.

Graham A, Higinbotham J, Allan D, Donaldson K, Beswick, PH (1999) Chemical differences between long and short amosite asbestos: differences in oxidation state and coordination sites of iron, detected by infrared spectroscopy. *Occup Environ Medicine* **56**, 606–611.

Gu ZW, Keane MJ, Ong TM, Wallace WE (2005) Diesel exhaust particulate matter dispersed in a phospholipid surfactant induces chromosomal aberrations and micronuclei but not 6-thioguanine-resistant gene mutation in V79 cells. *J Toxicol Environ Health A* **68**, 431–444.

Hills BA (2002) Surface-active phospholipid: a Pandora's box of clinical applications. Part II. Barrier and lubricating properties. *Intern Med J* **32**, 242–251.

Jiang L, Gao L, Sun J (2003) Production of aqueous colloidal dispersions of carbon nanotubes. *J Colloid Interface Sci* **260**, 89–94.

Kagan VE, Tyurina YY, Tyurin VA, Konduru NV, Potapovich AI, Osipov AN *et al.* (2006) Direct and indirect effects of single walled carbon nanotubes on RAW 264.7 macrophages: role of iron. *Toxicol Lett* **165**, 88–100.

Kim SH, Franses EI (2005) New protocols for preparing dipalmitoylphosphatidylcholine dispersions and controlling surface tension and competitive adsorption with albumin at the air/aqueous interface. *Colloids Surf B Biointerfaces* **43**, 256–266.

Kingma PS, Whitsett JA (2006) In defense of the lung: surfactant protein A and surfactant protein D. *Curr Opin Pharmacol* **6**, 277–283.

Kreyling W, Semmler M, Erbe F, Mayer P, Takenaka S, Oberdorster G *et al.* (2002) Minute translocation of inhlaed ultrafine insoluble iridium particles from lung epithelium to extrapulmonary tissues. *Ann Occup Hyg* **46**(Suppl. 1), 223–226.

Lam CW, James JT, McCluskey R, Hunter RL (2004) Pulmonary toxicity of single-wall carbon nanotubes in mice 7 and 90 days after intratracheal instillation. *Toxicol Sci* **77**, 126–134.

Li Q, Kinloch IA, Windle AH (2005) Discrete dispersion of single-walled carbon nanotubes. *Chem Commun (Camb)* 3283–3285.

Li Z, Hulderman T, Salmen R, Chapman R, Leonard SS, Young SH *et al.* (2007) Cardiovascular effects of pulmonary exposure to single-wall carbon nanotubes. *Environ Health Perspect* **115**, 377–382.

Lynch KM, Smith WA (1935) Pulmonary asbestosis. III. Carcinoma of lung in asbestos-silicosis. *Am J Cancer* **1935**, 56–64.

Maynard AD, Baron PA, Foley M, Shvedova AA, Kisin ER, Castranova V (2004) Exposure to carbon nanotube material: aerosol release during the handling of unrefined singlewalled carbon nanotube material. *J Toxicol Environ Health A* **67**, 87–107.

Maynard AD, Aitken RJ, Butz T, Colvin V, Donaldson K, Oberdorster G *et al.* (2006) Safe handling of nanotechnology. *Nature* **444**, 267–269.

Meyer KC, Zimmerman JJ (2002) Inflammation and surfactant. *Paediatr Respir Rev* **3**, 308–314.

Miller BG, Searl A, Davis JM, Donaldson K, Cullen RT, Bolton RE *et al.* (1999) Influence of fibre length, dissolution and biopersistence on the production of mesothelioma in the rat peritoneal cavity. *Ann Occup Hyg* **43**, 155–166.

Mossman BT, Bignon J, Corn M, Seaton A, Gee JB (1990) Asbestos: scientific developments and implications for public policy. *Science* **247**, 294–301.

Mossman BT, Churg A (1998) Mechanisms in the pathogenesis of asbestosis and silicosis. *Am J Respir Crit Care Med* **157**, 1666-1680.

Muller J, Huaux F, Moreau N, Misson P, Heilier JF, Delos M *et al.* (2005) Respiratory toxicity of multi-wall carbon nanotubes. *Toxicol Appl Pharmacol* **207**, 221–231.

Nel A, Xia T, Madler L, Li N (2006) Toxic potential of materials at the nanolevel. *Science* **311**, 622–627.

Nemmar A, Hoylaerts MF, Hoet PH, Nemery B (2004) Possible mechanisms of the cardiovascular effects of inhaled particles: systemic translocation and prothrombotic effects. *Toxicol Lett* **149**, 243–253.

Nii T, Ishii F (2004) Properties of various phosphatidylcholines as emulsifiers or dispersing agents in microparticle preparations for drug carriers. *Colloids Surf B Biointerfaces* **39**, 57–63.

Noh H, Vogler EA (2007) Volumetric interpretation of protein adsorption: competition from mixtures and the Vroman effect. *Biomaterials* **28**, 405–422.

Oberdorster G, Sharp Z, Elder AP, Gelein R, Kreyling W, Cox C (2004) Translocation of inhaled ultrafine particles to the brain. *Inhal Toxicol* **16**, 437–445.

O'Reilly KM, Mclaughlin AM, Beckett WS, Sime PJ (2007) Asbestos-related lung disease. *Am Fam Physician* **75**, 683–688.

Poland CA, Duffin R, Kinloch I, Maynard A, Wallace WA, Seaton A *et al.* (2008) Carbon nanotubes introduced into the abdominal cavity of mice show asbestos-like pathogenicity in a pilot study. *Nat Nanotechnol* **3**, 423–428.

Ramos-Nino ME, Testa JR, Altomare DA, Pass HI, Carbone M, Bocchetta M *et al.* (2006) Cellular and molecular parameters of mesothelioma. *J Cell Biochem* **98**, 723–734.

Renwick LC, Donaldson K, Clouter A (2001) Impairment of alveolar macrophage phagocytosis by ultrafine particles. *Toxicol Appl Pharmacol* **172**, 119–127.

Royal Society and Royal Academy of Engineering (2004) *Nanoscience and Nanotechnologies: Opportunities and Uncertainties*. Royal Society: London.

Selikoff IJ, Lee DHK (1978) *Asbestos and Disease*. Academic Press: New York.

Shvedova AA, Kisin ER, Mercer R, Murray AR, Johnson VJ, Potapovich AI *et al.* (2005) Unusual inflammatory and fibrogenic pulmonary responses to single-walled carbon nanotubes in mice. *Am J Physiol Lung Cell Mol Physiol* **289**, L698–L708.

Takagi A, Hirose A, Nishimura T, Fukumori N, Ogata A, Ohashi N *et al.* (2008) Induction of mesothelioma in p53+/- mouse by intraperitoneal application of multi-wall carbon nanotube. *J Toxicol Sci* **33**, 105–116.

Tran CL, Buchanan D, Cullen RT, Searl A, Jones AD, Donaldson K (2000) Inhalation of poorly soluble particles. II. Influence of particle surface area on inflammation and clearance. *Inhal Toxicol* **12**, 1113–1126.

Tweedale G (2002) Asbestos and its lethal legacy. *Nat Rev Cancer* **2**, 311–315.

Vaisman L, Wagner HD, Marom G (2006) The role of surfactants in dispersion of carbon nanotubes. *Adv Colloid Interface Sci* **128–130**, 37–46.

Veldhuizen EJ, Batenburg JJ, van Golde LM, Haagsman HP (2000) The role of surfactant proteins in DPPC enrichment of surface films. *Biophys J* **79**, 3164–3171.

Wagner JC, Sleggs CA, Marchand P (1960) Diffuse pleural mesothelioma and asbestos exposure in the North Western Cape Province. *Br J Ind Med* **17**, 260–271.

Warheit DB, Hartsky MA, Webb TR (2000) Biodegradability of inhaled p-aramid respirable fibre-shaped particulates: representative of other synthetic organic fibre-types? *Int Arch Occup Environ Health* **73**(Suppl.), S75–S78.

Warheit DB, Laurence BR, Reed KL, Roach DH, Reynolds GA, Webb TR (2004) Comparative pulmonary toxicity assessment of single-wall carbon nanotubes in rats. *Toxicol Sci* **77**, 117–125.

Wood WB, Gloyne SR (1930) Pulmonary asbestosis. *Lancet* **1**, 445–448.

World Health Organization (WHO) (1997) *Determination of Airborne Fibre Number Concentrations: a Recommended Method by Phase Contrast Optical Microscopy.* World Health Organization: Geneva.

Worle-Knirsch JM, Pulskamp K, Krug HF (2006) Oops they did it again! Carbon nanotubes hoax scientists in viability assays. *Nano Lett* **6**, 1261–1268.

5

Application of Zinc Oxide Quantum Dots in Food Safety

Tony Jin, Dazhi Sun, Howard Zhang and Hung-Jue Sue

5.1 Introduction

Outbreaks of foodborne pathogens continue to draw public attention to food safety. The Center for Disease Control (CDC, Atlanta, GA, USA) estimates that approximately 76 million cases of foodborne disease occur each year in the United States, and there are 325 000 hospitalizations and 5000 deaths related to foodborne diseases each year (Mead *et al.*, 1999). The most commonly recognized foodborne infections are those caused by bacteria including *Salmonella, Escherichia coli* O157:H7 and *Listeria monocytogenes*. Occurrences of egg-related outbreaks of *Salmonellosis* (CDC, 2001) and potential outbreaks from egg contamination of *L. monocytogenes* (Leasor and Foegeding, 1989; Moore and Madden, 1993) have heightened the concern for the safety of egg-related products. *E. coli* O157:H7 and *Salmonella* have been implicated in several outbreaks involving acidic fruits or fruit juices (Beuchat, 1996; CDC, 1999; Burnett and Beuchat, 2001; Luedtke and Powell, 2002*)*. Although no outbreaks of listeriosis have been associated with contaminated fruit juice, this ubiquitous pathogen has been recovered from unpasteurized apple juice (pH 3.78) and an unpasteurized apple-raspberry juice blend (pH 3.75) (Sado *et al.*, 1998). Therefore, there is a need to develop new antimicrobials to ensure food safety. The use of antimicrobial agents added directly to foods or through antimicrobial packaging is one effective approach.

In recent years, the use of inorganic antimicrobial agents in non-food applications has attracted interest for the control of microbes (Okouchi *et al.*, 1995; Wilczynski, 2000). The key advantages of inorganic antimicrobial agents are their improved safety and stability under high temperature treatment, compared with organic antimicrobial agents. Presently

Nanotoxicity: From In Vivo *and* In Vitro *Models to Health Risks*   Edited by Saura Sahu and Daniel Casciano
© 2009 John Wiley & Sons, Ltd

most antibacterial inorganic materials are TiO_2 (Shirashi *et al.*, 1999; Huang *et al.*, 2000) and ceramic-immobilized antimicrobial metals, such as silver and copper (Kourai, 1993; Wang *et al.*, 1995). Sawai *et al.* (1995) evaluated the antibacterial activity of 26 ceramic powders, and ten were found to inhibit bacterial growth. Among these active powders, MgO, CaO and ZnO exhibited strong antibacterial activity (Sawai *et al.*, 1995, 1998, 1999, 2000). It was found that the treatment with ZnO formulation caused a net reduction in bacterial cells of 78 % and 62 % in the case of treated cotton and cotton/polyester fabrics, while the net reduction in fungi was calculated to be 80.7 % and 32 %, respectively (Zohdy *et al.*, 2003). Antibacterial activities of metal oxide (ZnO, MgO and CaO) powders against *Staphylococcus aureus*, *E. coli* or fungi were quantitatively evaluated in culture media (Sawai, 2003; Sawai and Yoshikawa, 2004).

ZnO is one of five zinc compounds that are currently listed as 'generally recognized as safe', or GRAS, by the US Food and Drug Administration (21CFR182.8991). Zinc salt has been used for the treatment of zinc deficiency (Saldamli *et al.*, 1996; Lopes de Romana *et al.*, 2002). However, there are few data that demonstrate antimicrobial efficacy of ZnO in foods.

Nanotechnology has the potential to impact many aspects of food and agricultural systems. Food security, disease treatment delivery methods, new tools for molecular and cellular biology, new materials for pathogen detection, and protection of the environment are examples of the important links of nanotechnology to the science and engineering of agriculture and food systems (Weiss *et al.*, 2006). Nanoparticles have been reported for applications in nano-sensors and nanotracers (Moraru *et al.*, 2003). Although nanotechnology has the potential to provide new solutions to improve the safety and quality of food supply, currently there are few reports that focus on the application of nanoparticles in food safety.

Nisin is a bacteriocin produced by certain strains of *Lactococcus lactis*. It was affirmed as GRAS by the Food and Drug Administration (FDA) in 1988 (FDA, 1988), and is now used as a biopreservative in 57 countries around the world. Because it is nontoxic, heat stable, and does not contribute to off-flavors, nisin is used commercially in a variety of foods including dairy, eggs, vegetables, meat, fish, beverages and cereal-based products, to inhibit growths of food-borne pathogens including *Listeria monocytogenes* (Schillinger *et al.*, 1996). Nisin, at various concentrations alone and with other antimicrobial agents incorporated into polyethylene or other edible polymer films, was effective against various microorganisms, including *L. plantarum*, *L. monocytogenes*, *E. coli* and *Salmonella* spp. (Padgett *et al.*, 1998; Cutter *et al.*, 2001; Eswaranandam *et al.*, 2004).

Polylactic acid (PLA) is a biodegradable and compostable polymer that can be derived from renewable resources. PLA is of current interest not only because of the need to ultimately replace many fossil fuel-derived polymers, but also due to the growing global problems associated with plastic waste disposal (Plackett *et al.*, 2006). The use of PLA in food packaging has already received wide attention (Conn *et al.*, 1995; Sinclair, 1996; Haugaard *et al.*, 2002; Frederiksen *et al.*, 2003). There have been developments in Europe and in North America that have involved the use of PLA-based packaging for supermarket products, such as Biota™ PLA bottled water, Noble™ PLA bottled juices, and Dannon™ yogurts. The special characteristics of PLA, such as GRAS status, biodegradability and being a bio-resource, put PLA in a unique position for food applications. The results of our previous study (Jin and Zhang, 2008) suggested that the incorporation of nisin into the PLA polymer could provide a possible delivery system for improving the efficacy of nisin in food applications.

ZnO quantum dots (ZnO QDs) are nanoparticles of sizes smaller than 7 nm which exhibit the quantum confinement phenomenon. In our previous study, we investigated the antimicrobial activity of ZnO QDs against *L. monocytogenes, S.* Enteritidis and *E. coli* O157:H7 in growth media (Jin *et al.*, 2009). In the present study, the efficacy of ZnO alone or in combination with nisin in reducing *L. monocytogenes, S.* Enteritidis and *E. coli* O157:H7 in liquid egg white, apple cider and orange juice was investigated. Application of ZnO QDs to food systems (direct addition or slow release from a PLA coating) was also evaluated.

5.2 Materials and Methods

5.2.1 Preparation of ZnO QD Samples

5.2.1.1 ZnO QD Powder

Colloidal ZnO QDs were prepared and purified according to our previous report (Sun *et al.*, 2007). Briefly, 200 ml of methanol containing 0.08 M of KOH (99.99 %; Aldrich, Milwaukee, WI) was first heated at 60° C for 30 min to obtain a homogeneous solution. Powder of 0.008 mol $Zn(AC)_2 \cdot H_2O$ (99 %, Fluka, Castle Hill, Sweden) was then directly added into a basic methanol solution to make the final $[Zn^{2+}]$ of 0.04 M. This starting solution was allowed to react at 60° C for 2 hours with stirring and refluxing. After reaction, the ZnO solution was concentrated 10 times via rotary evaporation under vacuum at 40° C. Subsequently, hexane and isopropanol was added into the concentrated ZnO solution with a volume ratio of hexane:isopropanol:methanol of 5:1:1. The ZnO QDs precipitated immediately after adding hexane and isopropanol. This mixture was kept at 0° C overnight until the ZnO QDs were fully precipitated and settled to the bottom of the container. After removal of the supernatant, ZnO QDs were redispersed in methanol. The above operations were repeated at least twice to purify the ZnO nanoparticles. The purified ZnO QDs were then baked in an oven at 110° C for 2 hours and white ZnO QD powder was obtained.

Figure 5.1 shows the high-resolution transmission electron microscopy (HR-TEM) image of prepared ZnO QDs. The spherical particles were highly crystalline (inset of the TEM image) and had an average uniform size distribution of 5 nm. After concentrating and washing, zinc byproduct (zinc layered double hydroxide) and unwanted ions (K^+ and AC^-) were effectively removed (>99.8 % of K^+ can be removed in three washes) (Sun *et al.*, 2007). The prepared ZnO QDs were directly used to study the antimicrobial activity of ZnO QD powder after drying.

5.2.1.2 ZnO QD Coating

One gram of PLA resin (Natureworks, Minnetonka, MN) and 0.25 g of ZnO were accurately weighed and dispersed in 15 ml of methylene chloride (Fisher Scientific, Fairlawn, NJ). This mixture was stirred by a magnetic bar until the polymer was totally dissolved. The mixture was distributed to a 4 oz glass jar (Chases Scientific Glass, Rockwood, TN) that was horizontally rolling so the mixture could coat the inside wall of the jar. The methylene chloride was allowed to evaporate at room temperature during the jar's rolling under a chemical hood. The coated jars were dried in a vacuum oven for 24 h, and then fitted with lids and stored until time of use. For comparison, 0.25 g of nisin (2.5 % purity,

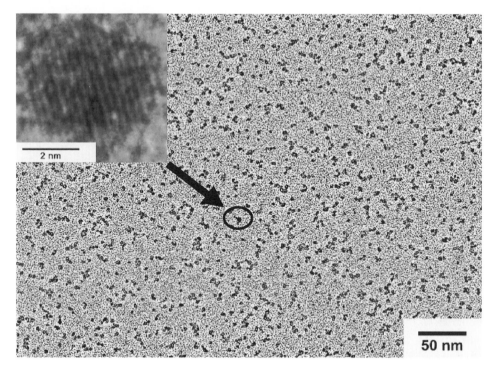

Figure 5.1 *HR-TEM image of ZnO QD powder. Reproduced from Jin et al. (2009), with permission of John Wiley & Sons Ltd.*

Sigma Chemical Co., St. Louis, MO) alone or combined with ZnO was added to the PLA/methylene chloride mixture.

5.2.2 Bacterial Inhibition Tests

5.2.2.1 Food Samples

Commercial pasteurized liquid egg white (LEW), apple cider (AC) and orange juice (OJ) without preservatives were purchased from a local grocery store. The pHs of LEW, AC and OJ were 8.2, 3.9 and 3.7, respectively.

5.2.2.2 Cultures

Listeria monocytogenes Scott A 724, *Escherichia coli* O157:H7 Oklahoma, and *Salmonella enterica serovar* Enteritidis ATCC 13076 were obtained from the culture collection of the US Department of Agriculture, Agricultural Research Service, Eastern Regional Research Center. Stock cultures were maintained at −80° C. The stains were propagated on Tryptic Soy Agar (TBA: Difco Laboratories, Detroit, MI) at 37° C and maintained at 0–2° C until use. Prior to the inoculum preparation, *E. coli* O157:H7 and *S.* Enteritidis cells were grown in tryptic soy broth (TSB: Remel, Inc., Lenexa, KS), and *L. monocytogenes* cells were grown in brain heart infusion broth (BHIB: Difco Laboratory, Detroit, MI) aerobically at 37° C for 16–18 h.

5.2.2.3   Antibacterial Test

For the direct addition test, a certain amount of ZnO powder was placed in a 4 oz glass jar with 100 ml of liquid medium. For the coating/release test, coated 4 oz glass jars were used and they contained 100 ml of liquid medium. For all the tests the liquid media were inoculated with overnight cultures of *L. monocytogenes, E. coli* O157:H7 or *S.* Enteritidis. The final cell density in a medium was approximately 1×10^4 cells per ml in each bacterial inhibition test. The jars were shaken at 50 rpm at 22° C, 10° C or 4° C. The inoculated media were sampled (1.0 ml) at certain time intervals. Specimens were serially diluted by sterile Butterfield's phosphate buffer (pH 7.2, Hardy Diagnostics, Santa Maria, CA), then pour plated onto BHI agar or tryptic soy agar. Plates were incubated at 37° C for 24 h. A ZnO-free inoculated medium served as a control. Room temperature (22° C) was selected for the incubation tests after considering the worst case scenario, in which food was left at room temperature, at which the pathogens grow much faster, rather than refrigerated. The incubation temperature was also tested at 10° C which is the 'abused' refrigeration temperature, and 4° C which is the ideal storage temperature according to manufacturers.

5.2.3   High-Resolution Transmission Electron Microscopy (HR-TEM)

The purified ZnO QDs were redispersed in methanol, diluted, and droplets of each solution were placed onto a 400-mesh carbon-coated copper grid. The grids were then dried in a desiccator for one day before imaging. HR-TEM of the above samples was carried out using a JEOL 2010 high-resolution transmission electron microscope operated at 200 kV.

5.2.4   Statistical Analysis

Antimicrobial experiments were conducted in triplicate. Data points were expressed as the mean ± SD. Data were analyzed using analysis of variance from SAS version 9.1 software (SAS Institute, Cary, NC). Duncan's multiple range tests were used to determine the significant difference of mean values. Unless stated otherwise, significance was expressed at the 5 % level.

5.3   Results and Discussion

5.3.1   Antibacterial Activity of ZnO QD against *Listeria monocytogenes* in Apple Cider and Liquid Egg White

Figure 5.2 illustrates the effect of ZnO QDs against growth of *L. monocytogenes* in AC during 5-day storage at 22° C. Three concentrations of ZnO QDs were used as the antimicrobial treatments in AC samples: 0.28, 0.56 and 1.12 mg/ml. Treatments of ZnO QDs at 0.56 mg/ml and 1.12 mg/ml significantly reduced *Listeria* cells and exhibited significant inhibitory effect on the growth of *L. monocytogenes* in AC during 5-day storage, as compared with the control. The treatment of ZnO QDs at 0.28 mg/ml had less inhibitory effect on the growth of *L. monocytogenes* after 2-day storage, although the treated AC sample had significantly lower *Listeria* cells than the control at day 3 and day 5. These data indicate that the inactivation of *Listeria* cells in AC was dependent on the concentration of the ZnO; the higher the concentration of ZnO used, the more efficacy was achieved.

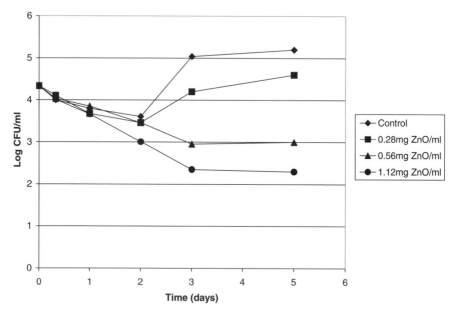

Figure 5.2   *Effect of ZnO nanoparticles on the growth of L. monocytogenes in apple cider at 22° C.*

The coating/release tests were used in this study to evaluate the potential application of antimicrobial packaging. Glass jars were coated with ZnO QDs, nisin, or their combination with a carrier of PLA. The growth of *L. monocytogenes* in LEW in the coated jars was investigated at 22° C, 10° C and 4° C, and the results are displayed in Figure 5.3 (a, b and c, respectively). The treatments of PLA+ZnO, PLA+nisin and PLA+ZnO+nisin coatings significantly inhibited the growth of *L. monocytogenes* in LEW at 22° C for 8 days, 10° C for 36 days, and 4° C for 86 days, respectively. Among them, the PLA+ZnO+nisin coating was the most effective in the reduction of *Listeria* cells. After an initial drop in population of *Listeria*, cell populations in controls increased to 7.2, 9.0 and 5.5 log CFU/ml by day 8, 36 and 86 at 22° C, 10° C and 4° C, respectively, while the treatment of PLA+ZnO+nisin coating reduced the cells to undetectable levels (detection limit 10 CFU/ml) by day 2 and remained at those levels throughout the rest of the storage period at each temperature. These data suggest that there was a synergistic anti-listerial activity in the combination of ZnO and nisin. The ZnO coating had less effectiveness against *Listeria* in LEW than the nisin coating at all three temperatures.

No significant differences ($P > 0.05$) between control and PLA coating in the growth of *L. monocytogenes* occurred at the three temperatures, which suggests that the treatment of PLA alone did not contribute to anti-listerial effects.

5.3.2   Antibacterial Activity of ZnO QD against *Salmonella* Enteritidis in Liquid Egg White and Orange Juice

Figure 5.4 (A and B) shows the growth of *S.* Enteritidis in LEW with or without the presence of ZnO QDs at 22° C and 4° C. The numbers of *Salmonella* cells in LEW samples

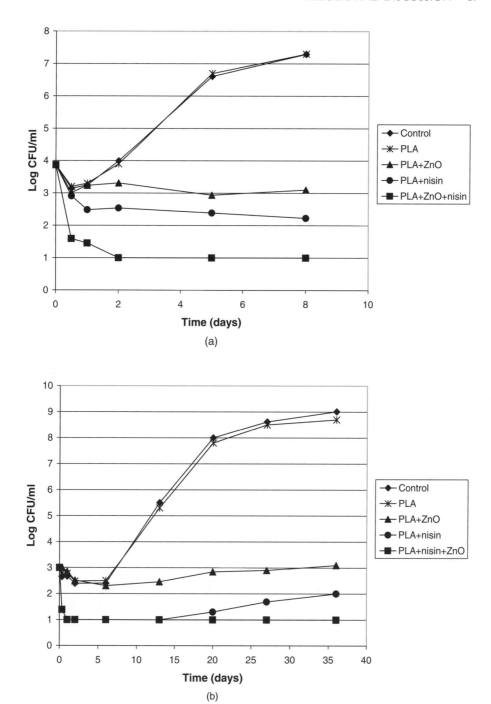

Figure 5.3 *Effect of ZnO, nisin and ZnO+nisin coatings on the growth of L. monocytogenes in liquid egg white at (a) 22° C, (b) 10° C and (c) 4° C.*

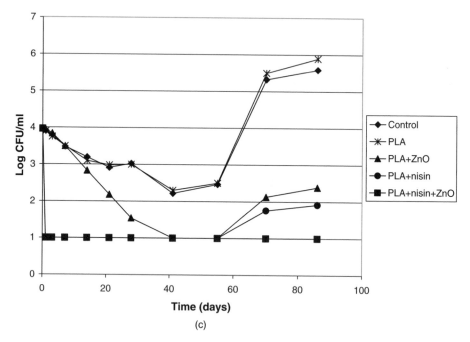

(c)

Figure 5.3   *(Continued)*

were significantly reduced in the presence of ZnO at both temperatures. These data are consistent with the *L. monocytogenes* data where the inhibitory effect of ZnO nanoparticles against *Salmonella* was dependent on the concentration of ZnO. As shown in Figure 5.4a, the numbers of *Salmonella* cells at day 8 were 5.5 and 3.5 log CFU/ml for 0.28 and 1.12 mg/ml, respectively, whereas the control had 9.7 log CFU/ml. In addition, when tested at refrigeration temperature (Figure 5.4b), *S.* Enteritidis cells in all samples grew at a much slower rate than those at room temperature. There were no significant differences in cell counts between control and ZnO-treated samples within 5-day storage at $4°$ C. After that, ZnO-treated samples had significantly lower *Salmonella* cells than the controls. Lowest cell counts were observed after 15 days for all samples at $4°$ C. The cell counts increased during further storage. After 41 days at $4°$ C, samples with ZnO had 3 and 4.7 log CFU/ml of *Salmonella* cells at 1.12 and 0.28 mg/ml, respectively. The *Salmonella* cells in the control samples (Figure 5.4b) increased to 7.4 log CFU/ml, which was similar to the samples stored at $24°$ C for 5 days (Figure 5.4a). These data suggest that the combination of ZnO treatment with low storage temperatures could significantly prevent the growth of *Salmonella* in LEW over an extended storage period.

Results for populations of *S.* Enteritidis in OJ stored at $22°$ C in the coating tests are shown in Figure 5.5. Populations in controls and PLA-coated jars increased from 3.8 log CFU/ml on day 0 to 5.9 log CFU/ml by day 13, whereas samples in the presence of ZnO coating or the nisin coating increased to only 4.6 and 4.5 log CFU/ml, respectively. Similar to *Listeria*, the PLA+ZnO+nisin coating exhibited more effectiveness against *Salmonella* in OJ than the individual ZnO coating or nisin coating. The PLA+ZnO+nisin

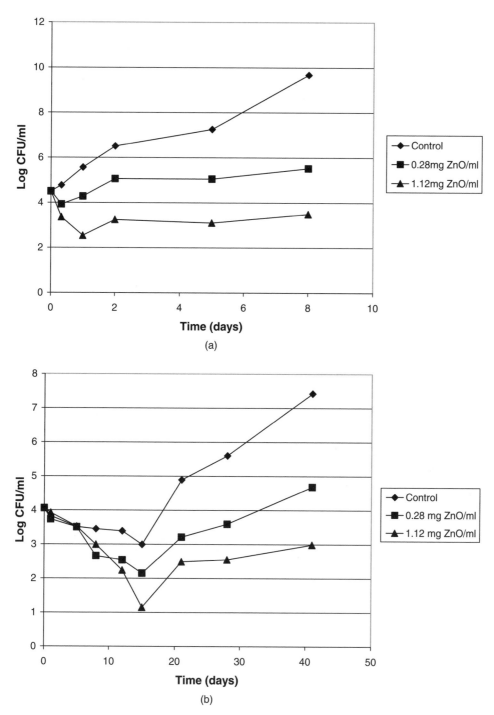

Figure 5.4 *Effect of ZnO nanoparticles on the growth of S. Enteritidis in liquid egg white at (a) 22° C, and (b) 4° C. Reproduced from Jin* et al. *(2009), with permission of John Wiley & Sons Ltd.*

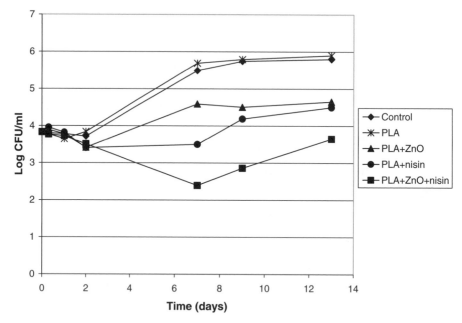

Figure 5.5   *Effect of ZnO, nisin and ZnO+nisin coatings on the growth of S. Enteritidis in orange juice at 22° C.*

coating reduced *Salmonella* from 3.9 log CFU/ml to 2.3 log CFU/ml by day 7, then slowly increasing to 3.6 log CFU/ml by day 13, which was 2.3 log CFU/ml less than in the control.

5.3.3   Antibacterial Activity of ZnO QD against *E. coli* O157:H7 in Apple Cider and Orange Juice

Figure 5.6 illustrates the effect of ZnO QD coatings on the growth of *E. coli* O157:H7 in AC at 22° C. A slight increase (ca. 0.5 log CFU/ml) in population occurred within 21-day storage in AC treated by ZnO and nisin coatings, whereas the control and PLA coating increased to 4.5 log CFU/ml by day 10, then slightly decreased to 4 log CFU/ml by day 21. A reduction of approximate 0.4 log CFU/ml was achieved by the PLA+ZnO+nisin coating at day 3. Throughout the rest of the 21-day storage period, the *E. coli* O157:H7 cell population in the PLA+ZnO+nisin coating samples remained at a constant level of around 2.8 log CFU/ml.

When 3.3 log CFU/ml of *E. coli* O157:H7 were inoculated in OJ, the cells in the PLA coating and control samples gradually decreased during storage at 22° C and declined to 2.2 log CFU/ml by day 21 (Figure 5.7). The treatments of PLA+ZnO, PLA+nisin and PLA+ZnO+nisin coatings significantly affect the survival of *E. coli* O157:H7 in OJ. Coatings containing PLA+ZnO, PLA+nisin and PLA+ZnO+nisin reduced the number of organisms to undetectable populations (<10 CFU/ml) at day 21, day 17 and day 10, respectively. Similar to *L. monocytogenes* and *S.* Enteritidis, when used in combination, ZnO QDs plus nisin had enhanced antimicrobial activity against *E. coli* O157: H7 in both AC and OJ.

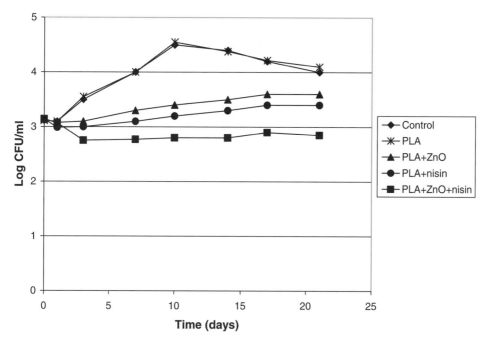

Figure 5.6 *Effect of ZnO, nisin and ZnO+nisin coatings on the growth of* E. coli *O157:H7 in apple cider at 22° C.*

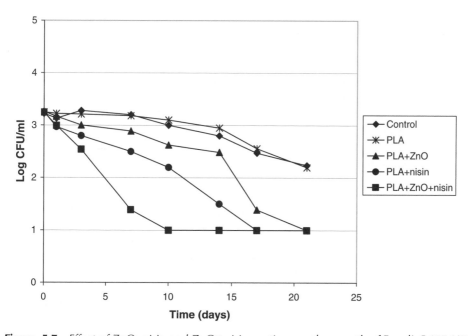

Figure 5.7 *Effect of ZnO, nisin and ZnO+nisin coatings on the growth of* E. coli *O157:H7 in orange juice at 22° C.*

These data indicate that *E. coli* O157:H7 can grow in AC (Figure 5.6) or survive well in OJ (Figure 5.7). Other researchers also reported that *E. coli* O157:H7 can survive or even grow in some acidic fruit juices, such as apple cider/juice (pH 3.5 to 4.1) (Zhao *et al.*, 1993; Miller and Kaspar, 1994; Beuchat, 1996; Semanchek and Golden, 1996; Ryu and Beuchat, 1998; Duffy and Schaffner, 2001) and OJ (pH 3.5 to 4.0) (Linton *et al.*, 1999). Thus, there would be a risk of food poisoning if AC or OJ becomes contaminated with *E. coli* O157:H7 and does not receive proper treatment before consumption. Using ZnO QDs or its combination with nisin is an effective approach to increase the safety of AC and OJ substantially.

The inhibition action of ZnO QDs may contribute to two modes of microbial inactivation: killing and growth suppression. In higher ZnO concentrations, ZnO treatments initially kill a sensitive subpopulation of cells and reduce the total microbial population, and then retard the growth of a resistant surviving subpopulation. In lower ZnO concentrations, ZnO treatments suppress the growth of bacteria and result in a 2-day or 8-day lag before a slow growth. Increasing ZnO concentration in a food system resulted in more killing of a sensitive subpopulation of cells, a longer lag, and less growth of a resistant surviving subpopulation. The mechanism for killing and growth suppression induced by ZnO nanoparticles has not been reported yet, and needs further investigation. It has been reported that zinc homeostasis is regulated through a number of specific and nonspecific membrane-bound uptake and efflux pumps (Blencowe and Morby, 2003), and zinc is capable of inhibiting nitric oxide formation (Abou-Mohamed *et al.*, 1998) and prevents sulfhydryl groups from oxidation (Lim *et al.*, 2004).

Nisin has been proven more effective against gram positive bacteria than gram negative bacteria because of the difference in cell wall structure (Hauben *et al.*, 1996). The same phenomena were also observed in this study. Nisin causes cell death as a result of the formation of relatively large pores in the membranes of sensitive cells, resulting in the release of ions, amino acids and adenosine triphosphate (ATP), and the subsequent collapse of proton motive force (Moll *et al.*, 1997). Although the mechanisms for the synergistic effect of ZnO+nisin against the three pathogens are not clear, it might be explained that the large pores caused by nisin treatment facilitate ZnO molecules to contact or penetrate into microbial cells to express their antibacterial activities. Further study is needed to confirm the above conjecture.

Antimicrobials can be added to food formulations directly or through slow release from packaging materials. Direct addition of antimicrobials into foods results in an immediate reduction of bacterial populations, but may not prevent the recovery of injured cells or the growth of cells that were not destroyed by direct addition, if residues of the antimicrobial are rapidly depleted (Zhang *et al.*, 2004). The application of antimicrobial films or coatings allows for the migration of the antimicrobial to the coating or film surface, and provides a continuous antimicrobial effect on the food during extended exposure. In our previous study (Jin and Zhang, 2008), we demonstrated that the retention of nisin activity occurred when nisin was incorporated into a PLA polymer, and the PLA/nisin polymer exhibits effective antibacterial activity against foodborne *L. monocytogenes*, *E. coli* O157:H7 and *S.* Enteritidis. Use of polymers as carriers of antimicrobials such as ZnO QDs and nisin not only controls release of these antimicrobials, but also prevents dramatic reductions in their antimicrobial activities due to its affinity for food particles and its inactivation by components in foods. Therefore, the use of antimicrobial packaging such as bottle/jar coatings can offer advantages compared with the direct addition of preservatives to the liquid foods, such that only low levels of preservative come into contact with the food.

5.4   Conclusions

ZnO nanoparticles possess antimicrobial effects against growths of *L. monocytogenes*, *S.* Enteritidis and *E. coli* O157:H7 in liquid foods. The inhibitory effect of ZnO was concentration-dependent, and synergistically enhanced by the combination with nisin. This study also demonstrates that a coating method using PLA as a carrier was an effective approach for incorporation of ZnO into a food system. The development of antimicrobial packaging with ZnO nanoparticles and its application in food safety appear to be promising.

Acknowledgements

The authors wish to thank Drs Joshua Gurtler and Bill Dantzer for their thoughtful reviews of this manuscript, and Anita Parameswaran for technical assistance.

References

Abou-Mohamed G, Papapetropoulos A, Catravas JD, Caldwell RW (1998) Zn^{+2} inhibits nitric oxide formation in response to lipopolysaccharides: implication in its anti-inflammatory activity. *Eur J Pharmacol* **341**, 265–272.

Beuchat LR (1996) Pathogenic microorganisms associated with fresh produce. *J Food Prot* **59**, 204–216.

Blencowe DK, Morby AP (2003) Zn(II) metabolism in prokaryotes. *FEMS Microbiol Rev* **27**, 291–311.

Burnett SL, Beuchat LR (2001) Human pathogens associated with raw produce and unpasteurized juices, and difficulties in decontamination. *J Ind Microbiol Biotechnol* **27**, 104–110.

Center for Disease Control (CDC) (1999) Outbreak of salmonella serotype muenchen infections associated with unpasteurized orange juice – United States and Canada, June 1999. *MMWR Morb Mortal Wkly Rep* **48**, 582–585.

Center for Disease Control (CDC) (2001) Outbreaks of multidrug-resistant *Salmonella* Typhimurium associated with veterinary facilities – Idaho, Minnesota, and Washington, 1999. *MMWR Morb Mortal Wkly Rep* **50**, 701–704.

Conn RE, Kolstad JJ, Brozelleca JF, Dixler DS, Filer LJ, LaDu BN Jr, Pariza MW (1995) Safety assessment of polylactide (PLA) for use as a food-contact. *Polym Food Chem Toxic* **33**, 273–283.

Cutter CN, Willett JL, Siragusa GR (2001) Improved antimicrobial activity of nisin-incorporated polymer film by formulation change and addition of food grade chelator. *Lett Appl Microbiol* **33**, 325–328.

Duffy S, Schaffner DW (2001) Modeling the survival of *Escherichia coli* O157:H7 in apple cider using probability distribution functions for quantitative risk assessment. *J Food Prot* **64**, 599–605.

Eswaranandam S, Hettiarachchy NS, Johnson MG (2004) Antimicrobial activity of citric, lactic, malic, or tartaric acids and nisin-incorporated soy, protein film against *Listeria monocytogenes, Escherichia coli* O157:H7, and *Salmonella gaminara*. *J Food Sci* **69**, 78–84.

Food and Drug Administration (FDA) (1988) Nisin preparation: affirmation of GTAS status as a direct human food ingredient. *Federal Register* **53**, 11247–11251.

Frederiksen CS, Haugaard VK, Poll L, Becker EM (2003) Light-induced quality changes in plain yogurt packaged in polylactate and polystyrene. *Eur Food Res Technol* **217**, 61–69.

Hauben KJA, Wuytack EY, Scootjens CCF, Michiels CW (1996) High pressure transient sensitization of *E. coli* to lysozyme and nisin by disruption of outer membrane permeability. *J Food Prot* **59**, 350–355.

Haugaard VK, Weber CJ, Danielsen B, Bertelsen G (2002) Quality changes in orange juice packaged in materials based on polylactate. *Eur Food Res Technol* **214**, 423–428.

Huang Z, Maness PC, Blakee DM, Wolfrum EJ, Smoliski SL, Jacoby WA (2000) Bacterial mode of titanium dioxide photocatalysis. *J Photochemery Photobiology A: Chemistry* **130**, 163–170.

Jin T, Zhang H (2008) Biodegradable polylactic acid polymer with nisin for use in antimicrobial food packaging. *J Food Sci* **73**, M127–134.

Jin T, Sun D, Su JY, Zhang H, Sue HJ (2009) Antimicrobial efficacy of zinc oxide quantum dots against *Listeria monocytogenes, Salmonella* Enteritidis and *Escherichia coli* O157:H7. *J Food Sci* **74**, M46–M52.

Kourai H (1993) Immobilized microbiocide. *J Antibact Antifungal Agents* **21**, 331–337 (in Japanese).

Leasor SB, Foegeding PM (1989) Listeria species in commercially broken raw liquid whole egg. *J Food Prot* **52**, 777–780.

Lim Y, Levy M, Bray TM (2004) Dietary zinc alters early inflammatory responses during cutaneous wound healing in weanling CD-1 mice. *J Nutr* **134**, 811–816.

Linton M, McClements JMJ, Patterson MF (1999) Survival of *Escherichia coli* O157:H7 during storage in pressure-treated orange juice. *J Food Prot* **62**, 1038–1040.

Lopes de Romana D, Brown KH, Guinard JX (2002) Sensory trial to assess the acceptability of zinc fortificants added to iron-fortified wheat products. *J Food Sci* **67**, 461–465.

Luedtke AN, Powell DA (2002) A review of North American apple cider-associated *E. coli* O157:H7 outbreaks, media coverage and a comparative analysis of Ontario apple cider producers' information sources and production practices. *Dairy Food Environ Sanit* **22**, 590–598.

Mead PS, Slutsker L, Dietz V, McCaig LF, Bresee JS, Shapiro C, Griffin PM, Tauxe RV (1999) Food-related illness and death in the United States. *Emerging Infectious Diseases* **5**, 607–625.

Miller LG, Kaspar CW (1994) *Escherichia coli* O157:H7 acid tolerance and survival in apple cider. *J Food Prot* **57**, 460–464.

Moll GN, Clark J, Chan WC, Bycroft BW, Roberts GCK, Konings WM, Driessen AJM (1997) Role of transmembrance pH gradient and membrane binding in nisin pore formation. *J Bacteriol* **179**, 135–140.

Moore J, Madden RH (1993) Detection and incidence of Listeria species in blended raw egg. *J Food Prot* **56**, 652–660.

Moraru CI, Panchapakesan CP, Huang Q, Takhistove P, Liu S, Kokini JL (2003) Nanotechnology: A new frontier in food science. *Food Technol* **57**(12), 24–29.

Okouchi S, Murata R, Sugita H, Moriyoshi Y, Maeda N (1995) Calorimetric evaluation of the antimicrobial activities of calcined dolomite. *J Antibact Antifungal Agents* **26**, 109–114 (in Japanese).

Padgett T, Han IY, Dawson PL (1998) Incorporation of food antimicrobial compounds into biodegradable packaging films. *J Food Prot* **61**, 1330–1335.

Plackett DV, Holm VK, Johansen P, Ndoni S, Nielsen PV, Sipilainen-Malm T, Sodergard A, Verstichel S (2006) Characterization of L-polylactide and L-polylactide-polycaprolactone copolymer film for use in cheese-packaging applications. *Packaging Technology and Science* **19**, 1–24.

Ryu JH, Beuchat LR (1998) Influence of acid tolerance responses on survival, growth, and thermal cross-protection of *Escherichia coli* O157:H7 in acidified media and fruit juices. *Int J Food Microbiol* **45**, 185–193.

Sado P, Jinneman KC, Husby GJ, Sorg SM, Omiecinski CJ (1998) Identification of *Listeria monocytogenes* from unpasteurized apple juice using rapid test kits. *J Food Prot* **61**, 1199–1202.

Saldamli I, Kokshel H, Ozboy O, Ozalp I, Kilic I (1996) Zinc-supplemented bread and its utilization in zinc deficiency. *Cereal Chem* **73**, 424–427.

Sawai J (2003) Quantitative evaluation of antibacterial activities of metallic oxide powders (ZnO, MgO and CaO) by conductimetric assay. *J Microbiological Methods* **54**, 177–182.

Sawai J, Igarashi H, Hashimoto A, Kokugan T, Shimizu M (1995) Evaluation of growth inhibitory effect of ceramic powder slurry on bacteria by conductance method. *J Chem Eng Jpn* **28**, 288–293.

Sawai J, Shoji S, Igarashi H, Hashimoto A, Kokugan T, Shimizu M, Kojima H (1998) Hydrogen peroxide as an antibacterial factor in zinc oxide powder slurry. *J Ferment Bioeng* **86**, 521–522.

Sawai J, Kojima H, Igarashi H, Hashimoto A, Shoji S, Shimizu M (1999) Bactericidal action of calcium oxide powder. *Trans Mater Res Soc Jpn* **24**, 667–670.

Sawai J, Kojima H, Igarashi H, Hashimoto A, Shoji S, Sawaki T, Hakoda A, Kawada E, Kokugan T, Shimizu M (2000) Antibacterial characteristics of magnesium oxide powder. *World J Microbiol Biotechnol* **16**, 187–194.

Sawai J, Yoshikawa T (2004) Quantitative evaluation of antifungal activity of metallic oxide powders (MgO, CaO and ZnO) by an indirect conductimetric assay. *J Appl Microbiol* **96**, 803–809.

Schillinger U, Geisen R, Holzapfel WH (1996) Potential of antagonistic microorganisms and bacteriocins for the biological preservation of foods. *Trends Food Sci Tech* **7**, 158–164.

Semanchek JJ, Golden DA (1996) Survival of *Escherichia coli* O157:H7 during fermentation of apple cider. *J Food Prot* **59**, 1256–1259.

Shirashi F, Toyoda K, Fukinbara S (1999) Photolytic smf photocatalytic treatment of an aqueous solution containing microbial cells and organic compounds in an annular-flow reactor. *Chem Eng Sci* **54**, 1547–1552.

Sinclair RG (1996) The case for polylactic acid as a commodity packaging plastic. *J Macromol Sci Pure Appl Chem* **33**, 585–597.

Sun D, Wong M, Sun L, Li Y, Miyatake N, Sue HJ (2007) Purification and stabilization of colloidal ZnO nanoparticles in methanol. *J Sol-Gel Sci Technol* **43**, 237–243.

Wang YL, Wan YZ, Dong XH, Cheng GX, Tao HM, Wen TY (1995) Preparation and characterization of antibacterial viscose-based activated carbon fiber supporting silver. *Carbon* **36**, 1567–1571.

Weiss J, Takhistov P, McClements DJ (2006) Functional materials in food nanotechnology. *J Food Sci* **71**(9), R107–R116.

Wilczynski M (2000) Anti-microbial porcelain enamels. *Ceram Eng Sci Proc* **21**, 81–83.

Zhang YC, Yam KL, Chikindas ML (2004) Effective control of *Listeria monocytogenes* by combination of nisin formulated and slowly released into a broth system. *Int J Food Microbiol* **90**, 15–22.

Zhao T, Doyle MP, Besser RE (1993) Fate of enterohemorrhagic *Escherichia coli* O157:H7 in apple cider with and without preservatives. *Appl Environ Microbiol* **59**, 2526–2530.

Zohdy MH, Abdel Kareem H, El-Naggar AM, Hassan MS (2003) Microbial detection, surface morphology, and thermal stability of cotton and cotton/polyester fabrics treated with antimicrobial formulations by a radiation method. *J Appl Polym Sci* **89**, 2604–2610.

6

Evaluation of Nanotoxicity of Foods and Drugs: Biological Properties of Red Elemental Selenium at Nano Size (Nano-Se) *In Vitro* and *In Vivo*

Jinsong Zhang

6.1 Introduction

The first description of selenium was published by Jacob Berzelius, who obtained a reddish precipitate in 1817 while burning pyrite in a sulfuric acid plant. Upon analysis this precipitate turned out to be a new element. He called the element after Selene, the Greek goddess of the moon. The human body metabolizes various selenium forms into selenide, which seems to be the common point for regulating selenium metabolism. Hydrogen selenide provides selenium for the synthesis of a number of selenoenzymes (Suzuki *et al.*, 2006; Ohta and Suzuki, 2008), including types 1, 2, 3 and 4 glutathione peroxidase; types 1, 2 and 3 thioredoxin reductase; and types 1, 2 and 3 iodothyronine deiodinase (Flohé *et al.*, 2000; Stadtman, 2000). Glutathione peroxidases catalyze the reduction of peroxides that can cause cellular damage (Brigelius-Flohé, 1999; Brigelius-Flohé *et al.*, 2001; Conrad *et al.*, 2007). Thioredoxin reductases provide reducing power for thioredoxin, which regulates multiple important biochemical processes and defends against oxidative stress. Additionally, thioredoxin reductase alone possesses a number of antioxidant functions, such as reducing selenium compounds, regenerating ascorbyl radical, lipoic acid and ubiquinone, and scavenging lipid peroxides and hydrogen peroxide (Arnér and Holmgren, 2000; Lillig and Holmgren, 2007; Papp *et al.*, 2007). Thyroid deiodinases function in the formation and regulation of active thyroid hormone (Beckett and Arthur, 2005; Köhrle

Nanotoxicity: From In Vivo *and* In Vitro *Models to Health Risks*   Edited by Saura Sahu and Daniel Casciano
© 2009 John Wiley & Sons, Ltd

et al., 2005). A direct deleterious consequence of selenium deficiency is compromised selenoenzymes. Consequently, severe selenium deficiency can result in the genesis of many diseases, such as muscular dystrophy and liver necrosis in animals (Oldfield, 1989; Schwarz and Foltz, 1999; Rederstorff *et al.*, 2006); Keshan disease (a cardiomiopathy with myocardial insufficiency) and Kashin–Beck disease (a type of osteoarthritis) in humans (Yang *et al.*, 1984; Moreno-Reyes *et al.*, 1998). In recent years selenium has attracted increasing interest as a chemical with potential capacity to reduce carcinogenesis (Schrauzer, 2000a; El-Bayoumy and Sinha, 2004; Brigelius-Flohé, 2008; Naithani, 2008; Zeng and Combs, 2008). Case-controlled epidemiological studies have shown an inverse relationship between selenium status and cancer risk (Rayman, 2005; Brinkman *et al.*, 2006). A randomized placebo-controlled Nutritional Prevention of Cancer study involving 1312 subjects showed that selenium supplementation reduced the overall cancer morbidity by nearly 50 %, and lowered the prevalence of developing prostate cancer by 63 % (Clark *et al.*, 1996). Another noticeable feature of selenium is its toxicity. Seleniferous plants can cause the hoofs of animals to drop off (O'Toole and Raisbeck, 1995). An endemic disease due to a selenium uptake of about 5 mg per day has been described; classic symptoms of selenium toxicity in humans are changes in skin, hair, nails, and nervous system function (Yang *et al.*, 1983).

Selenium is one micronutrient whose deficiency and toxic concentrations are close to each other. On the other hand, the cancer-preventive effects of selenium were mostly observed at high dose in experimental animals (El-Bayoumy and Sinha, 2004). The biological activity and toxicity of selenium are associated with chemical form (Ip, 1998; Whanger, 2002; Rayman *et al.*, 2008). For nutritional supplementation and cancer prevention, the optimal form of selenium is expected to possess high biological activity and low toxicity. Inorganic sodium selenite and organic selenomethionine and Se-methylselenocysteine have been extensively investigated for selenium supplementation (Whanger, 2002). There exists a long-held dogma that selenium in the redox state of zero (i.e. elemental selenium) is biologically inert due to the fact that the bioavailability of gray or red elemental selenium formed in some bacteria, or prepared via reducing sodium selenite by using glutathione or ascorbate as a reductant, is only 2–4 % of sodium selenite (Combs *et al.*, 1996). Red elemental selenium, formed in the redox system of sodium selenite and glutathione, is unstable and can further aggregate into gray or black elemental selenium if there are no controlling factors in the redox system. Protein existing in the redox system can control the aggregation of red elemental selenium; the resultant red elemental selenium particles are stable and at nano sizes, thereby being designated as Nano-Se (Zhang *et al.*, 2001). The present chapter outlines comparative studies between Nano-Se and other selenium compounds, namely sodium selenite, selenomethionine and Se-methylselenocysteine, in terms of bioavailability and toxicity.

6.2 Preparation and Characterization of Nano-Se

One ml sodium selenite (25 mM) was mixed with 4 ml glutathione (25 mM) containing 20 mg bovine serum albumin. The mixture was adjusted to pH 7.2 with 1.0 M sodium hydroxide to form Nano-Se (Zhang *et al.*, 2001). X-ray photoelectric energy spectra showed the binding energy of selenium 3d was 55.3 eV, indicating elemental selenium (Zhang *et al.*, 2001). High-resolution transmission electron microscopy (HR-TEM) showed that the sizes of Nano-Se were between 20–60 nm, the average size being 36 nm (Zhang *et al.*, 2001; Peng *et al.*, 2007). The interaction between bovine serum albumin and elemental selenium was examined by infrared spectroscopy; the shapes and peak intensities of native bovine serum albumin and bovine serum albumin in Nano-Se were similar. However, the wavenumbers of the transmittance peaks of native bovine serum albumin were slightly higher than those in Nano-Se, revealing that there are no covalent bonds, but there exist weak interactions between elemental selenium and the NH, C=O, COO$^-$, and C-N groups of bovine serum albumin (Gao *et al.*, 2002).

6.3 *In Vitro* Comparison with Sodium Selenite

Sodium selenite has been used to prevent selenium deficiency disorders in livestock and humans; it is also a reference selenium compound in selenium bioavailability studies (Levander, 1983). In addition, most animal studies dealing with the relationship of selenium to carcinogenesis used sodium selenite (Combs and Gray, 1998). Sodium selenite can easily be reduced to elemental selenium by reducing agents such as ascorbate and glutathione (Combs *et al.*, 1996). The reduction process can generate prooxidant intermediates (Spallholz, 1994; Seko and Imura, 1997). Presumably the reason sodium selenite causes cataracts in young rats (Schearer *et al.*, 1980) is associated with its prooxidant effect. To observe the biological activities and toxicity of Nano-Se, sodium selenite was firstly used as a reference.

6.3.1 Impact on Selenoenzymes

Human hepatoma HepG2 cells, after reaching 50 % confluence, were supplemented with selenium in the forms of Nano-Se and sodium selenite. Cells were harvested after 4 days incubation with selenium. Sodium selenite at the concentrations of 10, 20 and 50 nM dose-dependently and significantly increased type 1 glutathione peroxidase activity by 2.2, 4.3 and 8.2-fold, and type 4 glutathione peroxidase activity by 4.6, 9.3 and 19.7-fold, respectively (Zhang *et al.*, 2001). However, there were no significant differences between sodium selenite and Nano-Se at the same doses by ANOVA analysis, demonstrating that the two selenium forms have equal capacity for increasing the activities of the selenoenzymes.

6.3.2 Reaction with Glutathione

The molecular mechanism underlying selenium toxicity is still not completely understood. Toxicity or prooxidant activity of sodium selenite is associated with the interaction of selenite with glutathione to form reactive selenotrisulfides, leading to the production of toxic superoxide and hydrogen peroxide (Spallholz, 1994; Seko and Imura, 1997). After glutathione and selenium were mixed *in vitro*, the ratio of reacted glutathione responded linearly to increasing selenium concentrations. For Nano-Se, the linear range was 0.5–4.0 μg selenium/ml, y = 4.233x + 0.113 (r = 0.9998). For sodium selenite, the linear range was 0.05–0.6 μg selenium/ml, y = 51.837x − 0.0159 (r = 0.9958). The ratio of the two slope rates was 12.3, which suggests that sodium selenite is one order of magnitude more effective than Nano-Se in oxidizing glutathione (Zhang *et al.*, 2001). The reaction with glutathione implies that sodium selenite may have enhanced cytotoxicity as compared with Nano-Se. Thus the following experiments were carried out to test the hypothesis.

Firstly, the cytotoxic effect of sodium selenite and Nano-Se without the supplementation of glutathione was investigated. After seeding, HepG2 cells were allowed to adhere overnight in culture plates. Sodium selenite and Nano-Se were added to the medium and the cells were further cultured for 72 h before methyl thiazolyl tetrazolium (MTT) survival assay. Exposure to sodium selenite at the doses of 25, 50 and 100 μM significantly retarded cell growth in a dose-dependent fashion by 21, 44 and 66 % (Zhang *et al.*, 2001). However, there were no significant differences between sodium selenite and Nano-Se at the same doses by ANOVA analysis, indicating that the two selenium forms without the aid of extracellular glutathione equally inhibit HepG2 cell proliferation.

Secondly, the impact of added glutathione on the cytotoxic effects of sodium selenite and Nano-Se was evaluated. HepG2 cells were treated with 12.5 μM sodium selenite, 12.5 μM Nano-Se, 12.5 μM sodium selenite along with 250 μM glutathione, and 12.5 μM Nano-Se along with 250 μM glutathione for 72 h. Exposure to sodium selenite or Nano-Se or the co-treatment of Nano-Se and glutathione did not significantly retard cell growth compared with the non-selenium treatment. However, exposure to the co-treatment of sodium selenite and glutathione significantly retarded cell growth by 64 % compared with the non-selenium treatment. These data demonstrate that the cytotoxic effect of sodium selenite but not Nano-Se is enhanced by extracellular glutathione, being consistent with the property that sodium selenite is more potent in reacting with glutathione.

In contrast to the notion that selenium may preferentially or selectively kill transformed hepatocytes, it has been reported that the cytotoxic effect of sodium selenite on hepatocytes is greater than on hepatic carcinoma cells (Weiller *et al.*, 2004). Therefore the enhanced cytotoxic effect of sodium selenite observed in HepG2 hepatic carcinoma cells may also occur in liver tissue where extracellular glutathione and hepatocytes co-exist. Consequently sodium selenite would be more toxic than Nano-Se *in vivo*. To confirm the latter hypothesis, comparison studies between sodium selenite and Nano-Se in terms of acute, short-term and subchronic toxicities were carried out as shown below.

6.4   *In Vivo* Comparison with Sodium Selenite

Before the investigation of toxicity, it is paramount to know whether there is a difference in bioavailability between sodium selenite and Nano-Se *in vivo*, although this question has been resolved at the cellular level described above (Zhang *et al.*, 2001). To address this issue, Wistar rats were fed with selenium-deficient diet (<0.01 ppm selenium) for 6 weeks, then were divided into a selenium-deficient group, a Nano-Se group (selenium-deficient diet supplemented with 0.1 ppm selenium in the form of Nano-Se) and a sodium selenite group (selenium-deficient diet supplemented with 0.1 ppm selenium in the form of sodium selenite). The rats were sacrificed after 8 weeks of selenium supplementation. Sodium selenite significantly increased liver selenium by 8.8-fold, and liver glutathione peroxidase activity by 48.8-fold (Zhang *et al.*, 2001). With respect to these changes caused by the sodium selenite supplementation, no significant differences between sodium selenite and Nano-Se could be found by ANOVA analysis, demonstrating that the bioavailability of the two selenium forms is equal *in vivo*, being consistent with the observations *in vitro* (Zhang *et al.*, 2001).

6.4.1 Acute Toxicity

Kunming mice were randomly divided into 12 groups with ten mice in each group. A bolus of Nano-Se or sodium selenite at 1.43-fold dose escalation was orally administered and mortality was recorded over 14 days. Sodium selenite caused complete mortality at a dose of 25 mg selenium/kg. However, Nano-Se did not cause death at this dose. For sodium selenite the median lethal dose (LD_{50}) was 15.7 mg selenium/kg (95 % confidence limits: 13.4–18.5). For Nano-Se the LD_{50} was 113.0 mg selenium/kg (95 % confidence limits: 89.9–141.9). The acute toxicity of sodium selenite was 7.2-fold that of Nano-Se, based on selenium dose (Zhang *et al.*, 2001).

6.4.2 Short-term Toxicity

Selenium intoxication has been extensively studied in animals. Growth retardation has long been known to be a major characteristic in selenium-intoxicated animals. Selenium causes growth retardation through the reduction of growth hormone, somatomedin C (Thorlacius-Ussing *et al.*, 1987, 1988), insulin-like growth factor-binding proteins, and insulin-like growth factor-I (Grønbaek *et al.*, 1995). The US National Research Council concluded that growth inhibition might be the best indicator of toxic effects from selenium (Orskov and Flyvbjerg, 2000).

C57BL/6 mice were divided into three groups with eight mice in each group. The mice were orally administered saline as control, Nano-Se and sodium selenite at 4 mg selenium/kg each day for 4 weeks. Body weight in the sodium selenite group was 72–76 % of the control during the last two weeks (P all < 0.001), while body weight in the Nano-Se group remained not significantly different compared with the control throughout the period of 4 weeks (Zhang *et al.*, 2008a).

In addition to growth inhibition, sodium selenite also caused liver injury, predominantly in the form of inflammatory cell infiltration, while liver architecture in the Nano-Se group remained normal (Zhang *et al.*, 2008a). Glutathione S-transferase is responsible for detoxifying harmful compounds (Keen and Jakoby, 1978). In addition, it has an antioxidant effect (Hayes *et al.*, 2005). It is well known that oxidative stress is able to upregulate glutathione S-transferase as an adaptive response (Fiander and Schneider, 1999, 2000). Liver glutathione S-transferase activity was significantly increased by both selenium forms compared with the control, whereas glutathione S-transferase activity in sodium selenite treated mice was significantly lower than that in Nano-Se treated mice (Zhang *et al.*, 2008a). The higher activity of glutathione S-transferase in Nano-Se treated mice provides a mechanistic explanation for the observed lower toxicity. In addition to the difference in adaptive response as indicated by glutathione S-transferase activity, higher toxicity of sodium selenite may also be enhanced by its potent prooxidant capacity, since sodium selenite reacts with glutathione more efficiently as described above (Zhang *et al.*, 2001). Indeed, in Kunming mice, it was found that oral administration of sodium selenite at 6 mg selenium/kg for 12 days significantly increased liver malondialdehyde (a lipid peroxidation product formed as a consequence of excess reactive oxygen species) by 1.6-fold as compared with the normal control. In contrast, Nano-Se at the same selenium dose and regimen decreased liver malondialdehyde level by 27 % compared with the normal control. Furthermore, sodium selenite but not Nano-Se significantly suppressed liver superoxide dismutase and catalase activities (Zhang *et al.*, 2005). Therefore, the lower toxicity of Nano-Se compare with sodium selenite can be ascribed to at least two pathways: lower prooxidant reactions initiated by Nano-Se, and greater adaptive response induced by Nano-Se in mice.

6.4.3   Subchronic Toxicity

The subchronic toxicity of Nano-Se was compared with sodium selenite in rats. Groups of Sprague-Dawley rats (12 males and 12 females per group) were fed diets containing Nano-Se and sodium selenite at the concentrations of 0, 2, 3, 4 and 5 ppm selenium for 13 weeks. Clinical observations were made, and body weight and food consumption were recorded weekly. At the end of the study the rats were subjected to a full necropsy, and blood samples were collected for hematology and clinical chemistry determination. Histopathological examination was performed on the livers. At the two higher doses (4 and 5 ppm selenium), significant abnormal changes were found in body weight, hematology, clinical chemistry, relative organ weights, and histopathology parameters. However, the toxicity was more pronounced in the sodium selenite group than in the Nano-Se group. At the dose of 3 ppm selenium, significant growth inhibition and degeneration of liver cells were found in the sodium selenite group. No changes attributable to the administration of Nano-Se at the dose of 3 ppm selenium were found. Taken together, the 'no-observed-adverse-effect' level of Nano-Se and sodium selenite in male and female rats was considered to be 3 and 2 ppm selenium, respectively (Jia *et al.*, 2005). In conclusion, Nano-Se was less toxic than sodium selenite in the 13-week rat study.

6.5   *In Vivo* Comparison with Selenomethionine

Selenomethionine is the predominant chemical form of selenium in foodstuffs and selenium-enriched yeast. Numerous experimental studies have established that selenomethionine and selenium-enriched yeast are the most appropriate forms of selenium for use in nutritional supplements, due to their excellent bioavailability and lower toxicity among the various selenium forms (Schrauzer, 2000b, 2001, 2003). Both selenomethionine and Nano-Se possess the advantage of lower toxicity over selenite, but it remains unknown whether there are differences in terms of bioavailability and toxicity between Nano-Se and selenomethionine.

6.5.1   Acute Lethal Dose

Kunming mice were randomly divided into ten groups with ten mice per group. A bolus of selenomethionine at 1.25-fold dose escalation or Nano-Se at 1.43-fold dose escalation was administered orally. Cumulative mortality within 14 days after the treatment was used for the calculation of LD_{50}.

Selenomethionine caused 90 % mortality at a dose of 32 mg selenium/kg. However, Nano-Se caused only 10 % mortality at a dose of 36 mg selenium/kg and 70 % mortality at a dose as high as 150 mg selenium/kg. The respective LD_{50} of selenomethionine and Nano-Se were 25.6 mg selenium/kg (95 % confidence limits: 22.6–28.6) and 92.1 mg selenium/kg (95 % confidence limits: 71.1–131.1), which are consistent with the results reported earlier (Okuno *et al.*, 2001; Zhang *et al.*, 2001). Thus, the acute toxicity of selenomethionine was 3.6-fold that of Nano-Se based on selenium dose (Wang *et al.*, 2007).

6.5.2 Acute Toxicity at Nonlethal Dose

Kunming mice were randomly divided into three groups with six mice per group. They respectively received saline as control or a bolus of selenomethionine or Nano-Se at a dose of 10 mg selenium/kg orally. Mice in selenium-treated groups were sacrificed at 12 h.

Serum alanine aminotransferase activity increased by 23.8 and 0.9-fold in selenomethionine and Nano-Se treated mice, respectively ($P < 0.001$ between the two selenium forms). Serum aspartate aminotransferase activity increased by 4.7 and 0.7-fold in selenomethionine and Nano-Se treated mice, respectively ($P < 0.001$). Serum lactate dehydrogenase activity increased by 3.7 and −0.13-fold in selenomethionine and Nano-Se treated mice, respectively ($P < 0.001$). Liver malondialdehyde levels increased by 1.6 and −0.12-fold in selenomethionine and Nano-Se treated mice, respectively ($P < 0.01$ between the two selenium forms) (Wang *et al.*, 2007).

Taken together, these data clearly indicate that selenomethionine but not Nano-Se causes liver oxidative stress and robust acute liver injury. These results could provide an explanation for the large difference in LD_{50} values. Furthermore, it could be inferred that there would be a difference in toxicity if both selenium forms were given repeatedly to mice. Therefore, the following experiment was carried out to evaluate the difference in short-term toxicity.

6.5.3 Short-term Toxicity

Kunming mice were randomly divided into three groups with ten mice per group. The control mice were administered saline orally; the mice in the other groups were administered selenomethionine or Nano-Se orally at the dose of 5 mg selenium/kg for 7 consecutive days. Consistent with the notion that liver is the major target organ in mice under selenium toxicity, both selenium forms caused liver injury, but selenomethionine was more potent. Serum alanine aminotransferase activity increased by 4.9 and 1.9-fold in selenomethionine and Nano-Se treated mice, respectively ($P < 0.001$ between the two selenium forms). Serum aspartate aminotransferase activity increased by 1.8 and 0.9-fold in selenomethionine and Nano-Se treated mice, respectively ($P < 0.01$). Serum lactate dehydrogenase activity increased by 1.0 and 0.2-fold in selenomethionine and Nano-Se treated mice, respectively ($P < 0.001$). In addition, compared with normal liver architecture, selenomethionine caused irreversible and severe pathological change in the form of pyknosis, whereas Nano-Se caused hydropic degeneration which belongs to reversible and moderate pathological change. Overall, both selenium forms in this short-term toxicity study were toxic to mice as manifested by liver injury; however, the toxicity of Nano-Se was much lower (Wang *et al.*, 2007).

6.5.4 Bioavailability

Empiric deduction may lead one to suggest that Nano-Se has an impaired bioavailability due to the decrease in toxicity, especially because of its background as elemental selenium, which has been thought to be biologically inert for a long time. Therefore, a comparison of bioavailability was conducted by orally administering selenomethionine and Nano-Se to selenium-deficient mice once daily for 7 consecutive days at two nutritional doses (35 and 70 μg selenium/kg) to observe dose-dependent increases of selenium and selenoenzymes, and a supranutritional dose (1000 μg selenium/kg) to indicate saturated selenoenzymes, to observe selenium accumulation, and to induce phase 2 enzymes.

Both selenomethionine and Nano-Se increased selenium levels in whole blood, liver and kidney in a dose-dependent fashion. ANOVA analysis indicated that selenomethionine and Nano-Se equally increased selenium levels at the same nutritional dose, but at the supranutritional dose selenomethionine increased selenium level more efficiently than Nano-Se (the percentages of the selenium level in Nano-Se treated mice to the selenium level in selenomethionine treated mice were 55.5 % in whole blood, 30.3 % in liver, and 26.3 % in kidney, all $P < 0.001$ between the two selenium forms) (Wang *et al.*, 2007). Selenium accumulation includes selenocysteine in selenoenzymes with vital functions, and other forms whose functions are largely unknown. Among a number of retention mechanisms, that selenomethionine replaces the methionine of proteins is well elucidated. Hitherto, there is no evidence showing that this form of retention has a biologically beneficial function. On the contrary, excess substitution of methionine residues with selenomethionine may alter the physio-chemical properties of structural proteins, because the CH_3-Se-moiety of selenomethionine is more hydrophobic than the CH_3-S-moiety of methionine (Boles *et al.*, 1991). It has been found that selenomethionine-induced avian embryo malformation and embryo toxicity are at least in part due to excess substitution of methionine residue by selenomethionine (Spallholz and Hoffman, 2002). Thus the high accumulation of selenium at the supranutritional level in selenomethionine treated mice cannot be considered as a merit of selenomethionine; in contrast, it increases the risk of selenium intoxication.

Utilization of selenium, after its absorption, involves transformation to selenoenzymes such as glutathione peroxidase and thioredoxin reductase. Both selenomethionine and Nano-Se dose-dependently and significantly increased glutathione peroxidase activity in plasma, liver and kidney at the tested doses. Both selenomethionine and Nano-Se significantly increased liver and kidney thioredoxin reductase activities which had been saturated by the lowest tested dose, being consistent with the notion that thioredoxin reductase is more conserved under selenium deficiency and ranks higher in the hierarchy to utilize selenium compared with glutathione peroxidase (Berry, 2005). In general there were no significant differences between selenomethionine and Nano-Se in increasing these selenoenzymes activities by ANOVA analysis (Wang *et al.*, 2007).

The exact mechanism(s) by which selenium is a promising chemopreventive agent is not fully resolved. Induction of phase 2 enzymes is an effective and sufficient strategy for achieving protection against the toxic and neoplastic effects of many carcinogens (Talalay, 2000). Among the phase 2 enzymes, glutathione S-transferase plays a pivotal role in cellular protection against carcinogens by conjugating their electrophile metabolites with glutathione (Coles and Ketterer, 1990). It has been demonstrated that Nrf2 knockout mice, in which hepatic glutathione S-transferase activity is reduce by 50 % compared with wild-type mice, are more susceptible to benzo[a]pyrene carcinogenesis (Ramos-Gomez *et al.*, 2001); on the other hand, high expression of glutathione S-transferase has been shown to be protective against tumor development (Hayes *et al.*, 2005). For many cancer chemopreventive compounds, inducing glutathione S-transferase is a common attribute (Talalay, 2000). With different levels of efficacy, selenium compounds at supranutritional levels (550–1100 µg/kg selenium) administered to mice for 7 days can enhance glutathione S-transferase activity (El-Sayed *et al.*, 2006). At supranutritional doses, it was found that Nano-Se and selenomethionine increased glutathione S-transferase activity by 78 % and 22 %, respectively ($P < 0.001$) (Wang *et al.*, 2007).

In summary, both forms of selenium have comparable abilities to increase selenoenzyme activities; accumulation of selenium is higher in selenomethionine treated mice than in Nano-Se treated mice at supranutritional levels; Nano-Se is more efficient than selenomethionine in increasing glutathione S-transferase activity; and the toxicity of selenomethionine is higher than that of Nano-Se as indicated by LD_{50} values, acute liver injury, and short-term toxicity. The safety margin and potential toxic effects of selenium are important considerations for its role in supplementation. On the basis of the present results, Nano-Se can serve as an antioxidant with reduced risk of selenium toxicity, and as a potential chemopreventive agent if the induction of glutathione S-transferase by selenium is a crucial mechanism for its chemopreventive effect.

6.6   *In Vivo* Comparison with Se-methylselenocysteine

Se-methylselenocysteine, a precursor of methylselenol, is one of the most effective selenium compounds for chemoprevention (Medina *et al.*, 2001; Whanger, 2002). However, it is probably a highly toxic selenium compound for animals (Johnson *et al.*, 2008). Selenium compounds can activate glutathione S-transferase in cells ('t Hoen *et al.*, 2002). It has been found that methylselenol precursors were highly effective inducers of glutathione S-transferase in cells (Xiao and Parkin, 2006). Since glutathione S-transferase induction by chemopreventive agents plays an important role in cancer prevention, whether there is a difference in inducing glutathione S-transferase *in vivo* between Se-methylselenocysteine and Nano-Se is of interest. Given that Nano-Se and Se-methylselenocysteine have equivalent potency in inducing glutathione S-transferase, it would be important to know which one approaches closest to toxicity of selenium. Therefore, the following studies were carried out to compare the toxicity, glutathione S-transferase induction, and bioavailability of Nano-Se and Se-methylselenocysteine in mice.

6.6.1   Acute Lethal Dose

Kunming mice were randomly divided into ten groups with ten mice per group. A bolus of Se-methylselenocysteine at 1.33-fold dose escalation or Nano-Se at 1.43-fold dose escalation was administered orally. Cumulative mortality within 14 days after the treatment was used for the calculation of LD_{50}.

Se-methylselenocysteine caused 100 % mortality at a dose of 27.5 mg selenium/kg. However, Nano-Se caused only 10 % mortality at a dose of 36 mg selenium/kg and 70 %

mortality at a dose as high as 150 mg selenium/kg. The LD_{50} of Se-methylselenocysteine and Nano-Se were 14.6 mg selenium/kg (95% confidence limits: 13.1–16.2) and 92.1 mg selenium/kg (95 % confidence limits: 71.1–131.1), respectively. The acute toxicity of Se-methylselenocysteine was 6.3-fold that of Nano-Se based on selenium dose (Zhang *et al.*, 2008b).

6.6.2   Acute Toxicity at Nonlethal Dose and Survival Rate after Repetitive Administration

Kunming mice were randomly divided into three groups with six mice per group. They received saline as control, a bolus of Se-methylselenocysteine or Nano-Se, respectively, at the dose of 10 mg selenium/kg orally. Mice in selenium treated groups were sacrificed at 12 h.

Serum alanine aminotransferase activity increased by 3.2 and 0.9-fold in Se-methylselenocysteine and Nano-Se treated mice, respectively ($P < 0.01$ between the two selenium forms). Serum aspartate aminotransferase activity increased by 5.2 and 0.7-fold in Se-methylselenocysteine and Nano-Se treated mice, respectively ($P < 0.001$). Serum lactate dehydrogenase activity increased by 3.5 and -0.13-fold in Se-methylselenocysteine and Nano-Se treated mice, respectively ($P < 0.001$). Liver malondialdehyde levels increased by 3.7 and -0.12-fold in Se-methylselenocysteine and Nano-Se treated mice, respectively ($P < 0.001$) (Zhang *et al.*, 2008b).

Overall, compared with Nano-Se, acute treatment with Se-methylselenocysteine caused robust acute liver injury as displayed by serum alanine aminotransferase, aspartate aminotransferase, and lactate dehydrogenase as well as liver malondialdehyde. These results suggest that repetitive administration of such a dose of Se-methylselenocysteine could result in lethal consequence. To confirm this assumption, Kunming mice were orally administered Se-methylselenocysteine and Nano-Se daily for 7 days at the dose of 10 mg selenium/kg and then they were further observed for another 7 days. Se-methylselenocysteine caused 80 % death, whereas Nano-Se caused only 10 % death; there was a significant difference ($P < 0.01$) between the two selenium forms by log rank test of Kaplan-Meier survival analysis (Zhang *et al.*, 2008b).

6.6.3   Short-term Toxicity

Kunming mice were orally administered Se-methylselenocysteine and Nano-Se daily for 7 days at the dose of 5 mg selenium/kg to compare short-term toxicity. Both selenium forms caused liver injury, but Se-methylselenocysteine was more potent in this regard. As compared with normal liver architecture, Se-methylselenocysteine caused irreversible and severe pathological change in the forms of pyknosis and necrosis, whereas Nano-Se caused hydropic degeneration which is a reversible and moderate pathological change. Serum alanine aminotransferase activity increased by 4.0 and 1.7-fold in Se-methylselenocysteine and Nano-Se treated mice, respectively ($P < 0.05$ between the two selenium forms). Serum lactate dehydrogenase activity increased by 0.6 and 0.06-fold in Se-methylselenocysteine and Nano-Se treated mice, respectively ($P < 0.05$) (Zhang *et al.*, 2008b). Overall, both selenium forms in this short-term toxicity model were all toxic to mice as manifested by liver injury; however, the toxicity of Nano-Se was much lower.

6.6.4 Potential of Glutathione S-transferase Enhancement at Supranutritional Levels

Kunming mice were randomly divided into seven groups with six mice per group. Mice were treated with saline as control, Se-methylselenocysteine, and Nano-Se at the doses of 500, 1000 and 1500 µg selenium/kg for 7 days. Serum alanine aminotransferase and aspartate aminotransferase activities indicated that neither selenium compound caused liver toxicity at these doses. At 500 µg selenium/kg, Se-methylselenocysteine and Nano-Se only modestly increased glutathione S-transferase activity by 0.3 and 0.2-fold, respectively ($P > 0.05$ between the two selenium forms). At 1000 µg selenium/kg, Se-methylselenocysteine and Nano-Se significantly increased glutathione S-transferase activity by 1.0 and 0.9-fold, respectively ($P > 0.05$). At 1500 µg selenium/kg, Se-methylselenocysteine and Nano-Se significantly increased glutathione S-transferase activity by 1.3 and 1.2-fold, respectively ($P > 0.05$) (Zhang *et al.*, 2008b). Based on these data, it is concluded that supranutritional levels of Se-methylselenocysteine and Nano-Se as high as 1000 µg selenium/kg must be used to achieve a significant and marked increase of glutathione S-transferase activity in mice. Under the circumstances, Se-methylselenocysteine is more approaching toxicity of selenium. Therefore, Nano-Se can be considered as a chemopreventive agent with reduced risk of selenium toxicity, if the induction of glutathione S-transferase by selenium compounds is a crucial mechanism for cancer prevention.

The above claim may encounter the argument that Nano-Se with lower toxicity may also have compromised chemopreventive effect because both toxicity and carcinostatic activity of selenium are all involved in the oxidative catalysis of thiols (Spallholz, 1994). Obviously, this hypothesis largely depends upon the opinion that the cytotoxic mechanism is a central or sole mechanism for the chemopreventive effect of selenium. If this is the case, the risk of selenium toxicity seems to be unavoidable in the chemopreventive setting, unless the preferential selection of cancer cells is firmly established. Several investigators have reported that selenium compounds selectively inhibit growth and induce apoptosis in cancer cells compared with normal cells (Watrach *et al.*, 1984; Ghose *et al.*, 2001). In contrast, it was surprisingly found that the cytotoxic effect of either inorganic or organic selenium was more potent in normal hepatocytes compared with hepatic carcinoma cells (Weiller *et al.*, 2004). Consistent with this finding, nontumorigenic prostate cells are highly sensitive to selenium toxicity compared with prostate cancer cells at physiologically relevant concentrations (5–10 µM) (Rebsch *et al.*, 2006). Therefore the present evidence related to selectivity is largely unfavorable. Furthermore, the potent toxicity of Se-methylselenocysteine in mice revealed herein, and in rats and dogs reported very recently (Johnson *et al.*, 2008), also argue against a strategy of utilizing the cytotoxic mechanism as a central or sole mechanism for the chemopreventive effect of selenium. On the contrary, phase 2 enzyme induction represented by glutathione S-transferase herein appears to be a relatively safe and realistic strategy to identify optimal selenium forms for cancer prevention.

6.6.5 Bioavailability

Nutritional levels of selenium at the doses of 35 and 70 µg selenium/kg were used to compare bioavailability, and a supranutritional level of selenium at a dose of 1000 µg selenium/kg was used to indicate saturated activities of selenoenzymes.

Both Se-methylselenocysteine and Nano-Se dose-dependently and significantly increased glutathione peroxidase activity in plasma, liver and kidney at the tested doses. Both Se-methylselenocysteine and Nano-Se significantly increased liver and kidney thioredoxin reductase activities which had been saturated by the lowest tested dose, being consistent with the notion that thioredoxin reductase is more conserved under selenium deficiency and ranks higher in the hierarchy to utilize selenium compared with glutathione peroxidase (Berry, 2005). In general, there were no significant differences between Se-methylselenocysteine and Nano-Se in increasing these selenoenzymes activities (Zhang *et al.*, 2008b).

Both Se-methylselenocysteine and Nano-Se significantly and dose-dependently increased selenium in whole blood, red blood cells, liver and kidney compared with the control. At the nutritional levels, there were no significant differences in selenium accumulation between the two selenium forms; however, at the supranutritional level, selenium accumulation in Se-methylselenocysteine treated mice was significantly higher than that in Nano-Se treated mice (the percentages of the selenium level in Nano-Se treated mice to the selenium level in Se-methylselenocysteine treated mice were 55.5 % in whole blood, 36.7 % in red blood cells, 75.0 % in liver, and 72.4 % in kidney, all $P < 0.001$ between the two selenium forms) (Zhang *et al.*, 2008b). A similar phenomenon has also been observed in the comparison of selenomethionine with Nano-Se at supranutritional levels (Wang *et al.*, 2007). High accumulation of selenium in selenomethionine treated mice could be explained at least in part by its unspecific substitution of methionine. However, in the case of Se-methylselenocysteine, such a mechanism does not exist. It is inferred that the difference in selenium accumulation is probably associated with size. Compared with red elemental selenium particles of 36 nm, the Se-methylselenocysteine molecule is significantly smaller, thereby entering into cells more efficiently. The fact that in red blood cells the percentage of the selenium level in Nano-Se treated mice to the selenium level in Se-methylselenocysteine treated mice is only 36.7 % at a supranutritional level (Zhang *et al.*, 2008b) largely favors this hypothesis. To make the hypothesis more convincing, a demonstration that Nano-Se can size-dependently cause selenium accumulation is needed. Thus, size effects of Nano-Se *in vitro* and *in vivo* were further investigated as shown below.

6.7   Size Effects of Nano-Se *In Vitro* and *In Vivo*

Protein can control the aggregation of red elemental selenium; high protein concentration generates a small size of Nano-Se. To obtain Nano-Se with different sizes, 1 ml sodium selenite (25 mM) was mixed with 4 ml glutathione (25 mM) containing 200, 20 and 2 mg bovine serum albumin. The sizes of the resultant Nano-Se were 5–15 nm (small size), 20–60 nm (medium size) and 80–200 nm (large size), respectively (Zhang *et al.*, 2004).

6.7.1   Redox Activity *In Vitro*

Carbon-centered free radicals generated from 2,2'-azo-bis-(2-amidinopropane) hydrochloride, 1,1-diphenyl-2-picryhydrazyl free radicals, superoxide anion generated from the xanthine/xanthine oxidase system, and singlet oxygen generated by irradiated hemoporphyrin, can be scavenged by the three sizes of Nano-Se in a size-dependent fashion: namely, small size is more efficient (Huang *et al.*, 2003). The three sizes of Nano-Se also show size-dependent protective effects against the oxidation of DNA *in vitro* (Huang *et al.*, 2003). Moreover, the redox activities of Nano-Se vary linearly with the surface area of the particles (Mishra *et al.*, 2005). Thus, it can be concluded that a smaller size of Nano-Se has higher redox activity *in vitro*.

6.7.2   Upregulating Selenoenzyme *In Vitro* and *In Vivo*

HepG2 cells were supplemented with the three sizes of Nano-Se (small, medium and large) at the concentrations of 10, 20 and 50 nM for 4 days to compare their effects on inducing selenoenzymes. Type 1 and 4 glutathione peroxidase as well as type 1 thioredoxin reductase activities significantly and dose-dependently increased. Unexpectedly, the three sizes of Nano-Se at the same concentrations were equally efficient in all tested selenoenzymes (Zhang *et al.*, 2004). Thus it is interesting to know whether the size effect on selenoenzyme also disappears *in vivo*.

Selenium-deficient Kuming mice were orally administered with Nano-Se (small, medium and large) at the dose of 70 µg selenium/kg for 7 consecutive days. Liver glutathione peroxidase and thioredoxin reductase activities were significantly but equally increased by the three sizes of Nano-Se (Zhang *et al.*, 2004). Taken together, it can be concluded that Nano-Se ranging from 5–200 nm has no size effect in inducing selenoenzymes *in vitro* and *in vivo*.

Under conditions of selenium deficiency, the avidity of selenium uptake mechanisms may be increased to maintain the biosynthesis of selenoenzymes, which are fundamental for redox homeostasis. This increased avidity may override the potential advantage of small-size Nano-Se seen under selenium-replete conditions, thereby eliminating the size effect. It is inferred that once selenoenzymes have been saturated, selenium uptake mechanisms may downregulate; accordingly, the size effects of Nano-Se on the biomarkers sensitive to supranutritional levels of selenium can then appear. The next study was carried out to test this hypothesis.

6.7.3   Selenium Accumulation and Glutathione S-transferase Induction *In Vivo*

Selenium-deficient mice were administered either 36 or 90 nm Nano-Se at supranutritional doses, in both a short-term model and a single dose model. Under these conditions, Nano-Se indeed showed size effects on selenium accumulation and glutathione S-transferase activity. A size effect of Nano-Se was found in 15 out of 18 total comparisons between sizes at the same dose and time in the two models (Peng *et al.*, 2007).

Furthermore, it was found that the magnitude of the size effect was more prominent on selenium accumulation than on glutathione S-transferase activity (Peng *et al.*, 2007). Glutathione S-transferase is strictly regulated by transcriptional and translational mechanisms, so its increase in activity normally does not exceed 3-fold. In contrast, the homeostasis of selenium accumulation is not as tightly controlled. In the present experiments, glutathione S-transferase activity had reached or was approaching saturation, but liver selenium was far below saturation. Therefore, these results strongly suggest that the saturation profile of the tested biomarker has an impact on the size effect of Nano-Se, which explains why in previous studies (Zhang *et al.*, 2004) there was no size effect in increasing selenoenzymes, because they were easily saturated at nutritional selenium levels.

Size-dependent selenium accumulation has an important implication in interpreting the lower toxicity of Nano-Se. At the supranutritional level, in fact, approaching the toxic level of selenium, whereupon selenoenzymes have already been fully saturated, cells may in turn change to passive absorption of selenium. Under such circumstances, large size constitutes a barrier for selenium to enter into cells. Therefore, the remarkable attenuated selenium accumulation after ingesting Nano-Se at supranutritional levels in liver, kidney, whole blood, and red blood cells, compared with selenomethionine and Se-methylselenocysteine (Wang *et al.*, 2007; Zhang *et al.*, 2008b), may effectively delay the onset and development of selenium toxicity. In this regard, nanotechnology is an attractive strategy to utilize an appropriate size to control excess selenium accumulation and selenium toxicity, but reserve the capacities of increasing the activities of selenoenzyme and glutathione S-transferase.

6.8   Summary

Nanotechnology holds promise for medication and nutrition because material at nanometer dimensions exhibits novel properties different from those of both isolated atoms and bulk material. Elemental selenium in the redox state of zero in general is considered to be biologically inert. With bovine serum albumin as a disperser, neonatal elemental selenium atoms generated via reducing sodium selenite with glutathione can aggregate into Nano-Se. In contrast to biologically inert black elemental selenium at micrometer size, Nano-Se indeed manifests toxicity, which conforms to the notion that nanoparticles increase health risks. On the other hand, selenium is an essential trace element with a narrow margin between necessary intake amount and toxicity, whereas cancer-preventive effects of selenium were mostly observed at high doses in animals. For nutritional supplementation and cancer prevention, the optimal selenium form is expected to possess high biological activity and low toxicity. Compared with sodium selenite, selenomethionine and Se-methylselenocysteine, Nano-Se possesses equal efficacy in increasing the activities of selenoenzymes, but has much lower toxicity. These results suggest that Nano-Se can be considered as a novel selenium source with a reduced risk of selenium toxicity.

Acknowledgements

The author is grateful to Drs Xufang Wang, Huali Wang and Dungeng Peng for their work on Nano-Se in the author's laboratory.

References

Arnér ES, Holmgren A (2000) Physiological functions of thioredoxin and thioredoxin reductase. *Eur J Biochem* **267**, 6102–6109.
Beckett GJ, Arthur JR (2005) Selenium and endocrine systems. *J Endocrinol* **184**, 455–465.
Berry MJ (2005) Insights into the hierarchy of selenium incorporation. *Nat Genet* **37**, 1162–1163.

Boles JO, Cisneros RJ, Weir MS, Odom JD, Villafranca JE, Dunlap RB (1991) Purification and characterization of selenomethionyl thymidylate synthase from Escherichia coli: comparison with the wild-type enzyme. *Biochemistry* **30**, 11073–11080.

Brigelius-Flohé R (1999) Tissue-specific functions of individual glutathione peroxidases. *Free Radic Biol Med* **27**, 951–965.

Brigelius-Flohé R (2008) Selenium compounds and selenoproteins in cancer, *Chem Biodivers* **5**, 389–395.

Brigelius-Flohé R, Müller C, Menard J, Florian S, Schmehl K, Wingler K (2001) Functions of GI-GPx: lessons from selenium-dependent expression and intracellular localization. *Biofactors* **14**, 101–106.

Brinkman M, Buntinx F, Muls E, Zeegers MP (2006) Use of selenium in chemoprevention of bladder cancer. *Lancet Oncol* **7**, 766–774.

Clark LC, Combs GF, Turnbull BW, Slate EH, Chalker DK, Chow J, Davis LS, Glover RA, Graham GF, Gross EG, Krongrad A, Lesher JL, Park HK, Sanders BB, Smith CL, Taylor JR (1996) Effects of selenium supplementation for cancer prevention in patients with carcinoma of the skin. *J Am Med Assoc* **276**, 1957–1963.

Coles B, Ketterer B (1990) The role of glutathione and glutathione transferases in chemical carcinogenesis. *Crit Rev Biochem Mol Biol* **25**, 47–70.

Combs GF, Garbisu C, Yee BC, Yee A, Carlson DE, Smith NR, Magyarosy AC, Leighton T, Buchanan BB (1996) Bioavailability of selenium accumulated by selenite-reducing bacteria. *Biol Trace Elem Res* **52**, 209–225.

Combs GF, Gray WP (1998) Chemopreventive agents: selenium. *Pharmacol Ther* **79**, 179–192.

Conrad M, Schneider M, Seiler A, Bornkamm GW (2007) Physiological role of phospholipid hydroperoxide glutathione peroxidase in mammals. *Biol Chem* **388**, 1019–1025.

El-Bayoumy K, Sinha R (2004) Mechanisms of mammary cancer chemoprevention by organoselenium compounds. *Mutat Res* **551**, 181–197.

El-Sayed WM, Aboul-Fadl T, Lamb JG, Roberts JC, Franklin MR (2006) Effect of selenium-containing compounds on hepatic chemoprotective enzymes in mice. *Toxicology* **220**, 179–188.

Fiander H, Schneider H (1999) Compounds that induce isoforms of glutathione S-transferase with properties of a critical enzyme in defense against oxidative stress. *Biochem Biophys Res Commun* **262**, 591–595.

Fiander H, Schneider H (2000) Dietary ortho phenols that induce glutathione S-transferase and increase the resistance of cells to hydrogen peroxide are potential cancer chemopreventives that act by two mechanisms: the alleviation of oxidative stress and the detoxification of mutagenic xenobiotics. *Cancer Lett* **156**, 117–124.

Flohé L, Andreesen JR, Brigelius-Flohé R, Maiorino M, Ursini F (2000) Selenium, the element of the moon, in life on earth. *IUBMB Life* **49**, 411–420.

Gao X, Zhang J, Zhang L (2002) Hollow sphere selenium nanoparticles: Their in-vitro anti hydroxyl radical effect. *Adv Mater* **14**, 290–293.

Ghose A, Fleming J, El-Bayoumy K, Harrison PR (2001) Enhanced sensitivity of human oral carcinomas to induction of apoptosis by selenium compounds: involvement of mitogen-activated protein kinase and Fas pathways. *Cancer Res* **61**, 7479–7487.

Grønbaek H, Frystyk J, Orskov H, Flyvbjerg A (1995) Effect of sodium selenite on growth, insulin-like growth factor-binding proteins and insulin-like growth factor-I in rats. *J Endocrinol* **145**, 105–112.

Hayes JD, Flanagan JU, Jowsey IR (2005) Glutathione transferases. *Ann Rev Pharmacol Toxicol* **45**, 51–88.

Huang B, Zhang J, Hou J, Chen C (2003) Free radical scavenging efficiency of Nano-Se in vitro. *Free Radic Biol Med* **35**, 805–813.

Ip C (1998) Lessons from basic research in selenium and cancer prevention. *J Nutr* **128**, 1845–1854.

Jia X, Li N, Chen J (2005) A subchronic toxicity study of elemental Nano-Se in Sprague-Dawley rats. *Life Sci* **76**, 1989–2003.

Johnson WD, Morrissey RL, Kapetanovic I, Crowell JA, McCormick DL (2008) Subchronic oral toxicity studies of Se-methylselenocysteine, an organoselenium compound for breast cancer prevention. *Food Chem Toxicol* **46**, 1068–1078.

Keen JH, Jakoby WB (1978) Glutathione transferases. Catalysis of nucleophilic reactions of glutathione. *J Biol Chem* **253**, 5654–5657.

Köhrle J, Jakob F, Contempré B, Dumont JE (2005) Selenium, the thyroid, and the endocrine system. *Endocr Rev* **26**, 944–984.

Levander OA (1983) Considerations in the design of selenium bioavailability studies. *Fed Proc* **42**, 1721–1725.

Lillig CH, Holmgren A (2007) Thioredoxin and related molecules – from biology to health and disease. *Antioxid Redox Signal* **9**, 25–47.

Medina D, Thompson H, Ganther H, Ip C (2001) Se-methylselenocysteine: a new compound for chemoprevention of breast cancer. *Nutr Cancer* **40**, 12–17.

Mishra B, Hassan PA, Priyadarsini KI, Mohan H (2005) Reactions of biological oxidants with selenourea: formation of redox active nanoselenium. *J Phys Chem B* **109**, 12718–12723.

Moreno-Reyes R, Suetens C, Mathieu F, Begaux F, Zhu D, Rivera MT, Boelaert M, Nève J, Perlmutter N, Vanderpas J (1998) Kashin-Beck osteoarthropathy in rural Tibet in relation to selenium and iodine status. *N Engl J Med* **339**, 1112–1120.

Naithani R (2008) Organoselenium compounds in cancer chemoprevention. *Mini Rev Med Chem* **8**, 657–668.

Ohta Y, Suzuki KT (2008) Methylation and demethylation of intermediates selenide and methylselenol in the metabolism of selenium. *Toxicol Appl Pharmacol* **226**, 169–177.

Okuno T, Kubota T, Kuroda T, Ueno H, Nakamuro K (2001) Contribution of enzymic alpha, gamma-elimination reaction in detoxification pathway of selenomethionine in mouse liver. *Toxicol Appl Pharmacol* **176**, 18–23.

Oldfield JE (1989) Selenium in animal nutrition: the Oregon and San Joaquin Valley (California) experiences – examples of correctable deficiencies in livestock. *Biol Trace Elem Res* **20**, 23–29.

Orskov H, Flyvbjerg A (2000) Selenium and human health. *Lancet* **356**, 942–943.

O'Toole D, Raisbeck MF (1995) Pathology of experimentally induced chronic selenosis (alkali disease) in yearling cattle. *J Vet Diagn Invest* **7**, 364–373.

Papp LV, Lu J, Holmgren A, Khanna KK (2007) From selenium to selenoproteins: synthesis, identity, and their role in human health. *Antioxid Redox Signal* **9**, 775–806.

Peng D, Zhang J, Liu Q, Taylor EW (2007) Size effect of elemental selenium nanoparticles (Nano-Se) at supranutritional levels on selenium accumulation and glutathione S-transferase activity. *J Inorg Biochem* **101**, 1457–1463.

Ramos-Gomez M, Kwak MK, Dolan PM, Itoh K, Yamamoto M, Talalay P, Kensler TW (2001) Sensitivity to carcinogenesis is increased and chemoprotective efficacy of enzyme inducers is lost in nrf2 transcription factor-deficient mice. *Proc Natl Acad Sci USA* **98**, 3410–3415.

Rayman MP (2005) Selenium in cancer prevention: a review of the evidence and mechanism of action. *Proc Nutr Soc* **64**, 527–542.

Rayman MP, Infante HG, Sargent M (2008) Food-chain selenium and human health: spotlight on speciation. *Br J Nutr* **100**, 238–253.

Rebsch CM, Penna FJ, Copeland PR (2006) Selenoprotein expression is regulated at multiple levels in prostate cells. *Cell Res* **16**, 940–948.

Rederstorff M, Krol A, Lescure A (2006) Understanding the importance of selenium and selenoproteins in muscle function. *Cell Mol Life Sci* **63**, 52–59.

Schearer TR, McCormack DW, DeSart DJ, Britton JL, Lopez MT (1980) Histological evaluation of selenium induced cataracts. *Exp Eye Res* **31**, 327–333.

Schrauzer GN (2000a) Anticarcinogenic effects of selenium. *Cell Mol Life Sci* **57**, 1864–1873.

Schrauzer GN (2000b) Selenomethionine: a review of its nutritional significance, metabolism and toxicity. *J Nutr* **130**, 1653–1656.

Schrauzer GN (2001) Nutritional selenium supplements: product types, quality, and safety. *J Am Coll Nutr* **20**, 1–4.

Schrauzer GN (2003) The nutritional significance, metabolism and toxicology of selenomethionine. *Adv Food Nutr Res* **47**, 73–112.

Schwarz K, Foltz CM (1999) Selenium as an integral part of factor 3 against dietary necrotic liver degeneration. *Nutrition* **15**, 255.

Seko Y, Imura N (1997) Active oxygen generation as a possible mechanism of selenium toxicity. *Biomed Environ Sci* **10**, 333–339.

Spallholz JE (1994) On the nature of selenium toxicity and carcinostatic activity. *Free Radic Biol Med* **17**, 45–64.

Spallholz JE, Hoffman DJ (2002) Selenium toxicity: cause and effects in aquatic birds. *Aquat Toxicol* **57**, 27–37.

Stadtman TC (2000) Selenium biochemistry. Mammalian selenoenzymes. *Ann NY Acad Sci* **899**, 399–402.

Suzuki KT, Kurasaki K, Ogawa S, Suzuki N (2006) Metabolic transformation of methylseleninic acid through key selenium intermediate selenide. *Toxicol Appl Pharmacol* **215**, 189–197.

Talalay P (2000) Chemoprotection against cancer by induction of phase 2 enzymes. *Biofactors* **12**, 5–11.

't Hoen PA, Rooseboom M, Bijsterbosch MK, van Berkel TJ, Vermeulen NP, Commandeur JN (2002) Induction of glutathione-S-transferase mRNA levels by chemopreventive selenocysteine Se-conjugates. *Biochem Pharmacol* **63**, 1843–1849.

Thorlacius-Ussing O, Flyvbjerg A, Esmann J (1987) Evidence that selenium induces growth retardation through reduced growth hormone and somatomedin C production. *Endocrinology* **120**, 659–663.

Thorlacius-Ussing O, Flyvbjerg A, Jørgensen KD, Orskov H (1988) Growth hormone restores normal growth in selenium-treated rats without increase in circulating somatomedin C. *Acta Endocrinol (Copenh)* **117**, 65–72.

Wang H, Zhang J, Yu H (2007) Elemental selenium at nano size possesses lower toxicity without compromising the fundamental effect on selenoenzymes: comparison with selenomethionine in mice. *Free Radic Biol Med* **42**, 1524–1533.

Watrach AM, Milner JA, Watrach MA, Poirier KA (1984) Inhibition of human breast cancer cells by selenium. *Cancer Lett* **25**, 41–47.

Weiller M, Latta M, Kresse M, Lucas R, Wendel A (2004) Toxicity of nutritionally available selenium compounds in primary and transformed hepatocytes. *Toxicology* **201**, 21–30.

Whanger PD (2002) Selenocompounds in plants and animals and their biological significance. *J Am Coll Nutr* **21**, 223–232.

Xiao H, Parkin KL (2006) Induction of phase II enzyme activity by various selenium compounds. *Nutr Cancer* **55**, 210–223.

Yang GQ, Wang SZ, Zhou RH, Sun SZ (1983) Endemic selenium intoxication of humans in China. *Am J Clin Nutr* **37**, 872–881.

Yang GQ, Chen JS, Wen ZM, Ge KY, Zhu LZ, Chen XC, Chen XS (1984) The role of selenium in Keshan disease. *Adv Nutr Res* **6**, 203–231.

Zeng H, Combs GF (2008) Selenium as an anticancer nutrient: roles in cell proliferation and tumor cell invasion. *J Nutr Biochem* **19**, 1–7.

Zhang JS, Gao XY, Zhang LD, Bao YP (2001) Biological effects of a nano red elemental selenium. *Biofactors* **15**, 27–38.

Zhang J, Wang H, Bao Y, Zhang L (2004) Nano red elemental selenium has no size effect in the induction of seleno-enzymes in both cultured cells and mice. *Life Sci* **75**, 237–244.

Zhang J, Wang H, Yan X, Zhang L (2005) Comparison of short-term toxicity between Nano-Se and selenite in mice. *Life Sci* **76**, 1099–1109.

Zhang J, Wang H, Peng D, Taylor EW (2008a) Further insight into the impact of sodium selenite on selenoenzymes: high-dose selenite enhances hepatic thioredoxin reductase 1 activity as a consequence of liver injury. *Toxicol Lett* **176**, 223–229.

Zhang J, Wang X, Xu T (2008b) Elemental selenium at nano size (Nano-Se) as a potential chemopreventive agent with reduced risk of selenium toxicity: comparison with Se-methylselenocysteine in mice. *Toxicol Sci* **101**, 22–31.

7

Evaluation of Toxicity of Nanostructures in Biological Systems

Adam J. Gormley and Hamidreza Ghandehari

7.1 Introduction

Nanotechnology is emerging across disciplines in science and engineering as a dominating force. This is due to the unique properties of structures at the nanoscale, allowing scientists and engineers to manipulate the environment in new and unexplored ways. Nanotechnology may in the future have great impact in the fields of medicine and biotechnology. This is in part because many biological systems operate at the nanoscale (Figure 7.1). Proteins, the basic building blocks of biological systems, vary greatly in size, function and form. Human serum albumin, for example, the most abundant plasma protein, has a reported size of roughly 8 nm and its interactions with 5 nm nanoparticles have already been studied (Nayak and Shin, 2008). This ability of nanoparticles to imitate and interact with biological systems at the same scale of its parts is what has driven the recent flurry of nanobiotechnology research.

The application of nanobiotechnology in medicine, such as development of new devices and tools, clinical approaches and therapies, is termed *nanomedicine*. It is anticipated that research and development in nanomedicine will drastically improve healthcare across the globe. Although still in its infancy, application of nanotechnology in a number of consumer healthcare products such as sunscreens, toothpastes and wound dressings is already emerging. Applications in drug delivery, imaging, early disease detection and cancer treatment are just some of the anticipated uses of nanotechnology being explored today (Sahoo and Labhasetwar, 2003).

Like any new technology designed to interact with our environment, the proper assessment of the potential impact on our surroundings in terms of risk to benefit ratio needs

Nanotoxicity: From In Vivo *and* In Vitro *Models to Health Risks*   Edited by Saura Sahu and Daniel Casciano
© 2009 John Wiley & Sons, Ltd

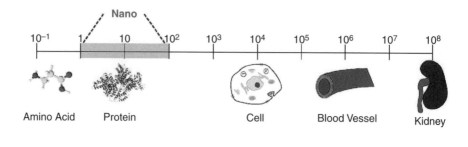

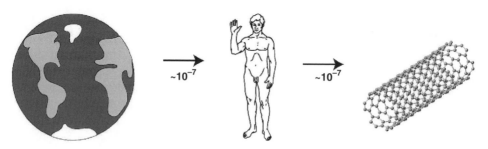

Figure 7.1   *Nanoscale bar showing the relative size of typical biological structures compared with nanosized constructs such as proteins and carbon nanotubes.*

specific attention. Classic examples of shortcomings in this regard include asbestos, chlorofluorocarbons (CFCs), radiation, and growth hormones in meat production. In all of these technologies, early warnings of their environmental impact were ignored, resulting in long term and often irreversible consequences (Harremoës *et al.*, 2001). This is particularly relevant to the field of nanotechnology, where the potential consequences of mass production and environmental exposure to such materials are largely unknown. Therefore, it is important to take the lessons learned from those earlier technologies and translate such experience to nanotech-related products, to avoid unnecessary risks to both our environment and the industry as a whole (Hansen *et al.*, 2008).

Adverse effects that nanoparticles may have to human health and the environment is undoubtedly the most significant consequence. This is particularly true in nanomedicine where the constructs under question are designed for interaction with humans and animals. Small enough to be engulfed by cell membranes by a variety of mechanisms, nanoconstructs may interact with the body in ways inaccessible to current drugs and biomedical devices.

Governing agencies such as the National Institutes of Health (NIH), National Science Foundation (NSF), Environmental Protection Agency (EPA), National Institute for Occupational Safety and Health (NIOSH) and others have responded to such safety concerns with appropriate measures. Basic scientific research aimed at better understanding how organisms respond to nanoconstructs has helped to shed light on the overall safety of such materials. Regulatory agencies such as the Food and Drug Administration (FDA) are also adapting to such concerns so that proper safety assessments can be made during the premarket approval process. As stated in their 2007 Nanotechnology Task Force Report,

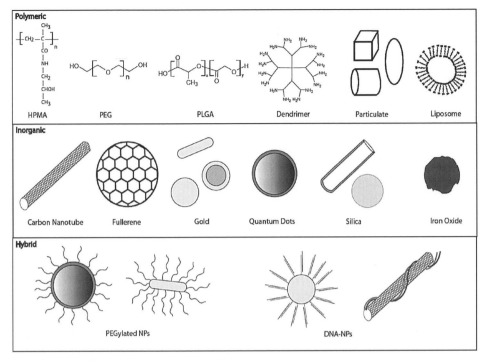

Figure 7.2   *Illustrations of various nanoparticle types. Nanoparticles can either be polymeric or inorganic in nature. Combinations of these two types create hybrid nanoconstructs.*

the FDA hopes to develop a transparent, consistent and predictable regulatory pathway for nanotechnology products (Food and Drug Administration, 2007).

Although these initial efforts have provided a solid foundation of information-gathering regarding the safety concerns of nanoconstructs, much more work needs to be done before regulations can be developed. The safe handling of nanotechnology-based products can only be assessed once a variety of important issues are addressed by sponsored research (Maynard *et al.*, 2006). The purpose of this chapter is to discuss concepts relating to nanoconstructs, how their physico-chemical properties may influence toxicity, and how to test for this *in vitro*. Specific emphasis will be on the use of nanoconstructs for biomedical applications.

7.2   Nanoconstructs

Nanomaterials, by definition, are any materials with one or more dimensions of the order of 100 nm or less (Scientific Committee on Emerging and Newly Identified Health Risks, 2007). They can either be polymeric, inorganic, or a hybrid of the two depending on the composition of their constituent parts (Figure 7.2). In all cases, however, each nanomaterial as dictated by their properties offer the scientist and engineer specific characteristics and advantages unique to their application (Table 7.1). These same characteristics, however, may also be harmful if not properly evaluated.

Table 7.1   *Examples of nanoconstructs and some of their medical applications.*

Nanoconstruct	Applications	References
Polymeric		
HPMA	Polymer-drug conjugates for cancer therapy Drugs: paclitaxel, camptothecin and doxorubicin	Vasey *et al.* (1999); Meerum *et al.* (2001); Wachters *et al.* (2004)
PEG	PEG-protein conjugates for disease therapy Proteins: asparaginase, interleukin 2, interferon-α	Zimmerman *et al.* (1989); Rajender *et al.* (2002); Graham (2003)
PAMAM dendrimer	Carrier for ibuprofen in A549 lung epithelial cells	Kolhe *et al.* (2006)
PLGA	Cisplatin loaded PEG-PLGA nanoparticles for prostate cancer	Dhar *et al.* (2008)
Inorganic		
Carbon nanotube	Biosensors: glucose, DNA, organophosphorous compound	Li *et al.* (2003); Wang *et al.* (2003); Lin Y *et al.* (2004)
	Drug delivery	Davis *et al.* (1998); Kam *et al.* (2004); Pantarotto *et al.* (2004)
Gold	Imaging and photothermal therapy	Lin A *et al.* (2004); El-Sayed *et al.* (2005); Huang *et al.* (2006)
	Gene/drug delivery	Paciotti *et al.* (2004); Tkachenko *et al.* (2004); Wijaya *et al.* (2009)
Quantum dots	Fluorescent imaging *in vitro* and *in vivo*	Bruchez *et al.* (1998); Akerman *et al.* (2002); Larson *et al.* (2003)
Silica	Drug loaded tubes for controlled release	Lee *et al.* (2002); Son *et al.* (2005); Sharma *et al.* (2005)

Polymeric nanoconstructs are increasingly being explored in biomedical applications such as drug delivery vehicles capable of homing bioactive agents to target tissues (Torchilin, 2008). Such carriers can be modified with a variety of functional groups to attach targeting moieties, imaging agents, as well as drug payloads. Once injected these systems will circulate in the blood stream, avoid immune detection, accumulate at the target site, and finally deliver their payload. In this way, drugs designed for specific tissues can act locally with controlled release characteristics. Evidence of success of this concept can be seen in multiple examples of anticancer agents attached or incorporated into polymers in both clinical trials and in the market today (Matsumura and Maeda, 1986; Wang *et al.*, 2002; Graham, 2003). With these polymeric systems, termed polymer therapeutics (Duncan, 2006), Paul Ehrlich's 'magic bullet' concept of targeted cancer treatment is slowly becoming a reality (Strebhardt and Ullrich, 2008).

Polymeric carriers can be divided into two different but related categories: water soluble and particulate. Such systems differ greatly in architecture depending on the monomeric units (building blocks) and the methods of synthesis and fabrication. For example, while advances in nanotechnology have enabled the fabrication of particulate polymeric nanoconstructs with well-defined sizes and shapes (Champion *et al.*, 2007; Gratton *et al.*, 2008), the size and shape of analogous water-soluble polymers greatly depend on the environmental conditions in which they reside. These characteristics greatly influence biodistribution and overall toxicity.

Water-soluble polymers can vary in structure, size, shape and form and are often described in terms of molecular weight (MW) and MW distribution. This is mostly due to the fact that in general, random polymeric nanoconstructs do not maintain specific shapes in solution. The size of these systems, therefore, is generally characterized by dynamic light scattering (DLS) and size exclusion chromatography (SEC). Polymeric systems are typically composed of repeating monomeric units that ultimately define their properties. Choice in monomers greatly influences these properties and creates different polymer types. For example, poly(N-isopropylacrylamide) (PNIPAAm) (Chilkoti *et al.*, 2002) and N-(2-hydroxypropyl)methacrylamide (HPMA) (Kopecek *et al.*, 2000) copolymers, two commonly used polymeric nanoconstructs in nanomedicine research, differ only slightly in monomer structure but have very different physico-chemical properties. The way these repeating units are connected by covalent bonds also defines the polymer's architecture. For example, polymeric nanoconstructs can be linear such as poly(ethylene glycol) (PEG), branched such as dendrimers, or self-assembled to form particulates such as micelles (Figure 7.2). The size and shape of water-soluble polymers greatly depends on environmental conditions such as pH, temperature, shear forces, as well as the presence or absence of ligand molecules. This property can often be controlled for in stimuli-sensitive polymers where the release of drugs may be desired in target environmental conditions.

Effect of size, MW, architecture, charge, hydrophobicity and surface functionality are some of the characteristics that need to be considered when addressing the toxicity of polymeric nanoconstructs (Table 7.2) (Alexis *et al.*, 2008). Polymer MW and hydrodynamic volume greatly influence circulation time and what tissue they reside in after leaving circulation. It was found, for example, that HPMA copolymers greater than 45 kDa in MW were above the glomerular filtration threshold of the kidneys and able to accumulate in organs of the reticuloendothelial system (RES) in a size-dependent manner (Seymour *et al.*, 1987).

The overall charge of the polymeric nanoconstructs, as well as their functional groups, also greatly influences their toxicity and biodistribution. Charge, most often described in terms of charge density or zeta potential, can range from highly negative to neutral to highly positive, depending on the polymer's functional groups. The presence of amine, sulfate, hydroxyl and carboxyl groups are the most common functionalities that dictate the polymer's zeta potential. Anionic micelles, for example, have been shown to have reduced liver and splenic uptake than those with a neutral charge (Yamamoto *et al.*, 2001). Similarly, the surface-bound functional groups also dictate the toxicity and biodistribution of polymeric carriers. HPMA copolymers conjugated with avidin and mesochlorin e_6 mono(N-2-aminoethylamide) (Mce$_6$), for example, decreased the immune response in mice compared with avidin alone (Hart *et al.*, 2000). If the surface chemistry is used to locally deliver highly toxic compounds such as chemotherapeutic agents (Minko *et al.*, 2000), non-specific tissue uptake of these polymeric systems with drug payloads could result in toxicity of otherwise healthy tissue.

Poly(amido amine) (PAMAM) dendrimers are one type of water-soluble polymers that have received significant attention for biomedical applications such as drug delivery and imaging (Table 7.1) (Florence, 2005). Due to their unique architecture (Figure 7.2), these

Table 7.2 *Physico-chemical characteristics known to influence toxicity and biodistribution of nanoconstructs.*

Physical
  Size
    Length, width, diameter
    Molecular weight
    Volume
    Surface area
    Aspect ratio
  Shape
    Sphere, rod, tube, wire, shell
    Linear, branched
  Polydispersity
  Material(s)
Chemical
  Hydrophobicity/hydrophilicity
  Charge
    Positive
    Negative
    Neutral
  Surface chemistry
    Bare
    Stabilizing agent
    Bound molecules (PEG)
  Functionality
    Drugs
    Targeting moieties

---

branched polymers have a well-defined shape with many surface groups available for conjugation and/or interaction with biological systems. Of particular interest is the ability of PAMAM dendrimers to translocate across the epithelial barrier of the gastrointestinal (GI) tract for oral drug delivery applications (Kitchens *et al.*, 2005; Kolhatkar *et al.*, 2008). The potential toxicity of dendrimers as a function of their size (generation), charge and surface functionality remains a concern, however. We have recently investigated the influence of size, charge, concentration and incubation time of PAMAM dendrimers on their toxicity towards Caco-2 human epithelial colorectal adenocarcinoma cells as models of intestinal epithelial cells. In the concentration range studied (0.1–10 mM) and over a 3.5 hour incubation period, hydroxyl-terminated PAMAM dendrimers did not show signs of toxicity on Caco-2 monolayers as measured by MTT (3-(4,5-Dimethylthiazol-2-yl)-2,5-diphenyltetrazolium bromide) and LDH (lactate dehydrogenase) assays (El-Sayed *et al.*, 2003). Carboxyl-terminated anionic dendrimers of generations G3.5 and G4.5 at 10 mM and 90 min incubation time did elicit size-dependent toxic effects in Caco-2 cells (El-Sayed *et al.*, 2003). However, compared with hydroxyl-terminated and carboxyl-terminated, cationic amine-terminated PAMAM dendrimers were significantly more toxic (Kitchens *et al.*, 2006). For example, while none of the hydroxyl-terminated or carboxyl-terminated dendrimers exhibited toxicity up to 1.0 mM for 2 hours, the G2 and G4 amine-terminated dendrimers did show signs of toxicity across all concentrations. Replacement of amine surface groups with acetyl groups significantly reduced these toxic effects depending on

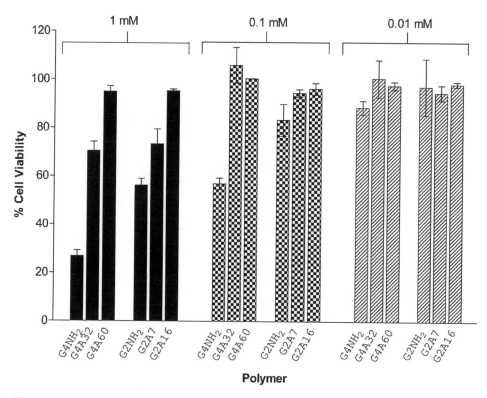

Figure 7.3 *Viability of Caco-2 cells after incubation for 3 hrs with dendrimers at (■) 1 mM, (▨) 0.1 mM and (▨) 0.01 mM. Dendrimers tested were of two different generations (G2 and G4) and either terminated with amine (NH$_2$) groups, or with n number of acetyl (A) groups (An). Results are expressed as mean ± SEM (n = 9). Reprinted from Kolhatkar et al. (2007), with permission from the American Chemical Society.*

the degree of acetylation, PAMAM generation and concentration (Figure 7.3) (Kolhatkar *et al.*, 2007). This decrease in toxicity had a linear correlation with surface density of PAMAM dendrimers.

Attachment of PEG to dendrimers (PEGylation) can increase their circulation time, decrease toxicity and facilitate conjugation of bioactive agents. In a recent study we conjugated PEG to G3.5 and G4.5 carboxyl-terminated PAMAM dendrimers to assess the impact of PEGylation on toxicity and transepithelial transport (Sweet *et al.*, 2009). Methoxy poly(ethylene glycol) (MePEG) (750 Da) was conjugated to carboxylic acid-terminated PAMAM dendrimers at feed ratios of 1, 2 and 4 PEG chains per dendrimer. In the concentration range 0.001 M to 0.1 M, PEGylation of anionic dendrimers did not significantly alter cytotoxicity measured by WST-1 (4-[3-(4-Iodophenyl)-2-(4-nitrophenyl)-2H-5-tetrazolio]-1,3-benzene disulfonate) assay (Figure 7.4). PEGylation of G3.5 dendrimers, however, significantly decreased cellular uptake and transepithelial transport, while PEGylation of G4.5 dendrimers led to a significant increase in uptake, but also a significant decrease in transport (Sweet *et al.*, 2009). Dendrimer PEGylation reduced the opening of tight junctions. Modulation of the tight junctional complex correlated well with changes in

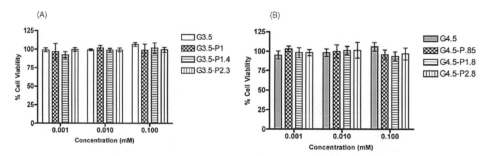

Figure 7.4 *Caco-2 cell viability in the presence of G3.5 (A) and G4.5 (B) native and PE-Gylated dendrimers after a 2-hour incubation time. n = 6, mean ± standard deviation. The number (n) of mPEG polymer chains conjugated to the dendrimers is represented here as G–Pn. Reprinted from Sweet et al. (2009), with permission from Elsevier.*

PEGylated dendrimer transport, and suggests that anionic dendrimers are also transported through the paracellular route. The examples used above illustrate that factors influencing the toxicity and biodistribution of water-soluble polymers include, but are not limited to, polymer charge, size, architecture and functional groups.

Particulate nanoconstructs are suspended in biological media, unlike water-soluble polymers that form solutions in aqueous environments. Traditional methods of nanoparticle fabrication led to systems with a distribution of sizes. Such polydispersity and heterogeneity influence biological fate. Advances in materials science and nanotechnology have led to the fabrication of monodisperse particles with well-defined geometries (Figure 7.2) (Gratton *et al.*, 2008). This promises a higher degree of control over material structure and the ability to systematically correlate architecture (sequence, length, shape, size, surface properties, etc.) with biodistribution, cellular internalization, toxicity and elimination. For example, a series of cubic and cylindrical nanoparticles derived from a mixture of trimethyloylpropane ethoxylate triacrylate, PEG monomethylether monomethacrylate and 2-aminoethylmethacrylate were recently prepared by PRINT (*Particle Replication In Non*-wetting *Templates*) technology (Gratton *et al.*, 2008). Other nanoparticles based on poly(lactic acid) (PLA), poly(glycolic acid) (PGA), poly(lactic acid-co-glycolic acid) (PLGA), poly(propylene oxide) (PPO) as well as latex were made into a variety of shapes including discs, ellipsoids, rods and spheres (Champion *et al.*, 2007). The internalization of monodisperse hydrogel particles fabricated by PRINT into HeLa cells as a function of size, shape, and surface charge was studied (Gratton *et al.*, 2008). The high-aspect-ratio particles with a diameter of 150 nm and height of 450 nm were internalized four times faster than the more symmetric, low-aspect-ratio particles ($d = 200$ nm, $h = 200$ nm). Cylindrical particles having a diameter of 100 nm were internalized to a lesser extent than the larger cylindrical particles having a diameter of 150 nm with the same aspect ratio. In other recent work, worm-like particles with very high aspect ratios (>20) were fabricated and it was shown that this shape exhibits negligible phagocytosis compared with conventional spherical particles of equal volume (Champion and Mitragotri, 2008).

Besides shape, differently sized polymeric nanoparticles have also been evaluated to study their effect on mechanisms of uptake and toxicity. For example, latex beads ranging from 50 nm to 1 μm in diameter were incubated with murine melanoma cells (B16-F10) for up to four hours in the presence of uptake inhibitors (Rejman *et al.*, 2004). It was found that particles smaller than 200 nm in diameter were internalized via clathrin-coated pits, but that larger particles up to 500 nm were internalized by caveolae-mediated endocytosis. Also, uptake of 50 and 100 nm latex beads was rapid, within 30 min, while other larger beads had much slower rates of internalization (several hours) (Rejman *et al.*, 2004). Similarly, it was found that HepG2 liver cancer cells were able to take up smaller MePEG-polycaprolactone (MePEG-PCL) nanoparticles with more efficiency than their larger counterparts (Hu *et al.*, 2007).

In another study, chitosan nanoparticles with varying MWs were evaluated for uptake and cytotoxicity against A549 lung epithelial cells, and it was found that changes in particle zeta potential was a greater contributor to uptake and toxicity than size (Huang *et al.*, 2004). The same was also found for PLGA nanoparticles modified with vitamin E succinated PEG 100, which showed significantly enhanced uptake compared with unmodified controls, indicating that surface functionality and charge have a significant impact on nanoparticle uptake (Yin Win and Feng, 2005). Factors that influence polymer degradation also need consideration since oligomeric degradation products can potentially be more toxic than the parent polymers. For example, it was found that oligo(lactic acid) degradation occurs by hydrolysis through backbiting (van Nostrum *et al.*, 2004), and that degradation products of lactic acid can result in toxicity (Taylor *et al.*, 1994).

Inorganic nanoparticles such as carbon nanotubes, gold nanorods and spheres, quantum dots, silica nanotubes and many others have also been investigated for applications such as disease detection, imaging and drug delivery (Table 7.1; Figure 7.2). The unique sizes, optical, conductive and surface reactive characteristics of these particles are exploited for specific applications. For example, gold nanoparticles have been used as early disease detection contrast agents (El-Sayed *et al.*, 2005) and drug delivery vehicles (Paciotti *et al.*, 2004), as well as photothermal therapy devices (Huang *et al.*, 2006). Therefore, it may be possible to use one system such as gold nanoparticles for combined disease detection and therapy strategies. Discussion of all the applications that inorganic nanoparticles have in modern medicine is beyond the scope of this chapter, but for further information the reader is encouraged to explore any number of the articles on this topic (Emerich and Thanos, 2003; Roco, 2003; Portney and Ozkan, 2006; Lewinski *et al.*, 2008). Here we discuss four of the most common inorganic nanoconstructs used in nanomedicine and the properties that dictate their toxicity (Table 7.2).

Carbon nanotubes (CNTs) and fullerenes are carbon-based structures trigonally bonded in graphite-like sheets of specific geometry such as tubes and spheres. These structures have unique mechanical, electrical and thermal characteristics. For example, the Young's modulus of multi-walled carbon nanotubes (MWCNTs) has been measured to be as high as 950 GPa with a strain to failure of 0.12, making these nanostructures particularly attractive for high-load engineering problems such as bridge cables or the imagined 'space elevator' (Yu *et al.*, 2000). Also, their ability to conduct both heat and electricity with extreme efficiency make them useful for circuit miniaturization (Collins and Avouris, 2000). CNTs and fullerenes are increasingly being explored as potential drug carriers by taking advantage of their unusually high aspect ratios and surface area. In addition to chemical functionalization of the outer walls of CNTs and fullerenes, such structures can also be loaded with biomolecules within their hollow interior for targeted delivery purposes (Bianco *et al.*, 2005). Other applications in biosensing, nano-guided assembly of biomedical devices, and imaging make these carbon-based materials particularly attractive in nanomedicine (Bekyarova *et al.*, 2005).

The overall toxicity of carbon structures dictated by their physico-chemical properties remains a concern (Jia *et al.*, 2005; Magrez *et al.*, 2006; Panessa-Warren *et al.*, 2006; Tian *et al.*, 2006; Grabinski *et al.*, 2007). For example, it was found that the toxicity of carbon-based nanoparticles (flakes, fibers and tubes) was aspect-ratio dependent, probably due to the presence of highly reactive dangling bonds found in carbon black with high density (Magrez *et al.*, 2006). Smaller, higher aspect ratio carbon nanomaterials such as single-walled carbon nanotubes (SWCNTs) and MWCNTs are more toxic than larger carbon-based nanomaterials such as carbon fibers, presumably due to increased interaction with the cellular membrane (Grabinski *et al.*, 2007). Similarly, smaller CNTs have been found to be taken up to a greater extent by fetal lung tissue (IMR90) cells than longer tubes, indicating a size dependency on uptake (Hobbie *et al.*, 2007). The surface functionalization of carbon nanomaterials also influences their cytotoxicity (Dumortier *et al.*, 2006; Magrez *et al.*, 2006; Nimmagadda *et al.*, 2006; Sayes *et al.*, 2006; Tian *et al.*, 2006). For example, it was found that replacement of stabilizing surfactant Pluronic F108 with phenyl-SO_3H significantly decreased the cytotoxicity of fullerenes (Sayes *et al.*, 2006). This is in part because the surfactants often used to stabilize these nanomaterials, including Triton X100 and sodium dodecylbenzene (SDBS), are highly toxic as detergents (Dong *et al.*, 2008). Finally, the overall dispersity and agglomeration of carbon nanotubes are known to influence their toxicity (Wick *et al.*, 2007). The CNT agglomerates may be larger and stiffer than well-dispersed CNTs, which may greatly impact upon the cell response to such bundles.

Quantum dots (QDs) are spherical particles typically with a cadmium selenide (CdSe) core and a zinc sulfide (ZnS) or cadmium sulfide (CdS) shell. Because of the semiconductor characteristics of these nanomaterials, QDs have size-dependent light emitting properties. The ability to fluoresce 10–100 times greater than typical fluorophores with little concern for photobleaching, coupled with the capacity to be functionalized with biomolecules on the particle's surface, makes these nanoconstructs particularly useful as *in vivo* and *in vitro* diagnostic markers (Smith *et al.*, 2008). Surface modifications with bioactive drugs may also make QDs capable of combined disease detection and therapy in response to internal or external stimuli. Concern over the toxicity of QDs due to the presence of heavy metal atoms at their core however remains a barrier. For example, toxicity of CdSe nanocrystals has been observed at very low concentrations (0.65 μM) due to the release of Cd^{2+} ions (Kirchner *et al.*, 2005). This toxicity was significantly reduced by the addition of ZnS or silica shells, but release of the Cd^{2+} ions could not be completely avoided as toxicity was still observed. PEGylating the surface of these QDs, however, further reduced the toxicity of these particles with no observed cytotoxicity at concentrations up to 30 μM (Kirchner *et al.*, 2005). These findings corroborate other studies investigating the toxicity of CdSe QDs which also found significant cytotoxicity at low concentrations due to Cd^{2+} ions (Derfus *et al.*, 2004). Therefore, although the size and charge of any nanomaterial is of importance when probing for toxicity, the inherent toxic elemental composition of QDs is of primary concern.

Gold nanoparticles have gained recent attention in nanomedicine due to their unique optical properties coupled with being relatively bio-inert. Able to absorb and scatter light, gold nanoparticles can emit light with size-dependent frequencies (Huang *et al.*, 2007). This size-dependent light emission and scattering creates the characteristic color spectrum long associated with colloidal gold. Such properties have potential in cancer detection and therapy as well as other biosensing applications (Jain et al., 2007). Additionally, the relatively inert nature of these nanoparticles makes them particularly attractive in drug delivery applications (Paciotti et al., 2004). They can be conjugated with a wide variety of biomolecules capable of targeted drug delivery while minimizing toxicity in non-targeted regions (Ghosh et al., 2008).

Gold is relatively bio-inert, but when observed, gold nanoparticle-induced toxicity most likely results from direct interactions of these particles with intracellular organelles. For example, internalized gold nanoparticles have been found to disorganize fibroblast cytoskeleton integrity and extracellular matrix (ECM) structure (Pernodet et al., 2006). This uptake has been found to be size dependent where 50 nm gold nanoparticles appear to be taken up more into cells than particles of other sizes (Chithrani et al., 2006; Jiang et al., 2008). Size probably dictates the mechanism of cell entry by gold nanoparticles due to geometric constraints at the receptor/nanoparticle interface (Rejman et al., 2004). Surface charge is also known to induce cytotoxicity where gold nanoparticles with positively charged surface modifications are more toxic than negatively charged particles (Goodman et al., 2004). Cetyl trimethylammonium bromide (CTAB), a highly cationic detergent used during gold nanorod synthesis, has very toxic effects (Takahashi et al., 2006). To mask this CTAB-induced toxicity, however, a commonly used technique is to functionalize the nanoparticles with long polymer chains such as PEG (Niidome et al., 2006).

Silica nanoparticles are another class of nanomaterials with applications in nanomedicine. These nanoconstructs are highly stable in most environments and are able to be conjugated with a variety of biomolecules, making them useful for drug delivery, biosensing and imaging applications (Tan et al., 2004). The potential toxicity of silica nanoparticles is a concern, however, as the inhalation of crystalline silica is well known to induce silicosis, a severe respiratory disease. In silicosis, inhaled silica particles are uptaken by macrophages which then stimulate an inflammatory response by triggering cytokine release and the production of reactive oxygen species (ROS) (Hamilton et al., 2008). It is thought that this response may actually be due to interactions of silica with the lysosomal membrane to which it is localized in the macrophage, and that permeation through this membrane may exude enzymes into the cytoplasm, thereby triggering apoptosis and cytokine release. Although silicosis is typically found in people who inhale silica dust in occupations such as mining, the potential to inhale synthetic nanosized silica particles is of concern. One study found similar responses to silica nanoparticles characterized by activation of macrophages and the production of pro-inflammatory cytokines and ROS (Park and Park, 2008). Despite findings such as these, little work investigating the toxicity of silica nanoparticles has been performed.

Silica nanotubes, created by template synthesis techniques (Figure 7.5), have attracted attention since their inner voids can be loaded with bioactive agents (Hillebrenner et al., 2006), their open ends gated for controlled drug release in response to environmental stimuli (Hillebrenner et al., 2007), and their surface functionalized to render biocompatibility. The toxicity of silica nanotubes dictated by their physico-chemical properties was recently investigated in an epithelial breast cancer cell line (MDA-MB-231) and a primary umbilical vein endothelial cell line (HUVEC) (Nan et al., 2008). Variations in size and charge of the silica nanotubes allowed for systematic study of the influence of these parameters on toxicity and uptake. Positively charged nanotubes were significantly more toxic than

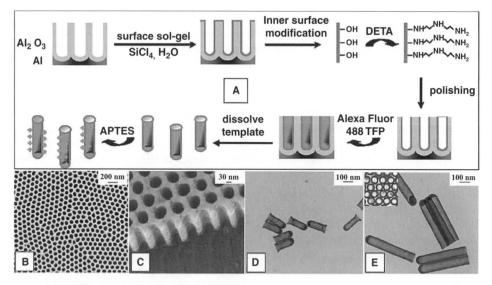

Figure 7.5 *(A) Schematic representation of template synthesis of silica nanotubes; (B,C) scanning electron microscope images of porous alumina template for SNT (diameter 50 nm, length 200 nm), (B) top view, (C) cross-section; (D,E) transmission electron microscope images of bare silica nanotubes (50 nm dia.) Length: (D) 200 nm, (E) 500 nm, with inset of TEM image for the cross-section of the SNTs bundle. Images confirm the monodisperse, well-defined dimensions of SNTs. Reprinted from Nan et al. (2008), with permission from the American Chemical Society.*

their bare counterparts as determined by the standard MTT assays (Figure 7.6), and the smaller, 200 nm nanotubes were significantly more toxic than the longer 500 nm nanotubes. These effects were also found to be cell-type dependent, potentially due to differences in metabolic activity or rate of proliferation between cancer and normal nonmalignant primary cells of the body. The influence of geometry of silica nanoparticles on toxicity, however, is unexplored.

7.3 Exposure

Nanosized constructs have been present in our environment since long before man evolved. One report suggests that CNTs and fullerenes were found in 10,000-year-old polar ice in Greenland (Esquivel and Murr, 2004). Many natural organisms such as viruses, and toxic proteins such as snake venom, also operate at the nanoscale. It is therefore not surprising that multicellular organisms such as human beings are equipped with evolved methods to remove or eradicate nanoscale foreign bodies. In the following discussions, a variety of considerations will be addressed when evaluating the toxicity of nanomaterials *in vivo*. Multicellular organisms are complex in nature, and therefore proper evaluation of the impact that nanoparticles have upon them is also complex. Factors that influence biodistribution, uptake and removal of nanosized constructs will be focused upon, as these variables dictate how the body is impacted by these particles (Table 7.2).

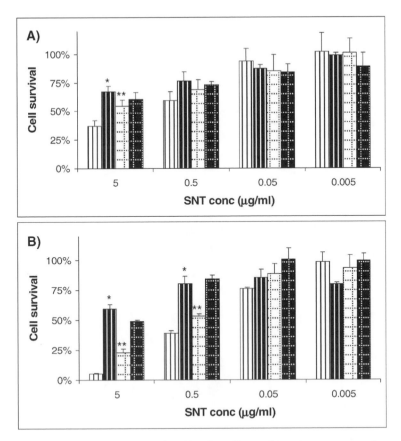

Figure 7.6 *MTT cytotoxicity assay showing the effects of varying concentrations of nanotubes on growth inhibition of (A) MDA-MB-231 breast cancer cells and (B) human umbilical vein endothelial cells (HUVEC) cultured in vitro. Increases in nanotube concentration resulted in decreased cell viability. At higher concentrations positively charged nanotubes were more toxic. At equal nanotube concentrations 200 nm nanotubes were more toxic than 500 nm. Legend: ⬜ 200 nm with positively charged APTS outer surface group; ■ 200 nm with no outer surface functionalization; ⬛ 500 nm with positively charged APTS outer surface group; ■ 500 nm with no outer surface functionalization. Results reported as mean ± SEM (n = 3). Statistically significant differences are indicated relative to 200 nm positively charged SNTs (*P < 0.002; **P < 0.02). Reprinted from Nan et al. (2008), with permission from the American Chemical Society.*

7.3.1 Absorption

Nanoparticle route of entry in nanomedicine is an important consideration when evaluating nanoparticle-induced toxicity. Routes of entry and administration include intravenous, subcutaneous or intraperitonial injection, as well as GI, pulmonary or dermal exposure (Figure 7.7). In all of these examples, it is likely that nanoparticles will first contact either epithelial or endothelial biological barriers before any toxicity can be observed. Therefore, mechanisms of nanoparticle transport across these barriers must be evaluated. Below are

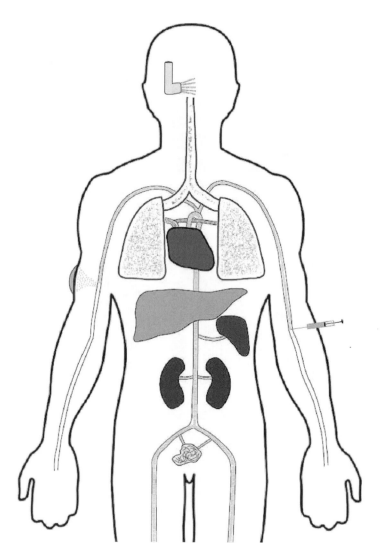

Figure 7.7   *Schematic of some routes of exposure and organs most likely to come in contact with nanoconstructs during circulation.*

three of the most common routes of absorption and some examples of factors that influence the transport and toxicity of these nanomaterials.

7.3.1.1   Skin

In the case of epithelial transport of nanoconstructs in drug delivery applications such as the skin, overcoming the robust natural barrier may be difficult. For skin in particular, transport across the stratum corneum (SC) followed by the epidermis, dermis and vascular endothelium may be difficult for nanosized constructs, as only small lipophilic molecules are known to pass freely through these layers (Barry, 2001). There are, however, reports that indicate the ability of nanoparticles to penetrate the skin, although most studies noted minimal penetration (Monteiro-Riviere and Inman, 2006; Baroli *et al.*, 2007; Zhang *et al.*, 2008). Factors that influence penetration include animal model choice, skin structure such as the presence of hair follicles, as well as nanoparticle physico-chemical properties (Table 7.2). For example, QDs of varying size, shape and surface chemistry have been applied to porcine skin to study the effects of these physico-chemical properties on skin penetration and toxicity (Ryman-Rasmussen *et al.*, 2006, 2007; Zhang *et al.*, 2008). It was found that PEGylated QDs, both spherical (35 nm) and ellipsoidal (6 nm minor axis, 12 nm major axis) were able to penetrate the SC and localize in the epidermal layers (Ryman-Rasmussen *et al.*, 2006). When coated with PEG-amine, the spherical QDs were able to penetrate further into the skin and localize in the deep dermal layers. However, this observation may also have been due to size, as the PEG-amine spheres were significantly smaller (15 nm). Another interesting observation was that while PEG-carboxyl spheres were able to penetrate the SC within 8 hours, PEG-carboxyl ellipsoids required 24 hours for penetration to the epidermal layers (Ryman-Rasmussen *et al.*, 2006). With roughly similar hydrodynamic diameters, these results may indicate a shape dependency on skin penetration, where spheres may have better penetration capabilities compared with ellipsoids. These results are supported by toxicity studies of the same QDs which indicate that amine- and carboxyl-coated QDs were toxic to primary human epidermal keratinocytes (HEKs), while regular PEG-coated QDs showed no signs of toxicity (Ryman-Rasmussen *et al.*, 2007). Carboxyl-coated QDs were also found to stimulate an inflammatory response unlike the PEG-amine or PEG-coated QDs, and these effects were dependent on size. Nail-shaped PEGylated QDs were also found to be cytotoxic but were unable to penetrate the SC (Zhang *et al.*, 2008). In another study, 10 nm metallic nanoparticles were able to penetrate the skin and reach the epidermis layers; interactions with the particle's charge and the different pH values of the individual skin layers may have influenced penetration depth (Baroli *et al.*, 2007). Taken together, results from these studies indicate that skin penetration, toxicity and inflammation may be influenced by properties such as size, shape and charge.

7.3.1.2 GI Tract

Translocation across the GI tract epithelial cell barriers may also be difficult for nanosized constructs due to tight gap junctions, although direct evidence both *in vitro* and *in vivo* indicates that uptake of these particles is possible (Kitchens *et al.*, 2005; Oberdorster *et al.*, 2005). Transport of these particles may occur by the paracellular pathway (through the gap junctions), by transcellular transport (endocytosis) or through Peyer's patches (Norris *et al.*, 1998). Additionally the characteristics that dictate these mechanisms are known to be the nanoparticle size, shape and charge (El-Sayed *et al.*, 2002, 2003). For example, it has been shown that 100 nm polystyrene nanoparticles have higher uptake in Caco-2 cells then other sizes tested (50, 200, 500 and 1000 nm) (Yin Win and Feng, 2005).

Transport through the intestinal mucus layer is also dependent on these physico-chemical properties (Norris and Sinko, 1997). It was found that transport through the mucus was heavily size-dependent, indicating that particles above 500 nm have limited transport, but that transport through the mucus may not be the rate-limiting barrier. Engineering nanoconstructs with mucoadhesive properties have been shown to increase nanoparticle transcytosis across the intestine (Hussain *et al.*, 1997). The concern over the toxicity

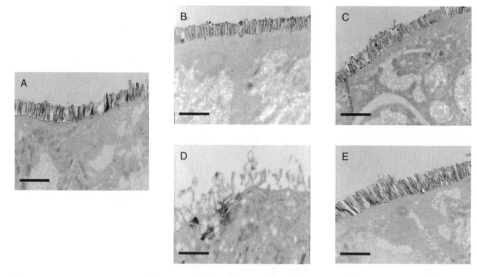

Figure 7.8   Transmission electron microscopy images of Caco-2 cell monolayers after treatment with PAMAM dendrimers (1 mM) for 2 hours. (A) control cells; (B) G2NH$_2$; (C) G1.5COOH; (D) G4NH$_2$; (E) G3.5COOH. The images display a generation-dependent effect of PAMAM dendrimers on Caco-2 microvilli (magnification ×12 500). Scale bars = 1 µm. Reprinted from Kitchens et al. (2007), with kind permission from Springer Science + Business Media.

of these nanoconstructs during uptake and transport has also been widely addressed. For example, thiolated chitosan nanoparticles have been studied in Caco-2 cells and the findings suggest that while size did have some impact, overall charge was found to be the greatest contributor towards toxicity (Loretz and Bernkop-Schnurch, 2007). It was found that while anionic and neutral chitosan nanoparticles have some toxic effects, cationic particles caused severe cytotoxicity. The results corroborate with findings on toxicity of dendrimers discussed earlier, which also characterize cationic nanoparticles as being toxic to Caco-2 cell monolayers (Kitchens *et al.*, 2006).

As discussed above, PAMAM dendrimers are one unique class of nanoconstructs with potential oral drug delivery applications. Their ability to be transported across the epithelial barrier of the GI tract has been evaluated *in vitro* (Kitchens *et al.*, 2005). Their transepithelial transport is dictated by their physico-chemical properties such as size (generation), charge and surface functionality. For example, smaller dendrimers (G0, G1 and G2) were found to be more permeable across Caco-2 cell monolayers than larger, higher-generation dendrimers (G3 and G4) (El-Sayed *et al.*, 2002). Also, it was found that cationic dendrimers were more permeable than their anionic counterparts, which exhibited a size-dependent permeability (El-Sayed *et al.*, 2003; Kitchens *et al.*, 2006). The mechanisms of transport were both para- and transcellular by endocytosis (Kitchens *et al.*, 2007). Transmission electron microscopy (TEM) images of the Caco-2 monolayers after exposure to dendrimers with varying size and charge indicates the dependence on these parameters for toxicity (Figure 7.8). It was found that larger, cationic dendrimers significantly disrupted monolayer morphology (Kitchens *et al.*, 2007).

7.3.1.3   Respiratory

The inhalation of particulate matter such as that found in cigarette smoke and pollution is known to cause a variety of respiratory diseases including asthma and cancer. Due to their unique size, it is anticipated that the inhalation of nanoconstructs may also induce toxicity depending on their physico-chemical properties. Several factors have to be considered when studying the toxicity of inhaled nanoconstructs, including deposition, transport and clearance (Borm and Kreyling, 2004). After inhalation, nanoconstructs will be deposited in regions throughout the respiratory system depending on size, shape, charge and surface functionality (Heyder *et al.*, 1986). Once deposited on the surface, translocation of these particles across the lung epithelial barrier provides these materials access to the blood stream. This mechanism of action is highly dependent on the described physico-chemical properties. For example, it was found that anionic dendrimers are transported by caveolae mediated endocytosis in lung epithelial cells, while cationic and neutral dendrimers are transported by a clathrin and caveolae independent mechanism (Perumal *et al.*, 2008).

Respiratory toxicity is also dependent on these parameters. For example, the ability of ultrafine TiO_2 nanoparticles to induce a host response and pulmonary inflammation has been found to be dependent on size due to prolonged pulmonary retention (Ferin *et al.*, 1992). Also, inhalation of 64, 202 and 535 nm polystyrene nanoparticles exhibited size-dependent toxic effects *in vivo* and *in vitro* (Brown *et al.*, 2001). It was found that while the 64 nm particles induced significant toxicity in lung lavage from rats instilled with 1 mg of the polystyrene particles, the other sizes did not induce significant toxicity. Expression of IL-8 was also size-dependent upon *in vitro* exposure, but interestingly the 64 nm particles were least able to stimulate an IL-8 response (Brown *et al.*, 2001).

7.3.2   Protein Adsorption

Upon contact with blood or other biological fluids such as tears, nanoconstructs will within seconds adsorb many different proteins that fully coat their surface. This phenomenon is true for all biomaterials and represents one of the greatest challenges for devices such as glucose sensors (Ratner *et al.*, 2004). Compared with most other biomaterials, this problem is particularly amplified at the nanoscale due to the surface reactivity characteristics of materials at this scale. Proteins such as fibronectin and albumin have high affinity for many biomaterial surfaces and very quickly convert their surface from a nonbiologically recognized material, to one that is. This is important because once adsorbed, these nanoparticles take on new surface characteristics that greatly dictate their biodistribution, clearance and toxicity. Therefore, proper characterization of how proteins adsorb to nanoparticle surfaces is required to fully understand their interaction with biological systems.

Protein adsorption on nanoparticles can result in immune recognition. Opsonization of antibodies such as IgG may activate humoral immunity against nanoparticles by stimulating natural killer (NK) cells and the complement system (Salvador-Morales *et al.*, 2006). Ultimately these particles will be phagocytosed by antigen-presenting cells and removed by the RES. Complete clearance from the body may not be possible once phagocytosed, due to the resistance of nanoparticles to degradation, and chronic immune activation may result in health-related pathologies such as tumorigenesis.

The adsorption of proteins is dependent upon the surface characteristics of particles such as charge and hydrophobicity (Lück *et al.*, 1998). For example, it has been found that latex nanoparticles with higher surface charge densities adsorb proteins to a greater extent than those with similar size and hydrophobicity (Gessner *et al.*, 2002). Similarly, modification of the latex nanoparticle's surface hydrophobicity, while maintaining constant size and charge, indicates that decreasing the surface hydrophobicity decreases the amount of adsorbed proteins (Gessner *et al.*, 2000). The influence of surface functionality was also examined, indicating that proteins such as albumin and IgG preferentially bind to latex nanoparticles with strongly basic groups, while other proteins such as Apolipoprotein (Apo) prefer those with acidic groups (Gessner *et al.*, 2003). Besides surface characteristics such as charge and hydrophobicity, other material parameters such as size and shape influence protein adsorption. For example, the curvature of silica nanoparticles was found to greatly influence the secondary structure of adsorbed proteins (Lundqvist *et al.*, 2004). Such alterations in the secondary structure of adsorbed proteins based on curvature dictate the overall stability and hence adsorption of these proteins onto the nanoparticle surface.

To study protein–nanoparticle interactions, one can characterize the nanoparticle surface before and after serum protein exposure. Studies investigating adsorption kinetics in either single protein solutions or in multifaceted complex protein solutions will provide evidence of the affinity that various proteins have with the nanoparticle under question. Some nanoparticles such as gold and QDs have very unique optical profiles, and shifts in the surface plasmon resonance (SPR) peak after protein exposure may provide evidence of nonspecific protein adsorption. Similarly, changes in hydrodynamic diameter as measured by DLS also provide evidence of protein adsorption. Although sometimes difficult due to low concentrations, it is ideal to perform these measurements from explanted nanoparticles to better understand protein adsorption *in vivo*.

One strategy commonly used is to coat the nanoparticle surfaces with long polymer chains such as PEG to prevent unwanted protein adsorption (Otsuka *et al.*, 2003; Niidome *et al.*, 2006). PEG's hydrophilicity and steric repulsion of other molecules prevents proteins such as fibronectin from adsorbing to biomaterial surfaces. These immune-evading water-soluble polymers can be added to otherwise immunogenic polymers to give them 'stealth-like' characteristics (Gref *et al.*, 2000). Similarly, other surface-immobilized biomolecules such as saccharides, lipids, natural proteins, and nucleic acids may be used to reduce nonspecific protein adsorption and therefore unwanted biorecognition of nanoconstructs.

7.3.3   Immune Response

The degree to which immune responses occur upon nanoparticle exposure is governed by the nanoparticle's physico-chemical properties (Table 7.2) (Vonarbourg *et al.*, 2006a). The mechanisms by which nanoconstructs trigger immunological responses is relatively unknown, and considerable evidence needs elucidation before strategies to reduce or avoid immune responses can be designed. Complete discussion of the potential immunogenicity of nanoparticles is beyond the scope of this chapter, although readers are encouraged to explore the review by Dobrovolskaia and McNeil (2007) for further information on this topic.

The immune system is divided into two different but interconnected systems termed innate and adaptive immunity (Abbas and Lichtman, 2005). Innate immunity is characterized as an early reaction to foreign bodies with very robust pathways capable of mounting quick responses against a large variety of antigens. If, however, such reactions are not completely effective and chronic foreign body exposure begins, adaptive immunity adapts to that specific antigen for more rigorous defense actions. It is expected that innate immunity is the principle mode of immune response action, as opsonization of antibodies is thought to be crucial for RES clearance (Gref *et al.*, 1994). For example, one study found immune recognition of C_{60} fullerenes due to IgG antibody adsorption (Chen *et al.*, 1998). After opsonization, neutrophils and macrophages are able to recognize the nanoparticles as foreign and activate for phagocytosis of the particles. Over-activation may result in phagocytic cell death and release of harmful products such as ROS, nitric oxide and lysosomal enzymes. Also, phagocytic cell activation results in release of pro-inflammatory cytokines, stimulating an inflammatory response and recruitment of more neutrophils and macrophages.

Complement activation is another pathway by which nanoparticles may activate the immune system. Such activation can occur by the classical or alternative pathway by either adsorbed antibodies or by adsorption of C3b and other complement proteins onto the nanoparticle surface. The influence of size, shape, charge and hydrophilicity on complement activation has been studied previously with PLA and PLGA nanoparticles, finding that modification of these parameters can reduce complement activation (Mosqueira *et al.*, 2001). In each pathway, after opsonization the complement cascade releases a cocktail of proinflammatory cytokines that result in further immune activation and phagocytosis of the particles. Similarly, the influence of liposome charge on complement activation has also been studied, indicating that negatively charged particles may activate complement by the classical pathway, while positively charged particles activate it through the alternative pathway (Chonn, 1991). Activation of both pathways by nanoconstructs such as CNTs has been reported (Salvador-Morales *et al.*, 2006). Surface modifications such as PEG coatings have been found to greatly reduce the extent of complement activation (Vonarbourg *et al.*, 2006b).

A number of studies have investigated the properties that influence immune activation, as these parameters dictate their overall clearance and circulation times (Vonarbourg *et al.*, 2006a). Nanoparticles made of polyalkylcyanoacrylate have been shown to induce the production of ROS in macrophages (Cruz *et al.*, 1997). CNTs show no sign of acute toxicity, but induce intracellular ROS in the presence of contaminants (Pulskamp *et al.*, 2007). The size and shape can also influence immune activation, as relative curvature dictates the degree of opsonization. This effect was observed where smaller liposomes had less opsonized proteins and less complement activation compared with large liposomes (Harashima *et al.*, 1994). Also, while negatively charged particles do not induce cytokine release from leukocytes, positively charged particles do (Dobrovolskaia and McNeil, 2007). Contaminants from synthesis (Pulskamp *et al.*, 2007) and nanoparticle degradation (Chellat *et al.*, 2005) have also been identified as being immunostimulatory.

7.3.4 Biodistribution

Once nanoconstructs have reached the bloodstream by any of the routes discussed earlier, they will circulate until they are absorbed into tissue or cleared by the immune system or filtration (Figure 7.7). Factors that determine which organs the nanoparticles accumulate in or are removed by are influenced by the physico-chemical characteristics of the nanoparticles. This is due to the fact that each organ differs in architecture and ultimately the potential modes of entry. For example, while some organs such as the liver are highly vascularized and populated with cells designed for uptake and metabolism of waste and foreign bodies, other organs such as the intervertebral disc lack a direct blood supply and are designed for mechanical strength without specific cellular function. Therefore, one would expect higher accumulation of nanoparticles in the liver in comparison to the intervertebral disc.

Nanoparticle size is of particular importance in biodistribution as the fenestrae, or pores, of organ endothelium differs by organ type and function. For example, the liver, spleen and bone marrow have fenestrae sizes between 50–100 nm and particles below these sizes pass through the endothelium and into the tissue (Alexis *et al.*, 2008). The same is also true for organs such as the kidneys which have much smaller pore sizes (roughly 4 nm in diameter) (Ohlson *et al.*, 2001; Deen *et al.*, 2001; Berne *et al.*, 2004). Tumor tissue also has large fenestrae due to their leaky vasculature and can be as large as 400 nm in size (Hashizume *et al.*, 2000). Because nanosized constructs typically have dimensions smaller than the fenestrae of these organs responsible for uptake and clearance of foreign objects, higher accumulation is observed in these organs (Alexis *et al.*, 2008). For example, intravenous injection of differently-sized bare gold nanoparticles resulted in a size-dependent biodistribution (De Jong *et al.*, 2008). It was found that while all sizes tested had widespread biodistribution profiles, the 10 nm particles were most pervasive. The effect of polymer MW on biodistribution has been extensively studied (Cartlidge *et al.*, 1987; Seymour *et al.*, 1987, 1995; Vexler *et al.*, 1994; Mitra *et al.*, 2004; Lammers *et al.*, 2005; Fang *et al.*, 2006). For example, HPMA copolymers of varying MWs, 4.5–800 kDa, were intravenously injected and the resulting biodistribution assessed (Noguchi *et al.*, 1998). It was found that while low MW HPMA copolymers were rapidly cleared through the kidneys and had the least tumor accumulation, high MW HPMA copolymers had long circulation times with enhanced tumor accumulation.

Nanoparticle charge and surface coating also influences biodistribution. For example, positively charged nanoparticles typically have reduced blood circulation times in comparison with negatively or neutrally charged particles. This may be due to several reasons including charge interactions with organs such as the liver and kidneys, as well as immune recognition specific to charge. For example, anionic PEG-poly(D,L-lactide) block copolymer micelles had reduced liver and spleen uptake compared with those with a neutral charge (Yamamoto *et al.*, 2001). Protein adsorption to nanoconstructs is also influenced by charge and surface coating, which also impacts biodistribution. Therefore, surface coating with PEG or other polymeric chains enhances blood circulation time and reduces uptake in organs such as the liver, spleen and kidneys (Niidome *et al.*, 2006).

Besides whole blood, the most common organs typically included in organ biodistribution include the liver, lungs, spleen and kidneys, as these organs are most likely to take up the particles due to their vasculature and fenestrae. Other potential organs of interest include the heart, stomach, brain, muscle, intestine and skin, as these organs may also absorb nanosized constructs (Wang *et al.*, 2004). If the nanoparticles are labeled with fluorophores or radio-isotopes, then standard counters may be used to assess nanoparticle concentrations in each organ. Various *in vivo* imaging systems can also be used here to obtain qualitative information on overall biodistribution of the nanoparticles. With this information, it may be possible to examine how and why nanoparticles accumulate in various organs and therefore which organs are particularly at risk of nanoparticle exposure.

7.3.5 Clearance

In some applications such as drug delivery, it is desirable to design a system that maximizes blood circulation time by minimizing renal filtration and hepatic uptake. Because rapid clearance can often be desired to avoid toxicity, it is important to discuss the factors that influence removal of nanoparticles. For a more in-depth discussion on this topic, readers are encouraged to refer to the review by Longmire *et al.* (2008).

As nanoparticles circulate, it is likely that they will pass through the complex vasculature of the renal system in the kidneys. Functioning as the primary organ to excrete unwanted molecules, proteins and foreign bodies, it is possible that many of the nanoparticles may be excreted into the urine. Such urinary excretion is dictated by the physico-chemical characteristics that define the nanoconstructs under question. For example, passive filtration is greatly dependent on size, as the pore sizes of the glomerular capillary wall are roughly 4–5 nm (Deen *et al.*, 2001; Ohlson *et al.*, 2001; Berne *et al.*, 2004). Although the fenestrae of the kidney's endothelial barrier is quite large, glomerular filtration represents the limiting barrier as its pore size is significantly smaller. Therefore, particles greater than 8 nm in size may not undergo mechanical filtration due to size constraints of the pores (Longmire *et al.*, 2008). Particle charge and surface chemistry also influences filtration, as membranes in the kidney are highly charged and selective. For example, the strong negative charge of the glomerular capillary wall makes filtration more selective towards positively charged particles compared with those with a neutral or negative charge (Deen *et al.*, 2001). The same is true for surface chemistry, as adsorbed proteins or surface-bound molecules will dictate passage into the urine.

Clearance by the liver is influenced by the same nanoparticle characteristics. While hepatocyte uptake may remove nanoparticles by biliary excretion into the GI tract, Kupffer cells of the RES do not have such mechanisms. In both modes of clearance, hepatocyte and Kupffer cell uptake, endocytosis and chemical- or enzyme-mediated degradation of the nanoparticles will follow. Uptake by hepatocytes and Kupffer cells can result in chronic accumulation and induce toxicity. Clearance by the liver is particularly influenced by adsorbed proteins, as opsonization of other proteins such as antibodies may trigger Kupffer cells to phagocytose the particles.

Immune activation represents the third main mechanism of clearance upon nanoconstruct exposure. This mechanism of clearance may occur locally (at the site of exposure such as in the lungs), systemically by free macrophages and monocytes, or in organs such as the liver. Refer to section 7.3.3 for information on the factors that influence immune clearance of nanoconstructs.

As mentioned earlier, nanoconstruct size and charge significantly impact upon blood circulation time and clearance. For example, the HPMA MW threshold for renal filtration was found to be 45 kDa, and polymers above this threshold have enhanced circulation times (Seymour *et al.*, 1987). Studies on dendrimers of varying size (generation) and charge have found that smaller, anionic dendrimers have longer circulation times compared with larger, more cationic dendrimers due to clearance from the liver (Malik *et al.*, 2000). Similarly, the influence of charge on liposome clearance indicates the importance of such parameters in biodistribution of nanoparticles *in vivo* (Levchenko *et al.*, 2002).

7.4   Evaluation of Toxicity *In Vitro*

Studies using epithelial and macrophage cell lines in the respiratory and GI tracts, as well as in the skin and vasculature, are among the most common *in vitro* models used in evaluating the toxicity of nanoparticles. The most notable examples of such toxicological studies are of the pulmonary response to inhaled asbestos (Mossman *et al.*, 1990). Much is now known about the development of pulmonary interstitial fibrosis and lung cancer in response to inhaled asbestos. Fiber characteristics such as concentration, type and size have been shown to be important factors in the development of the inflammatory response to inhaled asbestos, and have helped to develop more stringent requirements for regulation and importation of amphiboles (Mossman *et al.*, 1990).

Proper control of all variables involved is essential to understand fully the cell–nanoparticle interactions. Also, mimicking the *in vivo* environment as much as possible is crucial to obtaining representative results that may later be applied *in vivo*. Finally, the appropriate tests must be performed to measure accurately the cellular response under question. When these three factors are kept under consideration, *in vitro* analysis of nanotoxicity offers investigators a powerful tool for mechanistic and high throughput analysis. The following sections discuss the variables and proper control of study parameters associated with *in vitro* toxicological analysis of nanoparticles. Review of methods has been categorized by cellular function.

7.4.1   *In Vitro* Variables

Controlling the individual variables associated with *in vitro* nanotoxicological studies is a difficult task due to the inherently complex properties of nanoparticles (Schulze *et al.*, 2008). Variations in size, shape, surface area, surface chemistry and charge of the nanoparticles are just a small sample of the many important characteristics of these particles that have to be taken into account during study design (Table 7.2) (Zuin *et al.*, 2007). In addition to these properties, cell suspension variables and how the particles behave in this complex environment of proteins, amino acids and salts are important to consider. The following reviews a selection of some of these variables.

7.4.1.1   Serum/Media

As discussed above, the cytotoxicity of nanoparticles can be influenced by proteins present in serum. Because of the high surface reactivity of nanoparticles, protein absorption to the nanoparticle's surface is likely in the presence of serum. Some studies have shown that the cytotoxicity of nanoparticles is greatly reduced in the presence of serum, indicating that protein adsorption may mask nanoparticles from cells (Casey *et al.*, 2007a; Davoren *et al.*, 2007). For example, carbon nanotubes in the presence of albumin varied in dispersion and cytotoxicity compared with controls without albumin (Elgrabli, 2007). The presence of other media containing components such as phenol red may also influence cytotoxicity, as interactions with nanoparticles may influence absorbance during spectroscopy (Casey *et al.*, 2007a). Results from these studies indicate that proper controls, such as in the presence and absence of serum, must be used to ensure that there are no cell medium constituent interactions.

7.4.1.2   Stabilizing Surfactants

Because many nanoparticles are water-insoluble and have high binding affinities to one another, stabilizing agents are required to reduce agglomeration and increase aqueous dispersion. Many stabilizing agents currently used, however, have been shown to have strong cytotoxic effects. CTAB, used to synthesize and stabilize gold nanorods, is a strong detergent known to greatly increase nanorod cytotoxicity (Takahashi *et al.*, 2006). Similarly, the surfactants sodium dodecyl sulfate (SDS) and SDBS, used to stabilize CNTs, caused significant toxicity when compared with those stabilized using sodium cholate or single-stranded DNA (Dong *et al.*, 2008). Therefore, choice of surfactant for dispersion of nanoparticles must be considered carefully, and proper studies comparing toxicity in the presence of the surfactant with and without nanoparticles should be employed.

7.4.1.3   Concentration

Because of the unique sizes and shapes of nanoparticles, standardization of nanoparticle concentration remains a difficult task. Currently the most common expression of concentration in the literature is mass to volume ratio. Typically these reports are in terms of μg/ml, but large variations expand the range of concentration from ng/ml to mg/ml. Comparisons between studies using the current approach are not possible because variations in particle size, shape and surface chemistry produce non-uniform concentrations. To reduce these effects, it is suggested that investigators characterize and publish the size and shape of their particles, and that concentration be reported in terms of both mass to volume and number of particles per ml.

7.4.1.4   Incubation Time

Another significant variable that differs greatly from study to study is the time of incubation with nanoparticles. The majority of studies only investigate acute toxicity, generally 24 hrs, but longer incubation times are required to see the influence that nanoparticles have on cell cycle and proliferation. In general these studies should span at least two population doubling times to see the long-term, subcellular effects that particles have on cells.

7.4.2   Cell Lines

An investigator's choice in cell lines when designing *in vitro* experiments is pivotal to the ultimate goal of the study. In general, when evaluating nanotoxicity it is important to choose a cell line that best represents the most likely *in vivo* cell barrier exposed to such particles (Table 7.3). For example, if a certain nanoparticle is expected to be released in aerosol form, then alveolar epithelial and macrophage cell lines may be the best cell model to assess the potential toxicity when inhaled. Similarly, if concerns are being explored regarding nanoparticle content in topical creams such as suntan lotion, then studies using epidermal keratinocytes and dendritic cells would be the best model. Below are some of the most common cell lines used to date in nanotoxicology.

Table 7.3 *Common cell lines and assays used to study nanoconstruct toxicity* in vitro.

In vitro models	Description
Cell lines	
HeLa	Human cervical cancer
Macrophages	
RAW264.7	Mouse macrophage from leukemia virus-induced tumor
J774	Mouse BALB/c monocyte macrophage from tumor
NR8383	Normal rat alveolar macrophage
Epithelial Cells	
A549	Human carcinoma alveolar epithelial cell
HEK	Human epidermal keratinocytes from neonatal foreskin
Caco-2	Human epithelial colon cancer cell line
Fibroblasts	
NIH/3T3	Mouse embryo fibroblast
L929	Mouse fibroblast from normal mouse
Liver	
Hep G2	Human hepatocellular carcinoma
Kidney	
Cos-7	African green monkey kidney cell
Assays	
MTT	Measures metabolic activity of mitochondria by reduction of MTT tetrazolium salt
WST, MTS, XTT	Measures metabolic activity by reduction of tetrazolium salt at plasma membrane
LDH	Measures the release of cytosolic enzymes by reducing INT tetrazolium salt after cell lysis
Live/Dead	Stains live cells with green fluorescence and dead cells with red fluorescence
Neutral Red	Stains only viable cells red
TUNEL	Detects DNA fragmentation indicative of apoptosis

7.4.2.1   HeLa Cells

HeLa cells are among the most popular cell lines used in nanotoxicity studies (Tkachenko *et al.*, 2004; Tsoli *et al.*, 2005; Kam and Dai, 2005; Kam *et al.*, 2006). They are immortal cell lines derived from cervical cancer tissues and are used in toxicity studies because of their rapid population doubling time as well as their size. Cells that divide rapidly are useful in toxicity studies because the response to toxins in these cells is usually more rapid and evident than for other, slowly-dividing cells. For example, when exposed to Au_{55} clusters, 1.4 nm in diameter, HeLa cells exhibited strong indications of toxicity, probably due to interactions with the cell's DNA (Tsoli *et al.*, 2005). The large cytoplasm and nucleus of HeLa cells also offers enhanced visualization of nanoparticle internalization and subcellular trafficking by microscopy. For example, one study investigated the mechanism of uptake of CNTs in HeLa cells by microscopy, and found that such uptake is energy-dependent and most likely through clathrin-coated pits (Kam *et al.*, 2006). Also, after being used for more than 50 years, much is known about HeLa cells and how they function. Finally, because HeLa cells are an aggressive cancer cell line, lessons learned from these studies will help investigators to develop future technologies in cancer treatment.

7.4.2.2   Macrophages

Being the body's first line of defense against foreign bodies, it is no surprise that many studies investigate nanoparticle interactions with macrophages (Shukla *et al.*, 2005; Soto *et al.*, 2005; Hu *et al.*, 2006; Pulskamp *et al.*, 2007). Macrophages have the ability to phagocytose foreign objects and stimulate the foreign body response. Secretion of pro-inflammatory cytokines such as TNF-α and highly reactive chemicals such as ROS are also important measures of macrophage activation, providing input to the overall strength of the foreign body response to nanoparticles. For example, gold nanoparticles were tested with RAW264.7 macrophages, and while uptake of these nanoparticles was confirmed by various forms of microscopy, no signs of toxicity or immune activation were present (Shukla *et al.*, 2005). The same observations were also made for CNTs with NR8383 rat macrophages (Pulskamp *et al.*, 2007).

7.4.2.3   Epithelial Cells

Epithelial cells line the surfaces of organs such as lungs, skin and GI tract. It is likely that any exposure to nanoparticles from our surroundings will first come into contact with epithelial cells. As a consequence these cells are heavily studied for investigation of the fate of nanoparticles (Pulskamp *et al.*, 2007; Patra *et al.*, 2007). Often, immortalized carcinoma cell lines are used because they are easier to culture, have shorter doubling times and can be passaged many times. For example, one study found that gold nanoparticles were cytotoxic to A549 lung carcinoma epithelial cells, and that apoptosis was likely to be the primary mechanism of toxicity (Patra *et al.*, 2007).

7.4.2.4   Fibroblasts

Primarily responsible for synthesizing the ECM of tissue, fibroblasts are one of the most abundant cells in the body and are another class of cells most likely to come into contact with nanoparticles. For this reason, fibroblasts are widely studied in nanotoxicology (Tkachenko *et al.*, 2004; Cheng *et al.*, 2005; Kam and Dai, 2005; Nimmagadda *et al.*, 2006; Pernodet *et al.*, 2006; Zhang *et al.*, 2006; Lobo *et al.*, 2008). For example, incubation with carbon nanotubes indicated a dose-dependent cytotoxicity in 3T3 mouse fibroblasts (Nimmagadda *et al.*, 2006).

7.4.2.5   Liver and Kidney Cells

Acting as the body's natural filter of toxins, both the liver and the kidneys play essential roles in removing particulate debris such as nanoparticles. If these particles are highly toxic, however, then liver and/or kidney failure can result. Because of this, liver and kidney cell lines are frequently used in nanotoxicology (Olbrich *et al.*, 2001; Tkachenko *et al.*, 2004; Cheng *et al.*, 2005; Levi *et al.*, 2006; Patra *et al.*, 2007). One commonly used model here is the COS-7 monkey kidney cell line which has previously been used to evaluate Fe_3O_4 nanoparticles (Cheng *et al.*, 2005).

7.4.3   Common Cytotoxicity Assays

Colorimetric assays have widespread uses when studying the cytotoxic effects of nanoparticles in cell culture. The relatively low cost, ease of experimental design, preparation and implementation make these assays an attractive model. By using reagents easily reduced by very specific cellular biomolecules to produce changes in color, optical instruments such as spectrophotometers and flow cytometers can provide investigators with quantitative data that reflects specific cellular behavior. Colorimetric assays can be used to measure apoptosis, membrane necrosis, metabolic activity, cell cycle and many other cellular processes. Although very robust, it is important that the mechanism of action of these assays be well understood so that data are not misinterpreted (Table 7.3).

7.4.3.1   MTT

The most commonly used cytotoxicity assay in the literature is the MTT assay (Cheng *et al.*, 2005; Shukla *et al.*, 2005; Soto *et al.*, 2005; Tsoli *et al.*, 2005; Patra *et al.*, 2007; Pulskamp *et al.*, 2007; Zhang *et al.*, 2007, 2008; Lobo *et al.*, 2008). Acting as a measure of mitochondria activity, the MTT tetrazolium salt penetrates the cell's plasma membrane where it is reduced to water-insoluble purple formazan to produce a colored solution (Mosmann, 1983). Only being reduced in metabolically active cells, the intensity of the resulting colored solution is directly proportional to the number of viable cells. Therefore, measures of the optical density of the colored solution provide the investigator with an accurate representation of the number of viable cells.

7.4.3.2   WST, MTS and XTT

Other variations of the MTT assay performing a similar function include the WST-1 (Olbrich *et al.*, 2001; Pulskamp *et al.*, 2007), WST-8 (CCK-8), MTS (CellTiter 96) (Kam and Dai, 2005; Fu *et al.*, 2005; Wan *et al.*, 2006) and XTT tetrazolium salts. All four of these salts are reduced to water-soluble formazan products by metabolically active cells, producing measurable color changes proportional to the number of viable cells. Unlike the MTT assay which has a net positive charge mediating its passage through the plasma membrane, these tetrazolium salts have net negative charges making them membrane impermeable (Berridge *et al.*, 2005). Because of this, these dyes require an intermediate electron acceptor like 1-methoxy PMS (mPMS) at the surface of the membrane to help to reduce tetrazolium salt reduction. Therefore, it is important to understand when using these assays that reduction occurs only in the extracellular space.

7.4.3.3   LDH

The LDH assay also makes use of reduced tetrazolium salts to form colored formazan products, but unlike the other described assays which measure the number of viable cells, this assay measures the number of lysed cells (Olbrich *et al.*, 2001). LDH is a cytosolic enzyme that is only released upon damage of the plasma membrane. Once in the extracellular space, LDH oxidizes lactate to pyruvate which then oxidizes nicotinamide adenine dinucleotide (NAD) to NADH (reduced NAD). Finally, NADH reacts with tetrazolium salt iodonitrotetrazolium (INT) to form a formazan product. Because of this unique mechanism, the LDH assay provides a powerful tool to quantitatively measure cell membrane lysis in the presence of toxic materials.

7.4.3.4   Cell Stains

Besides colorimetric assays that depend on the reduction of tetrazolium salts to formazan, other popular assays stain cells for certain functions. Stains used for this purpose have a wide variety of structures, functions and mechanisms. The most popular and basic cell stain used in toxicology is known as the Live/Dead assay (Levi *et al.*, 2006; Yu *et al.*, 2006). In this assay the membrane-permeable calcein acetoxymethyl (calcein AM) is cleaved by intracellular esterases and retained within the cells. Cleavage of calcein AM within the cell results in intense fluorescence that can be brightly imaged with a green color indicating the presence of a live cell. Alternatively, cells with damaged membranes will allow ethidium homodimer-1 (EthD-1) to enter the cell where it will bind with nucleic acids and fluoresce red under the microscope. The contrast between these two stains provides a good representation of which cells are dead or alive for use in microscopy or flow cytommetry.

Similar to the Live/Dead assay, stains such as Trypan Blue (Nimmagadda *et al.*, 2006; Hauck *et al.*, 2008) or Neutral Red can be used to stain for live or dead cells. In normal, healthy cells, Trypan Blue is excluded from passing through the cell's plasma membrane. If the cell's membrane is damaged, then Trypan Blue is able to pass into the cytoplasm, staining the cell blue. Neutral Red, on the other hand, is a weak cationic dye that can pass through the plasma membrane and accumulate in lysosomes where the low pH environment turns the dye red in color. Disruption of normal cell function will not stain the cells red, making Neutral Red a dye that only stains viable cells.

7.4.3.5   Limitations

Although the *in vitro* toxicity assays described above provide easy methods of assessing cell viability, they provide little input in determining the mechanism of cellular toxicity and death. The majority of the colorimetric assays using tetrazolium salts measure cell viability as a function of metabolic function, but questions regarding why these cells stopped mitochondrial activity remains unanswered. Any number of deleterious consequences from nanoparticle exposure may contribute to these effects, such as membrane lysis, cell cycle arrest and apoptosis. The LDH assay is more specific to membrane lysis, but exact mechanisms of nanoparticle action cannot be elucidated from this assay. Similarly, while stains such as Live/Dead, Trypan Blue and Neutral Red differentiate dead from live cells, little information is gained regarding the mechanisms of cell death.

Another limitation of these colorimetric assays is the potential interactions of the nanoparticles with the color-generating dyes such as formazan crystals. Many of the nanoparticles being evaluated for toxicity have unique optical and chemical properties that may interfere with the normal function of these dyes. For example, it has been found that SWCNTs interact with MTT-formazan crystals (Worle-Knirsch *et al.*, 2006). This has been linked with the insoluble properties of these crystals that bind to the SWCNTs by physisorption through Van der Waals forces, thereby stabilizing their structure (Herzog *et al.*, 2007). Similarly, CNTs have also been shown to interact with other dyes such as Neutral Red, Alamar Blue, Commassie and WST, and produce conflicting results (Monteiro-Riviere and Inman, 2006; Casey *et al.*, 2007b). Other variables such as surfactants used to stabilize the particles have also been shown to interfere with the dyes being measured (Belyanskaya *et al.*, 2007). Results from these studies indicate that the validity of these methods needs to be assessed in the presence of nanoparticles before results can be interpreted as absolute.

7.4.4   Cell Function – Specific Methods

Although the methods described above act as a good starting point to begin to understand reactions to nanomaterials and their physico-chemical properties, other more detailed

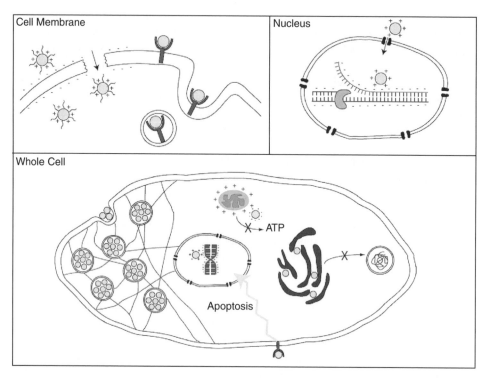

Figure 7.9   *Schematic of potential cell–nanoconstruct interactions. Nanoconstruct entry may be due to attachment of transduction peptides through which a pore can be created, or by an endocytotic mechanism. Interactions with the highly negatively charged DNA in the nucleus may prevent transcription or DNA replication. Also, interactions with the cytoskeleton, mitochondria, endoplasmic reticulum, or cell signalling cascades among other cellular processes, may induce cytotoxicity.*

in vitro studies must be performed to elucidate fully the complex mechanisms of cell toxicity in response to nanomaterials (Figure 7.9). Studies investigating the effects of nanoparticles on the plasma membrane, protein synthesis, cell cycle and division as well as overall cell function must be explored.

7.4.4.1   Plasma Membrane Function and Integrity

The plasma membrane is one of the most important components of cells. It protects the cell from extracellular toxins, controls the transport of molecules in and out of the cell, and is essential for signal transduction. The expression of a variety of proteins on the surface make the plasma membrane most susceptible to nanoparticle contact, making it the primary target of delivery technologies. Formed by an assembly of lipids constituting a bilayer, the plasma membrane maintains sufficient structural integrity to result in very specific mechanical, chemical, osmotic and electrical balances with respect to surrounding conditions. If, however, this function of the plasma membrane is disturbed by materials such as nanoparticles, then cellular functions may be disrupted, causing toxicity.

Because of the high surface reactivity and charge of many nanoparticles, the greatest concern associated with cellular toxicity is plasma membrane lysis. Interactions of nanoparticles with the lipid bilayer and associated proteins may cause instability associated with cell death. This is particularly true for positively charged particles that may enter cells due to their attraction to the negatively charged membrane. Binding of nanoparticles to the surface of the plasma membrane may also affect other essential cellular processes, such as ion transport and signal transduction. Bound nanoparticles may block or competitively bind signalling proteins or ion channels and isolate the cell from the extracellular space. Nutrient and water diffusion may also be affected by nanoparticles bound to the plasma membrane.

Several experimental methods are currently being used to investigate the effects of nanoparticles on the plasma membrane. Membrane lysis can be investigated colorimetrically by the LDH and Neutral Red assays described earlier. Another stain used to identify necrotic cells is propidium iodide (PI) (Pantarotto *et al.*, 2004; Ding *et al.*, 2005; Patra *et al.*, 2007; Pulskamp *et al.*, 2007). Normally membrane impermeant and generally excluded from viable cells, PI will stain necrotic cells red.

Another method that can be used to assess plasma membrane function and form is by atomic force microscopy (AFM) (Shukla *et al.*, 2005). With this method, cell membrane morphology can be visualized by measuring cantilever beam deflections as a function of position with a laser. Three-dimensional topographical maps of the membrane can therefore be visualized and assessed for health and injury. Nanoparticle presence on the surface of the membrane can also be visualized by this method. Other methods of membrane visualization such as bright field and electron microscopy, discussed later in detail, can be used to assess overall viability of the cell's plasma membrane in response to nanoparticle contact and exposure.

7.4.4.2   Cell Entry and Subcellular Localization

Besides maintaining mechanical, chemical, osmotic and electrical equilibrium of the cells relative to the outside environment, the plasma membrane also controls the translocation of specific molecules in and out of the cell's cytoplasm. Most membrane transport of molecules can be accomplished by either active or passive transport using protein-based channels or carriers depending on size, charge and specificity of the material or by simple diffusion. It is unlikely, however, that nanoparticles will be transported across the plasma membrane by these mechanisms due to their size.

Because nanoparticles are generally too big for passive or carrier-mediated transport, cellular uptake of these particles is most likely to occur by endocytosis. Endocytosis can occur by phagocytosis ('cellular eating'), pinocytosis ('cellular drinking'), clathrin-dependent receptor-mediated endocytosis, and clathrin-independent receptor-mediated endocytosis (Mukherjee *et al.*, 1997). Because these mechanisms are activated differently, it is important to understand which mechanisms are responsible for nanoparticle internalization. Particle characteristics such as size, charge, shape and surface coating influence mechanisms of internalization (Osaki *et al.*, 2004; Rejman *et al.*, 2004; Yin *et al.*, 2005; Magrez *et al.*, 2006; Chithrani *et al.*, 2006; Shenoy *et al.*, 2006; Pan *et al.*, 2007). The mechanisms of uptake of silica nanotubes by epithelial breast cancer cells (MDA-MB-231) was recently evaluated (Nan *et al.*, 2008). Confocal images of fluorescently labelled nanotubes incubated with MDA-MB-231 cells were taken in the presence and absence of known endocytosis

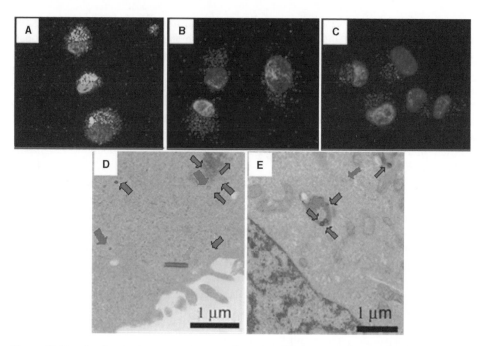

Figure 7.10 *Confocal microscope images of uptake of fluorescent silica nanotubes (SNTs) by MDA-MB-231 breast cancer cells. To elucidate endocytosis, uptake studies were carried out with 200 nm positively charged SNTs at 30° C (A), and in the presence of metabolic inhibitors of endocytosis, namely (B) sodium azide and (C) sucrose. Transmission electron microscope images showing uptake of 500 nm positively charged SNTs in MDA-MB-231 cells (D, E). Blue arrows indicate cross-sectional view and red arrows a horizontal view of nanotubes. Reprinted from Nan et al. (2008), with permission from the American Chemical Society. See colour plate section.*

inhibitors (Figure 7.10 A–C). The results suggest that silica nanotubes were taken up by endocytosis and that positively charged nanotubes had enhanced uptake in comparison with negatively charge, bare nanotubes. TEM images confirmed internalization within the cells (Figure 7.10 D–E).

Once internalized it is also important to track nanoparticles within the cell and investigate what organelles and cellular mechanisms are being affected. Interactions with mitochondria may reduce or alter cell metabolism, and binding to DNA may arrest cell cycle, division and protein synthesis. Similarly, attachment to the cytoskeleton may halt vesicular trafficking or cause mechanical instability and detachment from the ECM. If nanoparticles are able to escape the endosomes and lysosomes after entry, then almost unlimited numbers of cellular processes may be changed depending on subcellular location. Therefore, exact locations of particles within cells need to be observed to determine their toxic effects on cells.

To elucidate which mechanisms control the uptake of nanoparticles, numerous methods can be employed. The most common method to confirm endocytosis is by measuring nanoparticle uptake as a function of temperature (Shukla *et al.*, 2005; Kam *et al.*, 2006). Because endocytosis is an active, energy-consuming process, measurements of nanoparticle concentrations inside cells at both 37° C and 4° C can elucidate the mechanism of uptake. If endocytosis-driven, then uptake of nanoparticles at 4° C should be minimal if not completely absent. Concentration of nanoparticles within cells can be measured using a variety of techniques. Fluorescently labelled nanoparticles can be detected within cells by flow cytometry where the intensity of fluorescence is proportional to the concentration of nanoparticles within the cell. Similarly, changes in fluorescence and/or absorbance using a spectrophotometer enable the detection of changes in nanoparticle concentration within the supernatant. Another method used to measure nanoparticle uptake quantitatively is by inductively coupled plasma emission spectroscopy (ICP-AES) (Hu *et al.*, 2006; Hauck *et al.*, 2008; Zhang *et al.*, 2008). Originally developed to detect trace metals in materials such as rock, ICP-AES can be used to detect the presence of metallic nanoparticles such as gold, superparamagnetic iron oxide and QDs within cells.

Distinctions between clathrin dependent and independent pathways, such as caveolae and lipid-raft pathways, have been made by inhibiting well-defined mechanisms of action. Treatment with NaN_3, for example, disrupts adenosine 5'-triphosphate (ATP) and therefore the endocytotic pathway (Kam *et al.*, 2006). Similarly, treatment with sucrose or K^+ depleted medium disrupts formation of clathrin-coated vesicles. Cholesterol distribution on the plasma membrane can be inhibited by filipin and nystatin to block the clathrin-independent pathways (Kam *et al.*, 2006). These, along with many other methods, can be employed to determine the mechanism of endocytosis.

Visualization of nanoparticles in and around cells can be performed by either light or electron microscopy, each system offering its own advantages and disadvantages. Although limited in resolution due to the diffraction limit of light, light microscopes can image fluorescently labelled nanoparticles within cells. Confocal microscopy, for example, can create three-dimensional images of particles within cells (Fu *et al.*, 2005; Kam and Dai, 2005; Shukla *et al.*, 2005; Levi *et al.*, 2006; Patra *et al.*, 2007). Stains used to label various organelles such as the plasma membrane, nucleus, lysosomes and DNA can show colocalization of nanoparticles with site-specific regions of cells. Other methods such as video-enhanced contrast (VEC) and differential interference contrast (DIC) microscopy are used to provide information about cell shape, position and motility (Tkachenko *et al.*, 2004; Zhang *et al.*, 2008). New systems such as the CytoViva150 Ultrahigh Resolution Imaging (URI) microscope with 90 nm resolution have been used to look at uptake and agglomeration of nanoparticles (Skebo *et al.*, 2007).

Electron microscopy is another tool being used by investigators to visualize intracellular uptake of particles (Tkachenko *et al.*, 2004; Levi *et al.*, 2006; Nimmagadda *et al.*, 2006; Pulskamp *et al.*, 2007; Zhang *et al.*, 2007, 2008; Lobo *et al.*, 2008). With 0.1 nm resolution, techniques such as scanning election microscopy (SEM) and TEM do not require fluorescence to visualize nanoparticles. Although SEM does not offer the resolution that TEM can, SEM images do not require much sample preparation and may be able to visualize nanoparticles on the surface of cell membranes. Electron microscope images are also crucial for the proper characterization of the size and shape of nanoparticles. With these powerful techniques, investigators have the ability to visualize directly the presence and location of nanoparticles within cells.

7.4.4.3   Biomolecule Synthesis

Protein synthesis is one of the core functions of cells, and up and down regulation of genes in response to extra- and intra-cellular signals represents one of the main areas of study when evaluating toxicity. The secretion of ROS, for example, is one of the main responses that macrophages have when contacting foreign bodies such as nanoparticles. Production of ECM such as collagen or elastin in fibroblasts is another example of changes that cells make from such signals. The secretion of pro-inflammatory cytokines is one of the key associated changes in cell behavior representing a cell's response to nanoparticles.

Changes in the protein production cannot be linked with changes in just one system alone. Starting with changes occurring in the DNA and ending with the process of exocytosis, changes in protein synthesis due to nanoparticle interactions can occur in subcellular components such as the nucleus, endoplasmic reticulum, ribosomes, cytoskeleton and cytoplasm. Therefore, measurements of protein presence or absence or differences in concentration are not enough for full understanding of the toxic effects on cells. Studies investigating which biomolecular pathways were activated are necessary steps to ensure that the response being measured is representative of the mechanisms being initiated within the cells.

Because the pro-inflammatory response is a good representation of the negative effects that nanoparticles have on cells, a number of studies have actively measured the secretion of proteins such as IL-6, 8, 10, 1β and TNF-α (Shukla *et al.*, 2005; Pulskamp *et al.*, 2007; Zhang *et al.*, 2007, 2008). The most common methods for measurement of these proteins make use of the enzyme-linked immunosorbent assay (ELISA). Similarly, the Bio-Plex suspension array system using antibody-attached beads for specific analysis of cytokine presence has been used for this purpose (Zhang *et al.*, 2007, 2008). Western blots and immunohistochemistry are other tools commonly used by microbiologists to assess the up and down regulation of genes and proteins.

DNA microarray systems are other powerful methods used in nanotoxicology to assess the regulation of genes (Ding *et al.*, 2005; Zhang *et al.*, 2006; Hauck *et al.*, 2008). The Affymetrix HTA GeneChip System, for example, is a high-throughput technology of measuring gene expression by detecting specific pieces of mRNA on a chip. DNA microarray systems profile full spectrums of DNA expression for analysis. Because macrophages and their response when exposed to nanoparticles is important to understanding their toxicity, the secretion of ROS and reactive nitric oxide species (RNS) is another area of active research (Shukla *et al.*, 2005; Pulskamp *et al.*, 2007). For this purpose, the presence of these chemicals have been measured spectrophotometrically using Griess reagent or 2'7'-dichlorofluoroscein (DCF) dye.

7.4.4.4   Cell Cycle and Division

Control of the cell cycle, division and programmed cell death is a major, complex process necessary for proper function and control of biological systems. Without control over the cell cycle, pathological processes can initiate, with the most common example being cancer. Since it is well known that many toxins similar to nanoparticles such as asbestos are carcinogenic, it is important to study the effects that nanoparticles have on the cell cycle and division for proper risk assessment. Interactions with necessary molecular pathways during M, G_1, S and G_2 phases may influence how a cell divides and grows. Inhibition or activation of programmed cell death, known as apoptosis, is also a major focus of study when assessing the toxic effects that nanoparticles have on cells.

Although important, little work has been done to date investigating the effects that nanoparticles have on cell-cycle control. BrdU, a synthetic nucleoside that binds to newly synthesized DNA during S phase, has been used in cytotoxicity studies to investigate these effects (Ding *et al.*, 2005; Zhang *et al.*, 2006). Used in conjunction with PI stain, graphs indicating the percentage of cells in G_0/G_1, S and G_2/M phases can be made. By this method, changes in cell cycle in response to nanoparticles can be evaluated.

The ability of nanoparticles to initiate apoptosis in cells is the most studied effect on the cell cycle in the literature (Pantarotto *et al.*, 2004; Kam and Dai, 2005; Patra *et al.*, 2007; Pulskamp *et al.*, 2007). Whether directly due to nanoparticle presence or due to a natural cellular response to such foreign bodies, much is still unknown as to why cell contact with nanoparticles initiates the apoptotic mechanism. To study this, apoptotic assays are being used such as Annexin V-APC and poly(ADP-ribose) polymerase (PARP). During apoptosis Annexin V, which is conjugated to the fluorochrome adenomatous polyposis coli (APC), binds to phosphatidylserine (PS) causing fluorescence. Similarly, during apoptosis PARP is cleaved by caspase-2, -3 and -7. These along with several other methods will help investigators to understand the mechanisms of programmed cell death during nanoparticle contact.

Besides experiments to investigate effects on the cell cycle, simple cell proliferation assays can be performed. Used as simple cell counters, the viability assays can be used over two doubling times to investigate the effects that nanoparticles have on cell proliferation. Decreases in cell density compared with controls may be indicative of decreased cell division and growth. One study used the clonogenic assay commonly used in oncological research to evaluate the effects that carbon nanomaterials have on cell viability and proliferation (Herzog *et al.*, 2007). With this approach, a low-density, single-cell suspension culture is plated and observed over time. The number of clones produced in a given time from a single cell provides information on the cell's ability to proliferate. Although time-consuming, this assay does not require the use of dyes that may give confounding results.

7.4.4.5   Cytoskeleton and the Extracellular Matrix

The cell's cytoskeleton and surrounding ECM are important for maintaining mechanical support and intracellular trafficking. With proper distribution of actin, and intermediate filaments as well as microtubules, the cells would be highly susceptible to damage and necrosis would more than likely occur. Focus on the cytoskeleton in nanotoxicology is particularly important because nanoparticles are structural materials that may take up intracellular space and interfere with the cell's cytoskeleton. Large vacuoles formed from internalized nanoparticles have been implicated as interfering with the internal network of cells, potentially causing decreases in actin content within the cell (Pernodet *et al.*, 2006). Necessary to cell division, decreases in actin filaments may also reduce cell proliferation.

Investigations into the effects of nanoparticles on the cytoskeleton have mostly been performed using microscopy techniques mentioned earlier. Many different stains can be used to label the cytoskeleton, including Alexa Fluor 488 and FITC-labelled phalloidin for actin filament visualization among others (Pernodet *et al.*, 2006; Levi *et al.*, 2006). Chromatin staining with 4'-6-diamidino-2-phenylindole (DAPI), for example, has been used to examine the viability of cells, where normal, healthy cells have blue, highly organized intranuclear chromatin, whereas stressed cells fluoresce bright blue and condensed (Patra *et al.*, 2007).

Besides effects on the intracellular network of mechanical filaments, nanoparticles have the potential to negatively impact the ECM and cell adhesion. Physical or chemical modifications to the ECM may have irreversible adverse effects on a cellular and physiological basis. Decreased cell adhesion, migration and motility are all possible outcomes if the ECM becomes reduced or damaged in the presence of nanoparticles. Studies investigating these effects generally make use of microscopy to visualize the ECM. SEM images have been used to examine membrane projections and interactions with their neighboring cells and CNTs (Lobo *et al.*, 2008). Migration assays have also been used to assess the effects that nanoparticles have on cell migration and motility in agarose gels (Pernodet *et al.*, 2006).

7.5   Conclusion and Future Direction

In evaluating their toxicity it is essential to fully characterize the physico-chemical properties of nanoconstructs such as size, shape, charge, surface chemistry, composition, polydispersity, aggregation, optical fluorescence and absorbance. Route of entry, biodistribution, protein adsorption and immune activation are biological factors that influence the toxicity of nanomaterials. Mechanistic *in vitro* evaluation of nanoparticles as a function of physico-chemical properties can elucidate what cellular functions they may influence. This in turn can guide the modification of nanoconstructs to minimize toxicity and maximize the desired effects.

Acknowledgement

Financial support was in part provided by the National Institutes of Health (1R01DE019050) and the National Science Foundation Nanoscale Interdisciplinary Research Team (NSF-NIRT-0608906) Award.

Abbreviations

AFM	Atomic force microscopy
Apo	Apolipoprotein
ARC	Adenomatous polyposis coli
ATP	Adenosine 5'-triphosphate
Calcein AM	Calcein acetoxymethyl
CdS	Cadmium sulfide
CdSe	Cadmium selenide
CFC	Chlorofluorocarbons
CNT	Carbon nanotube
CTAB	Cetyl trimethylammonium bromide
DAPI	4'-6-diamidino-2-phenylindole
DCF	2'7'-dichlorofluorescein
DIC	Differential interference contrast

DLS	Dynamic light scattering
ECM	Extracellular matrix
ELISA	Enzyme-linked immunosorbent assay
EPA	Environmental Protection Agency
EthD-1	Ethidium homodimer-1
FDA	Food and Drug Administration (US)
GI	Gastrointestinal
HEK	Human epidermal keratinocyte
HPMA	N-(2-hydroxypropyl)methacrylamide
HUVEC	Human umbilical vein endothelial cells
ICP-AES	Inductively coupled plasma emission spectroscopy
IgG	Immunoglobulin G
IL-8	Interleukin-8
INT	Iodonitrotetrazolium
LDH	Lactate dehydrogenase
Mce_6	Mesochlorin e_6 mono(N-2-aminoethylamide)
MePEG	Methoxy poly(ethylene glycol)
mPMS	1-Methoxy PMS
MTT	3-(4,5-dimethylthiazol-2-yl)-2,5-diphenyltetrazolium bromide
MW	Molecular weight
MWCNT	Multi-walled carbon nanotube
NAD	Nicotinamide adenine dinucleotide
NADH	Reduced nicotinamide adenine dinucleotide
NIH	National Institutes of Health
NIOSH	National Institute for Occupational Safety and Health
NK	Natural killer
NSF	National Science Foundation
PAMAM	Poly(amido amine)
PARP	Poly(ADP-ribose) polymerase
PEG	Poly(ethylene glycol)
PGA	Poly(glycolic acid)
PI	Propidium iodide
PLA	Poly(lactic acid)
PLGA	Poly(lactic acid-co-glycolic acid)
PNIPAAm	Poly(N-isopropylacrylamide)
PPO	Poly(propylene oxide)
PS	Phosphatidylserine
PRINT	Particle Replication in Non-Wetting Templates
QD	Quantum dot
RES	Reticuloendothelial system
RNS	Reactive nitric oxide species
ROS	Reactive oxygen species
SC	Stratum corneum
SDBS	Sodium dodecyl benzene
SDS	Sodium dodecyl sulfate
SEC	Size exclusion chromatography

SEM	Scanning electron microscopy
SPR	Surface plasmon resonance
SWCNT	Single-walled carbon nanotube
TEM	Transmission electron microscopy
TNF-α	Tumor necrosis factor-α
URI	Ultrahigh Resolution Imaging
VEC	Video-enhanced contrast
WST-1	4-[3-(4-iodophenyl)-2-(4-nitrophenyl)-2H-5-tetrazolio]-1,3-benzene disulfonate
ZnS	Zinc sulfide

References

Abbas AK, Lichtman AH (2005) *Cellular and Molecular Immunology*. Elsevier Saunders: Philadelphia.

Alexis A, Pridgen E, Molnar LK, Farokhzad OC (2008) Factors affecting the clearance and biodistribution of polymeric nanoparticles. *Mol Pharm* **5**, 505–515.

Akerman ME, Chan WC, Laakkonen P, Bhatia SN, Ruoslahti E (2002) Nanocrystal targeting in vivo. *Proc Natl Acad Sci USA* **99**, 12617–12621.

Baroli B, Ennas MG, Loffredo F, Isola M, Pinna R, López-Quintela MA (2007) Penetration of metallic nanoparticles in human full-thickness skin. *J Invest Dermatol* **127**, 1701–1712.

Barry BW (2001) Novel mechanisms and devices to enable successful transdermal drug delivery. *Eur J Pharm Sci* **14**, 101–114.

Bekyarova E, Ni Y, Malarkey EB, Montana V, McWilliams JL, Haddon RC, Parpura V (2005) Applications of carbon nanotubes in biotechnology and biomedicine. *J Biomed Nanotechnol* **1**, 3–17.

Belyanskaya L, Manser P, Spohn P, Bruinink A, Wick P (2007) The reliability and limits of the MTT reduction assay for carbon nanotubes–cell interaction. *Carbon* **45**, 2643–2648.

Berne RM, M. Levy MN, Koeppen BM, Stanton BA (2004) *Physiology*. Mosby: St. Louis.

Berridge MV, Herst PM, Tan AS (2005) Tetrazolium dyes as tools in cell biology: new insights into their cellular reduction. *Biotechnol Annu Rev* **11**, 127–152.

Bianco A, Kostarelos K, Partidos CD, Prato M (2005) Biomedical applications of functionalised carbon nanotubes. *Chem Commun* **2005**, 571–577.

Borm PJ, Kreyling W (2004) Toxicological hazards of inhaled nanoparticles – potential implications for drug delivery. *J Nanosci Nanotechnol* **4**, 521–531.

Brown DM, Wilson MR, MacNee W, Stone V, Donaldson K (2001) Size-dependent proinflammatory effects of ultrafine polystyrene particles: A role for surface area and oxidative stress in the enhanced activity of ultrafines. *Toxicol Appl Pharmacol* **175**, 191–199.

Bruchez M Jr, Moronne M, Gin P, Weiss S, Alivisatos AP (1998) Semiconductor nanocrystals as fluorescent biological labels. *Science* **281**, 2013–2016.

Cartlidge SA, Duncan R, Lloyd JB, Kopeckova-Rejmanova P, Kopecek J (1987) Soluble, crosslinked N-(2-hydroxypropyl) methacrylamide copolymers as potential drug carriers. II: Effect of molecular weight on blood clearance and body distribution in the rat after intravenous administration. Distribution of unfractionated copolymer after intraperitoneal, subcutaneous or oral administration. *J Control Release* **4**, 253–264.

Casey A, Davoren M, Herzog E, Lyng FM, Byrne HJ, Chambers G (2007a) Probing the interaction of single walled carbon nanotubes within cell culture medium as a precursor to toxicity testing. *Carbon* **45**, 34–40.

Casey A, Herzog E, Davoren M, Lyng FM, Byrne HJ, Chambers G (2007b) Spectroscopic analysis confirms the interactions between single walled carbon nanotubes and various dyes commonly used to assess cytotoxicity. *Carbon* **45**, 1425–1432.

Champion JA, Katare YK, Mitragotri S (2007) Particle shape: a new design parameter for micro- and nanoscale drug delivery carriers. *J Control Release* **121**, 3–9.

Champion JA, Mitragotri S (2008) Shape induced inhibition of phagocytosis of polymer particles. *Pharm Res* **26**, 244–249.

Chellat F, Grandjean-Laquerriere A, Naour RL, Fernandes J, Yahia LH, Guenounou M, Laurent-Maquin D (2005) Metalloproteinase and cytokine production by THP-1 macrophages following exposure to chitosan-DNA nanoparticles. *Biomaterials* **26**, 961–970.

Chen BX, Wilson SR, Das M, Coughlin DJ, Erlanger BF (1998) Antigenicity of fullerenes: antibodies specific for fullerenes and their characteristics. *Proc Natl Acad Sci USA* **95**, 10809–10813.

Cheng FY, Su CH, Yang YS, Yeh CS, Tsai CY, Wu CL, Wu MT, Shieh DB (2005) Characterization of aqueous dispersions of Fe_3O_4 nanoparticles and their biomedical applications. *Biomaterials* **26**, 729–738.

Chilkoti A, Dreher MR, Meyer DE, Raucher D (2002) Targeted drug delivery by thermally responsive polymers. *Adv Drug Deliv Rev* **54**, 613–630.

Chithrani BD, Ghazani AA, Chan WC (2006) Determining the size and shape dependence of gold nanoparticle uptake into mammalian cells. *Nano Lett* **6**, 662–668.

Chonn A (1991) The role of surface charge in the activation of the classical and alternative pathways of complement by liposomes. *J Immunol* **146**, 4234–4241.

Collins PG, Avouris P (2000) Nanotubes for electronics. *Sci Am* **283**, 62–69.

Cruz T, Gaspar R, Donato A, Lopes C (1997) Interaction between polyalkylcyanoacrylate nanoparticles and peritoneal macrophages: MTT metabolism, NET reduction, and NO production. *Pharm Res* **14**, 73–79.

Davis JJ, Green MLH, Allen H, Hill O, Leung YC, Sadler PJ, Sloan J, Xavier AV, Chi Tsang S (1998) The immobilisation of proteins in carbon nanotubes. *Inorganica Chimica Acta* **272**, 261–266.

Davoren M, Herzog E, Casey A, Cottineau B, Chambers G, Byrne HJ, Lyng FM (2007) In vitro toxicity evaluation of single walled carbon nanotubes on human A549 lung cells. *Toxicol in Vitro* **21**, 438–448.

Deen WM, Lazzara MJ, Myers BD (2001) Structural determinants of glomerular permeability. *Am J Physiol Renal Physiol* **281**, F579–F596.

De Jong WH, Hagens WI, Krystek P, Burger MC, Sips A, Geertsma RE (2008) Particle size-dependent organ distribution of gold nanoparticles after intravenous administration. *Biomaterials* **29**, 1912–1919.

Derfus AM, Chan WCW, Bhatia SN (2004) Probing the cytotoxicity of semiconductor quantum dots. *Nano Lett* **4**, 11–18.

Dhar S, Gu FX, Langer R, Farokhzad OC, Lippard SJ (2008) Targeted delivery of cisplatin to prostate cancer cells by aptamer functionalized Pt (IV) prodrug-PLGA–PEG nanoparticles. *Proc Natl Acad Sci USA* **105**, 17356.

Ding L, Stilwell J, Zhang T, Elboudwarej O, Jiang H, Selegue JP, Cooke PA, Gray JW, Chen FF (2005) Molecular characterization of the cytotoxic mechanism of multiwall carbon nanotubes and nano-onions on human skin fibroblast. *Nano Lett* **5**, 2448–2464.

Dobrovolskaia MA, McNeil SE (2007) Immunological properties of engineered nanomaterials. *Nat Nano* **2**, 469–478.

Dong L, Joseph KL, Witkowski CM, Craig MM (2008) Cytotoxicity of single-walled carbon nanotubes suspended in various surfactants. *Nanotechnology* **19**, 255702.

Dumortier H, Lacotte S, Pastorin G, Marega R, Wu W, Bonifazi D, Briand JP, Prato M, Muller S, Bianco A (2006) Functionalized carbon nanotubes are non-cytotoxic and preserve the functionality of primary immune cells. *Nano Lett* **6**, 1522–1528.

Duncan R (2006) Polymer conjugates as anticancer nanomedicines. *Nat Rev Cancer* **6**, 688–701.

Elgrabli DAN (2007) Effect of BSA on carbon nanotube dispersion for in vivo and in vitro studies. *Nanotoxicology* **1**, 266–278.

El-Sayed IH, Huang X, El-Sayed MA (2005) Surface plasmon resonance scattering and absorption of anti-EGFR antibody conjugated gold nanoparticles in cancer diagnostics: applications in oral cancer. *Nano Lett* **5**, 829–834.

El-Sayed M, Ginski M, Rhodes C, Ghandehari H (2002) Transepithelial transport of poly (amido amine) dendrimers across Caco-2 cell monolayers. *J Control Release* **81**, 355–365.

El-Sayed M, Ginski M, Rhodes CA, Ghandehari H (2003) Influence of surface chemistry of poly(amido amine) dendrimers on Caco-2 cell monolayers. *J Bioact Compat Polym* **18**, 7–22.

Emerich DF, Thanos CG (2003) Nanotechnology and medicine. *Expert Opin Biol Ther* **3**, 655–663.

Esquivel EV, Murr LE (2004) A TEM analysis of nanoparticulates in a polar ice core. *Materials Characterization* **52**, 15–25.

Fang C, Shi B, Pei YY, Hong MH, Wu J, Chen HZ (2006) In vivo tumor targeting of tumor necrosis factor-a-loaded stealth nanoparticles: Effect of MePEG molecular weight and particle size. *Eur J Pharm Sci* **27**, 27–36.

Ferin J, Oberdorster G, Penney DP (1992) Pulmonary retention of ultrafine and fine particles in rats. *Am J Respir Cell Mol Biol* **6**, 535–542.

Florence AT (ed) (2005) Dendrimers: a versatile targeting platform. *Adv Drug Deliv Rev* **57**, 2101–2286.

Food and Drug Administration (US) (2007) *Nanotechnology: A Report of the US Food and Drug Administration Nanotechnology Task Force*. FDA: Silver Spring, MD.

Fu W, Shenoy D, Li J, Crasto C, Jones G, Dimarzio C, Sridhar S, Amiji M (2005) Biomedical applications of gold nanoparticles functionalized using hetero-bifunctional poly (ethylene glycol) spacer. *Mater Res Soc Symp Proc* **845**, AA5.4.1–AA5.4.6.

Gessner A, Waicz R, Lieske A, Paulke B, Mader K, Muller RH (2000) Nanoparticles with decreasing surface hydrophobicities: influence on plasma protein adsorption. *Int J Pharm* **196**, 245–249.

Gessner A, Lieske A, Paulke B, Muller R (2002) Influence of surface charge density on protein adsorption on polymeric nanoparticles: analysis by two-dimensional electrophoresis. *Eur J Pharm Biopharm* **54**, 165–170.

Gessner A, Lieske A, Paulke BR, Muller RH (2003) Functional groups on polystyrene model nanoparticles: Influence on protein adsorption. *J Biomed Mater Res A* **65**, 319–326.

Ghosh P, Han G, De M, Kim CK, Rotello VM (2008) Gold nanoparticles in delivery applications. *Adv Drug Deliv Rev* **60**, 1307–1315.

Goodman CM, McCusker CD, Yilmaz T, Rotello VM (2004) Toxicity of gold nanoparticles functionalized with cationic and anionic side chains. *Bioconjug Chem* **15**, 897–900.

Grabinski C, Hussain S, Lafdi K, Braydich-Stolle L, Schlager J (2007) Effect of particle dimension on biocompatibility of carbon nanomaterials. *Carbon* **45**, 2828–2835.

Graham ML (2003) Pegaspargase: a review of clinical studies. *Adv Drug Deliv Rev* **55**, 1293–1302.

Gratton SE, Ropp PA, Pohlhaus PD, Luft JC, Madden VJ, Napier ME, DeSimone JM (2008) The effect of particle design on cellular internalization pathways. *Proc Natl Acad Sci USA* **105**, 11613–11618.

Gref R, Minamitake Y, Peracchia MT, Trubetskoy V, Torchilin V, Langer R (1994) Biodegradable long-circulating polymeric nanospheres. *Science* **263**, 1600–1603.

Gref R, Luck M, Quellec P, Marchand M, Dellacherie E, Harnisch S, Blunk T, Muller RH (2000) 'Stealth' corona-core nanoparticles surface modified by polyethylene glycol (PEG): influences of the corona (PEG chain length and surface density) and of the core composition on phagocytic uptake and plasma protein adsorption. *Colloids Surf B Biointerfaces* **18**, 301–313.

Hamilton RF, Thakur SA, Holian A (2008) Silica binding and toxicity in alveolar macrophages. *Free Radic Biol Med* **44**, 1246–1258.

Hansen SF, Maynard A, Baun A, Tickner JA (2008) Late lessons from early warnings for nanotech-nology. *Nat Nanotechnol* **3**, 444–447.

Harashima H, Sakata K, Funato K, Kiwada H (1994) Enhanced hepatic uptake of liposomes through complement activation depending on the size of liposomes. *Pharm Res* **11**, 402–406.

Harremoës P, Gee D, MacGarvin M, Stirling A, Keys J, Wynne B, Vaz SG (2001) *Late Lessons from Early Warnings: the Precautionary Principle 1896–2000*. European Environment Agency: Copenhagen.

Hart PR, Kopeckova P, Omelyanenko V, Enioutina E, Kopecek J (2000) HPMA copolymer-modified avidin: immune response. *J Biomater Sci Polym Ed* **11**, 1–12.

Hashizume H, Baluk P, Morikawa S, McLean JW, Thurston G, Roberge S, Jain RK, McDonald DM (2000) Openings between defective endothelial cells explain tumor vessel leakiness. *Am J Pathol* **156**, 1363–1380.

Hauck TS, Ghazani AA, Chan WC (2008) Assessing the effect of surface chemistry on gold nanorod uptake, toxicity, and gene expression in mammalian cells. *Small* **4**, 153–159.

Herzog E, Casey A, Lyng FM, Chambers G, Byrne HJ, Davoren M (2007) A new approach to the toxicity testing of carbon-based nanomaterials – the clonogenic assay. *Toxicol Lett* **174**, 49–60.

Heyder J, Gebhart J, Rudolf G, Schiller CF, Stahlofen W (1986) Deposition of particles in the human respiratory tract in the size range 0.005–15 μm. *J Aerosol Sci* **17**, 811–825.

Hillebrenner H, Buyukserin F, Stewart JD, Martin CR (2006) Template synthesized nanotubes for biomedical delivery applications. *Nanomed* **1**, 39–50.

Hillebrenner H, Buyukserin F, Stewart JD, Martin CR (2007) Biofunctionalization and capping of template synthesized nanotubes. *J Nanosci Nanotechnol* **7**, 2211–2221.

Hobbie EK, Lacerda SH, Migler KB, Jakupciak JP (2007) Length-dependent uptake of DNA-wrapped single-walled carbon nanotubes. *Adv Mater* **19**, 939–945.

Hu F, K. Neoh KG, Cen L, Kang ET (2006) Cellular response to magnetic nanoparticles "PEGylated" via surface-initiated atom transfer radical polymerization. *Biomacromolecules* **7**, 809–816.

Hu Y, Xie J, Tong YW, Wang CH (2007) Effect of PEG conformation and particle size on the cellular uptake efficiency of nanoparticles with the HepG2 cells. *J Control Release* **118**, 7–17.

Huang M, Khor E, Lim LY (2004) Uptake and cytotoxicity of chitosan molecules and nanoparticles: Effects of molecular weight and degree of deacetylation. *Pharm Res* **21**, 344–353.

Huang X, El-Sayed IH, Qian W, El-Sayed MA (2006) Cancer cell imaging and photothermal therapy in the near-infrared region by using gold nanorods. *J Am Chem Soc* **128**, 2115–2120.

Huang X, Jain PK, El-Sayed IH, El-Sayed MA (2007) Gold nanoparticles: interesting optical properties and recent applications in cancer diagnostics and therapy. *Nanomed* **2**, 681–693.

Hussain N, Jani PU, Florence AT (1997) Enhanced oral uptake of tomato lectin-conjugated nanoparticles in the rat. *Pharm Res* **14**, 613–618.

Jain PK, I. El-Sayed IH, El-Sayed MA (2007) Au nanoparticles target cancer. *Nano Today* **2**, 18–29.

Jia G, Wang H, Yan L, Wang X, Pei R, Yan T, Zhao Y, Guo X (2005) Cytotoxicity of carbon nanomaterials: single-wall nanotube, multi-wall nanotube, and fullerene. *Environ Sci Technol* **39**, 1378–1383.

Jiang W, Kim BYS, Rutka JT, Chan WC (2008) Nanoparticle-mediated cellular response is size-dependent. *Nat Nanotechnol* **3**, 145–150.

Kam NWS, Jessop TC, Wender PA, Dai H (2004) Nanotube molecular transporters: internalization of carbon nanotube-protein conjugates into mammalian cells. *J Am Chem Soc* **126**, 6850–6851.

Kam NWS, Dai H (2005) Carbon nanotubes as intracellular protein transporters: generality and biological functionality. *J Am Chem Soc* **127**, 6021–6026.

Kam NWS, Liu Z, Dai H (2006) Carbon nanotubes as intracellular transporters for proteins and DNA: an investigation of the uptake mechanism and pathway. *Angew Chem Int Ed* **45**, 577–581.

Kirchner C, Liedl T, Kudera S, Pellegrino T, Javier AM, Gaub HE, Stolzle S, Fertig N, Parak WJ (2005) Cytotoxicity of colloidal CdSe and CdSe/ZnS nanoparticles. *Nano Lett* **5**, 331–338.

Kitchens KM, El-Sayed MEH, Ghandehari H (2005) Transepithelial and endothelial transport of poly (amido amine) dendrimers. *Adv Drug Deliv Rev* **57**, 2163–2176.

Kitchens KM, Kolhatkar RB, Swaan PW, Eddington ND, Ghandehari H (2006) Transport of poly(amido amine) dendrimers across Caco-2 cell monolayers: Influence of size, charge and fluorescent labeling. *Pharm Res* **23**, 2818–2826.

Kitchens KM, Foraker AB, Kolhatkar RB, Swaan PW, Ghandehari H (2007) Endocytosis and interaction of poly (amidoamine) dendrimers with Caco-2 cells. *Pharm Res* **24**, 2138–2145.

Kolhatkar RB, Kitchens KM, Swaan PW, Ghandehari H (2007) Surface acetylation of poly(amido amine) (PAMAM) dendrimers decreases cytotoxicity while maintaining membrane permeability. *Bioconjug Chem* **18**, 2054–2060.

Kolhatkar R, Sweet D, Ghandehari H (2008) Functionalized dendrimers as nanoscale drug carriers. In *Multifunctional Pharmaceutical Nanocarriers*, Torchilin V (ed.). Fundamental Biomedical Technologies 4. Springer: New York; 201–232.

Kolhe P, Khandare J, Pillai O, Kannan S, Lieh-Lai M, Kannan RM (2006) Preparation, cellular transport, and activity of polyamidoamine-based dendritic nanodevices with a high drug payload. *Biomaterials* **27**, 660–669.

Kopecek J, Kopecková P, Minko T, Lu ZR (2000) HPMA copolymer–anticancer drug conjugates: design, activity, and mechanism of action. *Eur J Pharm Biopharm* **50**, 61–81.

Lammers T, Kühnlein R, Kissel M, Subr V, Etrych T, Pola R, Pechar M, Ulbrich K, Storm G, Huber P (2005) Effect of physicochemical modification on the biodistribution and tumor accumulation of HPMA copolymers. *J Control Release* **110**, 103–118.

Larson DR, Zipfel WR, Williams RM, Clark SW, Bruchez MP, Wise FW, Webb WW (2003) Water-soluble quantum dots for multiphoton fluorescence imaging in vivo. *Science* **300**, 1434–1437.

Lee SB, Mitchell DT, Trofin L, Nevanen TK, Soderlund H, Martin CR (2002) Antibody-based bio-nanotube membranes for enantiomeric drug separations. *Science* **296**, 2198–2200.

Levchenko TS, Rammohan R, Lukyanov AN, Whiteman KR, Torchilin VP (2002) Liposome clearance in mice: the effect of a separate and combined presence of surface charge and polymer coating. *Int J Pharm* **240**, 95–102.

Levi N, Hantgan RR, Lively MO, Carroll DL, Prasad GL (2006) C 60-Fullerenes: detection of intracellular photoluminescence and lack of cytotoxic effects. *J Nanobiotechnology* **4**, 14–24.

Lewinski N, Colvin V, Drezek R (2008) Cytotoxicity of nanoparticles. *Small* **4**, 26–49.

Li J, Ng HT, Cassell A, Fan W, Chen H, Ye Q, Koehne J, Han J, Meyyappan M (2003) Carbon nanotube nanoelectrode array for ultrasensitive DNA detection. *Nano Lett* **3**, 597–602.

Lin A, Hirsch L, Lee MH, Barton J, Halas N, West J, Drezek R (2004) Nanoshell-enabled photonics-based imaging and therapy of cancer. *Technol Cancer Res Treat* **3**, 33–40.

Lin Y, Lu F, Wang J (2004) Disposable carbon nanotube modified screen-printed biosensor for amperometric detection of organophosphorus pesticides and nerve agents. *Electroanalysis* **16**, 145–149.

Lobo AO, Antunes EF, Machado AHA, Pacheco-Soares C, Trava-Airoldi VJ, Corat EJ (2008) Cell viability and adhesion on as grown multi-wall carbon nanotube films. *Materials Science & Engineering C* **28**, 264–269.

Longmire M, Choyke PL, Kobayashi H (2008) Clearance properties of nano-sized particles and molecules as imaging agents: considerations and caveats. *Nanomed* **3**, 703–717.

Loretz B, Bernkop-Schnurch A (2007) In vitro cytotoxicity testing of non-thiolated and thiolated chitosan nanoparticles for oral gene delivery. *Nanotoxicology* **1**, 139–148.

Lück M, Paulke BR, Schröder W, Blunk T, Müller RH (1998) Analysis of plasma protein adsorption on polymeric nanoparticles with different surface characteristics. *J Biomed Mater Res* **39**, 478–485.

Lundqvist M, Sethson I, Jonsson BH (2004) Protein adsorption onto silica nanoparticles: Conformational changes depend on the particles' curvature and the protein stability. *Langmuir* **20**, 10639–10647.

Magrez A, Kasas S, Salicio V, Pasquier N, Seo JW, Celio M, Catsicas S, Schwaller B, Forro L (2006) Cellular toxicity of carbon-based nanomaterials. *Nano Lett* **6**, 1121–1125.

Malik N, Wiwattanapatapee R, Klopsch R, Lorenz K, Frey H, Weener JW, Meijer EW, Paulus W, Duncan R (2000) Dendrimers: relationship between structure and biocompatibility in vitro, and preliminary studies on the biodistribution of ^{125}I-labelled polyamidoamine dendrimers in vivo. *J Control Release* **65**, 133–148.

Matsumura Y, Maeda H (1986) A new concept for macromolecular therapeutics in cancer chemotherapy: mechanism of tumoritropic accumulation of proteins and the antitumor agent smancs. *Cancer Res* **46**, 6387–6392.

Maynard AD, Aitken RJ, Butz T, Colvin V, Donaldson K, Oberdorster G, Philbert MA, Ryan J, Seaton A, Stone V, Tinkle SS, Tran L, Walker NJ, Warheit DB (2006) Safe handling of nanotechnology. *Nature* **444**, 267–269.

Meerum TJM, ten Bokkel HWW, Schellens JH, Schot M, Mandjes IA, Zurlo MG, Rocchetti M, Rosing H, Koopman FJ, Beijnen JH (2001) Phase I clinical and pharmacokinetic study of PNU166945, a novel water-soluble polymer-conjugated prodrug of paclitaxel. *Anticancer Drugs* **12**, 315–323.

Minko T, Kopeckova P, Kopecek J (2000) Efficacy of the chemotherapeutic action of HPMA copolymer-bound doxorubicin in a solid tumor model of ovarian carcinoma. *Int J Cancer* **86**, 108–117.

Mitra A, Nan A, Ghandehari H, McNeil E, Mulholland J, Line BR (2004) Technetium-99m-labeled N-(2-hydroxypropyl) methacrylamide copolymers: synthesis, characterization, and in vivo biodistribution. *Pharm Res* **21**, 1153–1159.

Monteiro-Riviere NA, Inman AO (2006) Challenges for assessing carbon nanomaterial toxicity to the skin. *Carbon* **44**, 1070–1078.

Mosqueira VCF, Legrand P, Gulik A, Bourdon O, Gref R, Labarre D, Barratt G (2001) Relationship between complement activation, cellular uptake and surface physicochemical aspects of novel PEG-modified nanocapsules. *Biomaterials* **22**, 2967–2979.

Mosmann T (1983) Rapid colorimetric assay for cellular growth and survival: application to proliferation and cytotoxicity assays. *J Immunol Methods* **65**, 55–63.

Mossman BT, Bignon J, Corn M, Seaton A, Gee JB (1990) Asbestos: scientific developments and implications for public policy. *Science* **247**, 294–301.

Mukherjee S, Ghosh RN, Maxfield FR (1997) Endocytosis. *Physiol Rev* **77**, 759–803.

Nan A, Bai X, Son SJ, Lee SB, Ghandehari H (2008) Cellular uptake and cytotoxicity of silica nanotubes. *Nano Lett* **8**, 2150–2154.

Nayak NC, Shin K (2008) Human serum albumin mediated self-assembly of gold nanoparticles into hollow spheres. *Nanotechnology* **19**, 265603.

Niidome T, Yamagata M, Okamoto Y, Akiyama Y, Takahashi H, Kawano T, Katayama Y, Niidome Y (2006) PEG-modified gold nanorods with a stealth character for in vivo applications. *J Control Release* **114**, 343–347.

Nimmagadda A, Thurston K, Nollert MU, McFetridge PS (2006) Chemical modification of SWNT alters in vitro cell–SWNT interactions. *J Biomed Mater Res A* **76**, 614–625.

Noguchi Y, Wu J, Duncan R, Strohalm J, Ulbrich K, Akaike T, Maeda H (1998) Early phase tumor accumulation of macromolecules: A great difference in clearance rate between tumor and normal tissues. *Cancer Science* **89**, 307–314.

Norris DA, Sinko PJ (1997) Effect of size, surface charge, and hydrophobicity on the translocation of polystyrene microspheres through gastrointestinal mucin. *J Appl Polym Sci* **63**, 1481–1492.

Norris DA, Puri N, Sinko PJ (1998) The effect of physical barriers and properties on the oral absorption of particulates. *Adv Drug Deliv Rev* **34**, 135–154.

Oberdorster G, Oberdorster E, Oberdorster J (2005) Nanotoxicology: an emerging discipline evolving from studies of ultrafine particles. *Environ Health Perspect* **113**, 823–839.

Ohlson M, Sorensson J, Haraldsson B (2001) A gel-membrane model of glomerular charge and size selectivity in series. *Am J Physiol Renal Physiol* **280**, 396–405.

Olbrich C, Bakowsky U, Lehr CM, Muller RH, Kneuer C (2001) Cationic solid-lipid nanoparticles can efficiently bind and transfect plasmid DNA. *J Control Release* **77**, 345–355.

Osaki F, Kanamori T, Sando S, Sera T, Aoyama Y (2004) A quantum dot conjugated sugar ball and its cellular uptake. On the size effects of endocytosis in the subviral region. *J Am Chem Soc* **126**, 6520–6521.

Otsuka H, Nagasaki Y, Kataoka K (2003) PEGylated nanoparticles for biological and pharmaceutical applications. *Adv Drug Deliv Rev* **55**, 403–419.

Paciotti GF, Myer L, Weinreich D, Goia D, Pavel N, McLaughlin RE, Tamarkin L (2004) Colloidal gold: a novel nanoparticle vector for tumor directed drug delivery. *Drug Deliv* **11**, 169–183.

Pan Y, Neuss S, Leifert A, Fischler M, Wen F, Simon U, Schmid G, Brandau W, Jahnen-Dechent W (2007) Size-dependent cytotoxicity of gold nanoparticles. *Small* **3**, 1941–1949.

Panessa-Warren BJ, Warren JB, Wong SS, Misewich JA (2006) Biological cellular response to carbon nanoparticle toxicity. *Journal of Physics: Condensed Matter* **18**, S2185–S2201.

Pantarotto D, Briand JP, Prato M, Bianco A (2004) Translocation of bioactive peptides across cell membranes by carbon nanotubes. *Chem Commun* **2004**, 16–17.

Park EJ, Park K (2008) Oxidative stress and pro-inflammatory responses induced by silica nanoparticles in vivo and in vitro. *Toxicol Lett* **184**, 18–25.

Patra HK, Banerjee S, Chaudhuri U, Lahiri P, Dasgupta AK (2007) Cell selective response to gold nanoparticles. *Nanomedicine* **3**, 111–119.

Pernodet N, Fang X, Sun Y, Bakhtina A, Ramakrishnan A, Sokolov J, Ulman A, Rafailovich M (2006) Adverse effects of citrate/gold nanoparticles on human dermal fibroblasts. *Small* **2**, 766–773.

Perumal OP, Inapagolla R, Kannan S, Kannan RM (2008) The effect of surface functionality on cellular trafficking of dendrimers. *Biomaterials* **29**, 3469–3476.

Portney NG, Ozkan M (2006) Nano-oncology: drug delivery, imaging, and sensing. *Anal Bioanal Chem* **384**, 620–630.

Pulskamp K, Diabaté S, Krug HF (2007) Carbon nanotubes show no sign of acute toxicity but induce intracellular reactive oxygen species in dependence on contaminants. *Toxicol Lett* **168**, 58–74.

Rajender RK, Modi MW, Pedder S (2002) Use of peginterferon alfa-2a (40 KD)(Pegasys) for the treatment of hepatitis C. *Adv Drug Deliv Rev* **54**, 571–586.

Ratner BD, Hoffman AS, Schoen FJ, Lemons JE (2004) *Biomaterials Science: An Introduction to Materials in Medicine*. Elsevier Academic Press: San Diego, CA.

Rejman J, Oberle V, Zuhorn IS, Hoekstra D (2004) Size-dependent internalization of particles via the pathways of clathrin- and caveolae-mediated endocytosis. *Biochem J* **377**, 159–169.

Roco MC (2003) Nanotechnology: convergence with modern biology and medicine. *Curr Opin Biotechnol* **14**, 337–346.

Ryman-Rasmussen JP, Riviere JE, Monteiro-Riviere NA (2006) Penetration of intact skin by quantum dots with diverse physicochemical properties. *Toxicol Sci* **91**, 159–165.

Ryman-Rasmussen JP, Riviere JE, Monteiro-Riviere NA (2007) Surface coatings determine cytotoxicity and irritation potential of quantum dot nanoparticles in epidermal keratinocytes. *J Invest Dermatol* **127**, 143–153.

Sahoo SK, Labhasetwar V (2003) Nanotech approaches to drug delivery and imaging. *Drug Discov Today* **8**, 1112–1120.

Salvador-Morales C, Flahaut E, Sim E, Sloan J, Green MLH, Sim RB (2006) Complement activation and protein adsorption by carbon nanotubes. *Mol Immunol* **43**, 193–201.

Sayes CM, Liang F, Hudson JL, Mendez J, Guo W, Beach JM, Moore VC, Doyle CD, West JL, Billups WE (2006) Functionalization density dependence of single-walled carbon nanotubes cytotoxicity in vitro. *Toxicol Lett* **161**, 135–142.

Schulze C, Kroll A, Lehr CM, Schafer UF, Becker K, Schnekenburger J, Isfort CS, Landsiedel R, Wohlleben W (2008) Not ready to use–overcoming pitfalls when dispersing nanoparticles in physiological media. *Nanotoxicology* **2**, 51–61.

Scientific Committee on Emerging and Newly Identified Health Risks (SCENIHR) (2007) *The scientific aspects of the existing and proposed definitions relating to products of nanoscience and nanotechnologies.* European Commission: Brussels.

Seymour LW, Duncan R, Strohalm J, Kopecek J (1987) Effect of molecular weight (Mw) of N-(2-hydroxypropyl) methacrylamide copolymers on body distribution and rate of excretion after subcutaneous, intraperitoneal, and intravenous administration to rats. *J Biomed Mater Res* **21**, 1341–1358.

Seymour LW, Miyamoto Y, Maeda H, Brereton M, Strohalm J, Ulbrich K, Duncan R (1995) Influence of molecular weight on passive tumour accumulation of a soluble macromolecular drug carrier. *Eur J Cancer* **31**, 766–770.

Sharma RK, Das S, Maitra A (2005) Enzymes in the cavity of hollow silica nanoparticles. *J Colloid Interface Sci* **284**, 358–361.

Shenoy D, Fu W, Li J, Crasto C, Jones G, DiMarzio C, Sridhar S, Amiji M (2006) Surface functionalization of gold nanoparticles using hetero-bifunctional poly(ethylene glycol) spacer for intracellular tracking and delivery. *Int J Nanomedicine* **1**, 51–57.

Shukla R, Bansal V, Chaudhary M, Basu A, Bhonde RR, Sastry M (2005) Biocompatibility of gold nanoparticles and their endocytotic fate inside the cellular compartment: a microscopic overview. *Langmuir* **21**, 10644–10654.

Skebo JE (2007) Assessment of metal nanoparticle agglomeration, uptake, and interaction using high-illuminating system. *Int J Toxicol* **26**, 135–141.

Smith AM, Duan H, Mohs AM, Nie S (2008) Bioconjugated quantum dots for in vivo molecular and cellular imaging. *Adv Drug Deliv Rev* **60**, 1226–1240.

Son SJ, Reichel J, He B, Schuchman M, Lee SB (2005) Magnetic nanotubes for magnetic-field-assisted bioseparation, biointeraction, and drug delivery. *J Am Chem Soc* **127**, 7316–7317.

Soto KF, Carrasco A, Powell TG, Garza KM, Murr LE (2005) Comparative in vitro cytotoxicity assessment of some manufactured nanoparticulate materials characterized by transmission electron microscopy. *J Nanopart Res* **7**, 145–169.

Strebhardt K, Ullrich A (2008) Paul Ehrlich's magic bullet concept: 100 years of progress. *Nat Rev Cancer* **8**, 473–480.

Sweet DM, Kolhatkar RB, Ray A, Swaan PW, Ghandehari H (2009) Transepithelial transport of PEGylated anionic poly(amido amine) dendrimers: implications for oral drug delivery. *J Control Release*, 2009 Apr 22 (Epub.).

Takahashi H, Niidome Y, Niidome T, Kaneko K, Kawasaki H, Yamada S (2006) Modification of gold nanorods using phosphatidylcholine to reduce cytotoxicity. *Langmuir* **22**, 2–5.

Tan W, Wang K, He X, Zhao XJ, Drake T, Wang L, Bagwe RP (2004) Bionanotechnology based on silica nanoparticles. *Med Res Rev* **24**, 621–638.

Taylor MS, Daniels AU, Andriano KP, Heller J (1994) Six bioabsorbable polymers: In vitro acute toxicity of accumulated degradation products. *J Appl Biomater* **5**, 151–157.

Tian F, Cui D, Schwarz H, Estrada GG, Kobayashi H (2006) Cytotoxicity of single-wall carbon nanotubes on human fibroblasts. *Toxicol in Vitro* **20**, 1202–1212.

Tkachenko AG, Xie H, Liu Y, Coleman D, Ryan J, Glomm WR, Shipton MK, Franzen S, Feldheim DL (2004) Cellular trajectories of peptide-modified gold particle complexes: comparison of nuclear localization signals and peptide transduction domains. *Bioconjug Chem* **15**, 482–490.

Torchilin VP (2008) Multifunctional pharmaceutical nanocarriers: development of the concept. In *Multifunctional Pharmaceutical Nanocarriers*, Torchilin VP (ed.). Fundamental Biomedical Technologies 4. Springer: New York; 1–32.

Tsoli M, Kuhn H, Brandau W, Esche H, Schmid G (2005) Cellular uptake and toxicity of Au55 clusters. *Small* **1**, 841–844.

van Nostrum CF, Veldhuis TFJ, Bos GW, Hennink WE (2004) Hydrolytic degradation of oligo (lactic acid): a kinetic and mechanistic study. *Polymer* **45**, 6779–6787.

Vasey PA, Kaye SB, Morrison R, Twelves C, Wilson P, Duncan R, Thomson AH, Murray LS, Hilditch TE, Murray T (1999) Phase I clinical and pharmacokinetic study of PK1 [N-(2-hydroxypropyl) methacrylamide copolymer doxorubicin]: first member of a new class of chemotherapeutic agents-drug-polymer conjugates. Cancer Research Campaign Phase I/II Committee. *Clin Cancer Res* **5**, 83–94.

Vexler VS, Clement O, Schmitt-Willich H, Brasch RC (1994) Effect of varying the molecular weight of the MR contrast agent Gd-DTPA-polylysine on blood pharmacokinetics and enhancement patterns. *J Magn Reson Imaging* **4**, 381–388.

Vonarbourg A, Passirani C, Saulnier P, Benoit JP (2006a) Parameters influencing the stealthiness of colloidal drug delivery systems. *Biomaterials* **27**, 4356–4373.

Vonarbourg A, Passirani C, Saulnier P, Simard P, Leroux JC, Benoit JP (2006b) Evaluation of pegylated lipid nanocapsules versus complement system activation and macrophage uptake. *J Biomed Mater Res A* **78**, 620–628.

Wachters FM, Groen HJ, Maring JG, Gietema JA, Porro M, Dumez H, de Vries EG, van Oosterom AT (2004) A phase I study with MAG-camptothecin intravenously administered weekly for 3 weeks in a 4-week cycle in adult patients with solid tumours. *Br J Cancer* **90**, 2261–2267.

Wan S, Huang J, Guo M, Zhang H, Cao Y, Yan H, Liu K (2006) Biocompatible superparamagnetic iron oxide nanoparticle dispersions stabilized with poly (ethylene glycol)-oligo (aspartic acid) hybrids. *J Biomed Mater Res A* **80**, 946–954.

Wang H, Wang J, Deng X, Sun H, Shi Z, Gu Z, Liu Y, Zhaoc Y (2004) Biodistribution of carbon single-wall carbon nanotubes in mice. *J Nanosci Nanotechnol* **4**, 1019–1024.

Wang SG, Zhang Q, Wang R, Yoon SF (2003) A novel multi-walled carbon nanotube-based biosensor for glucose detection. *Biochem Biophys Res Commun* **311**, 572–576.

Wang YS, Youngster S, Grace M, Bausch J, Bordens R, Wyss DF (2002) Structural and biological characterization of pegylated recombinant interferon alpha-2b and its therapeutic implications. *Adv Drug Deliv Rev* **54**, 547–570.

Wick P, Manser P, Limbach LK, Dettlaff-Weglikowska U, Krumeich F, Roth S, Stark WJ, Bruinink A (2007) The degree and kind of agglomeration affect carbon nanotube cytotoxicity. *Toxicol Lett* **168**, 121–131.

Wijaya A, Schaffer SB, Pallares IG, Hamad-Schifferli K (2009) Selective release of multiple DNA oligonucleotides from gold nanorods. *ACS Nano* **3**, 80–86.

Worle-Knirsch JM, Pulskamp K, Krug HF (2006) Oops they did it again! Carbon nanotubes hoax scientists in viability assays. *Nano Lett* **6**, 1261–1268.

Yamamoto Y, Nagasaki Y, Kato Y, Sugiyama Y, Kataoka K (2001) Long-circulating poly (ethylene glycol)–poly (d, l-lactide) block copolymer micelles with modulated surface charge. *J Control Release* **77**, 27–38.

Yin H, Too HP, Chow GM (2005) The effects of particle size and surface coating on the cytotoxicity of nickel ferrite. *Biomaterials* **26**, 5818–5826.

Yin Win K, Feng SS (2005) Effects of particle size and surface coating on cellular uptake of polymeric nanoparticles for oral delivery of anticancer drugs. *Biomaterials* **26**, 2713–2722.

Yu MF, Lourie O, Dyer MJ, Moloni K, Kelly TF, Ruoff RS (2000) Strength and breaking mechanism of multiwalled carbon nanotubes under tensile load. *Science* **287**, 637–640.

Yu WW, Chang E, Sayes CM, Drezek R, Colvin VL (2006) Aqueous dispersion of monodisperse magnetic iron oxide nanocrystals through phase transfer. *Nanotechnology* **17**, 4483–4487.

Zhang LW (2007) Biological interactions of functionalized single-wall carbon nanotubes in human epidermal keratinocytes. *Int J Toxicol* **26**, 103–113.

Zhang LW, Yu WW, Colvin VL, Monteiro-Riviere NA (2008) Biological interactions of quantum dot nanoparticles in skin and in human epidermal keratinocytes. *Toxicol Appl Pharmacol* **228**, 200–211.

Zhang T, Stilwell JL, Gerion D, Ding L, Elboudwarej O, Cooke PA, Gray JW, Alivisatos AP, Chen FF (2006) Cellular effect of high doses of silica-coated quantum dot profiled with high throughput gene expression analysis and high content cellomics measurements. *Nano Lett* **6**, 800–808.

Zimmerman RJ, Aukerman SL, Katre NV, Winkelhake JL, Young JD (1989) Schedule dependency of the antitumor activity and toxicity of polyethylene glycol-modified interleukin 2 in murine tumor models. *Cancer Res* **49**, 6521–6528.

Zuin S, Pojana G, Marcomini A (2007) Effect-oriented physicochemical characterization of nanomaterials. In *Nanotoxicology: Characterization, Dosing and Health Effects*, Monteiro-Riviere NA, Tran LC (eds). Informa Healthcare: New York; 19–58.

8

Developing Bioassay Methods for Evaluating Pulmonary Hazards from Nanoscale or Fine Quartz/Titanium Dioxide Particulate Materials

David B. Warheit, Kenneth L. Reed and Christie M. Sayes

8.1   Introduction

The development of new products using nanoparticles as components in the formulations is an emerging multidisciplinary technology that involves the synthesis of molecules in the nanoscale (i.e. 10^{-9} m) size range. From a material science and chemistry perspective, nanotechnology applications are exciting because of the potential versatility and new properties that are derived from the enhanced applications. For instance, as primary particle sizes decrease below the 100 nm range (i.e., as one moves down the nanoscale for a given material), completely new physical properties are likely to emerge. As an example, titanium dioxide particles become devoid of their 'white' pigment-like color to become colorless at size ranges < 50 nm, while retaining other properties such as ultraviolet ros attenuation as well as facilitating dispersion. This modification can have significant utility when developing cosmetic materials. Other particle types which have been utilized for electrical insulation applications can suddenly become conductive; or insoluble substances can become more soluble below 100 nm. As a consequence, these alterations in physical properties enhance the versatility of products and thus are likely to give rise to new industrial and medical applications as well as more eclectic products – and these possibilities have generated great interest in this new technology (Colvin, 2003).

Nanotoxicity: From In Vivo *and* In Vitro *Models to Health Risks*   Edited by Saura Sahu and Daniel Casciano
© 2009 John Wiley & Sons, Ltd

Apart from the natural environment, the existence of particulates in the nanoscale size range is not a new phenomenon. Some particulate types have been commercially produced for decades. In this regard, nanosized carbon black particles, utilized in the manufacture of rubber products (tires) and pigments, have been produced for more than a century. Likewise, nanosized fumed amorphous silica particulates and other metal oxide particles such as alumina, titania and zirconia have been produced as nanomaterials for over 50 years and used for applications such as thixotropic agents in pigments and cosmetics applications, and more recently in polishing powders in the microelectronics industry. Many of these nanoscale particle-types are produced using gas phase flame reactions carried out under very controlled conditions (Borm and Kreyling, 2004).

The determination of health risks from any substance or material is known to comprise two main factors: hazard assessments and exposure evaluations. In attempting to characterize the hazards of inhaling nanoscale materials, a pulmonary bioassay has been developed to assess the toxicity of fine and nanoscale particles delivered to the lungs via intratracheal instillation or following inhalation exposures. The results of two lung bioassay studies are presented in this brief review to demonstrate that strategies for assessing pulmonary hazards from nanomaterials can be developed, and that surface reactivity indices may play a more prominent role in producing toxicity compared with particle size characteristics. In the first study presented herein, four different titanium dioxide particle types were evaluated – three in the nanoscale size range and one in the fine size range. In the second set of studies, the pulmonary toxicity of four different quartz particle samples were assessed – two particle types in the nanoscale range and two in the fine size range.

Prior to the implementation of experimentation for hazard testing, it is always important to characterize accurately the physico-chemical properties of the nanoparticle types that are being evaluated. Accordingly, it has been strongly suggested to assess most if not all of the following characteristics listed below (Warheit, 2008):

- particle size and size distribution (wet state);
- surface area (dry state) in the relevant media being utilized – depending upon the route of exposure;
- crystal structure/crystallinity;
- aggregation status in the relevant media;
- composition/surface coatings;
- surface reactivity;
- method of nanomaterial synthesis and/or preparation including post-synthetic modifications (e.g. neutralization of ultrafine TiO_2 particle types);
- purity of sample.

8.2   A Typical Experimental Design for Pulmonary Bioassay Studies

Below we describe typical experimental designs for assessing the toxicity of nanoscale materials which have been utilized in several recently published pulmonary bioassays (Warheit *et al.*, 2007a,b) (see Table 8.1).

The fundamental components of this lung bioassay methodology are dose-response assessments and time-course assessments to determine the sustainability of any observed

Table 8.1   *General components of the pulmonary bioassay.*

Bronchoalveolar lavage assessments
    Lung inflammation and cytotoxicity
      • Cell differential analysis
      • BAL fluid lactate dehydrogenase (cytotoxicity)
      • BAL fluid alkaline phosphatase (epithelial cell cytotoxicity)
      • BAL fluid protein (lung permeability)

Lung tissue analysis
      • Lung weights
      • Lung cell proliferation (BrdU)
        • Parenchymal
        • Airway
      • Lung histopathology

effect following particulate exposures. Accordingly, the major endpoints of these studies are the following: (1) time course and dose/response potency of pulmonary inflammation and cytotoxicity; (2) airway and lung parenchymal cell proliferation; and (3) histopathological assessments of lung tissue.

For bronchoalveolar lavage fluid (BALF) studies, groups of rats (5 rats/group/dose/time point) were intratracheally instilled with single doses of either 1 or 5 mg/kg rutile-type ultrafine-TiO_2 particle-type 1 (uf-1); rutile ultrafine-TiO_2 particle-type 2 (uf-2); rutile F-1 fine-TiO_2 particles; anatase/rutile ultrafine uf-3 TiO_2 particle-types; or quartz (crystalline silica) particles (see Table 8.2). The intratracheal instillation method of exposure can be a reliable qualitative screen for assessing the pulmonary toxicity of inhaled particles (Warheit *et al.*, 2005). All particles were prepared in a volume of phosphate-buffered saline (PBS) solution and subjected to ultrasonic probe sonication for 15 minutes at 60 Hz. Groups of PBS-instilled rats served as controls. The lungs of PBS and particle-exposed rats were evaluated by BALF analyses at 24 hr, 1 week, 1 month and 3 months postexposure (pe).

For the lung tissue studies, additional groups of animals (4 rats/group/high dose/time period) were instilled with the particle types listed above plus the vehicle control (i.e. PBS).

Table 8.2   *Protocol for lung bioassay study with ultrafine and fine TiO_2 particles.*

Exposure groups
^ PBS (vehicle control)
Particle types (1 and 5 mg/kg)

• rutile-type uf-1 TiO_2
• rutile-type uf-2 TiO_2
• anatase/rutile-type uf-3 TiO_2
• rutile-type F-1 fine-TiO_2 (negative control)
• α-quartz particles (positive control)

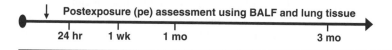

Postexposure (pe) assessment using BALF and lung tissue

  24 hr   1 wk   1 mo   3 mo

These studies and corresponding groups of rats were dedicated to lung tissue analyses, but only the high dose groups (5 mg/kg) and PBS controls were utilized in the morphology studies. These studies consisted of cell proliferation assessments and histopathological evaluations of the lower respiratory tract. Similar to the BAL fluid studies, the intratracheal instillation exposure period was followed by 24-hour, 1-week, 1-month and 3-month recovery periods (see Table 8.2).

8.2.1   Pulmonary Lavage

The lungs of sham and particulate-exposed rats were lavaged with a phosphate-buffered saline (PBS) solution as described previously. Methodologies for cell counts, differentials and pulmonary biomarkers in lavaged fluids were conducted as previously described (Warheit *et al.*, 1991, 1997). Briefly, the first 12 ml of lavaged fluids recovered from the lungs of PBS or particulate-exposed rats was centrifuged at 700 g, and 2 ml of the supernatant was removed for biochemical studies. All biochemical assays have been conducted on BALFs using a Roche Diagnostics (BMC)/Hitachi® 717 clinical chemistry analyzer using Roche Diagnostics (BMC)/Hitachi® reagents. Lactate dehydrogenase is a cytoplasmic enzyme and is used as a marker of cell injury. Alkaline phosphatase (ALP) activity can be a measure of Type II alveolar epithelial cell secretory activity, and increased ALP activity in BALFs may be an indicator of Type II lung epithelial cell toxicity. Increases in BALF micro-protein (MTP) concentrations indicate enhanced migration of vascular proteins into the alveolar regions, representing a breakdown in the integrity of the alveolar-capillary barrier.

8.2.2   Pulmonary Cell Proliferation Studies

Groups of particulate-exposed rats and corresponding controls were pulsed 24 hrs after instillation, as well as postexposure periods of 1 week, 1 and 3 months, with intraperitoneal injections of 100 mg/kg body weight of 5-bromo-2'deoxyuridine (BrdU) dissolved in a 0.5 N PBS solution at a dose of 100 mg/kg body weight. The animals were euthanized 6 hrs later by overdose pentobarbital injections. Following cessation of spontaneous respiration, the lungs were infused with neutral buffered formalin fixatives at airway pressures of 21 cm H_2O. Following 15 minutes of fixation, the tracheas were clamped, and the heart and lungs were carefully removed *en bloc* and immersion-fixed in formalin. In addition, a 1-cm piece of duodenum (which serves as a positive labelling tissue control) was removed and stored in formaldehyde. Subsequently, parasagittal sections from the right cranial and caudal lobes and regions of the left lung lobes, as well as the duodenal sections, were dehydrated in 70 % ethanol and sectioned for histology. The sections were embedded in paraffin, cut, and mounted on glass slides. The slides were stained with an anti-BrdU antibody, along with an AEC (3-amino-9-ethyl carbazole) marker, and counter-stained with aqueous hematoxylin. A minimum of 1000 cells/animal were counted in terminal airways as well as in alveolar regions. For each treatment group, immunostained nuclei in airways (i.e. terminal bronchiolar epithelial cells) or lung parenchyma (i.e. epithelia, interstitial cells or macrophages) were counted by light microscopy at ×1000 magnification (Warheit *et al.*, 1991, 1997).

8.2.3 Lung Histopathology Studies

The lungs of rats exposed to particulates or PBS controls were prepared for microscopy by airway infusion of formalin fixative under pressure (21 cm H_2O) at 24 hours, 1 week, 1 and 3 months postexposure. Sagittal sections of the left and right lungs were cut with a razor blade. Tissue blocks were dissected from left, right upper and right lower regions of the lung, and were subsequently prepared for light microscopy (paraffin-embedded, sectioned, and hematoxylin-eosin stained) (Warheit *et al.*, 1991, 1997).

8.2.4 Statistical Analyses

For analyses, each of the experimental values was compared with their corresponding sham control values for each time point. A one-way analysis of variance (ANOVA) and Bartlett's test were calculated for each sampling time. When the F test from ANOVA was significant, Dunnett's test was used to compare means from the control group to each of the groups exposed to particles. Significance vs. PBS controls was judged at the 0.05 probability level.

8.3 Results of Bioassays

8.3.1 Pulmonary Bioassay Studies of Fine and Nanoscale TiO$_2$ Particle Types

The objective of this bioassay study was to evaluate lung toxicities of newly developed, ultrafine or nanoscale TiO_2 particle types in rats, and compare them with other TiO_2 samples in two different size ranges or crystal structures. Following physico-chemical characterization of particle types (Table 8.3), groups of rats were exposed via intratracheal instillation with doses of 1 or 5 mg/kg of either two ultrafine/nano rutile TiO_2 particle types (termed for the purposes of the study uf-1 or uf-2); along with rutile, fine-sized TiO_2 (F-1); or an ultrafine TiO_2 particle type with a different crystal structure (i.e. 80/20 anatase/rutile ultrafine-TiO_2 particles (uf-3)); Min-U-Sil quartz particles served as positive benchmark control samples. In addition, PBS-instilled rats served as vehicle controls. Following exposures to the lungs, PBS and particle-exposed rats were assessed for lung inflammatory and cytotoxicity biomarkers, cell proliferation activity, and assessed for lung tissue effects at post-instillation exposure time points of 24 hrs, 1 week, 1 month and 3 months (see Table 8.2).

Table 8.3 *Physico-chemical characteristics of the TiO$_2$ particles studied (modelled after Warheit* et al., *2007a), with permission from Elsevier.*

Particle type	Crystal phase	Average particle size in nm		Surface area m²/g	pH in suspension		Surface reactivity
		H$_2$O	PBS		H$_2$O	PBS	
F-1	Rutile	382.0	2667.2	5.8	7.49	6.75	0.4
uf-1	Rutile	136.0	2144.3	18.2	5.64	6.78	10.1
uf-2	Rutile	149.4	2890.7	35.7	7.14	6.78	1.2
uf-3	80/20 anatase/rutile	129.4	2691.7	53.0	3.28	6.70	23.8

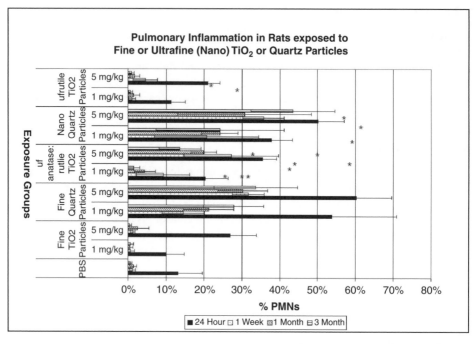

Figure 8.1 *Composite summary from two studies of pulmonary inflammation in particulate-exposed rats and controls as evidenced by % neutrophils (% PMNs) in BALFs at 24 hrs, 1 week, 1 month and 3 months postexposure (pe). Values given are means ± SD. Intratracheal instillation exposures of several particle types produced a short-term pulmonary inflammatory response, as evidenced by an increase in the percentages/numbers of BAL-recovered neutrophils, measured at 24 hrs pe. However, the exposures to fine and nanoquartz particles (1 and 5 mg/kg) produced sustained pulmonary inflammatory responses, as measured at 3 months pe (*P < 0.05 vs. PBS controls). Similarly high-dose exposures to anatase/rutile uf-3 TiO₂ particles produced significant inflammation by 1 week pe (Warheit et al., 2007a,b). Reprinted by permission of Oxford University Press.*

Results demonstrated that the toxicity range (in descending order of potency) of lung inflammation/cytotoxicity/cell proliferation and histopathological responses was quartz > uf-3 > F-1 = uf-1 = uf-2 (see Figure 8.1). The findings revealed that pulmonary exposure to quartz particle types (Figure 8.2) and, to a lesser degree, uf-3 anatase/rutile TiO₂ particles, resulted in sustained lung inflammation, cytotoxicity and corresponding adverse lung tissue effects. In contrast, exposure to F-1 fine-TiO₂ particles or to uf-1/uf-2 rutile ultrafine-TiO₂ particle types produced transient inflammatory responses (Figures 8.1 and 8.3) and no corresponding adverse lung tissue effects. The findings indicate that factors such as crystal structure, inherent pH of the particles, or surface chemical reactivity could account for the differential effects measured in rats exposed to anatase/rutile uf-3 TiO₂ particles but not the rutile uf-1 or uf-2 TiO₂ particle types. The data demonstrate that exposure to ultrafine-TiO₂ particle types can produce pulmonary hazard potentials of different potencies, based upon their composition, crystal structure and other surface characteristics (Warheit *et al.*, 2007a).

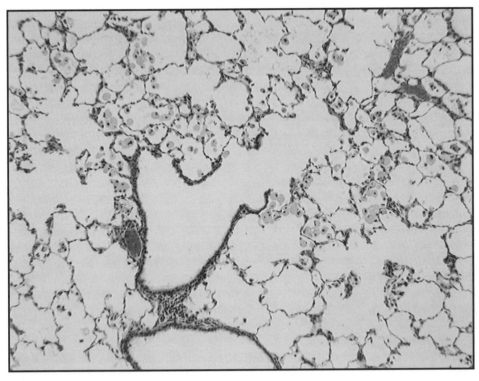

Figure 8.2 *Lung tissue response in a rat exposed to 5 mg/kg Min-U-Sil quartz particles, 3 months post-instillation exposure. Two common features of quartz-induced pulmonary responses are apparent: (1) the development of foamy macrophages filling alveolar airspaces; and (2) the early tissue thickening response leading to the progressive development of pulmonary fibrosis.*

8.3.2   Pulmonary Bioassay Studies of Fine and Nanoscale α-Quartz Particle Types

The aim of this study was to compare the pulmonary hazards of (1) synthetic 50 nm nanoquartz particle types (study 1); (2) synthetic 12 nm nanoquartz particulates (study 2); (3) synthetic fine quartz particles ($\sim$300 nm), and compare the lung effects with (mined) Min-U-Sil quartz particle types ($\sim$500 nm). A second aim was to evaluate the surface reactivities among the samples as a potential indicator of toxicity. Accordingly, these nanoscale and fine-sized particle samples were also tested for surface reactivity as evidenced by assessments of hemolytic potential. For the study design, groups of rats were intratracheally instilled with doses of either 1 or 5 mg/kg carbonyl iron particles (as a negative control particle type) or with the different α-quartz particle types in PBS solution (vehicle). The pulmonary effects were assessed in a dose-response/time-course protocol employing BALF biomarkers for inflammation and cytotoxicity, lung cell proliferation indices, and histopathological evaluation of lung tissue at 24 hrs, 1 week, 1 month and 3 months postexposure (see Table 8.4).

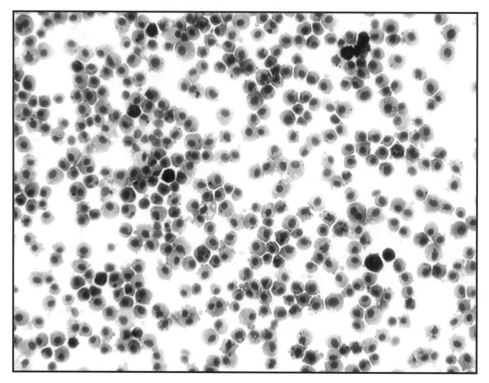

Figure 8.3 *Lung cells recovered by bronchoalveolar lavage 1 month after exposure to rutile-type uf TiO₂ particles. Notice that the inflammatory response has been resolved and the cytocentrifuge preparation contains mainly alveolar macrophages.*

Table 8.4 *Experimental design for nanoscale/fine quartz particle lung bioassay study.*

Exposure groups

- PBS (vehicle control)
- Particle-types (1 and 5 mg/kg)
  - carbonyl iron particles (negative control)
  - Min-U-Sil quartz particles (534 nm)
  - nano quartz II particles (12 nm)
  - fine quartz particles (300 nm)

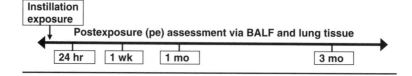

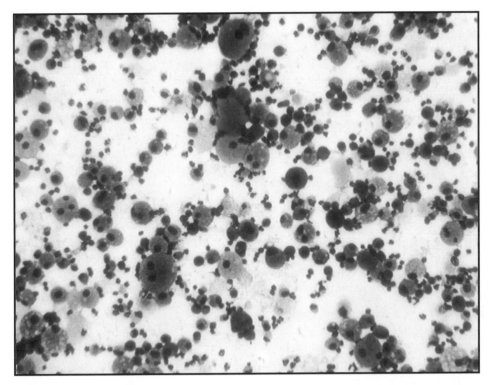

Figure 8.4 *Cytocentrifuge preparation demonstrating pulmonary inflammation recovered from bronchoalveolar lavage (BAL) fluids in rats following exposure to nano-sized quartz particles, at 3 months postexposure. The micrograph demonstrates that these pulmonary inflammatory responses remained sustained and progressive.*

Pulmonary exposures to the different α-quartz particle types resulted in different degrees of pulmonary inflammation and cytotoxicity (Figures 8.1 and 8.4), and these findings were not always consistent with particle size but correlated better with surface reactivity endpoints, in the form of erythrocyte hemolytic potential (see Table 8.5). Histopathological evaluations of lung tissues from three of the quartz samples demonstrated common quartz-related pulmonary effects – dose-dependent lung inflammatory 'foamy' macrophage

Table 8.5   *Summary of selected nanoscale and fine quartz particle endpoints.*

Endpoint	50 nm nano quartz	12 nm nano quartz	300 nm fine quartz	534 nm MinU-Sil
Particle size	++	+	+++	++++
Surface area	+++	++++	++	+
Hemolytic potential	+	+++	++	+++
Pulmonary inflammation	++	+++	++	+++
Cell injury indices	++	+++	+	+++

accumulation responses, along with early development of lung tissue thickening and corresponding pulmonary fibrosis (Figure 8.2). The qualitative aspects of α-quartz-related pulmonary effects were similar in form but with different potencies. The range of corresponding particle-related lung toxicities and tissue effects in order of greatest potency were nanoscale quartz 12 nm = Min-U-Sil quartz > fine quartz > nanoscale quartz 50 nm > carbonyl iron particles. The findings showed that the pulmonary toxicities of α-quartz particles were not associated with particle size *per se*, but correlated better with surface reactivity relative to particle size and/or surface area (Warheit *et al.*, 2007b).

8.4   Summary

We have described a methodology for conducting pulmonary bioassays with fine and nanoscale particle types. The four fundamental features of the bioassay are the following: (1) dose response characteristics; (2) time course experimental design; (3) appropriate physico-chemical characterization of the particle types being tested; and (4) utilization of appropriate benchmark control particle samples. Effective use of this bioassay can provide important pulmonary hazard information on the particulate materials of interest.

References

Borm PJ, Kreyling W (2004) Toxicological hazards of inhaled nanoparticles – potential implications for drug delivery. *J Nanosci Nanotechnology* **4**, 521–531.

Colvin VL (2003) The potential environmental impact of engineered nanomaterials. *Nat Biotechnol* **21**, 1166–1170.

Warheit DB (2008) How meaningful are the results of nanotoxicity studies in the absence of adequate material characterization? *Toxicol. Sci* **101**, 183–185.

Warheit DB, Carakostas MC, Hartsky MA, Hansen JF (1991) Development of a short-term inhalation bioassay to assess pulmonary toxicity of inhaled particles: Comparisons of pulmonary responses to carbonyl iron and silica. *Toxicol Appl Pharmacol* **107**, 350–368.

Warheit DB, Hansen JF, Yuen IS, Kelly DP, Snajdr S, Hartsky MA (1997) Inhalation of high concentrations of low toxicity dusts in rats results in pulmonary and macrophage clearance impairments. *Toxicol Appl Pharmacol* **145**, 10–22.

Warheit DB, Brock W, Lee KP, Webb TR, Reed KL (2005) Comparative pulmonary toxicity inhalation and instillation studies with different TiO$_2$ particle formulations: Impact of surface treatments on particle toxicity. *Toxicol Sci* **88**, 514–524.

Warheit DB, Webb TR, Reed KL, Frerichs S, Sayes CM (2007a) Pulmonary toxicity study in rats with three forms of ultrafine-TiO$_2$ particles: Differential responses related to surface properties. *Toxicology* **230**, 90–104.

Warheit DB, Webb TR, Colvin VL, Reed KL, Sayes CM (2007b) Pulmonary bioassay studies with nanoscale and fine quartz particles in rats: Toxicity is not dependent upon particle size but on surface characteristics. *Toxicol Sci* **95**, 270–280.

9

Nanoparticles: Is Neurotoxicity a Concern?

Jianyong Wang, Wenjun Sun and Syed F. Ali

9.1 Background

Nanotechnology, which is about controlling matter at near-atomic scales to produce unique or enhanced materials, products and devices, has become a major scientific endeavor in the last decade. As defined by the United States Nanotechnology Initiative (available at www.nano.gov), it is 'the understanding and control of matter at dimensions of roughly 1–100 nanometers, where unique phenomena enable novel applications'. Engineered nanomaterials have been used to create materials that exhibit novel physico-chemical properties and functions. The unusual physico-chemical properties are attributable to their small size (surface area and size distribution), chemical composition (purity, crystallinity, electronic properties), surface structure (surface reactivity, surface groups, inorganic or organic coatings), solubility, shape and aggregation (Nel *et al.*, 2006). These nanomaterials, including fullerenes, nanotubes, nanoparticles, liposomes, polymers and quantum dots, are increasingly being used for commercial purposes such as fillers, opacifiers, catalysts, semiconductors, cosmetics, microelectronics and drug carriers. More than 300 claimed nanotechnology products are already on the market (Maynard *et al.*, 2006), and it is estimated that the sale of products utilizing nanotechnology could become a $1 trillion market by 2015 (Xia *et al.*, 2009).

Rapid developments in nanotechnology have led, like other technological advances, to predictions of great benefits and also great dangers to humanity and the ecosystem (Seaton and Donaldson, 2005). With the fast growth of their new applications, nanomaterials are produced globally in massive quantities, and as a consequence, human exposure to these materials is inevitably and rapidly increasing. Although impressive from a

Nanotoxicity: From In Vivo *and* In Vitro *Models to Health Risks* Edited by Saura Sahu and Daniel Casciano
© 2009 John Wiley & Sons, Ltd

physico-chemical viewpoint, the novel properties of nanomaterials raise concerns about adverse effects on biological systems (Donaldson *et al.*, 2008; Lewinski *et al.*, 2008; Stern and McNeil, 2008). Nanoparticles and nanotubes are two major classes of nanomaterials that elicit foreseeable hazards. The medical problems associated with amphibole asbestos are well known, which is due to its accumulation in the lung, and thus hazard is explained by its thinness, long needle-like shape, and insolubility. This understanding allows us to predict pulmonary toxicity associated with exposure to nanotubes, which are also thin, long and insoluble (Seaton and Donaldson, 2005).

Many health concerns have been raised due to exposure to various nanoparticles, including potential adverse effects on respiratory and cardiovascular systems, blood, the gastrointestinal tract, skin, liver, and central nervous system (CNS). For example, traffic-derived nanoparticles are likely to be responsible for adverse cardiovascular effects, by three possible pathways (Seaton and Donaldson, 2005). Firstly, the inhaled nanoparticles may cause inflammation, changes in blood coagulability, and an increased risk of heart attack in people made vulnerable by coronary artery disease (Seaton *et al.*, 1995). Secondly, nanoparticles may translocate from the lungs to the blood, where they might interact with endothelium to promote thrombosis or destabilize atheromatous plaques (Nemmar *et al.*, 2004). Thirdly, some of the cardiovascular effects of particulate pollution suggest that a malfunctioning autonomic reflex may play a role (Liao *et al.*, 1999). Another example is that nanoparticles contained in some sunscreens and cosmetics may be able to penetrate the skin and potentiate ultraviolet damage. This chapter will focus on the discussion of potential neurotoxicity induced by nanoparticles.

There are a number of classes of nanoparticles, which were well summarized in a review by Medina *et al.* (2007). Liposomes are nanoparticles comprising lipid bilayer membranes surrounding an aqueous interior; emulsions comprise oil in water-type mixtures that are stabilized with surfactants to maintain size and shape. They have been engineered and used in drug delivery for improving the efficacy and safety of certain compounds (Sarker, 2005). Polymers and ceramic nanoparticles are also used as drug delivery systems. Polymers can form conjugates with proteins or drugs for certain desired properties, such as a plasma half-life increase, immunogenicity reduction, or permeability enhancement (Lee, 2006). Metal nanoparticles have been widely used in electronics, and the increasing exposure levels have elicited a serious concern over neurotoxicity (Hussain *et al.*, 2006). Fullerenes are novel carbon allotropes, which are characterized by having numerous points of attachment whose surfaces can also be functionalized for tissue binding (Bosi *et al.*, 2003). Quantum dots are semiconductor nanocrystals with unique optical and electrical properties currently applied in biomedical imaging and electronic industries. Crucial for biological applications, quantum dots must be covered with other materials allowing dispersion and preventing leakage of the toxic heavy metals (Hardman, 2006; Weng and Ren, 2006).

With the rapid growth of nano-industry and accordingly the increasing exposure levels to various nanoparticles, the need for assessing the risk of nanoparticles to human health is demanding; however, to date the research conducted to investigate the potential neurotoxicity induced by nanoparticles is very limited. Searching the PubMed database (January 2009) using the combination of 'nanoparticle' and 'neurotoxicity' gave only 21 hits; the combination of 'nanoparticle', 'CNS' and 'toxicity' gave 9 hits, and the combination of 'nanoparticle', 'brain' and 'toxicity' gave 93 hits. In addition, most of the studies were tested in cell cultures; only a few *in vivo* studies have been conducted.

9.2   How do Nanoparticles Gain Access to the CNS?

There are a few pathways through which nanoparticles can reach the CNS and elicit potentially toxic effects. Generally speaking, nanoparticles have unique ways in their translocation to the systemic circulation and CNS due to their small size and large surface area. However, it is important to emphasize that the biokinetics of different kinds of nanoparticles varies. It is largely dependent upon the characteristics of that particular nanoparticle, such as size, chemical composition, surface structure, solubility, shape, aggregation, and so on (Nel *et al.*, 2006). These parameters can modify cellular uptake, protein binding, translocation from portal of entry to the target site, and the possibility of causing tissue injury (Oberdörster *et al.*, 2005). The interactions between nanoparticles and cells, body fluids and proteins will significantly affect their ability to distribute throughout the body.

One of the most important portals of entry is the respiratory tract. Inhalation is probably the major route for nanoparticles in atmospheric pollutants, combustion-derived nanoparticles, and freely dispersible mineral or metal nanoparticles resulting from bulk manufacture and handing. Inhaled nanoparticles are efficiently deposited by diffusion mechanisms in all regions of the respiratory tract; however, the distribution in different regions is dependent upon the size of the nanoparticle. For inhaled 1 nm particles, 90 % of them are deposited in the nasopharyngeal compartment, 10 % in the tracheobronchial region, and essentially none in the alveolar region; for 5 nm particles, the deposition is approximately equal in all three regions; for 20 nm particles, the highest deposition is in the alveolar region (~50 %) (Oberdörster *et al.*, 2005). Once deposited, nanoparticles, in contrast to larger-sized particles, appear to translocate readily to extrapulmonary sites and reach other target organs, including the CNS, by different transfer routes and mechanisms. Normally the most prevalent mechanism for solid particle clearance in the alveolar region is mediated by alveolar macrophages, through phagocytosis of deposited particles. Interestingly, evidence has shown that nanoparticles could avoid normal phagocytic defense mechanisms in the respiratory system (Ferin and Oberdörster, 1992; Medina *et al.*, 2007). In several studies in which rats were exposed to different sized nanoparticles, after 24 hours only approximately 20 % of 15–20 nm or 80 nm nanoparticles could be lavaged with macrophages from the lung (Ferin *et al.*, 1991; Oberdörster *et al.*, 1992; Oberdörster, 2000). Possibly because of the inefficient alveolar macrophage phagocytosis, nanoparticles readily gain access to epithelial and interstitial sites; once nanoparticles enter the blood circulation or lymphatic pathway, they would be distributed throughout the body. A number of studies with different particle types and sizes have demonstrated the existence of this translocation pathway (Kreyling *et al.*, 2002; Oberdörster *et al.*, 2002; Heckel *et al.*, 2004).

Other routes to gain access to the systemic circulation include the gastrointestinal (GI) tract and skin. Nanoparticles that are cleared from the respiratory tract via the mucociliary escalator can subsequently be swallowed into the GI tract. Alternatively, nanoparticles can be ingested directly from food, water, cosmetics or drugs. However, a few studies that were conducted to investigate the uptake and disposition of nanomaterials by the GI tract did not show significant absorption and translocation. For example, C_{60} fullerenes were administered orally to rats; 98 % of the nanoparticles were cleared in the faeces within 48 hours (Yamago *et al.*, 1995). In another study, Jani *et al.* (1990) found a particle size-dependent uptake of polystyrene particles (ranging from 50 to 3000 nm) by the GI mucosa (6.6 % of the 50 nm nanoparticles; 5.8 % of the 100 nm nanoparticles). Dermal exposure to nanoparticles occurs regularly during the use of cosmetics and sunscreen products; nanoparticles include TiO_2 and ZnO, which are often coated to minimize their reactivity while maintaining their UV absorption properties (Nel *et al.*, 2006). In healthy skin, the epidermis provides excellent protection against particle spread to the dermis. However, flexing of normal skin – as in wrist movements – facilitates the penetration of micrometer-sized fluorescent beads to the dermis (Tinkle *et al.*, 2003). Damaged skin also allows micrometer-sized particles access to the dermis. Once in the dermis, lymphatic uptake is a major translocation route, probably facilitated by uptake in dendritic cells and macrophages. Kim *et al.* (2004) showed in mice and pig studies that intradermally injected quantum dots were localized to regional lymph nodes. Nanoparticles will eventually be translocated to systemic circulation via the lymphatic pathway.

There is another important route for nanoparticles to gain access to the CNS, although it is not a generally recognized mechanism: the uptake of nanoparticles by sensory nerve endings embedded in airway epithelia, by olfactory nerve endings in the nose, or by sensory nerve endings in the dermis, followed by axonal translocation to ganglionic and CNS structures. An inhalation study in rats has shown that after exposure to ^{13}C nanoparticles for 6 hours, ^{13}C concentration in the lung decreased after the exposure, while there was a significant and persistent increase in ^{13}C concentration in the olfactory bulb (Oberdörster *et al.*, 2004). In a study that was conducted to investigate pulmonary and systemic distribution of inhaled silver (Ag) nanoparticles in rats, Ag concentration in the olfactory region was much higher than those in the rest of the brain after one day of inhalation (Takenaka *et al.*, 2001). Another study in rats clearly showed that the olfactory neuronal pathway is efficient for translocating inhaled manganese oxide (MnO_2) nanoparticles to the CNS. After a 12-day exposure, the concentration of MnO_2 in the olfactory bulb, striatum, frontal cortex and cerebellum were all significantly increased (Elder *et al.*, 2006). When one nostril was occluded during a 6-hr exposure, Mn accumulation in the olfactory bulb was restricted to the side of the open nostril only (Feikert *et al.*, 2004). When larger-sized MnO_2 particles (1.3 and 18 μm mass median aerodynamic diameter) were administered to rats via inhalation for 15 days, no significant increases in olfactory Mn were found (Fechter *et al.*, 2002). This was to be expected given that the individual axons of the fila olfactoria (forming the olfactory nerve) are only 100–200 nm in diameter (Plattig, 1989). Translocation into deeper brain structures may possibly occur as well, as shown in the manganese accumulation in rats (Gianutsos *et al.*, 1997). Collectively, these studies demonstrate that the olfactory nerve pathway should also be considered a portal of entry to the CNS for humans under conditions of environmental and occupational exposure to airborne nanoparticles. However, this mechanism is not 'new' if one considers evidence from virology. Many studies have clearly demonstrated that viruses, such as the polio virus (approximately 30 nm), could gain entry to the CNS through the olfactory bulb and olfactory nerve (Oberdörster *et al.*, 2005). Human meningitis virus can move through olfactory and trigeminal neurons; herpes virus can move up and down through the trigeminal neuron to trigger outbreaks of herpes cold sores (Kennedy and Chaudhuri, 2002). Similarly, sensory nerve endings at other sites, such as the airway epithelia or dermis, would be expected to have a similar uptake and axonal translocation mechanism for nanoparticles.

Unlike most other organs, the brain has a unique barrier for protection – the blood–brain barrier (BBB), which strictly regulates the composition of the fluid microenvironment. Even a slight alteration in the brain fluid microenvironment in which neurons and glial cells are suspended would lead to altered CNS function (Sharma and Westman, 2004). Anatomically, the BBB resides in the endothelial cells of the cerebral microvessels that are connected with tight junctions. These CNS endothelial cells lack vesicular transport; therefore, the permeability of the BBB is comparable with that of plasma membrane. Accordingly, lipid-soluble substances permeate across the BBB easily, while water-soluble compounds are largely excluded (Sharma, 2007). The prevailing opinion is that small (<500 Da), compact hydrophobic molecules have the best chance of crossing the BBB by passive diffusion (Barnham et al., 2004). It should be noted that nanoparticles have significant advantages in crossing the BBB due to their unique physio-chemical properties, and therefore have been used to deliver drugs to the CNS (Lockman et al., 2002). There is also evidence showing that nanoparticles can disrupt the BBB and lead to brain edema formation (Sharma and Sharma, 2007). The influence of nanoparticles on brain function will largely depend upon their ability to modify BBB function.

9.3   Mechanisms for Potential Neurotoxicity Induced by Nanoparticles

The main characteristic of nanoparticles is their size, which falls in the transitional zone between individual atoms or molecules and the corresponding bulk materials. This can modify the physico-chemical properties of the material as well as create the opportunity for increased uptake and interaction with biological tissues. This combination of effects can generate adverse biological effects in living cells that would not otherwise be possible with the same material in larger form. Particle size and surface area are important material characteristics from a toxicological perspective. As the size of a particle decreases, its relative surface area increases and also allows a greater proportion of its atoms or molecules to be exposed on the surface rather than in the interior of the material. Surface molecules increase exponentially when particle size decreases <100 nm, reflecting the importance of surface area for increased chemical and biological activity of nanoparticles. The change in the physico-chemical and structural properties of engineered nanoparticles with a decrease in size could be responsible for a number of material interactions that could lead to toxicological effects: generate specific surface groups that could function as reactive sites. An example of how those surface properties can lead to toxicity is the interaction of electron donor or acceptor active sites (chemically or physically activated) with molecular dioxygen (O_2). Electron capture can lead to the formation of the superoxide radical (O_2^-), which through dismutation or Fenton chemistry can generate additional reactive oxygen species (ROS) (Oberdörster et al., 2005; Nel et al., 2006).

Due to their small size and enhanced reactivity, some nanoparticles readily travel throughout the body, deposit in target organs, penetrate cell membranes, lodge in sub-cellular compartments such as mitochondria, and may trigger injurious responses. It is evident that any intrinsic toxicity of the particle surface will be enhanced (Medina et al., 2007). Studies have shown that nanoparticles can induce bigger inflammatory responses and increased lung toxicity compared with larger particles with the same chemical composition at equivalent mass dose (Oberdörster et al., 2005; Donaldson et al., 2006). Studies have also shown that the high reactivity of copper (Cu) nanoparticles produces substantial toxicity differences in mice between the nanosize (23.5 nm) and much larger size (17 μm) particles. The LD_{50} in mice for nano-Cu particles and micro-Cu particles are 413 and >5000 mg/kg, respectively (Chen *et al.*, 2006; Meng *et al.*, 2007).

Nel *et al.* (2006) well summarized the experimental nanomaterial effects, including ROS generation, oxidative stress, mitochondrial perturbation, inflammation, protein denaturation and degradation, DNA damage, endothelial dysfunction, neoantigen generation, and altered cell cycle regulation, most of which are supported by experimental or clinical evidence. ROS generation and oxidative stress is currently the best-developed paradigm for nanoparticle toxicity. Under normal coupling conditions in the mitochondria, ROS are generated at low frequency and are easily neutralized by antioxidant defenses such as glutathione (GSH) and antioxidant enzymes. However, under conditions of excess ROS production during ambient or occupational nanoparticle exposure, the natural antioxidant defenses may be overwhelmed (Nel, 2005). Nanoparticles of various chemistries, such as C_{60} fullerenes and quantum dots, have been shown to create ROS, especially under concomitant exposure to light, ultra-violet radiation or transition metals. The exact mechanism by which nanoparticles cause ROS is not yet fully understood, but suggested mechanisms include: (1) photo-excitation of fullerenes causing intersystem crossing to create free electrons; (2) metabolism of nanoparticles creating redox active intermediates; and (3) inflammation responses *in vivo* that may cause oxyradical release by macrophages (Oberdörster *et al.*, 2005).

It has been demonstrated that nanoparticles of various sizes and chemical compositions preferentially mobilize to mitochondria (Rodoslav *et al.*, 2003). Because mitochondria are redox-active organelles, there is a likelihood of altering ROS production, and thereby overloading or interfering with antioxidant defenses. Furthermore, studies have shown that ambient nanoparticles perturb the mitochondrial permeability transition pore, which leads to the release of pro-apoptotic factors and programmed cell death (Li *et al.*, 2003). *In vitro* studies using different cell systems showed varying degrees of pro-inflammatory and oxidative-stress-related cellular responses after dosing with laboratory-generated or filter-collected ambient nanoparticles (Brown *et al.*, 2001; Li *et al.*, 2003). Inflammation and oxidative stress can be mediated by several primary pathways: (1) the particle surface causes oxidative stress resulting in increased intracellular calcium and gene activation; (2) transition metals released from particles result in oxidative stress, increased intracellular calcium, and gene activation; (3) cell surface receptors are activated by transition metals released from particles, resulting in subsequent gene activation; or (4) intracellular distribution of nanoparticles to mitochondria generates oxidative stress (Oberdörster *et al.*, 2005).

Oxidative stress is a well-known molecular mechanism for the pathogenesis of various neurodegenerative diseases, such as Alzheimer's disease, Parkinson's disease, Huntington's disease, and amyotrophic lateral sclerosis (Barnham *et al.*, 2004). By ROS production and inducing oxidative stress, exposure to various nanoparticles certainly elicits concerns for neurotoxicity. Although by consensus this is a major mechanism underlying the adverse effects of nanoparticles, many other mechanisms have been or will be elucidated due to the diversity and novel characteristics of nanoparticles. For instance, direct interaction between nanoparticles and intracellular protein may cause denaturation or accelerated degradation, which will lead to functional and structural changes of the protein. Our group has investigated the dopaminergic neurotoxicity induced by Mn, Ag and Cu nanoparticles using the dopaminergic neuronal cell line, PC12 (Wang *et al.*, 2008). By examining the expression alteration of a number of dopamine system-related genes, our data indicate that besides oxidative stress, enzyme alterations may also play a significant role in the dopamine depletion induced by metal nanoparticles.

9.4   Evidence for Neurotoxicity Induced by Various Nanoparticles

Although to date there is almost no direct evidence that links human neurodisease or CNS histopathology findings to the exposure to certain nanoparticles, a number of experimental studies have indicated that nanoparticles could initiate adverse biological responses that could lead to toxicological outcomes in various target systems/organs, including the CNS (Xia *et al.*, 2009). As stated above, most of the investigations were conducted using *in vitro* systems, which can hardly be used to extrapolate the biological effects to humans. Some significant studies that demonstrated evidence of nanoparticle-induced neurotoxicity are summarized below.

In a study conducted by Oberdörster (2004), largemouth bass were exposed to 0.5 ppm uncoated fullerenes (C_{60}). Significant lipid peroxidation was found in the brains of bass after 48 hours. GSH was also marginally depleted in gills of fish. This was the first study showing that uncoated fullerenes can cause oxidative damage in the brain and depletion of GSH *in vivo* in an aquatic species.

In a study conducted by Sharma and Sharma (2007), nanoparticles derived from metals (Cu, Ag or Al, approximately 50–60 nm) were administered to rats for one week. Animals exhibited mild cognitive impairment and cellular alterations in the brain. Subjection of these nanoparticle-treated rats to whole-body hyperthermia resulted in profound cognitive and motor deficits, exacerbation of blood-brain barrier disruption, edema formation and brain pathology, compared with naive animals.

Beckett *et al.* (2005) reported that inhalation of ZnO fumes in an occupational setting can cause metal fume fever, with clinical manifestations such as fatigue, chills, fever, myalgias, cough, dyspnea, leukocytosis, metallic taste, and salivation.

Studies by Calderon-Garciduenãs *et al.* (2002, 2003) demonstrated an interesting link between air pollution and CNS effects. These authors described significant inflammatory or neurodegenerative changes, such as degenerating neurons, non-neuritic plaques and neurofibrillary tangles, in the olfactory mucosa, olfactory bulb, and cortical and subcortical brain structures in dogs from a heavily polluted area in Mexico City, whereas these changes were not seen in dogs from a less-polluted rural control city. However, whether direct effects of airborne nanoparticles are the cause of these symptoms remains to be determined.

Campbell *et al.* (2005) demonstrated that exposure (4 h/day, 5 days per week for 2 weeks) to concentrated airborne particulate matter increases the levels of proinflammatory cytokines interleukin-1 alpha and tumor necrosis factor alpha in mice brain, which indicates that components of inhaled particulate matter may trigger a proinflammatory response in nervous tissue which could contribute to the pathophysiology of neurodegenerative diseases.

Veronesi *et al.* (2005) reported that subchronic exposure of concentrated ambient particulates produces neuropathological damage in the brains of Apo E-deficient mice, which are characterized by elevated levels of oxidative stress in the brain. In the study, neurons from the substantia nigral nucleus compacta were significantly reduced, supporting an environmental role for the development of neurodegenerative disease.

Hussain *et al.* (2006) showed that Mn nanoparticles (40 nm) induced the depletion of dopamine and its metabolites in a dopaminergic neuronal cell line, PC12; and such depletion was accompanied by an increase in ROS production. The study also showed that Ag nanoparticles (15 nm) moderately decreased dopamine content with an increase of ROS production. Further investigation (Wang *et al.*, 2008; Rahman *et al.*, 2009) showed that besides oxidative stress induced by Mn, Ag or Cu nanoparticles, enzymatic alterations may also play an important role in dopaminergic neurotoxicity; and the induced neurotoxicity may share some common mechanisms with neurodegeneration.

Tang *et al.* (2008) found that cadmium selenium quantum dots induce elevation of cytoplasmic calcium levels and impairment of sodium channels in rat primary cultured hippocampal neurons, and eventually neuron death. Long *et al.* (2007) reported that titanium dioxide nanoparticles stimulate ROS in microglia, and induce neuron death in primary cultures of rat striatum. Their data indicated the critical role of microglia in titanium dioxide nanoparticle-induced neurotoxicity. Block *et al.* (2004) showed that diesel exhaust particles (DEP) induce dose-dependent dopaminergic neuron death. This study also indicated that DEP must be phagocytized by microglia to induce oxidative insult. Microglia activation plays a critical role in ROS production and progressive dopaminergic neurotoxicity (Block and Hong, 2007); therefore its role in nanoparticle-induced neurotoxicity needs further evaluation.

9.5   Future Research Needed

Nanotechnology is promising to lead to substantial advances in many scientific fields, including electronics, consumer products, and medical diagnosis and treatment. With the massive production of nanoparticles, it is important to emphasize that development of novel nanoparticles must proceed in tandem with assessment of any toxicological and environmental effects, because it is likely that nanoparticles with novel physico-chemical properties may introduce new mechanisms of injury. Therefore, it is critical to understand the interactions happening at the nano-bio interface, and identify toxicological pathways and mechanisms of injury (Medina *et al.*, 2007; Xia *et al.*, 2009).

Maynard *et al.* (2006) proposed five grand challenges to stimulate future research that are needed in the field of nanotoxicology:

(1) Develop instruments to assess exposure to engineered nanomaterials in air and water, within the next 3–10 years.
(2) Develop and validate methods to evaluate the toxicity of engineered nanomaterials, within the next 5–15 years.
(3) Develop models for predicting the potential impact of engineered nanomaterials on the environment and human health, within the next 10 years.
(4) Develop robust systems for evaluating the health and environmental impact of engineered nanomaterials over their entire life, within the next 5 years.
(5) Develop strategic programs that enable relevant risk-focused research, within the next 12 months.

Thus far, most reported nanotoxicity studies have focused on *in vitro* cell culture studies. However, data obtained from such studies may not correspond to *in vivo* results. *In vivo* systems are extremely complicated and the interactions of the nanostructures with biological components, such as proteins and cells, could lead to unique biodistribution, clearance, immune response and metabolism. An understanding of the relationship between the physical and chemical properties of the nanostructure and their *in vivo* behavior would provide a basis for assessing toxic response, and more importantly could lead to predictive models for assessing toxicity (Fischer and Chan, 2007). Before conducting the toxicity study, physical characterization of the test nanoparticles is critical to ensure proper interpretation of the obtained data (Warheit, 2008; Murdock *et al.*, 2008).

Compared with other nanotoxicology fields, neuronanotoxicology research is even more limited (as indicated by the amount of literature in PubMed) and in great demand. Most nanoparticles are formed from transition metals, silver, copper, aluminum, silicon, carbon and metal oxides. These nanoparticles can either cross the BBB easily and/or produce damage to barrier integrity (Sharma, 2007). Once these nanoparticles reach the brain, ROS production, oxidative damage and neurodegeneration are the likely sequential events. Well-designed *in vivo* toxicity studies using appropriate administration methods and toxicity markers, supported by systemic and quantitative pharmacokinetics studies (absorption, distribution, metabolism and excretion of nanoparticles) are needed for neurotoxicity characterization and risk assessment.

Disclaimer: The views and conclusions in this report are those of the authors and do not necessarily represent the views of the US Food and Drug Administration.

References

Barnham KJ, Masters CL, Bush AI (2004) Neurodegenerative diseases and oxidative stress. *Nat Rev Drug Discov* **3**, 205–214.

Beckett WS, Chalupa DF, Pauly-Brown A, Speers DM, Stewart JC, Frampton MW, Utell MJ, Huang LS, Cox C, Zareba W, Oberdörster G (2005) Comparing inhaled ultrafine versus fine zinc oxide particles in healthy adults: a human inhalation study. *Am J Respir Crit Care Med* **171**, 1129–1135.

Block ML, Wu X, Pei Z, Li G, Wang T, Qin L, Wilson B, Yang J, Hong JS, Veronesi B (2004) Nanometer size diesel exhaust particles are selectively toxic to dopaminergic neurons: the role of microglia, phagocytosis, and NADPH oxidase. *FASEB J* **18**, 1618–1620.

Block ML, Hong JS (2007) Chronic microglial activation and progressive dopaminergic neurotoxicity. *Biochem Soc Trans* **35**, 1127–1132.

Bosi S, Da Ros T, Spalluto G, Prato M (2003) Fullerene derivatives: an attractive tool for biological applications. *Eur J Med Chem* **38**, 913–923.

Brown DM, Wilson MR, MacNee W, Stone V, Donaldson K (2001) Size-dependent proinflammatory effects of ultrafine polystyrene particles: a role for surface area and oxidative stress in the enhanced activity of ultrafines. *Toxicol Appl Pharmacol* **175**, 191–199.

Calderón-Garcidueñas L, Azzarelli B, Acuna H, Garcia R, Gambling TM, Osnaya N, Monroy S, DEL Tizapantzi MR, Carson JL, Villarreal-Calderon A, Rewcastle B (2002) Air pollution and brain damage. *Toxicol Pathol* **30**, 373–389.

Calderón-Garcidueñas L, Maronpot RR, Torres-Jardon R, Henríquez-Roldán C, Schoonhoven R, Acuña-Ayala H, Villarreal-Calderón A, Nakamura J, Fernando R, Reed W, Azzarelli B, Swenberg

JA (2003) DNA damage in nasal and brain tissues of canines exposed to air pollutants is associated with evidence of chronic brain inflammation and neurodegeneration. *Toxicol Pathol* **31**, 524–538.

Campbell A, Oldham M, Becaria A, Bondy SC, Meacher D, Sioutas C, Misra C, Mendez LB, Kleinman M (2005) Particulate matter in polluted air may increase biomarkers of inflammation in mouse brain. *Neurotoxicology* **26**, 133–140.

Chen Z, Meng H, Xing G, Chen C, Zhao Y, Jia G, Wang T, Yuan H, Ye C, Zhao F, Chai Z, Zhu C, Fang X, Ma B, Wan L (2006) Acute toxicological effects of copper nanoparticles in vivo. *Toxicol Lett* **163**, 109–120.

Donaldson K, Aitken R, Tran L, Stone V, Duffin R, Forrest G, Alexander A (2006) Carbon nanotubes: a review of their properties in relation to pulmonary toxicology and workplace safety. *Toxicol Sci* **92**, 5–22.

Donaldson K, Stone V, Tran CL, Kreyling W, Borm PJ (2008) Nanotoxicology. *Occup Environ Med* **61**, 727–728.

Elder A, Gelein R, Silva V, Feikert T, Opanashuk L, Carter J, Potter R, Maynard A, Ito Y, Finkelstein J, Oberdörster G (2006) Translocation of inhaled ultrafine manganese oxide particles to the central nervous system. *Environ Health Perspect* **114**, 1172–1178.

Fechter LD, Johnson DL, Lynch RA (2002) The relationship of particle size to olfactory nerve uptake of a non-soluble form of manganese into brain. *Neurotoxicology* **23**, 177–183.

Feikert T, Mercer P, Corson N, Gelein R, Opanashuk L, Elder A, Silva V, Carter J, Maynard A, Finkelstein J, Oberdorster G. 2004. Inhaled solid ultrafine particles (UFP) are efficiently translocated via neuronal naso-olfactory pathways. *Toxicologist* **78**(S1), 435–436.

Ferin J, Oberdörster G, Soderholm SC, Gelein R (1991) Pulmonary tissue access of ultrafine particles. *J Aerosol Med* **4**, 57–68.

Ferin J, Oberdörster G (1992) Translocation of particles from pulmonary alveoli into the interstitium. *J Aerosol Med* **5**, 179–187.

Fischer HC, Chan WC (2007) Nanotoxicity: the growing need for in vivo study. *Curr Opin Biotechnol* **18**, 565–571.

Gianutsos G, Morrow GR, Morris JB (1997) Accumulation of manganese in rat brain following intranasal administration. *Fundam Appl Toxicol* **37**, 102–105.

Hardman R (2006) A toxicologic review of quantum dots: toxicity depends on physicochemical and environmental factors. *Environ Health Perspect* **114**, 165–172.

Heckel K, Kiefmann R, Dorger M, Stoeckelhuber M, Goetz AE (2004) Colloidal gold particles as a new in vivo marker of early acute lung injury. *Am J Physiol Lung Cell Mol Physiol* **287**, L867–L878.

Hussain SM, Javorina AK, Schrand AM, Duhart HM, Ali SF, Schlager JJ (2006) The interaction of manganese nanoparticles with PC-12 cells induces dopamine depletion. *Toxicol Sci* **92**, 456–463.

Jani P, Halbert GW, Langridge J, Florence AT (1990) Nanoparticle uptake by the rat gastrointestinal mucosa: quantitation and particle size dependency. *J Pharm Pharmacol* **42**, 821–826.

Kennedy P, Chaudhuri A (2002) Herpes simplex encephalitis. *J Neurol Neurosurg Psychiatry* **73**, 237–238.

Kim S, Lim YT, Soltesz EG, De Grand AM, Lee J, Nakayama A, Parker JA, Mihaljevic T, Laurence RG, Dor DM, Cohn LH, Bawendi MG, Frangioni JV (2004) Near-infrared fluorescent type II quantum dots for sentinel lymph node mapping. *Nat Biotechnol* **22**, 93–97.

Kreyling WG, Semmler M, Erbe F, Mayer P, Takenaka S, Schulz H, Oberdörster G, Ziesenis A (2002) Translocation of ultrafine insoluble iridium particles from lung epithelium to extrapulmonary organs is size dependent but very low. *J Toxicol Environ Health* **65A**, 1513–1530.

Lee LJ (2006) Polymer nano-engineering for biomedical applications. *Ann Biomed Eng* **34**, 75–88.

Lewinski N, Colvin V, Drezek R (2008) Cytotoxicity of nanoparticles. *Small* **4**, 26–49.

Li N, Sioutas C, Cho A, Schmitz D, Misra C, Sempf J, Wang M, Oberley T, Froines J, Nel A (2003) Ultrafine particulate pollutants induce oxidative stress and mitochondrial damage. *Environ Health Perspect* **111**, 455–460.

Liao D, Creason J, Shy C, Williams R, Watts R, Zweidinger R (1999) Daily variation of particulate air pollution and poor cardiac autonomic control in the elderly. *Environ Health Perspect* **107**, 521–525.

Lockman PR, Mumper RJ, Khan MA, Allen DD (2002) Nanoparticle technology for drug delivery across the blood-brain barrier. *Drug Dev Ind Pharm* **28**, 1–13.

Long TC, Tajuba J, Sama P, Saleh N, Swartz C, Parker J, Hester S, Lowry GV, Veronesi B (2007) Nanosize titanium dioxide stimulates reactive oxygen species in brain microglia and damages neurons in vitro. *Environ Health Perspect* **115**, 1631–1637.

Maynard AD, Aitken RJ, Butz T, Colvin V, Donaldson K, Oberdörster G, Philbert MA, Ryan J, Seaton A, Stone V, Tinkle SS, Tran L, Walker NJ, Warheit DB (2006) Safe handling of nanotechnology. *Nature* **444**, 267–269.

Medina C, Santos-Martinez MJ, Radomski A, Corrigan OI, Radomski MW (2007) Nanoparticles: pharmacological and toxicological significance. *Br J Pharmacol* **150**, 552–558.

Meng H, Chen Z, Xing G, Yuan H, Chen C, Zhao F, Zhang C, Zhao Y (2007) Ultrahigh reactivity provokes nanotoxicity: explanation of oral toxicity of nano-copper particles. *Toxicol Lett* **175**, 102–110.

Murdock RC, Braydich-Stolle L, Schrand AM, Schlager JJ, Hussain SM (2008) Characterization of nanomaterial dispersion in solution prior to in vitro exposure using dynamic light scattering technique. *Toxicol Sci* **101**, 239–253.

Nel A (2005) Atmosphere. Air pollution-related illness: effects of particles. *Science* **308**, 804–806.

Nel A, Xia T, Mädler L, Li N (2006) Toxic potential of materials at the nanolevel. *Science* **311**, 622–627.

Nemmar A, Hoylaerts MF, Hoet PH, Nemery B (2004) Possible mechanisms of the cardiovascular effects of inhaled particles: systemic translocation and prothrombotic effects. *Toxicol Lett* **149**, 243–253.

Oberdörster E (2004) Manufactured nanomaterials (fullerenes, C_{60}) induce oxidative stress in the brain of juvenile largemouth bass. *Environ Health Perspect* **112**, 1058–1062.

Oberdörster G (2000) Toxicology of ultrafine particles: in vivo studies. *Philos Trans R Soc Lond A* **358**, 2719–2740.

Oberdörster G, Ferin J, Gelein R, Soderholm SC, Finkelstein J (1992) Role of the alveolar macrophage in lung injury: studies with ultrafine particles. *Environ Health Perspect* **97**, 193–197.

Oberdörster G, Sharp Z, Atudorei V, Elder A, Gelein R, Lunts A, Kreyling W, Cox C (2002) Extra-pulmonary translocation of ultrafine carbon particles following whole-body inhalation exposure of rats. *J Toxicol Environ Health* **65A**, 1531–1543.

Oberdörster G, Sharp Z, Atudorei V, Elder A, Gelein R, Kreyling W, Cox C (2004) Translocation of inhaled ultrafine particles to the brain. *Inhal Toxicol* **16**, 437–445.

Oberdörster G, Oberdörster E, Oberdörster J (2005) Nanotoxicology: an emerging discipline evolving from studies of ultrafine particles. *Environ Health Perspect* **113**, 823–839.

Plattig K-H (1989) Electrophysiology of taste and smell. *Clin Phys Physiol Meas* **10**, 91–126.

Rahman MF, Wang J, Patterson TA, Saini UT, Robinson BL, Duhart HM, Newport GD, Murdock RC, Schlager JJ, Hussain SM, Ali SF (2009) Expression of gene related to oxidative stress in the mouse brain after exposure to silver-25 nanoparticles. *Toxicol Lett* **187**, 15–21.

Rodoslav S, Laibin L, Eisenberg A, Dusica M (2003) Micellar nanocontainers distribute to defined cytoplasmic organelles. *Science* **300**, 615–618.

Sarker DK (2005) Engineering of nanoemulsions for drug delivery. *Curr Drug Deliv* **2**, 297–310.

Seaton A, MacNee W, Donaldson K, Godden D (1995) Particulate air-pollution and acute health-effects. *Lancet* **345**, 176–178.

Seaton A, Donaldson K (2005) Nanoscience, nanotoxicology, and the need to think small. *Lancet* **365**, 923–924.

Sharma HS (2007) Nanoneuroscience: emerging concepts on nanoneurotoxicity and nanoneuroprotection. *Nanomed* **2**, 753–758.

Sharma HS, Westman J (2004) *The Blood-Spinal Cord and Brain Barriers in Health and Disease*. Elsevier Academic Press: San Diego, CA.

Sharma HS, Sharma A (2007) Nanoparticles aggravate heat stress induced cognitive deficits, blood-brain barrier disruption, edema formation and brain pathology. *Prog Brain Res* **162**, 245–273.

Stern ST, McNeil SE (2008) Nanotechnology safety concerns revisited. *Toxicol Sci* **101**, 4–21.

Takenaka S, Karg E, Roth C, Schulz H, Ziesenis A, Heinzmann U, Schramel P, Heyder J (2001) Pulmonary and systemic distribution of inhaled ultrafine silver particles in rats. *Environ Health Perspect* **109**, 547–551.

Tang M, Xing T, Zeng J, Wang H, Li C, Yin S, Yan D, Deng H, Liu J, Wang M, Chen J, Ruan DY (2008) Unmodified CdSe quantum dots induce elevation of cytoplasmic calcium levels and impairment of functional properties of sodium channels in rat primary cultured hippocampal neurons. *Environ Health Perspect* **116**, 915–922.

Tinkle SS, Antonini JM, Rich BA, Roberts JR, Salmen R, DePree K, Adkins EJ (2003) Skin as a route of exposure and sensitization in chronic beryllium disease. *Environ Health Perspect* **111**, 1202–1208.

Veronesi B, Makwana O, Pooler M, Chen LC (2005) Effects of subchronic exposures to concentrated ambient particles. VII. Degeneration of dopaminergic neurons in Apo E-/- mice. *Inhal Toxicol* **17**, 235–241.

Wang J, Rahman MF, Duhart HM, Newport GD, Patterson TA, Murdock RC, Hussain SM, Schlager JJ, Ali SF (2008) Expression changes of dopaminergic system-related genes in PC12 cells induced by manganese, silver, or copper nanoparticles. *Toxicologist* **102**(S1), 214.

Warheit DB (2008) How meaningful are the results of nanotoxicity studies in the absence of adequate material characterization? *Toxicol Sci* **101**, 183–185.

Weng J, Ren J (2006) Luminescent quantum dots: a very attractive and promising tool in biomedicine. *Curr Med Chem* **13**, 897–909.

Xia T, Li N, Nel AE (2009) Potential health impact of nanoparticles. *Annu Rev Public Health* **30**, 130–150.

Yamago S, Tokuyama H, Nakamura E, Kikuchi K, Kananishi S, Sueki K, Nakahara H, Enomoto S, Ambe F (1995) In vivo biological behavior of a watermiscible fullerene: [14]C labeling, absorption, distribution, excretion and acute toxicity. *Chem Biol* **2**, 385–389.

10

Hepatotoxic Potential of Nanomaterials

Saura C. Sahu

10.1 Introduction

Nanotechnology is an emerging, rapidly developing new branch of modern technology. This technology deals with nanomaterials generally in the size range from 1 to 100 nm. Nanomaterials exist in different shapes and, therefore, they are commonly known as nanotubes, nanowires, as well as crystalline structures called quantum dots and fullerenes. They are developed for different uses. Nanomaterials exhibit unusual physical, chemical and biological properties that are quite different from their larger counterparts having larger mass and size. Extremely small size and high surface area of nanomaterials associated with their greater strength, stability, chemical and biological activity have led to a wide range of potential applications in modern society. They are used in rapidly increasing numbers of nanoproducts (i.e. nano-silver and nano-gold for diagnostic use, magnetic materials for high density data storage) and nanodevices (i.e. optoelecronics, cellular motors, biosensors), electronics, photovoltaics, diagnostics and drug delivery systems. They are present in a variety of consumer products such as foods, drugs, cosmetics, food color additives, food containers, paints and surface coatings.

This trend is expected to result in an ever-increasing presence of nanoparticles in the human environment from their production, use and disposal (Peter *et al.*, 2004; Oberdorster *et al.*, 2005). Because of their extremely small size, nanomaterials can potentially be readily inhaled, ingested and/or penetrate through the skin. Because of their greater stability it is anticipated that they may remain in the body and in the environment for long periods of time. However, it is not known at what dose level they can exhibit toxicity. Therefore, health effects of nanomaterials are of concern.

Nanotoxicity: From In Vivo *and* In Vitro *Models to Health Risks*   Edited by Saura Sahu and Daniel Casciano
© 2009 John Wiley & Sons, Ltd

The fact that nanotechnology is a rapidly emerging field means that very limited information is available on possible adverse health effects for humans. The limited information available from published literature does suggest potentially toxic effects. For example, silver nanomaterials have been used for treatment of a range of diseases including malaria, lupus, tuberculosis, typhoid, tetanus and cancer (Food and Drug Administration, 1999) and in emerging nanomedicines (Wagner *et al.*, 2006). However, their use has resulted in cases of argyria, a permanent blue-gray discoloration of the skin (Food and Drug Administration, 1999), and kidney damage (Hori *et al.*, 2002; White *et al.*, 2003). The potential health hazard of nanomaterials may be significant and, therefore, their risk assessment is important. For the last couple of years serious efforts have been made to assess these potential adverse effects, leading to the birth of a new discipline of toxicology known as 'nanotoxicology'.

The liver is the primary organ involved in the metabolism and detoxification of xenobiotics. Blood carrying the toxicants is filtered by the liver before being distributed to other parts of the body. The high rate of blood flow to the liver leads to delivery of high concentrations of the toxicants to this organ. The high levels of exposure and the high metabolic activity makes the liver a major target organ of toxicants. Therefore, the liver is prone to injury in spite of its capacity for regenerative growth. Once the nanomaterials, whether ingested, inhaled, absorbed through the skin or administered by intravenous injections and medical devices, reach the blood circulation, they will be translocated to the liver. Therefore, nanomaterials are potential hepatotoxicants. The purpose of this report is to put together the limited information available in the literature on the hepatotoxic potential of nanomaterials.

10.2   Liver: A Potential Target Organ of Nanomaterials

Most of the studies on health effects of nanomaterials have focused on toxicity due to inhalation exposure. However, whether nanoparticles enter the body by inhalation, ingestion, skin penetration or by intravenous injections and medical devices, they can be translocated to other organs of the body by the blood circulation system (Sadauskas *et al.*, 2007). Therefore, irrespective of the route of exposure, there can be potentially multiple target organs of nanomaterials in the body.

Nanomaterials can easily reach the liver by the blood circulation, and therefore it is a potential target organ (Patel, 1992; Stolnik *et al.*, 1995; McClean *et al.*, 1998; Oberdorster *et al.*, 2002; Chen *et al.*, 2006; Park *et al.*, 2006; Zhou *et al.*, 2006; Geze *et al.*, 2007; Kamruzzaman *et al.*, 2007; Sadauskas *et al.*, 2007; Wang *et al.*, 2007; Jain *et al.*, 2008). Oberdorster *et al.* (2002) observed significant amounts of ^{13}C-labelled carbon particles (22–30 nm in diameter) in the liver after 6 hours of inhalation exposure of rats to these nanoparticles at a concentration range 80 to 180 $\mu g/m^3$. It is not known whether they were eliminated from the liver. But there is a good possibility that these nanoparticles can enter the liver cells, resulting in cellular dysfunction and liver injury. Therefore, the effect of nanoparticles on the liver needs further study.

Gopee *et al.* (2007) examined the biodistribution of nanomaterials in mouse. They injected 9-week-old mice intradermally with poly(ethylene glycol)-coated quantum dots at a dose of 48 pmol/animal. The animals were euthanized at 24 h postexposure. They observed accumulation of quantum dots in the liver, kidney, spleen and lymph nodes. Studies with pig skin indicate some penetration of quantum dots and fullerenes into epidermis and dermis, but when tested in human skin the quantum dots do not penetrate beyond the stratum corneum (Kraeling, personal communication). There appears to be minimal permeation of nanomaterials in human skin, most of which appear to accumulate in the hair follicles.

The nanomaterials are taken up by the cells of the rethyculo-endothelial system (RES) mainly located in the liver and spleen (McClean *et al.*, 1998). It has been shown that amphiphilic ß-cyclodextrin nanospheres, developed as a colloidal drug carrier for intra-venous delivery, concentrate in the liver (Geze *et al.*, 2007). Self-assembled nanoparticles based on glycol chitosan bearing hydrophobic moieties as carriers for doxorubicin, used against a wide range of cancers such as breast, bladder, lung and thyroid, are found in the liver following intravenous injection (Park *et al.*, 2006). Nanosize colloidal carriers of site-specific clinical drug delivery systems are retained in the liver after intravenous ad-ministration (Patel, 1992; Stolnik *et al.*, 1995). A polyethylenimine polymer-based vesicle hybrid gene delivery system used for cancer treatment results in green fluorescent protein (GFP) transgene expression in the liver (Brownlie *et al.*, 2004). Solid lipid nanoparticles of mitoxantrone for local subcutaneous injection against breast cancer at a single dose of 15 mg/kg produce liver toxicity in mice in 14 days, as determined by histopathology (Lu *et al.*, 2006).

Superparamagnetic iron oxide nanoparticles (SPIONs) are widely used for cancer di-agnostics and treatment. Zhou *et al.* (2006) have shown that the SPIONs conjugated with luteinizing hormone-releasing-hormone (LHRH) accumulate in the breast tumors in mice treated intravenously once with a 250 mg/kg dose of nanoparticles, but most of the uncon-jugated SPIONs accumulate in the liver.

Colloidal gold is used for the treatment of rheumatoid arthritis (Mettier, 1946). Recent studies show that gold nanoparticles can be used for cancer diagnosis (Jain, 2005). But it has been shown that these nanomaterials accumulate in liver. Sadauskas et al. (2007) treated mice with a single intravenous injection of colloidal gold nanomaterials and observed the animals for 24 h. They observed the accumulation of these nanomaterials primarily in the Kupffer cells of the liver.

Magnetic resonance images obtained from rabbits injected with lactobionic acid (LA)-coated superparamagnetic magnetite nanoparticles show selective accumulation of LA-coated nanoparticles in hepatocytes (Kamruzzaman *et al.*, 2007).

The above published reports clearly demonstrate that the liver is a target organ for several nanomaterials which have been shown to concentrate there. They are likely to be retained in the liver and cause tissue injury (Wang *et al.*, 2007).

10.3 Hepatotoxicity of Nanomaterials *In Vitro*

The reported hepatotoxicity studies *in vitro* on nanomaterials are very limited. Few *in vitro* models have been evaluated for hepatotoxicity of nanomaterials. However, a couple of *in vitro* studies undertaken in the last few years demonstrate that nanomaterials are hepato-toxic *in vitro*. Hussain *et al.* (2005) used rat liver cell line BRL 3A as an *in vitro* model to assess toxicity of silver, aluminum, tungsten, molybdenum oxide, iron oxide, manganese oxide, cadmium oxide and titanium oxide nanoparticles. They used cytotoxicity, mitochon-drial dysfunction and oxidative stress as the endpoints of hepatotoxicity. They determined values of $EC_{50,}$ the concentration of nanoparticles that increased cytotoxicity by 50 %. They found the EC_{50} values for silver, molybdenum oxide and other oxides to be 24 μg/ml, 210 μg/ml and greater than 250 μg/ml, respectively. The study concluded that silver was highly toxic, molybdenum oxide moderately toxic, and other oxides were less toxic to the rat liver cells (Hussain *et al.*, 2005).

Sayes *et al.* (2005) examined toxicity of water-soluble fullerene aggregate nano-C_{60} in human liver carcinoma HepG2 cells. They observed that nano-C_{60} was cytotoxic to these cells at concentrations greater than 50 ppb. The nanoparticles induced lipid peroxidation in the cell membrane, as assessed by the thiobarbituric acid method. They concluded that the nano-C_{60} was toxic to HepG2 liver cells via cellular oxidative stress.

10.4   Hepatotoxicity of Nanomaterials *In Vivo*

The reported hepatotoxicity studies *in vivo* on nanomaterials are also very limited. However, the few *in vivo* studies undertaken in the last couple of years demonstrate that nanomaterials are hepatotoxic in rodents. Chen *et al.* (2006) administered copper nanoparticles to mice at varying doses by a single oral gavage and observed the animals for 72 h postexposure. The median lethal dose (LD_{50}) was determined to be 413 mg/kg. The histopathology and increased activity of the enzyme alkaline phosphatase (ALP) in serum were evaluated for liver injury. They observed dose-dependent liver injury caused by these nanoparticles.

Wang *et al.* (2007) exposed mice to titanium dioxide nanoparticles by a single oral dose of 5 g/kg by gavage. The animals were observed for two weeks; nanoparticles accumulated mainly in the liver. They were retained in the liver resulting in liver injury as evidenced by increased liver weight, hepatocyte necrosis by pathological examination of the tissue, and increased activities of liver enzymes alanine aminotransferase (ALT), aspartate aminotransferase (AST), alkaline phosphatase (ALP) and lactate dehydrogenase (LDH) in the serum.

Jain *et al.* (2008) studied the hepatotoxic effects of magnetic iron oxide nanoparticles, which are used as a magnetic resonance imaging (MRI) agent and a drug carrier system. They exposed rats to these nanoparticles with a single intravenous dose of 0.8 mg/kg. They evaluated the liver function by measuring the liver enzymes ALT, AST and ALP in the serum over 3 weeks following the nanoparticle administration. They measured the total iron levels in the serum and in the liver tissue during this period. Also they analyzed the liver tissues for oxidative stress and evaluated them histologically to determine biocompatibility of these nanoparticles. They observed serum iron levels gradually increasing for up to 1 week, but levels slowly declined thereafter. Biodistribution of iron in various body tissues changed with time, but a greater fraction of the injected iron was localized in the liver and spleen. Magnetization measurements of the liver and spleen samples showed a steady decrease over 3 weeks, suggesting particle degradation. Serum showed a transient increase in ALT, AST, ALP and iron levels over a period of 6–24 h following the nanoparticle injection. The increase in oxidative stress reached a peak at 3 days and then slowly declined thereafter. Histological analyses of liver, spleen and kidney samples collected at 1 and 7 days showed no apparent abnormal changes. In conclusion, their study did not find long-term changes in the liver enzyme levels or induced oxidative stress. The authors concluded that the iron oxide nanoparticles were safe for drug delivery and imaging applications (Jain *et al.*, 2008).

10.5 Mechanism of Nanomaterial Hepatotoxicity

It is difficult to discuss the mechanisms of hepatotoxicity of nanomaterials when only a few limited studies have been published in the last couple of years. However, these studies indicate that oxidative stress, which is known to play an important role in the toxicity of many chemicals and drugs, is involved in the hepatotoxicity of nanomaterials. Several nanomaterials such as carbon nanotubes, quantum dots and fullerenes have been shown to produce reactive oxygen species (ROS) both *in vivo* and *in vitro* (Li *et al.*, 2003; Nel *et al.*, 2006). Fullerenes are found to be model compounds for producing superoxides (Li *et al.*, 2003). Nanoparticles of various sizes and chemical compositions are able to localize preferentially in mitochondria, leading to oxidative stress and cellular damage. Fullerene and titanium oxide have been shown to induce oxidative damage in the liver cells (Hussain *et al.*, 2005; Sayes *et al.*, 2005).

Hoet *et al.* (2004) have reported that nanoparticles that enter the rat liver can induce oxidative stress locally. A single (1 day; 20 and 100 mg/kg) or repeated (14 days) intravenous administration of poly-isobutyl cyanoacrylate (PIBCA, a biodegradable particle) or polystyrene (PS, a nonbiodegradable particle) nanoparticles resulted in depletion of reduced glutathione (GSH) and oxidized glutathione (GSSG), as well as inhibition of superoxide dismutase (SOD) activity and a slight increase in catalase activity in the liver. These nanoparticles did not distribute in the hepatocytes, implying that the oxidative species most probably were produced by activated hepatic macrophages, after nanoparticle phagocytosis.

Uptake of polymeric nanoparticles by the Kupffer cells in rat liver induces modifications in hepatocyte antioxidant systems, probably due to the production of reactive oxygen species (Fernandez-Urrusuno *et al.*, 1997). This study demonstrated the depletion of the glutathione level in hepatocytes, but did not detect lipid peroxides measured by conjugated dienes. The report reasoned that the depletion of glutathione was not sufficient to initiate significant oxidative damage to hepatocytes. However, chronic depletion of the cellular antioxidant defense can lead to severe health problems. Therefore, long-term studies are needed to prove the safe use of these nanoparticles.

10.6 Conclusions

Nanomaterials are capable of entering the human body by inhalation, ingestion, skin penetration, or by intravenous injections and medical devices. Then they can be translocated to other parts of the body by blood circulation. It is possible that their distribution in the body may be a function of their size, and surface characteristics such as polarity, hydrophilicity, lipophilicity and catalytic activity. As the particle size decreases, its surface area per unit mass increases, and therefore the nanoparticles are expected to exhibit increased chemical and biological activity in the body (Chan, 2006).

The literature on rodent models *in vivo* strongly indicates that most nanomaterials tend to accumulate in the liver (Zhou *et al.*, 2006; Kamruzzaman *et al.*, 2007; Sadauskas *et al.*, 2007). They have been shown to be retained by the liver leading to tissue injury in mice (Wang *et al.*, 2007). Therefore there are obvious safety concerns, but information on their potential adverse health effects is very limited at the present time. More research on their toxicity, especially hepatotoxicity, is needed. It is anticipated that more and more attention will be given in the future to the hepatotoxic potential of nanomaterials present in day-to-day consumer products such as foods, cosmetics, drugs, drug delivery systems and other medical devices. More information is required on the characterization, biological interaction and health hazards of nanomaterials. *In vivo* and *in vitro* models need to be established for screening their toxic potential, including pharmacokinetics. Reliable biomarkers of their exposure, effects and susceptibility need to be identified *in vivo* as well as *in vitro*. Toxicokinetics and mechanisms of their toxicity need to be established. A set of high throughput and low-cost screening tests for these products should be identified. Finally, the safety analysis for potential adverse health effects of nanomaterials should be established.

Acknowledgements

I thank Drs Margaret Kraeling, Paddy Wiesenfeld and Robert Sprando for their critical review of this manuscript.

References

Brownlie A, Uchegbu IF, Schatziein AG (2004) PEI-based vesicle-polymer hybrid gene delivery system with improved biocompatibility. *International Journal of Pharmaceutics* **274**, 41–52.

Chan VSW (2006) Nanomedicine: An unresolved regulatory issue. *Regulatory Toxicology and Pharmacology* **46**, 218–224.

Chen Z, Meng H, Xing G, Zhao Y, Jia G, Yuan H, Ye C, Chai Z, Zhu C, Fang X, Ma B, Wan L (2006) Acute toxicological effects of copper nanoparticles in vivo. *Toxicology Letters* **163**, 1090–1120.

Fernandez-Urrusuno R, Fattal E, Feger J, Couvreur P, Therond P (1997) Evaluation of hepatic antioxidant systems after intravenous administration of polymeric nanoparticles. *Biomaterials* **18**, 511–517.

Food and Drug Administration (1999) FDA issues final ruling on OTC products containing colloidal silver. Available at http://www.fda.gov/bbs/topics/ANSWERS/ANS00971.html

Geze A, Chau LT, Choisnard L, Mathieu JP, Marti-Batlle D, Riou L, Putaux JL, Wouessidjewe D (2007) Biodistribution of intravenously administered amphiphiilic ß-cyclodextrin nanospheres. *International Journal of Pharmaceutics* **344**, 135–142.

Gopee NV, Roberts DW, Webb P, Cozart CR, Siitonen PH, Warbritton AR, Yu WW, Colvin VL, Walker NJ, Howard PC (2007) Migration of intradermally injected quantum dots to sentinel organs in mice. *Toxicological Sciences* **98**, 249–257.

Hoet PHM, Brüske-Hohlfeld I, Salata OV (2004) Nanoparticles – known and unknown health risks. *Journal of Nanobiotechnology* **2**, 12–23.

Hori K, Martin TG, Rainey P (2002) Believe it or not – silver still poison! *Veterinary and Human Toxicology* **44**, 291–291.

Hussain SM, Hess KL, Gearhart JM, Geiss KT, Schlager JJ (2005) In vitro toxicity of nanoparticles in BRL 3a rat liver cells. *Toxicology In Vitro* **19**, 975–983.

Jain KK (2005) Nanotechnology in clinical laboratory diagnostics. *Clinica Chimica Acta* **358**, 37–54.

Jain TK, Reddy MK, Morales MA, Leslie-Pelecky DL, Labhasetwar V (2008) Biodistribution, clearance, and biocompatibility of iron oxide magnetic nanoparticles in rats. *ASAP Mol. Pharmaceutics* **10**, 1021.

Kamruzzaman SKM, Ha YS, Kim SJ, Chang Y, Kim TJ, Ho Lee G, Kang IK (2007) Surface modification of magnetite nanoparticles using lactobionic acid and their interaction with hepatocytes. *Biomaterials* **28**, 710–716.

Li N, Sioutas C, Cho A, Schmitz D, Misra C, Sempf J, *et al.* (2003) Ultrafine particulate pollutants induce oxidative stress and mitochondrial damage. *Environ Health Perspect* **111**, 455–460.

Lu B, Xiong S, Yang H, Yin X, Chao R (2006) Solid nanoparticles of mitoxantrone for local injection against breast cancer and its lymph node metastases. *European Journal of Pharmaceutical Sciences* **28**, 86–95.

McClean S, Prosser E, Meehan E, O'Malley D, Clarke N, Ramtoola A, Brayden D (1998) Binding and uptake of biodegradable poly-DL-lactide micro- and nanoparticles in intestinal epithelia. *European Journal of Pharm Science* **6**, 153–163.

Mettier SR (1946) Classification and treatment of rheumatic diseases with special emphasis on infectious and rheumatoid arthritis. *American Journal of Orthodontics and Oral Surgery* **32**, A440–A444.

Nel A, Xia T, Li N (2006) Toxic potential of materials at the nanolevel. *Science* **311**, 622–627.

Oberdorster G, Sharp Z, Atudorei V, Elder A (2002) Extrapulmonary translocation of ultrafine carbon particles following whole-body inhalation exposure of rats. *J Toxicol Environ Hlth Part A* **65**, 1531–1543.

Oberdorster G, Oberdorster E, Oberdorster J (2005) Nanotoxicology: an emerging discipline evolving from studies of ultrafine particles. *Nanotoxicology* **113**, 823–839.

Park JH, Kwon S, Lee M, Chung H, Kim J, Kim Y, Park R, Kim I, Seo S, Kwon I, Jeong S (2006) Self-assembled nanoparticles based on glycol chitosan bearing hydrophobic moieties as carriers for doxorubicin: In vivo biodistribution and anti-tumor activity. *Biomaterials* **27**, 119–126.

Patel HM (1992) Serum opsonin and liposomes: their interaction and opsonophagocytosis. *Crit Rev Ther Drug Carrier Syst* **9**, 39–90.

Peter HH, Irene BH, Oleg VS (2004) Nanoparticles – known and unknown health risks. *Journal of Nanobiotechnology* **2**, 12.

Sadauskas E, Wallin H, Stoltenberg M, Vogel U, Doering P, Larsen A, Danscher G (2007) Kupffer cells are central in the removal of nanoparticles from the organism. *Particle and Fibre Toxicology* **4**, 10.

Sayes CM, Gobin AM, Ausman KD, Mendez J, West JL, Colvin VL (2005) Nano-C60 cytotoxicity is due to lipid peroxidation. *Biomaterials* **26**, 7587–7595.

Stolnik S, Illum L, Davis SS (1995) Long circulating microparticulate drug carriers. *Adv Drug Del Rev* **16**, 195–214.

Wagner V, Dullaart A, Bock A, Zweck A (2006) The emerging nanomedicine landscape. *Nature Biotechnology* **24**, 1211–1217.

Wang J, Zhou G, Chen C, Yu H, Ma Y, Jia G, Li B, Jiao F, Zhao Y, Chai Z (2007) Acute toxicity and distribution of different sized titanium dioxide in mice after oral administration. *Toxicology Letters* **168**, 176–185.

White JM, Powell AM, Brady K (2003) Severe generalized argyria secondary to ingestion of colloidal silver protein. *Clinical and Experimental Dermatology* **28**, 254–256.

Zhou J, Leuschner C, Kumar C, Hormes JF, Soboyejo WO (2006) Sub-cellular accumulation of magnetic nanoparticles in breast tumors and metastases. *Biomaterials* **27**, 2001–2008.

11

Nanotoxicity in Blood: Effects of Engineered Nanomaterials on Platelets

Jan Simak

11.1 Introduction

Different engineered nanomaterials have a profound impact on the development of diagnostic biosensors, drug delivery nanosystems, imaging nanoprobes for intravascular use, or other devices that come in contact with blood (Moghimi *et al.*, 2005; Buxton, 2007; Liu *et al.*, 2007; Morrow *et al.*, 2007; Nie *et al.*, 2007; Patel and Bailey, 2007; Pope-Harman *et al.*, 2007; Wei *et al.*, 2007; Zuo *et al.*, 2007; Banerjee and Verma, 2008). The critical task is to design nontoxic nanomaterials with optimum biocompatibility. Biocompatibility has been defined as the ability of a biomaterial to perform its desired function with respect to a medical therapy, without eliciting any undesirable local or systemic effects in the recipient or beneficiary of that therapy, but generating the most appropriate beneficial cellular or tissue response in that specific situation, and optimizing the clinically relevant performance of that therapy (Williams, 2008). While biocompatibility of engineered nanomaterials in a complex system of blood and vasculature is desirable and would represent the optimal situation, the first step is a screening of the toxicity of a nanomaterial. This review discusses the evaluation of potentially toxic effects of nanomaterials in blood, with a specific focus on the investigation of nanomaterial interactions with blood platelets.

Nanotoxicity: From In Vivo *and* In Vitro *Models to Health Risks*   Edited by Saura Sahu and Daniel Casciano
© 2009 John Wiley & Sons, Ltd

11.2   Toxicity of Engineered Nanomaterials in Blood

Any foreign material, unless optimally biocompatible, when exposed intravascularly, induces a complex cooperative defense response of several cell types and other blood components, vessel wall and other exposed tissues. The material recognized as a pathogenic factor (insult) elicits the defense process at the tissue level known as the acute inflammatory response. Principles and mechanisms of the inflammatory response have been excellently reviewed (Nathan, 2002, 2006; Pober and Sessa, 2007; Barton, 2008; Medzhitov, 2008; Serhan *et al.*, 2008). In general, the acute inflammatory response has the following goals: (1) to eliminate the pathogenic factor; (2) to remove irreversibly destroyed tissues; (3) to repair damaged tissues to restore their physiological functions; and finally (4) to resolve inflammatory process and achieve homeostasis. In the case of extensive tissue damage and/or when a pathogenic factor cannot be eliminated, the acute inflammation progresses to a chronic adaptive response, including replacement of damaged tissues with fibroblasts and other not fully functional cells.

Although the classic concept of inflammation is based on infectious triggers, inflammatory response can also be initiated by several types of exogenous or endogenous 'sterile' pathogenic factors of physical or chemical nature, or by metabolic malfunctions. Thus, we can expect that general mechanisms of inflammatory response to new types of nanomaterials will be similar to responses to previously studied pathogenic factors. The acute inflammatory response involves the coordinated delivery of participating cellular and noncellular components to the site of injury. Main components in the inflammatory interplay induced by intravascularly exposed nanomaterials would be endothelial cells of the vessel wall, platelets, white blood cells (particularly neutrophils, monocytes, tissue residential macrophages and mast cells), red blood cells, plasma coagulation system, and complement system.

Depending upon the level of biocompatibility, interactions of a biomaterial with blood components and vessel wall may or may not be clinically significant. While controlled inflammatory response is an essential physiological mechanism of tissue defense, if dysregulated it can become detrimental. In an attempt to eliminate a pathogenic factor and to remove damaged cells, the highly potent mediators, like reactive oxygen and nitrogen species (RONS), proteases and others, are released. This induces collateral damage at the site of inflammation. To minimize autoaggressive effects, the inflammatory processes have to be localized at the site of injury, and well controlled. When the action of the pathogenic factor is not locally restricted but systemic, which in our case would be nontargeted intravascular administration of a nanoparticulate material, the host defense becomes a systemic inflammatory response. Delocalization and dysregulation of inflammatory response lead to autoaggressive inflammation, where inflammatory cells and mediators damage the host's own tissues (Bone, 1992, 1996; Davies and Hagen, 1997; van Eeden and Hogg, 2002; Matsuda and Hattori, 2006). At a certain level of intensity, systemic inflammatory response leads to multiple organ dysfunction (MODS). MODS is associated with a failure of adequate perfusion of vital organs due to microcirculation disorders including edema, microthromboses, and other thrombohemorrhagic complications. Endothelial injury in systemic inflammation is a critical limiting factor of the reversibility of MODS and prognosis of the patient. Interactions of the main cellular and plasma systems of

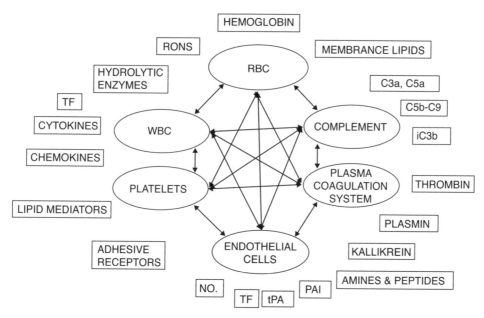

Figure 11.1 *Main intravascular components participating in the acute inflammatory response and examples of their potentially autoaggressive mediators. RBC, red blood cells; WBC, white blood cells; NO., nitric oxide; TF, tissue factor; tPA, tissue plasminogen activator; PAI, plasminogen activator inhibitor; iC3b, complement factor with opsonizing activity; C3a, C5a, complement factors anaphylatoxins; C5b-C9, complement membrane attack complex; RONS, reactive oxygen and nitrogen species.*

intravascular inflammatory response with examples of highly potent mediators is shown in Figure 11.1.

Discussion of the interplay between different intravascular components of inflammatory response is beyond the topic of this chapter. It is important to note, however, that an interaction of a foreign material *in vivo* with blood and vessel wall is always complex, although the effect on one specific component may be more apparently manifested in clinical symptoms. The evaluation of interactions of biomaterials with blood and vessel wall has been extensively studied; however, development and validation of specific clinically-relevant assays still remains a difficult task (Gorbet *et al.*, 1999; Gemmell, 2000, 2001; Sefton *et al.*, 2000, 2001; Sefton, 2001; Seyfert *et al.*, 2002; Spijker *et al.*, 2003; Gorbet and Sefton, 2004; McGuigan and Sefton, 2007). It is important to stress that most assays discussed below are evaluating gross effects of biomaterials on isolated blood components or tissue cultures under artificial static conditions. Thus, information obtained using these methods may be useful for screening purposes only, and cannot address all of the safety concerns, if any, associated with the biomaterial use in a specific clinical indication. A biomaterial safety can only be established in toxicity and biocompatibility studies in appropriate animal models, and by careful monitoring of adverse effects in well-designed clinical trials. General categories of recommended testing for different types of medical devices in contact with blood have been specified in the standard ISO 10993-4 (Association for the Advancement of Medical Instrumentation, 2002). Since nanomaterials may interfere with certain assays, and their evaluation requires specific approaches (e.g. quantitation of nanomaterials and preparation of sample for testing), several new international standard assays are currently under development (Dobrovolskaia and McNeil, 2007; Hall *et al.*, 2007; Dobrovolskaia *et al.*, 2008a,b).

Biocompatibility studies of biomaterials in contact with circulating blood are generally focused on three types of adverse effects: hemolysis, complement activation, and thrombogenicity (Dobrovolskaia *et al.*, 2008a). This practical approach is in principle addressing effects on the main intravascular systems participating in acute inflammatory response, as shown in Figure 11.1.

With regard to material effects on red blood cells (RBC), testing of *in vitro* hemolytic activity of materials has historically been considered as one of the key parameters of biocompatibility in blood. Indeed, massive acute intravascular hemolysis is the most severe effect that a biomaterial can induce on RBC. It is well known that hemolysis results in a decreased RBC count and thus decreased hemoglobin content in circulating blood, a disorder called hemolytic anemia, which leads to failure of oxygen delivery to tissues. However, it is important to realize that there are other serious consequences of intravascular hemolytic activities of biomaterials, significant at low levels of hemolysis, when oxygen delivery is not yet compromised. When RBC plasma membrane disintegrates, free hemoglobin in soluble form or in membrane vesicles and microvesicles can induce extensive oxidative damage intravascularly and during its filtration in renal tubuli, which may lead to renal failure (Minetti *et al.*, 2007; Cimen, 2008). Damaged RBCs and RBC membranes potentiate the injury by adhering to endothelium in microcirculation, as well as in large vessels. This results in prothrombotic and proinflammatory changes and characteristic microangiopathies. Disorder of hemocoagulation balance leads not only to thrombotic events, but at the same time, post-traumatic hemostasis is impaired and bleeding complications may occur. Thus, a biomaterial-induced intravascular hemolysis may trigger or potentiate the development of MODS (Scharte and Fink, 2003).

A standard test method for *in vitro* analysis of hemolytic properties of nanoparticles, ASTM E2524-08, has been developed (ASTM, 2008b; Dobrovolskaia *et al.*, 2008b). This test method assesses acute *in vitro* damage of red blood cells resulting in a release of hemoglobin assayed in plasma. Validation of this standard assay is in progress. Although the test of *in vitro* hemolytic activity is very useful for screening of material toxic effects, it should be pointed out that it provides information only about the acute *in vitro* disintegrating effect of material (direct or plasma mediated) on outer membrane of red blood cells. There are several other possible biochemical and physical activities of nanomaterials which may influence survival, redox and metabolic status, deformability, adhesivity, membrane vesiculation, and other properties of red blood cells *in vivo* (Nohl and Stolze, 1998; Scharte and Fink, 2003; Zhang and Guo, 2007).

Complement activation is an early and often primary event in the inflammatory response to foreign materials in blood (Gorbet and Sefton, 2004; Nilsson *et al.*, 2007). There are several positive feedback interactions in complement activation, in the crosstalk with other components of inflammatory response. Although complement plays an essential role in the defense against pathogens and foreign cells, similarly to other systems participating in inflammation, dysregulated activation of complement leads to tissue injury, including hemolysis and endothelial impairment. Damage to the host cells is mediated either directly by the complement membrane attacking complex C5b-C9, or indirectly by the anaphyla-toxic peptides C3a and C5a which stimulate monocytes/macrophages, neutrophils and mast cells. In addition, iC3b fragment plays an important role in the white blood cell (WBC) recruitment. Various assays have been employed to evaluate the complement activating effects of biomaterials, and standard tests for evaluation of solid materials are available (ASTM, 2000, 2006, 2008d; Gemmell, 2000; Bertholon *et al.*, 2006; Reddy *et al.*, 2007; Toda *et al.*, 2008). An *in vitro* screening assay optimized for evaluation of engineered nanoparticles has been proposed (Dobrovolskaia *et al.*, 2008a). The test is based on the Western blot detection of C3 cleavage products after incubation of the material in plasma. This protocol is designed for rapid and inexpensive assessment of complement activation. Nanomaterials found positive in this assay have to be evaluated further using quantitative assays aimed at specific complement activation pathways.

To understand interactions of a biomaterial with components of inflammatory response, as well as to predict a particulate biomaterial clearance and biodistribution, it is essential to evaluate adsorption of plasma proteins onto the material surface, known as opsoniza-tion (Owens and Peppas, 2006; Dobrovolskaia *et al.*, 2008a). Opsonization is a critical issue in biomaterial/blood biocompatibility. Any biomaterial will probably be instantly coated with various plasma proteins when exposed intravascularly. Depending upon the surface characteristics, immunoglobulins, albumin, complement proteins, kininogens and fibrinogen are usually found as major coating proteins. Binding of fibrinogen is particularly important. While in a soluble form this protein is not proinflammatory, but when adsorbed on the material surfaces, it prompts complex inflammatory response mainly by recruiting white blood cells onto the material surface (Hu *et al.*, 2001). A protocol for analysis of nanoparticle interaction with plasma proteins, using two-dimensional gel electrophoresis and mass spectrometry, has been developed (Owens and Peppas, 2006; Dobrovolskaia *et al.*, 2008a).

Different types of white blood cells play an essential role in the inflammatory response. Leukocyte biology is very complex due to its diversity. The evaluation of material effects on monocyte/macrophages and on neutrophils is the most important in terms of the intravascu-lar acute inflammatory response. Protocols for the evaluation of effects of nanoparticulate materials on macrophage and neutrophil *in vitro* functions were developed, including as-say of phagocytosis, cytokine induction (TNFalpha, IL-1beta, IL-6, IL-8, IL-10, IL12), chemotaxis, and oxidative burst (NO. production) (Dobrovolskaia *et al.*, 2008a). Although assays of induction of monocyte/macrophage surface expression of CD11b (a receptor for fibrinogen, iC3b and factor Xa) and CD142 (tissue factor – a main trigger of the plasma coagulation system) are not routinely used, these tests provide valuable information on biomaterial biocompatibility (Gorbet and Sefton, 2004; Sefton *et al.*, 2000). In addition, interaction of materials with lymphocytes, natural killer (NK) cells, dendritic cells, and biomaterial effects on cell differentiation during hemopoiesis, are critical issues in certain clinical applications. A standard test method for evaluation of the effects of nanoparticulate materials on the formation of mouse granulocyte-macrophage colonies has been proposed, and protocols for the evaluation of effects on leukocyte proliferation and cytotoxic activity of NK cells are available (ASTM, 2008a; Dobrovolskaia *et al.*, 2008a).

Thrombogenicity is defined as the ability of a material to induce or promote the forma-tion of thromboemboli. This definition, based on *in vivo* adverse effects, does not provide full information on what type of material activities should be evaluated *in vitro* (Sefton *et al.*, 2000). Generally, thrombotic and hemorrhagic symptoms are caused by dysregulation of hemostatic balance in the organism, which is maintained by the interplay of vascular endothelium, platelets and the plasma coagulation system (McGuigan and Sefton, 2007). As shown in Figure 11.1, such interplay is not isolated from other systems forming inflammatory response. Thus, hemostasis is an integral part of inflammation and has to be understood in this context. Dysregulation of hemostatic balance leads to a disorder with either thrombotic or hemorrhagic symptoms prevailing. It is important to note that in several types of hemocoagulation disorders, we may clinically observe thrombosis and bleeding at the same time (i.e. thromboembolic symptoms may be associated with failure of posttraumatic hemostasis). Therefore an understanding of underlying mechanisms of dysregulation of hemostatic balance is essential. In the evaluation of effects of a biomaterial on the hemocoagulation balance, the aim is to learn how the tested biomaterial interacts with, or affects functions of, three systems – endothelial cells, the plasma coagulation system, and platelets. Interaction of biomaterials with endothelium has been a focus of extensive research (McGuigan and Sefton, 2007). Several studies of effects of different nanomaterials on cultured endothelial cells *in vitro* have been published; however, these assays are not yet used as part of routine evaluations of biomaterial vascular biocompatibility profiles (Gelderman *et al.*, 2008; Gojova *et al.*, 2007; Lohbach *et al.*, 2006; Peters *et al.*, 2007; Stasko *et al.*, 2007; Yamawaki and Iwai, 2006). Most *in vitro* screening tests of biomaterial effects on hemostasis are directed to the plasma coagulation system (PCS). To evaluate material interaction with PCS for biocompatibility purposes, the first question is how the material affects the regulation of activities of the three most important final effector enzymes in PCS (thrombin, plasmin and kallikrein), and also how the biomaterial interferes with the conversion of soluble fibrinogen to fibrin polymer – a structural basis of the plasmatic clot. Thrombin is the most important enzyme in hemostasis, but also the most dangerous when dysregulated. Activities of thrombin are not restricted in PCS. Thrombin is a strong activator of platelets and complement. It also stimulates WBC and endothelial cells to proinflammatory status. Therefore, evaluation of the effects of biomaterials on the regulation of thrombin activity is a critical issue.

For screening purposes, effects of biomaterials on PCS are routinely evaluated using historically established plasma clotting assays like activated partial thromboplastin time (APTT). The tested material is preincubated with citrate anticoagulated plasma, and the activation of PCS is triggered via the contact pathway (intrinsic pathway, via FXII). In the APTT test, the coagulation is triggered by a contact activator (e.g. ellagic acid), phospholipids (cephalin) and Ca^{2+}, while in the PTT test (partial thromboplastin time), just phospholipids and Ca^{2+} are added. Negatively charged phospholipids with Ca^{2+} allow the formation of subsequent activation complexes, leading finally to the generation of thrombin activity. It was historically believed that materials exposed to blood induce thrombin generation via the contact pathway. It is more likely, however, that biomaterials *in vivo* induce tissue factor expression on monocytes. Thus the biomaterial-induced thrombin activity in blood is predominantly mediated by the tissue factor pathway (formerly extrinsic) and not by contact activation (Gorbet and Sefton, 2004). The tissue factor pathway is activated in the prothrombin time (PT) test. In the PT test, plasma coagulation is triggered by an excess amount of tissue factor (TF) with phospholipids (tissue thromboplastin) and Ca^{2+}. In general, the effects of a biomaterial on clotting-time based tests in plasma, like APTT, are very hard to interpret. A biomaterial may interact with coagulation factors and inhibitors at different stages of the activation process in PCS in coagulation and fibrinolytic pathway. These interactions may be in an antagonist manner. In addition, all clotting-time based assays are dependent upon conversion of fibrinogen to fibrin. The clotting time marks the moment when enough thrombin has acted for long enough to make fibrinogen clot (Hemker, 2007). It means that if the biomaterial interferes with fibrinogen to fibrin conversion, thrombin generation cannot be monitored by any clotting-based test. Direct effects of biomaterials on fibrinogen to fibrin conversion can be evaluated using thrombin time (TT) and reptilase time (RT) tests. Both these enzymes convert fibrinogen to fibrin. In contrast to thrombin, reptilase is not sensitive to heparin-antithrombin inhibition. Besides interference with fibrinogen to fibrin conversion, prolongation of reptilase and thrombin time may indicate activated fibrinolysis (i.e. plasmin activity). Plasmin splits fibrin, and when dysregulated also fibrinogen, to degradation products (FDP). High levels of FDP inhibit fibrinogen to fibrin conversion, and thus prolong reptilase and thrombin time. Direct measurements of generation of thrombin activity are not yet widely used (Hemker *et al.*, 2004). In some biocompatibility studies, surrogate markers of thrombin generation, like thrombin-antithrombin complex (TAT), and prothrombin fragment 1.2 (PF1.2), were analyzed using ELISA assays (Seyfert *et al.*, 2002). A protocol for screening assays of the effects of nanoparticles on plasma coagulation has been developed. It consists of four clotting assays: prothrombin time (PT), activated partial thromboplastin time (APTT), thrombin time (TT) and reptilase time (RT) (Dobrovolskaia *et al.*, 2008a).

Similarly to thrombin, plasmin is a major player in the inflammatory response. Besides its role in fibrinolysis, it interacts with all systems of inflammatory response shown in Figure 11.1. In addition, it has been suggested that plasminogen/plasmin plays a key role in the inflammatory response to biomaterials (Busuttil *et al.*, 2004). In most cases, intravascularly exposed biomaterials affect the generation of plasmin activity indirectly through stimulation of the release of plasminogen activators, like tPA, or inhibitors (PAI) from WBC and endothelial cells. Therefore the evaluation of fibrinolytic activity *in vitro* in plasma exposed to materials has a very limited use. *In vitro* whole blood experiments and better endothelial and macrophage tissue culture assays may be more useful.

Activation of prekallikrein is historically associated with a contact phase of plasma coagulation dependent on FXII. However, it has been shown recently that *in vivo* there is a more favorable proteolytic pathway of prekallikrein activation on endothelial cells and some other cell types, independent of contact activation and FXII (Schmaier and McCrae, 2007). Kallikrein has several activities within inflammatory interplay. Its most recognized activity is a cleavage of kininogens to liberate kinin peptides, particularly bradykinin. Bradykinin is a potent stimulator of endothelial cells to release vasodilating mediators (NO., PGI2) and also tissue plasminogen activator. Thus, the kallikrein-kinin system is involved in regulation of blood pressure, flow, and microvascular permeability, and other processes of inflammatory response. Methods for evaluation of the effects of biomaterials on the kallikrein-kinin system are not well established in biocompatibility studies, since the clinical importance of the kallikrein-kinin system in adverse effects of intravascularly exposed biomaterials remain to be elucidated. Nevertheless, evaluation of effects on the kallikrein-kinin system should not be overlooked in the recent development of various nanomaterials for intravascular use.

In this review, I will focus further on the evaluation of *in vitro* interactions of nanomaterials with blood platelets. This is one of the most challenging and clinically important areas of potential intravascular toxicity of engineered nanomaterials in blood.

11.3   Blood Platelets

Platelet physiology and pathophysiology has been extensively reviewed (Michelson, 2007). Only a brief introduction is provided to give basic information on blood platelets and their hemostatic functions. Platelets are small discoid cells, annucleated in mammals, of about 2–5 μm × 0.5 μm in size with a volume of 6–10 fl. Platelets are derived from megakaryocytes in the bone marrow (Kaushansky, 2005). One megakaryocyte yields about 5000–10 000 platelets. Under physiological conditions the platelet count in peripheral blood is $150–450 \times 10^9/l$, and they circulate for about 10 days. There is a pool of platelets stored in the spleen, which can be rapidly mobilized into circulation. Old platelets are removed by macrophages (RES) in the liver and spleen.

Platelets play a key role in hemostasis. In the blood stream, platelets circulate near the vessel wall maintaining the integrity of vasculature. At least 5×10^9 functional platelets/l are required to prevent spontaneous bleeding in non-traumatized thrombocytopenic patients. The clinical threshold for prophylactic platelet (PLT) transfusion is $10–20 \times 10^9/l$. Trauma or surgery requires a substantially higher count of functional platelets to prevent bleeding (Heal and Blumberg, 2004).

Platelets are very reactive cells, equipped with multiple surface receptors for various activation and inhibition agonists (Varga-Szabo *et al.*, 2008). Activation of different signalling pathways finally leads to an increase in the intracellular concentration of calcium and following changes in cytoskeletal organization. This results in platelet shape change from discoid to spherical, formation of pseudopodia, spreading, adhesion and aggregation. Platelet activation leads to the exposure of phosphatidyl serine (PS) on the plasma membrane, forming procoagulant surface for activation complexes of plasma coagulation system. Activated platelets also release procoagulant membrane microparticles from the plasma membrane and soluble mediators from different types of platelet granules. Thus, platelet activation results in the formation of primary hemostatic plug, which is enforced with fibrin network as a structural result of activation of the plasma coagulation system. Clot retraction is an additional function of platelets, dependent on functional platelet cytoskeleton. Activation signals are generated in any type of injury of endothelial intima of blood vessel, and exposure of subendothelial matrix proteins like collagen or subendothelial tissues.

Besides vessel wall injury, there are several other types of natural and artificial activation platelet agonist. Platelets can be activated by immunoglobulins, various microorganisms (bacterias, parasites) and their products, different drugs and various artificial surfaces and particles, including engineered nanomaterials.

Evaluation of activation or inhibition effects of biomaterials is usually based on material interference with the ability of platelets to perform their hemostatic functions: adhesion, aggregation, and release reaction. Besides their critical role in hemostasis, platelets play multiple roles in other different processes in inflammation. Platelets participate in the host defense against pathogens, regulation of vascular tone, in wound healing, regeneration and reparation of tissues. Platelets also play an important role in tumor growth and metastasis.

Platelets themselves release different growth factors and other mediators. As was discussed in the Introduction (Figure 11.1), platelets interact with all cellular and protein systems of inflammatory response with potentially adverse effects when this response is delocalized or dysregulated. Platelets are involved in recruitment of endothelial progenitor cells; they interact with monocytes and neutrophils in circulation and at the site of injury. Platelets induce activation of these cells leading to production of cytokines, proteolysis, oxidative burst and other processes (Freedman, 2008; May *et al.*, 2008). Under physiological conditions, healthy vascular endothelium prevents uncontrolled platelet activation in circulation by means of different inhibitory mediators. Interference of biomaterials with these processes may have dramatic clinical consequences.

11.4 Possible Adverse Effects of Nanomaterial Interaction with Platelets *In Vivo*

To evaluate possible clinical effects of nanomaterials on platelets *in vivo*, we have to consider different categories of nanomaterials coming into contact with platelets: (1) nanoparticles (NP, e.g. fullerenes, dendrimers, liposomes); (2) macrosurfaces of nanomaterials or nanoparticles containing composites (e.g. plastics with NP additives); (3) nanostructured macrosurfaces (filtration membranes, tissue scaffold). The type of exposure is also a critical issue. Basic types of intravascular and blood exposure are as follows. NP may be administered directly to the intravascular compartment by injection, or a nanomaterial-based device may be implanted intravascularly, or to extravascular tissues. Another type of direct exposure is the use of nanomaterials in surgery or trauma, use of nanomaterials in different extracorporeal circulation techniques (hemodialysis, apheresis, cardiopulmonary bypass (CPB), extracorporeal membrane oxygenation (ECMO)), and finally, exposure of nanomaterials to blood during collection, processing and storage of blood transfusion products. Besides direct exposure of nanomaterials to blood, traffic of NP to intravascular compartment or to bone marrow (BM) after other types or exposure (oral, inhalation, dermal) should also be considered. Type of exposure and a pharmacokinetic behavior of each specific nanomaterial in the given intended use will critically impact the clinical effects on platelets. In general, possible disorders in platelet count and in platelet functions and corresponding clinical manifestations are summarized in Figure 11.2. In the worst case scenario, platelet interaction with nanomaterial may lead to development of irreversible multiple organ dysfunction and death. It is important to note that in most cases, preclinical evaluation of adverse effects of biomaterials, particularly *in vivo* animal studies, are based on one insult model where a healthy organism is exposed to tested biomaterial. The same principle is also applied in human studies in healthy volunteers. In the real life of clinical practice, however, many patients in whom nanomaterials will be used for diagnostic or therapeutic purposes will already be severely ill. These patients will have certain levels of dysregulation of hemostasis and inflammatory response, and some level of dysfunction of multiple vital organs. It is likely that in these patients, much smaller effects of nanomaterials on platelets may have critical consequences, compared with testing in a healthy organism. A good example is the effect of platelets on lung injury. While intact platelets in healthy humans are not sequestered or cleared in the lungs, the activated or damaged platelets

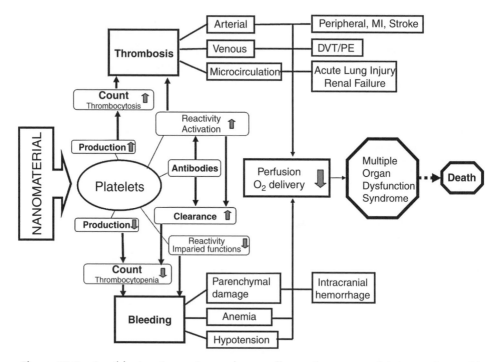

Figure 11.2 *Possible in vivo serious adverse effects of nanomaterial interactions with platelets. MI, myocardial infarction; DVT, deep venous thrombosis; PE, pulmonary embolism.*

adhere to pulmonary vasculature in healthy lungs. Moreover, in the injured pulmonary vasculature, platelets have been shown to stimulate strongly the recruitment of neutrophils and other leukocytes. Thus, platelets may play an important role in the propagation and possibly also in the initiation of inflammatory lung diseases, including acute lung injury (Tabuchi and Kuebler, 2008). Therefore, possible pulmonary adverse events should always be carefully monitored when testing materials interacting with platelets *in vivo*.

11.5   Methods for Evaluation of Effects of Nanomaterials on Platelets

11.5.1   *In Vitro* Testing

There are several types of assays for evaluating platelet hemostatic functions – adhesion, aggregation and release reaction, based on various principles. Methods for platelet testing have been reviewed extensively (Cardigan *et al.*, 2005; Harrison, 2005; Michelson *et al.*, 2006; Gurbel *et al.*, 2007; Harle, 2007). In the following review, for practical reasons, assays are grouped according to the instrumentation needed. There is no standard panel of platelet assays for nanomaterial evaluation available. Based on a review of published results, a list of assays used in evaluation of effects of nanomaterials on platelets is summarized in Table 11.1. The value of the different assays for this purpose is discussed below.

Table 11.1 *Methods used in studies of platelet interactions with engineered nanomaterials.*

Method	PLT sample	Assay	Nanomaterial	References
Cell counter	PRP	PLT-liposome aggregation	LIP NP	Reinish et al. (1988)
				Loughrey et al. (1990)
	PRP	PLT adhesion; counting of adhered PLT after release by trypsin	Au NP-PU film	Hsu et al. (2006)
Light microscopy phase contrast	PRP	PLT-NP aggregates	LIP NP	Zbinden et al. (1989)
	PRP			Loughrey et al. (1990)
	PLS			Male et al. (1992)
	PLS		LAT NP	Miyamoto et al. (1989)
	PRP		Poly NP	Cenni et al. (2008)
immunostaining	PRP	CD62P, CD41 immunostaining peroxidase detection	Nanopores	Ferraz et al. (2008)
Fluorescence microscopy	PLS	Immunodetection of MMP9	C NP	Radomski et al. (2005)
	PRP	PLT adhesion: acridine orange staining	DEN-SK film	Fernandes et al. (2006)
	PLS	Binding of NBD labelled liposomes to activated PLT	LIP NP	Zhu et al. (2007)
	PLS	Binding Alexa488 lab. NP to PLT	DEN NP	Kim et al. (2008)
	PLS	Intracellular free Ca^{2+} (FURA 2AM)	C NP	Semberova et al. (2008)
TEM		PLT morphology	C NP	Radomski et al. (2005)
			LIP NP	Male et al. (1992)
SEM	PRP	PLT morphology	C60 –PU film	Lin and Wu (1999)
	PLS	PLT aggregates	LAT NP	Miyamoto et al. (1989)
	PRP	PLT adhesion	Au NP-PU film	Hsu et al. (2006)
			Scaffold	Yim et al. (2007)
			Nanopores	Ferraz et al. (2008)

(Continued)

Table 11.1 (Continued)

Method	PLT sample	Assay	Nanomaterial	References
Aggregometry impedance	WB	PLT aggregation + ADP release luminescence assay	C NP	Radomski et al. (2005)
	PRP	PLT aggregation	C NP	Semberova et al. (2008)
	WB	PLT aggregation		Koziara et al. (2005)
	PRP	PLT aggregation	LIP NP	Zbinden et al. (1989)
light transmission	PRP, PLS		LIP NP	Juliano et al. (1983)
	PLS		DEN NP	Kim et al. (2008)
screen filtration pressure	Rab WB		C NP	Niwa and Iwai (2007)
Photometry	PRP	PLT aggregation	Gd NP	Oyewumi et al. (2004)
Flow cytometry	WB	CD62P, PAC-1, CD42b, CD61	C NP	Radomski et al. (2005)
	PRP	CD62P, CD63, MPs	C NP	Semberova et al. (2008)
	PRP	CD62P, CD63	DEN-SK film	Fernandes et al. (2006)
	PRP,WB	CD42b, FITC-liposomes	LIP NP	Constantinescu et al. (2003)
	WB	PAC-1		Koziara et al. (2005)
	PRP	CD41/CD61+ MP	Scaffold	Yim et al. (2007)
ELISA	PRP	PF4 release	Poly NP	Cenni et al. (2008)
	PRP	betaTG release	Nanopores	Ferraz et al. (2008)
Luminometry	PLS	ATP release: Luciferin-luciferase assay	LAT NP	Miyamoto et al. (1989)
Fluorimetry	PRP	HPTS-lab. Liposomes Phagocytosis – transfer to acidic pH	LIP NP	Male et al. (1992)
	PLS	PLT membrane integrity assay: FLDA		
Scintillation	PRP	[3H] and [14C] lab liposomes Phagocytosis, exocytosis, intracellular localization	LIP NP	Male et al. (1992)
SDS-PAGE zymography	PRP	Gelatinase MMP2, MMP9 activities	C NP	Radomski et al. (2005)

Legend: Au, gold; βTG, β-thromboglobulin; C, carbon; C60, fullerene; DEN, dendrimer; FLDA, fluorescein diacetate; FITC, fluorescein isothiocyanate; FURA 2AM, fura-2-acetoxymethyl ester; HPTS, 1-hydroxypyrene-3,6,8-trisulfonic acid; LAT, latex; LIP, liposome; MMP9, mitochondrial metaloprotease 9; MPs, membrane microparticles; NBD, nitrobenzodiazole; PF4, platelet factor 4; PLT, platelets; PLS, platelet suspension in buffer (Tyrode's); PRP, platelet-rich plasma; WB, anticoagulated whole blood; POLY, synthetic polymer; PU, polyurethane; SEM, scanning electron microscopy; SK, streptokinase; TEM, transmission electron microscopy.

11.5.1.1   Preparation of a Nanomaterial Sample for Testing

Standard preparation of a nanomaterial sample for testing, and its characterization, is a critical part of the study. Particularly, materials insoluble in aquatic solvents and unstable in suspension must be well characterized. For example, carbon nanotubes provided as a powder from the manufacturer need to be resuspended in a buffer, which is usually done with sonication. Sonication makes dramatic changes in material characteristics. Some studies are performed with nanomaterials resuspended in the presence of a detergent. Every additive or impurity turns the experiment into a different story. Another important issue is stability of the resuspended nanomaterial, including oxidative and other chemical changes. Strong advice is to involve a material scientist early in the project and work with a well-characterized nanomaterial. Otherwise, interpretation of results will be very problematic and value of the study compromised.

11.5.1.2   Platelets for Testing

To achieve maximum robustness of the platelet assays, it is critical to collect and process blood in a standard way with a minimum impact on platelets. Human blood should be used if possible, unless *ex vivo* studies on platelets exposed to a nanomaterial *in vivo* in an animal model are performed. Reactivity of human platelets from individual donors is very variable due to various reasons. Various drugs, including widely used nonsteroid anti-inflammatory agents, impact platelet functions for several days (Shen and Frenkel, 2007), therefore screening of donors in this respect is essential. Shear stress, inappropriate anticoagulation, temperature changes, and metabolic stress due to long storage are the main factors to consider during blood processing and platelet preparation (Harrison, 2005). For *in vitro* testing, nanomaterials are usually exposed to anticoagulated whole blood, platelet-rich plasma (PRP), or platelet suspension in the Tyrode's buffer.

11.5.1.3   Cell Counters

Cell counters based on the Coulter principle of electrical sensing zone method are routinely used to assay platelet count, size, volume, and % of aggregates (Briggs *et al.*, 2007). Cell counters have been used to assess PLT–nanoparticle aggregation (Loughrey *et al.*, 1990; Reinish *et al.*, 1988) and also to assess PLT adhesion to macrosurfaces, where adhered platelets were released by trypsin from the material surface and counted (Hsu *et al.*, 2006). In addition, a simple screening method of platelet aggregation by nanoparticulate materials may be performed using this instrument (Dobrovolskaia *et al.*, 2008a). Briefly, PRP is incubated with the material sample for 15 min at 37° C under static conditions and analyzed using a Coulter analyzer. Since platelet aggregates are not counted in the single platelet window, decrease of platelet count allows calculation of the percentage of aggregation. The main drawback of this assay is absence of any agitation, which would provide homogeneous contact of platelets with tested NP under low shear. Introduction of stirring, however, would be hard to standardize and would require additional instrumentation. This simple assay is easy to perform and may be useful for screening of nanomaterials during product development to identify NP with strong aggregation activity on platelets. Cell counters are also useful for any platelet assay to characterize whole blood, PRP or platelet suspension.

11.5.1.4   Microscopy

Different light and electron microscopic techniques have been widely used in studies on PLT–nanomaterial interactions. Phase contrast light was employed in several studies to observe PLT–NP aggregates (Miyamoto *et al.*, 1989; Zbinden *et al.*, 1989; Loughrey *et al.*, 1990; Male *et al.*, 1992; Cenni *et al.*, 2008). Platelet count and basic morphology are other useful parameters provided by phase contrast microscopy. PLT morphology may be evaluated using different scoring systems; for example, Kunicki's morphology score is frequently used in evaluation of platelet transfusion products (Kunicki *et al.*, 1975). Immunomicroscopy with peroxidase staining has been used for detection of platelet constitutive receptors and activation markers on PLT adhered to a nanoporous material (Ferraz *et al.*, 2008). This can be done with more advanced techniques of fluorescence microscopy, including immunodetection of MMP9 gelatinase after PLT treatment with NP (Radomski *et al.*, 2005). Fluorescence microscopy has various applications. It has been used for detection of adherent platelets labelled with nucleic-acid specific dyes like acridine orange (Fernandes *et al.*, 2006), or to study PLT adhesion of NP labelled with fluorescent dyes (Zhu *et al.*, 2007; Kim *et al.*, 2008). From other applications, we have used fluorescence microscopy of platelets loaded with the Calcium probe Fura 2AM for monitoring of intracellular free Ca^{2+} concentration after PLT exposure to nanoparticles (Semberova *et al.*, 2008). Transmission electron microscopy has been used to evaluate the detailed morphology of PLT exposed to NP (Male *et al.*, 1992; Radomski *et al.*, 2005), while scanning electron microscopy has been more utilized to visualize platelet adhesion to nanocomposite surfaces, scaffolds or nanoporous materials (Lin and Wu, 1999; Hsu *et al.*, 2006; Yim *et al.*, 2007; Ferraz *et al.*, 2008). Generally, microscopy techniques require well-trained experienced professionals and evaluate low numbers of platelets, which may lead to artefacts during sample processing. Although microscopic techniques are very valuable tools, they should be used as complementary to other methods like flow cytometry.

11.5.1.5   Aggregometry

Platelet aggregation can be detected using different physical principles. Light transmission aggregometry (LTA) is performed using PRP stirred in an aggregometer cuvette. Transmission of light through the PRP sample is set as 0 % aggregation, while a transmission through platelet-poor plasma (PPP) is set to 100 %. Thus, the aggregation response after addition of an agonist can be monitored in real time. A typical panel of agonists used in PLT aggregation studies includes thrombin receptor activating peptide (TRAP), collagen, ADP, epinephrine, arachidonic acid, and ristocetin (facilitates vWf binding to GPIb platelet receptor) (Jennings and McCabe White, 2007).

Platelet aggregation can also be studied in whole blood aggregometry. In this assay, whole blood is stirred in a cuvette and changes of impedance between two electrodes are measured. The increase in impedance is proportional to platelet aggregation, since platelet aggregates adhere to electrodes. Whole blood impedance aggregometry has several advantages over LTA, including no need for PRP preparation, so blood is minimally manipulated before testing, and smaller volumes are required for testing. However, presence of blood cells makes the analysis more complex. LTA and whole blood aggregometry (WBA) are considered gold standard assays for evaluation of platelet function in the diagnosis of hereditary and acquired platelet disorders. Simultaneously with aggregation in LTA or WBA, a release reaction may be monitored in a lumiaggregometer. ADP released from dense granules is converted to ATP, which reacts with luciferin-luciferase system to produce adenyl-luciferon. Luminescence light proportional to the ATP content is generated when adenyl-luciferon is oxidized. Both LTA (Juliano *et al.*, 1983; Kim *et al.*, 2008; Semberova *et al.*, 2008) and WBA (Zbinden *et al.*, 1989; Koziara *et al.*, 2005; Radomski *et al.*, 2005) have been utilized in different NP–PLT interaction studies. A different principle is used in the whole blood aggregometry by screen filtration. It measures resistance of flow of a whole blood sample through a screen of microsieve with 30 μm^2 openings, and provides pressure rate as an index of platelet aggregation (Ozeki *et al.*, 2001). This method is not widely used, but it has already been employed in one NP study (Niwa and Iwai, 2007). Platelet aggregometry assays are good candidates for development of a standard assay for characterization of nanomaterial effects on platelets. Firstly, however, a possible interference of different nanomaterials with LTA, impedance aggregometry, or other aggregometry methods needs to be evaluated. The quality of platelets used for aggregation experiments is a critical issue. Different drugs influence the ability of platelets to aggregate, and methods of blood collection or manipulation including shear stress or temperature changes have great impact on platelet function in aggregation assays. Also, platelets need to be used for aggregation experiments within 4 hrs after blood collection, since their reactivity changes over time. Appropriate positive and negative controls in aggregation experiments in each run are necessary. For characterization of nanomaterial effects on platelets, information that can be obtained includes: (1) direct platelet aggregating activity in the PRP and on washed platelets; (2) effects of different inhibitors like ASA, prostacyclin, NO. donors, calcium channel blockers, kinase inhibitors or ion chelators (EDTA); (3) release of ATP; (4) potentiating or inhibitory effects on platelet aggregation induced by standard agonists.

Simple assays recommended for *in vitro* evaluation of platelet transfusion products are hypotonic stress response and extent of shape change (Holme *et al.*, 1998). These assays are usually performed in a light transmission aggregometer, but a photometer with the capability of stirring the suspension in the cuvette can also be used. Hypotonic stress response (HSR) is based on the ability of platelets to extrude water and electrolytes after rapid swelling, which occurs when placed in hypotonic solution. The extent of shape change (ESC) is determined in platelets activated by ADP in the presence of EDTA to prevent platelet aggregation. While HSR relates to the metabolic status of platelets and correlates with ATP level, ESC correlates with morphology score. Both HSR and ESC were found to correlate with *in vivo* platelet recovery. Besides evaluation of platelet transfusion products, ESC and HSR could also serve as simple complementary tests for evaluation of PLT treated with nanomaterials. No data, however, are available in this respect.

11.5.1.6 Flow Cytometry

Flow cytometry is a multipurpose tool allowing a wide variety of platelet assays (Michelson *et al.*, 2007). With regard to studies on PLT–nanomaterial interactions, platelet flow cytometry may be used particularly for the following purposes: (1) to measure the activation state of circulating platelets and their reactivity by activation-dependent changes in platelet surface antigens, leukocyte-platelet aggregation, or platelet-derived microparticles; (2) assay of platelet intracellular changes after activation, induction of apoptosis, necrosis, or types of stimulation (intracellular free calcium concentration, mitochondrial potential). Flow cytometry is also valuable for evaluation of PLT–NP interactions *in vivo*, assay of binding of immunoglobulins to PLT for diagnostics of immune thrombocytopenias, or monitoring thrombopoiesis by counting young 'reticulated' platelets. When using nanoparticles conjugated to a fluorophor or naturally fluorescent NP, a binding of NP to PLT can be evaluated (Constantinescu *et al.*, 2003). In addition, PLT count as well as count of PLT aggregates may also be assayed. Flow cytometry allows analysis of several thousand platelets in one run. Platelets, cells and other events in the sample flow in a hydrodynamically focused stream of fluid subjected to one or more laser beams of specific wavelengths. The detector in line with the light beam provides for each passing particle a forward scatter (FSC) characteristic, which correlates with the particle volume or size. A detector perpendicular to the light beam records side scatter (SSC) characteristics, which depend on the inner complexity of the cell, like granularity, shape of the nucleus or membrane roughness. Thus platelets and platelet membrane microparticles form characteristic populations on a FSC/SSC plot. Additional detectors of fluorescence recognize the fluorescence signal from flowing events, such as platelets labelled with different monoclonal antibodies, each conjugated to a specific fluorophore emitting in a different fluorescence band. For immunodetection of platelets, monoclonal antibodies (Mabs) against major platelet surface receptors GPIIb-IIIa (CD41/CD61; CD41a) or GP Ib (CD42b) are usually used (Constantinescu *et al.*, 2003; Radomski *et al.*, 2005; Shin *et al.*, 2007; Yim *et al.*, 2007; Semberova *et al.*, 2008). Laboratory markers of platelet activation include activation-dependent conformational changes in the GPIIb-IIIa complex, exposure of granule membrane proteins, platelet surface binding of secreted platelet proteins, platelet and platelet–leukocyte aggregates, and development of a procoagulant surface (Michelson, 2006). An example of the first group of platelet activation markers is the Mab PAC-1 which is directed against the GPIIb-IIIa fibrinogen binding site exposed only after platelet activation. The most widely used platelet surface activation marker is P-selectin (CD62P). P-selectin is expressed on the membrane of platelet α-granules and it is exposed on the platelet surface after α-granule secretion. It is believed that the activation-dependent increase in platelet surface P-selectin exposure is not reversible over time *in vitro* (Michelson, 2006). However, P-selectin may be released from platelet surface in soluble or membrane microparticle associated form. Particularly, *in vivo* circulating degranulated platelets lose their surface P-selectin quickly, and therefore it is not a reliable marker for detection of circulating degranulated platelets. Another example of degranulation-dependent platelet surface markers is CD63, which in resting platelets resides on the membranes of lysosomes and dense granules (Israels and McMillan-Ward, 2007). Flow cytometry analysis of the PLT surface expression of CD62P and CD63, or PLT binding of PAC-1 antibody, have been performed in some studies on nanomaterial–PLT interactions (Koziara *et al.*, 2005; Radomski *et al.*, 2005; Fernandes *et al.*, 2006; Semberova *et al.*, 2008).

A useful marker of platelet activation *in vivo* seems to be analysis of leukocyte–platelet aggregates. It has been shown that circulating monocyte–platelet aggregates are a more sensitive marker of *in vivo* platelet activation, compared with platelet surface P-selectin (Michelson, 2006). Flow cytometry can also be used for analysis of platelet membrane microparticles (MPs). MPs are phopholipid vesicles of about 0.1–1 μm in size released from plasma membrane of stimulated platelets and other cell types (Simak and Gelderman, 2006). Platelet MPs expose various platelet membrane antigens, and the majority of platelet

MPs also expose accessible phosphatidylserin, which can be detected using fluorophor conjugated annexin V (in the presence of Ca^{2+}) or lactadherin. Platelet MPs are procoagulant, promoting activation of the plasma coagulation system, and may be prothrombotic *in vivo*. Therefore an assessment of whether a nanomaterial is inducing MP release from platelets is a useful complementary assay in *in vitro* as well as in *in vivo* testing.

11.5.1.7 Solid Phase Assays (ELISA)

Activation markers released to platelet supernatant (either true soluble or associated with different types of membrane vesicles) can be assayed using solid phase assays like ELISA (Harrison and Keeling, 2007). Analysis of different proteins like platelet factor 4 (PF4), β-thromboglobulin (βTG), 'soluble' forms of surface antigens or activation markers (sCD40L, sCD62P, GPV) can be performed to evaluate platelet activation. These assays have been used in research of *in vivo* and *in vitro* platelet activation including studies on nanomaterials (Cenni *et al.*, 2008; Ferraz *et al.*, 2008). Since these methods are prone to artefacts, variability of results, particularly for *in vivo* experiments, is usually very high. As for non-protein mediators, stable thromboxane metabolite thromboxan B_2 (TXB_2) or its metabolite 11–dehydro TXB_2 can be assayed in serum and in urine, respectively, to detect platelet activation *in vivo*.

The enzymatic assay for analysis of LDH release can be used for evaluation of cytotoxic activity of biomaterials on platelets (Massa *et al.*, 2007). Similarly, LDH release is a part of standard assay for evaluation of *in vitro* cytotoxicity of nanomaterials using cell lines (ASTM, 2008c).

11.5.1.8 Biochemical Status: pH, pO_2, pCO_2, HCO_3^-, Lactate, Glucose, ATP

Parameters of biochemical status of platelets usually assayed using clinical analyzers are used for evaluation of platelet transfusion products (Cardigan *et al.*, 2005). These assays will be useful for testing platelets after a long exposure to nanomaterials, for example in devices for processing and storage of transfusion products. Biochemical status of PRP or platelet suspension may be assayed for characterization of platelets before any testing to make sure that they are not in metabolic stress.

11.5.1.9 Dynamic Light Scattering (DLS)

Analysis of a hydrodynamic size of platelets and its temperature-dependent changes using DLS was proposed, together with DLS analysis of number of MPs, as a test of quality for platelet transfusion products (Maurer-Spurej and Chipperfield, 2007). Pilot studies look promising; however, so far only few researchers have applied DLS to platelet analysis. It would be beneficial to establish a DLS assay for evaluation of nanomaterial effects on platelets. A DLS instrument is usually available in a nanomaterial laboratory and DLS analysis is easy to perform. The main problem is that DLS does not work well for polydisperse particle suspension, or precipitating suspensions, which is exactly what platelets with MPs and plasma proteins are. Therefore interpretation of DLS results for a complex mixture like platelets in plasma or in other solutions should be done with great caution.

11.5.1.10   Platelet Function Analyzers and Thrombelastography

There are several types of analyzers based on different principles for clinical evaluation of platelet dysfunction, usually at the point of care (Harrison and Keeling, 2007). Some of them are based on a high shear platelet adhesion or aggregation, like PFA-100. Others are global hemostasis analyzers like thrombelastography. These analyzers have so far had very limited use in evaluation of effects of nanomaterials in blood (Jovanovic *et al.*, 2006; Murugesan *et al.*, 2006). Some of these analyzers, however, may be shown to be useful as screening or complementary tests in future.

11.5.1.11   Perfusion Chambers

Different assays based on perfusion chambers have been used for evaluation of platelet adhesion to biomaterial surfaces under specific shear stress conditions (Sakakibara *et al.*, 2002). Perfusion could be very useful for evaluating the impact of shear stress on platelet adhesion to nanomaterial-containing composites or nanostructured surfaces. There is, however, a need to develop a standard perfusion chamber assay for interlaboratory comparison of the results. In a classical Baumgartner type perfusion chamber, platelets flow along a subendothelial surface of a vessel segment (Baumgartner *et al.*, 1980). Such a chamber, or its modification using a collagen surface or even cultured endothelial cells, may be useful to assess adhesion functions of platelets treated with nanoparticles prior to or during perfusion.

11.5.1.12   Platelet Proteomic, Genomic and Transcriptome Analysis

Recent advances in technologies for proteomics and genomic analysis make these powerful tools widely available and routinely run in several laboratories focused on platelet research (Macaulay *et al.*, 2005). Although the role of proteomic analysis for testing of material effects on platelets has not been demonstrated so far, there is a good chance that a comparative proteomic analysis of treated and nontreated platelets may lead to identification of a panel of proteomic markers useful for evaluation of material biocompatibility or toxicity. Since platelets lack nuclei, it has been assumed that they have limited or no capacity for protein synthesis. Recent discoveries demonstrate that platelets synthesize proteins with important biological activities in response to stimulation. This process is termed signal-dependent translation, uses a constitutive transcriptome and specialized pathways, and can alter platelet phenotype and functions (Zimmerman and Weyrich, 2008). Genomic and transcriptomic analysis may therefore be useful for determination of different changes in platelet status induced by interaction with a nanomaterial. It is important to note that extensive controls excluding the possibility of contamination of platelet samples with blood cells is absolutely critical for these assays to avoid misleading results. We can expect that proteomic, genomic, and transcriptomic analysis will open a new chapter in our understanding of nanomaterial–platelet interactions and general principles of nanomaterial blood and vascular biocompatibility.

11.5.2   *In Vivo* Animal Models

Observations from *in vitro* and tissue culture studies of toxicity and other effects of nanomaterials may not correspond to *in vivo* results (Fischer and Chan, 2007). Therefore *in vivo* animal studies are essential in the preclinical investigation of nanomaterial effects on platelets. Besides general toxicity studies in animals, including careful histological examination of thrombosis in large vessels, in microcirculation, or signs of hemorrhage, the effects of nanomaterials on platelets *in vivo* can be evaluated in specific animal models designed according to intended use. Several animal models of arterial, venous or microvascular thrombosis have been developed (Badimon, 2001; Folts, 1991; Harker *et al.*, 1991; Rumbaut *et al.*, 2005; Verbeuren, 2006). Potentiating or inhibiting effects of nanoparticles can be studied in these models (Radomski *et al.*, 2005). Similarly, effects of nanomaterials on hemostatic efficacy of platelets may be evaluated in different models of hemorrhage or thrombocytopenia (Blajchman and Lee, 1997; James *et al.*, 2008; Majde, 2003). Platelet activation can also be studied in animal models of complement-mediated hypersensitivity reactions to nanomaterials and other *in vivo* models of a nanomaterial-induced inflammatory response (Szebeni *et al.*, 2007). In addition, as discussed above, the evaluation of nanomaterial interactions with platelets by administration to healthy animals is not usually the pathophysiological model most relevant to the intended clinical use of the nanomaterial. To mimic the clinical situation in preclinical testing of nanomaterials, 'second insult' animal models should be developed. Thus, nanomaterials could be evaluated in animals preconditioned with acute lung injury, sepsis, circulatory shock with hypoxia-reperfusion injury, or other types of multiple organ dysfunctions (Blajchman and Lee, 1997; Parker and Watkins, 2001; Majde, 2003; Lomas-Niera *et al.*, 2005; James *et al.*, 2008; Matute-Bello *et al.*, 2008). This approach will provide preclinical *in vivo* results more relevant to clinical use. The right selection of animal species is also important, since there are considerable differences between various animal species and humans in platelets and other systems participating in hemostasis, as well as in the inflammatory response in general.

11.5.2.1 Clinical trials

Finally, the evaluation of nanomaterial safety in a specific clinical indication, including investigation of clinical adverse effects, has to be done in appropriately designed clinical trials. Platelet–nanomaterial interactions should be investigated not only in bleeding and thromboembolic complications, but also in other adverse events like acute lung injury, renal failure, or other organ dysfunction caused by microcirculation disorders.

11.6 Studies on Effects of Different Types of Nanomaterials on Platelets

Relatively few studies investigating the effects of new nanomaterials on platelets have been published so far. There is a more extensive literature on liposomes, which are often particles at nanosize but have a long history of medical use, and thus liposomes are not typical representatives of new types of engineered nanoparticles. The main focus of the following review is to highlight different methodological approaches in individual assays for *in vitro* preclinical testing of effects on platelets. Absence of any standard assays makes it difficult to compare results from different studies and draw general conclusions. Unless an animal species is specified, the following studies were evaluating the effects of nanomaterials on human platelets.

11.6.1   Carbon Nanomaterials

Radomski *et al.* (2005) reported a key study on effects of carbon nanomaterials on platelets. Multiwalled (MWCNT) and single-walled (SWCNT) carbon nanotubes, C60 fullerenes (nC60) and mixed carbon nanoparticles (MCN, a mixture of amorphous carbon with approximately 7 % C60) (0.2–300 µg/ml) were investigated. Nanoparticles were compared with standard urban particulate matter (SRM1648, average size 1.4 µm). This study is a good example of a comprehensive methodological approach in evaluation of nanomaterial effects on platelets. Since all tested materials are not soluble in aquatic solvents, the critical information is how the suspensions were prepared for testing. In this study, particles were suspended in Tyrode's solution and sonicated and vortexed prior testing. Under these conditions, fullerene C60 probably formed, at least in part, a nanocrystalline suspension of stable aggregates (nC60). Therefore, material described as C60 SC in the original report is called nC60 in this review. nC60 refers to an unknown number 'n' of C60 molecules agglomerating to form the suspension. The formed clusters contain unmodified C60 in their centers, surrounded by partially hydroxylated C60 on the outside (Fortner *et al.*, 2005). Radomski *et al.* (2005) did not show size distribution of NP in the tested materials. Just for example, in our study on C60 fullerene (Gelderman *et al.*, 2008), we prepared nC60 nanocrystalline suspension by stirring in water for two weeks and finally spun the suspension at 4000 g to remove large agglomerates. Based on the DLS size distributions of particles present in our nC60 water suspension, the major water-'soluble' component was about 220 nm in hydrodynamic diameter. In the study of Radomski *et al.*, the size of agglomerate was probably larger and the surface less hydroxylated, since the suspension was not stirred for so long.

Radomski *et al.* (2005) performed aggregation studies using washed platelets resuspended in Tyrode's solution preincubated with various inhibitors (Prostacyclin, s-nitrosoglutathione, aspirin, 2-methylthio-AMP, phenanthroline, EDTA and Gö6976) prior to addition of nanomaterials. Platelet aggregation was assayed in a whole blood ionized calcium lumi-aggregometer. The release of ADP was measured by luciferin-luciferase assay. For microscopy, platelet aggregation was terminated at 20 % of maximal response, as determined by the aggregometer, and the samples were fixed with and evaluated by phase contrast microscopy or further processed for immunofluorescence microscopy or TEM. Flow cytometry of the platelet surface expression of activated GPIIb-IIIa (PAC-1), GPIb and CD62P, or beta3 integrin (GPIIIa, CD61) was also performed using nonfixed diluted samples. Zymography of gelatinase activities in the platelet releasate was performed using SDS-PAGE with copolymerized gelatin. Activities of platelet gelatinases MMP2 and MMP9 were identified by their molecular weight and quantified by reference standards.

The study reported that mixed carbon nanomaterials (MCN), carbon nanotubes (SWCNT and MWCNT) and SRM1648 caused activation and aggregation of human platelets (MCN>or=SWCNT>MWCNT>SRM1648). Interestingly, nC60 fullerene did not activate platelets. Compared with a strong platelet response to MCN, carbon nanotubes caused partial aggregation with little or no granular release. SRM1648, MCN, MWCNT and SWCNT induced activation of GPIIb-IIIa and treatment with EDTA completely inhibited aggregation, indicating the essential role of this integrin complex in the carbon NP-induced aggregation response. MCN caused maximum aggregation response and platelet degranulation associated with ADP release. In agreement with this result, MCN-induced aggregation was inhibited by P2Y12 receptor antagonist 1-MeSAMP. In addition, MCN, but not other carbon nanomaterials, induced increased surface expression of CD62P and decreased GPIb. All carbon NP-induced aggregation was inhibited by prostacyclin, and with less efficacy also by S-nitroso-glutathione (endogenous NO donor), but not by aspirin. A classical PKC isoform inhibitor Gö6976 did not affect MCN, SWCNT and MWCNT-induced aggregation. Phenanthroline, which inhibits the MMP gelatinase dependent pathway of aggregation, significantly reduced the aggregation response to carbon nanoparticles. Moreover, all carbon NP tested, including nonaggregating nC60, released cytosolic gelatinases proaggregatory MMP-2 and antiaggregatory MMP-9 from platelets. The authors suggest that the release of MMP-9 may be a sensitive index of cell exposure to nanoparticles.

Radomski *et al.* (2005) discussed different characteristics of carbon NP, which may impact their ability to activate and aggregate platelets. Size of particles, shape and surface characteristics are the main features to be considered. Tested carbon nanomaterials form a polydisperse suspension in aquatic solvents, forming agglomerates up to tens of μm in size. In this study, authors did not find a clear relationship between size of agglomerates and platelet-aggregating potency. Regarding shape, nanotubes may form interplatelet bridges, and thus promote aggregation, while spherical fullerenes may not. This is true for single molecules. However, with an assumption of formation of agglomerates, it would be interesting to compare shapes of agglomerates formed by carbon nanotubes and nC60. Also surface characteristics, including surface charge, are different in carbon nanotubes compared with nC60. Nanoparticle surface charge may play an important role in interaction of NP with platelet receptors like GPII-IIIa. Another important factor is the level of contaminants, particularly transition metals, present in different carbon NP preparations. Transition metals on the surface of NP can cause an increase in ROS generation. The authors also mentioned a difference in the physico-chemical nature of particles resulting from contamination with Ni and I (present in SWCNT only), or the structure of NP, being amorphous in the case of MCN, or crystal-like in the case of SWCNT and MWCNT.

Soluble contaminants were unlikely to affect NP aggregatory potency, since the aggregating ability of particles was greatly reduced following filtration through a 100 nm filter. The hypothesis that NPs containing amorphous carbon have higher potency to activate and aggregate platelets compared with crystal-like carbon NP is unlikely, since our pilot experiments showed that several CNT materials induced a higher PLT aggregation response compared with amorphous carbon black NP (Sigma Aldrich) (Semberova *et al.*, 2008). The activating effects of carbon NP on platelets observed by Radomski *et al.* (2005) *in vitro,* the authors confirmed in a rat thrombosis model *in vivo.* Very likely, platelet activation by carbon NP may be important in humans at systemic levels of NP achieved in therapeutic and diagnostic applications, and as a result of environmental pollution or other exposures. Although maximum aggregation response was observed *in vitro* at concentrations >100 μg/ml, the detection of activated GPIIb-IIIa or MMP-9 zymography was positive at concentrations 4–100 fold lower.

Studies on soluble C60 derivatives could further elucidate effects of fullerenes on platelets, eliminating problems with formation of different types of agglomerates. When a fullerene molecule is hydroxylated it becomes soluble in aquatic solvents. Niwa and Iwai (2007) studied the effects of a polyhydroxylated soluble derivative of C60 – fullerenol C60(OH)24 – compared with amorphous carbon black (CB) on rabbit platelets. They used a method of whole blood screen filtration pressure aggregometry (Ozeki *et al.*, 2001). The effects of CB or C60(OH)24 on ADP-induced platelet aggregation in whole blood was evaluated. CB and C60(OH)24 alone did not induce platelet aggregation. However, when whole blood was pretreated with C60(OH)24, ADP-induced aggregation threshold index values were elevated in a dose-dependent manner. In contrast, collagen- and thrombin-induced platelet aggregation was not affected by C60(OH)24. Although these results indicate that C60(OH)24 specifically facilitates ADP-induced platelet aggregation, we were not able to confirm this conclusion with our experiments using LTA of human PRP. In our hands, pretreatment of human PRP with C60(OH)24 had a significant inhibitory effect on ADP-induced PLT aggregation (unpublished results). These controversial results emphasize the role of experimental setting, and possibly also differences in functions of human and rabbit platelets.

Carbon nanomaterials are very attractive candidates as components of plastic composites, improving the mechanical properties of plastics for various biomedical applications. Lin and Wu (1999) studied platelet adhesion to a C60-grafted polyurethane (PU) surface. Samples of C60-grafted PU film preconditioned in buffer were incubated with platelet-rich plasma. After incubation, PRP was removed and samples were incubated with buffer for further development in adhered platelets. Samples were then fixed and processed for SEM analysis. The results showed significantly higher adhesion of platelets on the C60-grafted PU compared with the nontreated PU control. In addition, the degree of platelet activation on these C60-grafted surfaces was higher than the control. The authors concluded that this effect might be attributed to the synergistic effect of the grafted C60 molecules and the few residual amine functional groups which are left after the C60 grafting reaction. Testing of platelet adhesion on CNT-containing plastic composite surfaces is essential to avoid potential adverse effects of these materials in clinical use. Standardization of a perfusion chamber assay would be very useful for this purpose.

11.6.2 Metal and Metal Composite Nanomaterials

Gold NP have been studied to improve the mechanical properties and biostability of polyurethane polymers. Hsu *et al.* (2006) have prepared the nanocomposites of a polyether-type waterborne polyurethane (PU) incorporated with different amounts (17.4–174 ppm) of gold (Au) nanoparticles (approximately 5 nm). The nanocomposite containing a certain amount (43.5 ppm) of gold (PU-Au 43.5 ppm) was found to have optimal properties, and therefore its biocompatibility was evaluated. For evaluation of effects on platelets, NP were incubated in culture plates with PRP. After incubation the plates were washed and the adherent platelets were either detached by trypsin and counted by cell counter, or fixed on the surface and processed for SEM. Morphology score was evaluated and the average degree of platelet activation was calculated. The results showed that platelet adherence on PU-Ag nanocomposites was not significantly different compared with the original PU; however, the adhered platelets were evaluated as less activated in all PU-Au nanocomposites compared with PU. Interestingly, the concentration of gold in these seemed to make a difference to platelet compatibility. Nanocomposite PU-Au 43.5 ppm showed lower platelet adhesion and less activated platelet morphology compared with a nanocomposite with higher gold content, PU-Au 174 ppm. Counting of adhered platelets after trypsin detachment is not a widely used approach. It would be interesting to see how robust and reproducible this method is.

Gadolinium-containing NP (GdNP) have great importance as contrast agents in nuclear magnetic resonance (NMR) imaging. In addition, GdNP have also been studied for potential use in tumor-specific neutron capture therapy (NCT). The folate receptors were considered to be an effective strategy of targeting GdNP to tumors. Oyewumi *et al.* (2004) evaluated the effects of folate-coated Gadolinium NP on platelets by a simple assay using a general spectrophotometer. Various concentrations of folate-coated Gd nanoparticles were incubated at 37° C for 10 min with PRP under constant stirring. Platelet aggregation in the presence of various nanoparticle concentrations was measured by absorbance at 500 nm. The absorbance values of PRP and PPP and nanoparticle aqueous suspensions were measured as controls. Collagen and epinephrine were used as the positive controls. The study showed that, as assayed by this simple screening method of turbidity measurements, Gd nanoparticles in concentrations ranging from 50 to 300 µg/ml did not cause platelet aggregation. This study is an example of a simple screening of material effects on platelets, which is very useful during product development. More comprehensive studies are needed to obtain conclusive results.

11.6.3 Dendrimers

Dendrimer chemistry allows synthesis of a wide variety of NP of specific properties suited for the intended use. Fernandes *et al.* (2006) studied the effects of a bioconjugate of streptokinase (Sk) and a polyglycerol dendrimer (PGLD, generation 5) on platelets. In this study, polystyrene microplates coated with PGLD-Sk bioconjugate were incubated with PRP, and the surface expression of CD62P and CD63 was analyzed on nonfixed platelets using flow cytometry. This experiment showed that platelet contact with PGLD-Sk plates caused significantly less CD62P expression compared with PGLD-only coated plates and uncoated polystyrene plates. In addition, the authors studied platelet adhesion using microscopy imaging. After incubation with citrate anticoagulated whole blood, the plates were washed and adhered platelets were fixed, stained with acridine orange, and evaluated by epifluorescence microscopy. The authors demonstrated that uncoated polystyrene plates showed platelet adhesion and thrombus formation, while PGDL-Sk coated plates were almost free from platelet adhesion and thrombus formation.

11.6.4 Liposomes

Liposomes were found to interact rapidly with cellular components of blood. Constantinescu *et al.* (2003) studied the effects of various liposomes on platelets. Liposomes, labelled with carboxy fluorescein, or with membrane-sequestering R18 or FITC-labelled phospholipids, were mixed with washed platelets in buffer, PRP or whole blood. Platelets were labelled with CD42b Mab and analyzed by flow cytometry. Irrespective of composition, with or without poly(ethylene glycol), all types of liposomes were found to interact rapidly and dose-dependently with platelets and other blood cells. This took place equally in the presence or absence of plasma proteins. These results suggest that the interaction of liposomes with platelets is opsonization-independent.

The important question of whether liposomes interacting with platelets induce their activation or inhibition has been addressed in several studies. Juliano *et al.* (1983) investigated the effects of liposomes on *in vitro* platelet aggregation using light transmission aggregometry of human PRP and platelet suspension in buffer. Small unilamellar vesicles (SUV) and multilamellar vesicles (MLV) prepared from various conventional lipids and from a photopolymerizable phosphatidylcholine derivative (DPL, bis[1,2-(methacryloyloxy)dodecanoyl]-L-alpha-phosphatidylcholine) were used. None of the liposome preparations studied caused marked platelet aggregation in either plasma or buffer solution. However, positively charged vesicles containing stearylamine impaired the ability of platelets in plasma to aggregate in response to ADP. In contrast, negatively charged vesicles containing phosphatidyl glycerol impaired the ability of platelets in buffer to aggregate in response to thrombin. DPL vesicles had only modest effects on platelets in plasma or buffer. This study demonstrated that the surface charge of liposomes and availability of plasma components are critical factors in interactions of liposomes with platelets.

Similar results showing that some liposomes can inhibit ADP-induced platelet aggregation were also observed by Bonte *et al.* (1987). They studied the biocompatibility properties of polymerizable phosphatidylcholine bilayer membranes, in the form of liposomes, with a view toward the eventual utilization of such polymerized lipid assemblies in drug carrier systems or as surface coatings for biomaterials. The SH-based polymerizable lipid 1,2-bis[1,2-(lipoyl)dodecanoyl]-sn-glycero-3-phosphocholine (dilipoyl lipid, DLL) and the methacryl-based lipid 1,2-bis[(methacryloyloxy)dodecanoyl]-sn-glycero-3-phosphocholine (dipolymerizable lipid, DPL) were studied in comparison with 'conventional' zwitterionic or charged phospholipids. The authors examined effects of liposomes on platelet aggregation. Polymerized DPL liposomes and DLL liposomes, in polymerized or nonpolymerized form, were without substantial effect on platelet aggregation. However, DPL nonpolymerized vesicles, while not causing aggregation, did impair ADP-induced aggregation of platelets.

A comprehensive study of *in vitro* platelet compatibility of liposome NP with promising potential for clinical use was reported by Koziara *et al.* (2005). Effects of cetyl alcohol/polysorbate nanoparticles (E78 NP) and polyethylene glycol (PEG)-coated E78 NPs (PEG-E78 NPs) on activation and aggregation of human platelets was investigated using flow cytometry and whole blood impedance aggregometry. Both NP formulations E78 and PEG-E78 did not activate platelets as assayed by PAC Mab binding specific for activated GPIIb-IIIa, and did not cause platelet aggregation. Interestingly, both NP formulations very rapidly inhibited agonist-induced platelet activation and aggregation in a dose-dependent manner. Apparently these liposomes interact promptly with platelets without causing their activation. When high nanoparticle-to-platelet ratios are achieved, the liposomes interfere with platelet activation irrespective of agonists. The authors estimated that E78 NPs should not interfere with platelet functions *in vivo* at clinically relevant doses.

To further investigate the effect of surface charge on liposome interaction with platelets, different *in vivo* animal models have been employed. Reinish *et al.* (1988) studied the interaction of liposomes with platelets in rats. Rats were injected intravenously with liposomes of various compositions and sizes, and blood platelet count measured using the Coulter counter. It was found that negatively-charged liposomes induced a transient reduction in platelet count in the first 5 minutes after injection, which recovered by 60 minutes post-injection. This effect was most striking for multilamellar vesicles (MLVs) containing phosphatidylglycerol (PG). The authors showed that negatively charged liposomes temporarily associate with platelets *in vitro* as well as *in vivo*. Platelets *in vivo* are very likely transiently sequestered with liposomes in lungs and in the liver. *In vivo* study with radiolabelled platelets indicated that 60 minutes after liposome administration, the transiently sequestrated platelets reappear in circulation. Thus, the platelet count recovery is not due to mobilization of stored platelets from the spleen. It was not possible to prevent transient liposome platelet interaction by acetylsalicylic acid (ASA), heparin or ancrod, indicating that this interaction is not associated with induction of irreversible platelet aggregation. Interestingly, this transient thrombocytopenia was prevented by liposomal pretreatment, probably saturating binding sites on the endothelium. Compared with negatively charged liposomes, positively charged liposomes produced a less pronounced transient reduction in platelet count, while neutral liposomes caused only a mild, transient platelet decline. Besides rats, the study was also performed in rabbits with similar results (Doerschuk *et al.*, 1989), and also in a different laboratory using negatively charged liposomes (phosphatidyl choline/ phosphatidyl acid) on guinea pigs (Zbinden *et al.*, 1989).

In the following *in vitro* study, Loughrey *et al.* (1990) demonstrated that the transient interaction of negatively charged PG liposomes with rat platelets is mediated by complement factor, probably C3b. The aggregation of platelets with PG liposomes was monitored using phase contrast microscopy and also monitored as a reduction in platelet count using a Coulter counter. Interaction of platelets with PG liposomes did not occur in plasma depleted from C3b as well as in plasma treated with the cobra venom factor which specifically degrades C3 (confirmed by Western blotting). The interaction was also inhibited by EDTA. Thus, this study showed that the transient interaction of liposomes with rat platelets required the presence of negatively charged lipids in the liposome composition, divalent cations and intact complement factors in the environment. The authors suggested that the binding of C3b to rat platelets probably occurred due to CR1 receptor present on rodent platelets. In contrast, human platelets do not express CR1 receptors and accordingly, authors did not observe interaction of human platelets with PG liposomes in the presence of human plasma. This study is a good example of where rodent studies cannot be translated to humans.

After initial adhesion of NP like liposomes on platelet membrane, the second part of the story is what mechanism is further employed. Is it energy-independent sequestration in the open canallicular system, fusion with the plasma membrane, or active phagocytosis? This is very likely dependent on the size and surface characteristics of NP. Male *et al.* (1992) have shown that platelets are capable of phagocytosing liposomes rather than simply sequestering particles. The authors used [3H]-cholesterylhexadecyl ether as the lipid label and [3H]-inulin as a label of the aqueous phase. Incubation of human platelets with small neutral unilamellar liposomes (approximately 74 nm) resulted in uptake of the liposomes. While liposome lipids were retained, platelets rapidly released the liposome aqueous-phase components. Uptake of liposomes was proportional to the number of liposomes added and to the incubation time. Approximately 250 liposomes per platelet were taken up within a 5-hr incubation period. The thin-section electron microscopy confirmed that the uptake of the liposomes occurred through the open canallicular system. Changes in fluorescence of the pH-sensitive probe and hydrolysis of the cholesteryl [14C]oleate membrane marker indicated that liposomes were accumulated and degraded in acid- and esterase-containing vesicles. Uptake was inhibited by the addition of EDTA, indicating that divalent cations are involved in the phagocytosing process. Inhibitory effects of cytochalasin B showed that microfilaments and plasticity of the membrane play an important role in this process. The energy dependence of the liposome uptake has been confirmed by inhibitory effects of 2,4-dinitrophenol and iodoacetate, which interfere with glycolysis and oxidative phosphorylation. The platelet function or morphology was not affected by liposome loading and phagocytosis, as was confirmed by results from the serotonin release assay, micro-aggregation assay, fluorescein diacetate membrane integrity assay, and electron microscopy.

Finally, we have an example of a strategy in drug delivery using platelet-targeted liposomes. Biomimetic glycoliposomes for targeting P-selectin seems to be promising in this respect. Zhu *et al.* (2007) reported on glycocalyx-mimicking liposomes, prepared by incorporating a glycolipid of 3'-sulfo-Lewis a (SuLe(a))-PEG-DSPE with a headgroup of SuLe(a) and a spacer of polyethylene glycol (PEG) linked to two hydrophobic tails. This PEG-spaced structure was used to mimic the extended structure of P-selectin glycoprotein ligand 1 (PSGL-1) on activated leukocytes, in order to facilitate the specific binding of liposomes to the receptor of P-selectin expressed on activated platelets. The binding ability of liposomes to recombinant rat P-selectin coated on plates was assayed by ELISA. In addition, binding of liposomes labelled by nitrobenzodiazole-modified phophatidylcholine on activated platelets adsorbed on collagen-coated glass coverslips was evaluated by fluorescence microscopy. The results showed that SuLe(a)-PEG-DSPE can form stable, narrowly distributed liposomes with 1,2-distearoyl-sn-glycero-3-phosphocholine (DSPC) and cholesterol, with a vesicle size of about 113 nm. Binding ability of the resultant SuLe(a)-PEG-liposomes to P-selectin was 22 times higher compared with SuLe(a)-liposomes without a PEG spacer. Studies by fluorescence microscopy show that SuLe(a)-PEG-liposomes can bind to activated platelets *in vitro* effectively. It suggests that biomimetic SuLe(a)-PEG-liposomes could be used as nanocarriers to target activated platelets for drug delivery.

11.6.5   Other Engineered Nanomaterials

Miyamoto *et al.* (1989) investigated mechanisms of platelet aggregation induced by latex particles. Platelet suspension in Tyrode's buffer with 1 mM Ca^{2+} was stirred (1100 rpm) with particle suspension, and fixed. The platelet count was performed, and platelets and platelet aggregates were evaluated by SEM, aggregometry, and the assay of ATP release (luciferin-luciferase luminometry). Latex particles induced platelet aggregation associated with the release of ATP from the platelets. The smaller the diameter of particles having the same surface structure, the greater numbers or greater total surface area of particles were required for both reactions. The more hydrophobic and more negatively charged particles, having a diameter of about 0.3 μm, exhibited the highest platelet aggregation potency. Hydrophilic particles without high negative surface potential activated platelets only a little. Particles with high hydrophilicity and without high negative surface potential did not activate platelets at all. This study is very illustrative, showing that changes of just three parameters – size, surface charge, and hydrophobicity – in one type of material can greatly affect NP interaction with platelets.

Cenni *et al.* (2008) evaluated platelet compatibility of nanoparticles made of a conjugate of poly(D,L-lactide-co-glycolide) with alendronate about 200 nm in diameter. After agitation with PRP, a platelet count was assayed in a diluted aliquot of PRP by counting under the microscope using a hemocytometer, and platelet factor 4 was assayed using ELISA in the supernatant after mixing with platelet-inhibiting solution (EDTA, procain, 2-chloradenosine) and spin. Tested nanoparticles did not show any effect on PLT count (aggregation) or on PF4 concentration in supernatant (release reaction). There was, however, no positive control used in these experiments, which is always essential, since reactivity of platelets in individual experiments may be very variable due to multiple factors, as discussed previously.

11.7 Platelet Coating with Nanomaterials

Another area of interest is development of specific methods for assembly of nanostructures on the platelet surface. Platelet coating with different engineered nanoparticles has been studied (Ai *et al.*, 2002). Platelets were coated with 78 nm silica nanoparticles, 45 nm fluorescent nanospheres, or bovine IgG through layer-by-layer assembly by alternate adsorption with oppositely charged linear polyions. Sequential deposition on platelet surfaces of cationic poly(dimethyldiallylammonium chloride) and anionic poly(styrene sulfonate) was followed by adsorption of nanoparticles or immunoglobulins. This technique may serve to modify platelet surfaces for various purposes. It may be used to modulate platelet receptor-agonist interactions including aggregation or adhesion response, or, for example, antibody-coated platelets may be targeted to specific sites *in vivo* for diagnostic or therapeutic purposes.

11.8 Nanomaterials Designed to Inhibit or Activate Platelets

One promising approach is a targeted modification of the NP surface with agonists known to inhibit platelet functions. For example, activation of the A2A receptor, a G protein-coupled receptor (GPCR), by extracellular adenosine, is antiaggregatory in platelets and anti-inflammatory. This pathway of platelet inhibition was employed by Kim *et al.* (2008). Multiple copies of an A2A agonist, the nucleoside CGS21680, were coupled covalently to poly amidoamine (PAMAM) dendrimers. An Alexa 488 fluorescent PAMAM-CGS21680 conjugate 5 inhibited the aggregation of washed human platelets and was effectively internalized. Thus multivalent dendrimer conjugates may improve overall pharmacological profiles compared with the monovalent GPCR ligands.

Nanoparticles have also been employed to study structural requirements for platelet-activating activity of collagen-related peptides (Cejas *et al.*, 2007). Normally, collagen-related peptides (CRPs), even one as long as a 30-mer (10 Gly-Pro-Hyp (GPO) repeats), are unable to effectively express collagen's platelet-activating behavior. Two short CRPs, AcHN-(Gly-Pro-Hyp)nGly-OH with $n = 5$ and $n = 10$, were attached via the C-terminus to amino-functionalized latex nanoparticles to create a multimeric display of triple helical motifs. It was shown that a CRP with a 31-mer sequence can mimic the function of collagen when this peptide is displayed in a multivalent fashion by linking its C-terminus to a nanoparticle scaffold. These results indicate the importance of multiple triple-helical motives for robust stimulation of platelets by CRPs via the GPVI receptors.

11.9 Tissue Scaffolds

Progress in the field of tissue scaffolds for tissue engineering is greatly enhanced by the development of nanotechnology. Since many of these products are intended for intravascular use, investigation of effects of these materials on platelets is important. Interfacial polyelectrolyte complexation (PEC) fiber has been proposed as a biostructural unit and biological construct for tissue engineering applications, with its ability to incorporate proteins, drug molecules, DNA nanoparticles, and cells. The PEC is a process of self-assembly that occurs when two oppositely charged polyelectrolytes come together. The interaction between two naturally derived polyelectrolytes, chitosan and alginate, has been exploited to produce different biomaterials including tissue scaffolds. Yim *et al.* (2007) evaluated the biocompatibility and blood compatibility of PEC fiber in order to assess its potential for *in vivo* applications in tissue engineering. Although chitosan-alginate PEC fibrous scaffold was found to be thrombogenic, the blood compatibility of the scaffold could be significantly improved by incorporating a small amount of heparin in the polyelectrolyte solution during fiber formation. For testing of effects on platelets, the samples of different scaffolds were incubated with rabbit PRP under gentle agitation (60 rpm). After incubation, the material surfaces were washed, fixed and processed for SEM. The samples of PRP were labelled for detection of CD41/CD61, fixed, and analyzed by flow cytometry for evaluation of platelet MP. The platelet microparticle production and platelet adhesion on the chitosan-alginate-heparin fibrous scaffold were comparable to those on the resting control, indicating platelet compatible properties. Evaluation of MP release is a valuable complementary assay for evaluation of platelet-activating effects. It should always be accompanied by evaluation of other activation markers, like activation of GPIIb-IIIa (PAC-1), CD62P and CD63 on the surface of platelets.

11.10   Effect of Nanostructure Properties on Platelets

Ferraz *et al.* (2008) evaluated the influence of biomaterial nano-topography on platelet adhesion and activation. Nano-porous alumina membranes with pore diameters of 20 and 200 nm were studied. Blood collected in heparin (controls EDTA, citrate) was incubated in a polymethylacrylate chamber coated with heparin where the alumina membranes were placed covering the wells. The incubation was performed under vertical rotation (22 rpm) for 60 min at 37° C. The same experiment was also performed with heparin anticoagulated PRP. After incubation, samples of blood or PRP were mixed with EDTA or citrate. The EDTA samples were used for PLT count by Coulter counter, the citrate samples spun twice to get platelet-free plasma, frozen, and used for analysis of betaTG by ELISA. Membranes were fixed and processed either for SEM or for immunostaining of CD62P and CD41 using peroxidase detection for light microscopy. The results showed a slight drop in platelet numbers after incubation of both alumina membranes with whole blood but not with PRP. In the case of whole blood, slightly greater reduction was caused by 20 nm membrane compared with 200 nm membrane. After whole blood incubation, levels of betaTG were markedly elevated compared to 0 min control, however, no difference between 20 nm and 200 nm membrane was observed. The platelets found on the 20 nm membrane showed signs of activation such as spread morphology and protruding filipodia, as well as P-selectin expression. However, no microparticles were detected on this surface. CD41 and CD62P positive events were present at higher levels on the 200 nm membrane; with consideration of the SEM results, these objects were probably clusters of platelet MPs expressing CD62P. Despite the fact that very few platelets were found on the 200 nm alumina in contrast to the 20 nm membrane, many MPs were detected on this surface. Interestingly, all MPs were found inside circular-shaped areas of approximately 3 μm in diameter. This study demonstrates how nanotexture can influence platelet microparticle generation. The study highlights the importance of understanding the effects of nanotexture on molecular and cellular events in platelets when designing new biomaterials.

11.11 Conclusion

Thrombotic and hemorrhagic complications have a major impact on the mortality and morbidity of cancer patients, as well as patients with vascular and other diseases. Thus, understanding of the nanomaterial effects on platelets, other systems in blood, and vasculature is a critical safety issue. Few studies are available on the effects of different engineered nanomaterials on platelets. The results showed that some nanomaterials induce platelet activation, adhesion, aggregation and release reaction, while other nanomaterials may induce impairment of platelet functions and platelet damage. A panel of *in vitro* standard assays for (1) screening and (2) characterization of the platelet effects of (A) nanoparticles, and (B) macrodevices with nanostructured surfaces in contact with blood, needs to be developed. In addition, for *in vivo* evaluation of nanomaterial effects on platelets, standard animal models, including second-insult models where animals are preconditioned prior to nanomaterial exposure, are needed. Finally, well-designed clinical trials should address careful monitoring of possible adverse effects of nanomaterials on platelets, with consequent thrombotic or bleeding complications, or other platelet-mediated organ dysfunctions.

Disclaimer: The findings and conclusions in this article have not been formally disseminated by the Food and Drug Administration and should not be construed to represent any Agency determination or policy.

References

Ai H, Fang M, Jones SA, Lvov YM (2002) Electrostatic layer-by-layer nanoassembly on biological microtemplates: platelets. *Biomacromolecules* **3**, 560–564.

Association for the Advancement of Medical Instrumentation (2002) *ANSI/AAMI/ISO 10993-4:2002 Biological Evaluation of Medical Devices – Part 4: Selection of Tests for Interaction with Blood.* ANSI/AAMI/ISO.

ASTM (2000) *ASTM Standard F2065-00e, Standard practice for testing for alternative pathway complement activation in serum by solid materials.* ASTM International: West Conshohocken, PA. Available at www.astm.org.

ASTM (2006) *ASTM Standard F2567-06, Standard practice for testing for classical pathway complement activation in serum by solid materials.* ASTM International: West Conshohocken, PA. Available at www.astm.org.

ASTM (2008a) *ASTM Standard E2525-08, Standard test method for evaluation of the effect of nanoparticulate materials on the formation of mouse granulocyte-macrophage colonies.* ASTM International: West Conshohocken, PA. Available at www.astm.org.

ASTM (2008b) *ASTM Standard E 2524-08, Standard test method for analysis of hemolytic properties of nanomaterials.* ASTM International: West Conshohocken, PA. Available at www.astm.org.

ASTM (2008c) *ASTM Standard E 2526-08, Standard test method for evaluation of cytotoxicity of nanoparticulate materials in porcine kidney cells and human hepatocarcinoma cells.* ASTM International: West Conshohocken, PA. Available at www.astm.org.

ASTM (2008d) *ASTM Standard F1984-99, Standard practice for testing for whole complement activation in serum by solid materials.* ASTM International: West Conshohocken, PA. Available at www.astm.org.

Badimon L (2001) Atherosclerosis and thrombosis: lessons from animal models. *Thromb Haemost* **86**, 356–365.

Banerjee HN, Verma M (2008) Application of nanotechnology in cancer. *Technol Cancer Res Treat* **7**, 149–154.

Barton GM (2008) A calculated response: control of inflammation by the innate immune system. *J Clin Invest* **118**, 413–420.

Baumgartner HR, Turitto V, Weiss HJ (1980) Effect of shear rate on platelet interaction with subendothelium in citrated and native blood. II. Relationships among platelet adhesion, thrombus dimensions, and fibrin formation. *J Lab Clin Med* **95**, 208–221.

Bertholon I, Vauthier C, Labarre D (2006) Complement activation by core-shell poly(isobutylcyanoacrylate)-polysaccharide nanoparticles: influences of surface morphology, length, and type of polysaccharide. *Pharm Res* **23**, 1313–1323.

Blajchman MA, Lee DH (1997) The thrombocytopenic rabbit bleeding time model to evaluate the in vivo hemostatic efficacy of platelets and platelet substitutes. *Transfus Med Rev* **11**, 95–105.

Bone RC (1992) Toward an epidemiology and natural history of SIRS (systemic inflammatory response syndrome). *Jama* **268**, 3452–3455.

Bone RC (1996) Toward a theory regarding the pathogenesis of the systemic inflammatory response syndrome: what we do and do not know about cytokine regulation. *Crit Care Med* **24**, 163–172.

Bonte F, Hsu MJ, Papp A, Wu K, Regen SL, Juliano RL (1987) Interactions of polymerizable phosphatidylcholine vesicles with blood components: relevance to biocompatibility. *Biochim Biophys Acta* **900**, 1–9.

Briggs C, Harrison P, Machin SJ (2007) Platelet counting. In *Platelets*, Michelson AD (ed.). Elsevier: New York; 475–483.

Busuttil SJ, Ploplis VA, Castellino FJ, Tang L, Eaton JW, Plow EF (2004) A central role for plasminogen in the inflammatory response to biomaterials. *J Thromb Haemost* **2**, 1798–1805.

Buxton DB (2007) Nanotechnology in the diagnosis and management of heart, lung and blood diseases. *Expert Rev Mol Diagn* **7**, 149–160.

Cardigan R, Turner C, Harrison P (2005) Current methods of assessing platelet function: relevance to transfusion medicine. *Vox Sang* **88**, 153–163.

Cejas MA, Chen C, Kinney WA, Maryanoff BE (2007) Nanoparticles that display short collagen-related peptides. Potent stimulation of human platelet aggregation by triple helical motifs. *Bioconjug Chem* **18**, 1025–1027.

Cenni E, Granchi D, Avnet S, Fotia C, Salerno M, Micieli D, Sarpietro MG, Pignatello R, Castelli F, Baldini N (2008) Biocompatibility of poly(D,L-lactide-co-glycolide) nanoparticles conjugated with alendronate. *Biomaterials* **29**, 1400–1411.

Cimen MY (2008) Free radical metabolism in human erythrocytes. *Clin Chim Acta* **390**, 1–11.

Constantinescu I, Levin E, Gyongyossy-Issa M (2003) Liposomes and blood cells: a flow cytometric study. *Artif Cells Blood Substit Immobil Biotechnol* **31**, 395–424.

Davies MG, Hagen PO (1997) Systemic inflammatory response syndrome. *Br J Surg* **84**, 920–935.

Dobrovolskaia MA, McNeil SE (2007) Immunological properties of engineered nanomaterials. *Nat Nanotechnol* **2**, 469–478.

Dobrovolskaia MA, Aggarwal P, Hall JB, McNeil SE (2008a) Preclinical studies to understand nanoparticle interaction with the immune system and its potential effects on nanoparticle biodistribution. *Mol Pharm* **5**, 487–495.

Dobrovolskaia MA, Clogston JD, Neun BW, Hall JB, Patri AK, McNeil SE (2008b) Method for analysis of nanoparticle hemolytic properties in vitro. *Nano Lett* **8**, 2180–2187.

Doerschuk CM, Gie RP, Bally MB, Cullis PR, Reinish LW (1989) Platelet distribution in rabbits following infusion of liposomes. *Thromb Haemost* **61**, 392–396.

Fernandes EG, de Queiroz AA, Abraham GA, San Roman J (2006) Antithrombogenic properties of bioconjugate streptokinase-polyglycerol dendrimers. *J Mater Sci Mater Med* **17**, 105–111.

Ferraz N, Carlsson J, Hong J, Ott MK (2008) Influence of nanoporesize on platelet adhesion and activation. *J Mater Sci Mater Med* **19**, 3115–3121.

Fischer HC, Chan WC (2007) Nanotoxicity: the growing need for in vivo study. *Curr Opin Biotechnol* **18**, 565–571.

Folts J (1991) An in vivo model of experimental arterial stenosis, intimal damage, and periodic thrombosis. *Circulation* **83**, IV3–14.

Fortner JD, Lyon DY, Sayes CM, Boyd AM, Falkner JC, Hotze EM, Alemany LB, Tao YJ, Guo W, Ausman KD, Colvin VL, Hughes JB (2005) C60 in water: nanocrystal formation and microbial response. *Environ Sci Technol* **39**, 4307–4316.

Freedman JE (2008) Oxidative stress and platelets. *Arterioscler Thromb Vasc Biol* **28**, s11–16.

Gelderman MP, Simakova O, Clogston JD, Patri AK, Siddiqui SF, Vostal AC, Simak J (2008) Adverse effects of fullerenes on endothelial cells: fullerenol C60(OH)24 induced tissue factor and ICAM-I membrane expression and apoptosis in vitro. *Int J Nanomedicine* **3**, 59–68.

Gemmell CH (2000) Flow cytometric evaluation of material-induced platelet and complement activation. *J Biomater Sci Polym Ed* **11**, 1197–1210.

Gemmell CH (2001) Activation of platelets by in vitro whole blood contact with materials: increases in microparticle, procoagulant activity, and soluble P-selectin blood levels. *J Biomater Sci Polym Ed* **12**, 933–943.

Gojova A, Guo B, Kota RS, Rutledge JC, Kennedy IM, Barakat AI (2007) Induction of inflammation in vascular endothelial cells by metal oxide nanoparticles: effect of particle composition. *Environ Health Perspect* **115**, 403–409.

Gorbet MB, Yeo EL, Sefton MV (1999) Flow cytometric study of in vitro neutrophil activation by biomaterials. *J Biomed Mater Res* **44**, 289–297.

Gorbet MB, Sefton MV (2004) Biomaterial-associated thrombosis: roles of coagulation factors, complement, platelets and leukocytes. *Biomaterial* **25**, 5681–5703.

Gurbel PA, Becker RC, Mann KG, Steinhubl SR, Michelson AD (2007) Platelet function monitoring in patients with coronary artery disease. *J Am Coll Cardiol* **50**, 1822–1834.

Hall JB, Dobrovolskaia MA, Patri AK, McNeil SE. 2007. Characterization of nanoparticles for therapeutics. *Nanomed* **2**, 789–803.

Harker LA, Kelly AB, Hanson SR (1991) Experimental arterial thrombosis in nonhuman primates. *Circulation* **83**, IV41–55.

Harle CC (2007) Point-of-care platelet function testing. *Semin Cardiothorac Vasc Anesth* **11**, 247–251.

Harrison P (2005) Platelet function analysis. *Blood Rev* **19**, 111–123.

Harrison P, Keeling D (2007) Clinical tests of platelet functions. In *Platelets*, Michelson AD (ed.). Elsevier: New York; 445–474.

Heal JM, Blumberg N (2004) Optimizing platelet transfusion therapy. *Blood Rev* **18**, 149–165.

Hemker HC (2007) The initiation phase – a review of old (clotting-) times. *Thromb Haemost* **98**, 20–23.

Hemker HC, Al Dieri R, Beguin S (2004) Thrombin generation assays: accruing clinical relevance. *Curr Opin Hematol* **11**, 170–175.

Holme S, Moroff G, Murphy S (1998). A multi-laboratory evaluation of in vitro platelet assays: the tests for extent of shape change and response to hypotonic shock. Biomedical Excellence for Safer Transfusion Working Party of the International Society of Blood Transfusion. *Transfusion* **38**, 31–40.

Hsu SH, Tang CM, Tseng HJ (2006) Biocompatibility of poly(ether)urethane-gold nanocomposites. *J Biomed Mater Res A* **79**, 759–770.

Hu WJ, Eaton JW, Ugarova TP, Tang L (2001) Molecular basis of biomaterial-mediated foreign body reactions. *Blood* **98**, 1231–1238.

Israels SJ, McMillan-Ward EM (2007) Platelet tetraspanin complexes and their association with lipid rafts. *Thromb Haemost* **98**, 1081–1087.

James ML, Warner DS, Laskowitz DT (2008) Preclinical models of intracerebral hemorrhage: a translational perspective. *Neurocrit Care* **9**, 139–152.

Jennings LK, McCabe White M (2007) Platelet aggregation. In *Platelets*, Michelson AD (ed.). Elsevier: New York; 495–507.

Jovanovic AV, Flint JA, Varshney M, Morey TE, Dennis DM, Duran RS (2006) Surface modification of silica core-shell nanocapsules: biomedical implications. *Biomacromolecules* **7**, 945–949.

Juliano RL, Hsu MJ, Peterson D, Regen SL, Singh A (1983) Interactions of conventional or pho-topolymerized liposomes with platelets in vitro. *Exp Cell Res* **146**, 422–427.

Kaushansky K (2005) The molecular mechanisms that control thrombopoiesis. *J Clin Invest* **115**, 3339–3347.

Kim Y, Hechler B, Klutz AM, Gachet C, Jacobson KA (2008) Toward multivalent signaling across G protein-coupled receptors from poly(amidoamine) dendrimers. *Bioconjug Chem* **19**, 406–411.

Koziara JM, Oh JJ, Akers WS, Ferraris SP, Mumper RJ (2005) Blood compatibility of cetyl alcohol/polysorbate-based nanoparticles. *Pharm Res* **22**, 1821–1828.

Kunicki TJ, Tuccelli M, Becker GA, Aster RH (1975) A study of variables affecting the quality of platelets stored at "room temperature". *Transfusion* **15**, 414–421.

Lin JC, Wu CH (1999) Surface characterization and platelet adhesion studies on polyurethane surface immobilized with C60. *Biomaterials* **20**, 1613–1620.

Liu Y, Miyoshi H, Nakamura M (2007) Nanomedicine for drug delivery and imaging: a promising avenue for cancer therapy and diagnosis using targeted functional nanoparticles. *Int J Cancer* **120**, 2527–2537.

Lohbach C, Neumann D, Lehr CM, Lamprecht A (2006) Human vascular endothelial cells in primary cell culture for the evaluation of nanoparticle bioadhesion. *J Nanosci Nanotechnol* **6**, 3303–3309.

Lomas-Niera JL, Perl M, Chung CS, Ayala A (2005) Shock and hemorrhage: an overview of animal models. *Shock* **24** (Suppl. 1), 33–39.

Loughrey HC, Bally MB, Reinish LW, Cullis PR (1990) The binding of phosphatidylglycerol liposomes to rat platelets is mediated by complement. *Thromb Haemost* **64**, 172–176.

Macaulay IC, Carr P, Gusnanto A, Ouwehand WH, Fitzgerald D, Watkins NA (2005) Platelet genomics and proteomics in human health and disease. *J Clin Invest* **115**, 3370–3377.

Majde JA (2003) Animal models for hemorrhage and resuscitation research. *J Trauma* **54**, S100–105.

Male R, Vannier WE, Baldeschwieler JD (1992) Phagocytosis of liposomes by human platelets. *Proc Natl Acad Sci USA* **89**, 9191–9195.

Massa TM, McClung WG, Yang ML, Ho JY, Brash JL, Santerre JP (2007) Fibrinogen adsorption and platelet lysis characterization of fluorinated surface-modified polyetherurethanes. *J Biomed Mater Res A* **81**, 178–185.

Matsuda N, Hattori Y (2006) Systemic inflammatory response syndrome (SIRS): molecular patho-physiology and gene therapy. *J Pharmacol Sci* **101**, 189–198.

Matute-Bello G, Frevert CW, Martin TR (2008) Animal models of acute lung injury. *Am J Physiol Lung Cell Mol Physiol* **295**, L379–399.

Maurer-Spurej E, Chipperfield K (2007) Past and future approaches to assess the quality of platelets for transfusion. *Transfus Med Rev* **21**, 295–306.

May AE, Seizer P, Gawaz M (2008) Platelets: inflammatory firebugs of vascular walls. *Arterioscler Thromb Vasc Biol* **28**, s5–10.

McGuigan AP, Sefton MV (2007) The influence of biomaterials on endothelial cell thrombogenicity. *Biomaterials* **28**, 2547–2571.

Medzhitov R (2008) Origin and physiological roles of inflammation. *Nature* **454**, 428–435.

Michelson AD (2006) Evaluation of platelet function by flow cytometry. *Pathophysiol Haemost Thromb* **35**, 67–82.

Michelson AD (ed.) (2007) *Platelets*. Elsevier Inc.: New York.

Michelson AD, Frelinger AL 3rd, Furman MI (2006) Current options in platelet function testing. *Am J Cardiol* **98**, 4N–10N.

Michelson AD, Linden MD, Barnard MR, Furman MI, Frelinger AL 3rd (2007) Flow cytometry. In *Platelets*, Michelson AD (ed.). Elsevier: New York; 545–563.

Minetti M, Agati L, Malorni W (2007) The microenvironment can shift erythrocytes from a friendly to a harmful behavior: pathogenetic implications for vascular diseases. *Cardiovasc Res* **75**, 21–28.

Miyamoto M, Sasakawa S, Ozawa T, Kawaguchi H, Ohtsuka Y (1989) Platelet aggregation induced by latex particles. I. Effects of size, surface potential and hydrophobicity of particles. *Biomaterials* **10**, 251–257.

Moghimi SM, Hunter AC, Murray JC (2005) Nanomedicine: current status and future prospects. *Faseb J* **19**, 311–330.

Morrow KJ Jr, Bawa R, Wei C (2007) Recent advances in basic and clinical nanomedicine. *Med Clin North Am* **91**, 805–843.

Murugesan S, Park TJ, Yang H, Mousa S, Linhardt RJ (2006) Blood compatible carbon nanotubes – nano-based neoproteoglycans. *Langmuir* **22**, 3461–3463.

Nathan C (2002) Points of control in inflammation. *Nature* **420**, 846–852.

Nathan C (2006) Neutrophils and immunity: challenges and opportunities. *Nat Rev Immunol* **6**, 173–182.

Nie S, Xing Y, Kim GJ, Simons JW (2007) Nanotechnology applications in cancer. *Annu Rev Biomed Eng* **9**, 257–288.

Nilsson B, Ekdahl KN, Mollnes TE, Lambris JD (2007) The role of complement in biomaterial-induced inflammation. *Mol Immunol* **44**, 82–94.

Niwa Y, Iwai N (2007) Nanomaterials induce oxidized low-density lipoprotein cellular uptake in macrophages and platelet aggregation. *Circ J* **71**, 437–444.

Nohl H, Stolze K (1998) The effects of xenobiotics on erythrocytes. *Gen Pharmacol* **31**, 343–347.

Owens, DE 3rd, Peppas NA (2006) Opsonization, biodistribution, and pharmacokinetics of polymeric nanoparticles. *Int J Pharm* **307**, 93–102.

Oyewumi MO, Yokel RA, Jay M, Coakley T, Mumper RJ (2004) Comparison of cell uptake, biodistribution and tumor retention of folate-coated and PEG-coated gadolinium nanoparticles in tumor-bearing mice. *J Control Release* **95**, 613–626.

Ozeki Y, Sudo T, Toga K, Nagamura Y, Ito H, Ogawa T, Kimura Y (2001) Characterization of whole blood aggregation with a new type of aggregometer by a screen filtration pressure method. *Thromb Res* **101**, 65–72.

Parker SJ, Watkins PE (2001) Experimental models of gram-negative sepsis. *Br J Surg* **88**, 22–30.

Patel DN, Bailey SR (2007) Nanotechnology in cardiovascular medicine. *Catheter Cardiovasc Interv* **69**, 643–654.

Peters K, Unger RE, Gatti AM, Sabbioni E, Tsaryk R, Kirkpatrick CJ (2007) Metallic nanoparticles exhibit paradoxical effects on oxidative stress and pro-inflammatory response in endothelial cells in vitro. *Int J Immunopathol Pharmacol* **20**, 685–695.

Pober JS, Sessa WC (2007) Evolving functions of endothelial cells in inflammation. *Nat Rev Immunol* **7**, 803–815.

Pope-Harman A, Cheng MM, Robertson F, Sakamoto J, Ferrari M (2007) Biomedical nanotechnology for cancer. *Med Clin North Am* **91**, 899–927.

Radomski A, Jurasz P, Alonso-Escolano D, Drews M, Morandi M, Malinski T, Radomski MW (2005) Nanoparticle-induced platelet aggregation and vascular thrombosis. *Br J Pharmacol* **146**, 882–893.

Reddy ST, Van Der Vlies AJ, Simeoni E, Angeli V, Randolph GJ, O'Neil CP, Lee LK, Swartz MA, Hubbell JA (2007) Exploiting lymphatic transport and complement activation in nanoparticle vaccines. *Nat Biotechnol* **25**, 1159–1164.

Reinish LW, Bally MB, Loughrey HC, Cullis PR (1988) Interactions of liposomes and platelets. *Thromb Haemost* **60**, 518–523.

Rumbaut RE, Slaff DW, Burns AR (2005) Microvascular thrombosis models in venules and arterioles in vivo. *Microcirculation* **12**, 259–274.

Sakakibara M, Goto S, Eto K, Tamura N, Isshiki T, Handa S (2002) Application of ex vivo flow chamber system for assessment of stent thrombosis. *Arterioscler Thromb Vasc Biol* **22**, 1360–1364.

Scharte M, Fink MP (2003) Red blood cell physiology in critical illness. *Crit Care Med* **31**, S651–657.

Schmaier AH, McCrae KR (2007) The plasma kallikrein-kinin system: its evolution from contact activation. *J Thromb Haemost* **5**, 2323–2329.

Sefton MV (2001) Perspective on hemocompatibility testing. *J Biomed Mater Res* **55**, 445–446.

Sefton MV, Gemmell CH, Gorbet MB (2000) What really is blood compatibility? *J Biomater Sci Polym Ed* **11**, 1165–1182.

Sefton MV, Sawyer A, Gorbet M, Black JP, Cheng E, Gemmell C, Pottinger-Cooper E (2001) Does surface chemistry affect thrombogenicity of surface modified polymers? *J Biomed Mater Res* **55**, 447–459.

Semberova J, Holada K, Simakova O, Gelderman-Fuhrmann M, Simak J (2008) Carbon nanotubes activate platelets by facilitating extracellular Ca2+ influx. *Blood* **112**, 365–366 abs.

Serhan CN, Chiang N, Van Dyke TE (2008) Resolving inflammation: dual anti-inflammatory and pro-resolution lipid mediators. *Nat Rev Immunol* **8**, 349–361.

Seyfert UT, Biehl V, Schenk J (2002) In vitro hemocompatibility testing of biomaterials according to the ISO 10993-4. *Biomol Eng* **19**, 91–96.

Shen YM, Frenkel EP (2007) Acquired platelet dysfunction. *Hematol Oncol Clin North Am* **21**, 647–661, vi.

Shin SH, Ye MK, Kim HS, Kang HS (2007) The effects of nano-silver on the proliferation and cytokine expression by peripheral blood mononuclear cells. *Int Immunopharmacol* **7**, 1813–1818.

Simak J, Gelderman MP (2006) Cell membrane microparticles in blood and blood products: potentially pathogenic agents and diagnostic markers. *Transfus Med Rev* **20**, 1–26.

Spijker HT, Graaff R, Boonstra PW, Busscher HJ, van Oeveren W (2003) On the influence of flow conditions and wettability on blood material interactions. *Biomaterials* **24**, 4717–4727.

Stasko NA, Johnson CB, Schoenfisch MH, Johnson TA, Holmuhamedov EL (2007) Cytotoxicity of polypropylenimine dendrimer conjugates on cultured endothelial cells. *Biomacromolecules* **8**, 3853–3859.

Szebeni J, Alving CR, Rosivall L, Bunger R, Baranyi L, Bedocs P, Toth M, Barenholz Y (2007) Animal models of complement-mediated hypersensitivity reactions to liposomes and other lipid-based nanoparticles. *J Liposome Res* **17**, 107–117.

Tabuchi A, Kuebler WM (2008) Endothelium-platelet interactions in inflammatory lung disease. *Vascul Pharmacol* **49**, 141–150.

Toda M, Kitazawa T, Hirata I, Hirano Y, Iwata H (2008) Complement activation on surfaces carrying amino groups. *Biomaterials* **29**, 407–417.

van Eeden SF, Hogg JC (2002) Systemic inflammatory response induced by particulate matter air pollution: the importance of bone-marrow stimulation. *J Toxicol Environ Health A* **65**, 1597–1613.

Varga-Szabo D, Pleines I, Nieswandt B (2008) Cell adhesion mechanisms in platelets. *Arterioscler Thromb Vasc Biol* **28**, 403–412.

Verbeuren TJ (2006) Experimental models of thrombosis and atherosclerosis. *Therapie* **61**, 379–387.

Wei C, Wei W, Morris M, Kondo E, Gorbounov M, Tomalia DA (2007) Nanomedicine and drug delivery. *Med Clin North Am* **91**, 863–870.

Williams DF (2008) On the mechanisms of biocompatibility. *Biomaterials* **29**, 2941–2953.

Yamawaki H, Iwai N (2006) Cytotoxicity of water-soluble fullerene in vascular endothelial cells. *Am J Physiol Cell Physiol* **290**, C1495–1502.

Yim EK, Liao IC, Leong KW (2007) Tissue compatibility of interfacial polyelectrolyte complexation fibrous scaffold: evaluation of blood compatibility and biocompatibility. *Tissue Eng* **13**, 423–433.

Zbinden G, Wunderli-Allenspach H, Grimm L (1989) Assessment of thrombogenic potential of liposomes. *Toxicology* **54**, 273–280.

Zhang Y, Guo S (2007) The mechanical biocompatibility of nanoparticles. In *Nanotoxicology*, Zhao Y, Nalwa HS (eds). American Scientific Publishers: Stevenson Ranch; 201–248.

Zhu J, Xue J, Guo Z, Zhang L, Marchant RE (2007) Biomimetic glycoliposomes as nanocarriers for targeting P-selectin on activated platelets. *Bioconjug Chem* **18**, 1366–1369.

Zimmerman GA, Weyrich AS (2008) Signal-dependent protein synthesis by activated platelets: new pathways to altered phenotype and function. *Arterioscler Thromb Vasc Biol* **28**, s17–24.

Zuo L, Wei W, Morris M, Wei J, Gorbounov M, Wei C (2007) New technology and clinical applications of nanomedicine. *Med Clin North Am* **91**, 845–862.

12

Sources, Fate and Effects of Engineered Nanomaterials in the Aquatic Environment

David S. Barber, Nancy D. Denslow, R. Joseph Griffitt and Christopher J. Martyniuk

12.1 Introduction

As nanomaterials become a more common component of many industries in the high-tech sector, significant concern has been raised about the potential for these materials to produce adverse environmental effects (Moore, 2006). As with many chemicals, a likely sink for nanomaterials released into the environment will be aquatic systems. In this chapter, we focus specifically on engineered nanomaterials (ENMs) which are intentionally prepared materials having at least one dimension less than 100 nm. ENMs fall into the definition of colloids which are defined by IUPAC (1997) as materials with at least one dimension between 1 nm and 1 μm. There are many other colloids that are found in aquatic systems including natural and anthropogenic substances. Natural colloids include inorganic silicates, metal oxides and hydroxides, carbonates, phosphates and sulfides, as well as natural organic materials such as humic and fulvic acids. Incidental nanomaterials are released as byproducts of human activity, and include wear and corrosion products as well as combustion byproducts. Background levels of nanomaterials can vary widely, but typical values for surface waters range from mass concentrations of 10–500 μg/l and number concentrations of 10^6–10^7 particles/ml, with the highest numbers of particles having diameters below 100 nm (Rosse and Loizeau, 2003). As we consider the issue of nanomaterials in aquatic environments, it is important to remember that the presence of nanoparticles in aqueous systems is not new. However, the presence of ENMs prepared from potentially toxic materials, designed to be highly reactive and form stable dispersions, is novel.

Nanotoxicity: From In Vivo *and* In Vitro *Models to Health Risks*   Edited by Saura Sahu and Daniel Casciano
© 2009 John Wiley & Sons, Ltd

We will examine how ENMs may enter the aquatic environment, the fate of ENMs in the aquatic environment, and the interaction of ENMs with aquatic biota.

12.2   Entry of ENMs into the Aquatic Environment

The Project on Emerging Nanotechnologies at the Woodrow Wilson International Center for Scholars maintains an up-to-date listing of nanoparticle-containing consumer products. As of August 2008, there were 803 products from 21 countries on the list, the majority of which (502) were for health and fitness applications, including cosmetics (126) sunscreens (33), clothing (115) and personal care products (153). The database also lists 80 different products for use with food or beverages, and 56 in consumer electronics applications. While worldwide production of nanomaterials is not accurately known, estimates range from 500 metric tons per year for silver nanoparticles and carbon nanotubes, to 5000 metric tons per year for titanium dioxide particles (Muller and Nowack, 2008). This value is expected to rise in coming years, clearly providing potential environmental releases.

As indicated in Figure 12.1, nanomaterials can be released into the environment during manufacture, use or disposal of ENM-containing products. Releases during the initial nanoparticle (NP) production process has been demonstrated for carbon nanotubes. Maynard *et al.* (2004) observed that $<53\ \mu g/m^3$ single-walled carbon nanotubes (SWCNT) were released from a production facility into the surrounding air. Yeganeh *et al.* (2008) reported low levels of carbonaceous nanomaterials in a manufacturing facility, but found that handling of the particles inside a fume hood readily aerosolized carbonaceous NP,

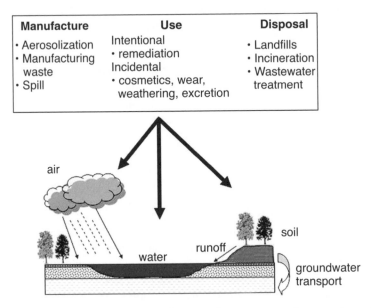

Figure 12.1 *Possible sources of ENM release and transport to the aquatic environment. ENMs may be released during manufacture, use, or disposal of products. Release may occur into air, water, or soil; however, deposition and transport processes can move ENM from air and soil into water.*

suggesting that stringent containment protocols during NP manufacturing are needed to prevent environmental release.

Nanomaterials may also be released intentionally or incidentally into the environment during use. Zero-valent iron nanoparticles degrade certain organic contaminants and are released directly into groundwater to assist in remediation efforts (Cundy *et al.*, 2008). The use of cosmetic products containing ENMs, such as sunscreens and skincare products, will release ENMs into the environment as they are washed off skin. In most cases, however, ENMs will be released by abrasion and weathering of ENM-containing products. Many nanotechnologies involve the use of composites in which nanomaterials, such as carbon nanotubes, are embedded in a matrix which limits release of nanomaterials, although frictional wear over the lifespan of these components can release the embedded ENMs. ENMs are increasingly used as fine powders and thin films that may release nanoparticles into the environment. Benn and Westerhoff (2008) demonstrated that silver NP contained in personal wear could be released into washing machine wastewater. Kaegi *et al.* (2008) showed that weathering of nanotitania-containing paints could release significant numbers of nanoparticulate TiO_2 into nearby streams through runoff following rain events. The use of ENMs in biomedical applications, such as targeted drug delivery and imaging, can be a source of environmental release as these particles are excreted, entering wastewater streams. Chen *et al.* (2008) found that in rats exposed to quantum dots, 57 % of the initial dose was released within 120 hours, with 33.3 % in the feces and 23.8 % in urine.

The final disposal method for products containing ENMs is likely to be either burial in a landfill or incineration, though some nanomaterials will also enter wastewater streams. For landfilled material, release into aquatic ecosystems will be a function of the long-term resilience of the component and the structural integrity of the landfills. Incineration is likely to release at least a portion of ENMs into the atmosphere, although the physical properties of those particles may be altered. For particles that enter wastewater, release will be a function of the efficiency of the wastewater treatment process in removing the particles. Recent work suggests that many nanoparticles are efficiently captured by adhesion to sludge, but capture efficiency can be markedly affected by particle charge and presence of surfactants (Limbach *et al.*, 2008).

Overall, the potential environmental risk from nanoparticle-containing products is largely determined by the product lifecycle. While these processes vary considerably across the spectrum of ENMs, several recent studies have estimated the potential impact of nanomaterial release into the environment (Blaser *et al.*, 2008; Muller and Nowack, 2008). Mueller and Nowack (2008) suggest that carbon nanotubes are not likely to pose an ecological threat, based on comparing environmental concentrations due to releases with those reported to cause toxicity. However, studies by Blaser *et al.* (2008) estimate that the use of nano-silver in biocidal applications could account for up to 15 % of the total silver released into water in Europe in 2010, reaching levels that could have adverse effects on freshwater organisms.

12.3   Fate of Engineered Nanomaterials in the Aquatic Environment

Once released, ENMs can enter the aquatic environment via airborne deposition, direct addition to water, terrestrial runoff and groundwater transport (Figure 12.1). Deposition and direct introduction to water will clearly contribute to loading of aquatic environments. The role of transport through soil is less clear. Intuitively, nanomaterials should be transported readily through soil due to their small size relative to pores; however, studies demonstrate that mobility in soils varies considerably (Lecoanet and Wiesner, 2004; Lecoanet et al., 2004). Studies have shown that typical groundwater chemistry promotes retention of iron oxide, TiO_2 and fullerene nanomaterials by soils (Guzman et al., 2006; Espinasse et al., 2007; Saleh et al., 2008a), suggesting that many nanomaterials may have limited mobility in soils.

12.3.1   Factors Influencing Nanomaterial Aggregation

Once in the aquatic environment, monodispersed nanomaterials will tend to remain suspended nearly indefinitely if the suspension is stable, giving nanomaterials the potential to become widely dispersed and persist for long durations in aquatic environments. Therefore, the fate of nanomaterials in the aquatic environment is determined largely by their rate of aggregation, transformation and dissolution (see Figure 12.2). ENMs may aggregate with themselves or with other colloids. The rate of aggregation will be affected by a variety of factors including ionic strength, presence of natural organic matter, and particle concentration. Aggregation is driven by the balance of attractive and repulsive forces between particles. While beyond the scope of this chapter, these forces are mathematically described by the extended Derjaguin-Landau-Verwey-Overbeck (DLVO) theory which has

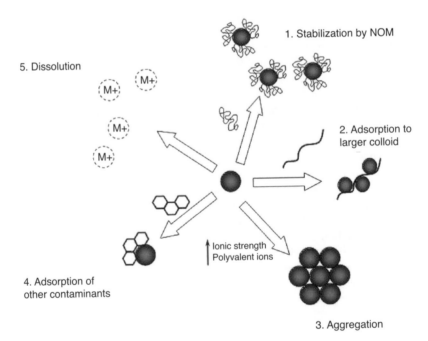

Figure 12.2 *Potential transformation of ENMs in aquatic environments. Once suspended in water, ENMs are likely to be modified by aggregation, adsorption and dissolution. The impact of each pathway will be influenced strongly by water chemistry, especially the identity and concentration of natural organic matter (NOM), pH, and ionic composition.*

been shown to be predictive for a number of nanomaterials (Kim *et al.*, 2008). However, it is uncertain if the DLVO theory will predict behavior of particles <20 nm (Kallay and Zalac, 2002) or of non-spherical particles, particles with unevenly distributed surface charges or roughness (Elimelech *et al.*, 1995; Li Y *et al.*, 2008). Additional information on models of particle interaction can be found in several recent reviews (Handy *et al.*, 2008a; Klaine *et al.*, 2008a).

Major factors that affect the rate of aggregation are ionic strength and composition of the receiving waters and the concentration of natural organic matter (see Figure 12.2). Particles are surrounded by a layer of charge at the surface and an electric field induced by the charged surface (diffuse layer) referred to as an electrical double layer (EDL). The EDL tends to prevent collision of similar particles by means of charge repulsion. However, addition of salts to the solution tends to screen surface charges and shrink (or collapse) the EDL. This allows particles to come into closer proximity to each other and increases the effect of attractive forces (e.g. van der Waal's forces), promoting particle aggregation. The higher the salt concentration, the more charge shielding occurs and the smaller the EDL. Di- and trivalent ions are very effective at charge shielding and typically have a large effect on particle aggregation. The effect of ionic strength on stability of nanoparticle suspensions has been demonstrated for a variety of nanoparticles including multi-walled carbon nanotubes (MWCNT) (Saleh *et al.*, 2008b), fullerenes (Brant *et al.*, 2005), and metallic nanoparticles (Tiraferri *et al.*, 2008). The importance of ionic composition on aggregation is evident in work by Zhang *et al.* (2008), in which monovalent ions had little effect on the stability of thioglycolate capped quantum dots, but even low concentrations of di- and trivalent ions caused aggregation. Based on these studies, most nanoparticles will aggregate rapidly in seawater.

The interactions of natural organic matter (NOM) with nanoparticles is relatively complex and the result of the interaction depends on nanoparticle composition, type of organic matter, and pH. A number of studies have demonstrated that organic matter (e.g. Suwannee River Humic Acid, SRHA) can stabilize nanoparticle suspensions (Chen and Elimelech, 2008; Diegoli *et al.*, 2008; Ghosh *et al.*, 2008). The exact mechanism by which stabilization occurs is unclear, but probably involves adsorption of organic matter to particle surfaces resulting in steric and electrostatic repulsion. However, other studies have demonstrated the opposite effect, where NOM increases the rate of aggregation of iron oxide nanoparticles (Baalousha *et al.*, 2008). In all cases, the effects are dependent upon NOM concentration, pH and ionic strength.

Another driving factor in aggregation is particle concentration. As particle number increases, the likelihood of collision and subsequent aggregation also increases. This has implications for aquatic toxicity tests, where wide ranges of particle concentrations may be used, leading to differences in particle size distributions and unexpected concentration changes. It is critical that careful characterization be conducted under actual exposure conditions to interpret results of exposures accurately. Additional information on nanomaterial characterization methods and strategies is available (Powers *et al.*, 2006).

Aggregation will have a significant impact on the distribution of ENMs in the aquatic system (Handy *et al.*, 2008a). Monodispersed nanoparticles and small clusters will remain within the water column and be available to pelagic organisms. However, aggregates will tend to be removed from the water column by sedimentation, increasing the concentration of nanomaterials in sediment and increasing exposure of benthic organisms. Due to the tendency toward aggregation, sediment is likely to be the ultimate sink for most nanomaterials in the aquatic environment. To date, relatively few studies have been conducted on benthic organisms, although there is a clear need for this information. Due to the processes involved in sedimenting nanomaterials, benthic organisms will be primarily exposed to aggregates, whose characteristics may differ substantially from those of the nanomaterial initially introduced into the environment. Again, careful thought regarding exposure scenarios and organism biology should go into experimental design of these types of toxicity tests.

12.3.2   Environmental Transformation of Nanomaterials

The role of environmental transformation on the fate of nanomaterials is unclear. Oxidation of many organic compounds by UV light and ozone plays an important role in their environmental fate. While photocatalytic reactions of fullerenes have been widely studied (Hotze *et al.*, 2008), the role of photodegradation of fullerenes is unclear. Studies do demonstrate the potential for dissolved ozone to oxidize fullerene suspensions, possibly leading to increased solubility and a spectrum of oxidation products that must be considered (Fortner *et al.*, 2007). Ozone also rapidly oxidizes the sidewalls of SWCNTs (Simmons *et al.*, 2006). Biochemical oxidation of organic molecules is also an important factor in determining their environmental fate. The ability of these processes to act on carbonaceous nanomaterials remains unclear, although studies have demonstrated that chemical models of cytochrome P450 can oxidize fullerenes (Hamano *et al.*, 1995). Other work has demonstrated enzymatic degradation of SWCNTs using horseradish peroxidase (Allen *et al.*, 2008). Because transformation of these nanomaterials is likely to modify their environmental fate and biological effects, more research is needed in this area.

Another variable that must be considered when examining the fate of nanomaterials is dissolution. While dissolution is negligible for some materials (e.g. fullerenes, nanotubes), significant dissolution of some metallic particles can occur (Franklin *et al.*, 2007; Griffitt *et al.*, 2007). Studies demonstrate that ionic species of metals are released from many metal and metal oxide nanoparticles as well as quantum dots. Ionic forms of many of these metals are well-described toxicants in aquatic systems and may impact upon the interpretation of results. In some cases, essentially all of the toxicity observed following exposure to zinc nanoparticles can be ascribed to release of soluble ions (Franklin *et al.*, 2007). Because exposures conducted with nanomaterials that dissolve will contain both soluble and particulate forms of the material, it can be very difficult to separate the effects caused by each form. However, by measuring the release of soluble species from nanoparticulates during exposures and comparing these levels to the concentration response of soluble forms of the material, it is possible to ascertain the role that dissolution plays in toxic responses elicited by nanoparticulate exposure (Griffitt *et al.*, 2008).

Due to their high surface area, nanoparticles can adsorb other contaminants onto their surface which can alter the bioavailability and fate of the adsorbed contaminant. This has been demonstrated in carp exposed to TiO_2 and arsenic (Sun *et al.*, 2008) or cadmium (Zhang *et al.*, 2007). Studies with benthic invertebrates showed that incorporation of SWCNTs into sediment influenced the bioavailability of a variety of hydrophobic organic compounds (Ferguson *et al.*, 2008). The effect of adsorption on availability of chemicals appears to vary considerably with chemical and nanomaterial, but this possibility must be considered in evaluating the environmental effects of nanomaterials.

The environmental fate of nanomaterials can be further impacted by their interactions with biota. Studies have demonstrated that polysaccharides, such as those produced by fungi and bacteria, can dramatically increase aggregation of nanomaterials (Espinasse *et al.*, 2007). Work by Roberts *et al.* (2007) demonstrated that carbon nanotubes coated with lysophosphatidylcholine were readily ingested by daphnia, which were able to digest the lipid coating, resulting in changes in water solubility of the nanomaterial.

Aqueous suspensions of nanomaterials are a dynamic system in which the characteristics of nanomaterials are affected by water chemistry, particle number, light, biota and the presence of other chemical contaminants. A significant challenge for ecotoxicologists working with nanomaterials is to design experiments that allow interpretation of results but maintain relevance to natural systems. The need to produce stable dispersions to understand the effects of particle properties such as size, must be balanced with the recognition that results obtained with monodispersed particles coated with a surfactant may not be indicative of what a particle will do in natural water when partially aggregated and coated with NOM.

12.4 Biotic Fate and Transport in Aquatic Organisms

Interaction of nanomaterials with aquatic biota will be a function of the physical properties of the particle as well as the biology of the target organism. Particle properties such as size, aggregation state, charge and composition may all affect the uptake, distribution and elimination of nanomaterials by aquatic biota. It is likely that factors such as physiology (e.g. osmoregulation, reproductive state), genetics, sex and age will impact the kinetics of NPs as they do for other aquatic contaminants (Duffy *et al.*, 2002; McClennan-Green *et al.*, 2007). The possible routes of exposure and mechanisms of uptake in vertebrates, invertebrates and unicellular organisms differ substantially. As indicated above, there may be differences in the characteristics of nanomaterials to which pelagic organisms are exposed compared with benthic organisms, due to aggregation and interaction with organic matter. This section summarizes our current knowledge on uptake, cellular and tissue distribution, and mechanisms of elimination for ENMs in aquatic organisms.

12.4.1 Uptake Mechanisms

For most macroscopic aquatic organisms, the two major routes of nanomaterial uptake are (1) across the gill epithelium and (2) across the epithelium of the digestive tract or the hepatopancreas, the site of nutrient absorption and storage in invertebrates. Several studies have demonstrated uptake of nanomaterials by the gill. Fluorescent latex nanoparticles (~39 nm in diameter) accumulated in the gills of see-through medaka (Kashiwada, 2006) exposed to 10 mg/l concentrations of particles. Similarly, carp exposed to 0.160 g of nano-TiO_2 in 16-litre tanks over a 25-day period accumulated up to 0.74 µg/g of TiO_2 in the gills (Zhang *et al.*, 2006). In blue mussel exposed to nanoparticles of glass wool (silicon dioxide), uptake of nanoparticles into gill was observed within 12 hours by electron microscopy (Koehler *et al.*, 2008). Zebrafish exposed to silver nanoparticles also appear to accumulate silver nanoparticles (Griffitt *et al.*, 2009). The mechanism by which nanoparticles are taken up by the gill epithelium is not clear. It is likely that the first step is adsorption of the nanoparticle to the mucus layer and its translocation to the epithelial surface

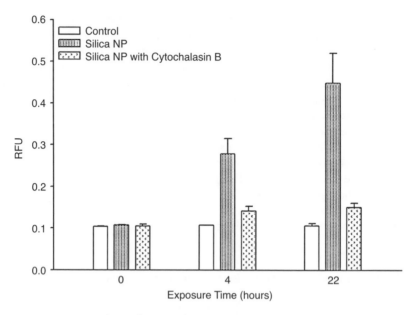

Figure 12.3   *Nanoparticles undergo endocytosis in fish gill cells. Uptake of 40 nm fluorescent silica nanoparticles by RTgill-W1 rainbow trout gill cells in the presence and absence of the endocytosis inhibitor, cytochalasin-B. Uptake was conducted in serum-free L-15 media. When used, cytochalasin was added at a concentration of 5 µg/ml 30 minutes prior to the addition of nanoparticles. Data is presented as relative fluorescence units (RFU) and values are mean ± SD of three independent experiments. Control cells had no fluorescent particles added, and values represent background autofluorescence.*

(Handy *et al.*, 2008b). Indeed, studies have demonstrated that SWCNTs are accumulated by gill mucus following exposure of rainbow trout (Smith *et al.*, 2007). Once in contact with the epithelial surface, particles are most likely taken up by endocytosis. This is corroborated by data in Figure 12.3, showing that the endocytosis inhibitor cytochalasin B blocks uptake of fluorescent silica nanoparticles by cultured rainbow trout gill epithelial cells. However, data from mussel suggest that some very small particles (less than 5 nm) may be able to enter gills by diffusion as they are not associated with endocytotic vesicles (Koehler *et al.*, 2008).

    The gut epithelium is another potential source of NP uptake in vertebrates and invertebrates. While most freshwater fish do not drink large amounts of water, waterborne NP can be ingested during stress-induced drinking or through inadvertent ingestion while eating. NP may also be ingested within normal prey organisms if they are taken up by lower organisms. The likelihood of nanoparticle ingestion by invertebrates may depend significantly on their feeding strategies. Filter-feeding invertebrates such as daphnia and mussels have been shown to ingest a variety of nanoparticles (Lovern and Klaper, 2006; Roberts *et al.*, 2007; Koehler *et al.*, 2008). Monodispersed nanoparticles may not be captured efficiently by these organisms, so the rate of ingestion may depend greatly on aggregation and adsorption to larger particulates. However, the effect of aggregation on ingestion probably differs with species, feeding strategy and nanomaterial, as illustrated in studies with daphnia and amphipods (Kennedy *et al.*, 2008).

The interaction of microorganisms with nanomaterials has received a great deal of attention due to biocidal activities of some nanomaterials. However, the ability of various types of microorganisms to internalize nanomaterials has been less studied. Bacteria do not have the capacity for endocytosis, and uptake of nanoparticles will require passage through specific transport mechanisms or through damaged cell walls. Studies indicate that particles <5 nm can indeed enter bacteria, although successful labelling depends on conjugation of nanomaterials with specific ligands such as adenine (Kloepfer *et al.*, 2005; Hirschey *et al.*, 2006). Nanomaterials clearly have the capacity to cause toxicity to algal cells (Griffitt *et al.*, 2008; Van Hoecke *et al.*, 2008; Aruoja *et al.*, 2009), but it is unclear whether toxicity is induced following internalization of particulates, or is due to release of soluble ions or extracellular generation of oxidative species. At least one study demonstrates that nanomaterials adhere to algae but are not internalized (Van Hoecke *et al.*, 2008). In any case, uptake or adhesion of nanomaterials to microorganisms may serve as a means of nanomaterial concentration and promote trophic transfer to higher organisms.

12.4.2   Diet and Trophic Transfer of ENMs

Ingestion of NPs is a route of uptake, and trophic transfer may contribute to the accumulation of NPs in the organism. In general, there are three potential outcomes of exposures through trophic transfer that depends upon the properties of the contaminant within the organism, the rate of uptake and elimination, and the position of the organism in the aquatic food chain. These include biomagnification, biominification (trophic dilution), or similar levels within prey and predator. Using a simplified food web, Holbrook *et al.* (2008) studied the trophic transfer of quantum dots (QD) between *Escherichia coli*, a ciliate *Tetrahymena pyriformes*, and a rotifer *Brachionus calyciflorus*. Using transmission electron microscopy, QD aggregates were observed to be attached to the outer membrane of *E. coli* but did not enter the organism. In contrast, ciliates accumulated both carboxylated and biotinylated QD from the water column. Trophic transfer of QD via ingestion of ciliates by rotifers was observed. After ingestion of prey, QD in the rotifers migrated from the digestive gut to the body cavity. However, biomagnification was not observed in this food chain due to rapid depuration of QD in rotifers. In another simplified food web, Bouldin *et al.* (2008) exposed freshwater algae to carboxylated QD for 96 hours and observed significant QD assimilation in algae. *Ceriodaphnia dubia*, allowed to feed on algae labelled with 55 ppb QD, had higher levels of fluorescence than those feeding on algae not labelled with QD. These studies provide evidence that the transfer of NPs from prey to predator can occur in food webs.

Further studies are needed to study the impacts of NPs in an aquatic food chain to determine whether this is a significant source of exposure for larger aquatic vertebrates such as predatory fish and marine mammals. To date, there are no studies demonstrating trophic transfer in higher-order predators. Additional studies are also needed to determine whether assimilation of NPs via food webs is a consideration for wildlife and human risk assessment decisions.

12.4.3   Tissue and Cellular Distribution

NPs entering the organism have the potential to be distributed to a number of tissues in aquatic organisms. In medaka exposed to 10 mg/l of non-ionized fluorescent latex particles for 7 days, nanomaterials accumulated in the blood, gills, intestine and gall bladder of adults (Kashiwada, 2006). However, brain, liver, kidney and testis also exhibited fluorescence and it is likely that NPs were absorbed and distributed through the blood to the major organs. In the same study, fertilized eggs incubated with 1 mg/l of non-ionized 39 nm fluorescent polystyrene microspheres for 24 hours showed deposition of nanoparticles in the chorion and oil droplets of the eggs, with lower fluorescence in the yolk. In embryos exposed to nanoparticles for three days and then placed in clean water until hatching, nanoparticles appeared to concentrate in the gall bladder and yolk sac of larval fish. Other studies demonstrate that exposure of medaka to selenium as selenium oxide nanoparticles results in higher levels of selenium in the liver, gills and muscle than in fish exposed to sodium selenite (Li H *et al.*, 2008). Zhang *et al.* (2006) reported bioaccumulation of nanosized TiO_2 in carp, with most material being present in the gills and viscera.

Like epithelial cells, other cell types appear to internalize nanomaterials primarily by endocytosis (Shi Kam *et al.*, 2004; Moore, 2006). NPs may enter via caveolae-mediated endocytosis or through fluid phase endocytosis (non-receptor mediated). In the cell, NPs appear to translocate from the plasma membrane to membrane-bound organelles such as the endoplasmic reticulum, golgi apparatus and lysosomes. In mammals, C60 fullerene preferentially localizes in cellular compartments that include the mitochondria, cytoplasm and nuclei (Porter *et al.*, 2006, 2007). However, comparatively fewer studies on cell and tissue distribution of NPs have been done in aquatic organisms. Canesi *et al.* (2008) performed an *in vitro* assay and incubated nanosized carbon black (Printex 90) at different concentrations (1, 5, 10 µg/ml) for 60 minutes in the mussel (*Mytilus galloprovincialis* LAM). Hemocytes, cells that play a role in the immune system of arthropods, were isolated from hemolymph. NPs appeared to be taken up by hemocytes via endocytosis, and NP aggregates were observed adhering to the cell surface.

Detection of nanomaterials in cells and tissues remains a significant challenge. Methods such as ICP-MS can detect the presence of metals in a tissue, but cannot determine whether the metal was particulate or ionic. Electron microscopy can identify nanoparticles, but can be confounded by the presence of other structures of similar size. This has limited our knowledge of the tissue and cellular distribution of nanomaterials, and our understanding of how the physio-chemical properties of NPs influence their distribution in a biological system. Moger *et al.* (2008) recently outlined criteria for successful localization of metal oxide NPs using imaging techniques. These criteria include sufficient spatial resolution and tissue penetration, 3D imaging capabilities of NP distribution within the tissue, and high contrast ability to distinguish between tissue and metal oxides. Other methods for the detection of NPs in tissue include fluorescent tracking which has been used successfully, but improved techniques in imaging will facilitate studies investigating the distribution of different types of NPs in multiple biological tissues.

12.4.4   Elimination

Absorbed NPs may be unique compared to more well studied aquatic organic contaminants (e.g. pharmaceuticals, pesticides, herbicides) because NPs have not yet been shown to undergo biotransformation through phase I and phase II detoxification enzymes. Therefore the metabolism of NPs by the liver is unlikely to play a significant role in elimination. It is likely that the major routes of elimination are renal and biliary excretion. In mammals, studies clearly demonstrate the renal excretion of QD and SWCNT (Singh *et al.*, 2006; Choi *et al.*, 2007). Renal excretion is size-dependent, and particles >6 nm in diameter, including adsorbed proteins, exhibit poor renal clearance (Choi *et al.*, 2007). It is also possible that elimination may occur via deposition of material into eggs during oogenesis and spawning. *In vitro* experiments with embryos suggest that NPs are able to enter the egg (Kashiwada, 2006; Zhu X *et al.*, 2008a), but evidence is lacking to suggest NPs enter developing embryos *in vivo*. While absorption and clearance of nanoparticles has been described in aquatic organisms, to date the mechanisms of elimination are unknown.

12.5 Toxicity of Nanomaterials in Aquatic Organisms

The number of studies examining toxicity of nanomaterials in aquatic organisms has increased dramatically in the past two years. Recent reviews covering these data extensively include Moore (2006), Klaine *et al.* (2008b) and Handy *et al.* (2008a). Examining these studies reveals that various nanomaterials have the potential to cause toxicity to microorganisms, invertebrates and vertebrates (Table 12.1). They also demonstrate that significant differences in effect of a given nanomaterial can be observed (Table 12.1). Differences may be due to particle properties, method of preparation or species sensitivity. There are clearly still significant gaps in our knowledge about nanomaterial toxicity; however, we examine some of the issues surrounding toxicity of carbonaceous and metallic nanoparticles below.

12.5.1 Carbonaceous Nanomaterials (Fullerenes, Carbon Nanotubes)

Fullerenes are among the most studied of nanomaterials. The first vertebrate study showing effects of C_{60} was a study by Oberdoster (2004) who treated largemouth bass with $nC_{60(THF)}$ at 0.5 and 1 mg/l for 48 hr and found evidence of lipid peroxidation in the brain. Subsequent studies have reported that fullerenes and modified fullerenes can cause toxicity in a variety of bacteria (Mashino *et al.*, 1999, 2003; Tsao *et al.*, 2002; Lyon *et al.*, 2005), algae (Weir *et al.*, 2008), freshwater crustaceans such as *Ceriodaphnia* (Lovern and Kaper, 2006; Zhu S *et al.*, 2006; Lovern *et al.*, 2007; Roberts *et al.*, 2007; Baun *et al.*, 2008; Kennedy *et al.*, 2008), and fish (Oberdorster, 2004; Zhu S *et al.*, 2006; Zhu X *et al.*, 2007; Cheng *et al.*, 2007; Henry *et al.*, 2007; Smith *et al.*, 2007). The primary mechanism of action appears to be induction of oxidative stress, perhaps through photocatalysis (Zhu X *et al.*, 2008b; Usenko *et al.*, 2008). However, other studies report little if any toxicity for fullerenes (Blaise *et al.*, 2008; Blickley *et al.*, 2008), and there are reports of antioxidant activity and protection from radiotoxicity in zebrafish using modified fullerenes (Daroczi *et al.*, 2006).

At least part of the discrepancy surrounding fullerene toxicity may be explained by differences in methods of producing the aqueous suspensions used for testing. Some studies have used tetrahydrofuran (THF) to dissolve fullerene and then conducted solvent exchange to prepare aqueous suspensions. Other studies have prepared aqueous suspensions of fullerenes by prolonged stirring, referred to as nC60. Studies in fish and daphnia suggest that fullerenes prepared with THF are much more toxic than those prepared by prolonged stirring (Oberdorster, 2004; Zhu S *et al.*, 2006; Henry *et al.*, 2007). Lovern and Klaper (2006) tested C_{60} prepared either by the THF method or by sonication in water on *Daphnia magna*, and found that the THF preparation was toxic with an LC_{50} of 0.46 mg/l compared

Table 12.1 *Selected toxicity studies on nanomaterials in freshwater organisms.*

Nanomaterial	Organism	Endpoint (concentration producing effect)	Reference
Fullerene (prepared as aqueous suspensions of nC60)	Bacteria (*E. Coli, B. subtilis*)	Growth inhibition (0.4 mg/l)	Fortner *et al.* (2007)
	Algae (*P. subcapitata*)	48-hr 30% growth inhibition (90 mg/l)	Tsao *et al.* (2002)
	Algae (*P. subcapitata*)	72-hr growth inhibition (>100 mg/l)	Blaise *et al.* (2008)
	Invertebrate *(Daphnia magna)*	LC_{50} (>35 mg/l)	Zhu S *et al.* (2006)
	Fish (*Fundulus heteroclitus*)	LC_{50} (>10 mg/l)	Blickley *et al.* (2008)
Fullerene (prepared with THF)	Invertebrate *(Daphnia magna)*	LC_{50} (460 µg/l)	Lovern and Klaper (2006)
	Fish (*Micropterus salmoides*)	Oxidative damage (0.5 mg/l)	Oberdorster (2004)
	Fish (*Danio rerio* embryo)	Delayed development, reduced hatching (1.5 mg/l)	Zhu X *et al.* (2007)
Single walled carbon nanotubes (SWCNTs)	Invertebrate *(Daphnia magna)*	96-hr LC_{100} (20 mg/l)	Roberts *et al.* (2007)
	Invertebrate (*Amphiascus tenuiremis*)	Mortality (10 mg/l)	Templeton *et al.* (2006)
	Fish (*Onchorhyncus mykiss*)	Gill pathology (0.5 mg/l)	Smith *et al.* (2007)
	Fish (*Danio rerio* embryo)	Hatching delay (120 mg/l)	Cheng *et al.* (2007)
Nano TiO$_2$	Algae (*Desmodemus subspicatus*)	50% growth inhibition (44 mg/l)	Hund-Rinke and Simon (2006)
	Invertebrate *(Daphnia magna)*	Nontoxic (no effect at 20 g/l)	Heinlaan *et al.* (2008)
	Invertebrate *(Daphnia magna)*	Nontoxic	Lovern and Klaper (2006)
	Invertebrate (*Ceriodaphnia dubia*)	48-hr LC_{50} (>10 mg/l)	Griffitt *et al.* (2008)
	Fish (Rainbow trout)	14-day gill pathology (1 mg/l)	Federici *et al.* (2007)

with an LC_{50} of 7.0 mg/l for the stirred preparation. The reason for the differences in toxicity of these preparations may result from differences in physical properties of the C60 clusters resulting from these preparations, including size and morphology of clusters. However, subsequent work suggests that much of the toxicity of fullerenes prepared with THF may be due to the presence of THF decomposition products (Henry *et al.*, 2007).

Perhaps more important are the surface properties of fullerenes, especially those that have been derivatized with different groups, yielding surfaces that are highly charged.

Fullerenes derivatized with cationic groups, including bis(N,N-dimethylpyrrolidinium io-dide and bis(N-methylpiperazinium iodide), were able to suppress the growth of *E. coli* probably through interference of energy metabolism and respiration (Mashino *et al.*, 1999). At high concentration, these particles were observed to produce H_2O_2 (Mashino *et al.*, 2003). In comparison, fullerenes derivatized with anionic groups such as dimalonic acid had no effect (Mashino *et al.*, 1999). While these properties can be exploited as antibacterial agents, release of derivatized C_{60} into the environment may also change natural bacterial populations.

Single and multi-walled carbon nanotubes (SWCNTs and MWCNTs, respectively) have also been tested. Smith *et al.* (2007) exposed rainbow trout to a range of concentrations between 0.1 and 0.5 mg/l SWCNTs dispersed in sodium dodecyl sulfate (final concentration of 0.15 mg/l in the tank) for ten days. The main route of exposure was through the gills and intestinal tract, confirmed by their finding of precipitated SWCNTs in these organs. In the gill, there was a dose-dependent increase in ventilation rate, gill pathologies including hyperplasia and oedema, mucous secretion and Na^+/K^+ ATPase activity. They also found a dose-dependent decrease in lipid peroxidation in the gill, brain and liver, and a dose-dependent increase in GSH in gills and liver. In another study, SWCNT toxicity was tested on zebrafish embryos from 4 to 96 hpf (Cheng *et al.*, 2007). At 120 mg/l a delay in hatching was observed. Double-walled nanotubes appeared less toxic, requiring a concentration of 240 mg/l to produce hatching delay, and carbon black in the same study caused no hatching delay.

MWCNTs that were prepared by sonication for 5 hr in water were tested on the unicellular protozoan *Stylonychia mytilus* (Zhu Y *et al.*, 2006). These investigators found a dose-dependent inhibition of growth at concentrations greater than 1 mg/l. The MWCNTs were found colocolized with mitochondria, suggesting that the effects on viability were due to mitochondrial dysfunction.

12.5.2 Metal/Metal Oxide Nanoparticles

A wide variety of metals have been used to synthesize nanoparticles. Studies have demonstrated toxicity as a result of exposure to copper, silver, titanium dioxide, silica, zinc oxide, nickel, indium, and others (Blaise *et al.*, 2008; Griffitt *et al.*, 2008; Van Hoecke *et al.*, 2008). While studies in aquatic organisms of some of these materials are limited, it is clear that metallic ENMs vary in their toxicities, depending on their size, core metal and biological species tested. The finding that some metallic ENMs are toxic is not terribly surprising since dissolved ions of many of the core metals cause substantial toxicity in biological systems. As indicated above, a critical question in studies with metallic nanomaterials is whether observed toxicity is due to the nanoparticle itself or release of dissolved ions. Most studies have not distinguished between the effects of the ENMs themselves and their dissolved metal ions, but the distinction must be made. For example, Morones *et al.* (2005) tested silver nanoparticles on several types of bacteria and found that particles with diameters of 1–10 nm bound to bacterial membranes, disrupting permeability and respiration, and also penetrated the bacteria. Once inside, the nanomaterials released silver ions, further increasing the bacteriocidal effects. In this case, toxicity appeared to be due to both the nanoparticles themselves and the released silver ions. In another study, Franklin *et al.* (2007) tested the effects of 30 nm ZnO particles on the freshwater microalga

Pseudokirchneriella subcapitata and found toxicity with an EC_{50} of 60 μg/l. They ascribed the toxicity to dissolved Zn ions, since they measured rapid dissolution of the ZnO particles at pH 7.6 using equilibrium dialysis.

More recent careful studies comparing toxicity of various nanometals suggest that copper NMs are more toxic to zebrafish than silver NMs, while TiO_2 appeared nontoxic (Griffitt *et al.*, 2008). Toxicity of NMs was much less than the soluble forms of the metals, but these studies determined that the toxicity of the particulate form was not totally attributable to the portion that dissolved (Griffitt *et al.*, 2007, 2008). Quantum dots containing a core of material that functions as a semiconductor (e.g. CdTe) have also been studied for their potential toxicity. Early studies indicate that toxicity may be due to leaching of heavy metals from the core (Tang *et al.*, 2008); however, most quantum dots used today are capped by ZnS or other materials and passified by pegylation or protein adsorption, making them considerably less toxic to biological systems (Cho *et al.*, 2007).

As seen with carbonaceous nanomaterials, a variety of factors are likely to affect the observed toxicity of metallic nanomaterials. Clearly size of nanomaterials is important; 25 nm TiO_2 particles were toxic to the green alga *Desmodesmus subspicatus* (EC_{50} 44 mg/l), whereas the 100 nm particles were not (Hund-Rinke and Simon, 2006). Both particles were toxic to *Daphnia magna* but at higher concentrations than for the alga. Duration of exposure is also an important issue. While most studies of TiO_2 find little acute toxicity, a recent study by Federici *et al.* (2007) found that TiO_2 in the range of 0.1 to 1 mg/l can cause sublethal effects after 14 days including gill pathologies, decreased Na^+/K^+ ATPase in the gills and intestine, and increased oxidative stress.

Substantial differences for the same nanomaterial can be observed across species (Table 12.1). Studies suggest that *B. subtilis* appears to be more sensitive than *E. coli* to ZnO (Adams *et al.*, 2006). Algae appear to be more sensitive than daphnia for some nanometals (Wang *et al.*, 2008) while daphnia are more sensitive than higher organisms such as fish (Griffitt *et al.*, 2008). The reasons for species sensitivity must be elucidated, but are likely to involve differences in the rate or extent of nanomaterial uptake, tissue distribution and elimination. It is clear that more detailed studies on the kinetics of nanoparticles in aquatic organisms are needed. There may also be inherent differences in susceptibility of a given species to nanomaterial toxicity, which will be largely based on the mechanism by which nanomaterials cause toxicity.

While data has begun to identify the gill and intestines as targets of nanomaterial action, to date, the mechanism by which nanomaterials cause toxicity remains unclear. There is substantial evidence that generation of reactive oxygen species is important for a variety of nanomaterials. Careful studies to elucidate the mechanisms by which various nanomaterials cause toxicity are clearly needed to allow extrapolation between particles and species. In these efforts, genomic technology is emerging as a method to examine sublethal toxicities and identify biochemical pathways that are affected by nanomaterial exposure (Oberdorster *et al.*, 2006; Griffitt *et al.*, 2007; Henry *et al.*, 2007).

12.6   Conclusions

As exploration of the ecotoxicity of nanomaterials moves forward, it is important that the differences between nanomaterials and traditional chemical toxicants are recognized. While many classic ecotoxicity tests are likely to work with nanomaterials, it is likely that they will need modification to deal with suspensions instead of solutions. The importance of particle characterization under actual exposure conditions cannot be overemphasized and must be included in study designs. Careful consideration of dosimetry is also required. Many studies in mammalian toxicology demonstrate that surface area is a more important variable than particle mass (Oberdorster, 1996; Oberdorster *et al.*, 2005), and similar results have been obtained for silica in algae (Van Hoecke *et al.*, 2008). Studies must consider particle number, surface area and mass when evaluating responses, especially when comparing results for a given nanomaterial that differs in size or shape. Finally, ENMs have enormous potential complexity due to the plethora of possible particles differing in size, shape, and surface functionalization that can be prepared from a single material. To deal with this complexity, it is critical that we develop an understanding of how particle physical properties influence particle fate and transport in the environment and in biota, as well as their toxicity.

References

Adams LK, Lyon DY, Alvarez PJ (2006) Comparative eco-toxicity of nanoscale TiO_2, SiO_2, and ZnO water suspensions. *Water Res* **40**, 3527–3532.

Allen BL, Kichambare PD, Gou P, Vlasova II, Kapralov AA, Konduru N, Kagan VE, Star A (2008) Biodegradation of single-walled carbon nanotubes through enzymatic catalysis. *Nano Lett* **8**, 3899–3903.

Aruoja V, Dubourguier HC, Kasemets K, Kahru A (2009) Toxicity of nanoparticles of CuO, ZnO and TiO(2) to microalgae Pseudokirchneriella subcapitata. *Sci Total Environ* **407**, 1461–1468.

Baalousha M, Manciulea A, Cumberland S, Kendall K, Lead JR (2008) Aggregation and surface properties of iron oxide nanoparticles: influence of pH and natural organic matter. *Environ Toxicol Chem* **27**, 1875–1882.

Baun A, Sorensen SN, Rasmussen RF, Hartmann NB, Koch CB (2008) Toxicity and bioaccumulation of xenobiotic organic compounds in the presence of aqueous suspensions of aggregates of nano-C(60). *Aquat Toxicol* **86**, 379–387.

Benn TM, Westerhoff P (2008) Nanoparticle silver released into water from commercially available sock fabrics. *Environ Sci Technol* **42**, 4133–4139.

Blaise C, Gagne F, Ferard JF, Eullaffroy P (2008) Ecotoxicity of selected nano-materials to aquatic organisms. *Environ Toxicol* **23**, 591–598.

Blaser SA, Scheringer M, Macleod M, Hungerbuhler K (2008) Estimation of cumulative aquatic exposure and risk due to silver: contribution of nano-functionalized plastics and textiles. *Sci Total Environ* **390**, 396–409.

Blickley TM, McClellan-Green P (2008) Toxicity of aqueous fullerene in adult and larval Fundulus heteroclitus. *Environ Toxicol Chem* **27**, 1964–1971.

Bouldin JL, Ingle TM, Sengupta A, Alexander R, Hannigan RE, Buchanan RA (2008) Aqueous toxicity and food chain transfer of Quantum DOTs in freshwater algae and Ceriodaphnia dubia. *Environ Toxicol Chem* **27**, 1958–1963.

Brant J, Lecoanet H, Hotze M, Wiesner M (2005) Comparison of electrokinetic properties of colloidal fullerenes (n-C60) formed using two procedures. *Environ Sci Technol* **39**, 6343–51.

Canesi L, Ciacci C, Betti M, Fabbri R, Canonico B, Fantinati A, Marcomini A, Pojana G (2008) Immunotoxicity of carbon black nanoparticles to blue mussel hemocytes. *Environ Int* **34**, 1114–1119.

Chen KL, Elimelech M (2008) Interaction of fullerene (C60) nanoparticles with humic acid and algi-nate coated silica surfaces: measurements, mechanisms, and environmental implications. *Environ Sci Technol* **42**, 7607–7614.

Chen LD, Liu J, Yu XF, He M, Pei XF, Tang ZY, Wang QQ, Pang DW, Li Y (2008) The biocompati-bility of quantum dot probes used for the targeted imaging of hepatocellular carcinoma metastasis. *Biomaterials* **29**, 4170–4176.

Cheng J, Flahaut E, Cheng SH (2007) Effect of carbon nanotubes on developing zebrafish (Danio rerio) embryos. *Environ Toxicol Chem* **26**, 708–716.

Cho SJ, Maysinger D, Jain M, Roder B, Hackbarth S, Winnik FM (2007) Long-term exposure to CdTe quantum dots causes functional impairments in live cells. *Langmuir* **23**, 1974–1980.

Choi HS, Liu W, Misra P, Tanaka E, Zimmer JP, Itty Ipe B, Bawendi MG, Frangioni JV (2007) Renal clearance of quantum dots. *Nat Biotechnol* **25**, 1165–1170.

Cundy AB, Hopkinson L, Whitby RL (2008) Use of iron-based technologies in contaminated land and groundwater remediation: a review. *Sci Total Environ* **400**, 42–51.

Daroczi B, Kari G, McAleer MF, Wolf JC, Rodeck U, Dicker AP (2006) In vivo radioprotection by the fullerene nanoparticle DF-1 as assessed in a zebrafish model. *Clin Cancer Res* **12**, 7086–7091.

Diegoli S, Manciulea AL, Begum S, Jones IP, Lead JR, Preece JA (2008) Interaction between man-ufactured gold nanoparticles and naturally occurring organic macromolecules. *Sci Total Environ* **402**, 51–61.

Duffy JE, Carlson E, Li Y, Prophete C, Zelikoff JT (2002) Impact of polychlorinated biphenyls (PCBs) on the immune function of fish: age as a variable in determining adverse outcome. *Mar Environ Res* **54**, 559–563.

Elimelech M, Gregory J, Jia X, Williams R (1995) *Particle Deposition and Aggregation: Measure-ment, Modelling and Simulation.* Butterworth-Heinemann: Woburn.

Espinasse B, Hotze EM, Wiesner MR (2007) Transport and retention of colloidal aggregates of C60 in porous media: effects of organic macromolecules, ionic composition, and preparation method. *Environ Sci Technol* **41**, 7396–7402.

Federici G, Shaw BJ, Handy RD (2007) Toxicity of titanium dioxide nanoparticles to rainbow trout (Oncorhynchus mykiss): gill injury, oxidative stress, and other physiological effects. *Aquat Toxicol* **84**, 415–430.

Ferguson PL, Chandler GT, Templeton RC, DeMarco A, Scrivens WA, Englehart BA (2008) Influence of sediment-amendment with single-walled carbon nanotubes and diesel soot on bioaccumulation of hydrophobic organic contaminants by benthic invertebrates. *Environ Sci Technol* **42**, 3879–3885.

Fortner JD, Kim DI, Boyd AM, Falkner JC, Moran S, Colvin VL, Hughes JB, Kim JH (2007) Reaction of water-stable C60 aggregates with ozone. *Environ Sci Technol* **41**, 7497–7502.

Franklin NM, Rogers NJ, Apte SC, Batley GE, Gadd GE, Casey PS (2007) Comparative toxicity of nanoparticulate ZnO, bulk ZnO, and ZnCl2 to a freshwater microalga (Pseudokirchneriella subcapitata): the importance of particle solubility. *Environ Sci Technol* **41**, 8484–8490.

Ghosh S, Mashayekhi H, Pan B, Bhowmik P, Xing B (2008) Colloidal behavior of aluminum oxide nanoparticles as affected by pH and natural organic matter. *Langmuir* **24**, 12385–12391.

Griffitt RJ, Weil R, Hyndman KA, Denslow ND, Powers K, Taylor D, Barber DS (2007) Exposure to copper nanoparticles causes gill injury and acute lethality in zebrafish (Danio rerio). *Environ Sci Technol* **41**, 8178–8186.

Griffitt RJ, Luo J, Gao J, Bonzongo JC, Barber DS (2008) Effects of particle composition and species on toxicity of metallic nanomaterials in aquatic organisms. *Environ Toxicol Chem* **27**, 1972–1978.

Griffitt RJ, Hyndman K, Denslow ND, Barber DS (2009) Comparison of molecular and histological changes in zebrafish gills exposed to metallic nanoparticles. *Toxicol Sci* **107**, 404–415.

Guzman KA, Finnegan MP, Banfield JF (2006) Influence of surface potential on aggregation and transport of titania nanoparticles. *Environ Sci Technol* **40**, 7688–7693.

Hamano T, Mashino T, Hirobe M (1995) Oxidation of C60 fullerene by cytochrome P450 chemical models. *J Chem Soc Chem Commun* 1537–1538.

Handy RD, von der Kammer F, Lead JR, Hassellov M, Owen R, Crane M (2008a) The ecotoxicology and chemistry of manufactured nanoparticles. *Ecotoxicology* 17, 287–314.

Handy RD, Owen R, Valsami-Jones E (2008b) The ecotoxicology of nanoparticles and nanomaterials: current status, knowledge gaps, challenges, and future needs. *Ecotoxicology* 17, 315–325.

Heinlaan M, Ivask A, Blinova I, Dubourguier HC, Kahru A (2008) Toxicity of nanosized and bulk ZnO, CuO and TiO_2 to bacteria Vibrio fischeri and crustaceans Daphnia magna and Thamnocephalus platyurus. *Chemosphere* 71, 1308–1316.

Henry TB, Menn FM, Fleming JT, Wilgus J, Compton RN, Sayler GS (2007) Attributing effects of aqueous C60 nano-aggregates to tetrahydrofuran decomposition products in larval zebrafish by assessment of gene expression. *Environ Health Perspect* 115, 1059–1065.

Hirschey MD, Han YJ, Stucky GD, Butler A (2006) Imaging Escherichia coli using functionalized core/shell CdSe/CdS quantum dots. *J Biol Inorg Chem* 11, 663–669.

Holbrook RD, Murphy KE, Morrow JB, Cole KD (2008) Trophic transfer of nanoparticles in a simplified invertebrate food web. *Nat Nanotechnol* 3, 352–355.

Hotze EM, Labille J, Alvarez P, Wiesner MR (2008) Mechanisms of photochemistry and reactive oxygen production by fullerene suspensions in water. *Environ Sci Technol* 42, 4175–4180.

Hund-Rinke K, Simon M (2006) Ecotoxic effect of photocatalytic active nanoparticles (TiO_2) on algae and daphnids. *Environ Sci Pollut Res Int* 13, 225–232.

IUPAC (1997) *Compendium of Chemical Terminology* (2nd edn), McNaught A, Wilkinson A (eds). Blackwell Scientific Publications: Oxford.

Kaegi R, Ulrich A, Sinnet B, Vonbank R, Wichser A, Zuleeg S, Simmler H, Brunner S, Vonmont H, Burkhardt M, Boller M (2008) Synthetic TiO_2 nanoparticle emission from exterior facades into the aquatic environment. *Environ Pollut* 156, 233–239.

Kallay N, Zalac S (2002) Stability of nanodispersions: a model for kinetics of aggregation of nanoparticles. *J Colloid Interface Sci* 253, 70–76.

Kashiwada S (2006) Distribution of nanoparticles in the see-through medaka (Oryzias latipes). *Environ Health Perspect* 114, 1697–1702.

Kennedy AJ, Hull MS, Steevens JA, Dontsova KM, Chappell MA, Gunter JC, Weiss CA Jr (2008) Factors influencing the partitioning and toxicity of nanotubes in the aquatic environment. *Environ Toxicol Chem* 27, 1932–1941.

Kim T, Lee CH, Joo SW, Lee K (2008) Kinetics of gold nanoparticle aggregation: experiments and modeling. *J Colloid Interface Sci* 318, 238–243.

Klaine SJ, Alvarez PJ, Batley GE, Fernandes TF, Handy RD, Lyon DY, Mahendra S, McLaughlin MJ, Lead JR (2008a) Nanomaterials in the environment: behavior, fate, bioavailability, and effects. *Environ Toxicol Chem* 27, 1825–1851.

Klaine SJ, Alvarez PJ, Batley GE, Fernandes TF, Handy RD, Lyon DY, Mahendra S, McLaughlin MJ, Lead JR (2008b) Nanomaterials in the environment: behavior, fate, bioavailability, and effects. *Environ Toxicol Chem* 27, 1825–1851.

Kloepfer JA, Mielke RE, Nadeau JL (2005) Uptake of CdSe and CdSe/ZnS quantum dots into bacteria via purine-dependent mechanisms. *Appl Environ Microbiol* 71, 2548–2557.

Koehler A, Marx U, Broeg K, Bahns S, Bressling J (2008) Effects of nanoparticles in Mytilus edulis gills and hepatopancreas – a new threat to marine life? *Mar Environ Res* 66, 12–14.

Lecoanet, HF, Wiesner MR (2004) Velocity effects on fullerene and oxide nanoparticle deposition in porous media. *Environ Sci Technol* 38, 4377–4382.

Lecoanet HF, Bottero JY, Wiesner MR (2004) Laboratory assessment of the mobility of nanomaterials in porous media. *Environ Sci Technol* 38, 5164–5169.

Li H, Zhang J, Wang T, Luo W, Zhou Q, Jiang G (2008) Elemental selenium particles at nano-size (Nano-Se) are more toxic to Medaka (Oryzias latipes) as a consequence of hyper-accumulation of selenium: a comparison with sodium selenite. *Aquat Toxicol* **89**, 251–256.

Li Y, Wang Y, Pennell KD, Abriola LM (2008) Investigation of the transport and deposition of fullerene (C60) nanoparticles in quartz sands under varying flow conditions. *Environ Sci Technol* **42**, 7174–7180.

Limbach LK, Bereiter R, Muller E, Krebs R, Galli R, Stark WJ (2008) Removal of oxide nanoparticles in a model wastewater treatment plant: influence of agglomeration and surfactants on clearing efficiency. *Environ Sci Technol* **42**, 5828–5833.

Lovern SB, Klaper R (2006) Daphnia magna mortality when exposed to titanium dioxide and fullerene (C60) nanoparticles. *Environ Toxicol Chem* **25**, 1132–1137.

Lovern SB, Strickler JR, Klaper R (2007) Behavioral and physiological changes in Daphnia magna when exposed to nanoparticle suspensions (titanium dioxide, nano-C60, and C60HxC70Hx). *Environ Sci Technol* **41**, 4465–4470.

Lyon DY, Fortner JD, Sayes CM, Colvin VL, Hughe JB (2005) Bacterial cell association and antimicrobial activity of a C60 water suspension. *Environ Toxicol Chem* **24**, 2757–2762.

Mashino T, Okuda K, Hirota T, Hirobe M, Nagano T, Mochizuki M (1999) Inhibition of E. coli growth by fullerene derivatives and inhibition mechanism. *Bioorg Med Chem Lett* **9**, 2959–2962.

Mashino T, Usui N, Okuda K, Hirota T, Mochizuki M (2003) Respiratory chain inhibition by fullerene derivatives: hydrogen peroxide production caused by fullerene derivatives and a respiratory chain system. *Bioorg Med Chem* **11**, 1433–1438.

Maynard AD, Baron PA, Foley M, Shvedova AA, Kisin ER, Castranova V (2004) Exposure to carbon nanotube material: aerosol release during the handling of unrefined single-walled carbon nanotube material. *J Toxicol Environ Health A* **67**, 87–107.

McClellan-Green P, Romano J, Oberdorster E (2007) Does gender really matter in contaminant exposure? A case study using invertebrate models. *Environ Res* **104**, 183–191.

Moger J, Johnston BD, Tyler CR (2008) Imaging metal oxide nanoparticles in biological structures with CARS microscopy. *Opt Express* **16**, 3408–3419.

Moore MN (2006) Do nanoparticles present ecotoxicological risks for the health of the aquatic environment? *Environ Int* **32**, 967–976.

Morones JR, Elechiguerra JL, Camacho-Bragado A, Holt K, Kouri JB, Ramirez JT, Yacaman MJ (2005) The bactericidal effect of silver nanoparticles. *Nanotechnology* **16**, 2346–2353.

Mueller NC, Nowack B (2008) Exposure modeling of engineered nanoparticles in the environment. *Environ Sci Technol* **42**, 4447–4453.

Oberdorster E (2004) Manufactured nanomaterials (fullerenes, C60) induce oxidative stress in the brain of juvenile largemouth bass. *Environ Health Perspect* **112**, 1058–1062.

Oberdörster E, Larkin P, Rogers J (2006) *Rapid Environmental Impact Screening for Engineered Nanomaterials: A Case Study using Microarray Technology*. In Project on Emerging Nanotechnologies. Wilson Center and The Pew Charitable Trust. Available at www.nanotechproject.org.

Oberdorster G (1996) Significance of particle parameters in the evaluation of exposure-dose-response relationships of inhaled particles. *Inhal Toxicol* **8** Suppl., 73–89.

Oberdorster G, Maynard A, Donaldson K, Castranova V, Fitzpatrick J, Ausman K, Carter J, Karn B, Kreyling W, Lai D, Olin S, Monteiro-Riviere N, Warheit D, Yang H (2005) Principles for characterizing the potential human health effects from exposure to nanomaterials: elements of a screening strategy. *Part Fibre Toxicol* **2**, 8.

Porter AE, Muller K, Skepper J, Midgley P, Welland M (2006) Uptake of C60 by human monocyte macrophages, its localization and implications for toxicity: studied by high resolution electron microscopy and electron tomography. *Acta Biomater* **2**, 409–419.

Porter AE, Gass M, Muller K, Skepper JN, Midgley P, Welland M (2007) Visualizing the uptake of C60 to the cytoplasm and nucleus of human monocyte-derived macrophage cells using energy-filtered transmission electron microscopy and electron tomography. *Environ Sci Technol* **41**, 3012–3017.

Powers KW, Brown SC, Krishna VB, Wasdo SC, Moudgil BM, Roberts SM (2006) Research strategies for safety evaluation of nanomaterials. Part VI. Characterization of nanoscale particles for toxicological evaluation. *Toxicol Sci* **90**, 296–303.

Roberts AP, Mount AS, Seda B, Souther J, Qiao R, Lin S, Ke PC, Rao AM, Klaine SJ (2007) In vivo biomodification of lipid-coated carbon nanotubes by Daphnia magna. *Environ Sci Technol* **41**, 3025–3029.

Rosse P, Loizeau J-L (2003) Use of single particle counters for the determination of the number and size distribution of colloids in natural surface waters. *Colloids and Surfaces A* **217**, 109–120.

Saleh N, Kim HJ, Phenrat T, Matyjaszewski K, Tilton RD, Lowry GV (2008a) Ionic strength and composition affect the mobility of surface-modified Fe0 nanoparticles in water-saturated sand columns. *Environ Sci Technol* **42**, 3349–3355.

Saleh NB, Pfefferle LD, Elimelech M (2008b) Aggregation kinetics of multiwalled carbon nanotubes in aquatic systems: measurements and environmental implications. *Environ Sci Technol* **42**, 7963–7969.

Shi Kam NW, Jessop TC, Wender PA, Dai H (2004) Nanotube molecular transporters: internalization of carbon nanotube-protein conjugates into Mammalian cells. *J Am Chem Soc* **126**, 6850–6851.

Simmons JM, Nichols BM, Baker SE, Marcus MS, Castellini OM, Lee CS, Hamers RJ, Eriksson MA (2006) Effect of ozone oxidation on single-walled carbon nanotubes. *J Phys Chem B* **110**, 7113–7118.

Singh R, Pantarotto D, Lacerda L, Pastorin G, Klumpp C, Prato M, Bianco A, Kostarelos K (2006) Tissue biodistribution and blood clearance rates of intravenously administered carbon nanotube radiotracers. *Proc Natl Acad Sci USA* **103**, 3357–3362.

Smith CJ, Shaw BJ, Handy RD (2007) Toxicity of single walled carbon nanotubes to rainbow trout (Oncorhynchus mykiss): respiratory toxicity, organ pathologies, and other physiological effects. *Aquat Toxicol* **82**, 94–109.

Sun H, Zhang X, Zhang Z, Chen Y, Crittenden JC (2008) Influence of titanium dioxide nanoparticles on speciation and bioavailability of arsenite. *Environ Pollut* **157**, 1165–1170.

Tang M, Xing T, Zeng J, Wang H, Li C, Yin S, Yan D, Deng H, Liu J, Wang M, Chen J, Ruan DY (2008) Unmodified CdSe quantum dots induce elevation of cytoplasmic calcium levels and impairment of functional properties of sodium channels in rat primary cultured hippocampal neurons. *Environ Health Perspect* **116**, 915–922.

Templeton RC, Ferguson PL, Washburn KM, Scrivens WA, Chandler GT (2006) Life-cycle effects of single-walled carbon nanotubes (SWNTs) on an estuarine meiobenthic copepod. *Environ Sci Technol* **40**, 7387–7393.

Tiraferri A, Chen KL, Sethi R, Elimelech M (2008) Reduced aggregation and sedimentation of zero-valent iron nanoparticles in the presence of guar gum. *J Colloid Interface Sci* **324**, 71–79.

Tsao N, Luh TY, Chou CK, Chang TY, Wu JJ, Liu CC, Lei HY (2002) In vitro action of carboxyfullerene. *J Antimicrob Chemother* **49**, 641–649.

Usenko CY, Harper SL, Tanguay RL (2008) Fullerene C60 exposure elicits an oxidative stress response in embryonic zebrafish. *Toxicol Appl Pharmacol* **229**, 44–55.

Van Hoecke K, De Schamphelaere KA, Van Der Meeren P, Lucas S, Janssen CR (2008) Ecotoxicity of silica nanoparticles to the green alga Pseudokirchneriella subcapitata: importance of surface area. *Environ Toxicol Chem* **27**, 1948–1957.

Wang J, Zhang X, Chen Y, Sommerfeld M, Hu Q (2008) Toxicity assessment of manufactured nanomaterials using the unicellular green alga Chlamydomonas reinhardtii. *Chemosphere* **73**, 1121–1128s.

Weir E, Lawlor A, Whelan A, Regan F (2008) The use of nanoparticles in anti-microbial materials and their characterization. *Analyst* **133**, 835–845.

Yeganeh B, Kull CM, Hull MS, Marr LC (2008) Characterization of airborne particles during production of carbonaceous nanomaterials. *Environ Sci Technol* **42**, 4600–4006.

Zhang XZ, Sun HW, Zhang ZY (2006) Bioaccumulation of titanium dioxide nanoparticles in carp. *Huan Jing Ke Xue* **27**, 1631–1635.

Zhang X, Sun H, Zhang Z, Niu Q, Chen Y, Crittenden JC (2007) Enhanced bioaccumulation of cadmium in carp in the presence of titanium dioxide nanoparticles. *Chemosphere* **67**, 160–166.

Zhang Y, Chen Y, Westerhoff P, Crittenden JC (2008) Stability and removal of water soluble CdTe quantum dots in water. *Environ Sci Technol* **42**, 321–325.

Zhu S, Oberdorster E, Haasch ML (2006) Toxicity of an engineered nanoparticle (fullerene, C60) in two aquatic species, Daphnia and fathead minnow. *Mar Environ Res* **62** Suppl., S5–9.

Zhu X, Zhu L, Li Y, Duan Z, Chen W, Alvarez PJ (2007) Developmental toxicity in zebrafish (Danio rerio) embryos after exposure to manufactured nanomaterials: buckminsterfullerene aggregates (nC60) and fullerol. *Environ Toxicol Chem* **26**, 976–979.

Zhu X, Zhu L, Duan Z, Qi R, Li Y, Lang Y (2008a) Comparative toxicity of several metal oxide nanoparticle aqueous suspensions to Zebrafish (Danio rerio) early developmental stage. *J Environ Sci Health A Tox Hazard Subst Environ Eng* **43**, 278–284.

Zhu X, Zhu L, Lang Y, Chen Y (2008b) Oxidative stress and growth inhibition in the freshwater fish Carassius auratus induced by chronic exposure to sublethal fullerene aggregates. *Environ Toxicol Chem* **27**, 1979–1985.

Zhu Y, Zhao Q, Li Y, Cai X, Li W (2006) The interaction and toxicity of multi-walled carbon nanotubes with Stylonychia mytilus. *J Nanosci Nanotechnol* **6**, 1357–1364.

13

Nanotoxicity of Metal Oxide Nanoparticles *in Vivo*

Weiyue Feng, Bing Wang and Yuliang Zhao

13.1 Introduction

Nanotechnology is considered to be able to reengineer the man-made world, molecule by molecule, sparking a wave of revolution in commercial products. There are many definitions of 'nanotechnology', but the most widely used is that of the US National Nanotechnology Initiative (NNI): 'Nanotechnology is the understanding, control and manipulation of matter at atomic and molecular levels at dimensions between approximately 1 and 100 nanometers, where unique phenomena enable novel applications' (www.nano.gov). According to the US National Science Foundation (NSF), the global market for nanotechnology products is estimated at \$700 billion in 2008 and will exceed \$1 trillion annually by 2015 (Roco, 2005).

Engineered nanoparticles (NPs) in general terms are defined as engineered nanostructures intentionally produced for specific uses in science, technology, medicine, industries and many other applications in daily life. In terms of morphology, NPs can be categorized into spheres, tubes, wires, fibers, rings and quantum dots (Oberdörster *et al.*, 2005a). They are distinguished from ambient ultrafine particles (UFPs), a term being intensely used by environmental scientists and toxicologists, originating from naturally occurring (volcanic emissions, sea spray and forest fires) or anthropogenic sources (combustion of fossil fuels, power plants, welding, smelting) (Kreyling, 2006).

Human beings have been being exposed to airborne ultrafine particles ($<100\,\text{nm}$) throughout all stages of human evolution, but it was only with the advent of the industrial revolution that exposure to UFPs from anthropogenic sources increased dramatically. The

Nanotoxicity: From In Vivo *and* In Vitro *Models to Health Risks*   Edited by Saura Sahu and Daniel Casciano
© 2009 John Wiley & Sons, Ltd

rapid development of nanotechnology is likely to become another source of UFPs for human exposure.

Nanotoxicology was proposed as a new branch of toxicology to address the adverse health effects caused by exposure of nanoparticles (Oberdörster *et al.*, 2005a). The phenomena intrinsic to the nanoscale, such as size confinement, dominance of interfacial phenomena, quantum mechanics, and so forth, are far beyond our prediction of what we have known at the macroscale. Therefore it is necessary to explore or to revaluate the toxicity of manufactured nanomaterials, even though the toxicity of the bulk material with the same chemical composition may already be known. So far there is no standard protocol for nanotoxicity testing; however, it will suffice to mention that three key elements of a toxicity screening strategy should be a full physico-chemical characterization of the nanomaterial, *in vitro* assays (cellular and noncellular), and *in vivo* studies (Oberdörster *et al.*, 2005b). Unlike the toxicity testing of traditional chemicals, the physico-chemical parameters, such as size, shape, crystal structure, solubility, surface area, surface charge, surface coating, and so on, may all contribute to toxicity. The exploration of nanomaterials *in vivo* includes their absorption, distribution, metabolism, excretion (ADME), particokinetics, and the interactions between the nanomaterials and biomolecules *in vivo*.

The most commercially important nanomaterials include metal oxides, such as titanium oxide (TiO_2), alumina oxide (Al_2O_3), iron oxides (Fe_2O_3, Fe_3O_4), zinc oxide (ZnO), cerium oxide (CeO_2) and zirconia (ZrO_2). These engineered metal-oxide nanoparticles are attractive for a large variety of applications such as catalysis, sensors, (photo)electronic materials, and environmental remediation, due to their unique crystal morphology and superior mechanical, electrical, magnetic, optical and adsorption properties (Klabunde *et al.*, 1996).

In this chapter, we will focus on the nanotoxicity of metal oxide NPs *in vivo*, including the existing sources of nanoparticles in the environment, the potential exposure routes, their ADME via various exposure routes, and the health effects caused by these nanoparticles. We hope the analysis and discussion of the nanotoxicity of metal oxide nanoparticles *in vivo* might provide some information and guidelines to nanoparticle-related occupational health and safety control.

13.2 Metal Oxide Nanoparticles in the Environment

Engineered nanoparticles are likely to enter the environment via several routes, either during the manufacture of nanomaterials or through the use and disposal of products containing nanoparticles. The industrial effluents and spillages from nanoparticle synthesis and manufacture probably contribute to environmental contamination (Nowack and Bucheli, 2007). Nanoparticles used in personal-care products such as cosmetics and sunscreens, industrial applications and healthcare technology are also most likely to enter the environment through washing them off or leaking during the use and disposal process (Roco, 2005; Aitken *et al.*, 2006).

Metal oxide nanoparticles are ubiquitous in the environment due to the wide applications in emerging nanotechnologies. For instance, titanium dioxide and zinc oxide are used on a large scale in pigments, sun-screens, and in polymers or tyres as stabilizers (Nohynek *et al.*, 2008). Iron oxide nanoparticles have been widely applied in industry, the environment, and *in vivo* biomedical applications due to their size-dependent superparamagnetism and catalytic abilities (Gupta *et al.*, 2007). Lanthanide oxide nanoparticles have been developed as platforms for sensitive bio-assays, including immunoassays and DNA assays (Nichkova *et al.*, 2007). In addition, other metal oxide nanoparticles such as zirconia and ceria oxide are rapidly growing ceramic nanomaterials with broad applications in catalysis and polishing, and as additives in polymers and dental materials due to their appealing physical properties. The global market for nanoparticles used in biomedical, pharmaceutical and cosmetic applications increased from $170.7 million in 2006 to $204.6 million in 2007, and should reach $684.4 million by 2012 (BCC Research, 2007). These data, on the other hand, infer that the potential application of metal oxide nanoparticles in various fields will undoubtedly increase human exposure and potential health risks.

13.3   Exposure Route of Nanoparticles

Nanoparticle exposure can be divided into occupational exposure at the workplace; incidental exposure from unintentional environmental release; or intentional application of consumer products or biomedical products. Occupational exposure to nanoparticles at the workplace may occur in research and development facilities in the nanotechnology sector, chemical and pharmaceutical companies, facilities where paints, cement and other products are manufactured, and during processes where nanoparticles are byproducts such as in baking, welding and polymer processing. Incidental exposure to nanoparticles may occur via engineered nanoparticles released into the environment; and nanoparticles are use in specific biomedical applications for diagnostic and therapeutic purposes. The potential routes for occupational or incidental exposure to nanoparticles in organisms include: respiratory and gastrointestinal tract, skin and soft tissue, and injection (Figure 13.1) (Oberdörster *et al.*, 2005a).

13.3.1   Respiratory Tract

Most of the toxicity research on NPs *in vivo* has been carried out in mammalian systems, with a focus on the respiratory system. Specific defense mechanisms may protect the mammalian organism from harmful materials at the portal of entry; however, these defenses may not always be as effective for NPs. When NPs are inhaled, their deposition, clearance, and translocation within the respiratory tract will be different from the larger particles.

13.3.1.1   Deposition of Inhaled Nanoparticles in the Respiratory Tract

According to the diffusion motion, exposure to airborne NPs via the inhalation route will deposit throughout the entire respiratory tract, starting from nasopharyngeal and tracheobronchial, down to the alveolar regions. Depending on the particle size, the deposition fractions of inhaled NPs in these regions show significant differences. For instance, 90 % of inhaled 1 nm particles are deposited in the nasopharyngeal compartment, only ~10 % in the tracheobronchial region, and essentially none in the alveolar region. On the other hand, 5 nm particles show about equal deposition of ~30 % of the inhaled particles in all three regions; 20 nm particles have the highest deposition efficiency in the alveolar region (~50 %), whereas in tracheobronchial and nasopharyngeal regions this particle size

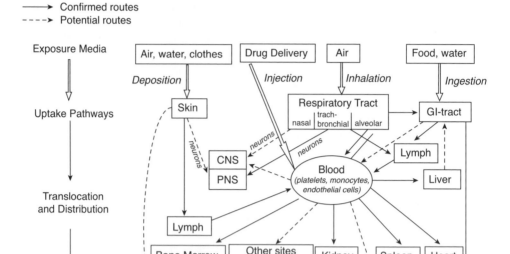

Figure 13.1   *Biokinetics of nanoparticles (Oberdörster et al., 2005a). Reproduced with permission from Environmental Health Perspectives*

deposits with ∼15 % efficiency (Figure 13.2) (ICRP, 1994). These different deposition efficiencies should have consequences for potential effects induced by inhaled NPs, as well as for their translocation beyond the respiratory tract.

13.3.1.2   Clearance Pathways

Pulmonary retention and clearance of particles has been studied for many years. Recent studies on occupational and environmental exposure of nanoparticles have generated a considerable amount of knowledge regarding the clearance of nanoparticles in the respiratory tract. The clearance of deposited particles in the respiratory tract is mainly by physical translocation to other sites or chemical clearance. Chemical dissolution in the upper or lower respiratory tract occurs for biosoluble nanoparticles in the intracellular or extracellular fluids. The solutes and soluble components can then undergo absorption and diffusion in other subcellular structures, or binding to proteins in cells, and may eventually be cleared into blood and lymphatic circulation. This clearance mechanism can happen at any location within the three regions of the respiratory tract, depending on the pH of local extracellular and intracellular compartments.

The clearance for insoluble or poorly soluble nanoparticles in the respiratory tract is basically via physical translocation. The efficiency of this clearance depends highly upon the site of deposition and particle size. The NPs trapped within the mucocillary escalator from the upper airways (nasopharyngeal and tracheobronchial) are expelled by pushing mucus towards the mouth; hence it is a relatively rapid process (Kreyling and Scheuch, 2000). For NPs within the alveolar regions, the most prevalent clearance is mediated by

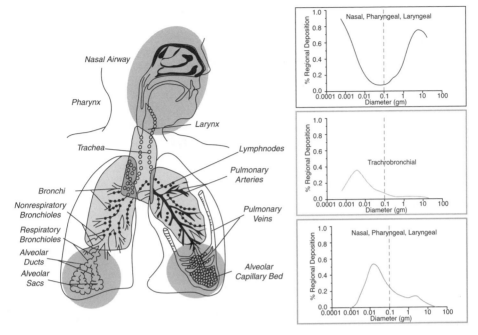

Figure 13.2 *Predicted fractional deposition of inhaled particles in the nasopharyngeal, tracheobronchial and alveolar regions of the human respiratory tract during nose-breathing (ICRP, 1994). Reproduced from ICRP Publication 66 (1994), with permission from the International Commission on Radiological Protection*

macrophage phagocytosis, which depends upon the ability of alveolar macrophages to 'sense' deposited particles, move to the site of their deposition, phagocytize them, and then move towards the mucociliary escalator (Kreyling and Scheuch, 2000).

Macrophage phagocytosis is a relatively slow process and involves several steps: specific receptors on the phagocyte membrane binding with specific molecules (ligands) localized on the surface of particle; internalization of the particle; fusing with lysosomes to form a phagolysosome; and digestion by lysosome enzymes. The physico-chemical parameters of particles, such as particle size, shape and surface charge, may interfere with the phagocytic ability of alveolar macrophages (Oberdörster *et al.*, 1994; Borm and Kreyling, 2004). Studies in rats following inhalation exposure to 20 nm and 250 nm TiO_2 particles have shown that the retention half-life of 250 nm TiO_2 in the alveolar regions was 117 days, while for 20 nm TiO_2 it was as long as 541 days (Oberdörster *et al.*, 1994). Furthermore, this mechanism of alveolar clearance is not perfect, as it allows smaller nanoparticles to penetrate the alveolar epithelium and reach the interstitial space from where nanoparticles may enter the circulatory and lymphatic systems and reach other sites throughout the body (Semmler-Behnke *et al.*, 2007).

13.3.1.3 Systemic Translocation of Nanoparticles

Because of the apparent inefficiency of alveolar macrophage phagocytosis of NPs, one might expect that these particles interact instead with epithelial cells. Indeed, results from several studies showed that NPs deposited in the respiratory tract readily gain access to epithelial and interstitial sites (Oberdörster *et al.*, 1992; Semmler-Behnke *et al.*, 2007). In a study evaluating the pulmonary inflammatory response of TiO_2 particle in rats, Oberdörster *et al.* (1992) reported that after intratracheal instillation of both nanoscale (20 nm) and larger (200 nm) TiO_2 particles, a highly increased interstitial access of the nanoscaled TiO_2 particles was observed. Such interstitial translocation represents a shift in target site away from the alveolar space to the interstitium.

Once the particles have reached pulmonary interstitial sites, uptake into the blood circulation and subsequent translocation into secondary target organs can occur (Oberdörster *et al.*, 2002; Takenaka *et al.*, 2006). In an earlier study, Berry *et al.* (1977) first confirmed the translocation of NPs across the alveolar epithelium into pulmonary capillaries using intratracheal instillations of 30 nm gold particles in rats. Recently, a number of studies based on different particle types ($^{59}Fe_2O_3$, ^{13}C, Ag) in rats also confirm the existence of this translocation pathway (Oberdörster *et al.*, 2002; Takenaka *et al.*, 2006; Zhu *et al.*, 2009). Zhu *et al.* (2009) reported that intratracheal-instilled nano-$^{59}Fe_2O_3$ could pass rapidly through the alveolar-capillary barrier into systemic circulation within 10 min, which is consistent with one-compartment model kinetics, and nano-$^{59}Fe_2O_3$ spread to the mononuclear phagocyte-rich organs, such as the liver, spleen, kidney and testicle (Zhu *et al.*, 2009).

However, evidence in humans for alveolar translocation of inhaled NPs into the blood circulation is still under debate. Nemmar *et al.* (2002) demonstrated a rapid 3–5 % appearance for humans inhaling ^{99m}Tc-labelled carbon particles (<100 nm) in the bloodstream within 1 minute post-exposure, and subsequent accumulation in the liver. However, some other studies in humans indicated that there was no significant translocation of inhaled ^{99m}Tc-labelled carbon NPs (35 nm, 100 nm) from the lung into the systematic circulation (Mills *et al.*, 2006; Wiebert *et al.*, 2006). The conflicting results from extrapulmonary translocation of experimental nanoparticles may be due to the differences in the particle chemical composition, particle size, surface characteristics, labelling materials, and the experimental models reported in the different studies. Taking together the evidence from all animal and human studies, alveolar translocation of NPs probably exists in humans; however, the extent of extrapulmonary translocation is highly dependent on characteristics of the nanoparticles.

Another translocation pathway for inhaled NPs involves uptake by olfactory neurons from the nasal lumen and subsequent axonal transport into the central nervous system (CNS). This translocation of NPs along olfactory neurons from the olfactory epithelium to the olfactory bulb was first reported by Bodian and Howe (1941) for the study of 30 nm polio virus in monkeys, and was later shown by nasally-deposited colloidal 50 nm gold particles moving into the olfactory bulb of squirrel monkeys as demonstrated by transmission electron microscopy (Delorenzo, 1957). Recent studies on inhalation of nanoparticles in rats showed that the inhaled ultrafine carbon (35 nm) and manganese dioxide (MnO_2) nanoparticles (30 nm) could translocate into the CNS via the olfactory neuronal pathway (Oberdörster *et al.*, 2004; Elder *et al.*, 2006). Furthermore, the intranasally-instilled TiO_2 (80 nm) and Fe_2O_3 (280 nm) nanoparticles in mice were observed transporting into the olfactory bulb along with different layers of the olfactory bulb, from the olfactory nerve layer (ON), granular cell layer (Gro), anterior olfactory nucleus, external part (AOE) to olfactory ventricle (OV), by synchrotron radiation X-ray fluorescence (Figure 13.3) (Wang

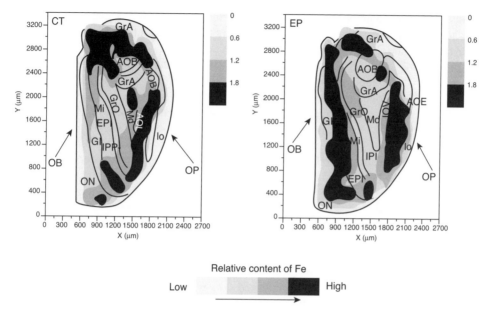

Figure 13.3 Fe distribution in the olfactory bulb section CT, control group (n = 3); EP, exposed group (n = 3); OB, olfactory bulb; OP, olfactory peduncle; ON, olfactory nerve; Gl, glomerular layer; Epl, external plexiform layer; Mi, mitral cell layer; Ipl, internal plexiform layer; GrO, granule cell layer of olfactory bulb; Md, medullary layer; GrA, granule cell layer of accessory olfactory bulb; AOB, accessory olfactory bulb; AOE, anterior olfactory nucleus external part; AOL, anterior olfactory nucleus, lateral part; Io, lateral tract. Reprinted from Wang et al. (2007), with kind permission from Springer Science + Business Media.

JX *et al.*, 2005, 2007). The above studies in animals point out that the olfactory neuronal pathway should be considered as a potential portal of airborne NP entry into the CNS of humans under conditions of environmental and occupational exposure.

13.3.2 Dermal Exposure

Nanomaterials, such as TiO_2 and ZnO nanoparticles, have been used in cosmetics and pharmaceuticals for many years. Therefore, dermal exposure has been another important route for NPs to gain the entry into the human body. The skin is composed of three layers: epidermis, dermis and subcutaneous. The outer portion of the epidermis, called the stratum corneum, is a 10 µm thick keratinized layer of dead cells and is difficult to pass through for ionic compounds or water-soluble molecules (Buzea *et al.*, 2007). The surface of the epidermis is highly microstructured, having a scaly appearance as well as pores for sweat, sebaceous glands, and hair follicle sites.

13.3.2.1 Absorption and Penetration of Nanoparticles in the Skin

Nanoparticle dermal penetration is still controversial. A current area under discussion is whether or not TiO_2 NPs in commercially available sunscreens penetrate the skin. Several studies in murine, porcine, and human skin confirmed that TiO_2 NPs remained on the skin surface or the outer layers of the skin, and had not penetrated into or through the living skin (Lademann *et al.*, 1999; Pflücker *et al.*, 2001; Schulz *et al.*, 2002). Schulz *et al.* (2002) and Pflücker *et al.* (2001) investigated the dermal absorption and penetration of different sizes (10–15 nm, 20 nm and 100 nm) and shapes (cubic or needles) of TiO_2 NPs in human volunteers. They found that none of these TiO_2 NPs penetrated beyond the outer layer of the stratum corneum. However, other evidence suggests that NPs may penetrate into the epidermis or dermis. Bennat and Müeller-Goymann (2000) applied TiO_2 NPs to human skin either as an aqueous suspension or oil-in-water emulsion, and evaluated skin penetration using the tape stripping method. They observed that TiO_2 NPs apparently penetrated deeper into human skin when applied as an oil-in-water emulsion, and that penetration was greater when applied to hairy skin, which suggests that TiO_2 NPs penetrate the surface through hair follicles or pores.

Nanoparticle penetration through the skin possibly also occurs in flexed and broken skin (Rouse *et al.*, 2007; Zhang and Monteiro-Riviere, 2008). Rouse *et al.* (2007) confirmed that repetitive motion of porcine skin could speed the passage of fullerene-derived amino-acid (Baa-Lys(FITC)-NLS) NPs through the skin. The confocal microscopy images depicted dermal penetration of the NPs at 8 h in flexed skin, whereas it had not penetrated into the dermis of unflexed skin until 24 h (Figure 13.4). Similarly, in a study of quantum dots (QD), Zhang and Monteiro-Riviere (2008) found the penetration of QD (655 and 565 nm) was primarily limited to the uppermost stratum corneum layers of intact skin, but abraded skin allowed QD to penetrate deeper into the dermal layers. These studies suggest that mechanical deformation or abrasion makes it easy to transport nanoparticles through the stratum corneum into the epidermis and dermis of skin.

13.3.2.2   Translocation of Nanoparticles via Dermal Exposure

The dermis has a rich supply of blood and macrophages, lymph vessels, dendritic cells and nerve endings (Oberdörster *et al.*, 2005a). Therefore, the interaction of penetrated nanoparticles within the dermis might lead to a translocation of NPs to other body sites. Currently, studies on the translocation of nanoparticles through the skin are still limited. A study by Kim *et al.* (2004) of intradermally-injected near-infrared QD (15–20 nm) in mice and pigs confirmed that NPs, once in the dermis, could localize to regional lymph nodes. The transport mechanism to the lymph nodes was suggested via skin macrophages and dendritic (Langerhans) cells mediated phagocytosis (Ohl *et al.*, 2004).

Neuronal transport of nanoparticles along sensory skin nerves may also be possible, since it has been demonstrated via the nasal and tracheobronchial regions of the respiratory tract (Oldfors and Fardeau, 1983). For example, an intra-muscular injection of ferritin (~112 nm), iron-dextran (11 or 21 nm) and gold protein (20–25 nm) nanoparticles showed rapid penetration through the basal lamina into the synaptic cleft of the neuromuscular junction, but this kind of penetration was restricted only to the smaller nanoparticles (Oldfors and Fardeau, 1983).

Overall, interpretation of *in vivo* skin penetration and translocation should use caution since the permeability of skin may vary widely depending both on the animal model and the substance being studied. Rats are the most common animal model, but rabbits, guinea

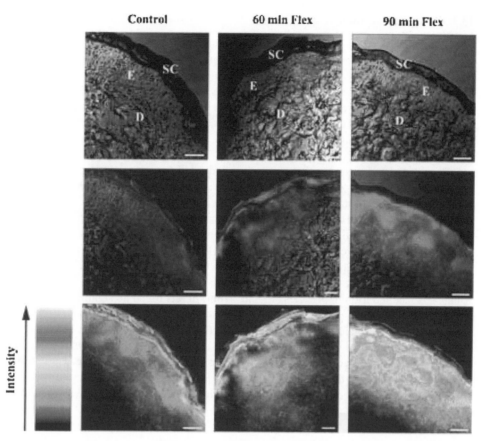

Figure 13.4 Confocal scanning microscopy images of intact skin dosed with Baa-Lys(FITC)-NLS for 8 h. Top row: confocal-DIC channel image shows an intact stratum corneum (SC) and underlying epidermal (E) and dermal layers (D). Middle row: Baa-Lys(FITC)-NLS fluorescence channel (green) and confocal-DIC channel shows fullerene penetration through the epidermal and dermal layers of skin. Bottom row: fluorescence intensity scan showing Baa-Lys(FITC)-NLS penetration. All scale bars represent 50 μm (Rouse et al., 2007). Reprinted with permission from the American Chemical Society, Copyright 2007. See color plate section.

pigs and pigs are also used to assess the absorption and penetration of NPs via the skin. In general, the penetration rate is in the order of rabbit > rat > pig > monkey > human, with the pig skin up to 4 times and the rat skin up to 9 times more permeable than human skin (Magnusson *et al.*, 2004). To assess the dermal penetration of nanoparticles, future studies will need to focus on (1) complete characterization of the nanomaterials being tested (e.g. size, specific surface area, surface charge, structure phase, monodispersed or aggregated), (2) the analytical method used to detect the nanoparticles (i.e. needs to be able to discriminate between epidermis, dermis, and hair follicles), and (3) the condition of the skin (e.g. intact, flexed, broken) before and after testing.

13.3.3   Gastrointestinal Tract

The gastrointestinal tract is one of the largest immunological organs of the body, containing more lymphocytes and plasma cells than the spleen, bone marrow and lymph nodes. It is considered that the exogenous sources of ingestion exposure primarily result from hand-to-mouth contact in the workplace. Alternatively, NPs can be ingested directly if contained in food or water or as drugs or drug delivery devices. In addition, NPs cleared from the respiratory tract *via* the mucociliary escalator can subsequently be ingested into the gastrointestinal (GI) tract. So, the gastrointestinal tract is considered an important portal of entry for NP exposure.

Previous studies based on microparticulates indicated that particulate uptake happens not only via the M-cells in the Peyer's patches (PP) and the isolated follicles of the gut-associated lymphoid tissue, but also the normal intestinal enterocytes in the gastrointestinal tract (Jani *et al.*, 1989; Hussain *et al.*, 2001). So far, only a few studies have investigated the uptake and translocation of NPs via gastrointestinal tract administration, and most have shown that NPs pass through the GI tract and are eliminated rapidly. In rats dosed orally with radio-labelled functionalized C60 fullerenes, 98 % were cleared in the feces within 48 hours, while the rest was eliminated via urine (Yamago *et al.*, 1995). Another study using ^{192}Ir-labelled iridium particles (15 and 80 nm) in rats found no significant uptake in the GI tract (Kreyling *et al.*, 2002).

In contrast, some other studies indicated that NP uptake via the GI tract enabled NPs to enter the lymphatic and capillary systems and subsequently other organs. The earlier studies in rats found that the larger TiO_2 particles (150–500 nm) were taken up into the blood and transported to the liver (Jani *et al.*, 1994). Recently, an oral gavage study of TiO_2 NPs (25 and 80 nm) according to the OECD procedure found that NPs could be transported

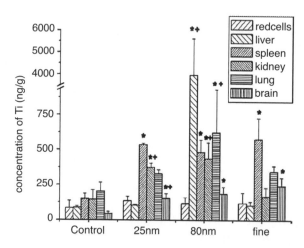

Figure 13.5   *Contents of titanium in each tissue of female mice 2 weeks post-exposure to different-sized TiO_2 particles by a single oral administration. *Represents significant difference from the control group (Dunnett's, P < 0.05); + represents significant difference from the fine group (Student's, P < 0.05) (Wang JX et al., 2007). Reprinted with permission from Elsevier*

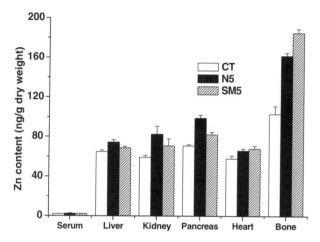

Figure 13.6 *Contents of Zn in serum and organic tissues of mice exposed to ZnO nanoparticles at the dose of 5 g/kg body weight on 14 days post-oral administration. CT, control; N5, 5 g/kg body weight 20 nm ZnO group; SM5, 5 g/kg body weight 120 nm ZnO group (Wang et al., 2008a). Reprinted with kind permission from Springer Science + Business Media. Copyright 2008*

to liver, spleen, kidneys and lung tissues (Figure 13.5) (Wang JX *et al.*, 2007). Another study in mice showed that orally administrated ZnO NPs (20 and 120 nm) could be retained in bone and pancreas and excreted via kidney (Figure 13.6) (Wang *et al.*, 2008).

Actually, the inconsistent results of absorption in the GI tract are probably due to the different particle composition, particle size and particle surface chemistry, which offers possibilities of site-specific targeting to different regions of the GI tract. A study of polystyrene particles with sizes between 50 nm and 3 μm indicated that the uptake decreased with increasing particle size from 6.6 % for 50 nm, 5.8 % for 100 nm, 0.8 % for 1 μm, to 0 % for 3 μm particles (Jani *et al.*, 1990). The kinetics of particles in the GI tract depend strongly on the charge of the particles. Positively charged latex particles were trapped in the negatively charged mucus, while negatively charged latex nanoparticles diffused across the mucus layer, interacted with epithelial cells and transported to other organs (Jani *et al.*, 1989).

13.3.4 Injection

Injection of NPs has been studied in drug delivery systems. Injection is the administration of a fluid into the subcutaneous tissue, muscle, blood vessels, or body cavities. The translocation of NPs following injection depends on the site of injection: intravenously injected NPs quickly spread throughout the circulatory system, with subsequent translocation to organs (Pereverzeva *et al.*, 2008); intradermal injection leads to lymph nodes uptake; while intramuscular injection is followed by neuronal and lymphatic system uptake (Kim *et al.*, 2004). For example, the intramuscular injection of magnetic nanoparticles smaller than 100 nm into the tongues and facial muscles of mice resulted in synaptic uptake (Olsson and Kristensson, 1981).

Elucidating the *in vivo* pharmacological profiles of intravenous administration of NPs is considered very important in the context of the underlying medical debate regarding the safety of novel nanomaterials. Once NPs have entered the blood circulation, they can be distributed throughout the body. The liver is the major distribution site via uptake by Kupffer cells, followed by the spleen as another organ of the reticulo-endothelial system. Studies in rats indicated that dextran-coated Fe_3O_4 nanoparticles were cleared rapidly from serum with a short half-life and accumulated dominantly in the liver and spleen with a total percentage of more than 90 % after intravenous injection (Ma *et al.*, 2008). Distributions of NPs to heart, kidney and immune-modulating organs (spleen, bone marrow) following intravenous injection have also been found. The intravenously-injected NPs, including iron oxide, fullerenes and gold with sizes ranging from 10–240 nm, showed localization in different organs, such as liver, spleen, bone marrow, lymph nodes, brain and lungs (Corot *et al.*, 2006; Jeong *et al.*, 2007; De Jong *et al.*, 2008; Sonavane *et al.*, 2008).

The distribution of NPs in the body is associated with their surface characteristics. It was found that coating iron-oxide NPs with polyethylene glycol or other polymers almost completely prevented hepatic and splenetic localization; therefore, in drug delivery or therapy applications, a target-specific purpose could be achieved (Park *et al.*, 2008). NPs modified by specific ligands (e.g. apolipoprotein, transferrin) on the surface might be transported into the CNS across the tight blood-brain barrier (Kreuter, 2001, 2004; Kreuter *et al.*, 2002). Therefore, the highly desirable properties of NPs must be carefully considered against potential adverse responses, and a rigorous toxicological assessment is mandatory.

13.4   Metal Oxide Nanoparticles and Toxicity

Several *in vivo* and *in vitro* studies have clearly shown that some NPs can pass through different membranes or barriers, transport into the body and accumulate in organs such as lung, liver, spleen, bone and brain. Once entering the organism, NPs may interact with biological systems. In the following section, the health effects induced by some metal oxide nanoparticles *in vivo* are discussed.

13.4.1   Iron Oxide

Iron oxide-based NPs have been suggested for many potential applications in industry, environment and biomedicine, such as photocatalysis, bioremediation, biomedical imaging, and so forth (Penpolcharoen *et al.*, 2001; Horányi and Kálmán, 2004). Currently, the iron oxide NPs available in the market mainly include ferric oxide (α-Fe_2O_3, γ-Fe_2O_3), magnetite (Fe_3O_4) and biocompatible superparamagnetic iron-oxide (SPIO) or ultrasmall superparamagnetic iron-oxide (USPIO) nanoparticles that consist of an iron oxide core (magnetite Fe_3O_4, maghemite or other insoluble ferrites) coated with polymers such as dextran. So far, some SPIO contrast agents including lumirem, Abdoscan, Feridex and Ferumoxide have been approved by the US Food and Drug Administration, which are extensively used for imaging of GI tract, liver and spleen.

The uncoated iron oxide NPs, such as ferric oxide (Fe_2O_3) and ferrous oxide (Fe_3O_4), are widely used in environmental catalysis, pigment, coating, magnetic storage, and so forth. Iron compounds are also highly exposed in mining areas, manufacturing and occupational workplaces for magnetopneumography users and welders. Zhu *et al.* (2008) investigated the size, dose and time dependence of Fe_2O_3 particles (22 nm and 280 nm) on pulmonary and coagulation systems in rats following intratracheal instillation. Both nano- and submicron-sized Fe_2O_3 particle intratracheal exposure could initiate acute lung injury. The long-term overload of particles in alveolar macrophages and epithelium cells could even cause pathological symptoms of lung emphysema and the pro-sign of lung fibrosis (Figure 13.7). Comparing with the submicron-sized Fe_2O_3 particle, the nano-sized Fe_2O_3 particle may increase microvascular permeability and cell lysis in lung epitheliums and disturb blood coagulation parameters significantly (Figure 13.7). Further, Wang *et al.* (2009) investigated the neurotoxicity and size effects of repeated low-dose (130 µg) intranasal exposure of these two sizes of Fe_2O_3 particles on mice. After the nano- and submicron-sized Fe_2O_3 particle treatment, the activities of glutathione peroxidase (GSH-Px), superoxide dismutase (Cu,Zn-SOD), and constitutive nitric oxide synthases (cNOS) were significantly elevated; meanwhile, the total glutathione (GSH) and GSH/GSSG ratio significantly decreased in the olfactory bulb and hippocampus. Furthermore, some ultrastructural alterations in the mice brain, such as neurodendron degeneration, membranous structure disruption and lysosome increase in the olfactory bulb, slight dilation in the rough endoplasmic reticulum and lysosome increase in the hippocampus, were induced by the nano-sized Fe_2O_3 particles (Wang *et al.*, 2009).

For functional iron oxide nanoparticles, several studies have investigated the biodistribution, clearance and biocompatibility of polymer-coated Fe_3O_4-based magnetic NPs, such as dextran-Fe_3O_4, oleic acid (OA)-Pluronic-Fe_3O_4, alginate-Fe_3O_4, and Polyvinyl alcohol (PVA)-Fe_3O_4 in rats after intravenous administration (Gamarra *et al.*, 2008; Jain *et al.*, 2008). The studies in rats via tail vein injection of oleic acid (OA)-Pluronic-Fe_3O_4 nanoparticles (11 ± 2 nm) (10 mg Fe/kg body weight) showed that greater fractions of the injected iron oxide nanoparticles were localized in the liver and spleen than in the brain, heart, kidney and lung tissues (Jain *et al.*, 2008). There was a transient increase in serum alanine aminotransferase (ALT), aspartate aminotransferase (AST), alkaline phosphatase (AKP), and total iron-binding capacity (TIBC) over a period of 6–24 h following nanoparticle injection; however these indices returned to normal levels within 3 days post-injection. Histological analyses of liver, spleen and kidney samples collected at 1 and 7 days post-exposure showed no apparent abnormal changes. Thus, it may indicate that the functional iron oxide nanoparticles may safely be used for drug delivery and imaging applications.

13.4.2   Titanium Oxide

TiO_2 nanoparticles are widely used as white pigments, cosmetics, and photocatalysts in air and water cleaning. They are regarded as poorly soluble or insoluble particulates and commonly found in three crystalline forms: rutile, anatase and brookite. Anatase and rutile are the most common polymorphs of synthetic TiO_2 NPs. The commercially available DeGussa P25 consists of $\sim$25 nm particles of about 80 % anatase and 20 % rutile.

Most studies based on TiO_2 NPs focus their effects on the respiratory system. The first example involving studies of ultrafine and fine TiO_2 particles showed that ultrafine anatase TiO_2 (20 nm), when intratracheally instilled into rats and mice, induced a much greater pulmonary-inflammatory neutrophil response (determined in the lung lavage 24 h after dosing) than fine anatase TiO_2 (250 nm) at the same instilled mass dose (Figure 13.8a).

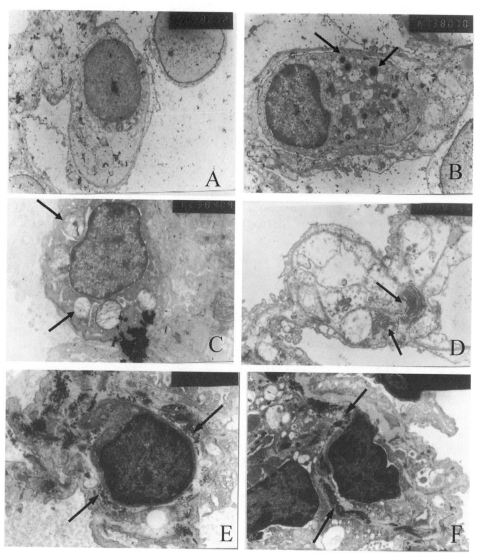

Figure 13.7 *TEM images of the lungs at days 7 and 30 after intratracheal instillation of particles or saline in rats (magnification ×8000). (A) Control group. (B) Day 7 after instillation of 20 mg/kg body weight 22 nm-Fe₂O₃. Lysosome increased in lung type II epithelium. (C) and (D) Day 30 after instillation of 20 mg/kg body weight 280 nm-Fe₂O₃. Mitochondria tumefaction (C) and organelles dissolution (D) (thin arrow) were shown. The increased amount of collagen formation could be observed. (E) 1 day after instillation of 20 mg/kg body weight 22 nm-Fe₂O₃. Particles translocated into the lung epitheliums (thin arrow). (F) Day 7 after instillation of 20 mg/kg body weight 22 nm-Fe₂O₃. Particles translocated into the lung epitheliums (thin arrow) (Zhu et al., 2008). Reprinted with permission from Elsevier*

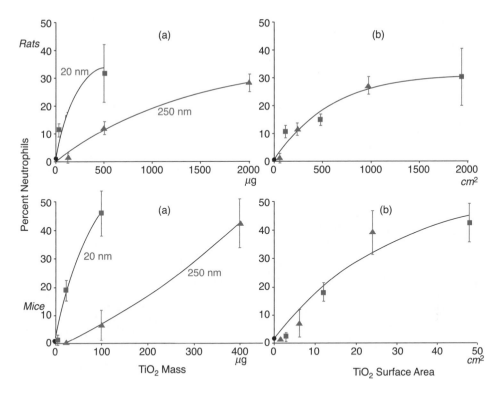

Figure 13.8 *Percentage of neutrophils in lung lavage of rats and mice as indicators of inflammation, 24 hrs after intratracheal instillation of different mass doses of 20 nm and 250 nm TiO$_2$ particles in rats and mice. (a) The steeper dose–response of nanosized TiO$_2$ is obvious when dose is expressed as mass. (b) The same dose–response relationship but with dose expressed as particle surface area, indicating that particle surface area seems to be a more appropriate dosemetric for comparing effects of different-sized particles, provided they are of the same chemical structure (anatase TiO$_2$ in this case) (Oberdörster, 2000). Reprinted with permission of the Royal Society*

However, when the instilled dose was expressed as particle surface area it became obvious that the neutrophil response in the lung for both ultrafine and fine TiO$_2$ fitted the same dose-response curve (Figure 13.8b), suggesting that particle surface area for particles of different sizes but of the same chemical structure is a better dosemetric than particle mass or particle number (Oberdörster, 2000).

Another study in mice with acute and subacute inhalation exposures (4 hr/day for 10 days) of TiO$_2$ NPs (2–5 nm, anatase) in a whole body exposure chamber showed no adverse effects of the animals, while modest pulmonary inflammatory response and pathological damage were observed following subacute exposure (Figure 13.9) (Grassian *et al.*, 2007). A further toxic mechanism study in mice indicated that the pulmonary toxicity of intratracheally instilled TiO$_2$ NPs (rutile, 19–21 nm), such as pulmonary emphysema, macrophages accumulation, extensive disruption of alveolar septa, type II pneumocyte

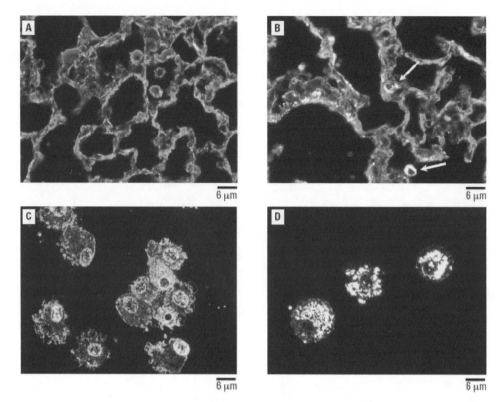

Figure 13.9 Dark field micrographs of lung tissue with H&E staining (A, B) and alveolar macrophages prepared by cytospinning and H&E staining (C, D). (A, C) Sentinels and (B, D) animals subacutely exposed to TiO₂ nanoparticles with a primary particle size of 2–5 nm and necropsied immediately after the last exposure. Arrows point to TiO₂ nanoparticle-laden macrophages (Grassian et al., 2007). Reproduced with permission from Environmental Health Perspectives. See color plate section

hyperplasia, and epithelial cell apoptosis, might be attributed to up-regulation of placenta growth factor (PlGF) and other chemokine (CXCL1, CXCL5 and CCL3) expressions (Chen *et al.*, 2006).

So far, few studies have investigated the health effects of TiO₂ NPs via injection, oral and skin exposure. Fabian *et al.* (2008) investigated the toxicity of intravenously administered TiO₂ nanoparticles in rats (5 mg/kg body weight, <100 nm) and found no obvious toxic effects, immune responses or changes in functions of organs throughout the study. In an acute oral toxicity study in mice with different sizes of TiO₂ NPs (25, 80 and 155 nm), Wang JX *et al.* (2007) reported that nanosized TiO₂ particles (25 nm and 80 nm) showed significant changes in serum biochemical parameters (ALT/AST, lactate dehydrogenase (LDH)) and pathology (hydropic degeneration around the central vein and spotty necrosis of hepatocytes) of liver compared with fine TiO₂ particles (155 nm). In addition, after 25 nm and 80 nm TiO₂ particle treatment, pathological changes of kidney, such as the renal tubules filled with protein liquids, were also observed. Compared with the control group, a significant increase in serum LDH and alpha-hydroxybutyrate dehydrogenase (alpha-HBDH) in 25 and 80 nm groups indicated myocardial damage as well. To sum up, the nanosized TiO_2 particles showed more severe acute oral toxicity than the fine particles.

13.4.3   Other Metal Oxide Nanoparticles

In addition to iron oxide and titanium oxide NPs, other metal oxide NPs such as zinc oxide (ZnO) and nickel oxide (NiO) also have various applications in industry and the environment. Engineered ZnO NPs, as a piezoelectric as well as semiconductive material, have been widely applied in electronics, optoelectronics, gas sensors and environmental remediation (Lee et al., 2002; Angappane et al., 2006; Maity and Chaudhuri, 2008). NiO NPs with a uniform size and good dispersion are desirable for many applications, such as designing ceramic, magnetic, electrochromic and heterogeneous catalytic materials (Tao and Wei, 2004). Furthermore, Zn, Mn, Cu and Ni were found to be the most abundant metals after Fe and Ti, in the standard reference material (SRM) of urban and diesel engine particulate matter (National Institute of Standards and Technology, Gaithersburg, MD) (Huggins et al., 2000). Therefore, during combustion and industrial processes, such as brass founding and welding or cutting galvanized sheet metal, high concentrations of ZnO, MnO_2 and NiO nanoparticles may be produced (Zimmer et al., 2002; Pagan et al., 2003).

In a human inhalation study, 12 healthy adults inhaled 500 µg/m^3 of ultrafine ZnO (40.4 ± 2.7 nm), fine ZnO (291.2 ± 20.2 nm), and filtered air while at rest for 2 hours. At this level of exposure, no differences in leukocyte surface markers, hemostasis and cardiac electrophysiology were detected between any of the three exposure conditions (Beckett et al., 2005). Wang et al. (2008), referring to the guidelines of OECD (2001) for Testing of Chemicals (No. 420), evaluated the acute oral toxicity of 20 and 120 nm ZnO particles at doses of 1, 2, 3, 4 and 5 g/kg body weight. According to the Globally Harmonized Classification System (GHS) for the classification of chemicals, both 20 and 120 nm ZnO were classified as nontoxic grade. The pathological results showed that liver, spleen, heart, pancreas and bone are the target organs for 20 and 120 nm ZnO oral exposure, and the two sizes of ZnO particles displayed opposite dose-effect relationships in pathological damage (Wang et al., 2008).

Studies in rats exposed to ultrafine MnO_2 particles (30 nm, ∼ 500 µg/m^3) with either nostrils patent or the right nostril occluded showed that the olfactory neuronal pathway was efficient for translocating inhaled MnO_2 NPs to the CNS, and the translocated MnO_2 NPs could result in inflammatory changes in the olfactory bulb such as significant increases in tumor necrosis factor-α mRNA (∼8-fold) and protein (∼30-fold), macrophage inflammatory protein-2, glial fibrillary acidic protein, and neuronal cell adhesion molecule mRNA (Figure 13.10) (Elder et al., 2006). For NiO NPs, Oyabu et al. (2007) demonstrated that the elimination half-time of NiO NPs (139 ± 12 nm) in the lung of rats was 62 days following inhalation exposure (1.0 ± 0.5 × 105 particles/cm^3, 6 h/day for 4 weeks) (Figure 13.11). However, the, pulmonary pathological findings indicated only mild inflammatory cell responses at 4 days after the 4-week inhalation exposure, but not severe throughout the observation time (Oyabu et al., 2007).

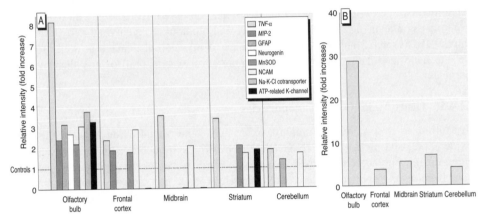

Figure 13.10   *Gene and protein expression changes, by brain region, from pooled samples after 11 days of inhalation exposure to ultrafine Mn oxide aerosols. (A) Gene expression changes represented as relative intensities (fold increase over normalized control, dashed line). (B) TNF-α protein expression changes (relative intensities, fold increase over normalized control). Abbreviations: MIP-2, macrophage inflammatory protein-2; MnSOD, manganese superoxide dismutase; NCAM, neuronal cell adhesion molecule (Elder et al., 2006). Reproduced with permission from Environmental Health Perspectives*

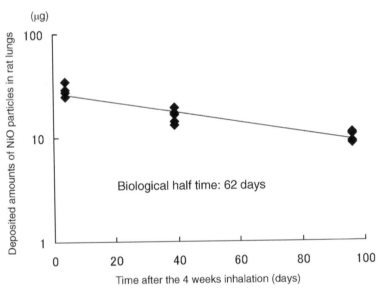

Figure 13.11   *Temporal change of NiO amounts in rat lungs in the observation time (Oyabu et al., 2007). Reproduced from Environmental Health Perspectives*

13.5   Conclusions and Prospects

Human exposure to NPs from natural and anthropogenic sources has occurred since ancient times. However, there is heightened concern today over whether the development of nanotechnology will have an impact on public health, and it is indisputable that engineered nanoparticles are a source of nanoparticle pollution when they have not been safely manufactured, handled, and disposed of. A large body of research exists regarding nanoparticle toxicity, comprising epidemiological, animal, human, and cell culture studies. Research on humans and animals indicates that some NPs are able to enter the body, and rapidly migrate to the organs via the circulatory and lymphatic systems. Toxic effects of NPs have already been documented on the lung, heart, liver, kidney and brain tissues.

Although more and more scientists are involved in the study of the health effects of NPs, our current knowledge about the toxic effects of NPs *in vivo* is still relatively limited, and only a glimpse of some toxic paradigms have been provided. Several important questions and research directions should be addressed in the near future by the scientific community involving in the study of NP sciences:

1. Advanced analysis of the physical and chemical characteristics of NPs will continue to be essential in revealing the relationship between their size, composition, crystallinity, morphology and their electromagnetic response properties, reactivity, aggregation, and kinetics.
2. Further studies on kinetics and biochemical interactions of NPs within organisms are also imperative.
3. How the interactions of NPs with biomolecules or other chemicals in the organism impact upon the toxicity of NPs should be pursued.
4. The special toxicological mechanisms of NPs *in vivo* should be clarified.
5. Research should also be directed toward finding ways to reduce the toxicity of nanomaterials.

We conclude that the study of the health effects of NPs has increased our awareness of environmental particulate pollution generated from natural, anthropogenic and nanotechnology sources. With increased knowledge and ongoing study, we are more likely to understand the causes and mechanisms of nanoparticle-associated toxicities, and will find ways to prevent them.

References

Aitken RJ, Chaudhry MQ, Boxall ABA, Hull M (2006) Manufacture and use of nanomaterials: current status in the UK and global trends. *Occup Med* **56**, 300–306.

Angappane S, John NS, Kulkarni GU (2006) Pyramidal nanostructures of zinc oxide. *J Nanosci Nanotechnol* **6**, 101–104.

BCC Research (2007) *Nanostructured Materials for the Biomedical, Pharmaceutical, & Cosmetic Markets*. Available at http://www.bccresearch.com/report/NAN017D.html.

Beckett WS, Chalupa DF, Pauly-Brown A, Speers DM, Stewart JC, Frampton MW, Utell MJ, Huang LS, Cox C, Zareba W, Oberdörster G (2005) Comparing inhaled ultrafine versus fine zinc oxide particles in healthy adults: a human inhalation study. *Am J Respir Crit Care Med* **171**, 1129–1135.

Bennat C, Mueller-Goymann CC (2000) Skin penetration and stabilization of formulations containing microfine titanium dioxide as physical UV filter. *Int J Cosmet Sci* **22**, 271–283.

Berry JP, Arnoux B, Stanislas G, Galle P, Chretien J (1977) A microanalytic study of particles transport across alveoli: role of blood-platelets. *Biomedicine* **27**, 354–357.

Bodian D, Howe HA (1941) The rate of progression of poliomyelitis virus in nerves. *Bull Johns Hopkins Hosp* **69**, 79–85.

Borm PJA, Kreyling W (2004) Toxicological hazards of inhaled nanoparticles-potential implications for drug delivery. *J Nanosci Nanotechnol* **4**, 521–531.

Buzea C, Pacheco II, Robbie K (2007) Nanomaterials and nanoparticles: sources and toxicity. *Biointerphases* **2**, MR17–MR71.

Chen HW, Su SF, Chien CT, Lin WH, Yu SL, Chou CC, Chen JJW, Yang PC (2006) Titanium dioxide nanoparticles induce emphysema-like lung injury in mice. *FASEB J* **20**, 2393–2395.

Corot C, Robert P, Idee JM, Port M (2006) Recent advances in iron oxide nanocrystal technology for medical imaging. *Adv Drug Deliver Rev* **58**, 1471–1504.

De Jong WH, Hagens WI, Krystek P, Burger MC, Sips A, Geertsma RE (2008) Particle size-dependent organ distribution of gold nanoparticles after intravenous administration. *Biomaterials* **29**, 1912–1919.

Delorenzo AJ (1957) Electron microscopic observations of the olfactory mucosa and olfactory nerve. *J Biophys Biochem Cyto* **3**, 839–850.

Elder A, Gelein R, Silva V, Feikert T, Opanashuk L, Carter J, Potter R, Maynard A, Finkelstein J, Oberdorster G (2006) Translocation of inhaled ultrafine manganese oxide particles to the central nervous system. *Environ Health Perspect* **114**, 1172–1178.

Fabian E, Landsiedel R, Ma-Hock L, Wiench K, Wohlleben W, van Ravenzwaay B (2008) Tissue distribution and toxicity of intravenously administered titanium dioxide nanoparticles in rats. *Arch Toxicol* **82**, 151–157.

Gamarra LF, Pontuschka WM, Amaro E, Costa-Filho AJ, Brito GES, Vieira ED, Carneiro SM, Escriba DM, Falleiros AMF, Salvador VL (2008) Kinetics of elimination and distribution in blood and liver of biocompatible ferrofluids based on Fe_3O_4 nanoparticles: An EPR and XRF study. *Mater Sci Eng C-Biomim Supramol Syst* **28**, 519–525.

Grassian VH, O'Shaughnessy PT, Adamcakova-Dodd A, Pettibone JM, Thorne PS (2007) Inhalation exposure study of titanium dioxide nanoparticles with a primary particle size of 2 to 5 nm. *Environ Health Perspect* **115**, 397–402.

Gupta AK, Naregalkar RR, Vaidya VD, Gupta M (2007) Recent advances on surface engineering of magnetic iron oxide nanoparticles and their biomedical applications. *Nanomedicine* **2**, 23–39.

Horányi G, Kálmán E (2004) Anion specific adsorption on Fe_2O_3 and AlOOH nanoparticles in aqueous solutions: comparison with hematite and gamma-Al_2O_3. *J Colloid Interface Sci* **269**, 315–319.

Huggins FE, Huffman GP, Robertson JD (2000) Speciation of elements in NIST particulate matter SRMs 1648 and 1650. *J Hazard Mater* **74**, 1–23.

Hussain N, Jaitley V, Florence AT (2001) Recent advances in the understanding of uptake of microparticulates across the gastrointestinal lymphatics. *Adv Drug Deliver Rev* **50**, 107–142.

ICRP (1994) Publication 66: Human respiratory model for radiological protection. *Ann ICRP* **24**, 1–482.

Jain TK, Reddy MK, Morales MA, Leslie-Pelecky DL, Labhasetwar V (2008) Biodistribution, clearance, and biocompatibility of iron oxide magnetic nanoparticles in rats. *Mol Pharm* **5**, 316–327.

Jani P, Halbert GW, Langridge J, Florence AT (1989) The uptake and translocation of latex nanospheres and microspheres after oral-administration to rats. *J Pharm Pharmacol* **41**, 809–812.

Jani P, Halbert GW, Langridge J, Florence AT (1990) Nanoparticle uptake by the rat gastrointestinal mucosa: quantitation and particle-size dependency. *J Pharm Pharmacol* **42**, 821–826.

Jani PU, McCarthy DE, Florence AT (1994) Titanium-dioxide (rutile) particle uptake from the rat Gi tract and translocation to systemic organs after oral-administration. *Int J Pharm* **105**, 157–168.

Jeong U, Teng XW, Wang Y, Yang H, Xia YN. 2007. Superparamagnetic colloids: Controlled synthesis and niche applications. *Adv. Mater.* **19**: 33–60.

Kim S, Lim YT, Soltesz EG, De Grand AM, Lee J, Nakayama A, Parker JA, Mihaljevic T, Laurence RG, Dor DM, Cohn LH, Bawendi MG, Frangioni JV (2004) Near-infrared fluorescent type II quantum dots for sentinel lymph node mapping. *Nat Biotechnol* **22**, 93–97.

Klabunde KJ, Stark J, Koper O, Mohs C, Park DG, Decker S, Jiang Y, Lagadic I, Zhang DJ (1996) Nanocrystals as stoichiometric reagents with unique surface chemistry. *J Phys Chem* **100**, 12142–12153.

Kreuter J (2001) Nanoparticulate systems for brain delivery of drugs. *Adv Drug Deliver Rev* **47**, 65–81.

Kreuter J (2004) Influence of the surface properties on nanoparticle-mediated transport of drugs to the brain. *J Nanosci Nanotechnol* **4**, 484–488.

Kreuter J, Shamenkov D, Petrov V, Ramge P, Cychutek K, Koch-Brandt C, Alyautdin R (2002) Apolipoprotein-mediated transport of nanoparticle-bound drugs across the blood-brain barrier. *J Drug Target* **10**, 317–325.

Kreyling WG (2006) Translocation and accumulation of nanoparticles in secondary target organs after uptake by various routes of intake. *Toxicol Lett* **164**, S34–S34.

Kreyling W, Scheuch G (2000) Clearance of particles deposited in the lungs. In *Particle-Lung Interactions*, Gehr P, Heyder J (eds). Marcel Dekker: New York; 323–376.

Kreyling WG, Semmler M, Erbe F, Mayer P, Takenaka S, Schulz H, Oberdörster G, Ziesenis A (2002) Translocation of ultrafine insoluble iridium particles from lung epithelium to extrapulmonary organs is size dependent but very low. *J Toxicol Environ Health A* **65**, 1513–1530.

Lademann J, Weigmann HJ, Rickmeyer C, Barthelmes H, Schaefer H, Mueller G, Sterry W (1999) Penetration of titanium dioxide microparticles in a sunscreen formulation into the horny layer and the follicular orifice. *Skin Pharmacol Appl Skin Physiol* **12**, 247–256.

Lee CJ, Lee TJ, Lyu SC, Zhang Y, Ruh H, Lee HJ (2002) Field emission from well-aligned zinc oxide nanowires grown at low temperature. *Appl Phys Lett* **81**, 3648–3650.

Ma HL, Xu YF, Qi XR, Maitani Y, Nagai T (2008) Superparamagnetic iron oxide nanoparticles stabilized by alginate: Pharmacokinetics, tissue distribution, and applications in detecting liver cancers. *Int J Pharm* **354**, 217–226.

Magnusson BM, Anissimov YG, Cross SE, Roberts MS (2004) Molecular size as the main determinant of solute maximum flux across the skin. *J Invest Dermatol* **122**, 993–999.

Maity AB, Chaudhuri S (2008) Gas sensing properties of ZnO nanowires. *Trans Indian Ceram Soc* **67**, 1–15.

Mills NL, Amin N, Robinson SD, Anand A, Davies J, Patel D, de la Fuente JM, Cassee FR, Boon NA, MacNee W, Millar AM, Donaldson K, Newby DE (2006) Do inhaled carbon nanoparticles translocate directly into the circulation in humans? *Am J Respir Crit Care Med* **173**, 426–431.

Nemmar A, Hoet PHM, Vanquickenborne B, Dinsdale D, Thomeer M, Hoylaerts MF, Vanbilloen H, Mortelmans L, Nemery B (2002) Passage of inhaled particles into the blood circulation in humans. *Circulation* **105**, 411–414.

Nichkova M, Dosev D, Gee SJ, Hammock BD, Kennedy IM (2007) Multiplexed immunoassays for proteins using magnetic luminescent nanoparticles for internal calibration. *Anal Biochem* **369**, 34–40.

Nohynek GJ, Dufour EK, Roberts MS (2008) Nanotechnology, cosmetics and the skin: is there a health risk? *Skin Pharmacol Physiol* **21**, 136–149.

Nowack B, Bucheli TD (2007) Occurrence, behavior and effects of nanoparticles in the environment. *Environ Pollut* **150**, 5–22.

Oberdörster G (2000) Toxicology of ultrafine particles: in vivo studies. *Philos Trans R Soc Lond Series A Math Phys Eng Sci* **358**, 2719–2739.

Oberdörster G, Ferin J, Gelein R, Soderholm SC, Finkelstein J (1992) Role of the alveolar macrophage in lung injury: studies with ultrafine particles. *Environ Health Perspect* **97**, 193–199.

Oberdörster G, Ferin J, Lehnert BE (1994) Correlation between particle-size, in-vivo particle persistence, and lung injury. *Environ Health Perspect* **102**, 173–179.

Oberdörster G, Sharp Z, Atudorei V, Elder A, Gelein R, Lunts A, Kreyling W, Cox C (2002) Extrapulmonary translocation of ultrafine carbon particles following whole-body inhalation exposure of rats. *J Toxicol Environ Health A* **65**, 1531–1543.

Oberdörster G, Sharp Z, Atudorei V, Elder A, Gelein R, Kreyling W, Cox C (2004) Translocation of inhaled ultrafine particles to the brain. *Inhal Toxicol* **16**, 437–445.

Oberdörster G, Maynard A, Donaldson K, Castranova V, Fitzpatrick J, Ausman K, Carter J, Karn B, Kreyling W, Lai D, Olin S, Monteiro-Riviere N, Warheit D, Yang H (2005a) Principles for characterizing the potential human health effects from exposure to nanomaterials: elements of a screening strategy. *Part Fibre Toxicol* **2**, 1–35.

Oberdörster G, Oberdörster E, Oberdörster J (2005b) Nanotoxicology: an emerging discipline evolving from studies of ultrafine particles. *Environ Health Perspect* **113**, 823–839.

OECD (2001) *OECD Guidelines for Testing of Chemicals. No. 425: Acute oral toxicity-up-and-down procedure.* Organisation for Economic Co-operation and Development: Paris.

Ohl L, Mohaupt M, Czeloth N, Hintzen G, Kiafard Z, Zwirner J, Blankenstein T, Henning G, Forster R (2004) CCR7 governs skin dendritic cell migration under inflammatory and steady state conditions. *Immunity* **21**, 279–288.

Oldfors A, Fardeau M (1983) The permeability of the basal lamina at the neuromuscular junction. An ultrastructural study of rat skeletal-muscle using particulate tracers. *Neuropathol Appl Neurobiol* **9**, 419–432.

Olsson T, Kristensson K (1981) Neuronal uptake of iron: somatopetal axonal-transport and fate of cationized and native ferritin, and iron-dextran after intramuscular injections. *Neuropathol Appl Neurobiol* **7**, 87–95.

Oyabu T, Ogami A, Morimoto Y, Shimada M, Lenggoro W, Okuyama K, Tanaka I (2007) Biopersistence of inhaled nickel oxide nanoparticles in rat lung. *Inhal Toxicol* **19**, 55–58.

Pagan I, Costa DL, McGee JK, Richards JH, Dye JA, Dykstra MJ (2003) Metals mimic airway epithelial injury induced by in vitro exposure to Utah Valley ambient particulate matter extracts. *J Toxicol Environ Health A* **66**, 1087–1112.

Park JY, Daksha P, Lee GH, Woo S, Chang YM (2008) Highly water-dispersible PEG surface modified ultra small superparamagnetic iron oxide nanoparticles useful for target-specific biomedical applications. *Nanotechnology* **19**, 365603–365609.

Penpolcharoen M, Amal R, Brungs M (2001) Degradation of sucrose and nitrate over titania coated nano-hematite photocatalysts. *J Nanopart Res* **3**, 289–302.

Pereverzeva E, Treschalin I, Bodyagin D, Maksimenko O, Kreuter J, Gelperina S (2008) Intravenous tolerance of a nanoparticle-based formulation of doxorubicin in healthy rats. *Toxicol Lett* **178**, 9–19.

Pflücker F, Wendel V, Hohenberg H, Gartner E, Will T, Pfeiffer S, Wepf R, Gers-Barlag H (2001) The human stratum corneum layer: An effective barrier against dermal uptake of different forms of topically applied micronised titanium dioxide. *Skin Pharmacol Appl Skin Physiol* **14**, 92–97.

Roco MC (2005) Environmentally responsible development of nanotechnology. *Environ Sci Technol* **39**, 106A–112A.

Rouse JG, Yang JZ, Ryman-Rasmussen JP, Barron AR, Monteiro-Riviere NA (2007) Effects of mechanical flexion on the penetration of fullerene amino acid-derivatized peptide nanoparticles through skin. *Nano Lett* **7**, 155–160.

Schulz J, Hohenberg H, Pflucker F, Gartner E, Will T, Pfeiffer S, Wepf R, Wendel V, Gers-Barlag H, Wittern KP (2002) Distribution of sunscreens on skin. *Adv Drug Deliver Rev* **54**, S157–S163.

Semmler-Behnke M, Takenaka S, Fertsch S, Wenk A, Seitz J, Mayer P, Oberdörster G, Kreyling WG (2007) Efficient elimination of inhaled nanoparticles from the alveolar region: Evidence for interstitial uptake and subsequent reentrainment onto airway epithelium. *Environ Health Perspect* **115**, 728–733.

Sonavane G, Tomoda K, Makino K (2008) Biodistribution of colloidal gold nanoparticles after intravenous administration: effect of particle size. *Colloids Surf B-Biointerfaces* **66**, 274–280.

Takenaka S, Karg E, Kreyling WG, Lentner B, Moller W, Behnke-Semmler M, Jennen L, Walch A, Michalke B, Schramel P, Heyder J, Schulz H (2006) Distribution pattern of inhaled ultrafine gold particles in the rat lung. *Inhal Toxicol* **18**, 733–740.

Tao DL, Wei F (2004) New procedure towards size-homogeneous and well-dispersed nickel oxide nanoparticles of 30 nm. *Mater Lett* **58**, 3226–3228.

Wang B, Feng WY, Wang M, Shi JW, Zhang F, Ouyang H, Zhao YL, Chai ZF, Huang YY, Xie YN, Wang HF, Wang J (2007) Transport of intranasally instilled fine Fe_2O_3 particles into the brain: Micro-distribution, chemical states, and histopathological observation. *Biol Trace Elem Res* **118**, 233–243.

Wang B, Feng WY, Wang M, Wang TC, Gu YQ, Zhu MT, Ouyang H, Shi JW, Zhang F, Zhao YL, Chai ZF, Wang HF, Wang J (2008) Acute toxicological impact of nano- and submicro-scaled zinc oxide powder on healthy adult mice. *J Nanopart Res* **10**, 263–276.

Wang B, Feng WY, Zhu MT, Wang Y, Wang M, Gu YQ, Ouyang H, Wang HJ, Li M, Zhao YL, Chai ZF, Wang HF (2009) Neurotoxicity of low-dose repeatedly intranasal instillation of nano- and submicron-sized ferric oxide particles in mice. *J Nanopart Res* **11**, 41–53.

Wang JX, Chen CY, Sun J, Yu HW, Li YF, Li B, Xing L, Huang YY, He W, Gao YX, Chai ZF, Zhao YL (2005) Translocation of inhaled TiO_2 nanoparticles along olfactory nervous system to brain studied by synchrotron radiation X-ray fluorescence. *High Ener Phys Nucl Phys* **29**, 76–79.

Wang JX, Zhou GQ, Chen CY, Yu HW, Wang TC, Ma YM, Jia G, Gao YX, Li B, Sun J, Li YF, Jiao F, Zhao YL, Chai ZF (2007) Acute toxicity and biodistribution of different sized titanium dioxide particles in mice after oral administration. *Toxicol Lett* **168**, 176–185.

Wiebert P, Sanchez-Crespo A, Falk R, Philipson K, Lundin A, Larsson S, Moller W, Kreyling WG, Svartengren M (2006) No significant translocation of inhaled 35-nm carbon particles to the circulation in humans. *Inhal Toxicol* **18**, 741–747.

Yamago S, Tokuyama H, Nakamura E, Kikuchi K, Kananishi S, Sueki K, Nakahara H, Enomoto S, Ambe F (1995) In-vivo biological behavior of a water-miscible fullerene-C-14 labeling, absorption, distribution, excretion and acute toxicity. *Chem Biol* **2**, 385–389.

Zhang LW, Monteiro-Riviere NA (2008) Assessment of quantum dot penetration into intact, tape-stripped, abraded and flexed rat skin. *Skin Pharmacol Phys* **21**, 166–180.

Zhu MT, Feng WY, Wang B, Wang TC, Gu YQ, Wang M, Wang Y, Ouyang H, Zhao YL, Chai ZF (2008) Comparative study of pulmonary responses to nano- and submicron-sized ferric oxide in rats. *Toxicology* **247**, 102–111.

Zhu MT, Feng WY, Wang Y, Wang B, Wang M, Ouyang H, Zhao YL, Chai ZF (2009) Particokinetics and extrapulmonary translocation of intratracheally instilled ferric oxide nanoparticles in rats and the potential health risk assessment. *Toxicol Sci* **107**, 342–351.

Zimmer AT, Baron PA, Biswas P (2002) The influence of operating parameters on number-weighted aerosol size distribution generated from a gas metal arc welding process. *J Aerosol Sci* **33**, 519–531.

14

In Vivo Hypersensitive Pulmonary Disease Models for Nanotoxicity

Ken-ichiro Inoue and Hirohisa Takano

14.1 Introduction

Epidemiological studies have demonstrated a correlation between exposure to air pollutant particles at the concentrations currently found in major metropolitan areas, and mortality and morbidity (Samet *et al.*, 2000). The concentration of particulate matter (PM) with a mass median aerodynamic diameter (a density-dependent unit of measure used to describe the diameter of the particle) of less than or equal to 2.5 μm ($PM_{2.5}$) is more closely associated with both acute and chronic respiratory effects and subsequent mortality than larger particles of ≤10 μm (PM_{10}) (Peters *et al.*, 1997). Moreover, one intriguing aspect of the epidemiological data is that health effects of $PM_{2.5}$ are primarily seen in subjects with predisposing factors, including pneumonia, asthma, chronic obstructive pulmonary disease, compromised immune systems, or aged over 65 years old (Dockery *et al.*, 1993). Consistent with the epidemiological studies, we have experimentally demonstrated that diesel exhaust particles (DEPs), major contributors to environmental $PM_{2.5}$ in urban areas, exhibit respiratory toxicity with or without predisposing factors *in vivo* (Ichinose *et al.*, 1995, 1997; Takano *et al.*, 1997, 2002).

To date, nanoparticles of less than 0.1 μm in mass median aerodynamic diameter have been shown to be increasing in ambient air (Cyrys *et al.*, 2003). Recent measurements indicate that nanoparticle numbers in ambient air range from $2 \times 10^4/cm^3$ to $2 \times 10^5/cm^3$, with mass concentrations of more than 50 μg/m^3 near major highways (Zhu *et al.*, 2002; Timonen *et al.*, 2004). Also, nanoparticles have been implicated in cardiopulmonary system effects (Utell *et al.*, 2000). Furthermore, compared with larger particles, nanoparticles have a higher deposition rate in the peripheral lung; they can cross the pulmonary epithelium

Nanoxicity: From In Vivo *and* In Vitro *Models to Health Risks* Edited by Saura Sahu and Daniel Casciano
© 2009 John Wiley & Sons, Ltd

and reach the interstitium (Oberdorster, 2001), and may thus be systemically distributed in the bloodstream (Seaton *et al.*, 1995). Nanoparticles have an enhanced capacity to produce reactive oxygen species, and consequently have a widespread toxicity (Brown *et al.*, 2001; Dick *et al.*, 2003; Li *et al.*, 2003).

On the other hand, development of nanotechnology has increased the risk of other types of particles than combustion-derived particles in the environment, namely, engineered nanomaterials (Oberdorster *et al.*, 2005). Considering the variety of their sizes and natures, it can be imagined that exposure to nanolevel (<100 nm) materials might also lead to adverse health effects, including cardiorespiratory ones. Indeed, health toxicity induced by nanomaterials is being researched worldwide. In particular, nanomaterial exposure reportedly induces several patterns of lung inflammation (Warheit *et al.*, 2004, 2006, 2007; Shvedova *et al.*, 2005; Chen *et al.*, 2006). Besides their toxic effects on health, therefore, it should be ascertained whether they also aggravate preexisting pathological conditions. Nonetheless, few ideal animal models to test facilitating effects of nanoparticles/materials have, if any, been developed.

14.2 Effects of Nanoparticles on Acute Lung Inflammation Induced by Bacterial Endotoxin

A glycolipid of gram-negative bacteria, known as endotoxin or lipopolysaccharide (LPS), stimulates host cells via innate immunity (Vincenti *et al.*, 1992). In animal models, intratracheal administration of LPS causes lung cytokine production, neutrophil recruitment and lung injury (Ulich *et al.*, 1991). LPS is found in bronchoalveolar lavage (BAL) fluid of patients with pneumonia (Flanagan *et al.*, 2001) and acute respiratory distress syndrome (Martin *et al.*, 1997), which sometimes results in a fatal outcome. In addition, LPS is a significant constituent of many air pollutant particles and has accordingly been implicated in the adverse effects of PM (Becker *et al.*, 2002). In accordance with the close links among LPS, lung inflammation (injury), and PM, we have previously shown that intratracheal administration of DEPs and their components facilitates lung inflammation induced by LPS (Takano *et al.*, 2002; Yanagisawa *et al.*, 2003) and subsequent systemic inflammation with coagulatory disturbance (Inoue *et al.*, 2006a). On the other hand, we have also shown that inhaled diesel exhaust (DE: peak particle size of 110 nm) at a soot concentration of 0.3, 1.0 or 3.0 mg/m^3 does not exacerbate the lung inflammation (Inoue *et al.*, 2006b). It is noteworthy, however, that in reality, the size of diesel engine-derived particles has been getting smaller (400 nm: Takano *et al.* (1998); 110 nm: Inoue *et al.* (2006b)).

We examined the effects of pulmonary exposure to nanoparticles (by an intratracheal instillation technique) on lung inflammation related to LPS in mice. The vehicle, two sizes (14 and 56 nm) of carbon black nanoparticles (4 mg/kg), LPS (2.5 mg/kg), or LPS + nanoparticles was administered intratracheally, and parameters of lung inflammation and coagulation were evaluated. Nanoparticles alone induced slight lung inflammation and significant pulmonary edema as compared with the vehicle. Fourteen nm nanoparticles intensively aggravated LPS-elicited lung inflammation and pulmonary edema, which was concomitant with the enhanced lung expression of interleukin (IL)-1β, macrophage inflammatory protein (MIP)-1α, macrophage chemoattractant protein (MCP)-1, MIP-2, and keratinocyte chemoattractant (KC) in overall trend, whereas 56 nm nanoparticles did not show apparent effects. Immunoreactivity for 8-hydroxyguanosine (OHdG), a proper marker for oxidative stress, was more intense in the lung from the LPS + 14 nm nanoparticle group than in that from the LPS group. The circulatory fibrinogen level was higher in the LPS + 14 nm nanoparticle group than in the LPS group. Taken together, nanoparticles can aggravate lung inflammation related to bacterial endotoxin, which is more prominent with smaller particles. The enhancing effect may be mediated, at least partly, by the increased local expression of proinflammatory cytokines and via the oxidative stress. Furthermore, nanoparticles can promote coagulatory disturbance accompanied by lung inflammation (Inoue et al., 2006c).

Furthermore, we confirmed that this model is suitable for testing of nanomaterials. In brief, ICR male mice were divided into 8 experimental groups that intratracheally received vehicle, three sizes (15, 50, 100 nm) of TiO$_2$ nanomaterials (8 mg/kg), LPS (2.5 mg/kg), or LPS plus nanomaterials. Twenty-four hours after the treatment, both nanomaterials exacerbated the lung inflammation and vascular permeability elicited by LPS, with an overall trend of amplified lung expressions of cytokines such as IL-1β, MCP-1, and KC. LPS plus nanomaterials, especially with size less than 50 nm, elevated circulatory levels of fibrinogen, IL-1β, MCP-1, KC, and von Willebrand factor compared with LPS alone. The enhancement tended overall to be greater with the smaller nanomaterials than with the larger ones. cDNA microarray analyses revealed that gene expression pattern was not different between the LPS group and the LPS + nanomaterial groups. These results suggest that nanomaterials exacerbate lung inflammation related to LPS with systemic inflammation and coagulatory disturbance, and the exacerbation is more prominent with smaller nanomaterials than with larger ones (Inoue et al., 2008).

Our next study was performed to determine whether the inhalation of diesel engine-derived nanoparticles also exacerbates the model. ICR mice were exposed for 5 h to clean air or diesel engine-derived nanoparticles at a concentration of 15, 36 or 169 µg/m^3 after intratracheal challenge with 125 µg/kg of LPS or vehicle, and were sacrificed for evaluation 24 h after the intratracheal challenge. Nanoparticles alone did not induce lung inflammation. Nanoparticle inhalation exaggerated LPS-elicited inflammatory cell recruitment in the BAL fluid and lung parenchyma, compared with clean air inhalation, in a concentration-dependent manner. Lung homogenates derived from the LPS + nanoparticle groups tended to have an increased tumor necrosis factor-α level and chemotaxis activity for polymorphonuclear leukocytes as compared to those from the LPS group or the corresponding nanoparticle groups. Nanoparticle inhalation did not significantly increase the lung expression of proinflammatory cytokines or facilitate systemic inflammation and coagulatory disturbance. Isolated alveolar macrophages from nanoparticle-exposed mice showed a greater production of IL-1β and KC stimulated with ex vivo LPS challenge than those from clean air-exposed mice, although the differences did not reach significance. These results suggest that acute exposure to diesel nanoparticles exacerbates lung inflammation induced by LPS (Inoue et al., 2008). Taken together, nanoparticle exposure exacerbates acute lung inflammation related to bacterial endotoxin (Figure 14.1). Furthermore, this in vivo model may be suitable to test the enhancing effects of nanoparticles/materials on innate immunity.

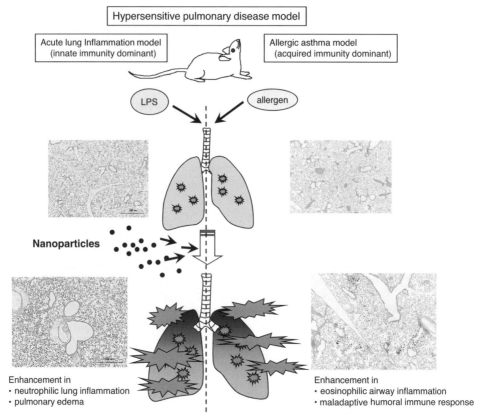

Figure 14.1 *Proposed model for enhancement of nanoparticles/materials on hypersensitive lung disease.*

14.3   Effects of Nanoparticles on Allergic Airway Inflammation

Bronchial asthma has been recognized as chronic airway inflammation that is characterized by an increase in the number of activated lymphocytes and eosinophils. A number of studies have shown that various particles including carbon black can enhance allergic sensitization (Maejima *et al.*, 1997; Lambert *et al.*, 1999; Inoue *et al.*, 2007a). Carbon black has been demonstrated to enhance the proliferation of antibody-forming cells and both IgE and IgG levels (Lovik *et al.*, 1997; Lambert *et al.*, 2000). Ultrafine particles (PM and carbon black) reportedly exaggerate allergic airway inflammation *in vivo* (van Zijverden *et al.*, 2000; Last *et al.*, 2004; Al-Humadi *et al.*, 2002). However, all studies have failed to examine the size of particles they used. Therefore, no research has addressed the size effects of particles or nanoparticles on allergic airway inflammation *in vivo*. Given the hypothesis, we investigated the effects of carbon black nanoparticles with a diameter of 14 nm or 56 nm on antigen-related airway inflammation. ICR mice were divided into six experimental groups. The vehicle, two sizes of carbon nanoparticles, ovalbumin (OVA), and OVA + nanoparticles were administered intratracheally. The cellular profile of BAL fluid, lung histology, expression of cytokines, chemokines, 8-OHdG, and immunoglobulin production were studied. Nanoparticles with a diameter of 14 nm or 56 nm aggravated the antigen-related airway inflammation characterized by the infiltration of eosinophils, neutrophils and mononuclear cells, and by an increase in the number of goblet cells in the bronchial epithelium. Nanoparticles with antigen increased protein levels of IL-5, IL-6 and IL-13, eotaxin, MCP-1, and regulated on activation and normal T cells expressed and secreted (RANTES) in the lung as compared with the antigen alone. The formation of 8-OHdG was moderately induced by nanoparticles or antigen alone, and was markedly enhanced by antigen plus nanoparticles as compared with either nanoparticles or antigen alone. The aggravation was more prominent with 14 nm nanoparticles than with 56 nm particles with overall trend. Particles with a diameter of 14 nm exhibited adjuvant activity for total IgE and antigen-specific IgG and IgE. Nanoparticles can aggravate antigen-related airway inflammation and immunoglobulin production, which becomes more prominent with smaller particles. The enhancement may be mediated, at least partly, by the increased local expression of IL-5 and eotaxin, and also by the modulated expression of IL-13, RANTES, MCP-1 and IL-6 (Inoue *et al.*, 2005).

In ongoing reports, nanoparticles alone or OVA alone moderately enhanced cholinergic airway reactivity, as assessed by total respiratory system resistance (R) and Newtonian resistance (R_n). All the parameters of lung responsiveness, such as R, compliance, elastance, R_n, tissue damping, and tissue elastance, were worse in the nanoparticle + OVA groups than in the vehicle group, the corresponding nanoparticle groups, or the OVA group. The lung mRNA level for Muc5ac was significantly higher in the OVA group than in the vehicle group, and further increased in the nanoparticle + OVA groups than in the OVA or nanoparticle groups. These data suggest that carbon nanoparticles can enhance lung hyperresponsiveness, especially in the presence of antigen. The effects may be mediated, at least partly, through the enhanced lung expression of Muc5ac (Inoue *et al.*, 2007b). Taken together, nanoparticle-exposure exacerbates allergic asthma (Figure 14.1). Furthermore, this *in vivo* model may be suitable to test the enhancing effects of nanoparticles on adaptive immunity.

Acknowledgements

Special thanks to Prof. Takamichi Ichinose, Dr Rie Yanagisawa and Dr Eiko Koike for greatly contributing to the work in the manuscript. The authors are supported by NIES grants.

References

Al-Humadi NH, Siegel PD, Lewis DM, *et al.* (2002) The effect of diesel exhaust particles (DEP) and carbon black (CB) on thiol changes in pulmonary ovalbumin allergic sensitized Brown Norway rats. *Exp Lung Res* **28**, 333–349.

Becker S, Fenton MJ, Soukup JM (2002) Involvement of microbial components and toll-like receptors 2 and 4 in cytokine responses to air pollution particles. *Am J Respir Cell Mol Biol* **27**, 611–618.

Brown DM, Wilson MR, MacNee W, Stone V, Donaldson K (2001) Size-dependent proinflammatory effects of ultrafine polystyrene particles: a role for surface area and oxidative stress in the enhanced activity of ultrafines. *Toxicol Appl Pharmacol* **175**, 191–199.

Chen HW, Su SF, Chien CT, *et al.* (2006) Titanium dioxide nanoparticles induce emphysema-like lung injury in mice. *Faseb J* **20**, 2393–2395.

Cyrys J, Stolzel M, Heinrich J, *et al.* (2003) Elemental composition and sources of fine and ultrafine ambient particles in Erfurt, Germany. *Sci Total Environ* **305**, 143–156.

Dick CA, Brown DM, Donaldson K, Stone V (2003) The role of free radicals in the toxic and inflammatory effects of four different ultrafine particle types. *Inhal Toxicol* **15**, 39–52.

Dockery DW, Pope CA, 3rd, Xu X, *et al.* (1993) An association between air pollution and mortality in six US cities. *N Engl J Med* **329**, 1753–1759.

Flanagan PG, Jackson SK, Findlay G (2001) Diagnosis of gram negative, ventilator associated pneumonia by assaying endotoxin in bronchial lavage fluid. *J Clin Pathol* **54**, 107–110.

Ichinose T, Furuyama A, Sagai M (1995) Biological effects of diesel exhaust particles (DEP). II. Acute toxicity of DEP introduced into lung by intratracheal instillation. *Toxicology* **99**, 153–167.

Ichinose T, Yajima Y, Nagashima M, Takenoshita S, Nagamachi Y, Sagai M (1997) Lung carcinogenesis and formation of 8-hydroxy-deoxyguanosine in mice by diesel exhaust particles. *Carcinogenesis* **18**, 185–192.

Inoue K, Takano H, Yanagisawa R, Sakurai M, Ichinose T, Sadakane K, Yoshikawa T (2005) Effects of nano particles on antigen-related airway inflammation in mice. *Respir Res* **6**, 106.

Inoue K, Takano H, Sakurai M, *et al.* (2006a) Pulmonary exposure to diesel exhaust particles enhances coagulatory disturbance with endothelial damage and systemic inflammation related to lung inflammation. *Exp Biol Med (Maywood)* **231**, 1626–1632.

Inoue K, Takano H, Yanagisawa R, Sakurai M, Ueki N, Yoshikawa T (2006b) Effects of diesel exhaust on lung inflammation related to bacterial endotoxin in mice. *Basic Clin Pharmacol Toxicol* **99**, 346–352.

Inoue K, Takano H, Yanagisawa R, *et al.* (2006c) Effects of airway exposure to nanoparticles on lung inflammation induced by bacterial endotoxin in mice. *Environ Health Perspect* **114**, 1325–1330.

Inoue KI, Takano H, Yanagisawa R, Hirano S, Kobayashi T, Fujitana Y, Shimada A, Yoshikawa T (2007a) Effects of inhaled nanoparticles on acute lung injury induced by lipopolysaccharide in mice. *Toxicology* **238**, 99–110.

Inoue K, Takano H, Yanagisawa R, Sakurai M, Abe S, Yoshino S, Yamaki K, Yoshikawa T (2007b) Effects of nanoparticles on lung physiology in the presence or absence of antigen. *Int J Immunopathol Pharmacol* **20**, 737–744.

Inoue K, Takano H, Ohnuki M, Yanagisawa R, Sakurai M, Shimada A, Mizushima K, Yoshikawa T (2008) Size effects of nanomaterials on lung inflammation and coagulatory disturbance. *Int J Immunopathol Pharmacol* **21**, 197–206.

Lambert AL, Dong W, Winsett DW, Selgrade MK, Gilmour MI (1999) Residual oil fly ash exposure enhances allergic sensitization to house dust mite. *Toxicol Appl Pharmacol* **158**, 269–277.

Lambert AL, Dong W, Selgrade MK, Gilmour MI (2000) Enhanced allergic sensitization by residual oil fly ash particles is mediated by soluble metal constituents. *Toxicol Appl Pharmacol* **165**, 84–93.

Last JA, Ward R, Temple L, Pinkerton KE, Kenyon NJ (2004) Ovalbumin-induced airway inflammation and fibrosis in mice also exposed to ultrafine particles. *Inhal Toxicol* **16**, 93–102.

Li N, Sioutas C, Cho A, *et al.* (2003) Ultrafine particulate pollutants induce oxidative stress and mitochondrial damage. *Environ Health Perspect* **111**, 455–460.

Lovik M, Hogseth AK, Gaarder PI, Hagemann R, Eide I (1997) Diesel exhaust particles and carbon black have adjuvant activity on the local lymph node response and systemic IgE production to ovalbumin. *Toxicology* **121**, 165–178.

Maejima K, Tamura K, Taniguchi Y, Nagase S, Tanaka H (1997) Comparison of the effects of various fine particles on IgE antibody production in mice inhaling Japanese cedar pollen allergens. *J Toxicol Environ Health* **52**, 231–248.

Martin TR, Rubenfeld GD, Ruzinski JT, *et al.* (1997) Relationship between soluble CD14, lipopolysaccharide binding protein, and the alveolar inflammatory response in patients with acute respiratory distress syndrome. *Am J Respir Crit Care Med* **155**, 937–944.

Oberdorster G (2001) Pulmonary effects of inhaled ultrafine particles. *Int Arch Occup Environ Health* **74**, 1–8.

Oberdorster G, Oberdorster E, Oberdorster J (2005) Nanotoxicology: an emerging discipline evolving from studies of ultrafine particles. *Environ Health Perspect* **113**, 823–839.

Peters A, Wichmann HE, Tuch T, Heinrich J, Heyder J (1997) Respiratory effects are associated with the number of ultrafine particles. *Am J Respir Crit Care Med* **155**, 1376–1383.

Samet JM, Dominici F, Curriero FC, Coursac I, Zeger SL (2000) Fine particulate air pollution and mortality in 20 US cities, 1987–1994. *N Engl J Med* **343**, 1742–1749.

Seaton A, MacNee W, Donaldson K, Godden D (1995) Particulate air pollution and acute health effects. *Lancet* **345**, 176–178.

Shvedova AA, Kisin ER, Mercer R, *et al.* (2005) Unusual inflammatory and fibrogenic pulmonary responses to single-walled carbon nanotubes in mice. *Am J Physiol Lung Cell Mol Physiol* **289**, L698–708.

Takano H, Yoshikawa T, Ichinose T, Miyabara Y, Imaoka K, Sagai M (1997) Diesel exhaust particles enhance antigen-induced airway inflammation and local cytokine expression in mice. *Am J Respir Crit Care Med* **156**, 36–42.

Takano H, Ichinose T, Miyabara Y, *et al.* (1998) Inhalation of diesel exhaust enhances allergen-related eosinophil recruitment and airway hyperresponsiveness in mice. *Toxicol Appl Pharmacol* **150**, 328–337.

Takano H, Yanagisawa R, Ichinose T, *et al.* (2002) Diesel exhaust particles enhance lung injury related to bacterial endotoxin through expression of proinflammatory cytokines, chemokines, and intercellular adhesion molecule-1. *Am J Respir Crit Care Med* **165**, 1329–1335.

Timonen KL, Hoek G, Heinrich J, *et al.* (2004) Daily variation in fine and ultrafine particulate air pollution and urinary concentrations of lung Clara cell protein CC16. *Occup Environ Med* **61**, 908–914.

Ulich TR, Watson LR, Yin SM, *et al.* (1991) The intratracheal administration of endotoxin and cytokines. I. Characterization of LPS-induced IL-1 and TNF mRNA expression and the LPS-, IL-1-, and TNF-induced inflammatory infiltrate. *Am J Pathol* **138**, 1485–1496.

Utell MJ, Frampton MW (2000) Acute health effects of ambient air pollution: the ultrafine particle hypothesis. *J Aerosol Med* **13**, 355–359.

van Zijverden M, Van Der Pijl A, Bol M, *et al.* (2000) Diesel exhaust, carbon black, and silica particles display distinct Th1/Th2 modulating activity. *Toxicol Appl Pharmacol* **168**, 131–139.

Vincenti MP, Burrell TA, Taffet SM (1992) Regulation of NF-kappa B activity in murine macrophages: effect of bacterial lipopolysaccharide and phorbol ester. *J Cell Physiol* **150**, 204–213.

Warheit DB, Laurence BR, Reed KL, Roach DH, Reynolds GA, Webb TR (2004) Comparative pulmonary toxicity assessment of single-wall carbon nanotubes in rats. *Toxicol Sci* **77**, 117–125.

Warheit DB, Webb TR, Sayes CM, Colvin VL, Reed KL (2006) Pulmonary instillation studies with nanoscale TiO_2 rods and dots in rats: toxicity is not dependent upon particle size and surface area. *Toxicol Sci* **91**, 227–236.

Warheit DB, Webb TR, Colvin VL, Reed KL, Sayes CM (2007) Pulmonary bioassay studies with nanoscale and fine quartz particles in rats: toxicity is not dependent upon particle size but on surface characteristics. *Toxicol Sci* **95**, 270–280.

Yanagisawa R, Takano H, Inoue K, *et al.* (2003) Enhancement of acute lung injury related to bacterial endotoxin by components of diesel exhaust particles. *Thorax* **58**, 605–612.

Zhu Y, Hinds WC, Kim S, Sioutas C (2002) Concentration and size distribution of ultrafine particles near a major highway. *J Air Waste Manag Assoc* **52**, 1032–1042.

15

In Vivo and *In Vitro* Models for Nanotoxicology Testing

Rosalba Gornati, Elena Papis, Mario Di Gioacchino, Enrico Sabbioni,
Isabella Dalle-Donne, Aldo Milzani and Giovanni Bernardini

15.1 Nanotechnology: Very Promising, But a Little Scary

Nanotechnology is perceived as one of the key technologies of this century. It offers the promise of generating new products that will revolutionize diverse areas of our life. Nanotechnology will change the way we diagnose and treat diseases, protect the environment, produce energy, protect our body from cold and heat and our skin from sun, preserve our food, build our houses, memorize our data, speed up our communications, and . . . nobody knows what more. Nanotechnology is then expected to become increasingly important in the very near future and to have a strong impact on the economy, creating new jobs and improving welfare. Such great promise makes nanotechnology a priority for public administrations of countries wishing to maintain their scientific and technological leadership. Nanotechnology is already becoming the subject of public debate, although survey researches suggest that most people are still unfamiliar with the term (Cobb and Macoubrie, 2004). However, concern about the possible drawbacks of this technology exists, and social science scholars tell us that it would diminish if measures were taken to enhance laypeople's trust in governmental agencies (Siegrist *et al.*, 2007). Therefore, there is a need for collecting and disseminating accurate, reliable and unbiased information on the risks and benefits of this new technology. Public perception and the consequent attitudes toward nanotechnology are likely to be influenced by the correctness of the provided information on risks and benefits, and, in turn, is likely to influence the realization of further technological advances. But is the needed information on risks of this emerging technology really available? Although the answer to this question is probably negative, some efforts have

Nanotoxicity: From In Vivo *and* In Vitro *Models to Health Risks*   Edited by Saura Sahu and Daniel Casciano
© 2009 John Wiley & Sons, Ltd

been made to study the toxicology of nanomaterials and, in this context, a new discipline, nanotoxicology, is emerging (Fisher and Chan, 2007).

15.1.1 Nano-terminology

The birth of nanotoxicology is accompanied by the emergence of many 'nano-terms' such as nanobiology, nanoboxes, nanocage, nanocluster, nanocrystals, nanocubes, nanofibers, nanofilm, nanoframe, nanomaterials, nanoparticles (NPs), nanopowder, nanosheet, nanoshell, nanospheres, nanostars, nanostructures, nanotubes, and so on. Such terminological cornucopia risks, however, generating confusion in the readers' minds: the nanoworld needs internationally accepted definitions not only for the sake of clarity, but also for regulatory purposes. It is generally accepted that a nanomaterial is a material having at least one dimension on the scale of 100 nanometers or less. Nanomaterials can be nanoscale in one dimension (e.g. films), two dimensions (e.g. fibers and tubes) or three dimensions (e.g. particles).

15.1.2 Nanofilms, Nanofibers and Nanoparticles

Nanofilms have a great use in biotechnology for encapsulating macromolecules; polypeptide nanofilms, for example, have recently been employed in the attempt to create artificial red blood cells (Palath *et al.*, 2006). A mat of nanofibers can mimic extracellular matrix to support cell attachment, proliferation and migration, and this is quite promising for tissue engineering (Venugopal *et al.*, 2007), while carbon nanotubes can be bio-functionalized for bioelectronics and for specifically delivering peptides or nucleic acid to cells (Yang *et al.*, 2007).

NPs of a particular shape can be better defined as nanospheres, nanocubes, nanoboxes, nanocages, nanoframes or nanostars. Delivery of the antioxidant vitamin E mediated by polyethylene glycol nanospheres may be a useful adjunct for antioxidant therapy in Alzheimer's disease (Shea *et al.*, 2005). Gold nanocages, obtained using silver nanocubes as a template, have been tailored for use in optical coherence tomography and in photothermal destruction of cancer cells (Skrabalak *et al.*, 2007). Iron oxide NPs have found their role as a contrast agent in magnetic resonance imaging (Patel *et al.*, 2008) and represent a promising possibility for magnetic targeting (Chertok *et al.*, 2008).

15.2 Is Shape Important?

The large fraction of surface atoms in comparison with the bulk influences the chemical and physical properties of NPs, so that they can be considered in an intermediate state between the solid and the molecular. Any decrease in size leads to a decrease in the cohesive energy and hence in the thermal stability and melting temperature, properties that are also affected by shape (Barnard *et al.*, 2005) and by clustering (Attarian-Shandiz *et al.*, 2008). Therefore, it becomes obvious that not only the size, but also the shape of NPs, as well their capacity to aggregate, might be correlated with their activity. The antibacterial properties of silver NPs, for example, depends on their shape (Pal *et al.*, 2007). Then, NP size, shape and aggregation state (which may drastically change depending on the surrounding environment, i.e. culture media, body fluids and intracellular milieu) need to be determined as fully as possible.

15.2.1 Optical Microscopy for Determining Size and Shape

Unfortunately, conventional light microscopy is of little use in investigating NPs because of its inadequate resolution. In fact, according to Abbe's theory of image formation, the resolution of an optical microscope is diffraction-limited, i.e. limited by the spreading out of each image point due to diffraction. Nevertheless, some methods have been proposed to detect and analyze metal NPs by far-field optical microscopy (van Dijk *et al.*, 2005).

Finer resolution limits can be obtained, instead, by resorting to near-field optical microscopy, a very recent technique which takes advantage of the properties of evanescent light which decays exponentially and usually goes undetected. The scanning near-field optical microscope (SNOM) is a microscope without lenses that builds an image by detecting light intensity for each point of the scan (Edidin, 2001; Hecht *et al.*, 2000). SNOM is in its infancy, and several technical obstacles still have to be overcome to render this technique available to a wider group of scientists. Apertureless SNOM, however, has already been used with success for characterizing metallic NPs (Pack *et al.*, 2003; Kim and Leone, 2008).

15.2.2 Electron Microscopy for Determining Size and Shape, Structure and Composition

In a conventional transmission electron microscope (TEM, cf. Reimer, 1997), a thin specimen is irradiated with an electron beam of uniform current density with an acceleration voltage of 60 to 120 kV in routine instruments, and reaching 3 MV in high voltage electron microscopes. Electrons are emitted from the electron gun and illuminate the specimen through a two- or three-stage condenser lens system. The electron intensity distribution behind the specimen is magnified with a three- or four-stage lens system and viewed on a fluorescent screen. The image can be recorded by direct exposure of a photographic emulsion or digitally by a charge coupled device (CCD) camera.

Specimens such as biological thin-sections (Figure 15.1) and surface replicas are mounted on metal grids with different mesh sizes, while others such as particles, viruses and macromolecules need a supporting film resistant to electron irradiation with a low atomic number to reduce diffraction, with electrical conductivity to avoid charging and with low granularity for high resolution. For medium magnification, grids covered with carbon-coated formvar films are ideal (Figure 15.2).

The high vacuum and the electron bombardment to which the NPs are subjected inside the TEM could themselves be a source of artefacts. Dramatic modifications of the appearance of iron oxide NPs have been recorded during observation. Such changes have been attributed to structural rearrangement of the NPs as a result of electron beam exposure (Latham *et al.*, 2006).

TEM can operate with small electron probes in various microanalytical modes with a good spatial resolution, and this is quite useful for NP characterization. The excitation of the atoms of the specimen by the electron beam causes the emission of X-rays with element-specific energies. If the microscope is equipped with an X-ray detector, reliable quantitative information on elemental composition of selected areas of the specimen can be obtained. Information similar to that obtained with the X-ray microanalysis can be obtained

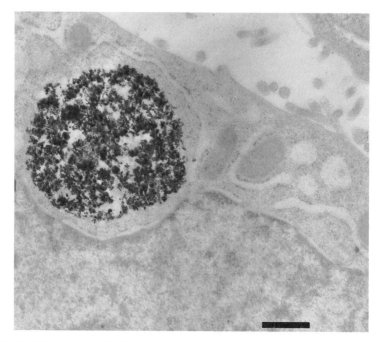

Figure 15.1   *TEM picture of a thin-section of culture cells exposed for 72 h to cobalt oxide NPs. Note the endocytic vesicle full of NPs. Scale bar is 0.5 μm.*

with electron energy loss spectroscopy (EELS), collecting the spectrum of the energy losses by the inelastic interaction of the electron beam with the atoms of the specimen.

Information on the lattice structure of a crystal is usually obtained by X-ray diffraction. The application of this technique to isolated NPs suffers a strong disadvantage as it inherently averages over a relatively large region, the size of the X-ray beam, and produces

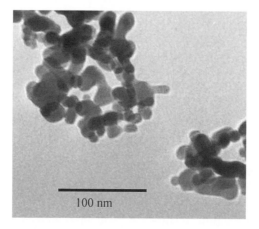

Figure 15.2   *TEM picture of cobalt oxide NPs. NPs were deposited on a grid covered with a carbon-coated formvar film.*

low signal-to-noise ratios for very small particles. TEM, instead, is better suited in that it delivers a very narrow high-intensity beam to the sample for analysis. Some instruments permit electron diffraction patterns to be collected from areas as small as 1 nm in diameter. This is fine enough to allow the possibility of determining the structure of different areas of composited NPs (Lu *et al.*, 2007).

In the scanning electron microscope (SEM), electrons emitted by an electron gun are focused into a fine beam (2–10 nm in diameter) by a set of condenser lenses. The electron beam is deflected by the scan coils in a controlled pattern so that it raster-scans a rectangular area where the sample is placed. The primary electrons (the incident beam) interact with the sample, producing secondary electrons (SE), backscattered electrons (BSE), Auger electrons, X-rays and photons. BSE result from elastic scattering (i.e. with little or no change in energy) of the primary electrons, with the positive nuclei of the specimen. SE, instead, are predominantly produced by the interactions of the beam electrons with the weakly bonded conduction-band electrons in metals, and therefore only a small amount of kinetic energy can be transferred to the secondary electrons; their energy in fact ranges from 2 to 5 eV. A comparison of images obtained with SE and with BSE can be extremely useful to detect intracellular uptake of NPs (Figure 15.3A and B). As for TEM, if the microscope is equipped with a microanalysis system, the X-ray fluorescence generated from the atoms of the specimen can be collected to produce a full quantitative analysis showing the sample composition.

To prevent buildup of electrical charges, to increase thermal conduction and to improve SE emission, samples prepared for SEM are usually sputtered with gold, and this layer, although kept as thin as possible, slightly modifies the sample topography. Moreover, the 'orange peel effect', often observed at high resolution, does not correspond to the true surface of the NP, but is an artefact caused by metal coating during sample preparation (Liu *et al.*, 2007). Therefore, caution should be used when determining exact measurements such as surface area and porosity with SEM. Several properties of NPs (e.g. their catalytic activity) depend upon their surface area and porosity, and these characteristics can be better determined by gas sorption (adsorption and desorption) methods; the resulting nitrogen isotherms are analyzed generally using BET theory to obtain the specific surface area (m^2/g). BET consists of the first initials of the family names of Stephen Brunauer, Paul Hugh Emmett, and Edward Teller, who extended the Langmuir theory from monolayer to multilayer adsorption (Brunauer *et al.*, 1938).

15.2.3   The Atomic Force Microscope is Ideal for Nanomaterials

The atomic force microscope (AFM) can analyze the sample as it is in air or under liquid without preliminary complex preparative work. An AFM is a mechanical instrument that works like our fingers, which feel an object when we cannot see it. By using a finger to 'visualize' an object, our brain is able to deduce its topography while touching it. The AFM consists of a cantilever with a sharp 'finger' (probe) at its end which is used to scan the specimen surface. The probe is positioned close enough to the sample so that it can interact with the force-fields associated with the sample surface. Such interaction leads to a deflection of the cantilever which is recorded to reconstruct a true three-dimensional surface profile. An AFM can easily visualize NPs and characterize their size and shape (Figure 15.4). AFM is less expensive and to some extent easier to use than TEM or SEM.

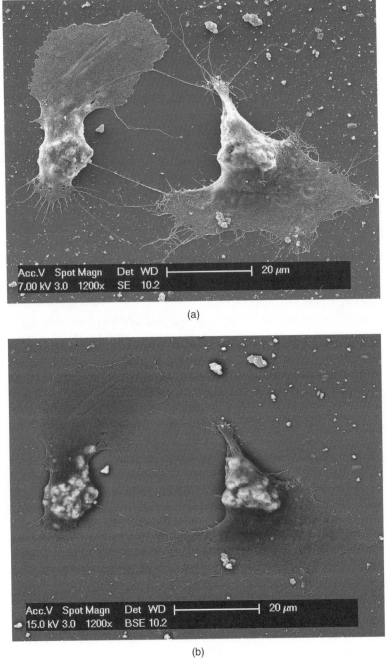

(a)

(b)

Figure 15.3 *SEM images of culture cells exposed for 72 h to cobalt oxide NPs. The image in panel A was obtained collecting the secondary electrons, while that in panel B with back-scattered electrons. Note the strong signal from metallic NPs present in panel B.*

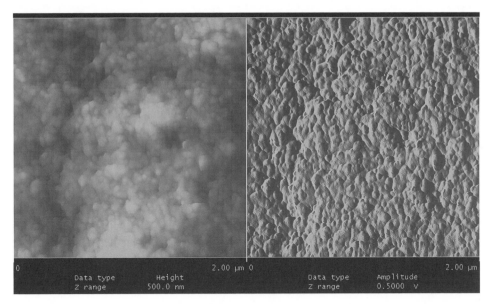

Figure 15.4 *AFM image of cobalt oxide NPs with its corresponding shading image in the right panel.*

The main difficulty, however, is developing methods for distributing and fixing NPs on the holder surface. In fact, if NPs do not firmly adhere to the holder, the probe can push them away or pick them up, creating in both cases artefactual images.

As we have seen, each technique has its own drawbacks. There is no single 'best technique' for all situations. Moreover, a combination of different techniques can guarantee the absence of artefacts, which in microscopy are always around the corner.

15.3   Dynamic Light Scattering and Nanoparticle Tracking Analysis

Electron microscope techniques are certainly powerful, but have several disadvantages: instrumentation is complex and expensive, sample preparation is time-consuming and might generate artefacts, observation has to be carried out under a vacuum, and expertise in the field of microscopy is necessary. AFM is also complex, and a deep understanding of the technique is necessary. Extensive experience, however, is not required for dynamic light scattering (DLS), which is currently one of the most popular methods used to determine particle size; measurements, in fact, are quick and relatively inexpensive, and the commercial instruments are almost completely automated for simple routine use. In a typical DLS measurement, a monochromatic and coherent light (a laser) is made to hit a suspension of NPs. As NPs are quite small compared with the wavelength, the light scatters in all directions in a Rayleigh scattering fashion. If a frosted glass screen were used to view the sample cell, a speckle pattern resulting from the constructive and destructive interferences would be seen. As the system is not stationary because of the Brownian motion of the NPs, the speckle pattern will fluctuate: the smaller the particles, the higher their average velocity, the faster the fluctuation. A detector, usually positioned at 90°, records the light intensity fluctuations. Their autocorrelation function can be related to the diffusion coefficient and, resorting to the Stokes-Einstein equation, the equivalent hydrodynamic diameter can be calculated.

Alternatively the diffusion coefficient can be determined by simultaneously but individually tracking NPs undergoing Brownian motion by nanoparticle tracking analysis (NTA). As for DLS, a laser is made to hit a suspension of NPs. In this case, a video of 20–60 s duration of the moving particles is taken by a conventional optical microscope aligned normally to the beam. The recording is analyzed by software that identifies each NP and tracks its trajectory. From these data the diffusion coefficient and hence the hydrodynamic diameter can be calculated.

The advantages of NTA over DLS include that NTA does not need calibration, avoids the DLS bias towards larger particles, and is better suited for samples containing particles of different dimensions (Carr *et al.*, 2008). DLS, in contrast, samples many more particles and allows a better detection limit. The increasing interest in nanotechnology and the consequent need for fast, easy and reliable NP characterization is driving novel implementations of DLS and NTA, such as differential dynamic microscopy (Cerbino and Trappe, 2008), which will boost the potentiality of these techniques.

15.4   Zeta Potential

The ability of NPs to stay in solution or conversely that to stick together (aggregate) or to proteins, cells and substrates of different natures, may be related to their ζ (zeta) potential. An electrically charged particle in a liquid is surrounded by an inner shell called the Stern layer where the ions are strongly bound to the particle, and by an outer region (called the diffuse layer) where the ions are less tightly associated. This external layer is delimited by a notional boundary called the slipping plane, which is the plane of hydrodynamic shear. The potential at this boundary is the ζ potential, and its value can be related to the stability of NP dispersions. When the ζ potential is high (either negative or positive) NPs will tend to repel each other; when it is low, NPs will tend to aggregate.

ζ potential is derived from measuring the average velocity of NPs as they are subjected to an electric field. This velocity can be determined by measuring the doppler shift of laser light scattered by the moving particles. The electrokinetic properties of a particle can also be measured by resorting to electroacoustic methods. A high-frequency electric field is applied to the sample and moves electrophoretically the NPs which generate an alternating acoustic wave from which ζ potential is derived. The estimates of ζ potential obtained by these two techniques, however, might differ slightly under some experimental conditions (Jailani *et al.*, 2008).

Measurements of ζ potential are usually performed to study the stability of NP suspensions, as this is important in several industrial processes. Interestingly, however, electrostatic interactions have been shown to play an important role in protein adsorption and cellular uptake of NPs (Patil *et al.*, 2007); moreover, ζ potential can be used as a marker to study interactions of NPs with normal and cancer cells under different experimental conditions (Zhang *et al.*, 2008).

15.5   Dissolution

Metal NPs do dissolve in some conditions. Quantifying dissolution is extremely important in nanotoxicology, to determine whether the observed effects are due to the NP itself or to the metal ion made available by NP dissolution. Therefore, dissolution has to be checked in water, in biological fluids, in the different cellular compartments as well as in the different conditions available in the environment. Quantification of the dissolved metals can be performed by atomic absorption spectroscopy (AAS) or inductively coupled plasma mass spectrometry (ICP-MS).

In AAS, the sample is atomized with a flame or a graphite furnace. The atoms are illuminated with light of a given wavelength (in general, each wavelength corresponds to only one element) and make the transition to higher electron energy levels. The analyte concentration is determined from the amount of absorption after calibrating the instrument with standards of known concentration.

ICP-MS makes use of an inductively coupled plasma to atomize and ionize the sample, and a mass spectrometer to separate and detect the ions. ICP-MS spectrometers can accept solid as well as liquid samples. Solid samples are introduced into the ICP by way of a laser ablation system, while liquids are nebulized. The samples move into the argon plasma torch body, whose extreme temperature causes the atomization and ionization of the injected sample. Ions flow through a small orifice into a mass spectrometer, maintained in a vacuum so that the ions are free to move without collisions, where the ions are separated on the basis of their mass-to-charge ratio and quantified.

Unlike AAS, which can only measure a single element at a time, ICP-MS can scan for all elements simultaneously.

15.6   *In Vitro* and *In Vivo* Models are Both Necessary for Nanotoxicology Research

In spite of increasing potential exposure (occupational and public) to nanomaterials, information on their toxicological properties remains lacking, precluding a reliable assessment of their human health impact (Kandlikar *et al.*, 2007). In this context, there is a need to implement a strategy to provide hazard identification data on the widest possible range of nanomaterials, to derive conclusions on the spectrum of toxicological effects associated with such materials (Oberdoster *et al.*, 2005). To reach this goal, significant basic nanotoxicology research is urgently needed in all the areas that demand human epidemiological and clinical data, as well as information derived from *in vivo* and *in vitro* systems (Donaldson *et al.*, 2004). Among other things, it is expected that nanotoxicology research will lead to a considerable increase in the number of animals used, raising ethical questions due to the EC Directive 86/609/EEC (Anon, 1986), the debate on animal use in nanotoxicology research being at present a more emotional than a rational question. In particular, the development of *in vitro* alternative (non-animal) methods is important not only for research purposes, but also to evaluate the potential toxicity of any NPs (screening) produced in the future, to provide guidance on their safe industrial production, and for legal purposes when they are introduced into the market. However, it is not completely correct to say that progress in nanotoxicology research will be wholly the result of animal research, nor is it correct to say

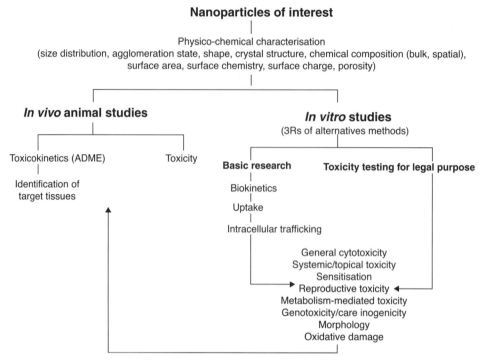

Figure 15.5   In vivo *and* in vitro *studies for nanotoxicology research.*

that animal research will contribute nothing of value. The truth is that *in vivo*, *in vitro* and epidemiological/clinical studies would each provide pieces of the research puzzle, and each would contribute to important toxicological information (Stephens, 1987). Thus, progress in nanotoxicology will depend on these three branches of investigation, interdependent upon other in a complementary and multidisciplinary approach. In particular, at this early stage of understanding of potential nanotoxicity, and due to the little knowledge of the factors influencing toxicity of NPs, animal models would be particularly useful to study aspects which cannot be mimicked by *in vitro* systems, such as toxicokinetics (absorption, distribution, metabolism, elimination: ADME) and the identification of target tissues. The *in vitro* systems would mainly be used to identify specific characteristics of nanomaterials which can be used as indicators of toxicity, in order to establish a ranking of NP toxicity and for mechanistic studies (Figure 15.5).

15.7   3Rs: An Ethical Basis for Nanotoxicology

In vitro nanotoxicology finds its main justification in the fact that it will make the toxicology of nanomaterials a more scientifically-based practice. The first aim is to provide the nanotoxicologist the opportunity to apply new protocols for his research and practice. These protocols should take advantage of existing molecular and cellular models of modern biology, including isolated perfused organs, organotypic culture, primary cells or established cell line cultures, and animal and human cells. From this wide spectrum of options, information of great relevance that is difficult to obtain in a classical toxicity study, such as time-dependence of the effect (immediate or progressive?), the degree of reversibility, the dose-effect relationship, its universality or its specificity toward animal species and the role of metabolic activation, can be obtained. The *in vitro* methods are also ideal in nanotoxicology research because of their ability to produce accurate results rapidly and inexpensively without the use of animals. In addition, *in vitro* nanotoxicology takes into account the substantial ethical, political, financial and regulatory pressures to reduce, refine and replace animal toxicology with alternative assays (3Rs approach) (Sabbioni, 2005). An alternative method is any method that can be used to replace, reduce or refine the use of animal experiments in biomedical research, testing or education. The 'alternatives' concept is attributed to Russel and Burch (1959) who defined three types of alternatives ('Three Rs'): reduction alternatives, which obtain a comparable level of information from the use of fewer animals, or more information from the same number of animals; refinement alternatives, which alleviate or minimise potential pain, suffering or distress; and replacement alternatives, which permit a given purpose to be achieved without using living vertebrate animals (Hartung, 2005). Although *in vitro* experiments are not a substitute for whole animal studies, the use of simple *in vitro* models with specific and relatively simple endpoints as markers of nanotoxicity in specialised cells (e.g. leakage of enzymes (LDH assay), neutral red (vital staining for lysosomes), mitochondrial integrity (MTT assay), apoptosis, necrosis) and that enable the understanding of general mechanisms of nanotoxicity (apoptosis, necrosis, oxidative stress, DNA damage) can be the basis for assessing the hazards of engineered NPs. *In vitro* nanotoxicology thus contributes to a strengthening of the scientific basis of nanotoxicology, by giving access to sound and scientifically justifiable data. This is of particular importance, taking into account the specific properties of the nanoscale materials conferred by a huge surface area/mass ratio and by their small size which can facilitate their uptake by cells and the passage across epithelial or endothelial barriers into the blood or the lymph circulation, from where they can reach internal organs. In particular, great advantages can be sought by using *in vitro* systems which are considered for validation at the European Centre of the Validation of Alternative Methods (ECVAM) of the European Commission (Hartung *et al.*, 2004) and which cover topical and systemic toxicological areas, such as reproductive toxicity, potential carcinogenicity and barrier functions – models which reflect the various routes of human exposure to NPs such as skin, gastro-intestinal epithelial barrier, blood-brain barrier, human vascular endothelial cells, lung epithelial macrophages and other phagocytic cells (Worth and Balls, 2002).

15.8 Particle Characteristics Affecting Nanoparticle Toxicity: *In Vivo* and *In Vitro* Studies

The purpose of this section is to assess the existing *in vivo/in vitro* evidence from the literature concerning particle characteristics that may affect nanotoxicity. In particular, relevant parameters considered are: dissolution, chemical composition, size, shape, agglomeration state, crystal structure, specific surface and surface changes and coating. In addition, the literature has been analysed with a view to establishing whether the present toxicological information available is enough to draw conclusions on the most appropriate dosemetric for NPs, the mechanisms of NP toxicity, the toxicokinetics and the systematic translocation, deposition and the fate in the body. The analysis of the literature hereafter considered focuses mainly on inorganic-based NPs, but carbon-based NPs were also considered.

15.8.1   *In Vivo* Studies

Most of the *in vivo* experimental research on metal-based nanoparticles has been carried out using traditional animal models, mainly rats (Ag, Au, CdO, Ir, Ni, TiO_2) and mice (Au, Fe_2O_3, Cu, Gd). Very few studies concern other animal species such as monkey (Mn_2O_3) or hamster (TiO_2). Although the potential human routes of exposure to nanoparticles are inhalation, ingestion, skin contact and intravenous injection, the predominant route of exposure of the existing *in vivo* studies is largely by inhalation and intratracheal instillation (Ag, Au, C, CdO, Fe, Mn_2O_3, Ir, Ni, TiO_2 and carbon nanotubes). This is not surprising because most of the toxicological knowledge related to *in vivo* research on engineered nanomaterials is drawn from the extensive research on ambient ultrafine particles (UFPs), nanotoxicology being considered an emerging discipline evolving from studies of UFPs (Oberdörster, 2005). A limited number of studies concern the intravenous route of administration which is relevant for nanoparticles of interest in nanomedicine, such as Co, $CoFe_2O_4$, Gd, Fe_2O_3, and oral administration of Au, Cu, Zn and ZnO. One study involves an intraperitoneal exposure (SiO_2), and one a nonphysiological route such as intramuscular and subcutaneous exposure of rats in an experiment on the carcinogenic potential of Co, Ni, SiO_2 and TiO_2.

15.8.1.1   Size

Size is a determining factor for the uptake and translocation of nanoparticles, as shown by TiO_2 particle translocation across the rat alveolar epithelium (Oberdörster *et al.*, 1994) and nanoferrite targeting tumors in athimic mice (Natarajan *et al.*, 2008); translocation of Au nanoparticles across mice gastrointestinal (GI) tract (Hillyer and Albrecht, 2001); and translocation of Ir particles from rat lung to the GI tract and feces (Kreyling *et al.*, 2002). The dependence of nanotoxicity on size has been observed in tissues of rats acutely exposed to Cu nanoparticles (Chen *et al.*, 2006), and in rats after exposure to TiO_2 particles (Oberdörster *et al.*, 1994). However, in rats exposed to three types of quartz, Warheit *et al.* (2007a) observed a differential degree of pulmonary inflammation and cytotoxicity not always consistent with particle size, but correlated with surface activity.

15.8.1.2   Surface Area

Rats exposed to TiO_2 anatase nanoparticles developed high pulmonary inflammatory reaction, epithelial effects, and impairment of alveolar macrophage function, which correlated with the surface area of the particles (Oberdörster *et al.*, 1994).

15.8.1.3   Crystal Structure

Exposure of rats to ultrafine particles of TiO_2 as rutile or anatase-rutile produced differential pulmonary effects, based on their composition and crystal structure (Warheit *et al.*, 2006).

15.8.1.4 Agglomeration State

Rapid clearance from the lung and translocation to other tissues of Ag nanoparticles inhaled by rats was found, compared with Ag aggregates which were retained in the lung (Takenaka *et al.*, 2001).

15.8.2 *In Vitro* Studies

Most of the alternative (non-animal) methods to study nanoparticle toxicity concern the use of cell lines. Only a few studies concern the use of primary cells (CoCr) or explant (TiO_2).

15.8.2.1 Size

Nanosize- and concentration-dependent toxicity (decrease of mitochondrial function and lack of ATP, effect on membrane integrity and increase of ROS generation) was induced by Ag nanoparticles in rat alveolar macrophages (Carlsson, 2006). A higher cytotoxicity of large Ni nanoparticles was found in mouse neuroblastoma Neuro2A cells, compared with small nanoparticles (Yin *et al.*, 2005). Chithrani *et al.* (2006) observed a size-dependent uptake of Au in HeLa cells. SiO_2 particles penetrated in a size-dependent way into the nuclei of human epithelial HEp-2 cells, inducing aberrant nucleoplasmic protein aggregates (Chen and von Mikecz, 2005). A size-dependent uptake of CeO_2 particles occurred in human lung fibroblasts MRC-9 (Limback *et al.*, 2005). Differential cytotoxicity according to particle size was established in primary human dermal fibroblasts exposed to CoCr nanoparticles (Papageorgiou *et al.*, 2007). The scavenging of free radicals induced by three different nanosized Se nanoparticles in solution were size-dependent (Huang *et al.*, 2003). Ultrafine TiO_2 nanoparticles ($<20\,nm$), but not TiO_2 ($>200\,nm$) increase the micronuclei and apoptosis in Syrian hamster embryo (SHE) fibroblasts (Rahman *et al.*, 2002).

15.8.2.2 Shape

Chen *et al.* (2006) have found that the uptake of Au nanoparticles (spherical and rod-shaped) is size- and shape-dependent. Cha and Myung (2007) showed that different sizes of Si microparticles did not affect cytotoxicity in human brain, liver, stomach and lung cells.

15.8.2.3 Solubility

The solubility of metal-oxide and calcium phosphate nanoparticles influences the cytotoxic response (Brunner *et al.*, 2006).

15.8.2.4 Aggregates

Cerium dioxide nanoparticles penetrate into cytoplasm of human lung epithelial cells (BEAS-2B), being located in the peri-region of the nucleus as aggregates (Park *et al.*, 2008). The clustering of cobalt nanoparticles was found to be less pronounced in the medium for the growth of mouse embryo fibroblasts (Balb/3T3 cells) than in pure water (Sabbioni *et al.*, 2006).

15.8.2.5 Surface Coatings

Nanosurface coating is a powerful means of reducing nanotoxicity. Goodman *et al.* (2004) observed a moderate cytotoxicity compared with the nontoxic effect induced in Cos-1 cells by Au nanoparticles functionalised with cationic and anionic side-chains, respectively. Meanwhile, Sayes *et al.* (2004, 2006) showed a strong dependence of cellular activity upon surface functionalization of four different surface-modified C_{60} derivatives (fullerenes) in human liver carcinoma cells (HepG2), the cytotoxicity of the surface-coated C60(OH)$_{24}$ being lower by seven orders of magnitude compared with the C60. In rat alveolar macrophages NR8383, Al_2O_3-coated nanoparticles displayed a greater cytotoxicity and a decrease in phagocytic ability compared with uncoated nanoparticles (Wagner *et al.*, 2007). The toxicity of Cd-Se quantum dots to hepatocytes as a result of the release of free Cd^{2+} from crystalline lattice is prevented by ZnS, polyethylene glycol or other coatings (Derfus *et al.*, 2004). The massive influence of the surface chemistry of particles on their cytotoxic effect has also been shown by using CdSe and CdSe/ZnS coated with mercaptopropionic acid (Kirchner *et al.*, 2004).

15.9 NPs and Gene Expression

The course of normal cellular development as well as pathological changes are believed to be driven by changes in gene expression. Similarly, modifications of the environment where cells and organisms live will modulate gene expression profiles. It is therefore reasonable to think that NPs might also be capable of influencing gene expression. The study of these modifications can help to gain insights into the mechanisms of action of NPs and, conversely, genes whose expression is modified by exposure to NPs can be used as biomarkers (indicators of a biological state) of exposure. Among the innumerable kinds of biomarkers, nucleotide sequences are highly specific, and easy to detect and quantify by polymerase chain reaction (PCR) and real-time PCR.

In both cases, screening for differentially expressed genes is the straightforward approach. In the last 15 years, several techniques have been developed for identifying genes differentially expressed among biological samples, including differential display polymerase chain reaction (DD; Liang and Pardee, 1992), serial analysis of gene expression (SAGE; Velculescu *et al.*, 1995), suppressive subtractive hybridization (SSH; Robert *et al.*, 2001), and microarrays (Schena *et al.*, 1995).

The original DD, invented in 1992 by Arthur Pardee and Peng Liang (Liang and Pardee, 1992), consisted of cDNA production from subsets of mRNA that had been reversely transcribed with different primers anchored to the polyadenylated tail, followed by amplification of cDNAs with arbitrary primers, with incorporation of a radioactive label and electrophoresis in polyacrylamide. With this method, two or more samples deriving from different tissues or developmental stages, or which are exposed and not exposed to a drug or to a toxicant, can be compared. The technique has proven to be versatile and sensible, but it has the handicap of a high rate of false positives. False positives depend on several factors, such as the quality of reagents and enzymes, the type and purity of primers, the integrity, concentration and purity of RNA used, as well as the systems being compared, experimental design, the choice of appropriate internal controls, criteria for picking bands, the reaction setup, type of PCR reaction tubes, and training and experience of the researcher.

Several modifications of the original DD protocols have been reported in attempts to minimize false-positive signals (Luce and Burrows, 1998; Miele *et al.*, 1999; Nagel *et al.*, 1999), to increase the efficiency using different primer designs (Wang *et al.*, 1998; Zhao *et al.*, 1998), and to eliminate the need for radioactive markers (Ripamonte *et al.*, 2005; Cho *et al.*, 2006).

DD has been used to search, in culture cells, for transcripts whose expression is modified by cobalt NPs with the dual aim of gaining insights into the mechanisms of action of NPs, and to obtain reliable biomarkers of exposure (Papis *et al.*, 2007a). Several transcripts were identified and some were confirmed and quantified by real-time PCR; such biomarkers, however, were not able to distingush between NPs and ions (Papis *et al.*, 2007b).

SAGE, a modification of DD, allows the quantitative and simultaneous analysis of a large number of transcripts, giving an overview of a cell's gene activity. Because SAGE does not require a preexisting clone, it can be used to identify and quantitate new genes as well as known genes.

SSH can be used for generating both subtracted cDNA or genomic DNA libraries. The method is based on a suppression PCR effect and combines normalization and subtraction in a single procedure. The normalization step equalizes the abundance of DNA fragments within the target population, and the subtraction step excludes sequences that are common to the populations being compared. This dramatically increases the probability of obtaining low-abundance differentially expressed cDNA or genomic DNA fragments, and simplifies the analysis of the subtracted library (Rebrikov *et al.*, 2004).

DNA microarrays consist of an arrayed series of thousands of microscopic spots of DNA oligonucleotides, each containing picomoles of a specific DNA sequence used as a probe to assay, through hybridization under high-stringency conditions, the presence of complementary DNA in a sample. Hybridization is usually detected and quantified by fluorescence-based detection of a fluorophore-labelled target to determine the relative abundance of each specific sequence in the sample. DNA microarrays, although relatively expensive, are easy to use, allow the quantitation of thousands of genes from many samples, and generate large amounts of data in little time. However, they are not appropriate for some applications, such as detecting unknown sequences. Microarrays have found their use to measure responses to toxicologically relevant genes, to identify biomarkers of toxicity, and also to predict the modes-of-action of toxicants (Eun *et al.*, 2008). These potentialities can easily be extended to NP research.

Moreover, microarrays offer an efficient means of interrogating the entire genome for variations such as amplification, deletions, insertions and rearrangements (Gresham *et al.*, 2008). As it is reasonable to think that some NPs could cause variations in the genome, microarrays could again be used for NP risk assessment. Microarray technology, however, is not limited to nucleic acids. Protein microarrays, in fact, are now becoming very popular in biomedicine and for biomarker discovery (Spisak *et al.*, 2007). Therefore, they could soon find their use in nanotoxicology.

15.10   Nanoparticle Cytotoxicity and Oxidative Stress

Apart from acute toxicity of NPs, the toxicity of degradation products, stimulation of cells with subsequent release of inflammatory mediators, and other toxic effects have to be seriously considered. A first indication of the potential cytotoxicity can be obtained by studying cell cultures after incubation with NPs (Lewin *et al.*, 2000; Dodd *et al.*, 2001). The cytotoxicity is usually much higher *in vitro* compared with *in vivo*. This may be explained by the fact that degradation products responsible for the toxicity are eliminated continuously from the application site *in vivo*. Therefore, cytotoxicity tests conducted *in vitro* may have limited application. However, some patterns are emerging for the more studied NPs. The major toxicological concern is the fact that some NPs pass across cell membranes, especially those into mitochondria (Foley *et al.*, 2002), and are redox active (Colvin, 2003).

In human and rat alveolar macrophages, after exposure to titanium dioxide (TiO_2) NPs, the level of reactive oxygen species (ROS) increased (Rahman *et al.*, 1997). ROS are both physiologically necessary and potentially destructive. Moderate levels of ROS play specific roles in the modulation of several cellular events, including signal transduction, proliferative response, gene expression, and protein redox regulation (Halliwell and Gutteridge, 2007; Dalle-Donne *et al.*, 2007). High ROS levels (oxidative stress) can cause damage to key cellular components such as lipids, proteins and nucleic acids, possibly leading to subsequent cell death by necrosis or apoptosis (Dalle-Donne *et al.*, 2006). Oxidative stress associated with TiO_2 NPs, for example, results in early inflammatory responses such as an increase in polymorphonuclear cells, impaired macrophage phagocytosis, and/or fibroproliferative changes in rodents (Bermudez *et al.*, 2004). TiO_2 NPs have also been shown to cause proinflammatory effects in human endothelial cells (Peters *et al.*, 2004).

There is evidence that carbon NPs can have detrimental effects on living organisms. For instance, carbon NPs have been shown to induce oxidative stress in fish brain cells and pulmonary inflammation in rats (Oberdörster, 2004; Warheit *et al.*, 2004), and exposure of human keratinocytes to insoluble carbon NPs was associated with oxidative stress and apoptosis (Shvedova *et al.*, 2003).

Silver NPs have been shown to be a promising antimicrobial material (Sondi and Salopek-Sondi, 2004). Furthermore, a recent study illustrated the potentially huge impact of nanosilver on the fight against AIDS (Elechiguerra *et al.*, 2005), demonstrating the ability of silver NPs (1–10 nm range) to attack HIV-1, preventing interaction of the virus with host cells. However, silver NPs cause high ROS levels, depletion of glutathione, and reduction of mitochondrial membrane potential in the BRL 3A immortal rat liver cell line, suggesting that oxidative stress is likely to contribute to silver NP cytotoxicity (Hussain *et al.*, 2005). It is possible that the loss of glutathione may compromise cellular antioxidant defences and lead to the accumulation of ROS that are generated as byproducts of normal cellular function.

Magnetic NPs can be used in a wide variety of biomedical applications (Figure 15.6), ranging from magnetic labelling, cell isolation, deterioration of cancer cells via hyperthermia, controlled drug release, to contrast agents for magnetic resonance imaging (MRI) (Xu and Sun, 2007). Most of these biomedical applications require well-defined and controllable interactions between the magnetic NPs and living cells. MRI is made possible by the combination of high spatial resolution with the high specificity of magnetic markers, such as iron oxide particles. To achieve this aim, cells must be selectively labelled with magnetic markers either *in vitro* or *in vivo*. The suitability of magnetic markers for MRI is determined by this specificity and by the change in the magnetic resonance signal caused by alterations in magnetic susceptibility. Magnetic NPs based on a core consisting of iron oxides coated with biocompatible materials have proven to fulfil these needs.

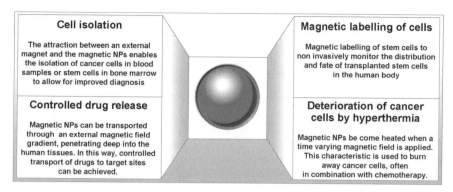

Figure 15.6 *Magnetic NPs.*

Magnetic NPs are usually prepared by micro-emulsion, or alkaline co-precipitation of appropriate ratios of Fe^{3+} and Fe^{2+} salts in water (Hyeon, 2003). The disadvantages of these water-based methods are that the size uniformity and crystallinity of the magnetic NPs are rather poor, and NP aggregation is commonly observed. However, a procedure that enables the synthesis of very monodisperse and highly crystalline magnetic NPs with sizes between 3 nm and 20 nm, without showing any signs of NP aggregation, was recently developed (Sun *et al.*, 2004; Xu and Sun, 2007; Hyeon, 2003).

Several tests on animals have shown that, with a large dosage of iron-based NPs, no side-effects occurred after a 7-day treatment, according to the histology and serological blood tests (Weissleder *et al.*, 1989; Lacava *et al.*, 1999a,1999b). However, cytotoxicity of every product for *in vivo* applications should be examined carefully. The suitability of iron oxide NPs for MRI and cellular labelling has been investigated by several studies both *in vitro* and *in vivo* (Lewin *et al.*, 2000; Bulte *et al.*, 2001, 2002). NPs are incorporated by the cells via endocytosis, giving an intracellular magnetic marker, resulting in a contrast in MRI. The network of blood and lymphatic vessels provides natural routes for the clearing of unwanted materials, and delivery of therapeutic agents. However, this network appears to provide little in the way of controlled and specific access to tissues, and the knowledge of these processes is scant. Regardless of these limitations, nanoparticulate systems provide possibilities for access to cell populations and body compartments. When injected intravenously, NPs are cleared rapidly from the circulation, predominantly by the liver (Kupffer cells) and the spleen (marginal zone and red pulp) macrophages (Moghimi *et al.*, 2001).

Normally, magnetic NPs are surrounded by different coatings to prevent the presence of free iron oxide, but the coating might be metabolized after some time (Schulze *et al.*, 1995). Certain polymeric constituents used in NP design and engineering act as inhibitors of P-glycoprotein efflux pumps expressed in polarized endothelial cells that form the exterior of the blood-brain barrier (Batrakova *et al.*, 2003), and could potentially interfere with transport of a number of modulators and homeostatic mediators in the central nervous system (King *et al.*, 2001). Some polymeric micelles, depending on the nature of their monomer constituents, can induce cell death via apoptosis and/or necrosis (Savic *et al.*, 2003; Moghimi *et al.*, 2004). Degradation products arising from poly(L-lactic acid) particles show cytotoxicity, at least to immune cells (Lam *et al.*, 1993).

Among the possible mechanisms of NP-induced toxicity, increased levels of ROS remains an important consideration. Thus, ROS generation by NPs could lead to possible adverse biological effects through an oxidant injury mechanism. It has recently been demonstrated that the incubation of rat macrophages with citrate-coated iron oxide NPs results in a highly significant increase in oxidative stress, indicated by an augmentation of malonyldialdehyde and protein carbonyls (Stroh *et al.*, 2004). The decrease in oxidative stress when compared with control levels, 24 h after incubation, indicated that the initial increase in oxidative stress is transient and closely linked to incubation of the cells with NPs. The incorporated NPs do not seem to affect cellular proliferation to a significant extent, and the population of rat macrophages does not seem to lose the iron oxide NP marker over a long time. The results of this work demonstrate that the *in vitro* magnetic labelling of macrophages with iron oxide NPs minimally affects cell growth and viability. However, because transient oxidative stress occurs, it is suggested that this stress be prevented by using antioxidants or iron chelators (Stroh *et al.*, 2004).

Possible mechanisms of NP-induced oxidative stress could include the dissolution of NPs and subsequent release of iron, which catalyzes ROS generation with formation of $O_2^{\cdot-}$ from H_2O_2 via the Fenton reaction. Furthermore, inert material does not give rise to spontaneous ROS production, yet is capable of inducing ROS production under biological conditions, based on the ability of the NPs to target mitochondria. For example, cationic polystyrene nanospheres have been shown to induce lysosomal leakage, $O_2^{\cdot-}$ production, and mitochondrial damage, which can eventually lead to apoptosis of murine macrophages (Xia *et al.*, 2006).

15.11　Conclusions

Although nanotechnology is developed sufficiently to enable scientists to make ready use of NPs, there are still fundamental questions about these materials that must be answered if nanotechnology is ultimately to have a significant impact extending beyond the laboratory, into everyday life. It goes without saying that before implementing the various applications of NPs, it is necessary to investigate their potential toxicological impacts. This task, however, is not so straightforward as it may appear. In fact, existing evidence on the factors that affect the uptake, the translocation to other tissues, the intracellular trafficking and the nanotoxicity of metal-based NPs reveals a paucity of data, with more uncertainties than certainties. What it is taken for granted is that it is not always so scientifically sound. For example, we have reasonable evidence that the size and surface coating influence the nanotoxic response *in vitro,* while the corresponding reliable *in vivo* data are mainly still limited to the study of Oberdörster (2004). Furthermore, no data are in practice available on the effects of fundamental physical factors that could influence nanotoxicity, including NP composition, shape, surface area, crystal structure and porosity, which largely determine the functional and toxicological impact of nanomaterials. Therefore, there is a clear need for a systematic approach to generating basic data not only on the toxicological effects of nanomaterials, but also on the relations between the factors that can affect their nanotoxicity. In addition, the dissolution of NPs in biological fluids and the determination of the possible contaminants of the nanomaterials tested are two other key aspects which would be considered in experimental *in vivo* and *in vitro* nanotoxicology research.

Acknowledgements

AFM and SEM images were obtained from a collaborative project with Professor Mario Raspanti, whom we thank. Supported by a grant from FO.CO.VA. Foundation.

References

Anon (1986) Directive 86/609/EEC of 24 November 1986 on the approximation of laws, regulations and administrative provisions of the Member States regarding the protection of animals used for experimental and other scientific purposes. *Official Journal of the European Communities* **L358**, 1–29

Attarian-Shandiz M, Safaei A, Sanjabi S, Barber ZH (2008) Modeling the cohesive energy and melting point of nanoparticles by their average coordination number. *Solid State Commun* **145**, 432–437.

Barnard AS, Lin XM, Curtiss LA (2005) Equilibrium morphology of face-centered cubic gold nanoparticles >3 nm and the shape changes induced by temperature. *J Phys Chem B* **109**, 24465–24472.

Batrakova EV, Li S, Alakhov VY, Miller DW, Kabanov AV (2003) Optimal structure requirements for pluronic block copolymers in modifying P-glycoprotein drug efflux transporter activity in bovine brain microvessel endothelial cells. *J Pharmacol Exp Ther* **304**, 845–854.

Bermudez E, Mangum JB, Wong BA, Asgharian B, Hext PM, Warheit DB, Everitt JI, Moss OR (2004) Pulmonary responses of mice, rats, and hamsters to subchronic inhalation of ultrafine titanium dioxide particles. *Toxicol Sci* **77**, 347–357.

Brunauer S, Emmett PH, Teller E (1938) Adsorption of gases in multimolecular layers. *J Am Chem Soc* **60**, 309–319.

Brunner TJ, Wick P, Manser P, Spohn P, Grass RN, Limbach LK, Bruinink A, Stark WJ (2006) In vitro cytotoxicity of oxide nanoparticles: comparison to asbestos, silica, and the effect of particle solubility. *Environ Sci Technol* **40**, 4374–4381.

Bulte JW, Douglas T, Witwer B, Zhang SC, Strable E, Lewis BK, Zywicke H, Miller B, van Gelderen P, Moskowitz BM, Duncan ID, Frank JA (2001) Magnetodendrimers allow endosomal magnetic labeling and in vivo tracking of stem cells. *Nat Biotechnol* **19**, 1141–1147.

Bulte JW, Duncan ID, Frank JA (2002) In vivo magnetic resonance tracking of magnetically labeled cells after transplantation. *J Cereb Blood Flow Metab* **22**, 899–907.

Carlsson C (2006) *In Vitro Toxicity Assessment of Silver Nanoparticles in Rat Alveolar Macrophages*. MSc dissertation, Wright State University, Pharmacology and Toxicology.

Carr R, Smith J, Nelson P, Hole P, Malloy A, Weld A, Warren J (2008) The real-time visualisation and size analysis of nanoparticles in liquids-nanoparticle tracking analysis. *J Nanoparticle Res*.

Cerbino R, Trappe V (2008) Differential dynamic microscopy: probing wave vector dependent dynamics with a microscope. *Phys Rev Lett* **100**, 188102.

Cha KE, Myung H (2007) Cytotoxic effects of nanoparticles assessed in vitro and in vivo. *J Microbiol Biotechnol* **17**: 1573–1578.

Chen M, von Mikecz A (2005) Formation of nucleoplasmic protein aggregates impairs nuclear function in response to SiO2 nanoparticles. *Exp Cell Res* **305**, 51–62.

Chen Z, Meng H, Xing G, Chen C, Zhao Y, Jia G, Wang T, Yuan H, Ye C, Zhao F, Chai Z, Zhu C, Fang X, Ma B, Wan L (2006) Acute toxicological effects of copper nanoparticles in vivo. *Toxicol Lett* **163**, 109–120.

Chertok B, Moffat BA, David AE, Yu F, Bergemann C, Ross BD, Yang VC (2008) Iron oxide nanoparticles as a drug delivery vehicle for MRI monitored magnetic targeting of brain tumors. *Biomaterials* **29**, 487–496.

Chithrani BD, Ghazani AA, Chan WC (2006) Determining the size and shape dependence of gold nanoparticle uptake into mammalian cell. *Nano Lett* **6**, 662–668.

Cho YJ, Stein S, Jackson RS Jr, Liang P (2006) Saturation screening for p53 target genes by digital fluorescent differential display. *Methods Mol Biol* **317**, 179–192.

Cobb MD, Macoubrie J (2004) Public perceptions about nanotechnology: Risks, benefits and trust. *J Nanoparticle Res* **6**, 395–405.

Colvin VL (2003) The potential environmental impact of engineered nanomaterials. *Nat Biotechnol* **21**, 1166–1170.

Dalle-Donne I, Rossi R, Colombo R, Giustarini D, Milzani A (2006) Biomarkers of oxidative damage in human disease. *Clin Chem* **52**, 601–623.

Dalle-Donne I, Rossi R, Giustarini D, Colombo R, Milzani A (2007) *S*-Glutathionylation in protein redox regulation. *Free Radic Biol Med* **43**, 883–898.

Derfus AM, Chan WC, Bhatia SN (2004) Probing the cytotoxicity of semiconductor quantum dots. *Nano Lett* **4**, 11–18.

Dodd CH, Hsu HC, Chu WJ, Yang P, Zhang HG, Mountz JD Jr, Zinn K, Forder J, Josephson L, Weissleder R, Mountz JM, Mountz JD (2001) Normal T-cell response and in vivo magnetic resonance imaging of T cells loaded with HIV transactivator-peptide-derived superparamagnetic nanoparticles. *J Immunol Methods* **256**, 89–105.

Donaldson K, Stone V, Tran CL, Kreyling W, Borm PJA (2004) Nanotoxicology. A new frontier in particle toxicology relevant to both the workplace and general environment and to consumer safety. *Occup Environ Med* **61**, 727–728.

Edidin M (2001) Near-field scanning optical microscopy, a siren call to biology. *Traffic* **2**, 797–803.

Elechiguerra JL, Burt JL, Morones JR, Camacho-Bragado A, Gao X, Lara HH, Yacaman MJ (2005) Interaction of silver nanoparticles with HIV-1. *J Nanobiotechnology* **3**, 6.

Eun JW, Ryu SY, Noh JH, Lee MJ, Jang JJ, Ryu JC, Jung KH, Kim JK, Bae HJ, Xie H, Kim SY, Lee SH, Park WS, Yoo NJ, Lee JY, Nam SW (2008) Discriminating the molecular basis of hepatotoxicity using the large-scale characteristic molecular signatures of toxicants by expression profiling analysis. *Toxicology* **249**, 176–183.

Fisher HC, Chan WCW (2007) Nanotoxicity: the growing need for *in vivo* study. *Curr Opin Biotechnol* **18**, 565–571.

Foley S, Crowley C, Smaihi M, BonWls C, Erlanger B, Seta P, Larroque C (2002) Cellular localisation of a water-soluble fullerene derivative. *Biochem Biophys Res Commun* **294**, 116–119.

Goodman CM, McCusker CD, Yilmaz T, Rotello VMB (2004) Toxicity of gold nanoparticles functionalized with cationic and anionic side chains. *Bioconjug Chem* **15**, 897–900.

Gresham D, Dunham MJ, Botstein D (2008) Comparing whole genomes using DNA microarrays. *Nat Rev Genet* **9**, 291–302.

Halliwell B, Gutteridge JMC (2007) *Free Radicals in Biology and Medicine* (4th edn). Oxford University Press, Inc: New York.

Hartung T (2005) ECVAM's Progress in implementing the 3Rs in Europe. *Altex* **22**, 18–25.

Hartung T, Bremer S, Casati S, Coecke S, Corvi R, Fortaner S, Gribaldo L, Halder M, Janusch Roi A, Prieto P, Sabbioni E, Worth A, Zuang V (2004) ECVAM's response to the changing political environment for alternatives: consequences of the European Union chemicals and cosmetics policies. *Atla* **31**, 473–481

Hecht B, Sick B, Wild UP, Deckert V, Zenobi R, Martin JFO, Pohl DW (2000) Scanning near-field optical microscopy with aperture probes: fundamentals and applications. *J Chem Phys* **112**, 7761–7774.

Hillyer JF, Albrecht RM (2001) Gastrointestinal persorption and tissue distribution of differently sized colloidal gold nanoparticles. *J Pharm Sci* **90**, 1927–1936.

Huang B, Zhang J, Hou J, Chen C (2003) Free radical scavenging efficiency of Nano-Se in vitro. *Free Radic Biol Med* **35**, 805–813.

Hussain SM, Hess KL, Gearhart JM, Geiss KT, Schlager JJ (2005) In vitro toxicity of nanoparticles in BRL 3A rat liver cells. *Toxicol in Vitro* **19**, 975–983.

Hyeon T (2003) Chemical synthesis of magnetic nanoparticles. *Chem Commun* **8**, 927–934.

Jailani S, Franks GV, Healy TW (2008) ζ Potential of nanoparticle suspensions: effect of electrolyte concentration, particle size, and volume fraction. *J Am Ceram Soc* **91**, 1141–1147.

Kandlikar M, Ramachandran G, Maynard A, Murdock W (2007) Health risk assessment for nanoparticles: A case for using expert judgment. *J Nanoparticle Research* **9**, 137–156.

Kim ZH, Leone SR (2008) Polarization-selective mapping of near-field intensity and phase around gold nanoparticles using apertureless near-field microscopy. *Opt Express* **16**, 1733–1741.

King M, Su W, Chang A, Zukerman A, Pasternack GW (2001) Transport of opioids from the brain to the periphery by P-glycoprotein: peripheral actions of central drugs. *Nat Neurosci* **4**, 268–274.

Kirchner C, Liedl T, Kudera S, Pellegrino T, Javier AM, Gaub HE, Stolzle S, Fertig N, Parak WJ (2004) Cytotoxicity of colloid CdSe and CdSe/ZnS nanoparticles. *Nano Lett* **5**, 331–338.

Kreyling WG, Semmler M, Erbe F, Mayer P, Takenaka S, Schulz H, Oberdörster G, Ziesenis A (2002) Translocation of ultrafine insoluble iridium particles from lung epithelium to extrapulmonary organs is size dependent but very low. *J Toxicol Environ Health A* **65**, 1513–1530.

Lacava ZGM, Azevedo RB, Lacava LM, Martins EV, Garcia VAP, Rébula CA, Lemos APC, Sousa MH, Morais PC, Tourinho FA, Da Silva MF (1999a) Toxic effects of ionic magnetic fluids in mice. *J Magn Magn Mater* **194**, 90–95.

Lacava ZGM, Azevedo RB, Martins EV, Lacava LM, Freitas MLL, Garcia VAP, Rébula CA, Lemos APC, Sousa MH, Tourinho FA, Da Silva MF, Morais PC (1999b) Biological effects of magnetic fluids: toxicity studies. *J Magn Magn Mater* **201**, 431–434.

Lam KH, Schakenraad JM, Esselbrugge H, Feijen J, Nieuwenhuis P (1993) The effect of phagocytosis of poly(Llactic acid) fragments on cellular morphology and viability. *J Biomed Mat Res* **27**, 1569–1577.

Latham AH, Wilson MJ, Schiffer P, Williams ME (2006) TEM-induced structural evolution in amorphous Fe oxide nanoparticles. *J Am Chem Soc* **128**, 12632–12633.

Lewin M, Carlesso N, Tung CH, Tang XW, Cory D, Scadden DT, Weissleder R (2000) Tat peptide-derivatized magnetic nanoparticles allow in vivo tracking and recovery of progenitor cells. *Nat Biotechnol* **18**, 410–414.

Liang P, Pardee AB (1992) Differential display of eukaryotic messenger RNA by means of the polymerase chain reaction. *Science* **257**, 967–971.

Limbach LK, Li Y, Grass RN, Brunner TJ, Hintermann MA, Muller M, Gunther D, Stark WJ (2005) Oxide nanoparticle uptake in human lung fibroblasts: effects of particle size, agglomeration, and diffusion at low concentrations. *Environ Sci Technol* **39**, 9370–9376.

Liu Y, Jehanathan N, Yang H, Laeng J (2007) SEM observation of the "orange peel effect" of materials. *Mater Lett* **61**, 1433–1435.

Lu Q, Yao K, Xi D, Liu Z, Luo X, Ning Q (2007) A comparative study on the selected area electron diffraction patterns of Fe oxide/Au core-shell structured nanoparticles. *J Mater Sci Technol* **23**, 189–192.

Luce MJ, Burrows PD (1998) Minimizing false positives in differential display. *Biotechniques* **24**, 766–770.

Miele G, Slee R, Manson J, Clinton M (1999) A rapid protocol for the authentication of isolated differential display RT-PCR CDNAs. *Prep Biochem Biotechnol* **29**, 245–255.

Moghimi SM, Hunter AC, Murray JC (2001) Long-circulating and target-specific nanoparticles: theory to practice. *Pharmacol Rev* **53**, 283–318.

Moghimi SM, Hunter AC, Murray JC, Szewczyk A (2004) Cellular distribution of nonionic micelles. *Science* **303**, 626–627.

Nagel A, Fleming JT, Sayler GS (1999) Reduction of false positives in prokaryotic mRNA differential display. *Biotechniques* **26**, 641–648.

Natarajan A, Gruettner C, Ivkov R, Denardo GL, Mirick G, Yuan A, Foreman A, Denardo SJ (2008) Nanoferrite particle based radioimmunonanoparticles: binding affinity and in vivo pharmacokinetics. *Bioconjug Chem* **19**, 1211–1218.

Oberdörster E (2004) Manufactured nanomaterials (fullerenes, C60) induce oxidative stress in the brain of juvenile largemouth bass *Environ Health Perspect* **112**, 1058–1062.

Oberdörster G (2005) Principles for characterizing the potential human health effects from exposure to nanomaterials:elements of a screening strategy. *Part Fibre Toxicol* **2**, 8.

Oberdörster G, Ferin J, Lehnert BE (1994) Correlation between particle size, in vivo particle persistence, and lung injury. *Environ Health Persp* **102** (Suppl. 5), 173–179.

Oberdörster G, Oberdörster E, Oberdörster J (2005) Nanotoxicology: an emerging discipline evolving from studies of ultrafine particles. *Environ Health Persp* **113**, 823–839.

Pack A, Grill W, Wannemacher R (2003) Apertureless near-field optical microscopy of metallic nanoparticles. *Ultramicroscopy* **94**, 109–123.

Pal S, Tak YK, Song JM (2007) Does the antibacterial activity of silver nanoparticles depend on the shape of the nanoparticle? A study of the gram-negative bacterium Escherichia coli. *Appl Environ Microbiol* **73**, 1712–1720.

Palath N, Bhad S, Montazeri R, Guidry CA, Haynie DT (2006) Polypeptide multilayer nanofilm artificial red blood cells. *J Biomed Mater Res B Appl Biomater* **81**, 261–268.

Papageorgiou I, Brown C, Schins R, Singh S, Newson R, Davis S, Fisher J, Ingham E, Case CP (2007) The effect of nano- and micron-sized particles of cobalt-chromium alloy on human fibroblasts in vitro. *Biomaterials* **28**, 2946–2958.

Papis E, Gornati R, Prati M, Ponti J, Sabbioni E, Bernardini G (2007a) Gene expression in nanotoxicology research: analysis by differential display in BALB3T3 fibroblasts exposed to cobalt particles and ions. *Toxicol Lett* **170**, 185–192.

Papis E, Gornati R, Ponti J, Prati M, Sabbioni E, Bernardini G (2007b) Gene expression in nanotoxicology: A search for biomarkers of exposure to cobalt particles and ions. *Nanotoxicology* **1**, 197–202.

Park EJ, Choi J, Park YK, Park K (2008) Oxidative stress induced by cerium oxide nanoparticles in cultured BEAS-2B cells. *Toxicology* **245**, 90–100.

Patel D, Moon JY, Chang Y, Kim TJ, Lee GH (2008) Poly(d,l-lactide-co-glycolide) coated superparamagnetic iron oxide nanoparticles: Synthesis, characterization and in vivo study as MRI contrast agent. *Colloids Surf A Physicochem Eng Asp* **313**, 91–94.

Patil S, Sandberg A, Heckert E, Self W, Seal S (2007) Protein adsorption and cellular uptake of cerium oxide nanoparticles as a function of zeta potential. *Biomaterials* **28**, 4600–4607.

Peters K, Unger RE, Kirkpatrick CJ, Gatti AM, Monari E (2004) Effects of nano-scaled particles on endothelial cell function in vitro: studies on viability, proliferation and inflammation. *J Mater Sci Mater Med* **15**, 321–325.

Rahman Q, Norwood J, Hatch G (1997) Evidence that exposure of particulate air pollutants to human and rat alveolar macrophages leads to differential oxidative response. *Biochem Biophys Res Commun* **240**, 669–672.

Rahman Q, Lohani M, Dopp E, Pemsel H, Jonas L, Weiss DG, Schiffmann D (2002) Evidence that ultrafine titanium dioxide induces micronuclei and apoptosis in Syrian hamster embryo fibroblasts. *Environ Health Perspect* **110**, 797–800.

Rebrikov DV, Desai SM, Siebert PD, Lukyanov SA (2004) Suppression subtractive hybridization. *Methods Mol Biol* **258**, 107–134.

Reimer L (1997) *Transmission Electron Microscopy: Physics of Image Formation and Microanalysis* (4th edn). Springer: Berlin.

Ripamonte P, Krempel MG, Watanabe FY, Caetano AR, Meirelles FV (2005) Development and optimization of a fluorescent differential display PCR system for studying bovine embryo development in vitro. *Genet Mol Res* **4**, 726–733.

Robert C, Gagne D, Bousquet D, Barnes FL, Sirard MA (2001) Differential display and suppressive subtractive hybridization used to identify granulosa cell messenger RNA associated with bovine oocyte developmental competence. *Biol Reprod* **64**, 1812–1820.

Russel WMS, Burch RL (1959) The Principles of Humane Experimental Technique. Methuen: London.

Sabbioni E (2005) *In vitro toxicology of nanoparticles.* Proceedings of the Workshop: Research Needs on Nanoparticles, Brussels, 25–26 January 2005: 22.

Sabbioni E, Ponti J, Del Torchio R, Farina M, Fortaner S, Munaro B, Sasaki T, Rossi F (2006) Recherche in vitro sur la toxicologie des nanoparticules au Joint Research Center. Médecine Nucléaire, Imagerie fonctionnelle et métabolique, 2006, 30, no. 1. Available at http://dossier.univ-st-etienne.fr/lbti/www/acomen/revue/2006/pdf1/ROSSI.pdf

Savic R, Luo LB, Eisenberg A, Maysinger D (2003) Nanocontainers distribute to defined cytoplasmic organelles. *Science* **300**, 615–618.

Sayes CM, Fortner JD, Guo W, Lyon D, Boyd AM, Ausman KD, Tao YJ, Sitharaman B, Wilson LJ, Hughes JB, West JL, Colvin VL (2004) The differential cytotoxicity of water-soluble fullerenes. *Nano Letters* **4**, 1881–1887.

Sayes CM, Liang F, Hudson JL, Mendez J, Guo W, Beach JM, Moore VC, Doyle CD, West JL, Billups WE, Ausman KD, Colvin VL (2006) Functionalization density dependence of single-walled carbon nanotubes cytotoxicity in vitro. *Toxicol Lett* **161**, 135–142.

Schena M, Shalon D, Davis RW, Brown PO (1995) Quantitative monitoring of gene expression patterns with a complementary DNA microarray. *Science* **270**, 467–470.

Schulze E, Ferrucci JR, Poss K, Lapointe L (1995) Cellular uptake and trafficking of prototypical magnetic iron oxide label in vitro. *Invest Radiol* **30**, 604–610.

Shea TB, Ortiz D, Nicolosi RJ, Kumar R, Watterson AC (2005) Nanosphere-mediated delivery of vitamin E increases its efficacy against oxidative stress resulting from exposure to amyloid beta. *J Alzheimers Dis* **7**, 297–301.

Shvedova AA, Castranova V, Kisin ER, Schwegler-Berry D, Murray AR, Gandelsman VZ, Maynard A, Baron P (2003) Exposure to carbon nanotube material: assessment of nanotube cytotoxicity using human keratinocyte cells. *J Toxicol Environ Health* **66**, 1909–1926.

Siegrist M, Keller C, Kastenholz H, Frey S, Wick A (2007) Laypeople's and experts' perception of nanotechnology hazards. *Risk Anal* **27**, 59–69.

Skrabalak SE, Au L, Lu X, Li X, Xia Y (2007) Gold nanocages for cancer detection and treatment. *Nanomed* **2**, 657–668.

Sondi I, Salopek-Sondi B (2004) Silver nanoparticles as antimicrobial agent: A case study on E. coli as a model for Gram-negative bacteria. *J Colloid Interface Sci* **275**, 177–182.

Spisak S, Tulassay Z, Molnar B, Guttman A (2007) Protein microchips in biomedicine and biomarker discovery. *Electrophoresis* **28**, 4261-4273.

Stephens MS (1987) The significance of alternative techniques in biomedical research: an analysis of Nobel prize awards. In *Advances in Animal Welfare Science*, Fox MW, Mickley LD (eds). Milhoff: Boston.

Stroh A, Zimmer C, Gutzeit C, Jakstadt M, Marschinke F, Jung T, Pilgrimm H, Grune T (2004) Iron oxide particles for molecular magnetic resonance imaging cause transient oxidative stress in rat macrophages. *Free Radic Biol Med* **36**, 976–984.

Sun S, Zeng H, Robinson DB, Raoux S, Rice PM, Wang SX, Li G (2004) Monodisperse MFe_2O_4 (M=Fe, Co, Mn) nanoparticles. *J Am Chem Soc* **126**, 273–279.

Takenaka S, Karg E, Roth C, Schulz H, Ziesenis A, Heinzmann U, Schramel P, Heyder J (2001) Pulmonary and systemic distribution of inhaled ultrafine silver particles in rats. *Environ Health Perspect* **4**, 547–551.

van Dijk MA, Lippitz M, Orrit M. 2005. Far-field optical microscopy of single metal nanoparticles. *Acc Chem Res* **38**, 594–601.

Velculescu VE, Zhang L, Vogelstein B, Kinzler KW (1995) Serial analysis of gene expression. *Science* **270**, 484–487.

Venugopal J, Low S, Choon AT, Ramakrishna S (2007) Interaction of cells and nanofiber scaffolds in tissue engineering. *J Biomed Mater Res B Appl Biomater* **84**, 34–48.

Wagner AJ, Bleckmann CA, Murdock RC, Schrand AM, Schlager JJ, Hussain SM (2007) Cellular interaction of different forms of aluminum nanoparticles in rat alveolar macrophages. *J Phys Chem B* **111**, 7353–7359.

Wang X, Li X, Feuerstein GZ (1998) Use of novel downstream primers for differential display RT-PCR. *Biotechniques* **24**, 382–384.

Warheit DB, Laurence BR, Reed KL, Roach DH, Reynolds GA, Webb TR (2004) Comparative pulmonary toxicity assessment of single-wall carbon nanotubes in rats. *Toxicol Sci* **1**, 117–125.

Warheit DB, Webb TR, Sayes CM, Colvin VL, Reed KL (2006) Pulmonary instillation studies with nanoscale TiO2 rods and dots in rats: toxicity is not dependent upon particle size and surface area. *Toxicol Sci* **91**, 227–236.

Warheit DB, Webb TR, Colvin VL, Reed KL, Sayes CM (2007a) Pulmonary bioassay studies with nanoscale and fine-quartz particles in rats: toxicity is not dependent upon particle size but on surface characteristics. *Toxicol Sci* **95**, 270–280.

Warheit DB, Webb TR, Reed KL, Frerichs S, Sayes CM (2007b) Pulmonary toxicity study in rats with three forms of ultrafine-TiO$_2$ particles: differential responses related to surface properties. *Toxicology* **230**, 90–100.

Weissleder R, Stark DD, Engelstad BL, Bacon BR, Compton CC, White DL, Jacobs P, Lewis J (1989) Superparamagnetic iron oxide: pharmacokinetics and toxicity. *Am J Roentgenol* **152**, 167–173.

Worth AP, Balls M (2002) Alternative (non-animal) methods for chemical testing: current status and future prospects. *Atla* **30**, suppl. 1.

Xia T, Kovochich M, Brant J, Hotze M, Sempf J, Oberley T, Sioutas C, Yeh JI, Wiesner MR, Nel AE (2006) Comparison of the abilities of ambient and manufactured nanoparticles to induce cellular toxicity according to an oxidative stress paradigm. *Nano Lett* **6**, 1794–1807.

Xu C, Sun S (2007) Monodisperse magnetic nanoparticles for biomedical applications. *Polymer International* **56**, 821–826.

Yang W, Thordarson P, Gooding JJ, Ringer SP, Braet F (2007) Carbon nanotubes for biological and biomedical applications. *Nanotechnology* **18** , 412001–412012.

Yin H, Too HP, Chow GM (2005) The effects of particle size and surface coating on the cytotoxicity of nickel ferrite. *Biomaterials* **26**, 5818–5826.

Zhang Y, Yang M, Portney NG, Cui D, Budak G, Ozbay E, Ozkan M, Ozkan CS (2008) Zeta potential: a surface electrical characteristic to probe the interaction of nanoparticles with normal and cancer human breast epithelial cells. *Biomed Microdevices* **10**, 321–328.

Zhao S, Molnar G, Zhang J, Zheng L, Averboukh L, Pardee AB (1998) 3'-end cDNA pool suitable for differential display from a small number of cells. *Biotechniques* **24**, 842–850.

16

In Vitro and *In Vivo* Toxicity Study of Nanoparticles

Jayoung Jeong, Wan-Seob Cho, Seung Hee Kim and Myung-Haing Cho

16.1 *In Vitro* Toxicity Study of Nanoparticles

Nanoparticles have a higher probability of successfully travelling through an organism than other materials or larger particles, thus holding great promise for use in nanomedicinal applications. However, the various interactions of nanoparticles with fluids, cells and tissues, and their biological responses at each of these sites, need to be considered prior to use in the field of nanomedicine. Therefore, nanoparticle biocompatibility must be tested using the appropriate methods to assess the safety of these particles for use in medicinal applications. Unfortunately, there have only been a limited number of studies that have examined the *in vitro* mechanistic toxicology of nanoparticles. Prior to the clinical use of nanoparticles, *in vitro* evaluations of nanoparticles must demonstrate a high degree of biocompatibility, with minimal negative effects on cell viability, immune function and blood components (Hall *et al.*, 2007).

16.1.1 Sterility

The sterility of nanoparticles, which may be used in pharmaceuticals, is an important part of nano-safety. Guidelines for testing sterility in pharmaceuticals has been addressed in various worldwide pharmacopeias (e.g. USP, EP, KP and JP), including Section 21 of the Code of Federal Regulations (CFR), International Conference on Harmonization (ICH) and Food and Drug Administration *Points to Consider* (PTC) documents. These documents provide a brief summary of the test methods and sample requirements used for the most common types of pharmaceutical products. In addition, it might be recommended

Nanotoxicity: From In Vivo *and* In Vitro *Models to Health Risks*   Edited by Saura Sahu and Daniel Casciano

that nanobio products pass sterility tests that are outlined under the good manufacturing practice (GMP) guidelines.

The sterility of nanoparticles should be evaluated before initiating *in vitro* and/or *in vivo* tests. Endotoxin levels, bacteria, yeast, mold or even mycoplasma contamination should be tested. Therefore, this chapter describes the work that has been done to examine nanoparticle sterility such as the detection of endotoxins and bacteria/yeast/mold.

16.1.1.1   Endotoxins

The association of endotoxins (lipopolysaccharide, LPS), which can trigger an immune response that complicates the interpretation of a toxic response in toxicological tests, is a concern in terms of nanoparticle safety, in addition to the release of endotoxins intimately associated with septic shock (Gottlieb, 1993; Munford, 2006). A new study shows that gold nanospheres, with a diameter of 7 nm and produced in conventional laboratory surroundings, activate human antigen presenting dendritic cells (DCs) to induce the proliferation of peripheral blood mononuclear cells (PBMC), when the particles were mixed with either allergenic or autologous DCs. This effect was found to be due to endotoxin contamination of the nanoparticles. However, when the particles were produced under controlled conditions, endotoxin contamination was eliminated and the activation of DCs did not take place (Vallhov *et al.*, 2006).

In addition, the quantitative detection of gram-negative bacterial endotoxins in nanoparticle preparations using an end-point Limulus Amebocyte Lysate (LAL) assay has been examined (Darkow *et al.*, 1999). Gram-negative bacterial endotoxins catalyze the activation of proenzymes in the Limulus Amebocyte Lysate. The activated enzyme then catalyzes the splitting of p-nitroanilin from the colorless substrate. The released p-nitroanilin was measured photometrically at 405 nm after the reaction was terminated with a stop reagent. The concentration of endotoxins in the sample was directly proportional to the absorbance and was calculated from a standard curve. The authors of this study recommended using an endotoxin-free environment during particle production, and that the bio-nanoparticles should be sterilized with gamma radiation before use in medical applications.

LPS contamination is probably avoided in a GLP (good laboratory practice) laboratory when particles are produced for biomedical applications. However, LPS levels should always be analyzed in nanoparticle preparations if they are to be tested on living cells or animals. Furthermore, LPS contamination should be avoided in nanoparticles used for other applications, since there is a risk of an unwanted immune response if particles enter the body via various routes.

16.1.1.2   Bacteria/yeast/mold

Microbial contamination tests should be carried out to ascertain the quality and safety of manufactured nanoparticles that will be used in nano-medicine. Biological products manufactured under GMP conditions require that sterility testing be performed under the GMP guidelines. There are two common methods to test sterility: immersion (direct inoculation) and membrane filtration. The pharmacopeias and 21CFR 610.12 recommend using two media for both the immersion and membrane filtration methods (Khuu *et al.*, 2004). In the case of nanoparticles, the immersion (direct inoculation) method may be preferred since there are currently no filters small enough for nanoparticles (Zhao *et al.*, 2008). Generally in both test methods the test article or membrane is incubated for 14 days in the test media. However, protocols can be customized according to the characteristics and properties of the nanoparticles, such as size, shape and surface modifications.

16.1.2 Blood Contact Properties

Blood compatibility is an important property for the intended *in vivo* functions of most bionanoparticles. The lack of blood compatibility results in the adsorption of plasma proteins, platelet adhesion, triggering coagulation, activating complement cascades and clot formation. The coagulation of nanoparticles is driven by intermicellar collisions. Therefore, the blood contact properties of nanoparticles should always be evaluated prior to clinical use.

16.1.2.1 Hemolysis

The erythrocyte is a highly specialized cell with a limited metabolic repertoire. As an oxygen shuttle, it must continue to perform this essential task while being exposed to a wide range of environments in the vascular circuit, and to a variety of xenobiotics througout its lifetime. Therefore, the red blood cell is susceptible to a variety of toxic effects from xenobiotics as well as nanoparticles. Toxicity in erythrocytes is associated with alterations in cell shape, which leads to a loss of cellular deformability (Piomelli, 1981). Therefore, exposure to nanoparticles may affect the basic rheological properties of erythrocytes, which can cause problems with erythrocyte deformability such as hemolysis (Leach *et al.*, 2002; Shi *et al.*, 2005).

Hemolysis can be assayed by quantitatively determining the amount of hemoglobin in the whole blood (total blood hemoglobin, TBH) and the amount of hemoglobin released into the plasma (plasma-free hemoglobin, PFH) by calorimetry when blood is exposed to nanoparticles. In this assay, the erythrocytes are collected by centrifugation at 1500 rpm for 15 min, and then washed three times with the appropriate buffer, such as phosphate buffered saline (PBS) buffered at pH 7.4. The nanoparticle dispersions are prepared in a PBS buffer at the appropriate concentration. Then 100 μl of the stock dispersion is added to 1 ml of the nanoparticle dispersion. The solutions are mixed and incubated for 4 h at 37° C in an incubator shaker. The percentage of hemolysis is measured by UV-vis analysis of the supernatant at an absorbance of 394 nm after centrifugation at 13 000 rpm for 15 min. One milliliter of PBS is used as the negative control with 0 % hemolysis, and 1 ml of double-distilled (DD) H_2O is used as the positive control with 100 % hemolysis. Hemoglobin and its derivatives, except sulfhemoglobin, are oxidized to methemoglobin by ferricyanide in the presence of alkali. Cyanmethemoglobin is then formed from the methemoglobin through its reaction with cyanide (Drabkin's solution). The cyanmethemoglobin can then be detected by a spectrophotometer that is set at 540 nm. The hemoglobin standard is used to build a standard curve that covers the concentration range of 0.025 to 0.80 mg/ml and to prepare quality control samples at low (0.0625 mg/ml), mid (0.125 mg/ml) and high (0.625 mg/ml) concentrations to monitor assay performance. The required sample volume is 300 μl (i.e. 100 μl per test-replicate). The results are expressed as a percentage of hemolysis and are used to evaluate the acute *in vitro* hemolytic properties of the nanoparticles.

The characterization and evaluation of a nanomaterial's blood contact properties is very important in relation to nanoparticle safety (De Jong and Borm, 2008). The hemolysis assay with mammalian erythrocytes is very easy to perform as a screening test of nanoparticle toxicity (Dobrovolskaia *et al.*, 2008; Lin *et al.*, 2008). The possible hemolytic factors of nanoparticles are enzymatic modification, oxidant damage of cell membranes, changes in the basic rheological properties of the cells, osmotic stability, and endotoxin and/or microbial contamination.

16.1.2.2 Platelet Aggregation

There is some evidence that microscale particles made from carbon can affect vascular hemostasis and precipitate thrombosis (Delfino *et al.*, 2005). In fact, it has been shown that personal exposure to ambient airborne particulate matter (PM) 10 nm or less in aerodynamic diameter (PM10) increased plasma fibrinogen levels (Seaton *et al.*, 1999). Furthermore, peripheral thrombosis in experimental animals has been detected following tracheal inhalation of diesel exhaust particles (Nemmar *et al.*, 2003). However, no studies have examined the effects of engineered or combustion-derived nanoparticles on platelet function or vascular thrombosis. Therefore, the effects of nanoparticles on human platelet function *in vitro* and vascular thrombosis in experimental animals should be evaluated.

The effect of nanoparticles on platelet function can be determined by obtaining blood from healthy volunteers who have not taken any drugs known to affect platelet function for two weeks prior to the study. Washed platelets must then be isolated and resuspended in Tyrode's solution (2.5×10^8 platelets/ml), as previously described (Radomski and Moncada, 1983). Platelets are preincubated for 2 min at $37°$ C in a whole blood ionized calcium lumi-aggregometer (Chronolog) prior to the addition of particles. Prostacyclin (PGI2), S-nitrosoglutathione (GSNO), aspirin, 2-methylthio-AMP, phenanthroline, ethylenediamine tetraacetic acid (EDTA) and Go6976 are preincubated with platelets for 1 min before the addition of nanoparticles. Platelet aggregation is studied for 8 min and analyzed using a Aggro-Link data reduction system (Radomski *et al.*, 2001, 2002; Jurasz *et al.*, 2003). The release of ATP is measured by luciferin-luciferase using a lumi-aggregometer as previously described (Sawicki *et al.*, 1997). Platelet-rich plasma (PRP) is obtained from human whole blood that was freshly pooled and incubated with the control or test sample for 15 min at a nominal temperature of $37°$ C. After that PRP is analyzed using a Z2 particle counter and size analyzer to determine the number of active platelets. Percentage aggregation is calculated by comparing the number of active platelets in the test sample to the number of active platelets in the control baseline tube.

16.1.2.3 Coagulation

The clotting time of the nano-treated sample can be compared with the clotting time of the pooled normal plasma or whole blood to provide a standard measurement of the sample's hemostatic status. The plasma coagulation is assayed in four tests: prothrombin time (PT), activated clotting time (ACT), activated partial thromboplastin time (APTT) and thrombin time (TT). As described below, such clotting assays are commonly used as screening tests to evaluate both the samples' intrinsic and extrinsic coagulation systems.

Prothrombin time test. The PT test measures the tissue factor-induced coagulation time of blood or plasma. It is used as a screening test to evaluate the integrity of the extrinsic coagulation pathway and is sensitive to coagulation factors I, II, V, VII and X (Glassock, 2007). The test is performed by adding thromboplastin and Ca^{2+} to a blood sample and measuring the time for clot formation. A prolonged clotting time suggests the presence of an inhibitor to, or a deficiency in, one or more of the coagulation factors of the extrinsic pathway. However, a prolonged PT clotting time can also occur in patients on warfarin therapy or for those with vitamin K deficiency (Zelis *et al.,* 2008) or liver dysfunction (Butenas and Mann, 2002; Tripodi *et al.,* 2007). The PT test can provide an assessment of the extrinsic coagulation pathway, and is widely used to monitor the effects of nanoparticle-treated blood.

Activated clotting time test. The ACT is a screening test that resembles the activated partial thromboplastin time (APTT) test, but is performed using fresh whole blood samples. The ACT is used primarily to monitor the status of blood coagulation in connection with clinical procedures that involve the administration of high doses of heparin (Galli and Palatnik, 2005). It is important to monitor the blood response to heparin during such procedures, because underdosing can result in pathological thrombus formation, whereas overdosing can lead to serious hemorrhagic complications (Kher *et al.,* 1997). Therefore, interactions between nanoparticles and heparin-treated blood should also be evaluated prior to use in *in vivo* applications.

Activated partial thromboplastin time test. The APTT test is used to evaluate the intrinsic coagulation pathway, which includes factors I, II, V, VIII, IX, X, XI and XII. The test is performed using a plasma sample, in which the intrinsic pathway is activated by the addition of phospholipid, an activator (ellagic acid, kaolin, or micronized silica), and Ca^{2+} (Ray *et al.,* 1992; Tripodi *et al.,* 1993; Dragoni *et al.,* 2001). Formation of the Xase and prothrombinase complexes on the surface of the phospholipid enables prothrombin to be converted into thrombin, which subsequently promotes clot formation (Pickering *et al.,* 2004). The APTT is used to assess the overall competence of a patient's coagulation system, as a preoperative screening test for bleeding tendencies, and as a routine test for monitoring heparin therapy. Since nanoparticles may affect the intrinsic coagulation pathway, nanoparticle treatment should be subject to the APTT prior to use in nanomedical application (Jiao *et al.,* 2002; Cenni *et al.,* 2008).

Thrombin time test. The TT test measures the rate of blood clot formation after thrombin treatment and compares this rate with that of a normal plasma control. The test is performed by adding a standard amount of thrombin to plasma that has been depleted of platelets, and measuring the time required for a clot to form. This test has been used as an aid in the diagnosis of disseminated intravascular coagulation (DIC) and liver disease (Ragni *et al.* 1982; Ho *et al.* 1998).

16.1.2.4 Complement Activation

Intravenous injection of some micelles and other lipid-based nanoparticles can cause acute hypersensitivity reactions (HSRs) in a high percentage (up to 45 %) of patients, with hemodynamic, respiratory and cutaneous manifestations (Szebeni *et al.,* 2007). The phenomenon can be explained by the activation of the complement (C) system on the surface of the lipid particles, which leads to anaphylatoxin (C5a and C3a) liberation and subsequent release reactions of mast cells, basophils and possibly other inflammatory cells in the blood (Vedhachalam *et al.*, 2007).

The total activation of the complement system in samples treated with nanoparticles may be qualitatively determined by Western blot (Nagayama *et al.*, 2007). The complement system represents an innate arm of the immune defense and the antibody-mediated immune response. Three major pathways that lead to complement activation have been described (Fujita, 2005). They are the classical pathway, the alternative pathway and the lectin pathway. The classical pathway is activated by immune (antigen-antibody) complexes (Lu *et al.*, 2008). Activation of the alternative pathway is antibody-independent (Toapanta and Ross, 2006). The lectin pathway (Gadjeva *et al.*, 2004) is initiated by plasma protein mannose-binding lectin. The complement system is composed of several components (C1, C2 ... C9) and factors (B, D, H, I and P). Activation of any one of the three pathways mentioned above results in the cleavage of the C3 component of complement. The effect of nanoparticles on the complement system can be determined by exposing the test nanoparticles to human plasma and subsequently analyzing the human plasma by polyacrylamide gel electrophoresis (PAGE) followed by Western blot with anti-C3 specific antibodies (Jiang *et al.*, 2008). These antibodies recognize both the native C3 component of the complement and its cleaved products. Native C3 and minor amounts of C3 cleavage products can be visualized by Western blot in control human plasma. When test compounds or positive controls (cobra venom factor) induce activation of the complement system, the majority of the C3 component is cleaved and the appearance of C3 cleavage products will be observed by Western blot.

16.1.3   Cell-Based Assays (Cytotoxicity)

Based on the unusual physical and chemical properties of nanoscale materials and their increasing use in medical and commercial applications, it is critical to develop and validate *in vitro* toxicity assays to aid in the safe implementation of these emerging nanotechnologies. Upon intentional or unintentional introduction into the human body, the nanoparticle properties will determine the mechanism of cellular uptake and the toxicity profile. Current cytotoxicity studies are largely limited to assays that measure average cell viability before and after exposure to nanoparticles, instead of analyzing the effects of these particles on individual cell function (Dey *et al.*, 2008). Assays that can detect the latter effect are pertinent because live cells that have accumulated nanoparticles are likely to be functionally compromised. Cytotoxicity has been measured using different methods, including tetrazolium dye assays (e.g. MTT, XTT, WST-1), viability markers (e.g. trypan blue, neutral red, thymidine uptake), apoptotic and necrotic markers with immunocytochemistry and flow cytometry staining (e.g. propidium iodide), cell growth by counting cells and subpassages and population doubling levels.

16.1.3.1   Necrosis

Before the efficacy of nanoparticles can be determined, cytotoxicity must be assessed in order to ensure that the desired effects of the nanoparticles are appropriately interpreted.

Many studies have measured the release of the cytoplasmic protein lactate dehydrogenase (LDH). This protein leaks into the extracellular media either as a consequence of membrane defects that lead to death, or due to cell death, which results in membrane breakdown. Therefore, the level of LDH enzyme activity in the cell culture medium of exposed cells should be directly proportional to the level of cell death. The 3-(4,5-Dimethyl-2-thiazolyl)-2,5-diphenyl-2H-tetrazolium bromide (MTT) assay is widely used as a measure of cell viability (Sayes *et al.*, 2007). The MTT molecule is converted to a blue formazan product by the mitochondrial enzyme succinate dehydrogenase, and so a decrease in the metabolic competence of the cells is used as an indicator of viability. Annexin V combined with propidium iodide staining provides a reliable and relatively easy technique to quantify viability and can also be used to distinguish between cell death via apoptosis and necrosis (Kodavanti *et al.*, 2006). The technique involves quantifying immunofluorescently stained cells by flow cytometry or fluorimetry. It is also possible to image the cell by fluorescence or confocal microscopy to observe changes in individual cells within a culture. Another assay, the live/dead assay, has also been used in some studies (Sayes *et al.*, 2006). This assay involves treating the cells with calcein acetoxymethylester (AM), which diffuses into the cells due to the lipophilic nature of AM. Once inside viable cells, cytoplasmic esterase enzymes will cleave the dye to release fluorescent calcein. In addition, the cells are treated with an ethidium homodimer, which can only enter cells if the cell membrane is compromised and hence are dead. The ability of the nanoparticles to interfere with fluorescence quantification must be taken into consideration in any of the assays that use fluorescence as a marker; for instance, TiO_2 can reflect UV light, while carbon black absorbs light and can alter the background fluorescence (Choi *et al.*, 2007c).

16.1.3.2 Apoptosis

Apoptosis is programmed cell death, which is an integral part of normal development and is involved in disease and infection. Apoptosis is a morphologically distinct form of cell death that is implemented and preceded by a well-conserved biochemical mechanism involving caspases, a family of systeine proteases that cleave proteins at aspartic acid residues. They are the main effectors of apoptosis or programmed cell death, and their activation leads to characteristic morphological changes of the cell such as shrinkage, chromatin condensation, DNA fragmentation and plasma membrane blebbing.

There are two main pathways to apoptotic cell death. The first pathway, which is referred to as the death receptor pathway, involves the interaction of a death receptor, such as the tumor necrosis factor (TNF) receptor-1 or the Fas receptor, with its ligand (Akazawa and Gores, 2007). The second pathway, which is referred to as the mitochondrial pathway, depends upon the participation of mitochondria, proapoptotic and antiapoptotic members of the Bcl-2 family (Siegel and Lenardo, 2002). The end result of both pathways is caspase activation and the cleavage of specific cellular substrates, resulting in the biochemical and morphological changes associated with the apoptotic phenotype (Zimmermann and Green, 2001). Monitoring apoptosis in porcine proximal tubule cells (LLC-PK1) and human hepatocarcinoma cells (Hep G2) that were treated with nanoparticles has been used to study the *in vitro* NCL preclinical characterization cascade. The protocol utilizes a fluorescent method to determine the degree of caspase-3 activation (Fox and Aubert, 2008).

16.1.3.3 Leukocyte Proliferation

To assess the effect of nanoparticle formulation on basic immunological function, leukocyte proliferation should be considered. This can be done by isolating lymphocytes from pooled human blood that has been anti-coagulated with Li-heparin using the Ficoll-Paque Plus solution. The isolated cells can then be incubated with or without phytohemaglutinin (PHAM) in the presence or absence of nanoparticles (Hoshino *et al.*, 2004). Therefore, this assay can be used to measure the ability of nanoparticles to induce a proliferative response in human lymphocytes or to suppress the response induced by PHAM.

16.1.3.4 Activity of NK Cells

Natural killer, or NK, cells were discovered in the 1970s and constitute up to 15 % of the total lymphocyte population in normal healthy subjects. They are capable of killing a broad range of tumor and viral infected cells. Depressed NK cell activity and populations are associated with the development and rapid progression of cancer, hepatitis, AIDS, chronic fatigue syndrome, various immunodeficiency syndromes, and certain autoimmune diseases (Kwak-Kim and Gilman-Sachs, 2008; Schmitt *et al.*, 2008; Vivier *et al.*, 2008).

Due to their capability of killing a broad range of infected cells both without prior sensitization and without major histocompatibility, NK cells represent an immune surveillance mechanism independent of classic T cell-mediated immunity. Studies with human and animal subjects have revealed that NK cells are the first line of defense against viruses, bacteria, parasites and tumors. Compromised or absent natural immunity is associated with acute and chronic viral infections such as AIDS, chronic fatigue immune dysfunction syndrome (CFIDS), psychiatric depression, and various immunodeficiency syndromes (Ojo-Amaize *et al.*, 1994). Although the exact nature of CFIDS is still controversial, it has been generally accepted that CFIDS is characterized by debilitating fatigue and by a number of immunological abnormalities, the most consistent being a significant depression of NK cell activity. Therefore, the effects of nanoparticles on NK cell cytotoxicity should be examined to determine whether there is an association between NK cell activity and nanoparticle treatment (Wilson *et al.*, 2007).

16.1.3.5 Phagocytosis

Phagocytic cells are typically neutrophils or macrophages, the latter being most common. Macrophages are found in all tissues of the body and can take up particles from anywhere they may accumulate. However, the principal organs in which they are normally found are the spleen, liver and lymph nodes. Because of the architecture of these tissues, the spleen tends to be involved in clearing larger particles (>250 nm diameter) that have become trapped (Fernandez-Urrusuno *et al.*, 1996). In contrast, the liver is more involved in clearing smaller particles, and the site-specific macrophages (Kupffer cells) will engulf nanoparticles (Fernandez-Urrusuno *et al.*, 1997).

16.1.4 Genotoxicity Assays

Genetic toxicity needs to be addressed early in the safety assessment process of chemicals for regulatory purpose. This usually involves a tiered strategy that includes *in vitro* assays

(ICH, 1998). The two main endpoints that are usually investigated are gene mutations and chromosome damage, which are detected by alterations in chromosome number (polyploidy and aneuploidy) or structure (breaks, deletions, rearrangements). Current genotoxicity tests include assays that detect DNA breaks, bacterial mutations, chromosomal aberrations, sister chromatid exchanges, micronuclei, unscheduled DNA synthesis, as well as the formation of 8-OHdG and DNA adducts (Auffan *et al.*, 2006; Colognato *et al.*, 2008; Fenech, 2008). Investigations of genotoxicity and cellular interactions as a result of engineered nanomaterials and nanoparticles manufactured on the nanometer scale have been limited, and the majority of these studies have screened only for cytotoxicity.

16.2 *In Vivo* Toxicity Study of Nanoparticles

Advances in the engineering of nanostructures with exquisite size and shape control made nanotechnology an exciting research area, owing to their unique properties and broad range of applications (Medintz *et al.*, 2005; Caruthers *et al.*, 2007). Engineered nanostructures are used as probes for ultrasensitive molecular sensing and diagnostic imaging, agents for photodynamic therapy, actuators for drug delivery, triggers for photothermal treatment, precursors for building solar cells, and electronic and light-emitting diodes (Akerman *et al.*, 2002; Gao *et al.*, 2004; Medintz *et al.*, 2005; Caruthers *et al.*, 2007). As transitions from academic findings to industrial products are emerging, toxicities of nanostructures have been issued (Maynard *et al.*, 2006; Tsuji *et al.*, 2006). To date, a complete understanding of the biological effects of size, shape, composition and surface chemistry has been limited, and thus it is unclear whether the exposure of humans, animals, insects and plants to engineered nanostructures could produce harmful biological responses (Colvin, 2003). Several studies have shown that nanoparticles have different biological responses when compared with micron-sized or larger materials. From these studies a new subdiscipline of nanotechnology called nanotoxicology has emerged (Oberdörster *et al.*, 2005, 2007). Because nanotechnology is an emerging science and widely used in various fields, an enormous interest in nanotoxicology research has been generated. *In vivo*, these nanostructures could be metabolized or altered, and since it is well known that engineered nanostructures have properties that are related to their size, shape and/or composition, the effects of the metabolized/altered nanostructures on biological systems are difficult to predict. As a result of this uncertainty, in combination with an incomplete understanding of the interactions of nanostructures with biological systems, regulatory agencies and the general public have raised questions with regard to nanotechnology-based products.

16.2.1 Pharmacokinetics (ADME)

A systemic analysis of the pharmacokinetics (absorption, distribution, metabolism and excretion; PK) of nanoparticles is a crucial component in understanding the activity of nanoparticles and for risk assessment. Since PK data give quantitative information about the behavior of nanoparticles in biological systems and their toxic effects, the data obtained from PK can help to improve the design of nanoparticles for biomedical applications such as diagnosis and therapeutics, and provide a better understanding of the behavior of nanoparticles in different tissues and cell types. In addition, PK data are crucial for making a risk assessment because specific PK parameters provide information about the potential toxicity of nanoparticles in the target organ. Even though PK is critical for the evaluation of nanoparticle characteristics or toxicity, very few classical pharmacokinetic studies have been conducted with nanoparticles (Table 16.1).

In general, nanoparticles enter the body via six principal routes: inhalation, intravenous, intraperitoneal, oral, dermal, and subcutaneous. Administration routes are divided into direct parenchymal routes (intravenous or intra-arterial), where the drug does not meet cellular barriers before reaching the circulatory system, and indirect routes, where the drug does meet barriers before it reaches the circulatory system. When the nanoparticles enter the body, absorption can occur through interactions with biological components. Afterwards they can be distributed to various organs in the body where they may remain structurally the same, become modified, or metabolized (Borm *et al.*, 2006). When nanoparticles are distributed to various organs, they can enter the cells of the organ and reside in the cells for an unknown amount of time before moving to other organs or being excreted. Excretion may also occur after the distribution steps. A number of these studies have focused on traditional toxic materials, such as asbestos and carbon black, that have a wide size distribution (some of which are nanometer in scale) (Oberdörster *et al.*, 2005; Tsuji *et al.*, 2006; Poland *et al.*, 2008).

16.2.1.1   Absorption

Absorption is the movement of the nanoparticle from the administration site directly to the therapeutic site or to the circulatory system. Absorption is determined by both the physico-chemical properties of the chemical substance or macromolecule, and the biological properties of the administration site. During the absorption process, the nanoparticles may interact with many biological components such as immune molecules processed by opsonization. Many of the serum proteins that interact with nanostructures, such as complement (Reddy *et al.*, 2007) and immunoglobulins, are immunoactive. In addition, absorbed nanoparticles can change the structure and function of protein, which can affect the fate of the nanoparticles (Lundqvist *et al.*, 2004; De *et al.*, 2007). Therefore, in PK studies of nanoparticles, one should focus more on protein binding and immune components. This type of data can yield relevant information about the structure–activity relationship (nanoparticle–protein interaction) and is important for evaluating the toxicity of nanoparticles (Reddy *et al.*, 2007).

16.2.1.2   Distribution

After absorption into the circulatory system, nanoparticles can subsequently be distributed to all parts of the body. However, the distribution of the nanoparticles would be uneven because of differences in blood flow, tissue binding, regional pH and permeability of endothelial membranes of the blood vessels. In the distribution process, the density of blood vessels is a key factor that defines the speed at which equilibrium and organ-specific concentrations are reached, where highly vascularized areas reach equilibrium more rapidly than poorly vascularized areas. The biodistribution of nanoparticles can only be determined by quantitatively mapping the location of nanoparticles administered at

Table 16.1 Pharmacokinetics studies.

Study	Species/ strain	Sex	Nanoparticles, dose and duration of exposure	Route of exposure	Duration of study	Main findings	References
PK	BALB/c nude mice	Female	SPIO 20 nm, NanoFerrite 30 nm, and 100 nm radioimmunonanoparticles	iv	5 min, and 1, 2, 4, 24 and 48 h	1. Blood and body clearance of the nanoparticles and mean concentrations in lung, kidney and lymph node were similar to ^{111}In-ChL6-NP. 2. Mean tumor uptakes (%ID/g ± SD) of each SPIO 20 nm, NanoFerrite 30 nm, and 100 nm radioimmunonanoparticles at 48 h were 9.00 ± 0.8 (20 nm), 3.0 ± 0.3 (30 nm), and 4.5 ± 0.8 (100 nm), respectively. 3. The ranges of tissue uptakes were liver (16–32 ± 1–8), kidney (7.0–15 ± 1), spleen (8–17 ± 3–8), lymph nodes (5–6 ± 1–2), and lung (2.0–4 ± 0.1–2).	Natarajan *et al.* (2008)
PK	A/J mice	Male	siRNA DOTAP/DOPE complexes (250 nm) siRNA RGD-PEG-PEI complexes (130 nm)	iv	5, 15, 30, 60 min	1. Complexation of siRNA with either DOTAP/DOPE or RGD-PEG-PEI did not affect siRNA blood levels at any of the time-points. 2. Complexes distributed mainly in liver and kidney, suggesting a rapid renal clearance by glomerular filtration. 3. DOTAP/DOPE: highest tissue levels were found in the liver, lung and kidney (16, 26, and 4.0% at 15 min, respectively). 4. RGD-PEG-PEI: accumulated in the liver, lung, kidney, and spleen (36, 7.0, 3.6, 3.4% at 15 min, respectively).	de Wolf *et al.* (2007)

(Continued)

Table 16.1 (Continued)

Study	Species/strain	Sex	Nanoparticles, dose and duration of exposure	Route of exposure	Duration of study	Main findings	References
PK	BALB/c mice	Female	DTPA-CNT with radiotracer [^{111}In]	iv	24 h	Functionalized SWNTs are not retained in any of the reticuloendothelial system organs (liver or spleen) and are rapidly cleared from systemic blood circulation through the renal excretion route.	Singh et al. (2006)
Distribution	Wister rats	Male	10 nm (77 µg/ml), 50 nm (96 µg/ml), 100 nm (89 µg/ml), 250 nm (108 µg/ml) gold nanoparticles	iv	24 h	1. 10 nm particles were present in various organ systems including blood, liver, spleen, kidney, testis, thymus, heart, lung and brain 2. Whereas, the larger particles were only detected in blood, liver and spleen	De Jong et al. (2008)
Distribution	ICR mice	Male	50 nm MNP@SiO$_2$ (RITC)	ip	4 wks	The particles were distributed in all organs and the distribution pattern was time-dependent	Kim et al. (2006a)
Distribution	ICR mice	Male	Water-soluble hydroxylated SWCNTs with radioactive ^{125}I atoms	ip		Accumulated in the liver and kidneys and was excreted in the urine in 18 days	Wang et al. (2004)
Clearance	SD rats	Male	CdSe/ZnS quantum dots with DHLA (anionic), cysteamine (cationic), cysteine (zwitterionic), and DHLA-PEG (neutral) coatings	iv	4 h	A final hydrodynamic diameter <5.5 nm resulted in rapid and efficient urinary excretion and elimination 1. Molecular weight and charge contribute to final hydrodynamic diameter in vivo and that pure charge (anionic and cationic) is associated with unexpected serum protein adsorption. This absorption did not appear to affect solubility but increased hydrodynamic diameter by almost 15 nm and prevented renal excretion.	Choi et al. (2007a)

different time points and at different doses. Unfortunately, only a few studies regarding the *in vivo* biodistribution of engineered nanoparticles have been conducted. Therefore, at this point we cannot make a general conclusion as to how the size, shape, aggregation and surface chemistry will affect nanostructure biodistribution. In a previous study, inorganic and organic nanoparticles were found to be mainly sequestered in the liver and spleen, and the surface chemistry was determined to be one of the most important factors influencing tissue biodistribution (Fischer *et al.*, 2006; Kim *et al.*, 2006a; Liu *et al.*, 2007; Yang *et al.*, 2007). In addition, lymph nodes and bone marrow were shown to be target organs of nanoparticles (Natarajan *et al.*, 2008). All of these organs contain large concentrations of macrophages, which are part of the reticuloendothelial system (RES). As a result of these findings it was concluded that nanoparticles were taken up by phagocytic cells. The RES, now called the mononuclear phagocyte system (MPS), is part of the immune system and consists of a collection of monocytes and macrophages that are involved in the uptake and metabolism of foreign molecules and particulates (Saba, 1970). Surface chemistry such as coating with polyethylene glycol and ammonium/chelator functional groups have been shown to prevent RES uptake (Paciotti *et al.*, 2004; Singh *et al.*, 2006). Aside from surface chemistry, the core nanostructure could also impact upon biodistribution behavior.

16.2.1.3 Metabolism

Metabolism is mostly performed in the hepatocytes (mainly liver cell type) of the liver. Drug metabolism pathways usually occur in two apparent phases: phase I and phase II. Phase I reactions involve forming a new or changed functional group or a cleavage (oxidation, reduction, hydrolysis) to ensure a higher rate of reactivity. Phase II reactions involve conjugation with an endogenous compound such as glucuronic acid, glycine or sulfate to ensure higher water solubility. The metabolites of these processes have a higher polarity and are excreted through the kidneys (urine) or the liver (bile) at a much higher rate than the original drug molecule. The most important enzyme system of phase I metabolism is the microsomal family of isoenzymes called cytochrome P450, which can transfer electrons supplied by certain flavoproteins (NADPH-cytochrome P450 reductase) to catalyze drug oxidation. The cytochrome P450 superfamily is grouped into 14 families and 17 subfamilies based on gene sequence identity. The isoenzymes are named by a root symbol CYP, followed by a number/letter combination. The most important CYP isoenzymes for drug metabolism are CYP1A2, CYP2C9, CYP2C19, CYP2D6 and CYP3A4. The most important phase II reaction is glucuronidation (conjugation to glucuronic acid), which occurs in the liver microsomal enzyme system and results in hydrophilic metabolites that are suitable for renal or biliary excretion. The other important phase II reactions are amino acid conjugation with glutamine or glycine, acetylation, methylation, or sulfoconjugation of phenolic acids such as cysteine. Until recently, the metabolism of nanoparticles had not been fully addressed. Polymer-based nanostructures and super paramagnetic iron oxide nanostructures for MRI contrast agents were shown to degrade; however, QDs, fullerenes and silica nanoparticles clearly did not degrade *in vivo* (Ballou *et al.*, 2004; Khan *et al.*, 2005; Singh *et al.*, 2006; Yang *et al.*, 2007). The breakdown of nanoparticles may elicit unique molecular responses that are not predictable, and thus the understanding and cataloging of what, when and how such nanostructures degrade is extremely important.

16.2.1.4   Elimination

Elimination is the total amount of drug that is lost from the body, which is mainly due to metabolism (chemical alteration) and excretion (elimination without further chemical change). The main routes of nanoparticle excretion are through the kidney (Singh *et al.*, 2006) or the liver/bile duct (Hardonk *et al.*, 1985; Renaud *et al.*, 1989), and the minor routes of excretion are the saliva, sweat, breast milk and lungs. Previous studies have shown that hydroxl-functionalized single-walled carbon nanotubes (SWCNTs), when administered intraperitoneally, accumulated in the liver and kidneys and were excreted in the urine in 18 days (Wang *et al.*, 2004). However, ammonium-functionalized SWCNTs, when administered intravenously, did not accumulate in the liver and had much faster renal excretion (Singh *et al.*, 2006). However, size and surface chemistry dependent elimination mechanisms for QDs have been published, where QDs smaller than 5.5 nm in diameter that were cysteine-coated were shown to be excreted in the urine (Choi *et al.*, 2007a). In addition, siRNA DOTAP/DOPE complexes (250 nm) and siRNA RGD-PEG-PEI complex (130 nm) showed a rapid renal clearance by glomerular filtration (de Wolf *et al.*, 2007).

16.2.2   Acute Toxicity Study

Acute toxicity study is one of the most important to evaluate the toxic characteristics of nanoparticles. This will provide information on health hazards that are likely to arise from short-term exposure for different routes of administration such as oral, dermal, and inhalation. Data from an acute toxicity study may serve as a basis for establishing a dosage regimen and may provide additional information on the mode of toxic action of a substance. Acute toxicity from oral/dermal/inhalation administration is the sum of all adverse effects that are caused by a substance following a single uninterrupted exposure through these routes of administration over a short period of time (24 hours or less). The LC_{50} (median lethal concentration) is the statistically derived concentration of a substance that is expected to cause death during exposure or within a fixed time in 50 % of animals that are exposed to the substance over a defined time.

Few studies have examined the systemic toxicity of nanoparticles, and of these, most have been cursory acute toxicity studies (Table 16.2) that have not identified the target organs or extensively characterized the test material. This lack of characterization is an important point, since without characterization it is impossible to compare studies and recognize parameters that influence toxicity. In many of the studies, the LD_{50} could not even be estimated due to the lack of lethality observed at the doses administered. As a result of the small size of nanoparticles, it has been generally assumed that nanoparticles cannot be detected by the immune systems. However, several studies have shown that most nanoparticles were entrapped by the reticuloendothelial system (RES), suggesting that the main target organ of these particles is the liver and spleen. Since nanoparticles that are used in biomedical applications are commonly coated to reduce opsonization and to avoid the RES, it is expected that these nanoparticles have target organs that are not associated with the RES. Aside from the primary RES-related organs, the kidney may also be a common target organ of toxicity for nanoparticles. The kidney has been identified as the primary clearance route for many nanoparticles, including CNTs, water-soluble

Table 16.2 *Acute toxicology studies.*

Species/strain	Sex	Nanoparticles	Dose and duration of exposure	Route of exposure	Duration of study	Toxicity	Other findings	References
BALB/c nude mice	Both	1. Noncovalently pegylated SWNTs (SWNT PEG) 2. Covalently functionalized SWNTs (SWNT O PEG)	1. SWNT PEG: 151 mg 2. SWNT O PEG: 47 mg 3. 2 times on day 0 and 7	iv	4 months	No evidence of toxicity		Schipper et al. (2008)
CD1 mice	Male	MWCNTs and CNx MWCNTs	1, 2.5, 5 mg/kg	Nasal, po, ip	1, 2, 3, 7, 30 days	No evidence of toxicity		Carrero-Sanchez et al. (2006) Mori et al. (2006)
Sprague-Dawley rats	Both	Fullerene (C60)	2000 mg/kg Single treatment	po	14 days	1. No deaths observed 2. No effects on body weights		
Sprague-Dawley rats	Female	Polyalkylsulfonated C60	po: 2500 mg/kg ip: 500, 750, 1000 mg/kg iv: 10, 100 mg/kg	po, ip, iv	2 wks	1. po: no deaths observed 2. ip: LD_{50} was 600 mg/kg Decrease in body weights and spleen weights 3. ip and iv: Kidney: phagolysosomal nephropathy		Chen et al. (1998)

(Continued)

Table 16.2 *(Continued)*

Species/strain	Sex	Nanoparticles	Dose and duration of exposure	Route of exposure	Duration of study	Toxicity	Other findings	References
ICR mice	Male	Polyhydroxylated fullerence (C60)	0.01, 0.1, 0.5 and 1 g/kg	ip	2 wks	1. LD_{50} was 1.2 g/kg 2. Dose-dependent increases in liver/body weight noted		Ueng et al. (1997)
Swiss mice	Both	Fullerene (C60)	2.5–5.0 g/kg	ip	14 days	1. No deaths observed 2. Hypertrophy and hyperplasia of hepatic satellite cells	1. Fullerene was absorbed, localized in spleen and liver	Moussa et al. (1996)
ICR mice	Both	58 nm Zn	5 g/kg	po	2 wks	1. Lethargy, vomiting, diarrhea at the beginning 2. Elevated ALT, ALP and LDH 3. Kidney, tubular dilation, casts 4. Liver, hydropic degeneration		Wang et al. (2006)
ICR mice	Both	25 nm Cu	108, 158, 232, 341, 501, 736, 1080 mg/kg	po	2 days	1. LD_{50} was 413 mg/kg 2. Kideny: proximal tubular necrosis 3. Liver: steatosis 4. Spleen: atrophy		Chen et al. (2006)

Animal	Sex	Nanoparticle	Dose	Route	Duration	Findings	Reference
ICR mice	Both	25, 80 and 155 nm TiO_2	5 g/kg	po	2 wks	1. Kidney, glomerular swelling 2. Liver, hydropic degeneration, spotty necrosis	Wang et al. (2007)
Sprague-Dawley rats	Male	Iron oxide magnetic nanoparticle Core: 11 nm DLS: 193 nm	10 mg/kg	iv	0, 6 h, 24 h; 3, 7, 21 d	1. Transient increase in ALT, AST, AKP, and total iron binding capacity (TIBC) 2. Oxidative stress peaked at ~3 days 3. No histopathological changes	Jain et al. (2008) Serum iron levels peaked 1 wk after treatment
BALB/c mice	Both	Dextran-coated magnetic nanoparticles	2250, 3150, 4500, 6300, 9000 mg/kg	sc	2 wks	LD_{50} of single treatment was 4409.61 ± 514.93 mg/kg	Yu et al. (2008)
ICR mice	Female	30 nm core-shell silica	3.3×10^{-7}, 3.3×10^{-8} M	iv	1, 3, 7, 21, 60 days	No toxicity was found	Choi et al. (2007b)

polyalkylsulfonated fullerenes, and dendrimers (Roberts *et al.*, 1996; Chen *et al.*, 1998; Nigavekar *et al.*, 2004; Wang *et al.*, 2004; Lee *et al.*, 2005). In another study with rats, the kidney was determined to be both the primary organ for particle removal and the target organ of toxicity for polyalkylsulfonated fullerenes (Chen *et al.*, 1998). Since many nanoparticles are opsonized and phagocytosed by the immune system, lysosomal disorders may be a common side-effect of nanoparticle exposure (Shukla *et al.*, 2005; Bottini *et al.*, 2006). In addition, dendrimer and phosphohydroxylated fullerenes can inhibit enzyme activity (Ueng *et al.*, 1997; Shcharbin *et al.*, 2006). A more thorough understanding of the potential mechanisms of nanoparticle toxicity is essential for hazard assessment and identification of exposure biomarkers.

16.2.3   Subacute Toxicity Study

In the assessment and evaluation of the toxic characteristics of a chemical, determining the oral/dermal/inhalation toxicity at repeated doses should be carried out after the effects of the chemical on acute toxicity has been determined. These studies would provide information on the possible health hazards that are likely to arise from repeated exposure over a relatively limited period of time. The duration of exposure is normally 28 days, although a 14-day study may be appropriate in certain circumstances; however, justification for use of a 14-day exposure period should be provided. A few subacute toxicity studies have been carried out with nanoparticles (Table 16.3). Silver nanoparticles that were 60 nm in diameter produced slight liver damage when the dosage was over 300 mg/kg (Kim *et al.*, 2008). However, fullerene and 50 nm magnetic nanoparticles coated with silica showed no toxicity in experimental animals (Gharbi *et al.*, 2005; Kim *et al.*, 2006b).

16.2.4   Inhalation Toxicity Study

An understanding of the exposure and effects of discrete nanoparticles is important to fully understand nanoscale materials. However, because of the broad applications of nanotechnology, simply focusing on discrete nanoparticles is not adequate. Inhalation is thought to be an important route of nanoparticle exposure, since nanoparticles can travel great distances in air by Brownian diffusion, and are respirable. This process would result in depositing the nanoparticles within the alveolar regions of the lung (Bailey, 1994). However, in the case of engineered nanoparticles, very few airborne exposure studies have been conducted. Often a confounding factor in airborne nanoparticle exposure studies has been the high level of incidental background nanoparticles.

An improved fundamental understanding of the behavior of airborne nanomaterials is critical to assessing accurately the effects of exposure. There are considerable data available on aerosol generation, distribution and deposition based on the aerodynamic diameter of small particles (Table 16.4). Cohesive forces cause nanoparticle aggregates and agglomerates, which markedly affects their propensity to become airborne, as well as their aerodynamic diameter. The use of nanomaterials in liquids and composites may severely limit or preclude airborne exposure. Deaggregation, deagglomeration, and dissolution in biological fluids are important factors that could potentially contribute to a complete understanding of the fate of nanoparticles. Although in some cases inhalation is a critical route

Table 16.3 Subacute toxicology studies.

Species/strain	Sex	Nanoparticles	Dose and duration of exposure	Route of exposure	Duration of study	Toxicity	Other findings	References
Sprague-Dawley rats	Both	60 nm silver	30, 300, 1000 mg/kg for 28 consecutive days	po	28 days	1. No significant changes in body weights 2. Dose-dependent changes in the ALP and cholesterol values in both sexes 3. Slight liver damage (over 300 mg/kg)	Silver contents in female kidneys were higher than male kidneys	Kim et al. (2008)
ICR mice	Male	50 nm MNP@SiO$_2$ (RITC)	25, 50, 100 mg/kg for 4 wks	ip	4 wks	No toxicity observed		Kim et al. (2006a)
Wistar rats	Male	Fullerene (C60)	0.25, 0.50, 2.00 g/kg	ip	14 days	No toxicity observed	Liver protection from free-radical damage	Gharbi et al. (2005)
Sprague-Dawley rats	Female	Polyalkylsulfonated C60	0.6, 6, 60 mg/kg for 12 consecutive days	ip	13 days	1. Less food and water consumption 2. Loss of body weights 3. Kidney: phagolysosomal nephropathy		Chen et al. (1998)

Table 16.4 Pulmonary toxicology studies.

Species	Nanoparticle	Administration route/ study duration	Dose	Adverse effects/lesions	References
Sprague-Dawley rats	11.80–14.82 nm silver	Inhalation/28 days	Particles/cm^3 Low: 1.73×10^4 Mid: 1.27×10^5 High: 1.32×10^6	1. No significant changes in body weights 2. No significant changes in hematology and blood chemistry	Ji et al. (2007)
Sprague-Dawley rats	18 nm silver	Inhalation/90 days	Particles/cm^3 Low: 0.7×10^6 Mid: 1.4×10^6 High: 2.9×10^6	1. No statistically significant differences in the cellular differential counts, the inflammation measurements of bronchoalveolar lavage 2. Dose-dependent increases in lesions related to silver nanoparticle exposure, such as infiltrate mixed cell and chronic alveolar inflammation, including thickened alveolar walls and small granulomatous lesions 3. Decreases in the tidal volume and minute volume and other inflammatory responses	Sung et al. (2008)
ICR mice	50 nm fluorescent magnetic nanoparticle	Inhalation/28 days	Particles/cm^3 Low: 4.89×10^5 High: 9.34×10^5	1. Fluorescent magnetic nanoparticles were distributed in various organs, including the liver, spleen, testis, lung and brain	Kwon et al. (2008)
Sprague-Dawley rats	22 nm, 280 nm ferric oxide	Intratracheal instillation/ 1, 7, 30 days	0.8 mg/kg and 20 mg/kg	1. 22 nm, 280 nm: induce oxidative stress 2. Inflammatory reaction including inflammatory and immune cells increase 3. Lung: follicular hyperplasia, protein effusion, pulmonary capillary vessel hyperaemia and alveolar lipoproteinosis	Zhu et al. (2008)

Animal	CNT type	Exposure route/duration	Dose	Findings	Reference
CD1 mice	MWCNTs	Intratracheal instillation/ 1, 2, 3, 7 and 30 days	1, 2.5 and 5 mg/kg	1. High mortality: 30, 60, 90 % for 1, 2.5, 5 mg/kg, respectively 2. Severe clinical signs 3. Atypical hyperplasia in bronchioles 4. Multiple granulomas 5. Goblet cell hyperplasia	Carrero-Sanchez et al. (2006)
CD1 mice	CNx MWCNTs	Intratracheal instillation/ 1, 2, 3, 7 and 30 days	1, 2.5 and 5 mg/kg	1. Tissue invasion 2. Granulomatous inflammation 3. Reactive fibrosis 4. Extensive papillomatous hyperplasia	Carrero-Sanchez et al. (2006)
Guinea pig	MWCNTs	Intratracheal instillation/ 4 weeks	25 mg/animal	1. No evidence of inflammation 2. No perturbation of lung function	Huczko et al. (2001)
Guinea pig	MWCNTs	Intratracheal instillation/ 90 days	15 mg/animal	1. Nonspecific desquamative interstitial pneumonia-like reaction 2. Increased lung resistance	Huczko et al. (2005)
Sprague-Dawley rats	MWCNTs	Intratracheal instillation/ 1 and 2 months	0.5, 2 and 5 mg/animal	1. Inflammation and dose-dependent fibrosis 2. Bronchiolar granulomatous lesions	Muller et al. (2005)
C57BL/6 mice	SWCNTs	Pharyngeal aspiration/ 1, 3, 7, 28 and 60 days	40 µg/animal	1. Transient inflammatory response 2. Dose-dependent epithelioid granulomas and interstitial fibrosis 3. Decreased bacterial clearance, and dose-dependent loss of pulmonary function	Shvedova et al. (2005)
ICR mice	SWCNTs	Intratracheal instillation/ 3, 14 days	0.5 mg/kg	1. Alveolar macrophage activation 2. Chronic inflammatory responses 3. Severe pulmonary granuloma	Chou et al. (2008)
CD rats	SWCNTs	Intratracheal instillation/ 24 h, 1 week, 1 and 3 months	1 and 5 mg/kg	1. Deaths in high dose group b 2. Transient inflammatory and cell injury responses 3. Nonprogressive, non-dose-dependent multifocal granulomas	Warheit et al. (2004)
B6C3F1 mice	SWCNTs	Intratracheal instillation/ 7 and 90 days	0.1 and 0.5 mg/animal	1. Deaths in high dose group 2. Progressive, dose-dependent multifocal epithelial granulomas 3. Interstitial inflammation 4. Peribronchial inflammation and necrosis	Lam et al. (2004)

of exposure, a full understanding of the effects of exposure and the fate of the nanoparticles will require an assessment of the technology applications and physical state of the nano-materials. However, very little is known about the disposition and fate of nanoparticles (<100 nm) in the body. Aerosolization of primary nanoparticles is probably limited to combustion and spray dispersion processes. Mechanical grinding processes produce low yields of particles that are on the nanoscale, although the aerosolization of micron-sized aggregates or agglomerates is possible. Cohesive forces (between nanoparticles) and ad-hesive forces (with surrounding media) strongly influence the state and fate of the primary particles in gas, liquid or solids. Airborne discrete nanoparticles predominantly move by convection and diffusion. Particles in this size range are deposited in the respiratory tract predominantly by diffusion (James *et al.*, 1991).

Once deposited, nanoparticles may cross biological membranes and access tissue that would not normally be exposed to larger particles. Nanoscale titanium dioxide (TiO_2) particles can translocate into lung interstitium, and iridium can translocate to secondary organs (Ferin and Oberdorster, 1992; Semmler *et al.*, 2004). Nanoparticle aggregates would likely be subjected to normal macrophage clearance mechanisms. Ultrafine TiO_2 particles were phagocytosed by alveolar macrophages through intratracheal instillation, which prevented both the pulmonary inflammatory reaction and interstitial access of the ultrafine particles (Oberdörster *et al.*, 1992). However, in some cases, inflammation has been observed in the respiratory tract due to the relatively large surface area. Inhalation of 10 mg/m^3 of ultrafine TiO_2 for 13 weeks resulted in pulmonary overload in rats and mice, with inflammation similar to that seen with higher mass doses of fine TiO_2 (Bermudez *et al.*, 2004).

Effects beyond the respiratory tract may also occur from exposure to nanomaterials. Advances in this area have been made from studies that have examined the effects of ultrafine particles that are associated with air pollution. Respiration of ultrafine particles can cause pro-inflammatory and oxidative stress-related cellular responses in cardiovascular systems (Oberdörster *et al.*, 2005).

16.2.5   Irritation

Skin is a complex dynamic organ that has several functions, the primary one being to act as a barrier to the external environment. The skin is the largest organ of the body and serves as a primary route of environmental and/or occupational exposure; it is one of the principal portals of entry by which environmental toxins or nanomaterials can enter the body. Currently, there is no information on whether nanoparticles can be absorbed across the stratum corneum barrier or whether systemically administered particles can accumulate in the dermal tissue. Skin is unique because it provides an environment within the avascular epidermis where particles could potentially lodge and not be susceptible to removal by phagocytosis. The ability of nanomaterials to traverse the skin is a primary determinant of their dermatotoxic potential. That is, nanomaterials or nanoparticles must penetrate the stratum corneum in order to exert toxicity in lower cell layers. The quantita-tive prediction of the rate and extent of percutaneous penetration (into skin) and absorption (through skin) of topically applied nanomaterials is complicated because the processes that drive nanoparticles into the skin may be different from those governing chemicals. The cutaneous toxicity of nanoparticles is summarized in Table 16.5. Anatomically,

Table 16.5 Cutaneous toxicology studies.

Species	Nanoparticle	Study design/duration	Dose	Adverse effects/lesions	References
BALB/c mice	Dextran-coated magnetic nanoparticles	Skin irritation, 24, 48, 72 h	0.1, 0.2, 0.3 ml % (W/V) injection on flank	1. Hemangiectasia and leukocyte infiltration were seen in subcutaneous tissues 2. These phenomena almost disappeared 72 h later	Yu et al. (2008)
SKH-1 mice	Carboxylated QD	Skin irritation with or without UVR exposure/8 and 24 h	~3 pmol/cm^2	1. Low levels of penetration were seen in both the non-UVR exposed mice and the UVR exposed mice 2. Qualitatively higher levels of penetration were observed in the UVR exposed mice	Mortensen et al. (2008)
Wistar rats	Carboxylated QD655 and QD565	Apply to flexed, tape-stripped and abraded rat skin/8 and 24 h	1 uM	1. Nonflexed skin did not show QD penetration at 8 or 24 h 2. Flexed skin showed an increase in QD on the surface of skin but no penetration at 8 and 24 h 3. Tape-stripped skin depicted QD only on the surface of the viable epidermis 4. QD655 penetrated into the viable dermal layers of abraded skin at both 8 and 24 h, while QD565 was present only at 24 h	Zhang and Monteiro-Riviere (2008)
Rabbit	Ultrafine TiO$_2$-C	Patch test/4 h Scored dermal irritation after 60 min, 24, 48 and 72 h after test substance removal	Single 0.5 g dermal dose	1. No dermal irritation 2. No clinical signs of toxicity were observed, and no body weight loss occurred.	Warheit et al. (2007)

(Continued)

Table 16.5 (Continued)

Species	Nanoparticle	Study design/duration	Dose	Adverse effects/lesions	References
SKH-1 mice	PEG coated QD, 37 nm diameter	Injected intradermally on the right dorsal flank	48 pmol QD	1. QD moved from the injection sites apparently through the lymphatic duct system to regional lymph nodes	Gopee et al. (2007)
Human	CNT soot	Patch test/96 h	Unknown, aqueous suspension	No irritation or signs of allergic response	Huczko and Lange (2001)
Human	Fullerene soot	Patch test/96 h	Unknown, aqueous suspension	No irritation or signs of allergic response	Huczko et al. (1999)
Rabbit	Fullerene soot	Ocular irritation (modified Draize test)/24, 48 and 72 h	Unknown, aqueous suspension	No irritation or signs of allergic response	Huczko et al. (1999)
Wistar rats	'Hat-stacked' carbon nanofibers	Subcutaneous implantation/1 and 4 weeks	Unknown	Foreign body granuloma No tissue necrosis No severe inflammation	Yokoyama et al. (2005)
Wistar rats	CNTs	Subcutaneous implantation/1 and 4 weeks	0.1 mg	Foreign body granuloma No tissue necrosis No severe inflammation	Sato et al. (2005)

nanomaterial absorption may occur through several routes, since the majority of lipid-soluble particles may move through the intercellular lipid pathway between the stratum corneum cells, through the cells, or through hair follicles or sweat ducts. *In vivo* studies have shown that carboxylated QD penetrate the skin at low levels, and the levels were elevated by UV irradiation (Mortensen *et al.*, 2008). With regard to the condition of skin, nonflexed and flexed skin did not show QD penetration and tape-stripped skin depicted QD only on the surface of the viable epidermis. However, QD can penetrate the dermal layer by abrasion of skin (Zhang and Monteiro-Riviere, 2008). By injection of nanoparticles into dermal layers of skin, nanoparticles can move to regional lymph nodes via the lymphatic duct system (Gopee *et al.*, 2007).

In conclusion, nanoparticles cannot penetrate normal skin. However, several conditions such as UV irradiation, tape-stripping and abrasion may influence the penetration efficacy. In addition, when nanoparticles penetrate into the dermal layer, nanoparticles can move to regional lymph nodes and several pathological phenomena can be seen.

Acknowledgement

This work was supported by a grant (08161KFDA541) from the Korean Food and Drug Administration in 2008.

References

Akazawa Y, Gores GJ (2007) Death receptor-mediated liver injury. *Semin Liver Dis* **27**, 327–338.

Akerman ME, Chan WC, Laakkonen P, Bhatia SN, Ruoslahti E (2002) Nanocrystal targeting in vivo. *Proc Natl Acad Sci USA* **99**, 12617–12621.

Auffan M, Decome L, Rose J, Orsiere T, De Meo M, Briois V, Chaneac C, Olivi L, Berge-Lefranc JL, Botta A, Wiesner MR, Bottero JY (2006) In vitro interactions between DMSA-coated maghemite nanoparticles and human fibroblasts: a physicochemical and cyto-genotoxical study. *Environ Sci Technol* **40**, 4367–4373.

Bailey MR (1994) The new ICRP model for the respiratory tract. *Radiat Prot Dosimetry* **53**, 107–114.

Ballou B, Lagerholm BC, Ernst LA, Bruchez MP, Waggoner AS (2004) Noninvasive imaging of quantum dots in mice. *Bioconjug Chem* **15**, 79–86.

Bermudez E, Mangum JB, Wong BA, Asgharian B, Hext PM, Warheit DB, Everitt JI (2004) Pulmonary responses of mice, rats, and hamsters to subchronic inhalation of ultrafine titanium dioxide particles. *Toxicol Sci* **77**, 347–357.

Borm P, Klaessig FC, Landry TD, Moudgil B, Pauluhn J, Thomas K, Trottier R, Wood S (2006) Research strategies for safety evaluation of nanomaterials, part V: role of dissolution in biological fate and effects of nanoscale particles. *Toxicol Sci* **90**, 23–32.

Bottini M, Cerignoli F, Dawson MI, Magrini A, Rosato N, Mustelin T (2006) Full-length single-walled carbon nanotubes decorated with streptavidin-conjugated quantum dots as multivalent intracellular fluorescent nanoprobes. *Biomacromolecules* **7**, 2259–2263.

Butenas S, Mann KG (2002) Blood coagulation. *Biochemistry (Mosc)* **67**, 3–12.

Carrero-Sanchez JC, Elias AL, Mancilla R, Arrellin G, Terrones H, Laclette JP, Terrones M (2006) Biocompatibility and toxicological studies of carbon nanotubes doped with nitrogen. *Nano Lett* **6**, 1609–1616.

Caruthers SD, Wickline SA, Lanza GM (2007) Nanotechnological applications in medicine. *Curr Opin Biotechnol* **18**, 26–30.

Cenni E, Granchi D, Avnet S, Fotia C, Salerno M, Micieli D, Sarpietro MG, Pignatello R, Castelli F, Baldini N (2008) Biocompatibility of poly(D,L-lactide-co-glycolide) nanoparticles conjugated with alendronate. *Biomaterials* **29**, 1400–1411.

Chen HH, Yu C, Ueng TH, Chen S, Chen BJ, Huang KJ, Chiang LY (1998) Acute and subacute toxicity study of water-soluble polyalkylsulfonated C60 in rats. *Toxicol Pathol* **26**, 143–151.

Chen Z, Meng H, Xing G, Chen C, Zhao Y, Jia G, Wang T, Yuan H, Ye C, Zhao F, Chai Z, Zhu C, Fang X, Ma B, Wan L (2006) Acute toxicological effects of copper nanoparticles in vivo. *Toxicol Lett* **163**, 109–120.

Choi HS, Liu W, Misra P, Tanaka E, Zimmer JP, Itty Ipe B, Bawendi MG, Frangioni JV (2007a) Renal clearance of quantum dots. *Nat Biotechnol* **25**, 1165–1170.

Choi J, Burns AA, Williams RM, Zhou Z, Flesken-Nikitin A, Zipfel WR, Wiesner U, Nikitin AY (2007b) Core-shell silica nanoparticles as fluorescent labels for nanomedicine. *J Biomed Opt* **12**, 064007.

Choi JH, Nguyen FT, Barone PW, Heller DA, Moll AE, Patel D, Boppart SA, Strano MS (2007c) Multimodal biomedical imaging with asymmetric single-walled carbon nanotube/iron oxide nanoparticle complexes. *Nano Lett* **7**, 861–867.

Chou CC, Hsiao HY, Hong QS, Chen CH, Peng YW, Chen HW, Yang PC (2008) Single-walled carbon nanotubes can induce pulmonary injury in mouse model. *Nano Lett* **8**, 437–445.

Colognato R, Bonelli A, Ponti J, Farina M, Bergamaschi E, Sabbioni E, Migliore L (2008) Comparative genotoxicity of cobalt nanoparticles and ions on human peripheral leukocytes in vitro. *Mutagenesis* **23**, 377–382.

Colvin VL (2003) The potential environmental impact of engineered nanomaterials. *Nat Biotechnol* **21**, 1166–1170.

Darkow R, Groth T, Albrecht W, Lutzow K, Paul D (1999) Functionalized nanoparticles for endotoxin binding in aqueous solutions. *Biomaterials* **20**, 1277–1283.

De M, You CC, Srivastava S, Rotello VM (2007) Biomimetic interactions of proteins with functionalized nanoparticles: a thermodynamic study. *J Am Chem Soc* **129**, 10747–10753.

De Jong WH, Borm PJ (2008) Drug delivery and nanoparticles: applications and hazards. *Int J Nanomedicine* **3**, 133–149.

De Jong WH, Hagens WI, Krystek P, Burger MC, Sips AJ, Geertsma RE (2008) Particle size-dependent organ distribution of gold nanoparticles after intravenous administration. *Biomaterials* **29**, 1912–1919.

Delfino RJ, Sioutas C, Malik S (2005) Potential role of ultrafine particles in associations between airborne particle mass and cardiovascular health. *Environ Health Perspect* **113**, 934–946.

de Wolf HK, Snel CJ, Verbaan FJ, Schiffelers RM, Hennink WE, Storm G (2007) Effect of cationic carriers on the pharmacokinetics and tumor localization of nucleic acids after intravenous administration. *Int J Pharm* **331**, 167–175.

Dey S, Bakthavatchalu V, Tseng MT, Wu P, Florence R, Grulke EA, Yokel R, Dhar SK, Yang HS, Chen Y, St Clair DK (2008) Interactions between SIRT1 and AP-1 reveal a mechanistic insight into the growth promoting properties of alumina (Al_2O_3) nanoparticles in mouse skin epithelial cells. *Carcinogenesis* **29**, 1920–1929.

Dobrovolskaia MA, Clogston JD, Neun BW, Hall JB, Patri AK, McNeil SE (2008) Method for analysis of nanoparticle hemolytic properties in vitro. *Nano Lett* **8**, 2180–2187.

Dragoni F, Minotti C, Palumbo G, Faillace F, Redi R, Bongarzoni V, Avvisati G (2001) As compared to kaolin clotting time, silica clotting time is a specific and sensitive automated method for detecting lupus anticoagulant. *Thromb Res* **101**, 45–51.

Fenech M (2008) The micronucleus assay determination of chromosomal level DNA damage. *Methods Mol Biol* **410**, 185–216.

Ferin J, Oberdörster G (1992) Translocation of particles from pulmonary alveoli into the interstitium. *J Aerosol Med* 179–187.

Fernandez-Urrusuno R, Fattal E, Rodrigues JM Jr, Feger J, Bedossa P, Couvreur P (1996) Effect of polymeric nanoparticle administration on the clearance activity of the mononuclear phagocyte system in mice. *J Biomed Mater Res* **31**, 401–408.

Fernandez-Urrusuno, R, Fattal E, Feger J, Couvreur P, Therond P (1997) Evaluation of hepatic antioxidant systems after intravenous administration of polymeric nanoparticles. *Biomaterials* **18**, 511–517.

Fischer HC, Liu L, Pang KS, Chan WCW (2006) Pharmacokinetics of nanoscale quantum dots: in vivo distribution, sequestration, and clearance in the rat. *Adv Funct Mater* **16**, 1299–1305.

Fox R, Aubert M (2008) Flow cytometric detection of activated caspase-3. *Methods Mol Biol* **414**, 47–56.

Fujita T (2005) Activation pathway of complement (classical, alternative, lectin). *Nippon Rinsho* **63** (Suppl. 4), 269–273.

Gadjeva M, Thiel S, Jensenius JC (2004) Assays for the mannan-binding lectin pathway. *Curr Protoc Immunol* **Chapter 13**, Unit 13, 6.

Galli A, Palatnik A (2005) What is the proper activated clotting time (ACT) at which to remove a femoral sheath after PCI? What are the best "protocols" for sheath removal? *Crit Care Nurse* **25**, 88–92, 94–95.

Gao X, Cui Y, Levenson RM, Chung LW, Nie S (2004) In vivo cancer targeting and imaging with semiconductor quantum dots. *Nat Biotechnol* **22**, 969–976.

Gharbi N, Pressac M, Hadchouel M, Szwarc H, Wilson SR, Moussa F (2005) [60]Fullerene is a powerful antioxidant in vivo with no acute or subacute toxicity. *Nano Lett* **5**, 2578–2585.

Glassock RJ (2007) Prophylactic anticoagulation in nephrotic syndrome: a clinical conundrum. *J Am Soc Nephrol* **18**, 2221–2225.

Gopee NV, Roberts DW, Webb P, Cozart CR, Siitonen PH, Warbritton AR, Yu WW, Colvin VL, Walker NJ, Howard PC (2007) Migration of intradermally injected quantum dots to sentinel organs in mice. *Toxicol Sci* **98**, 249–257.

Gottlieb T (1993) Hazards of bacterial contamination of blood products. *Anaesth Intensive Care* **21**, 20–23.

Hall JB, Dobrovolskaia MA, Patri AK, McNeil SE (2007) Characterization of nanoparticles for therapeutics. *Nanomed* **2**, 789–803.

Hardonk MJ, Harms G, Koudstaal J (1985) Zonal heterogeneity of rat hepatocytes in the in vivo uptake of 17 nm colloidal gold granules. *Histochemistry* **83**, 473–477.

Ho CH, Hou MC, Lin HC, Lee SD, Liu SM (1998) Can advanced hemostatic parameters detect disseminated intravascular coagulation more accurately in patients with cirrhosis of the liver? *Zhonghua Yi Xue Za Zhi (Taipei)* **61**, 332–338.

Hoshino A, Hanaki K, Suzuki K, Yamamoto K (2004) Applications of T-lymphoma labeled with fluorescent quantum dots to cell tracing markers in mouse body. *Biochem Biophys Res Commun* **314**, 46–53.

Huczko A, Lange H, Calko E (1999) Fullerenes: experimental evidence for a null risk of skin irritation and allergy. *Fuller Sci Tech* **7**, 935–939.

Huczko A, Lange H (2001) Carbon nanotubes: Experimental evidence for a null risk of skin irritation and allergy. *Fuller Sci Tech* **9**, 247–250.

Huczko A, Lange H, Calko E, Grubek-Jaworska H, Droszcz P (2001) Physiological testing of carbon nanotubes: Are they asbestos like? *Fuller Sci Tech* **9**, 251–254.

Huczko A, Lange H, Bystrzejewski P, Grubek-Jaworska H, Nejman P, Przybylowski T, Czuminska K, Glapinski J, Walton DRM (2005) Pulmonary toxicity of 1-D nanocarbon materials. *Fuller Nanotub Carbon Nanostruct* **13**, 141–145.

International Conference on Harmonization (ICH) (1998) *ICH harmonized tripartite guideline S2B. Genotoxicity: a standard battery for genotoxicity testing of pharmaceuticals.* Available at http://www.ich.org/cache/compo/502-272-1.html

Jain TK, Reddy MK, Morales MA, Leslie-Pelecky DL, Labhasetwar V (2008) Biodistribution, clearance, and biocompatibility of iron oxide magnetic nanoparticles in rats. *Mol Pharm* **5**, 316–327.

James JM, Herman KJ, Lloyd JJ, Shields RA, Testa HJ, Church S, Stretton TB (1991) Evaluation of 99Tcm Technegas ventilation scintigraphy in the diagnosis of pulmonary embolism. *Br J Radiol* **64**, 711–719.

Ji JH, Jung JH, Kim SS, Yoon JU, Park JD, Choi BS, Chung YH, Kwon IH, Jeong J, Han BS, Shin JH, Sung JH, Song KS, Yu IJ (2007) Twenty-eight-day inhalation toxicity study of silver nanoparticles in Sprague-Dawley rats. *Inhal Toxicol* **19**, 857–71.

Jiang Z, Huang W, Li J, Li M, Liang A, Zhang S, Chen B (2008) Nanogold catalysis-based immunoresonance-scattering spectral assay for trace complement component 3. *Clin Chem* **54**, 116–123.

Jiao Y, Ubrich N, Marchand-Arvier M, Vigneron C, Hoffman M, Lecompte T, Maincent P (2002) In vitro and in vivo evaluation of oral heparin-loaded polymeric nanoparticles in rabbits. *Circulation* **105**, 230–235.

Jurasz P, Alonso D, Castro-Blanco S, Murad F, Radomski MW (2003) Generation and role of angiostatin in human platelets. *Blood* **102**, 3217–3223.

Khan MK, Nigavekar SS, Minc LD, Kariapper MS, Nair BM, Lesniak WG, Balogh LP (2005) In vivo biodistribution of dendrimers and dendrimer nanocomposites – implications for cancer imaging and therapy. *Technol Cancer Res Treat* **4**, 603–613.

Kher A, Al Dieri R, Hemker HC, Beguin S (1997) Laboratory assessment of antithrombotic therapy: what tests and if so why? *Haemostasis* **27**, 211–218.

Khuu HM, *et al.* (2004) Comparison of automated culture systems with a CFR/USP-compliant method for sterility testing of cell-therapy products. *Cytotherapy* **6**, 183–195.

Kim JS, Yoon TJ, Yu KN, Kim BG, Park SJ, Kim HW, Lee KH, Park SB, Lee JK, Cho MH (2006a) Toxicity and tissue distribution of magnetic nanoparticles in mice. *Toxicol Sci* **89**, 338–347.

Kim TH, Jin H, Kim HW, Cho MH, Cho CS (2006b) Mannosylated chitosan nanoparticle-based cytokine gene therapy suppressed cancer growth in BALB/c mice bearing CT-26 carcinoma cells. *Mol Cancer Ther* **5**, 1723–1732.

Kim YS, Kim JS, Cho HS, Rha DS, Kim JM, Park JD, Choi BS, Lim R, Chang HK, Chung YH, Kwon IH, Jeong J, Han BS, Yu IJ (2008) Twenty-eight-day oral toxicity, genotoxicity, and gender-related tissue distribution of silver nanoparticles in Sprague-Dawley rats. *Inhal Toxicol* **20**, 575–583.

Kodavanti UP, Schladweiler MC, Ledbetter AD, Ortuno RV, Suffia M, Evansky P, Richards JH, Jaskot RH, Thomas R, Karoly E, Huang YC, Costa DL, Gilmour PS, Pinkerton KE (2006) The spontaneously hypertensive rat: an experimental model of sulfur dioxide-induced airways disease. *Toxicol Sci* **94**, 193–205.

Kwak-Kim J, Gilman-Sachs A (2008) Clinical implication of natural killer cells and reproduction. *Am J Reprod Immunol* **59**, 388–400.

Kwon JT, Hwang SK, Jin H, Kim DS, Minai-Tehrani A, Yoon HJ, Choi M, Yoon TJ, Han DY, Kang YW, Yoon BI, Lee JK, Cho MH (2008) Body distribution of inhaled fluorescent magnetic nanoparticles in the mice. *J Occup Health* **50**, 1–6.

Lam CW, James JT, McCluskey R, Hunter RL (2004) Pulmonary toxicity of single-wall carbon nanotubes in mice 7 and 90 days after intratracheal instillation. *Toxicol Sci* **77**, 126–134.

Leach JK, Hinman A, O'Rear EA (2002) Investigation of deformability, viscosity, and aggregation of mPEG-modified erythrocytes. *Biomed Sci Instrum* **38**, 333–338.

Lee CC, MacKay JA, Frechet JM, Szoka FC (2005) Designing dendrimers for biological applications. *Nat Biotechnol* **23**, 1517–1526.

Lin S, Du F, Wang Y, Ji S, Liang D, Yu L, Li Z (2008) An acid-labile block copolymer of PDMAEMA and PEG as potential carrier for intelligent gene delivery systems. *Biomacromolecules* 9, 109–115.

Liu Z, Cai W, He L, Nakayama N, Dhen K, Sun X, Chen X, Dai H (2007) In vivo biodistribution and highly efficient tumour targeting of carbon nanotubes in mice. *Nat Nano* 2, 47–52.

Lu JH, Teh BK, Wang L, Wang YN, Tan YS, Lai MC, Reid KB (2008) The classical and regulatory functions of C1q in immunity and autoimmunity. *Cell Mol Immunol* 5, 9–21.

Lundqvist M, Sethson I, Jonsson BH (2004) Protein adsorption onto silica nanoparticles: conformational changes depend on the particles' curvature and the protein stability. *Langmuir* 20, 10639–10647.

Maynard AD, Aitken RJ, Butz T, Colvin V, Donaldson K, Oberdorster G, Philbert MA, Ryan J, Seaton A, Stone V, Tinkle SS, Tran L, Walker NJ, Warheit DB (2006) Safe handling of nanotechnology. *Nature* 444, 267–269.

Medintz IL, Uyeda HT, Goldman ER, Mattoussi H (2005) Quantum dot bioconjugates for imaging, labelling and sensing. *Nat Mater* 4, 435–446.

Mori T, Takada H, Ito S, Matsubayashi K, Miwa N, Sawaguchi T (2006) Preclinical studies on safety of fullerene upon acute oral administration and evaluation for no mutagenesis. *Toxicology* 225, 48–54.

Mortensen LJ, Oberdorster G, Pentland AP, Delouise LA (2008) In vivo skin penetration of quantum dot nanoparticles in the murine model: the effect of UVR. *Nano Lett* 8, 2779–2287.

Moussa F, Trivin F, Colin R, Hadchouel M, Sizaret PY, Greugny V, Fabre C, Rassat A, Szwarc H (1996) Early effects of C_{60} administration in Swiss mice: a preliminary account for *in vivo* C_{60} toxicity. *Fullerenes, Nanotubes and Carbon Nanostructures* 4, 21–29.

Muller J, Huaux F, Moreau N, Misson P, Heilier JF, Delos M, Arras M, Fonseca A, Nagy JB, Lison D (2005) Respiratory toxicity of multi-wall carbon nanotubes. *Toxicol Appl Pharmacol* 207, 221–231.

Munford RS (2006) Severe sepsis and septic shock: the role of gram-negative bacteremia. *Annu Rev Pathol* 1, 467–496.

Nagayama S, Ogawara K, Fukuoka Y, Higaki K, Kimura T (2007) Time-dependent changes in opsonin amount associated on nanoparticles alter their hepatic uptake characteristics. *Int J Pharm* 342, 215–221.

Natarajan A, Gruettner C, Ivkov R, DeNardo GL, Mirick G, Yuan A, Foreman A, DeNardo SJ (2008) NanoFerrite particle based radioimmunonanoparticles: binding affinity and in vivo pharmacokinetics. *Bioconjug Chem* 19, 1211–1218.

Nemmar A, Hoet PH, Dinsdale D, Vermylen J, Hoylaerts MF, Nemery B (2003) Diesel exhaust particles in lung acutely enhance experimental peripheral thrombosis. *Circulation* 107, 1202–1208.

Nigavekar SS, Sung LY, Llanes M, El-Jawahri A, Lawrence TS, Becker CW, Balogh L, Khan MK (2004) 3H dendrimer nanoparticle organ/tumor distribution. *Pharm Res* 21, 476–483.

Oberdörster G, Ferin J, Gelein R, Soderholm SC, Finkelstein J (1992) Role of the alveolar macrophage in lung injury: studies with ultrafine particles. *Environ Health Perspect* 97, 193–199.

Oberdörster G, Oberdörster E, Oberdörster J (2005) Nanotoxicology: an emerging discipline evolving from studies of ultrafine particles. *Environ Health Perspect* 113, 823–839.

Oberdörster G, Stone V, Donaldson K (2007) Toxicology of nanoparticles: a historical perspective. *Nanotoxicology* 1, 2–25.

Ojo-Amaize EA, Conley EJ, Peter JB (1994) Decreased natural killer cell activity is associated with severity of chronic fatigue immune dysfunction syndrome. *Clin Infect Dis* 18 (Suppl. 1), S157–159.

Paciotti GF, Myer L, Weinreich D, Goia D, Pavel N, McLaughlin RE, Tamarkin L (2004) Colloidal gold: a novel nanoparticle vector for tumor directed drug delivery. *Drug Deliv* 11, 169–183.

Pickering W, Gray E, Goodall AH, Ran S, Thorpe PE, Barrowcliffe TW (2004) Characterization of the cell-surface procoagulant activity of T-lymphoblastoid cell lines. *J Thromb Haemost* **2**, 459–467.

Piomelli S (1981) Chemical toxicity of red cells. *Environ Health Perspect* **39**, 65–70.

Poland CA, Duffin R, Kinloch I, Maynard A, Wallace WA, Seaton A, Stone V, Brown S, MacNee W, Donaldson K (2008) Carbon nanotubes introduced into the abdominal cavity of mice show asbestos-like pathogenicity in a pilot study. *Nature Nanotechnology* **20**, 423–428.

Radomski M, Moncada S (1983) An improved method for washing of human platelets with prostacyclin. *Thromb Res* **30**, 383–389.

Radomski A, Stewart MW, Jurasz P, Radomski MW (2001) Pharmacological characteristics of solid-phase von Willebrand factor in human platelets. *Br J Pharmacol* **134**, 1013–1020.

Radomski A, Jurasz P, Sanders EJ, Overall CM, Bigg HF, Edwards DR, Radomski MW (2002) Identification, regulation and role of tissue inhibitor of metalloproteinases-4 (TIMP-4) in human platelets. *Br J Pharmacol* **137**, 1330–1338.

Ragni MV, Lewis JH, Spero JA, Hasiba U (1982) Bleeding and coagulation abnormalities in alcoholic cirrhotic liver disease. *Alcohol Clin Exp Res* **6**, 267–274.

Ray M, Carroll P, Smith I, Hawson G (1992) An attempt to standardize APTT reagents used to monitor heparin therapy. *Blood Coagul Fibrinolysis* **3**, 743–748.

Reddy ST, Van Der Vlies AJ, Simeoni E, Angeli V, Randolph GJ, O'Neil CP, Lee LK, Swartz MA, Hubbell JA (2007) Exploiting lymphatic transport and complement activation in nanoparticle vaccines. *Nat Biotechnol* **25**, 1159–1164.

Renaud G, Hamilton RL, Havel RJ (1989) Hepatic metabolism of colloidal gold-low-density lipoprotein complexes in the rat: evidence for bulk excretion of lysosomal contents into bile. *Hepatology* **9**, 380–392.

Roberts JC, Bhalgat MK, Zera RT (1996) Preliminary biological evaluation of polyamidoamine (PAMAM) Starburst dendrimers. *J Biomed Mater Res* **30**, 53–65.

Saba TM (1970) Physiology and physiopathology of the reticuloendothelial system. *Arch Intern Med* **126**, 1031–1052.

Sato Y, Yokoyama A, Shibata K, Akimoto Y, Ogino S, Nodasaka Y, Kohgo T, Tamura K, Akasaka T, Uo M, Motomiya K, Jeyadevan B, Ishiguro M, Hatakeyama R, Watari F, Tohji K (2005) Influence of length on cytotoxicity of multi-walled carbon nanotubes against human acute monocytic leukemia cell line THP-1 in vitro and subcutaneous tissue of rats in vivo. *Mol Biosyst* **1**, 176–182.

Sawicki G, Salas E, Murat J, Miszta-Lane H, Radomski MW (1997) Release of gelatinase A during platelet activation mediates aggregation. *Nature* **386**, 616–619.

Sayes CM, Liang F, Hudson JL, Mendez J, Guo W, Beach JM, Moore VC, Doyle CD, West JL, Billups WE, Ausman KD, Colvin VL (2006) Functionalization density dependence of single-walled carbon nanotubes cytotoxicity in vitro. *Toxicol Lett* **161**, 135–142.

Sayes CM, Reed KL, Warheit DB (2007) Assessing toxicity of fine and nanoparticles: comparing in vitro measurements to in vivo pulmonary toxicity profiles. *Toxicol Sci* **97**, 163–180.

Schipper ML, Nakayama-Ratchford N, Davis CR, Kam NWS, Chu P, Liu Z, Sun X, Dai H, Gambhir SS (2008) A pilot toxicology study of single-walled carbon nanotubes in a small sample of mice. *Nat Nano* **3**, 216–221.

Schmitt C, Ghazi B, Bensussan A (2008) NK cells and surveillance in humans. *Reprod Biomed Online* **16**, 192–201.

Seaton A, Soutar A, Crawford V, Elton R, McNerlan S, Cherrie J, Watt M, Agius R, Stout R (1999) Particulate air pollution and the blood. *Thorax* **54**, 1027–1032.

Semmler M, Seitz J, Erbe F, Mayer P, Heyder J, Oberdorster G, Kreyling WG (2004) Long-term clearance kinetics of inhaled ultrafine insoluble iridium particles from the rat lung, including transient translocation into secondary organs. *Inhal Toxicol* **16**, 453–459.

Shcharbin D, Jokiel M, Klajnert B, Bryszewska M (2006) Effect of dendrimers on pure acetyl-cholinesterase activity and structure. *Bioelectrochemistry* **68**, 56–59.

Shi HZ, Gao NN, Li YZ, Yu JG, Fan QC, Bai GE, Xin BM (2005) Effects of L.F04, the active fraction of Lycopus lucidus, on erythrocytes rheological property. *Chin J Integr Med* **11**, 132–135.

Shukla R, Bansal V, Chaudhary M, Basu A, Bhonde RR, Sastry M (2005) Biocompatibility of gold nanoparticles and their endocytotic fate inside the cellular compartment: a microscopic overview. *Langmuir* **21**, 10644–10654.

Shvedova AA, Kisin ER, Mercer R, Murray AR, Johnson VJ, Potapovich AI, Tyurina YY, Gorelik O, Arepalli S, Schwegler-Berry D, Hubbs AF, Antonini J, Evans DE, Ku BK, Ramsey D, Maynard A, Kagan VE, Castranova V, Baron P (2005) Unusual inflammatory and fibrogenic pulmonary responses to single-walled carbon nanotubes in mice. *Am J Physiol Lung Cell Mol Physiol* **289**, L698–708.

Siegel RM, Lenardo MJ (2002) Apoptosis signaling pathways. *Curr Protoc Immunol* **Chapter 11**, Unit 11, 9C.

Singh R, Pantarotto D, Lacerda L, Pastorin G, Klumpp C, Prato M, Bianco A, Kostarelos K (2006) Tissue biodistribution and blood clearance rates of intravenously administered carbon nanotube radiotracers. *Proc Natl Acad Sci USA* **103**, 3357–3362.

Sung JH, Ji JH, Yoon JU, Kim DS, Song MY, Jeong J, Han BS, Han JH, Chung YH, Kim J, Kim TS, Chang HK, Lee EJ, Lee JH, Yu IJ (2008) Lung function changes in Sprague-Dawley rats after prolonged inhalation exposure to silver nanoparticles. *Inhal Toxicol* **20**, 567–574.

Szebeni J, Alving CR, Rosivall L, Bunger R, Baranyi L, Bedocs P, Toth M, Barenholz Y (2007) Animal models of complement-mediated hypersensitivity reactions to liposomes and other lipid-based nanoparticles. *J Liposome Res* **17**, 107–117.

Toapanta FR, Ross TM (2006) Complement-mediated activation of the adaptive immune responses: role of C3d in linking the innate and adaptive immunity. *Immunol Res* **36**, 197–210.

Tripodi A, Chantarangkul V, Arbini AA, Moia M, Mannucci PM (1993) Effects of hirudin on activated partial thromboplastin time determined with ten different reagents. *Thromb Haemost* **70**, 286–288.

Tripodi A, Caldwell SH, Hoffman M, Trotter JF, Sanyal AJ (2007) Review article: the prothrombin time test as a measure of bleeding risk and prognosis in liver disease. *Aliment Pharmacol Ther* **26**, 141–148.

Tsuji JS, Maynard AD, Howard PC, James JT, Lam CW, Warheit DB, Santamaria AB (2006) Research strategies for safety evaluation of nanomaterials, part IV: risk assessment of nanoparticles. *Toxicol Sci* **89**, 42–50.

Ueng TH, Kang JJ, Wang HW, Cheng YW, Chiang LY (1997) Suppression of microsomal cytochrome P450-dependent monooxygenases and mitochondrial oxidative phosphorylation by fullerenol, a polyhydroxylated fullerene C60. *Toxicol Lett* **93**, 29–37.

Vallhov H, Qin J, Johansson SM, Ahlborg N, Muhammed MA, Scheynius A, Gabrielsson S (2006) The importance of an endotoxin-free environment during the production of nanoparticles used in medical applications. *Nano Lett* **6**, 1682–1686.

Vedhachalam C, Duong PT, Nickel M, Nguyen D, Dhanasekaran P, Saito H, Rothblat GH, Lund-Katz S, Phillips MC (2007) Mechanism of ATP-binding cassette transporter A1-mediated cellular lipid efflux to apolipoprotein A-I and formation of high density lipoprotein particles. *J Biol Chem* **282**, 25123–25130.

Vivier E, Tomasello E, Baratin M, Walzer T, Ugolini S (2008) Functions of natural killer cells. *Nat Immunol* **9**, 503–510.

Wang B, Feng WY, Wang TC, Jia G, Wang M, Shi JW, Zhang F, Zhao YL, Chai ZF (2006) Acute toxicity of nano- and micro-scale zinc powder in healthy adult mice. *Toxicol Lett* **161**, 115–123.

Wang H, Wang J, Deng X, Sun H, Shi Z, Gu Z, Liu Y, Zhao Y (2004) Biodistribution of carbon single-wall carbon nanotubes in mice. *J Nanosci Nanotechnol* **4**, 1019–1024.

Wang J, Zhou G, Chen C, Yu H, Wang T, Ma Y, Jia G, Gao Y, Li B, Sun J, Li Y, Jiao F, Zhao Y, Chai Z (2007) Acute toxicity and biodistribution of different sized titanium dioxide particles in mice after oral administration. *Toxicol Lett* **168**, 176–185.

Warheit DB, Laurence BR, Reed KL, Roach DH, Reynolds GA, Webb TR (2004) Comparative pulmonary toxicity assessment of single-wall carbon nanotubes in rats. *Toxicol Sci* **77**, 117–125.

Warheit DB, Hoke RA, Finlay C, Donner EM, Reed KL, Sayes CM (2007) Development of a base set of toxicity tests using ultrafine TiO2 particles as a component of nanoparticle risk management. *Toxicol Lett* **171**, 99–110.

Wilson KD, Raney SG, Sekirov L, Chikh G, deJong SD, Cullis PR, Tam YK (2007) Effects of intravenous and subcutaneous administration on the pharmacokinetics, biodistribution, cellular uptake and immunostimulatory activity of CpG ODN encapsulated in liposomal nanoparticles. *Int Immunopharmacol* **7**, 1064–1075.

Yang RS, Chang LW, Wu JP, Tsai MH, Wang HJ, Kuo YC, Yeh TK, Yang CS, Lin P (2007) Persistent tissue kinetics and redistribution of nanoparticles, quantum dot 705, in mice: ICP-MS quantitative assessment. *Environ Health Perspect* **115**, 1339–1343.

Yokoyama A, Sato Y, Nodasaka Y, Yamamoto S, Kawasaki T, Shindoh M, Kohgo T, Akasaka T, Uo M, Watari F, Tohji K (2005) Biological behavior of hat-stacked carbon nanofibers in the subcutaneous tissue in rats. *Nano Lett* **5**, 157–161.

Yu Z, Xiaoliang W, Xuman W, Hong X, Hongchen G (2008) Acute toxicity and irritation of water-based dextran-coated magnetic fluid injected in mice. *J Biomed Mater Res A* **85**, 582–587.

Zelis M, Zweegman S, Van Der Meer FJ, Kramer MH, Smulders YM (2008) The interaction between anticoagulant therapy with vitamin K-antagonists and treatment with antibiotics: a practical recommendation. *Ned Tijdschr Geneeskd* **152**, 1042–1046.

Zhang LW, Monteiro-Riviere NA (2008) Assessment of quantum dot penetration into intact, tape-stripped, abraded and flexed rat skin. *Skin Pharmacol Physiol* **21**, 166–180.

Zhao Y, Sugiyama S, Miller T, Miao X (2008) Nanoceramics for blood-borne virus removal. *Expert Rev Med Devices* **5**, 395–405.

Zhu MT, Feng WY, Wang B, Wang TC, Gu YQ, Wang M, Wang Y, Ouyang H, Zhao YL, Chai ZF (2008) Comparative study of pulmonary responses to nano- and submicron-sized ferric oxide in rats. *Toxicology* **247**, 102–111.

Zimmermann KC, Green DR (2001) How cells die: apoptosis pathways. *J Allergy Clin Immunol* **108**, S99–103.

17

In Vitro and *In Vivo* Models for Nanotoxicity Testing

Kyung O. Yu, Laura K. Braydich-Stolle, David M. Mattie, John J. Schlager and Saber M. Hussain

17.1 Introduction

Nanotechnology is an emerging area focused on the use of particles of a minute size for creating unique devices at a nanoscale level. These particles possess distinctive physical and chemical properties, which have yielded the production of materials with novel functionalities. Although many engineered nanoparticles (NP) and nanodevices offer benefits to society, the toxicity and environmental fate of these materials has not been addressed (Donaldson *et al.*, 2004; Service, 2005; Vinardell, 2005; Hardman, 2006). Nanoparticles are manufactured from diverse soluble and insoluble materials, and are created with various sizes, shapes and surface modifications (Ryman-Rasmussen *et al.*, 2006). These properties of the nanoparticle enable them to enter the body via different routes such as inhalation through the respiratory tract (Lam *et al.*, 2004; Warheit *et al.*, 2004; Oberdörster *et al.*, 2005), oral ingestion via the gastrointestinal tract (Zhou and Yokel, 2005), and passage through skin (Monteiro-Riviere *et al.*, 2005a,b). Engineered nanomaterials are often classified as either carbon-based or metal-based. In this chapter the carbon-based materials subcategories of single- and multiwalled carbon nanotubes will be discussed in detail. Metal-based nanoparticles such as titanium dioxide, silver, silicon dioxide and zinc oxide, listed in Table 17.1, will also be discussed. Additionally, this chapter will focus on *in vivo* and *in vitro* studies that address the three main routes of exposure with these different nanomaterials.

Nanotoxicity: From In Vivo and In Vitro Models to Health Risks   Edited by Saura Sahu and Daniel Casciano
© 2009 John Wiley & Sons, Ltd

Table 17.1   Commonly used nanomaterials and their abbreviations.

Nanomaterial	Abbreviation	Consumer products/uses
Carbon nanotubes (single or multiwalled)	CNT (SWCNT, MWCNT)	Biosensors, cell delivery systems
Silver	Ag	Antimicrobial in food packaging, band aids, socks, etc.
Silicon dioxide	SiO_2	Drug delivery, biosensing, cancer detection
Titanium dioxide	TiO_2	Paint, sunscreen, cosmetics, waste water treatments
Zinc oxide	ZnO	Sunscreen, antimicrobial

17.2   Inhalation Exposure

The most common form of nanoparticles is those ultrafine particles present in polluted air. As a result of dangerous airborne nanoparticles, which are easily inhaled and absorbed systemically, the majority of nanotoxicity studies performed have focused on lung exposure. Nanoparticles can easily enter into the nasal, oral and tracheal/bronchial regions of the respiratory tract. Further, studies have shown that nanoparticles are capable of penetrating the alveoli deep in the lung (Oberdörster *et al.*, 2005). Once inhaled, the nanoparticles can interact with internal fluids and agglomerate, which affects the overall size of the particle and its final deposition site within the lungs. Furthermore, clearance from the different nasal regions depends upon where the particles accumulate. For example, particles that are soluble in mucus may enter the blood and be distributed throughout the body systemically and ultimately secreted as waste, while insoluble particles could be swallowed (Oberdörster *et al.*, 2005). Studies have also shown that nanoparticles inhaled through the nose and deposited in olfactory nerve endings can cross the blood-brain barrier (Chen *et al.*, 2003; Borm and Kreyling, 2004) and enter into the olfactory bulb of the brain (Oberdörster *et al.*, 2005; Nel *et al.*, 2006).

17.2.1   *In Vivo* Studies

In toxicity studies, two methods exist for nanoparticle deposition into the lungs: inhalation and intratracheal instillation. In the first method, inhalation studies are performed in an exposure chamber, where the potentially toxic particles are inhaled by the subjects via the respiratory tract. However, maintenance of a constant concentration of particles in the exposure chamber is not easily achieved and can be costly (Driscoll *et al.*, 2000). In addition, not all nanoparticles can be delivered via an inhalation route. For instance, during a whole body exposure, an animal's hair can become coated with NPs, which leads to inhalation exposure as well as oral ingestion when the animal licks their hair (Ballantyne *et al.*, 2006). Toxic effects of nanoparticles have been shown to be closely related to the size of the particle. One study evaluated titanium dioxide NPs (size 2–5 nm) via inhalation exposure. Since characterization of the NPs is critical for toxicity evaluations (Murdock *et al.*, 2008), before exposing the mice to the particles, the authors (Grassian *et al.*, 2007) measured particle sizes by different methods such as power X-ray diffraction, transmission electron microscopy (TEM), and surface area using the Brauner, Emmer and Teller (BET) measurement. The aerosol formed in the exposure chamber was measured to be 5 nm with a surface area of $210 \pm 10\,m^2/g$, and all of the titanium oxide particles were anatase crystals with no rutile crystalline form present. Three weeks after exposure, it was found that mice exposed to $0.77\,mg/m^3$ NPs showed minimal lung toxicity, while those exposed to $7.22\,mg/m^3$ NPs revealed significant toxicity with moderate inflammation. Similarly, Ji *et al.* (2007) reported that there were no significant changes in health effects when rats were exposed for 28 days at a concentration near the current American Conference of Governmental Industrial Hygienists (ACGIH) silver dust limit ($100\,\mu g/m^3$). However, when rats were subchronically exposed for 90 days with silver NPs at a concentration twice that of the 28-day study, reports showed that there were lung function changes due to inflammation. The authors noted that doubling the dosage increased the surface area five-fold, which does not reflect normal exposure concentrations. Health effects in this study were related to surface area of silver NPs, while no effects were seen from the change in mass (Sung *et al.*, 2008).

The second method for examining toxic effects from inhalation studies is instillation. Intratracheal instillation is a method which provides an inexpensive manner to deliver precise doses of nanoparticles easily to the lungs in an effort to determine toxicity *in vivo* (Henderson *et al.*, 1995). A syringe with a feeding needle is inserted into the trachea to deliver nanoparticles to the lungs. However, this route bypasses the upper respiratory tract, which could be an important target organ for NP toxicity studies.

Pulmonary toxicity of intratracheally instilled single-walled carbon nanotubes (SWCNTs) was investigated in mice (Lam *et al.,* 2004). Due to insolubility of the test materials in water and dispersing agents, SWCNTs were sonicated before dosing. Although nanotubes induced dose-dependent interstitial inflammation in mice exposed for 7 days, toxicity levels increased dramatically when mice were exposed to 0.5 mg/ml for 90 days. Studies showed peribroncheal inflammation and necrosis extended in alveolar regions due to high levels of carbon nanotubes, compared with little toxicity seen with carbon black (negative control) or quartz (positive control) when delivered at similar ratios to the nanotubes. Ultrafine particles (carbon black, 14.3 nm (UCB) and TiO_2, 29 nm (UTiO$_2$)) induced inflammation and epithelial damage more than corresponding larger particles (carbon black, 290 nm (CB) and TiO_2 250 nm (TiO$_2$)). In this study, UCB was more toxic than UTiO$_2$ with regard to inflammation and epithelial damage.

When nanoparticles were used for inhalation or intratracheal studies, inflammatory responses were found to be size-dependent and/or surface area-dependent. In comparing anatase TiO_2 bulk scale materials (250 nm) and anatase TiO_2 nanoscale particles (20 nm) intratracheally instilled in rats and mice (Oberdörster, 2000), studies revealed that smaller size nanoparticles caused more inflammatory neutrophil response, as seen with lung lavage 24-h postexposure. Additionally, when the toxic effect (neutrophil) was normalized by surface area, and not mass, a good fit in a dose-response curve between the two sizes was generated. Osier *et al.* (1997) used intratracheal instillation and intratracheal inhalation methods to test two different primary sizes of TiO_2 (fine and ultrafine) at similar doses. The two methods showed different responses following measurement of the bronchoalvi-olar lavage (BAL) fluid for levels of two cytokines: macrophage inflammatory protein-2 (MIP-2) and tumor necrosis factor alpha (TNF-α). The authors reported that increased BALF MIP-2 protein levels seemed to increase the inflammatory response, while TNF-α did not. Furthermore, Li *et al.* (2007) investigated the acute pulmonary toxicity caused by 3 and 20 nm TiO_2 instilled intratracheally into lungs of mice. Three days after dosing at 0.4, 4 and 40 mg/kg, the biochemical parameters in BAL fluid and pathological examinations were employed to measure toxicity levels. Contrary to the general belief that small NPs induce a higher degree of toxicity, the authors in this study concluded that the 3 nm TiO_2 produced the same pulmonary toxicity as the 20 nm particles over all dosing levels. Further, it is believed that the pH levels of TiO_2 in the medium caused these adverse effects, while the size and surface area of the particles did not induce a response. Overall, there are limited inhalation studies in animal models. There is a great need to establish appropriate inhalation protocols that are relevant to realistic exposure to humans.

17.2.2   *In Vitro* Studies

Recently a number of papers have been published that describe *in vitro* toxicity of NPs exposed to cells of the respiratory system. *In vitro* studies using human bronchial epithelial cells have shown that SWCNTs generate reactive oxygen species (ROS), indicating that oxidative stress was the predominant mechanism of acute toxicity (Shvedova *et al.*, 2004a,b). Similarly, TiO_2 particles induced oxidative damage in a human bronchial epithelial cell line in a size-dependent manner (Gurr *et al.*, 2005). Additionally, A549 cells, which are alveolar basal epithelial cells derived from a carcinoma, have been used in a wide range of nanotoxicity studies. One study examining the size-dependent effects of crystalline silicon dioxide particles (15 nm and 46 nm) demonstrated that the toxicity was not size-dependent, but occurred in a dose-dependent manner along with increased levels of oxidative stress (Lin *et al.*, 2006). Furthermore, 47 nm TiO_2 particles have been shown to be nontoxic, while silica nanospheres, 80 nm Al, and <5 μm quartz particles induce toxicity (Erdos *et al.*, 2005). In comparison, nano-sized silver has demonstrated toxic effects related to particle size in rat alveolar macrophages, and the production of ROS was identified as the mechanism of toxicity (Carlson *et al.,* 2008).

17.3   Dermal Exposure

Skin is the first line of defense for the human body against the environment, yet it provides a significant surface area for potential contamination. The skin is a complex organ that comprises the epidermis, dermis and hypodermis. The epidermis, the most external layer, does not contain any blood vessels but is composed of several cell types including keratinocytes, melanocytes, Langerhans cells, and Merkels cells. The outer layer of the skin consisting of keratinocytes is called the stratum corneum. Between the epidermis and the dermis there is a very thin membrane, the basement membrane, which attaches the epidermis firmly, though not rigidly, to the dermis. The dermis is the second layer of skin, which consists of connective tissue, blood vessels, nerves, hair roots,and sweat glands. Below the dermis lies the hypodermis, which contains fibroblasts, adipose cells, and macrophages, as well as subcutaneous fat, larger blood vessels and nerves. The subcutaneous fat lies on the muscles and bones, to which the whole skin structure is attached by connective tissues. With their use in cosmetics, sunscreens, band aids, clothing, and so on, the incidence of dermal nanoparticle exposure is rapidly increasing, and this section will focus on the *in vivo* and *in vitro* studies that have evaluated dermal exposure, absorption and cytotoxicity.

17.3.1   *In Vivo* Studies

Titanium dioxide is used in a wide range of commercial products that come into contact with skin, which led investigators to study the dermal interaction with this nanometal. Pflucker *et al.* (1999, 2001) studied dermal uptake, using porcine skin, of titanium dioxide (TiO$_2$) particles averaging between 20–50 nm. Examination of the nanoparticles on the skin was performed using transmission electron microscopy (TEM), tape stripping and subsequent scanning electron microscopy (SEM) with energy dispersive X-ray analysis (EDXA). These techniques illustrated that TiO$_2$ was located exclusively in the stratum corneum. Recent *in vitro* and *in vivo* studies have also revealed the importance of hair follicles for dermal interaction with nanoparticles (Toll *et al.*, 2004; Lademann *et al.*, 2005). Open hair follicles can be an efficient reservoir for penetration of topically applied materials, and can contribute to the systemic circulation of particles. Tsuji *et al.* (2006) reported that nanoparticles such as TiO$_2$ or ZnO can penetrate the human epidermis and enter into some hair follicles, but these particles do not travel into deep layers of the skin when they are applied topically. Furthermore, percutaneous absorption of TiO$_2$ particles is size-dependent and chemically specific (Hoet *et al.*, 2004).

Tinkle *et al.* (2003) reported that 0.5 and 1.0 μm fluorospheres on the wrist with motion could penetrate the intact stratum corneum of human skin and reach the epidermis, the anatomical location of the Langerhans cells. The authors reported that once the particles are in the lipid channel of the stratum corneum (SC), they move into the skin and are phagotcytosed by the Langerhans cells. Hair follicle sizes and potential follicular reservoirs are different in various parts of the body, leading to differences in absorption rates based upon the thickness of SC (Otberg *et al.*, 2004). Since hair follicles and sweat glands only account for up to 0.1 % of the total body surface, they were not considered to contribute to the transdermal penetration route.

17.3.2   *In Vitro* Studies

Human skin fibroblasts have been used as an *in vitro* model to assess nanotoxicity where zinc oxide (50–70 nm) and titanium dioxide (less than 150 nm) have been shown to be toxic in these cells. This cytotoxicity occurs in a time-dependent manner, with zinc oxide inducing higher levels of toxicity than titanium dioxide (Dechsakulthorn *et al.*, 2007). Our laboratory has also shown that silicon dioxide and titanium dioxide nanoparticles can cause cytotoxicity and ROS production in mouse keratinocytes (Yu *et al.*, 2009; Braydich-Stolle *et al.*, 2008). In the case of the silica nanoparticles, Yu *et al.* (2009) reported that this toxicity occurs in a size-dependent manner (Figures 17.1, 17.2 and 17.3), while the titanium dioxide nanoparticles did not show a size-dependent effect. However, the mechanism of cell death varies with the crystal structure of the titanium dioxide nanoparticles (Braydich-Stolle *et al.*, 2008). Additionally, TiO$_2$ nanoparticles were capable of localizing in the lysosomes as well as the mitochondria (Figure 17.4) (Braydich-Stolle *et al.*, 2008). Furthermore, *in vitro* cultures of primary human epidermal keratinocytes treated with unmodified multiwalled carbon nanotubes (MWCNTs) demonstrate cytoxicity (Monteiro-Riviere *et al.*, 2005a). Nel *et al.* (2006) reported that treating keratinocytes with high doses of SWCNTs results in ROS generation and oxidative stress. Moreover, our laboratory has shown that mouse keratinocytes treated with MWCNTs and SWCNTs exhibit ROS generation due to impurities in iron (Grabinski *et al.*, 2007), which is consistent with data demonstrating

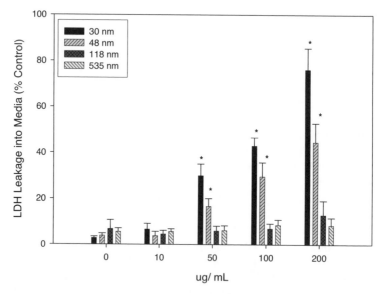

Figure 17.1   *Effect of SiO₂ on LDH leakage into media (% control). The HEL-30 cells were dosed with different sizes (30, 48, 118 and 535 nm) and various concentrations of silica (0, 10, 50, 100 and 200 µg/ml) for 24 h. Size- and dose-dependent LDH leakage was observed for 30 and 48 nm silica nanoparticles. Large size (118 and 535 nm) particles showed less toxicity compared with smaller nanoparticles. Three independent experiments (n = 3) were carried out and data are represented as mean ± SD. *Significantly different from control at P < 0.05. Reprinted from Yu et al. (2009)* Journal of Nanoparticle Research **11**, *15–24, with kind permission from Springer Science + Business Media. Copyright 2009*

that MWCNTs and SWCNTs induce cytotoxicity in human epidermal keratinocytes (Allen *et al.*, 2000; Shedova *et al.*, 2003; Monteiro-Riviere *et al.*, 2005b).

17.4   Oral Exposure/Gastrointestinal Tract

The gastrointestinal (GI) tract is composed of the upper and lower GI tract, which differs substantially from animal to animal and from animals to humans. Based on their uses in food, food packaging, drug delivery and cosmetics, nanoparticles have the potential to be ingested and pass through to the GI tract; therefore, this last section discussing nanoparticle exposure will focus on toxicity in the gastrointestinal tract.

17.4.1   *In Vivo* Studies

Early research has shown that particulate matter can be taken up by the M-cells in the Peyer's patches of the small intestines, the follicles associated with lymphoid tissue and the normal intestinal enterocytes (Florence and Hussain, 2001; Hussain *et al.*, 2001). Once the particles reach the submucosal tissue they are able to enter the lymphatic system and capillaries, which gives them access to the entire body via the circulatory system, thereby

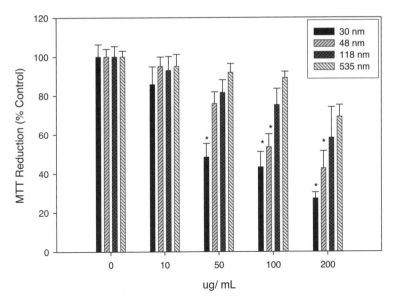

Figure 17.2 *Effect of silica nanoparticles on MTT reduction (% control). The HEL-30 cells were dosed with different sizes (30, 48, 118 and 535 nm) and various concentrations of silica (0, 10, 50, 100 and 200 μg/ml) for 24 h. Size- and dose-dependent MTT reduction for 35 and 51 nm NPs at higher concentrations (200 μg/ml) produced significant toxicity when compared with large size particles (118 and 42 nm). Three independent experiments (n = 4) were carried out and data are represented as mean ± SD. *Significantly different from control at P < 0.05. Reprinted from Yu et al. (2009)* Journal of Nanoparticle Research **11**, *15–24, with kind permission from Springer Science + Business Media. Copyright 2009*

having the ability to induce systemic immune responses (Hoet *et al.*, 2004). Furthermore, studies using gold (Hillyer and Albrecht, 2001) showed that colloidal gold uptake is dependent upon particle size; smaller particles are shown to be absorbed more, mainly through the small intestines. Reports on latex uptake (Jani *et al.*, 1989) were also size-dependent and demonstrated that surface modifications play a role in absorption, whereby uptake of non-ionic latex particles was higher than that for ionic carboxylated beads. Hillery *et al.* (1994) investigated the uptake of 60 nm polystyrene particles through lymphoid and non-lymphoid tissue of the small and large GI tract. Uptake by non-lymphatic tissue was less than that seen with lymphatic tissue. The majority of particles were absorbed through the lymphoid section of the large intestine.

Wang *et al.* (2006) investigated the acute oral toxicity of Zn powder in male and female mice at a dose level of 5 g/kg, using two sizes (58 nm, nanoscale, and 1.08 μm, microscale) of Zn. Severe symptoms of digestive illness were observed in mice treated with nanoscale Zn, but not in those treated with the microscale Zn. Furthermore, slight differences in blood biochemistry between the two zinc-treated mice groups were noted. Although the doses used in this study were much higher than real exposure levels, the authors wanted to evaluate the oral toxicity of nanoscale Zn powder according to the OECD (Organization for Economic Co-operation and Development) guidelines, which are international guidelines

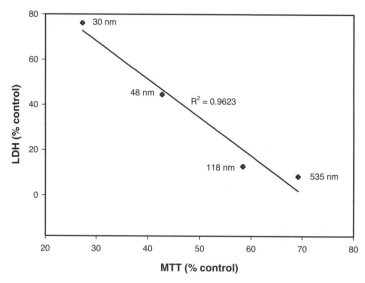

Figure 17.3 *Correlation between LDH and MTT for various sizes of silica nanoparticles at 200 μg/ml. The graph was generated using two data points in the highest exposure condition (200 μg/ml) in Figures 17.1 and 17.2. Reprinted from Yu et al. (2009)* Journal of Nanoparticle Research **11**, *15–24, with kind permission from Springer Science + Business Media. Copyright 2009*

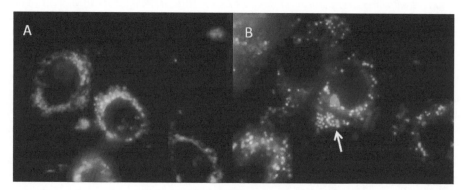

Figure 17.4 *Uptake and localization of titanium dioxide nanoparticles in mouse keratinocytes. (A) Confocal image of mouse keratinocytes treated with TiO$_2$-Ru labelled nanoparticles and exposed for 24 h. The lysosomes were stained green and the nanoparticles fluoresced red; yellow indicated co-localization. The majority of nanoparticles were localizing in the lysosomes, further suggesting endocytosis as the mechanism of uptake. However, there were nanoparticles that were not in the lysosomes. (B) Confocal image of mouse keratinocytes treated with TiO$_2$-Ru labelled nanoparticles and exposed for 24 h. The mitochondria were stained green and the nanoparticles fluoresced red; yellow indicated co-localization. The mitochondria demonstrated less localization of nanoparticles when compared to the lysosomes; however, there were instances of nanoparticles localizing in the mitochondria (as shown by the white arrow), indicating that mitochondrial function can be altered by the presence of nanoparticles in a cell. See color plate section.*

currently used for toxicity testing of chemicals and materials. In addition, titanium dioxide particles were studied for acute oral toxicity and biodistribution (Wang *et al.*, 2007). The adult mice were dosed with nano-sized (25 and 80 nm) and fine (155 nm) TiO_2. Similar to the ZnO study, the large dose of 5 g/kg was administered as a single dose according to the OECD protocol. For oral ingestion of 80 nm TiO_2, the particles were observed to be retained in the liver, while the 20 nm particles did not localize in the liver, but instead were found in the spleen, kidneys and lung tissues. Female mice showed higher organ to body weight ratios for liver, spleen and kidneys, expressed as milligrams (wet weight of tissues)/gram (body weight). No significant changes were observed in the males, while the ratio of liver in female mice was significantly different ($P < 0.05$). Although small sizes of NPs are generally more toxic than fine-size particles, the authors in this study found gender-dependent effects rather than size-dependent effects. Furthermore, this study shows that TiO_2 was translocated into systemic circulation and deposited in different tissues. Gender-related toxicity was also reported by Kim *et al.* (2008). They investigated the subchronic oral toxicity of silver in rats dosed with 0, 30, 300 and 1000 mg/kg for 28 days. Some dose-dependent toxicity was observed in alkaline phosphatase and cholesterol values in male and female mice, indicating liver damage for dose levels above 300 mg/kg. The tissue distribution of silver NPs revealed dose-dependent accumulation of particles in the kidneys, resulting in a two-fold increase in female kidneys compared with males. These studies show that NPs, after oral ingestion, can enter systemic circulation via the GI tract with distribution in other organs.

17.4.2 *In Vitro* Studies

Very limited information is available on *in vitro* studies for oral exposure based on the lack of a representative *in vitro* model. However, des Rieux and colleagues have established a co-culture system using Caco-2 and Raji cells, where the Raji cells induce the Caco-2 cells to differentiate into follicle associated eptithlium (FAE)-like cells, which can be used as a model for M-cells in the Peyer's patches of the small intestine to study nanoparticle transport (des Rieux *et al.*, 2005, 2007). The few studies that have been performed with this model have focused on polystyrene nanoparticles, which would be used in drug delivery. The initial findings report that transport across this epithelium is dependent upon concentration, temperature and size (des Rieux *et al.*, 2005) and that coating the nanoparticles with adhesion molecules, expressed by the epithelium, facilitates transport (Gullberg *et al.*, 2006).

17.5 *In Vivo–In Vitro* Comparison

While nanotoxicity is a relatively new field, a standard practice for testing the toxicity of these materials has not been established. *In vitro* assays receive criticism for their lack of reproducibility, and a recent study by Sayes and colleagues has shown that in the case of lung exposure, *in vitro* models did not correlate well with *in vivo* models (Sayes *et al.*, 2007). In this case, for the *in vitro* study, fine particles or nanoparticles (90–500 nm) were dosed to rat L2 lung epithelial cells, macrophages (MAC), and a co-culture of L2 cells and MAC for *in vitro* screening (dosing levels of 0.005–520 mg/cm^2). For the *in vivo* comparison, rats were intratracheally instilled with carbonyl iron, crystalline or amorphous silica, nano or fine zinc oxide (1–5 mg/kg). *In vitro* cytotoxicity was evaluated by examining MTT, LDH, MIP-2, TNF-α and IL-6, while the *in vivo* study examined inflammation (neutrophil response) and cytotoxic endpoints of BAL fluid LDH levels. According to Sayes and colleagues, the variables between the *in vitro* and *in vivo* studies, which could explain the poor correlation, are the following:

(1) Are the cell types used *in vivo* and *in vitro* systems comparable?
(2) Are exposure times comparable between the *in vivo* and *in vitro* systems?
(3) What are the best cytotoxicity endpoints between the *in vitro* and *in vivo* systems?

Another study by Lim and colleagues attempted to correlate *in vitro* and *in vivo* methods using silica nanoparticles. When silica nanoparticles were intratracheally instilled at various concentrations (0–40 mg/ml), they induced acute inflammation resulting in chronic fibrosis. For the *in vitro* comparison, A549 cells were dosed with various concentrations of silica NPs (0–50 μg/cm^2). A dose-dependent apoptosis was observed in both the *in vitro* and *in vivo* systems and was responsible for cytotoxicity, as measured through histopathology and cytotoxic endpoints like cell viability and BAL fluid analysis. Aside from these few studies, there has not been much of an effort to investigate the correlation between *in vitro* and *in vivo* studies systematically.

Despite the criticism that *in vitro* models have received, they are a valuable tool for understanding cytotoxic effects. In order for an *in vitro* model to be used as a predictor of toxicity, the cells need to be validated as an accurate model for representing that tissue. There is no argument that single cell cultures used during *in vitro* studies do not provide a good representation of complex tissues such as the lungs, skin, testes, and so on. However, single cell cultures are easily generated and are useful for providing a preliminary foundation for generation of useful measurement techniques, assessment of dosing ranges, and investigation of probable mechanisms of toxicity, before progressing to costly *in vivo* studies.

Furthermore, *in vivo* nanotoxicity studies have also yielded conflicting data. For example, Lam's SWCNT studies established that SWCNTs were more toxic than quartz in a dose-dependent manner in a mouse model (Lam *et al.*, 2004), while Warheit *et al.* (2004) demonstrated dose-dependent responses with quartz and non-dose-dependent responses with SWCNTs in a rat model. In addition, studies using rats, which have examined the toxicity of nanosized TiO$_2$, yielded conflicting results as to the nanoproperty that is responsible for generating a toxic response. One study concluded that ultrafine TiO$_2$ particles (29 nm) increased inflammation and altered macrophage chemotactic responses in rat lungs, when compared with TiO$_2$ particles of 250 nm (Renwick *et al.*, 2004). However, another study showed that rats exposed to nanoscale TiO$_2$ rods/dots produced inflammatory responses that were similar to pulmonary effects seen with larger TiO$_2$ particles based upon surface properties (Warheit *et al.*, 2006, 2007a,b). These studies used the same delivery method, but varied drastically in the nanoparticle source, amount of nanoparticles exposed, and the time of exposure. Furthermore, all of these studies display very little characterization of the nanomaterials prior to, during or after exposure. The need for characterization is clearly illustrated by Murdock *et al.* (2008) who have shown that the presence of serum in media not only alters the agglomerate size in solution, but also the relative toxicity of the nanoparticles, probably through nanomaterial–protein interactions. Based on this, questionable study designs and lack of characterization are the most likely source of the non-reproducible and potentially controversial results. Therefore, for future studies, there needs to be an optimization and validation of *in vitro* testing protocols and a strong focus needs to be placed on the characterization of these materials prior to, during and after exposure to cells.

In summary, for *in vitro* and *in vivo* exposure settings, characterization of NPs prior to, during and after exposure are very important since particle sizes change due to agglomeration. Although there is not a good correlation between *in vitro* and *in vivo* methods, *in vitro* testing is a valuable tool to investigate cytotoxic effects. *In vitro* systems can produce useful measurement techniques, finding of dosing ranges, and elucidation of mechanisms of toxicity before costly *in vivo* testing. For *in vivo* exposure, NPs can enter into the body via inhalation and via the GI tract, and can be transferred into systemic circulation and could result in a human health risk. NPs can cross the blood-brain barrier and can be used as drug delivery system to the brain. However, NPs could produce health risks when they show toxic effects in the brain. Dermal exposure of NPs shows less penetration through the epidermis. Size-dependent and chemical specific toxicity were observed by these routes, and high surface area plays an important role due to their reactivity compared with bulk materials. Specific surface area, not total mass, shows better prediction of toxicity determined by inflammatory and oxidative stress responses. Toxicity testing of newly synthesized NPs should be evaluated individually when health risks are suspected, since there is no testing protocol representative of all NPs.

References

Allen DG, Riviere JE, Monteiro-Riviere NA (2000) Identification of early biomarkers of inflammation produced by keratinocytes exposed to jet fuels Jet A, JP-8, and JP-8(100). *J Biochem Mol Toxicol* **14**, 231–237.

Ballantyne B, Snelling WM, Norris JC (2006) Respiratory peripheral chemosensory irritation, acute and repeated exposure toxicity studies with aerosols of triethylene glycol. *J Appl Toxicol* **26**, 387–396.

Borm PJ, Kreyling W (2004) Toxicological hazards of inhaled nanoparticles – potential implications for drug delivery. *J Nanosci Nanotechnol* **4**, 521–531.

Braydich-Stolle LK, Schaeublin NM, Murdock RC, Jiang J, Biswas P, Schlager JJ, Hussain SM (2008) Crystal structure mediates mode of cell death in TiO$_2$ nanotoxicity. *J Nano Research*, online only.

Carlson C, Hussain SM, Schrand AM, Braydich-Stolle LK, Hess K, Schlager JJ (2008) Unique cellular interaction of silver nanoparticles: Size dependent generation of reactive oxygen species. *J Phys Chem* **112**, 13608–13619.

Chen Y, Xue Z, Zheng D, Xia K, Zhao Y, Liu T, Long Z, Xia J (2003) Sodium chloride modified silica nanoparticles as a non-viral vector with a high efficiency of DNA transfer into cells. *Curr Gene Ther* **3**, 273–279.

Dechsakulthorn F, Hayes A, Bakand S, Joeng L, Winder C (2007) In vitro cytotoxicity assessment of selected nanoparticles using human skin fibroblasts. *AATEX* **14** (spec. issue), 397–400.

des Rieux A, Ragnarsson EG, Gullberg E, Preat V, Schneider YJ, Artursson P (2005) Transport of nanoparticles across an in vitro model of the human intestinal follicle associated epithelium. *Eur J Pharm Sci* **25**, 455–465.

des Rieux A, Fievez V, Theate I, Mast J, Preat V, Schneider YJ (2007) An improved in vitro model of human intestinal follicle-associated epithelium to study nanoparticle transport by M cells. *Eur J Pharm Sci* **30**, 380–391.

Donaldson K, Stone V, Tran CL, Kreyling W, Borm PJ (2004) Nanotoxicology. *Occup Environ Med* **61**, 727–728.

Driscoll KE, Costa DL, Hatch G, Henderson R, Oberdorster G, Salem H, Schlesinger RB (2000) Intratracheal instillation as an exposure technique for the evaluation of respiratory tract toxicity: uses and limitations. *Toxicol Sci* **55**, 24–35.

Erdos GW, Moraga D, Palazuelos M, Powers K, Kelley K (2005) Cellular response to nanoparticle exposure. *NSTI-Nanotech* **6**, 78–80.

Florence AT, Hussain N (2001) Transcytosis of nanoparticle and dendrimer delivery systems: evolving vistas. *Adv Drug Deliv Rev* **50** (Suppl. 1), S69–S89.

Grabinski CM, Hussain SM, Lafdi K, Braydich-Stolle L, Schlager JJ (2007) Effect of particle dimension on biocompatibility of carbon nanomaterials. *Carbon* **45**, 2828–2835.

Grassian VH, O'Shaughnessy PT, Mcakova-Dodd A, Pettibone JM, Thorne PS (2007) Inhalation exposure study of titanium dioxide nanoparticles with a primary particle size of 2 to 5 nm. *Environ Health Perspect* **115**, 397–402.

Gullberg E, Keita AV, Salim SY, Andersson M, Caldwell KD, Soderholm JD, Artursson P (2006) Identification of cell adhesion molecules in the human follicle-associated epithelium that improve nanoparticle uptake into the Peyer's patches. *J Pharmacol Exp Ther* **319**, 632–639.

Gurr JR, Wang AS, Chen CH, Jan KY (2005) Ultrafine titanium dioxide particles in the absence of photoactivation can induce oxidative damage to human bronchial epithelial cells. *Toxicology* **213**, 66–73.

Hardman R (2006) A toxicologic review of quantum dots: Toxicity depends on physicochemical and environmental factors. *Environ Health Perspect* **114**, 165–172.

Henderson RF, Driscoll KE, Harkema JR, Lindenschmidt RC, Chang IY, Maples KR, Barr EB (1995) A comparison of the inflammatory response of the lung to inhaled versus instilled particles in F344 rats. *Fundam Appl Toxicol* **24**, 183–197.

Hillery AM, Jani PU, Florence AT (1994) Comparative, quantitative study of lymphoid and non-lymphoid uptake of 60 nm polystyrene particles. *J Drug Target* **2**, 151–156.

Hillyer JF, Albrecht RM (2001) Gastrointestinal persorption and tissue distribution of differently sized colloidal gold nanoparticles. *J Pharm Sci* **90**, 1927–1936.

Hoet PH, Bruske-Hohlfeld I, Salata OV (2004) Nanoparticles – known and unknown health risks. *J Nanobiotechnology* **2**, 12.

Hussain N, Jaitley V, Florence AT (2001) Recent advances in the understanding of uptake of microparticulates across the gastrointestinal lymphatics. *Adv Drug Deliv Rev* **50**, 107–142.

Jani P, Halbert GW, Langridge J, Florence AT (1989) The uptake and translocation of latex nanospheres and microspheres after oral administration to rats. *J Pharm Pharmacol* **41**, 809–812.

Ji JH, Jung JH, Kim SS, Yoon JU, Park JD, Choi BS, Chung YH, Kwon IH, Jeong J, Han BS, Shin JH, Sung JH, Song KS, Yu IJ (2007) Twenty-eight-day inhalation toxicity study of silver nanoparticles in Sprague-Dawley rats. *Inhal Toxicol* **19**, 857–871.

Kim YS, Kim JS, Cho HS, Rha DS, Kim JM, Park JD, Choi BS, Lim R, Chang HK, Chung YH, Kwon IH, Jeong J, Han BS, Yu IJ (2008) Twenty-eight-day oral toxicity, genotoxicity, and gender-related tissue distribution of silver nanoparticles in Sprague-Dawley rats. *Inhal Toxicol* **20**, 575–583.

Lademann L, Otberg N, Jacobi U, Hoffman RM, Blume-Peytavi U (2005) Follicular penetration and targeting. *J Investig Dermatol Symp Proc* **10**, 301–303.

Lam CW, James JT, McCluskey R, Hunter RL (2004) Pulmonary toxicity of single-wall carbon nanotubes in mice 7 and 90 days after intratracheal instillation. *Toxicol Sci* **77**, 126–134.

Li J, Li Q, Xu J, Li J, Cai X, Liu R, Li Y, Ma J, Li W (2007) Comparative study on the acute pulmonary toxicity induced by 3 and 20nm TiO$_2$ primary particles in mice. *Environ Toxicol Pharmacol* **24**, 239–244.

Lin W, Huang YW, Zhou XD, Ma Y (2006) In vitro toxicity of silica nanoparticles in human lung cancer cells. *Toxicol Appl Pharmacol* **217**, 252–259.

Monteiro-Riviere NA, Inman AO, Wang YY, Nemanich RJ (2005a) Surfactant effects on carbon nanotube interactions with human keratinocytes. *Nanomedicine* **1**, 293–299.

Monteiro-Riviere NA, Nemanich RJ, Inman AO, Wang YY, Riviere JE (2005b) Multi-walled carbon nanotube interactions with human epidermal keratinocytes. *Toxicol Lett* **155**, 377–384.

Murdock RC, Braydich-Stolle L, Schrand AM, Schlager JJ, Hussain SM (2008) Characterization of nanomaterial dispersion in solution prior to in vitro exposure using dynamic light scattering technique. *Toxicol Sci* **101**, 239–253.

Nel A, Xia T, Madler L, Li N (2006) Toxic potential of materials at the nanolevel. *Science* **311**, 622–627.

Oberdörster G (2000) Toxicology of ultrafine particles: in vivo studies. *Phil Trans R Soc A* **358**, 2719–2740.

Oberdörster G, Oberdörster E, Oberdörster J (2005) Nanotoxicology: an emerging discipline evolving from studies of ultrafine particles. *Environ Health Perspect* **113**, 823–839.

Osier M, Baggs RB, Oberdörster G (1997) Intratracheal instillation versus intratracheal inhalation: influence of cytokines on inflammatory response. *Environ Health Perspect* **105** (Suppl. 5), 1265–1271.

Otberg N, Richter H, Schaefer H, Blume-Peytavi U, Sterry W, Lademann J (2004) Variations of hair follicle size and distribution in different body sites. *J Invest Dermatol* **122**, 14–19.

Pflucker F, Hohenberg H, Holzle E, Will T, Pfeiffer S, Wepf R, Diembeck W, Wenck H, Gers-Barlag H (1999) The outermost stratum corneum layer is an effective barrier against dermal uptake of topically applied micronized titanium dioxide. *Int J Cosmet Sci* **21**, 399–411.

Pflucker F, Wendel V, Hohenberg H, Gartner E, Will T, Pfeiffer S, Wepf R, Gers-Barlag H (2001) The human stratum corneum layer: an effective barrier against dermal uptake of different forms of topically applied micronised titanium dioxide. *Skin Pharmacol Appl Skin Physiol* **14** (Suppl. 1), 92–97.

Renwick LC, Brown D, Clouter A, Donaldson K (2004) Increased inflammation and altered macrophage chemotactic responses caused by two ultrafine particle types. *Occup Environ Med* **61**, 442–447.

Ryman-Rasmussen JP, Riviere JE, Monteiro-Riviere NA (2006) Penetration of intact skin by quantum dots with diverse physicochemical properties. *Toxicol Sci* **91**, 159–165.

Sayes CM, Reed KL, Warheit DB (2007) Assessing toxicity of fine and nanoparticles: comparing in vitro measurements to in vivo pulmonary toxicity profiles. *Toxicol Sci* **97**, 163–180.

Service RF (2005) Nanotechnology. Calls rise for more research on toxicology of nanomaterials. *Science* **310**, 1609.

Shvedova AA, Castranova V, Kisin ER, Schwegler-Berry D, Murray AR, Gandelsman VZ, Maynard A, Baron P (2003) Exposure to carbon nanotube material: assessment of nanotube cytotoxicity using human keratinocyte cells. *J Toxicol Environ Health A* **66**, 1909–1926.

Shvedova AA, Kisin E, Keshava N, Murray AR, Gorelik AO, Arepalli S, Gandelsman VZ, Castranova V (2004a) Cytotoxic and genotoxic effects of single wall carbon nanotube exposure on human keratinocytes and bronchial epithelial cells. 227th ACS National Meeting, March 28–April 1, 2004, Anaheim, CA.

Shvedova A, Kisin E, Murray A, Schwegler-Berry D, Gandelsman V, Baron P, Maynard A, Gunter M, Castranova V (2004b) Exposure of human bronchial cells to carbon nanotubes caused oxidative stress and cytotoxicity. Proceedings of the Meeting of the SFRR Europe, June 26–29, 2003.

Sung JH, Ji JH, Yoon JU, Kim DS, Song MY, Jeong J, Han BS, Han JH, Chung YH, Kim J, Kim TS, Chang HK, Lee EJ, Lee JH, Yu IJ (2008) Lung function changes in Sprague-Dawley rats after prolonged inhalation exposure to silver nanoparticles. *Inhal Toxicol* **20**, 567–574.

Tinkle SS, Antonini JM, Rich BA, Roberts JR, Salmen R, DePree K, Adkins EJ (2003) Skin as a route of exposure and sensitization in chronic beryllium disease. *Environ Health Perspect* **111**, 1202–1208.

Toll R, Jacobi U, Richter H, Lademann J, Schaefer H, Blume-Peytavi U (2004) Penetration profile of microspheres in follicular targeting of terminal hair follicles. *J Invest Dermatol* **123**, 168–176.

Tsuji JS, Maynard AD, Howard PC, James JT, Lam CW, Warheit DB, Santamaria A (2006) Research strategies for safety evaluation of nanomaterials, Part IV: risk assessment of nanoparticles. *Toxicol Sci* **89**, 42–50.

Vinardell MP (2005) *In vitro* cytotoxicity of nanoparticles in mammalian germ-line stem cell. *Toxicol Sci* **88**, 285–286.

Wang B, Feng WY, Wang TC, Jia G, Wang M, Shi JW, Zhang F, Zhao YL, Chai ZF (2006) Acute toxicity of nano- and micro-scale zinc powder in healthy adult mice. *Toxicol Lett* **161**, 115–123.

Wang J, Zhou G, Chen C, Yu H, Wang T, Ma Y, Jia G, Gao Y, Li B, Sun J, Li Y, Jiao F, Zhao Y, Chai Z (2007) Acute toxicity and biodistribution of different sized titanium dioxide particles in mice after oral administration. *Toxicol Lett* **168**, 176–185.

Warheit DB, Laurence BR, Reed KL, Roach DH, Reynolds GA, Webb TR (2004) Comparative pulmonary toxicity assessment of single-wall carbon nanotubes in rats. *Toxicol Sci* **77**, 117–125.

Warheit DB, Webb TR, Sayes CM, Colvin VL, Reed KL (2006) Pulmonary instillation studies with nanoscale TiO_2 rods and dots in rats: toxicity is not dependent upon particle size and surface area. *Toxicol Sci* **91**, 227–236.

Warheit DB, Webb TR, Colvin VL, Reed KL, Sayes CM (2007a) Pulmonary bioassay studies with nanoscale and fine-quartz particles in rats: toxicity is not dependent upon particle size but on surface characteristics. *Toxicol Sci* **95**, 270–280.

Warheit DB, Webb TR, Reed KL, Frerichs S, Sayes CM (2007b) Pulmonary toxicity study in rats with three forms of ultrafine-TiO_2 particles: differential responses related to surface properties. *Toxicology* **230**, 90–104.

Yu K, Grabinski C, Schrand A, Murdock C, Wang W, Gu B, Schlager J, Hussain S (2009) Toxicity of amorphous silica nanoparticles in mouse keratinocytes. *J Nanopart Res* **11**, 15–24.

Zhou Y, Yokel RA (2005) The chemical species of aluminum influences its paracellular flux across and uptake into Caco-2 cells, a model of gastrointestinal absorption. *Toxicol Sci* **87**, 15–26.

18

In Vitro Models
for Nanotoxicity Testing

Yinfa Ma

18.1 Introduction

Nanotechnology and applications of nanomaterials have become one of the leading technologies since the early 1990s, due to the unique physical and chemical characteristics of nanoparticles (Stix, 2001). Nanomaterials are increasingly important in many industrial products, with applications such as catalysts, pigments, resin, cosmetics and electronic devices. The use of nanomaterials has been extended to the biomedical and biotechnological fields in recent years for the purposes of drug/gene delivery, disease diagnosis, recognition, diagnosis and tracking of tumor cells, and imaging. Nanomaterials include several different nanostructures; this chapter mainly focuses on nanoparticles, although other nanostructures may be mentioned briefly in context.

It is well known that, with a decrease in particle size, the specific surface area (ratio of total surface area to mass of nanomaterials) increases rapidly and allows a higher proportion of its atoms or molecules to be displayed on the surface of the material rather than in the interior. An inverse relationship between particle size and the number of molecules displayed on its particle surface is clearly shown in Figure 18.1 (Oberdörster *et al.*, 2005). An increase in the specific surface area of a particle means that there are more reactive groups on the particle surface. Consequently, nanoparticles become very reactive in the cellular environment, and may impose dramatically different biological effects than those of their bulk forms (Kipen and Laskin, 2005; Nel *et al.*, 2006). As a result, the level of toxicity of nanomaterials is likely to be different from that of the bulk forms. Accordingly, the toxicology data on nanoparticles (NPs) need to be carefully assessed, and

Nanotoxicity: From In Vivo *and* In Vitro *Models to Health Risks*   Edited by Saura Sahu and Daniel Casciano
© 2009 John Wiley & Sons, Ltd

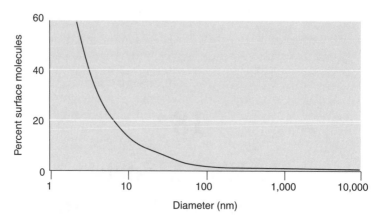

Figure 18.1 *Surface molecules as a function of particle size. It is clearly seen that the proportion of surface molecules increases exponentially when particle size decreases <100 nm, reflecting the importance of surface area for increased chemical and biological activity. Reproduced from Oberdörster* et al. *(2005), with permission from Environmental Health Perspectives*

the applications of nanoparticles may be limited because of concerns about their toxicity (Service, 2004).

Currently, there is a common assumption that the small size of nanomaterials allows them easily to enter tissues, cells, organelles, and functional biomolecular structures. A corollary is that the entry of nanomaterials into vital biological systems could cause damage, which could subsequently result in harm to human health. Many people can be exposed to nanostructures in various ways, including researchers who manufacture nanomaterials, patients who are injected with a nanomaterial, or people who use products containing nanomaterials. In all cases, there are unique routes of exposure that will dictate the specific fate of the nanomaterials (Fischer Hans and Chan Warren, 2007). Unfortunately, there is insufficient information available about the impact on human health and the environmental implications of manufactured nanomaterials.

Although the long-term effects of nanoparticles on human health remain unclear, many preliminary studies have demonstrated their toxic effects, selectively including carbon nanomaterials (carbon black, nanotubes, C_{60} and fullerene) (Shvedova *et al.*, 2003; Oberdorster, 2004; Warheit, 2004), silica and metal oxides (Lin *et al.*, 2006a,b, 2008), quantum dots (Green and Howman, 2005; Cho *et al.*, 2007), ultrafine polystyrene (Brown *et al.*, 2001), nanometer-sized diesel exhaust particles (Block *et al.*, 2004), and ultrafine particulate pollutants (Li *et al.*, 2003). Toxicity studies are normally conducted either *in vitro* (using cells) or *in vivo* (using animals), except in the case of ecotoxicity studies of NPs (Adams *et al.*, 2006; Hund-Rinke and Simon, 2006). In many cases, an *in vitro* study is conducted first to understand the mechanism of cytotoxicity of the nanoparticles before an *in vivo* experiment is designed. Since this chapter focuses on *in vitro* models for nanotoxicity testing, the *in vivo* models for nanotoxicity study will not be described.

In vitro nanotoxicity assessment of nanoparticles is actually a quite complicated issue because many factors are involved for a valid assessment, such as particle parameters, cell line selections, and biomarker detection. Consequently, interpretation of the toxicity data is frequently difficult because of the variations in one or more of these parameters. Careful consideration and good experimental designs are essential before conducting an *in vitro* study to compare the cytotoxicity of nanoparticles, and to avoid comparing apples with oranges.

In vitro studies, in general, monitor the following parameters to determine the levels of cytotoxicity: cell morphology, viability, proliferation, inflammation, oxidative stress response, mitochondrial damage, DNA and protein damage, cellular dysfunction, cellular uptake, and gene expression.

To assist readers in designing their *in vitro* experiments for a nanotoxicity study, the following key areas will be described in this chapter:

- characterization of nanoparticles;
- cell line selections;
- biomarker analysis;
- cell imaging;
- dosimetry in *in vitro* study;
- effect of nanoparticle morphology on nanotoxicity testing;
- results analysis.

18.2 Characterization of Nanoparticles

In conducting *in vitro* nanotoxicity testing, one needs to realize that the physico-chemical and structural properties of nanoparticles play a major role in their interactions with cells, and that could lead to quite different toxicological effects (Donaldson and Tran, 2002; Oberdörster *et al.*, 2005; Kreyling *et al.*, 2006; Lewinski *et al.*, 2008). The major parameters of nanoparticles to be considered, before conducting *in vitro* testing, include sizes of the particles, size distribution, surface area, porosity, charge density on the surface, solubility, morphology, functional groups on the surface, purity, stability, and so on.

As described earlier, particle size plays one of the most important roles in causing cytotoxicity of NPs, even though other parameters may also contribute significantly. For instance, shrinkage in size may create discontinuous crystal planes that increase the number of structural defects, as well as disrupting the well-structured configuration of the nanoparticle, thereby altering its electrical properties (Donaldson and Tran, 2002; Kipen and Laskin, 2005; Oberdörster *et al.*, 2005). This could establish specific surface groups that could function as reactive sites. Therefore, the sizes and size distributions of the nanoparticles must be accurately measured before an *in vitro* study is conducted. Several techniques that can be used to determine the sizes and size distributions of nanoparticles include atomic force microscopy (AFM), transmission electron microscopy (TEM), scanning electron microscopy (SEM), light scattering (static or dynamic), and zeta-meter. The most common techniques used by nanotoxicity researchers for size measurements are AFM and TEM, even though their use depends on the availability of these instruments. Conventional TEM instruments can provide two-dimensional images of NPs as shown in Figure 18.2 (a) and (b) (Lin *et al.*, 2009). The TEM software allows one to calculate the size distribution of the NPs by using the imaging data. AFM not only provides particle size information, but also provides particle height and volume because it directly produces three-dimensional images, which is the unique advantage of AFM over SEM and TEM.

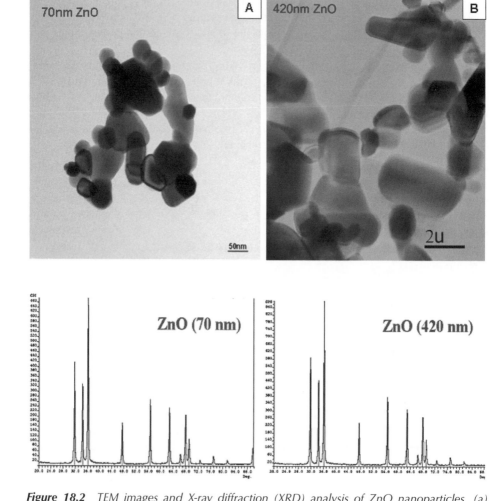

Figure 18.2 *TEM images and X-ray diffraction (XRD) analysis of ZnO nanoparticles. (a) 70 nm ZnO particles; (b) 420 nm ZnO particles. XRD analysis indicated that both particles were hexagonal. Reprinted from Lin et al. (2009), with kind permission from Springer Science + Business Media. Copyright 2009*

The specific surface area (SSA, m^2/g) of the nanoparticles is another important factor that can affect their cytotoxicity significantly because the interactions of NPs with cells may increase greatly as the SSA increases. Even though the SSA is inversely proportional to the nanoparticle size, the nanopores on the nanoparticles may vary from particle to particle. The SSA of NPs is normally measured through using the Brunauer, Emmett and Teller (BET) technique. In addition, the BET technique can also provide particle porosity information, even though other techniques can be used to measure porosity, such as the porosity analyzer. The fundamental principles of the BET technique for measuring the SSA can be found generally in the literature.

Table 18.1 Metal impurity levels in 70 nm and 420 nm ZnO particles tested in an ICP-MS system. The method detection limit is 0.04 ppm*. Reprinted with permission from Lin et al. (2009).

Elements	Metal impurity levels (ppm)	
	70 nm ZnO	420 nm ZnO
Na	45.4	7.5
K	44.9	4.4
Cu	37.9	9.3
Se	18.2	18.7
Ca	8.6	1.7
As	5.5	5.7
Pb	2.9	4.6
Mg	2.4	0.5
Cd	1.4	2.8
Ga	0.8	0.7
Al	0.6	0.3
Sb	0.3	0.2
Ti	0.2	ND
Ni	0.1	ND
Ag	0.1	0.1
Ba	0.1	ND
Rb	0.07	ND
Cr	0.04	0.05
Be, Fe, V, Mn, Co, Sr, Mo, Cs, Ti, U	ND	ND
Total	169.5	56.6

*One part per million (ppm) of a specific metal is defined as 1 gram of that metal per 10^6 gram of ZnO. ND, not detected.

Many studies have demonstrated that different particle structures (amorphous versus crystal structures, or different crystal structures) show different levels of cytotoxicity. Therefore, pertinent information on particle structures, or crystal structures, must be known before conducting a cytotoxicity study. X-ray diffraction (XRD) is a commonly used instrumental technique for this purpose. A representative XRD analysis of ZnO nanoparticles is shown in Figure 18.2.

At the nanoscale, the impurities in a particle or on a particle surface can significantly affect the cytotoxicity of nanoparticles. Sometimes they may be the major contribution to the cytotoxicity of the NPs, which may provide misleading information about the level of cytotoxicity of the NPs themselves. Therefore, the impurities in nanoparticles should be determined before an *in vitro* cytotoxicity study is undertaken. For example, inductively coupled plasma mass spectrometry (ICP-MS) is the most commonly used technique to quantitatively measure the metal impurities in many nanoparticles, because of its high sensitivity, but the instrumental conditions may vary from instrument to instrument. The impurities in one type of silica nanoparticle are shown in Table 18.1 (Lin *et al.*, 2009).

Due to their nanoscale size and surface properties, nanoparticles tend to aggregate or precipitate in suspensions. Because this process can affect the levels of cytotoxicity of NPs, the average hydrodynamic size and distribution of NPs in suspension must be determined. Dynamic laser scattering (DLS) and zeta-meter (or zeta-sizer) have commonly been used to determine the hydrodynamic sizes of nanoparticles. A zeta-meter can also provide zeta potential and surface charge information, which is essential to know before an *in vitro* study is conducted. To acquire accurate data on the hydrodynamic sizes of nanoparticles, as well as the charge density on the surface, it is recommended that measurements be taken in the cell media used for an *in vitro* study.

In summary, the parameters discussed above should be known as much as possible before an *in vitro* study is conducted. Otherwise, the cytotoxicity data acquired through the *in vitro* study may be hard to explain or be difficult to compare with on levels of cytotoxicity.

18.3   Cell Line Selections for *In Vitro* Nanotoxicity Study

What cell line should be chosen for an *in vitro* study? What can be learned from an *in vitro* study? These are the common questions that researchers always need to answer before an *in vitro* experiment is designed. A systematic and thorough analysis of *in vitro* study results can lead to improvements in design of nanoparticles for diagnostic and therapeutic applications. In addition, a better understanding of the interactions of nanoparticles with different types of tissues and cell lines, as well as assessments of NP distribution in cells compared with cellular damage, can serve as the basis for determining their toxicity and the direction of future investigations.

During the past decade, a majority of nanotoxicity research has been performed using cell culture systems. A better understanding of the relationship between the physical and chemical properties of nanomaterials and their *in vitro* behavior would provide a basis for assessing cytotoxic response and, more importantly, could lead to predictive models for assessing nanotoxicity (Fischer Hans and Chan Warren, 2007). The overall advantages of *in vitro* studies with cells are as follows:

(1) It is easy to delineate the mechanism of toxicity in the absence of the physiological and compensatory factors that confound the interpretation of whole animal studies.
(2) It allows examination of the effects of target cells in the absence of the secondary effects caused by inflammation.
(3) It is efficient, cost-effective, and is helpful in improving the design of subsequent expensive whole animal studies. The examination of nanotoxicity in cultured cells is both feasible and more physiologically relevant.

Actually, *in vitro* toxicity studies of NPs using different cell lines have been increasingly reported. Both animal- and human-derived cells have been widely used as an *in vitro* model for cytotoxicity studies (Oberdörster *et al.*, 2005). However, there is a lack of consensus in the published literature on nanoparticle toxicity due to variable cell lines. Researchers use different cell lines when conducting nanotoxicity studies based on several factors, such as cell line availability, specific areas they are interested in, and endpoints to evaluate the level of toxicity. Table 18.2 summarizes the *in vitro* systems that have been studied for potential organ targets, cell lines selected, and endpoints tested on metal nanoparticles, quantum dots, and carbon nanoparticles (including C_{60}, SWCNTs and MWCNTs). For each class of nanoparticle, a variety of cell lines, different experimental conditions and endpoints tested have been adopted. For example, the cytotoxicity of iron oxide nanoparticles has been studied using at least 16 different cell lines, as shown in Table 18.3. An outstanding

Table 18.2 *Available* in vitro *systems for potential organ target, cell lines, and endpoints tested.*

Potential Target	Cell Line	Test or Endpoint
Lung	Macrophages	Trypan blue, LDH, apoptosis, Chemotaxis assay
	Immune cells	Cytokine profile, adjuvant effects
	A549	SRB, LDH, GSH/GSSG, WST, MMP
	H1299	MTT, Live/Dead, LDH
	NR8383	MTT, WST
	IMR-90	YO-PRO1, PI, Brd U, Microarray
	H596	MTT
Skin	HEK	MTT, Neutral red,
	HDF	MTT, Live/Dead, LDH
	HaCaT	Alamar blue, GSH
	MSTO-211H	MTT
	HSF42	Hoechst
Liver	HepG2	MTT, LDH, GSH
	BRL-3A	MTT
Breast	MDA-MB-231	Cell titer96
	5K-BR-3	Live/Dead
	MDA-MB-435S	Live/Dead
	SKBr3	Live/Dead
	MCF-7	MTT
Kidney	HEK293	MTT, Western blot, Flow cytometry, Microarray
	COS-7	MTT
	COS-1	MTT
Bone	hBMSC	WST, Rt-PCR, Flow cytometry
Vein	HUVEC	LDH, WST, Microarray
Brain	NHA	MTT, Live/Dead, LDH
	EC219	MTT
Others	Hela	LDH
	K562	MTT
	RAW264.7	MTT
	Vero cells	WST, MTT
	KB cells	Microacopy
	Human fibroblasts	MTT, Live/Dead
	Melanoma cells	MTT
	Htert-BJ1	MTT
	MSTO-211H	MTT
	NIH 3T3	MTT
	PC12M	Live/Dead
	PC12	MTT
	WTK1	Comet, flow cytometry, MTT
	B16F10	Microscopy
	SH-SY5Y	MTT, flow cytometry

comprehensive overview of cytotoxicity of nanoparticles is available as a reference for interested researchers (Lewinski *et al.*, 2008).

The human bronchoalveolar carcinoma-derived cell line (A549) is one of the commonly used cell lines for cytotoxicity study (Lin *et al.*, 2006a, 2008, 2009) because the level of oxidative stress of cells can be evaluated by monitoring several key parameters, including

Table 18.3 Cytotoxicity studies on Fe_3O_4 nanoparticles. Reproduced from Lewinski et al., with permission. Copyright Wiley-VCH Verlag GmbH & Co. KGaA.

Cell line	Surface coating	Exposure conditions	NP concentration (average size)	Test	Exposure duration	Toxicity
COS-7 cells	none	3×10^4 cells/well, 24-well plates	0.92–23.05 mM (d = 9 nm)	MTT	4 h	No significant difference between control and exposed
Human fibroblasts	MA-PEG	10,000 CeLLS mL^{-1} in 24-well plates	0–1000 µg mL^{-1} (d = 50 nm)	MTT, Live/ Dead	24 h	250 µg mL^{-1}: 25–50% viability decrease for bare; 1 mg mL^{-1}: 99% viable for PEG-coated
Melanoma cells	PVA (amino, caroxyl, thiol)	N/A	0.25 mg mL^{-1} (core = −9 nm, d = 19.54 nm)	MTT	2 h	No cytotoxicity after 2 h for all polymer/iron ratio; after 24 hr, cytotoxicity at high polymer conc.
Primary human fibroblasts, hTERT-BJ1	pullulan	10^4 cells/well in 96-well plates	0–2 mg mL^{-1} (d = 40–50 nm)	MTT	24 h	Plain SPION showed significant decrease in viability, pn-SPION showed no cytotoxicity w/ >92% viability
Rat liver cells, BRL 3A	none	confluent in 6 or 24-well plates	0–250 µg mL^{-1} (d = 30, 47 nm)	LDH, MTT, GSH	24 h	EC50 > 250 µg mL^{-1}
Human mesothelioma MSTO-211H, rodent 3T3 fibroblast cells	none	N/A	3.75–15 ppm (d = 12–50 nm)	MTT	3, 6 days	3T3 cells viable w/ up to 30 ppm; MSTO cell viability decrease at 3.75 ppm, free radicals via Fenton rxn

Cells	Coating	Cell density	Concentration (size)	Assay	Time	Results
Rat brain-derived endothelial EC219; murine N9 & N11 microglial cells	PVA, amino PVA, caroxylPVA, ThiolPVA	96 or 48-well plates	2.5 µL NPs mL^{-1}, 11.3 µg iron mL^{-1} (core = 8–12 nm, d = 30 nm)	MTT	48 h	Only aminoPVA-SPION uptaken by N11
Mouse macrophages, RAW 264.7	P(PEGMA)	105 cells mL^{-1}	0.2 mg mL^{-1} (d = 6.2 ± 0.7 nm)	ratio of treated/control cells	1, 4 days	Cytotoxicity dose dependent; decrease w/ time attributed to cell division
Human breast carcinoma SK-BR-3 cells, human dermal fibroblasts	PMAO-PEG	Confluent	10–400 nm (9.6 nm)	Live/Dead	1, 24, 48 h	91% viability at 400 nm after 48 h in HDF cells
Human monocytemacrophages	dextran	1–2 ×10^6 cells/well plates & 0.5 ×10^6 cells/well in 48-well plates	0.0001–10 mg mL^{-1} (d = 30 nm)	MTT, Neutral Red	24, 48, 72 h, 4 days + 14 days grow	1 mg mL^{-1}: not toxic after 72 h; 10 mg mL^{-1}: mildly toxic; viability similar over 2 wks
rat pheochromocytoma cell line PC12M	DMSA	20,000 cells mL^{-1} in 6 or 12 well plates	15, 1.5 mm and 150 µm (d = 5–12 nm)	Live/Dead	2, 4, 6 days	SPION exposure reduced PC12 ability to respond to nerve growth factors
OCTY mouse cells	MPEG–Asp3-NH2, MPEG-PAA, PAA	104 cells/well in 96-well plates	0–400 µg mL^{-1} (d = 14 nm)	Cell Titer 96	72 h	MPEG–Asp3-NH2 almost no cytotoxicity; MPEG–PAA- and PAA-coated decrease cell viability

reactive oxygen species (ROS), lactate dedydrogenase (LDH) activity, cellular levels of GSH/GSSG, concentration of malondialdehyde (MDA), and cellular α-tocopherol level. In addition, MCF-7, the human breast cancer cell line, is also commonly used for cellular oxidative stress studies.

Gold nanoparticles constitute one of the most popularly studied nanoparticle groups. Monkey kidney cells, HelG2 cells, 3T3/NIH, hEPG2, and human leukemia (K562) cells have been used to study the nuclear targeting ability of gold nanoparticles, both alone and with full-length peptide attached (Thomas *et al.*, 2003; Tkachenko *et al.*, 2003, 2004; Connor *et al.*, 2005). The effects of gold nanoparticles on cell proliferation, nitric oxide, and reactive oxygen species production has been tested using macrophage cells (Shukla *et al.*, 2005). In addition, other cell lines, including breast cancer (MDA-MB-231) cells, human dermal fibroblast cells, SKBR2 breast-cancer cells, Vero cells and KB cells have also been used in the cytotoxicity study of gold NPs (Hirsch *et al.*, 2003; Salem *et al.*, 2003; Goodman *et al.*, 2004; Loo *et al.*, 2004, 2005; Fu *et al.*, 2005; Niidome *et al.*, 2006; Pernodet *et al.*, 2006; Shenoy *et al.*, 2006; Huff *et al.*, 2007; James *et al.*, 2007; Su *et al.*, 2007).

Based upon information provided by past studies, there are no standards for selecting cells for nanotoxicity studies. However, the same cell line must be chosen in order to allow comparison of levels of cytotoxicity. Several groups that studied the effects of carbon nanoparticles under various experimental conditions with different cell lines obtained different results. The most significant factor influencing cytotoxicity in this class of carbon nanoparticles seems to be cell-type specific. Macrophage cell lines (Sayes *et al.*, 2004), human epidermal keratinocytes (HEK) (Panessa-Warren *et al.*, 2006), human dermal fibroblasts, HL60, Jurkat T cells (Kam *et al.*, 2004; Manna *et al.*, 2005; Sayes, 2006), and other cell lines have been used as test models to evaluate the toxicity of SWCNTs, MWCNTs, and other carbon NPs. The testing results vary quite significantly from one model to another (Shvedova *et al.*, 2003; Cherukuri *et al.*, 2004; Kam *et al.*, 2004; Pantarotto *et al.*, 2004; Sayes *et al.*, 2004; Cui *et al.*, 2005; Ding *et al.*, 2005; Jia *et al.*, 2005; Manna *et al.*, 2005; Murr *et al.*, 2005; Sato *et al.*, 2005; Bellucci *et al.*, 2006; Chlopek *et al.*, 2006; Fiorito *et al.*, 2006; Flahaut *et al.*, 2006; Magrez *et al.*, 2006; Monteiro-Riviere and Inman, 2006; Nel *et al.*, 2006; Porter Alexandra *et al.*, 2006; Rouse *et al.*, 2006; Sayes, 2006; Witzmann and Monteiro-Riviere, 2006; Worle-Knirsch *et al.*, 2006; Yamawaki and Iwai, 2006; Helland *et al.*, 2007; Pulskamp *et al.*, 2007).

In summary, many cell lines have been used for cytotoxicity studies based on the factors described in this section, making a comparison of cytotoxicity among these different studies difficult. Therefore, standardization in the choice of cell line models is essential in order to be able to compare results from different studies and to conduct further mechanism studies (Wick *et al.*, 2007). In addition, human cells should probably be used to better predict toxicity levels for humans, and cell types used should also be kept consistent.

18.4   Biomarker Analysis

While the exact mechanism whereby nanoparticles induce adverse effects on human health remains unknown, it is suggested that nanoparticles of various compositions create reactive oxygen species (ROS), and thereby modulate intracellular calcium concentrations,

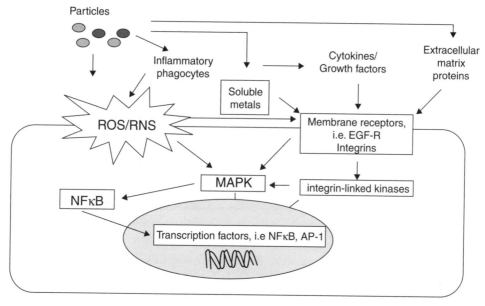

Figure 18.3 *Possible pathways of nanoparticle-induced cell proliferation and cell damage. Reproduced from Knaapen et al. by permission of John Wiley & Sons Inc*

activate transcription factors, and induce cytokine production. Reactive oxygen species (ROS) have been shown to damage cells by peroxidizing lipids, altering proteins, disrupting DNA, interfering with signalling functions, and modulating gene transcription (Xu *et al.*, 2004; Oberdörster *et al.*, 2005). Oxidative stress generation, inflammation and DNA damage are among the events in our current understanding of these complex cellular mechanisms after exposure to nanoparticles, as shown in Figure 18.3 (Knaapen *et al.*, 2004). Advances in life science and analytical techniques have identified some targets as indicators or biomarkers of such physiological status. In this section, some fundamentals and several classes of biomarker analyses for an *in vitro* nanotoxicity study are presented for quantitative evaluation of oxidative stress.

18.4.1   Intracellular ROS Measurement

Reactive oxygen species (ROS) and other free radicals are essential for life because they are important in cell signalling and are also used by phagocytes to fight against bacteria. However, nonessential production of ROS is harmful. Nevertheless, antioxidants are ready to scavenge ROS. When the balance between antioxidants and free radicals is broken after exposure to nanoparticles, oxidative stress takes place. The measurement of cellular ROS accumulation can be conducted using the 2′,7′-dichloroflurescin diacetate (DCFH2-DA) method (Wang and Joseph, 1999). Briefly, DCFH2-DA, a nonfluorescent compound, can passively enter cells and interact with ROS to form a highly fluorescent compound, DCF. The DCF fluorescence is measured at 485 nm excitation and 520 nm emission (Lin *et al.*,

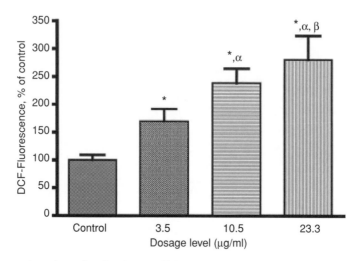

Figure 18.4 *The relative levels of intracellular reactive oxygen species (ROS) after 24 h exposure to 3.5, 10.5 or 23.3 μg/ml of 20 nm CeO₂ nanoparticles. Reprinted with permission from Lin et al. (2006b).*

2006a,b, 2007). A higher level of intracellular ROS will give a higher fluorescence, as shown in Figure 18.4.

18.4.2   LDH Measurement

Determination of lactate dehydrogenase (LDH) release is widely used in nanotoxicity studies. ROS generated during exposure to nanoparticles very often will cause lipid peroxidation, and subsequently the lipid peroxidation will cause cell membrane damage and disturb the integrity of cellular membranes, leading to the leakage of cytoplasmic enzymes such as lactate dehydrogenase (LDH) into the media. It is clear that the more severe the damage to the cell membrane by the ROS, the higher the level of LDH would be in the cell media. A LDH kit is available for LDH assay. In this assay, the LDH released from damaged cells catalyzes the oxidation of lactate to pyruvate with simultaneous reduction of NAD^+ to NADH. The rate of NAD^+ reduction is measured as an increase in absorbance at 340 nm. The rate of NAD^+ reduction is directly proportional to LDH activity in the cell medium, while the amount of LDH released is proportional to the number of cells damaged (Haslam *et al.*, 2000; Sayes *et al.*, 2004; Muller *et al.*, 2005; Lin *et al.*, 2006a,b, 2007), as shown in Figure 18.5.

18.4.3   GSH and GSSG Measurement

As one of the most important biomarkers of oxidative stress, glutathione (GSH) exists abundantly and is widely distributed in living cells where it acts as an important intracellular antioxidant. The tripeptide GSH (γ-glutamyl-cysteinyl-glycine) is involved in many important cellular functions, ranging from the control of physico-chemical properties of cellular proteins and peptides (and thus modulating the activity of many enzymes) to detoxification of xenobiotics and scavenging free radicals. It also protects cells against

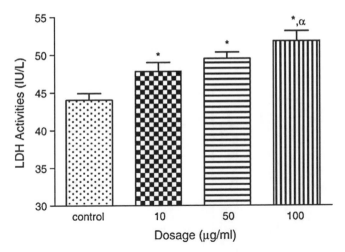

Figure 18.5 *The LDH activities in the cell culture medium after 48 h exposure to 10 μg/ml, 50 μg/ml or 100 μg/ml of 15 nm SiO$_2$ nanoparticles. Reprinted with permission from Lin et al. (2006a).*

the toxic effects of oxygen by reacting directly with reactive oxygen intermediates and, less directly, by maintaining other compounds which have antioxidant activity, such as ascorbate and α-tocopherol, in reduced forms. The antioxidant and reducing activity are exhibited by a reduced form of glutathione (GSH), whose concentration over the oxidized forms is relatively high. The intracellular level of glutathione in most mammalian cells is in the millimolar range, and is kept so by continuous synthesis. When oxidative stress occurs, cellular GSH levels will decrease, as shown in Figure 18.6. In contrast, the level of oxidized glutathione (GSSG) will increase. Therefore, the ratio of GSH and GSSG has been used as an indicator of cellular oxidative stress.

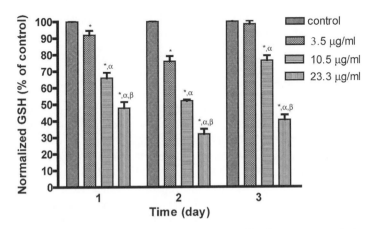

Figure 18.6 *Normalized cellular GSH levels of A549 cells after 24, 48 or 72 h exposure to 3.5, 10.5 or 23.3 μg/ml of 20 nm CeO$_2$. Reprinted with permission from Lin et al. (2006b).*

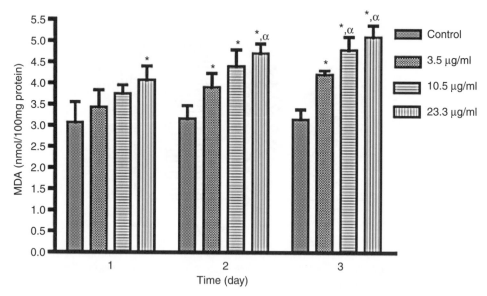

Figure 18.7   Cellular MDA levels of A549 cells after 24, 48 or 72 h exposure to 3.5, 10.5 or 23.3 µg/ml of 20 nm CeO₂. Reprinted with permission from Lin et al. (2006b).

To measure cellular GSH levels, frozen cells (at $-70°$ C) are homogenized in a serine borate buffer (100 mM Tris-HCl, 10 mM boric acid, 5 mM L-serine, 1 mM DETAPAC, pH 7.5). Twenty microliters of homogenate are added to 230 µl of HPLC grade H_2O and 750 µl of N-(1-pyrenyl)maleimide (NMP) solution (1 mM in acetonitrile). The resulting suspensions were incubated for 5 min at room temperature, then 5 µl of 2 M HCl are added to stop the reaction. The samples are filtered through a 0.2 µm Whatman Puradisc syringe filter, and an aliquot of 20 µl is injected for analysis using a high-performance liquid chromatography (HPLC) system with fluorescence detection (excitation at 330 nm, emission at 375 nm). A C_{18} column (250 × 4.6 mm) is normally used for the separation. An existing protocol can be found in several publications (Winters *et al.*, 1995; Lin *et al.*, 2006a,b).

18.4.4   MDA Measurement

Malondialdehyde (MDA) is a typical marker for measuring the degree of lipid peroxidation. MDA levels increase during oxidative stress due to free radical attacks on cell membrane lipids, as shown in Figure 18.7. The structure of MDA is shown in Figure 18.8. MDA formation has been studied since the 1940s, and the determination of cellular concentration of MDA has been well described (Draper *et al.*, 1993). The ROS, generated by exposure to nanoparticles, produce oxidative stress, as reflected by elevated production of MDA, which is a steady indicator of liquid peroxidation and membrane damage (Haslam *et al.*, 2000; Lin *et al.*, 2006a,b, 2007). A HPLC with the UV detection method is generally used for determination of cellular MDA concentrations. Detailed experimental conditions can be found in many publications (Draper *et al.*, 1993; Lin *et al.*, 2006b).

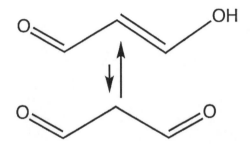

Figure 18.8   Structure of malondialdehyde (MDA).

18.4.5   α-Tocopherol Measurement

α-tocopherol is also a kind of antioxidant, and its presence in a cell membrane has been proven to represent the major defense system against peroxidation of membrane lipids, which are highly susceptible to peroxidative degradation. As cellular oxidative stress progresses, the α-tocopherol levels decrease accordingly, as shown in Figure 18.9. A method for extracting cellular α-tocopherol was developed by Lang, Gohil and Packer (Lang et al., 1986). The extracted sample solution can be analyzed by using the HPLC fluorescence method with an excitation wavelength of 292 nm and an emission wavelength of 324 nm.

18.4.6   Oxidative DNA Damage Measurement (Single Cell Gel Electrophoresis)

ROS are capable of causing oxidative damage to DNA, producing a variety of lesions, such as single- and double-strand breaks, and resulting in chemical changes to both pyrimidine and purine bases. Therefore, the level of oxidative DNA damage can be used as a major indicator to evaluate the extent of cellular damage caused by nanoparticles. The single cell gel electrophoresis (comet assay) has been commercially developed and a comet assay kit

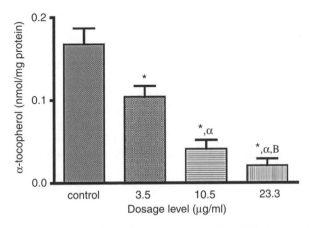

Figure 18.9   Cellular α-tocopherol levels of A549 cells after 72 h exposure to 3.5, 10.5 or 23.3 μg/ml of 20 nm CeO_2. Reprinted with permission from Lin et al. (2006b).

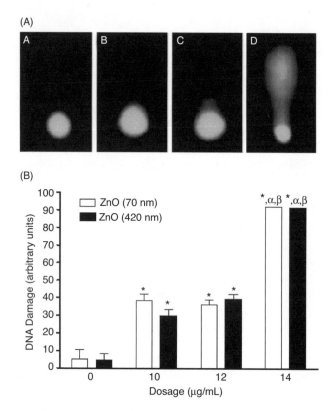

Figure 18.10   (A) The representative fluorescence images of nuclei treated in the comet assay after 24 h exposure to (A) 0 μg/ml, (B) 10 μg/ml, (C) 12 μg/ml, or (D) 14 μg/ml of 70 nm ZnO nanoparticles. (B) DNA damage was observed at all dosages. Reprinted from Lin et al. (2009), with kind permission from Springer Science + Business Media. Copyright 2009

is available for this purpose. Detailed procedures for conducting comet assay can be found in the literature. The key to being successful in this experiment is that all steps of single cell electrophoresis must be conducted under yellow light to prevent additional DNA damage. After the electrophoresis, slides are viewed using an epifluorescence microscope equipped with a fluorescein filter. Observations are normally made at a magnification of ×400. In general, about 30 randomly selected cells per experimental point are imaged and analyzed by using the software that comes with the microscope. Results are presented as tail DNA percentage, a parameter describing the number of migrated fragments, represented by the fluorescence intensity in the tail, as shown in Figure 18.10A (Lin *et al.*, 2009). In addition, the percentages of tail fluorescence intensity can be used to evaluate quantitatively the degree of oxidative DNA damage, as shown in Figure 18.10B.

18.4.7   Protein Assay

To normalize the level of each biomarker to the same level, the total cellular protein concentration must be determined so that each biomarker concentration can be presented

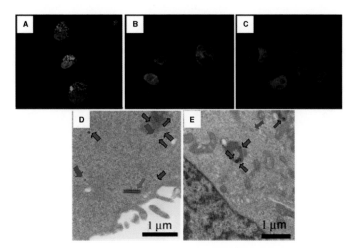

Figure 7.10 *Confocal microscope images of uptake of fluorescent silica nanotubes (SNTs) by MDA-MB-231 breast cancer cells. To elucidate endocytosis, uptake studies were carried out with 200 nm positively charged SNTs at 30° C (A), and in the presence of metabolic inhibitors of endocytosis, namely (B) sodium azide and (C) sucrose. Transmission electron microscope images showing uptake of 500 nm positively charged SNTs in MDA-MB-231 cells (D, E). Blue arrows indicate cross-sectional view and red arrows a horizontal view of nanotubes. Reprinted with permission from the American Chemical Society. Copyright 2008.*

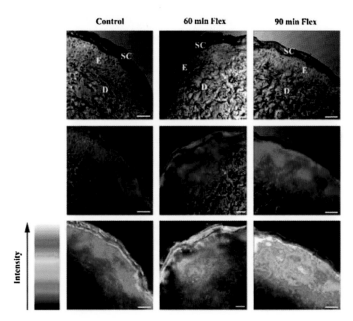

Figure 13.4 *Confocal scanning microscopy images of intact skin dosed with Baa-Lys(FITC)-NLS for 8 h. Top row: confocal-DIC channel image shows an intact stratum corneum (SC) and underlying epidermal (E) and dermal layers (D). Middle row: Baa-Lys(FITC)-NLS fluorescence channel (green) and confocal-DIC channel shows fullerene penetration through the epidermal and dermal layers of skin. Bottom row: fluorescence intensity scan showing Baa-Lys(FITC)-NLS penetration. All scale bars represent 50 μm (Rouse et al., 2007). Reprinted with permission from the American Chemical Society Copyright 2007.*

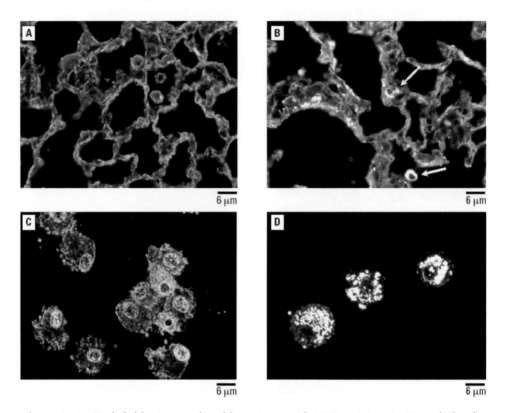

Figure 13.9 *Dark field micrographs of lung tissue with H&E staining (A, B) and alveolar macrophages prepared by cytospinning and H&E staining (C, D). (A, C) Sentinels and (B, D) animals subacutely exposed to TiO$_2$ nanoparticles with a primary particle size of 2–5 nm and necropsied immediately after the last exposure. Arrows point to TiO$_2$ nanoparticle-laden macrophages (Grassian et al., 2007). Reproduced with permission from Environmental Health Perspectives*

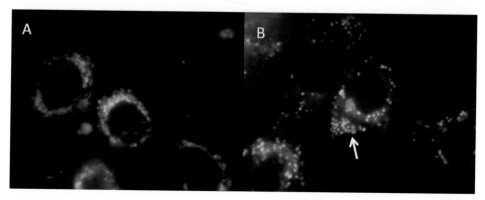

Figure 17.4 Uptake and localization of titanium dioxide nanoparticles in mouse keratinocytes. (A) Confocal image of mouse keratinocytes treated with TiO$_2$-Ru labelled nanoparticles and exposed for 24 h. The lysosomes were stained green and the nanoparticles fluoresced red; yellow indicated co-localization. The majority of nanoparticles were localizing in the lysosomes, further suggesting endocytosis as the mechanism of uptake. However, there were nanoparticles that were not in the lysosomes. (B) Confocal image of mouse keratinocytes treated with TiO$_2$-Ru labelled nanoparticles and exposed for 24 h. The mitochondria were stained green and the nanoparticles fluoresced red; yellow indicated co-localization. The mitochondria demonstrated less localization of nanoparticles when compared to the lysosomes; however, there were instances of nanoparticles localizing in the mitochondria (as shown by the white arrow), indicating that mitochondrial function can be altered by the presence of nanoparticles in a cell.

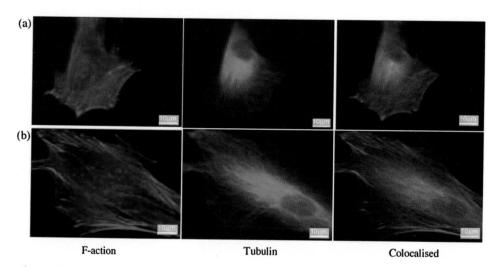

F-action   Tubulin   Colocalised

Figure 18.13 Human fibroblasts stained for F-actin (red), tubulin (green) and the nucleus (blue). The right-hand images show the signals combined into one image. (a) Control sample. (b) Treated with gelatin nanoparticles. Reprinted with permission from Gupta et al. (2004).

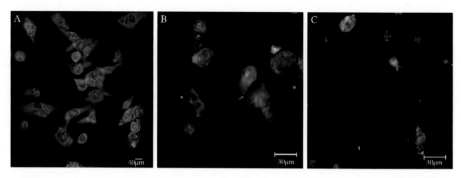

Figure 18.14 *Human hepatocellular carcinoma cell line HCCLM6 transfected with GFP using three different kinds of quaternized chitosan/pDNA nanoparticles as the transfection agent. (A) 60 % trimethylated chitosan oligomer; (B) chitosan (43–45 KDa, 87 %); and (C) chitosan (230 KDa, 90 % as vectors). Nuclei were stained red with PI. Reprinted with permission from Zheng* et al. *(2007).*

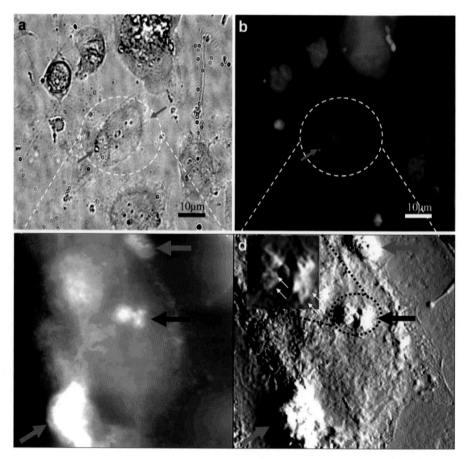

Figure 18.17 *Human ovarian cancer cells exposed to cicplatin-containing nano-liposomes. Red arrows indicate liposomes on the surface, and black arrows indicate liposomes that have been internalized. The images show (a) light microscopy (b) fluorescence microscopy and (c, d) two modes of AFM. Reprinted with permission from Ramachandran* et al. *(2006).*

as nmol/mg protein. The Bradford method is commonly followed by using a protein assay kit. Bovine serum albumin is routinely used as a standard.

18.5   Cell Imaging for *In Vitro* Nanotoxicity Analysis

An indispensable tool in evaluating nanoparticle toxicity has been *in vitro* cell imaging. Central to many areas of biological studies, cell imaging is nearly as old as the microscope itself. Advances in recent years, including the discovery of fluorescing nanoparticles, have particularly suited this discipline for determining the mechanisms of nanomaterial toxicity.

There are multiple ways in which cells may be imaged, and different information may be gained from each. One of the most universal methods used today is phase contrast microscopy, which utilizes the light of one phase passing through a sample and the light of a destructively interfering reference beam of a different phase to produce an image with exaggerated topographical differences. Differences in composition and density will also readily appear using this technique. It is important due to the fact that most monolayer cells and single cell organisms are only dimly visible, if not nearly transparent, under normal bright-field conditions. Phase contrast allows cell and nuclear membranes and some other subcellular structures (such as large vesicles) to be clearly seen. For these reasons, this has been a 'front line' observation technique used in NP toxicity studies to detect cell morphology changes. Figure 18.11 shows primary hepatocytes without and with the presence of CdSe nanocrystals (quantum dots), respectively (Derfus *et al.*, 2004). The morphology changes in the cells clearly indicate severe toxicity.

Differential interference contrast microscopy (DIC) is similar to phase contrast microscopy in that it renders nearly transparent samples easy to view without any form of staining. Also called Nomarski interference contrast microscopy (NIC), this technique

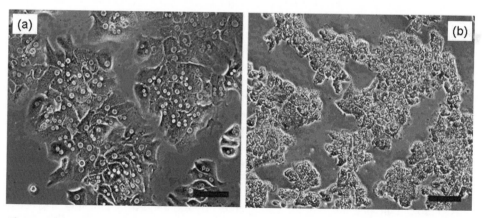

Figure 18.11   *Liver culture model exposed to CdSe quantum dots. (a) Control cells with their characteristic features of clearly visible cell boundaries, polygonal shape and clearly visible nuclei. (b) Cells exposed to CdSe nanoparticles, showing granular cytoplasm, blistering and unclear nuclei. The scale bar represents 100 μm. Reprinted from Derfus et al. (2004), with permission from the American Chemical Society. Copyright 2004*

Figure 18.12 *M21 human melanoma cells with internalized cholera toxin B-conjugated quantum dots. From left to right: DIC, confocal fluorescence, and overlaid images. Reprinted from Chakraborty* et al. *(2007), with permission from the American Chemical Society.*

uses slightly more complicated optics. Suffice to say that two polarizing filters and two Wollaston prisms are used to generate two beams of polarized light which are delivered to a cell sample, and recombined in such a way that there is interference between them. The result is that the cell sample gains a three-dimensional appearance, as though light were shining upon it at an oblique angle. This effect is particularly useful for showing details of cell surface morphology. Figure 18.12 shows a DIC image of M21 human melanoma cells, a confocal fluorescence image of quantum dots conjugated with cholera toxin B, and the two images combined (Chakraborty *et al.*, 2007).

In vitro wide field fluorescence imaging has yielded, to date, the greatest nanomaterial toxicology knowledge of all the imaging techniques. This is largely due to the diverse selection of fluorescent probes available today, yet it is due, in part, to the sensitivity of this technique and its relatively low cost. Fluorescence microscopy relies on the delivery of excitation light to the fluorescent probe(s) in the sample and the collection of the emitted fluorescent signal and its recording by a suitably sensitive camera. Care must be taken to avoid ultra-violet or excessively intense excitation light, as these will induce toxic effects of their own. Developed over the years for numerous purposes in cell biology, the selection of available fluorescent probes is both varied and selective. Organic dyes constitute the largest single collection, and allow a researcher to track and distinguish dead versus live cells; monitor cell membrane potential; measure ion concentrations, such as Ca^{2+} and Mg^{2+}, over time; view cell cytoskeletal proteins such as actin and tubulin; label DNA; monitor ATP release; monitor pH; and specifically stain and monitor organelles, such as nuclei, nucleoli, mitochondria, golgi apparatus, lysozymes, endoplasmic reticulum and other cell features such as lipid rafts. Very specific molecular cell elements have also been fluorescently labelled using immuno-labelling, where antibodies are pre-labelled with organic dyes. The antibodies then bind specifically to their antigens, effectively achieving molecular recognition in the placement of the fluorophores. By the fluorescent labelling of specific cell structures and imaging by fluorescent microscopy, it opens the door to identifying parts of cells not otherwise visible using techniques such as phase contrast microscopy or DIC. Figure 18.13 shows control and gelatin nanoparticle-exposed human fibroblasts. The cells were stained for F-actin, tubulin, and the nucleus in order to investigate 'cell adhesion, morphology, and cytoskeletal organization' (Gupta *et al.*, 2004). The selective labelling

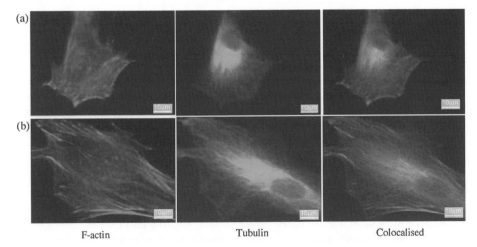

| F-actin | Tubulin | Colocalised |

Figure 18.13 *Human fibroblasts stained for F-actin (red), tubulin (green) and the nucleus (blue). The right-hand images show the signals combined into one image. (a) Control sample. (b) Treated with gelatin nanoparticles. Reprinted from Gupta* et al. *(2004), with permission from Elsevier. See color plate section.*

of these cell structures and the monitoring of these processes enables the visualization of changes in living cells caused by the presence of nanomaterials.

Beyond the scope of typical organic fluorescent dyes, there are valuable cell labels that can be used in fluorescence microscopy. One that has been most helpful is green fluorescent protein (GFP). This fluorescent marker is a protein subunit capable of emitting green fluorescence. It is produced by transfecting a population of cells with the DNA for GFP, so that it is added on to the end of a desired protein when that protein is expressed by the cell. Figure 18.14 shows the novel combination of using nanoparticles as the transfection vectors for GFP DNA (Zheng *et al.*, 2007).

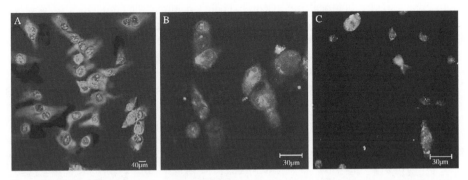

Figure 18.14 *Human hepatocellular carcinoma cell line HCCLM6 transfected with GFP using three different kinds of quaternized chitosan/pDNA nanoparticles as the transfection agent. (A) 60 % trimethylated chitosan oligomer; (B) chitosan (43–45 KDa, 87 %); and (C) chitosan (230 KDa, 90 % as vectors). Nuclei were stained red with PI. Reprinted from Zheng* et al. *(2007), with permission from Elsevier. See color plate section.*

By far the most fascinating fluorescent labels in the study of nanomaterial toxicity are the luminescent nanoparticles. To date, these particles have generally been made by one of three methods. The first method is the doping of a NP with fluorescent organic dyes. Silica NPs are commonly used for this purpose, as they are relatively transparent and serve to encapsulate the dye molecules while allowing the transmission of excitation and emission photons. The detection of a single fluorescent nanoparticle is limited only by the size of the dye-doped nanoparticles desired by the researcher and the sensitivity of the detection device used. Photobleaching – that is, the photodecomposition of a fluorophore – is reduced in dye-doped NPs, as the dye molecules are typically stabilized after being locked in the material matrix of the NP. The second method for producing luminescent NPs involves the use of semiconductor nanocrystals such as ZnSe, CdSe and CdTe. Termed 'quantum dots' shortly after their discovery, these particles are typically from 1.5 to 13.5 nm in diameter for visible wavelength emissions, while the wavelength of the emitted fluorescence is directly dependent upon the size of the nanocrystal. A blinking phenomenon characterizes their fluorescence. These fluorescent particles have several advantages over traditional organic dyes, including greater emission intensity, resistance to photobleaching, and a continuous excitation absorbance wavelength range, as opposed to the limited excitation range of organic dyes. Solubility and stability in water were at one time a problem, but those problems have been solved by coating the particles in a 'cap' layer, such as ZnS or CdS, and with a final coating layer of SiO_2 or a polymer. Such outer layers also prevent toxicity arising from the heavy metal particle cores. The third method of producing luminescent NPs involves the resonant scattering of visible or near-infrared light when a particle's surface plasmon oscillation is excited. This scattering is quite intense compared with fluorescence emissions, and the particles are not subject to photobleaching. For these experiments, gold nanoparticles are a prime example. Figure 18.15 shows noncancerous and cancerous cells after incubation with anti-epidermal growth factor receptor (EGFR)-labelled gold nanoparticles. The antibody binds to EGFR, which leads to an approximately 600 % greater labelling rate for the cancerous cells, and holds potential for cancer detection (El-Sayed *et al.*, 2005).

In vitro fluorescence imaging may also be performed on a specialized fluorescence microscope called a confocal microscope. There are three standard types of confocal microscopes, but they all deliver high-quality images using the same central technology. The essential difference between the wide field and confocal fluorescence microscopes

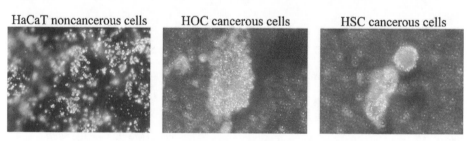

Figure 18.15 *One noncancerous and two cancerous types of cells exposed to anti-EGFR antibody-labelled gold nanoparticles. The gold nanoparticles exhibit surface plasmon light scattering. Reprinted from El-Sayed et al. (2005), with permission from the American Chemical Society.*

lies in the elimination of background noise. A confocal system limits excitation light to a single point on the sample at any given time, minimizing the excitation of fluorophores not in focus. Likewise, the emission light passes through a pinhole that serves to eliminate all light originating above or below the sample focal plane. The resulting images typically have both high sensitivity and resolution. Figure 18.12 shows a confocal fluorescence image of cholera toxin B-labelled quantum dots. Note the high definition of the signal compared with some of the wide field fluorescence images in Figure 18.13. Confocal systems were traditionally limited by speed, as images were obtained by scanning a sample point by point and row by row. As a result, samples were usually fixed, permeabilized and dyed. Later systems, incorporating spinning-disk geometries and especially laser scanning capabilities, allowed live cell imaging as well.

Beyond fluorescence and light scattering techniques, there are others ways to investigate the toxicity of nanomaterials with cells. One of these techniques is scanning electron microscopy (SEM). While SEM can deliver exceptionally detailed images of many different types of surface, it requires that the samples be fixed (dead). Figure 18.16 shows SEM images of Chang cells following exposure to 0.005 % benzalkonium chloride (BAC) (a positive control), chitosan NPs, and a control (de Campos *et al.*, 2004). Note how the NPs have not changed the morphology of the cells as the positive control has done.

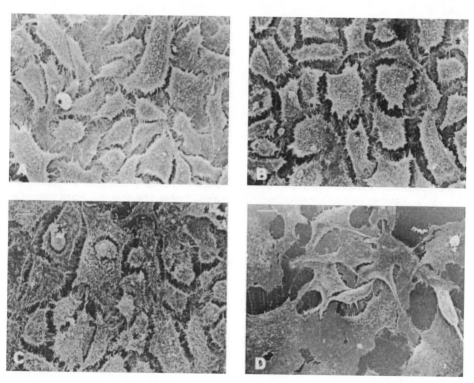

Figure 18.16 *SEM of Chang cells: (A) negative control; (B) 0.25 mg/ml chitosan nanoparticles; (C) 1 mg/ml chitosan nanoparticles; and (D) 0.005 % BAC as positive control. Reprinted from de Compos* et al. *(2004), with kind permission from Springer Science + Business Media.*

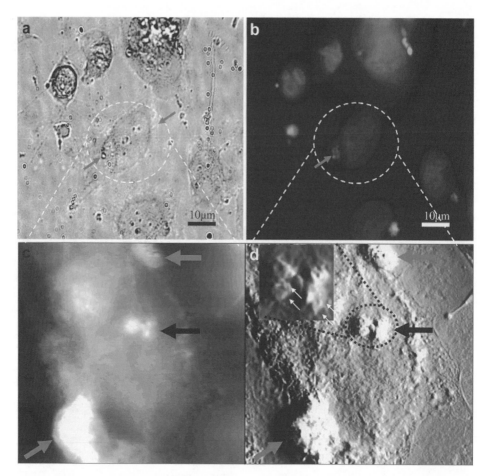

Figure 18.17   *Human ovarian cancer cells exposed to cicplatin-containing nano-liposomes. Red arrows indicate liposomes on the surface, and black arrows indicate liposomes that have been internalized. The images show (a) light microscopy; (b) fluorescence microscopy; and (c, d) two modes of AFM. Reprinted from Ramachandran et al. (2006), with permission from the American Chemical Society. See color plate section.*

Another fixed-cell alternative to fluorescence microscopy is atomic force microscopy (AFM). Its results are somewhat similar to SEM in that they are able to determine surface contours with excellent resolution. Figure 18.17 shows human ovarian cancer cells' internalization of small liposomes containing the anticancer drug cisplatin (Ramachandran *et al.*, 2006). The nanosized capsules aided in the delivery of the drug to the cells, as it has traditionally suffered from poor bioavailability. Note how Figure 18.16 combines the imaging advantages of (a) light microscopy, (b) fluorescence microscopy, and (c,d) AFM.

In closing, it is important to note that this is indeed only a brief overview of the progress made in utilizing *in vitro* cell imaging techniques for studying nanotoxicity. There remain a host of further examples and combinations of techniques that have shed light upon this important area of study. Besides the mammalian cell types presented here, studies have also been made of bacteria, yeast, and even plant species such as microalgae. The blossoming field of nanomaterial toxicology will undoubtedly rely heavily upon cell imaging in the future, as it makes great strides towards understanding the biological behaviors of nanomaterials.

18.6   Dosimetry for *In Vitro* Study

The expression of a dosimetry unit in nanotoxicity is a complicated issue. Traditionally, chemical dosimetry *in vitro* contains two factors: amount and time. Since time can be easily controlled, the amount of chemicals is the key factor for dosimetry. Also, most chemicals in solution have known physico-chemical properties and remain unchanged during the study process. Therefore, molarity is the most commonly used concentration unit in conducting many chemical reactions. In contrast, molarity may not be an appropriate concentration unit in conducting nanotoxicity studies, because nanoparticles are not really dissolved in solution. Dosimetry for nanoparticles *in vitro* is a problem of multiple dimensions, including not only the amount and time, but also particle characteristics, such as size, shape, density, agglomeration state and surface charge. The factors affecting a delivered dose *in vitro* from particles and media characteristics are shown in Table 18.4. More detailed information on dosimetry considerations in nanotoxicity testing can be found in published articles (Teeguarden *et al.*, 2007; Lison *et al.*, 2008).

Many preliminary histochemical analyses indicate that cell uptake of particles is a small fraction of the applied quantity. Moreover, at the suspension used, most nanoparticles do not precipitate from stock solutions over a long period of time (such as a 24 h dosing period). In this case, since the cells are exposed to a uniform suspension of nanoparticles, the use of μg/ml units to express the dosage does not seem inappropriate. Actually, this dosage unit has been used by many nanoparticle-related cytotoxicity studies (Gurr *et al.*, 2005; Hussain

Table 18.4   *The particle and media factors affecting delivered dose* in vitro. *Reproduced from Teeguarden* et al. *(2007) by permission of Oxford University Press*

	Size (nm)		Effect on Nanoparticle Transport	
	< 1000	1000	Diffusion	Gravitational settling
Material property				
Size	±	+	↓ with ↑ diameter	↑ With square of diameter
Shape	±	+	Uncertain	Spheres most efficient
Density	±	+	—	↑ With density
Surface chemistry	+	+	Agglomeration[a]	Agglomeration
Zeta potential[b]	+	+	Agglomeration	Agglomeration
Concentration	+	+	Agglomeration	Agglomeration
Media property				
Density	±	+	—	↓ with ↑ media density
Viscosity	±	+	↓ with ↑ viscosity	↓ with ↑ media viscosity

[a] Agglomeration refers to affects on diffusion and gravititational settling that are secondary to changes in size and shape due to agglomeration.
[b] A measure of particle charge.

et al., 2005; Sayes, 2006). Of course, the specific contributions of particle mass, particle number and particle specific surface area to the biological effects of nanoparticles are still under vigorous debate, and require both theoretical modelling and experimental validation.

18.7   Effect of Nanoparticle Morphology on the Results of Nanotoxicity Study

Morphology in materials science describes particle size, shape, texture, and phase distribution in physical objects. Researchers have often named nanoparticles after the real-world shapes that they might represent, such as nanospheres, nanoreefs, nanoboxes, nanotubes, and others. Many nano-factors can make nanomaterials very toxic, both medically and environmentally, among which nanoparticle morphology plays an essential role. Different nanosizes or crystal types are not only able to present different surface-to-volume ratios and surface active sites, but lead to an aggregation at a different level. All of these features can make the particles very reactive or catalytic during their interactions with the biological system.

The particle surface and interfaces are important components of nanoscale materials. As the morphology changes, the proportion of atoms at the surface may also change, relative to the proportion inside its volume. This may result in more reactive groups on a particle surface, and is likely to change the biological effects (more or less cytotoxic). In addition, it should be noted that, from a toxicological perspective, two different nanoparticle types may not be biologically equivalent even though the chemical components are the same. A typical example is that the cytotoxicity caused by amorphous silica is significantly different from that of crystal silica (Lin *et al.*, 2006a). The morphology differences can be variations in crystal structures, nanosizes, and aggregation status. These morphology differences are manifested in different pulmonary inflammatory and cytotoxic effects, ranging from benign to severe health impacts.

Although the morphology of particles, including particle size, is not likely to change a lot under cell culture conditions, one fact should not be ignored: nanoparticle toxicity assessments are affected by their solution dynamics; the nanoparticles tend to settle, diffuse and agglomerate over time in solution. These processes may change the nature of the particles and their transport to cells. These processes are hard to control because they are affected by the properties of the particles themselves, as well as the solution (viscosity, density, ion strength, and presence of peptides and proteins, etc.). For instance, particles of different morphology settle or diffuse at different rates, and these differences may lead to variations in transport to adherent cells in cell cultures. Consequently, these differences will result in variations in cytotoxicity. For this reason, before *in vitro* toxicity studies are conducted, particle morphology in solution needs to be accurately characterized as a prerequisite for implementing a study of nanoparticle toxicity. In addition, the nanoparticle morphology status should be considered when comparisons are made of the cytotoxicity of the different nanoparticles, to avoid misleading conclusions.

18.8   Data Analysis

In conducting an *in vitro* cytotoxicity study of nanoparticles, data analysis is one of the most important steps, especially when data are used to evaluate whether the two groups of cells or levels of biomarkers are significantly different. Due to the nature of cell culture systems, triplicate or even quadruplicate experiments should be designed for each experimental condition. After experimental data are acquired, statistical data analysis should be carried out. In general, every group of data is expressed as the mean $\pm$ standard deviation (SD) of three or four experiments. To compare whether two groups of data are significantly different, one-tailed unpaired Student's t test is often used for significance testing, using a P value of 0.05 or 0.01, based on the level of confidence being set. Detailed information on data analysis can be found in many books on statistics.

References

Adams LK, Lyon DY, Alvarez PJJ (2006) Comparative eco-toxicity of nanoscale TiO_2, SiO_2, and ZnO water suspensions. *Water Research* **40**, 3527–3532.

Bellucci S, Balasubramanian C, Bergamaschi A, Bottini M, Magrini A, Mustelin T (2006) Biomedical applications of carbon nanotubes and the related cellular toxicity. *NSTI Nanotech 2006, NSTI Nanotechnology Conference and Trade Show, Boston, MA, May 7–11, 2006* **2**, 217–220.

Block ML, Wu X, Pei Z, Li G, Wang T, Qin L, Wilson B, Yang J, Hong JS, Veronesi B (2004) Nanometer size diesel exhaust particles are selectively toxic to dopaminergic neurons: the role of microglia, phagocytosis, and NADPH oxidase. *FASEB Journal* **18**, 1618–1620, 1610 1096/fj 1604-1945fje.

Brown DM, Wilson MR, MacNee W, Stone V, Donaldson K (2001) Size-dependent proinflammatory effects of ultrafine polystyrene particles: a role for surface area and oxidative stress in the enhanced activity of ultrafines. *Toxicology and Applied Pharmacology* **175**, 191–199.

Chakraborty SK, Fitzpatrick JAJ, Phillippi JA, Andreko S, Waggoner AS, Bruchez MP, Ballou B (2007) Cholera toxin B conjugated quantum dots for live cell labeling. *Nano Lett* **7**, 2618–2626.

Cherukuri P, Bachilo SM, Litovsky SH, Weisman RB (2004) Near-infrared fluorescence microscopy of single-walled carbon nanotubes in phagocytic cells. *J Am Chem Soc* **126**, 15638–15639.

Chlopek J, Czajkowska B, Szaraniec B, Frackowiak E, Szostak K, Beguin F (2006) In vitro studies of carbon nanotubes biocompatibility. *Carbon* **44**, 1106–1111.

Cho SJ, Maysinger D, Jain M, Roeder B, Hackbarth S, Winnik FM (2007) Long-term exposure to CdTe quantum dots causes functional impairments in live cells. *Langmuir* **23**, 1974–1980.

Connor EE, Mwamuka J, Gole A, Murphy CJ, Wyatt MD (2005) Gold nanoparticles are taken up by human cells but do not cause acute cytotoxicity. *Small* **1**, 325–327.

Cui D, Tian F, Ozkan CS, Wang M, Gao H (2005) Effect of single wall carbon nanotubes on human HEK293 cells. *Toxicology Letters* **155**, 73–85.

de Campos AM, Diebold Y, Carvalho ELS, Sanchez A, Alonso MJ (2004) Chitosan nanoparticles as new ocular drug delivery systems: in vitro stability, in vivo fate, and cellular toxicity. *Pharm Res* **21**, 803–810.

Derfus AM, Chan WCW, Bhatia SN (2004) Probing the cytotoxicity of semiconductor quantum dots. *Nano Letters* **4**, 11–18.

Ding L, Stilwell J, Zhang T, Elboudwarej O, Jiang H, Selegue JP, Cooke PA, Gray JW, Chen FF (2005) Molecular characterization of the cytotoxic mechanism of multiwall carbon nanotubes and nano-onions on human skin fibroblast. *Nano Letters* **5**, 2448–2464.

Donaldson K, Tran CL (2002) Inflammation caused by particles and fibers. *Inhal Toxicol* **14**, 5–27.

Draper HH, Squires EJ, Mahmoodi H, Wu J, Agarwal S, Hadley M (1993) A comparative evaluation of thiobarbituric acid methods for the determination of malondialdehyde in biological materials. *Free Radical Biology & Medicine* **15**, 353–363.

El-Sayed IH, Huang X, El-Sayed MA (2005) Surface plasmon resonance scattering and absorption of anti-EGFR antibody conjugated gold nanoparticles in cancer diagnostics: Applications in oral cancer. *Nano Lett* **5**, 829–834.

Fiorito S, Serafino A, Andreola F, Bernier P (2006) Effects of fullerenes and single-wall carbon nanotubes on murine and human macrophages. *Carbon* **44**, 1100–1105.

Fischer Hans C, Chan Warren CW (2007) Nanotoxicity: the growing need for in vivo study. *Current Opinion Biotechnology* **18**, 565–571.

Flahaut E, Durrieu MC, Remy-Zolghadri M, Bareille R, Baquey C (2006) Investigation of the cytotoxicity of CCVD carbon nanotubes towards human umbilical vein endothelial cells. *Carbon* **44**, 1093–1099.

Fu W, Shenoy D, Li J, Crasto C, Jones G, Dimarzio C, Sridhar S, Amiji M (2005) Biomedical applications of gold nanoparticles functionalized using hetero-bifunctional poly(ethylene glycol) spacer. *Materials Research Society Symposium Proceedings* **845**, 223–228.

Goodman CM, McCusker CD, Yilmaz T, Rotello VM (2004) Toxicity of gold nanoparticles functionalized with cationic and anionic side chains. *Bioconjugate Chemistry* **15**, 897–900.

Green M, Howman E (2005) Semiconductor quantum dots and free radical induced DNA nicking. *Chemical Communications*, 121–123.

Gupta AK, Gupta M, Yarwood SJ, Curtis ASG (2004) Effect of cellular uptake of gelatin nanoparticles on adhesion, morphology and cytoskeleton organization of human fibroblasts. *J Controlled Release* **95**, 197–207.

Gurr JR, Wang AS, Chen CH, Jan KY (2005) Ultrafine titanium dioxide particles in the absence of photoactivation can induce oxidative damage to human bronchial epithelial cells. *Toxicology* **213**, 66–73.

Haslam G, Wyatt D, Kitos PA (2000) Estimating the number of viable animal cells in multi-well cultures based on their lactate dehydrogenase activities. *Cytotechnology* **32**, 63–75.

Helland A, Wick P, Koehler A, Schmid K, Som C (2007) Reviewing the environmental and human health knowledge base of carbon nanotubes. *Environmental Health Perspectives* **115**, 1125–1131.

Hirsch LR, Stafford RJ, Bankson JA, Sershen SR, Rivera B, Price RE, Hazle JD, Halas NJ, West JL (2003) Nanoshell-mediated near-infrared thermal therapy of tumors under magnetic resonance guidance. *Proc Nat Acad Sci USA* **100**, 13549–13554.

Huff TB, Hansen MN, Zhao Y, Cheng J-X, Wei A (2007) Controlling the cellular uptake of gold nanorods. *Langmuir* **23**, 1596–1599.

Hund-Rinke K, Simon M (2006) Ecotoxic effect of photocatalytic active nanoparticles on algae and daphnids. *Environmental Sci Pollution Res Inter* **13**, 225–232.

Hussain SM, Hess KL, Gearhart JM, Geiss KT, Schlager JJ (2005) In vitro toxicity of nanoparticles in BRL 3A rat liver cells. *Toxicology in Vitro* **19**, 975–983.

James WD, Hirsch LR, West JL, O'Neal PD, Payne JD (2007) Application of INAA to the build-up and clearance of gold nanoshells in clinical studies in mice. *J Radioanalytical Nuclear Chem* **271**, 455–459.

Jia G, Wang H, Yan L, Wang X, Pei R, Yan T, Zhao Y, Guo X (2005) Cytotoxicity of carbon nanomaterials: single-wall nanotube, multi-wall nanotube, and fullerene. *Environmental Sci Tech* **39**, 1378–1383.

Kam NWS, Jessop TC, Wender PA, Dai H (2004) Nanotube molecular transporters: internalization of carbon nanotube-protein conjugates into mammalian cells. *J Am Chem Soc* **126**, 6850–6851.

Kipen HM, Laskin DL (2005) Smaller is not always better: Nanotechnology yields nanotoxicology. *Am J Physiology* **289**, L696–L697.

Knaapen AM, Borm PJA, Albrecht C, Schins RPF (2004) Inhaled particles and lung cancer. Part A: mechanisms. *Int J Cancer* **109**, 799–809.

Kreyling WG, Semmler-Behnke M, Moeller W (2006) Health implications of nanoparticles. *J Nanoparticle Res* **8**, 543–562.

Lang JK, Gohil K, Packer L (1986) Simultaneous determination of tocopherols, ubiquinols, and ubiquinones in blood, plasma, tissue homogenates, and subcellular fractions. *Analytical Biochemistry* **157**, 106–116.

Lewinski N, Colvin V, Drezek R (2008) Cytotoxicity of nanoparticles. *Small* **4**, 26–49.

Li N, Sioutas C, Cho A, Schmitz D, Misra C, Sempf J, Wang M, Oberley T, Froines J, Nel A (2003) Ultrafine particulate pollutants induce oxidative stress and mitochondrial damage. *Environ Health Perspect* **111**, 455–460.

Lin W, Huang Y-W, Zhou X-D, Ma Y (2006a) In vitro toxicity of silica nanoparticles in human lung cancer cells. *Toxicology and Applied Pharmacology* **217**, 252–259.

Lin W, Huang Y-W, Zhou X-D, Ma Y (2006b) Toxicity of cerium oxide nanoparticles in human lung cancer cells. *Int J Toxicology* **25**, 451–457.

Lin W, Huang Y-W, Zhou X-D, Ma Y (2007) In vitro toxicity of silica nanoparticles in human lung cancer cells. *Toxicology and Applied Pharmacology* **217**, 252–259. Response from Dr Hoet. *Toxicology and Applied Pharmacology* **220**, 226.

Lin W, Stayton I, Huang Y-W, Zhou X-D, Ma Y (2008) Cytotoxicity and cell membrane depolarization induced by aluminum oxide nanoparticles in human lung epithelial cells A549. *Toxicological and Environmental Chemistry* **90**, 983–996.

Lin W, Xu Y, Huang C-C, Ma Y, Shannon KB, Chen D-R, Huang Y-W (2009) Toxicity of nano- and micro-sized ZnO particles in human lung epithelial cells. *Journal of Nanoparticle Research* **11**, 25–39.

Lison D, Thomassen LCJ, Rabolli V, Gonzalez L, Napierska D, Seo JW, Kirsch-Volders M, Hoet P, Kirschhock CEA, Martens JA (2008) Nominal and effective dosimetry of silica nanoparticles in cytotoxicity assays. *Toxicological Sciences* **104**, 155–162.

Loo C, Lin A, Hirsch L, Lee M-H, Barton J, Halas N, West J, Drezek R (2004) Nanoshell-enabled photonics-based imaging and therapy of cancer. *Technology in Cancer Research & Treatment* **3**, 33–40.

Loo C, Lowery A, Halas N, West J, Drezek R (2005) Immunotargeted nanoshells for integrated cancer imaging and therapy. *Nano Letters* **5**, 709–711.

Magrez A, Kasas S, Salicio V, Pasquier N, Seo JW, Celio M, Catsicas S, Schwaller B, Forro L (2006) Cellular toxicity of carbon-based nanomaterials. *Nano Letters* **6**, 1121–1125.

Manna SK, Sarkar S, Barr J, Wise K, Barrera EV, Jejelowo O, Rice-Ficht AC, Ramesh GT (2005) Single-walled carbon nanotube induces oxidative stress and activates nuclear transcription factor-kB in human keratinocytes. *Nano Letters* **5**, 1676–1684.

Monteiro-Riviere NA, Inman AO (2006) Challenges for assessing carbon nanomaterial toxicity to the skin. *Carbon* **44**, 1070–1078.

Muller J, Huaux F, Moreau N, Misson P, Heilier J-F, Delos M, Arras M, Fonseca A, Nagy JB, Lison D (2005) Respiratory toxicity of multi-wall carbon nanotubes. *Toxicology and Applied Pharmacology* **207**, 221–231.

Murr LE, Garza KM, Soto KF, Carrasco A, Powell TG, Ramirez DA, Guerrero PA, Lopez DA, Venzor J III (2005) Cytotoxicity assessment of some carbon nanotubes and related carbon nanoparticle aggregates and the implications for anthropogenic carbon nanotube aggregates in the environment. *Int J Environmental Research Public Health* **2**, 31–42.

Nel A, Xia T, Maedler L, Li N (2006) Toxic potential of materials at the nanolevel. *Science* **311**, 622–627.

Niidome T, Yamagata M, Okamoto Y, Akiyama Y, Takahashi H, Kawano T, Katayama Y, Niidome Y (2006) PEG-modified gold nanorods with a stealth character for in vivo applications. *Journal of Controlled Release* **114**, 343–347.

Oberdörster E (2004) Manufactured nanomaterials (fullerenes, C60) induce oxidative stress in the brain of juvenile largemouth bass. *Environmental Health Perspectives* **112**, 1058–1062.

Oberdörster G, Oberdörster E, Oberdörster J (2005) Nanotoxicology: an emerging discipline evolving from studies of ultrafine particles. *Environmental Health Perspectives* **113**, 823–839.

Panessa-Warren BJ, Warren JB, Wong SS, Misewich JA (2006) Biological cellular response to carbon nanoparticle toxicity. *Journal of Physics: Condensed Matter* **18**, S2185–S2201.

Pantarotto D, Briand J-P, Prato M, Bianco A (2004) Translocation of bioactive peptides across cell membranes by carbon nanotubes. *Chemical Communications* 16–17.

Pernodet N, Fang X, Sun Y, Bakhtina A, Ramakrishnan A, Sokolov J, Ulman A, Rafailovich M (2006) Adverse effects of citrate/gold nanoparticles on human dermal fibroblasts. *Small* **2**, 766–773.

Porter Alexandra E, Muller K, Skepper J, Midgley P, Welland M (2006) Uptake of C60 by human monocyte macrophages, its localization and implications for toxicity: studied by high resolution electron microscopy and electron tomography. *Acta Biomaterialia* **2**, 409–419.

Pulskamp K, Diabate S, Krug HF (2007) Carbon nanotubes show no sign of acute toxicity but induce intracellular reactive oxygen species in dependence on contaminants. *Toxicology Letters* **168**, 58–74.

Ramachandran S, Quist AP, Kumar S, Lal R (2006) Cisplatin nanoliposomes for cancer therapy: AFM and fluorescence imaging of cisplatin encapsulation, stability, cellular uptake, and toxicity. *Langmuir* **22**, 8156–8162.

Rouse JG, Yang J, Barron AR, Monteiro-Riviere NA (2006) Fullerene-based amino acid nanoparticle interactions with human epidermal keratinocytes. *Toxicology in Vitro* **20**, 1313–1320.

Salem AK, Searson PC, Leong KW (2003) Multifunctional nanorods for gene delivery. *Nature Materials* **2**, 668–671.

Sato Y, Yokoyama A, Shibata K-I, Akimoto Y, Ogino S-I, Nodasaka Y, Kohgo T, Tamura K, Akasaka T, Uo M, Motomiya K, Jeyadevan B, Ishiguro M, Hatakeyama R, Watari F, Tohji K (2005) Influence of length on cytotoxicity of multi-walled carbon nanotubes against human acute monocytic leukemia cell line THP-1 in vitro and subcutaneous tissue of rats in vivo. *Molecular BioSystems* **1**, 176–182.

Sayes CM (2006) The bio-nano interface: examining the interactions between water-soluble nanoparticles and cellular systems. Unpublished PhD thesis, Rice University.

Sayes CM, Fortner JD, Guo W, Lyon D, Boyd AM, Ausman KD, Tao YJ, Sitharaman B, Wilson LJ, Hughes JB, West JL, Colvin VL (2004) The differential cytotoxicity of water-soluble fullerenes. *Nano Letters* **4**, 1881–1887.

Service RF (2004) Nanotoxicology. Nanotechnology grows up. *Science* **304**, 1732–1734.

Shenoy D, Fu W, Li J, Crasto C, Jones G, DiMarzio C, Sridhar S, Amiji M (2006) Surface functionalization of gold nanoparticles using hetero-bifunctional poly(ethylene glycol) spacer for intracellular tracking and delivery. *Int J Nanomedicine* **1**, 51–57.

Shukla R, Bansal V, Chaudhary M, Basu A, Bhonde RR, Sastry M (2005) Biocompatibility of gold nanoparticles and their endocytotic fate inside the cellular compartment: a microscopic overview. *Langmuir* **21**, 10644–10654.

Shvedova A, Castranova V, Kisin E, Schwegler-Berry D, Murray A, Gandelsman V, Maynard A, Baron P (2003) Exposure to carbon nanotube material: assessment of nanotube cytotoxicity using human keratinocyte cells. *Journal of Toxicology and Environmental Health, Part A* **66**, 1909–1926.

Stix G (2001) Little big science. Nanotechnology. *Sci Am* **285**, 32–37.

Su C-H, Sheu H-S, Lin C-Y, Huang C-C, Lo Y-W, Pu Y-C, Weng J-C, Shieh D-B, Chen J-H, Yeh C-S (2007) Nanoshell magnetic resonance imaging contrast agents. *J Am Chem Soc* **129**, 2139–2146.

Teeguarden JG, Hinderliter PM, Orr G, Thrall BD, Pounds JG (2007) Particokinetics in vitro: dosimetry considerations for in vitro nanoparticle toxicity assessments. [Erratum to document cited in CA146:136487]. *Toxicological Sciences* **97**, 614.

Thomas ME, Blodgett DW, Hahn DV, Kaplan SG (2003) Characterization and modeling of the infrared properties of GaP and GaAs. *Proceedings of SPIE – The International Society for Optical Engineering* **5078**, 159–168.

Tkachenko AG, Xie H, Coleman D, Glomm W, Ryan J, Anderson MF, Franzen S, Feldheim DL (2003) Multifunctional gold nanoparticle-peptide complexes for nuclear targeting. *J Am Chem Soc* **125**, 4700–4701.

Tkachenko AG, Xie H, Liu Y, Coleman D, Ryan J, Glomm WR, Shipton MK, Franzen S, Feldheim DL (2004) Cellular trajectories of peptide-modified gold particle complexes: comparison of nuclear localization signals and peptide transduction domains. *Bioconjugate Chemistry* **15**, 482–490.

Wang H, Joseph JA (1999) Quantifying cellular oxidative stress by dichlorofluorescein assay using microplate reader. *Free Radical Biology & Medicine* **27**, 612–616.

Warheit DB (2004) Nanoparticles: health impacts? *Materials Today* **7**, 32–35.

Wick P, Manser P, Limbach LK, Dettlaff-Weglikowska U, Krumeich F, Roth S, Stark WJ, Bruinink A (2007). The degree and kind of agglomeration affect carbon nanotube cytotoxicity. *Toxicology Letters* **168**, 121–131.

Winters RA, Zukowski J, Ercal N, Matthews RH, Spitz DR (1995) Analysis of glutathione, glutathione disulfide, cysteine, homocysteine, and other biological thiols by high-performance liquid chromatography following derivatization by N-(1-pyrenyl)maleimide. *Analytical Biochemistry* **227**, 14–21.

Witzmann FA, Monteiro-Riviere NA (2006) Multi-walled carbon nanotube exposure alters protein expression in human keratinocytes. *Nanomedicine* **2**, 158–168.

Worle-Knirsch JM, Pulskamp K, Krug HF (2006) Oops they did it again! Carbon nanotubes hoax scientists in viability assays. *Nano Letters* **6**, 1261–1268.

Xu XH, Brownlow WJ, Kyriacou SV, Wan Q, Viola JJ (2004) Real-time probing of membrane transport in living microbial cells using single nanoparticle optics and living cell imaging. *Biochemistry* **43**, 10400–10413.

Yamawaki H, Iwai N (2006) Cytotoxicity of water-soluble fullerene in vascular endothelial cells. *Am J Physiology* **290**, C1495–C1502.

Zheng F, Shi X-W, Yang G-F, Gong L-L, Yuan H-Y, Cui Y-J, Wang Y, Du Y-M, Li Y (2007) Chitosan nanoparticle as gene therapy vector via gastrointestinal mucosa administration: Results of an in vitro and in vivo study. *Life Sci* **80**, 388–396.

19

In Vitro Human Lung Cell Culture Models to Study the Toxic Potential of Nanoparticles

Fabian Blank, Peter Gehr and Barbara Rothen-Rutishauser

19.1 Introduction

The rapid expansion of nanotechnology has resulted in the production of a variety of nanoparticles of different sizes, shapes, charge, chemistry, coating and solubility, and the number of products containing nanoparticles is increasing continuously. Despite the bright outlook for nanotechnology, there is an increasing concern that intentional or unintentional human exposure to some types of nanoparticles may lead to significant adverse health effects (Oberdörster *et al.*, 2007). The question about the size effects of nanoparticles is important, since the potential for exposure will increase as the quantity and types of nanoparticles used in the society grow. As the outcome of all these debates and concerns, nanotoxicology as a branch in toxicology research has emerged with the aim to investigate possible harmful effects of exposure to nanomaterials (Oberdörster *et al.*, 2005b; Nel *et al.*, 2006).

Nanoparticles are defined as particles with lengths in two or three dimensions greater than 1 nm and smaller than 100 nm, which is the same size covered by mainly combustion-derived ambient particles which have been termed ultrafine particles. Ultrafine particles and engineered nanoparticles have several characteristics in common, but there are also significant differences. Although there are obvious differences between ultrafine particles, which are polydispersed and have a chemically complex nature, and nanoparticles, which are in contrast monodispersed with precise chemically engineered characteristics, the same toxicological principles have been assumed (Oberdörster *et al.*, 2005b).

Nanotoxicity: From In Vivo *and* In Vitro *Models to Health Risks* Edited by Saura Sahu and Daniel Casciano
© 2009 John Wiley & Sons, Ltd

19.2   Interaction of Nanoparticles with Biological Systems

There is a lot of ongoing research on the mechanism of toxic action of nanoparticles. One important question is how the reactive surface of nanoparticles interacts with the environment in the body. A number of recent *in vitro* and *in vivo* studies on nanoparticle uptake and intracellular trafficking demonstrate that there is no single common uptake mechanism for nanoparticles (Unfried *et al.*, 2007; Muhlfeld *et al.*, 2008a). Furthermore, *in vitro* studies performed with cell cultures have confirmed the increased ability of nanoparticles to produce intracellular free radicals, which can then cause cellular damage. Generation of intracellular reactive oxygen species upon exposure to particles is nowadays considered relevant for possible nanoparticle-induced toxicity (Donaldson *et al.*, 2005). During recent years the number of papers on ongoing work in nanotoxicology has increased exponentially. However, it is difficult to draw any common conclusions about how nanoparticles interact with biological systems. This is in particular true because many studies include a broad range of nanoparticle concentrations and exposure times, making it difficult to determine whether the cytotoxicity observed is physiologically relevant. In addition, groups working with *in vitro* models use various cell lines as well as culturing conditions, which make direct comparisons between the available results difficult. There is a need to develop standardized *in vitro* test protocols, defining the most relevant particle doses and exposure scenarios for the nanomaterials. Several strategies to screen for the toxicity of nanoparticles have been proposed by investigators (Oberdörster *et al.*, 2005a; Borm *et al.*, 2006; Nel *et al.*, 2006).

19.3   Barriers to Nanoparticle Distribution

Nanoparticles may enter the human body via the skin, the gastrointestinal tract and the lung, despite the fact that these biological compartments act as barriers to the passage of nanosized materials. Because the lung is considered by far the most important portal of entry for nanoparticles into the human body (see below), we will mainly focus on the lung as a potential barrier for inhaled nanoparticles.

A series of structural and functional barriers protects the respiratory system against harmful and innocuous particulate material (Nicod, 2005). This is important as the internal surface area of the lungs is vast (alveoli and airways approximately total $150\,m^2$) (Gehr *et al.*, 1978a), facilitating broad access to the lung tissue. However, a number of epidemiological studies have shown that ambient particulate matter causes adverse health effects associated with increased pulmonary and cardiovascular morbidity and mortality (Pope *et al.*, 1995; Peters *et al.*, 1997; Schulz *et al.*, 2005). Recent studies indicate a specific toxicological role for inhaled combustion-derived ultrafine particles (Borm and Kreyling, 2004; Araujo *et al.*, 2008). Although there are contradictionary results about the translocation of particles into the blood circulation (Kreyling *et al.*, 2002; Nemmar *et al.*, 2002; Mills *et al.*, 2006; Wiebert *et al.*, 2006), it is currently accepted that the degree to which inhaled ultrafine particles and nanoparticles translocate to the circulation is rather small; however, evidence of cumulative effects of this translocation are lacking so far. These small particles can not only cross the blood–air barrier, but also, as shown in recent studies, have the ability to cross cellular membranes (Rothen-Rutishauser *et al.*, 2006, 2007). Hence they are able to enter different compartments of the cell like the endoplasmic reticulum, the mitochondria and even the nucleus. Once inside the cells, nanosized particles may therefore cause several biological responses including the generation of reactive oxygen species (Gonzalez-Flecha, 2004; Li *et al.*, 2008), the enhanced expression of pro-inflammatory cytokines (Muller *et al.*, 2005), and DNA strand breaks (Vinzents *et al.*, 2005; Schins and Knaapen, 2007).

There is a strong need for *in vitro* test systems to assess the toxicity of particulate matter and especially nanoparticles (Ayres *et al.*, 2008). Many lung cell culture models exist, providing an alternative to animal exposures for analysing the effects of different types of particles.

19.3.1 The Respiratory Tract: Main Portal of Entry for Ambient Particulate Matter

The respiratory tract has a large internal surface area and a very thin air–blood tissue barrier (Gehr *et al.*, 1978a,b), both of which are essential for an optimal gas exchange between air and blood by diffusion.

The respiratory tract can be subdivided into functionally and structurally distinct regions (Ochs and Weibel, 2008). Most proximal is the extrathoracic region, consisting of the nasal cavity, mouth, pharynx and larynx, followed by the tracheobronchiolar region, consisting of trachea, main bronchi, bronchi and bronchioles including terminal bronchioles. The main tasks of this part are air conditioning and air conduction. Ambient air, which is usually of lower temperature and humidity than the air in the lung, is efficiently modified and cleansed of much of the larger particulate material by mucociliary activity (fast particle clearance) before being conducted deeply into the lungs. The tracheobronchiolar region is followed by the proximal part of the alveolar-interstitial region, consisting of the respiratory bronchioli with only a few alveoli apposed, the task of which is air conduction, some gas exchange and slow clearance of particulate material. The following distal part of the alveolar-interstitial region consists of the most peripheral airways, the alveolar ducts with their 'walls' completely covered with alveoli (i.e. alveolar entrances), and the alveolar sacs (i.e. alveolar ducts with alveoli closing the end of the terminal ducts), including the interstitial connective tissue. The main task of this region is the gas exchange; particles are cleared only very slowly from this region (Gehr, 1994).

The deposition of particles in the lung is size-dependent. Besides the geometry of the airways and the breathing pattern, the particle size is important for deposition and clearance studies in the respiratory tract. Significant amounts of ultrafine particles are deposited in the most peripheral region of the lung, in the alveoli, but to a considerable extent also in the extrathoracic airways (Oberdörster *et al.*, 2005b). The deeper the particles are deposited in the lung, the longer it takes to clear them and the higher is the probability of adverse health effects due to particle–tissue and particle–cell interactions.

19.3.2 Particle Doses

Characterization of a dose of nanomaterials for toxicological investigation is very complex. With nanoparticles a variety of other material attributes have to be considered, including size and size distribution, shape, crystallinity, porosity, surface roughness, solubility, surface area, state of dispersion, surface chemistry and many other physico-chemical properties. Among all these, four parameters are defined to be relevant for nanoparticles or ultrafine particles when assessing their toxicity: particle number, length, surface area and mass concentration (Maynard and Aitken, 2007).

19.4   Models to Study Particle–Cell Interactions

Understanding the functional and pathological disorders induced in the respiratory tract by particles requires the investigation of the direct effects of these particulate pollutants on the state and activity of lung cells. So far, three approaches have been used: animal experiments, *ex vivo* studies of cells of isolated lung or biopsies, and *in vitro* cell culture systems to study the effect of pollutants under controlled conditions (Aufderheide, 2005).

The exposure of animals, isolated lungs or cell cultures to particles requires knowledge about several methodological problems, which are described below. To study the interactions of particles with cells, the interplay between cells upon particle exposure, as well as the final toxicity potential of the particles, the use of *in vivo* as well as *in vitro* models can be very powerful (Rothen-Rutishauser *et al.*, 2008a). However, the advantages and disadvantages of each model type need to be considered.

19.4.1   Animal Models

Study of the inhalation of aerosolized particles can be performed by whole body, head/nose/mouth-only or lung-only exposure (Sakagami, 2006; Muhlfeld *et al.*, 2008b). Whole body exposures are closest to environmental exposure and are less stressful for the animals than other exposure types, and are therefore recommended for chronic exposures. Head/nose/mouth-only exposure guarantees more efficient and controllable doses and allows testing of parameters (e.g. blood samples) during exposure. But it causes stress to the animals and cuts them off from their food and water supply for the time of exposure. The technically demanding lung-only exposures by instillation using intubation or tracheotomy provide the most precise dosage but may evoke inflammatory responses in animal models like rats (Brown *et al.*, 2001; Hohr *et al.*, 2002; Dick *et al.*, 2008; Warheit *et al.*, 2008) and mice (Andre *et al.*, 2006; Stoeger *et al.*, 2006). Furthermore, the lack of systemic and autonomous nervous system reactions may limit the significance of the results (Gardner and Kennedy, 1993; Pauluhn and Mohr, 2000; Pauluhn, 2005).

Although artificial, specialized techniques such as intratracheal instillation containing fluid are frequently used. However, they should not be regarded as an adequate substitute for inhalation studies as they do not represent the actual event of exposure (Gardner and Kennedy, 1993).

19.4.2   *Ex Vivo* Models

In one of the most commonly used methods, the isolated perfused lung is kept separated from the body in an artificial thorax system under experimental conditions (Sakagami, 2006). This model has been used to study the transport of pharmaceutically relevant compounds across the air–blood barrier. The disadvantage of such *ex vivo* models is the limited life span of only a few hours. Moreover, these models are afflicted with a very high grade of complexity (Steimer *et al.*, 2005; Sakagami, 2006). For example, particle translocation across the alveolar capillary barrier has been evaluated in an isolated perfused rabbit lung model, where it has been found that polystyrene nanoparticles (24, 110 or 190 nm) do not diffuse through the barrier. This result, which is different to *in vivo* findings, has been explained by the absence of hemodynamic factors, the absence of the lymph flow and inflammatory cells (Nemmar *et al.*, 2005).

19.4.3   Cell Culture Models of the Lung

19.4.3.1   General Aspects

Although animal models are essential to study the effects of ultrafine particles or nanoparticles in a whole organism, it is often very difficult to focus on or elucidate one specific mechanism of particle–cell interaction when working with animals, especially when an experiment has to be conducted in a standardized and well-characterized environment according to the needs of the researcher. During recent years there have been sustained efforts to complement or even replace animal experiments by cell culture approaches in many scientific fields. One of the major advantages of *in vitro* research is that, compared with animal models, cellular and subcellular functions (i.e. cell growth, cell interactions or metabolism), as well as the underlying molecular pathways can be studied with more ease in a simplified, direct biological model system, provided that the *in vitro* model is established appropriately to focus exactly upon the mechanism of interest. Furthermore, *in vitro* models provide the possibility of investigating toxic effects on human cells in extensive studies, which cannot be conducted *in vivo*. Cultured human and animal cells can be better controlled and therefore yield more reproducible data than *in vivo* systems, but require high standardization to maximize reproducibility. Guidelines for good cell culture practice are required, including control of the starting material, i.e. the cultured cells, the culture medium, and the culture substratum (Gstraunthaler and Hartung, 2002; Gruber and Hartung, 2004). However, one has to bear in mind that cell cultures are systems isolated from the *in vivo* microenvironment and may behave in a way which is different from the *in vivo* situation.

The cells may be used from freshly isolated tissue (primary cultures) or may stem from an immortalized cell line (secondary cultures). Both systems have advantages and disadvantages. Primary cultures isolated from animal tissue represent a heterogeneous population of different cell types, which show a relatively high differentiation level, which can become very similar to the *in vivo* situation. But even cells from primary cultures may show an altered phenotype due to the *in vitro* microenvironment, which lacks a complete pattern of stimuli (i.e. specific cytokines, growth factors, interactions with other cells and the extra-cellular matrix) affecting the cells in a particular *in vivo* microenvironment. Furthermore, primary cultures face limitations such as the lack of availability of normal human airway tissue, the limited number of cells which can be received during each isolation, and a certain donor variation (Gstraunthaler and Hartung, 2002).

Cell lines are homogeneous and more stable, and hence more reproducible. The disadvantage of cell lines, however, is that they retain little phenotypic differentiation. Therefore, a certain cell line should be selected very carefully with respect to the requirements needed for a particular study (Rothen-Rutishauser *et al.*, 2008a). Nevertheless, if cell cultures are used properly, they represent a sophisticated and reproducible system which can be used as a first step towards understanding how an agent will react in the body.

19.4.3.2   Cell Culture Models

Selected cell culture models. The complex nature of the lung architecture, which includes very distinct anatomical regions, presupposes the use of different *in vitro* models. In a nanoparticle exposure scenario, an *in vitro* model should highlight the most important characteristics of its corresponding region in the respiratory tract (see also section 19.3.1). We will focus our discussion on *in vitro* models which consist of human primary cells and cell lines derived from the airway- and alveolar epithelium, because of the high frequency of their use in nanoparticle cytotoxicity studies (Table 19.1). In addition, we describe some first nanotoxicology assays performed with immune cells (i.e. macrophages and dendritic cells) as well as with fibroblasts and mesothelial cells (Table 19.1).

Table 19.1   *Human cell culture models for nanotoxicological studies.*

Cell culture type	References
Primary cultures • Monocyte-derived macrophages • Monocyte-derived dendritic cells	Waldman *et al.* (2007) Blank *et al.* (2007); Rothen-Rutishauser *et al.* (2005, 2007)
Cell lines Airway epithelial cells • Calu-3 • 16HBE14o- • BEAS-2B	 Bivas-Benita *et al.* (2004); Grenha *et al.* (2007); Rotoli *et al.* (2008) Brzoska *et al.* (2004); Holder *et al.* (2008b) Herzog *et al.* (2007); Jang *et al.* (2006); Park *et al.* (2007); Veranth *et al.* (2007)
Alveolar epithelial cells • A549 • Immortalized human alveolar type 2 cells with alveolar type 1 phenotype	 Duffin *et al.* (2007); Park *et al.* (2007); Stearns *et al.* (2001) Kemp *et al.* (2008)
Macrophages • THP-1	 Chen *et al.* (2006); Ece *et al.* (2008); Goulaouic *et al.* (2008); Wottrich *et al.* (2004)
Fibroblasts • MRC-9	 Limbach *et al.* (2005)
Mesothelial cells • MSTO-211H	 Kaiser *et al.* (2008); Wick *et al.* (2007)
3D cultures • 3D aggregates of A549 cells • Triple cell co-culture model (epithelial cells, macrophages, dendritic cells)	 Carterson *et al.* (2005) Rothen-Rutishauser *et al.* (2007, 2008a, 2008b)

Primary cell cultures. Currently, there are few data available from nanotoxicity studies using human primary cultures, such as lung epithelial cells. One reason might be that the use of primary cell cultures has a number of disadvantages, which are summarized in section 19.4.3.1. One major problem might be the limited availability of human material. In addition, the establishment of a primary culture, which usually requires a longer culture time and the addition of a complex mixture of agents like extracellular matrix proteins, cytokines and growth factors, often emerges as too laborious for an extended nanotoxicity study, which is done much more quickly using the easily controllable and reproducible cell lines. Also, for screening studies the cell number available for primary cell cultures is too small.

However, there is a growing body of literature on the effects of nanoparticles on human blood monocyte-derived and *in vitro* differentiated cells like macrophages and dendritic cells, which represent the cells of the immune system. During recent years, several methods for generating dendritic cells and macrophages have been developed. In 1994, Sallusto and Lanzavecchia published their findings about the generation of human dendritic cells by culturing dendritic cell precursors such as peripheral blood monocytes in the presence of two cytokines (i.e. GM-CSF and IL-4) (Sallusto and Lanzavecchia, 1994). This has since become the most widely accepted protocol (Sallusto *et al.*, 1995). Macrophages can be generated from the same precursors but without supplementation of the growth factors.

Using this method of *in vitro* cell differentiation, monocyte-derived macrophages were used together with endothelial cells to demonstrate that carbon nanoparticles with surface-bound iron promote an inflammatory response in the macrophages, which induces an inflammatory phenotype in endothelial cells (Waldman *et al.*, 2007). In another study, monocyte-derived macrophages were used to visualize the uptake and subsequent in-tracellular location of C60 using energy-filtered transmission electron microscopy and electron tomography (Porter *et al.*, 2007). Furthermore, we are using monocyte-derived macrophages and dendritic cells in monocultures (Figure 19.1) or in an *in vitro* model of the human airway wall to study nanoparticle-cell interactions (Rothen-Rutishauser *et al.*, 2005, 2007; Blank *et al.*, 2007), which will be described in more detail later in this chapter.

Airway epithelial cell lines. Among the most popular human airway epithelial cell lines are the Calu-3 and 16HBE14o lines. The Calu-3 cell line is of human origin, commercially available from the American Type Culture Collection, and displays epithelial morphology as well as adherent growth. The presence of tight junctions and secretory activity makes the Calu-3 cell line a promising tool for pulmonary drug absorption studies (Steimer *et al.*, 2005), but it is frequently used for nanotoxicity studies with different nanoparticles like carbon nanotubes (Rotoli *et al.*, 2008), chitosan nanoparticles (Grenha *et al.*, 2007), and poly (D,L-lactide-co-glycolide) nanoparticles (Bivas-Benita *et al.*, 2004).

16HBE14o, immortalized and transformed by the SV40 large T-antigen bronchial epithelial cell line, is a normal human airway epithelial cell line and only available as a gift from Dieter Gruenert (Cardiovascular Research Institute, University of California, San Francisco). The cells form polarized monolayers with extensive tight junctional belts (Forbes, 2000; Steimer *et al.*, 2005), and when they are grown on collagen supports at an air–liquid interface they retain important properties of differentiated airway epithelial cells (Cozens *et al.*, 1994). This cell line has recently been used for toxicity studies with diesel exhaust ultrafine particles (Holder *et al.*, 2008a) and biodegradable nanoparticles (Brzoska *et al.*, 2004).

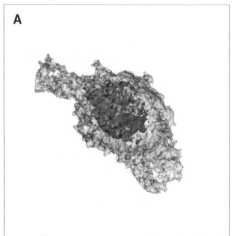
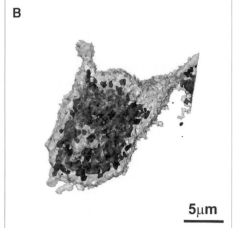

Figure 19.1 *Nanoparticle uptake by monocyte-derived macrophages and dendritic cells visualized by laser scanning microscopy. Primary cultures of monocyte-derived macrophages (A) or dendritic cells (B) were grown for 7 days and then incubated for 24 h with fluorescently labelled polystyrene particles (50 nm). After fixation the cells were stained for CD14 (macrophages (A)) or CD86 (dendritic cells (B)), and visualization of particles was done using a deconvolution algorithm. When the surface rendering (white) is made transparent the intracellular particles (black) can be seen.*

The BEAS-2B cell line was derived from normal human epithelial cells immortalized using the adenovirus 12-simian virus 40 hybrid virus (Reddel *et al.*, 1988) and is available from the American Type Culture Collection. The cells have often been used to study airway epithelial structure and function; however, they do not form tight junctions (for a review see Forbes, 2000). BEAS-2B cells have recently been used for cytotoxicity studies using cerium oxide nanoparticles (Park *et al.*, 2007), carbon nanomaterials (Herzog *et al.*, 2007), nanosized metal oxide and soil dust particles (Veranth *et al.*, 2007), and secondary organic aerosols coated on magnetic nanoparticles (Jang *et al.*, 2006).

Alveolar epithelial cell lines. The cell line A549, which originates from human lung carcinoma (Lieber *et al.*, 1976), belongs to the most well-characterized and widely used *in vitro* model (Foster *et al.*, 1998). It is also available from the American Type Culture Collection, and it has been shown that the A549 cells have many important biological properties of alveolar epithelial type II cells (e.g. membrane-bound inclusions, which resemble lamellar bodies of type II cells) (Shapiro *et al.*, 1978). Other ultrastructural characteristics common to type II cells have also been described, such as distinct polarization, tight junctions and extensive cytoplasmic extensions (Stearns *et al.*, 2001). A549 cells are very frequently used for nanoparticle cell interaction (Figure 19.2) and nanotoxicity studies. The biological effects of metallic nanoparticles (Stearns *et al.*, 2001; Duffin *et al.*, 2007; Limbach *et al.*, 2007; Park *et al.*, 2007) on A549 cells are described in many studies, and we have only cited some of the most recent publications.

Just recently the immortalization of human type II cells for the use in nanotoxicity studies has been reported by Kemp and co-workers. This new cell line exhibits a type

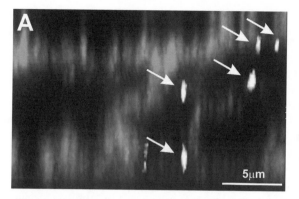 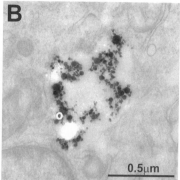

Figure 19.2   *Intracellular nanoparticle localisation in A549 cells. A549 cells were grown for 7 days and incubated for 24 h with fluorescently labelled polystyrene nanoparticles (50 nm) (A) or titanium dioxide nanoparticles (30 nm) (B). The polystyrene nanoparticles in (A) (white arrows) could be visualized in A549 cells stained for F-Actin (grey) by laser scanning microscopy combined with digital image restoration (Rothen-Rutishauser et al., 2007). The titanium dioxide nanoparticles were visualized by transmission electron microscopy and identified by electron energy loss spectroscopy (Rothen-Rutishauser et al., 2007, 2008b). The circles mark the region where the element analysis was performed.*

I-like phenotype, no longer expresses alkaline phosphatase and pro-surfactant protein C, but shows enhanced levels of caveolin-1 and RAGE (receptor for advanced glycation endproducts). The uptake of latex particles has been studied with these cells, and the cell line is postulated to be important for particle translocation studies (Kemp *et al.*, 2008).

Macrophage cell lines. Although the use of primary human macrophages derived from human-blood monocytes is more common for nanotoxicology studies, there are some studies described using the human monoctye/macrophage cell line THP-1 to test the toxicity of nanoparticles (Chen *et al.*, 2006; Ece *et al.*, 2008) or ultrafine particles (Wottrich *et al.*, 2004; Goulaouic *et al.*, 2008). This cell line is available from the American Type Culture Collection.

Fibroblast cell lines. The human lung fibroblast cell line MRC-9, also available from the American Type Culture Collection, has been shown to be a suitable cell type to study the uptake of cerium dioxide nanoparticles (Limbach *et al.*, 2005).

Mesothelial cell lines. Since the discovery of carbon nanotubes (Iijima, 1991), the fullerene-related graphite cylinders have initiated rapidly growth in product development in nanotechnology. However, a recent *in vivo* study has shown that when exposing the mesothelial lining of the body cavity to long multiwalled carbon nanotubes, this results in asbestos-like pathogenic behavior (Poland *et al.*, 2008). There are also initial *in vitro* studies with the human mesothelial-derived cell line MSTO-211H (available from the American Type Culture Collection), which has been used to study carbon nanotube toxicity *in vitro* (Wick *et al.*, 2007; Kaiser *et al.*, 2008).

19.4.3.3   Air–Liquid Cultures

In air-exposed cell cultures the interaction of particles with cells can be studied in an environment that, compared with immersed cultures, more closely mimics the *in vivo* situation. A fact of particular importance is that the cells in the lung are covered by a very thin liquid lining layer with a molecular surfactant film at the air–liquid interface. Surfactant plays an important role in particle displacement and retention (Gehr *et al.*, 1990, 1996; Schürch *et al.*, 1990). As an alternative to just adding the particles suspended in medium to the immersed cells, particles can be nebulised (as a suspension or as a dry powder) over the air-exposed system using a spraying device (e.g. a Microsprayer®) (Blank *et al.*, 2006) or an exposure chamber with an integrated aerosol generator (Tippe *et al.*, 2002). Another experimental approach working with an exposure device based on a cell exposure system is the CULTEX system, which was established by Aufderheide and co-workers (Aufderheide and Mohr, 2000). The cultivation of epithelial cells on permeable supports allows the culture medium to be kept on either side of the cultured epithelium separate, which leads to an increased differentiation of the cultured cells (Handler *et al.*, 1989). Furthermore, the medium can be removed from the upper side to expose the cells to air on one side and to make them fed from the medium in the chamber underneath (Voisin *et al.*, 1977a,b). The air–liquid culture technique has been described in different cell culture models (Aufderheide and Mohr, 2000; Ehrhardt *et al.*, 2002; Mathia *et al.*, 2002; Ritter *et al.*, 2004; Blank *et al.*, 2006). In recent particle–cell interaction studies where air-exposed A549 cells were used (Aufderheide *et al.*, 2003; Ritter *et al.*, 2004), the cells were exposed to air just at the time when test substances were applied. However, it has been shown that air-exposed monolayers of A549 cells, as type II-like cells, secrete surfactant in the liquid lining layer covering the cells. The released surfactant lowered the surface tension of the hypophase within 24 h down to a level of about 28 mN/m (Blank *et al.*, 2006), a value that is very close to *in vivo* values determined on the tracheal wall of anesthetised dogs, sheep or horses (Im Hof *et al.*, 1997; Schürch *et al.*, 1990). A549 cells exposed to air for at least 24 h are therefore a powerful tool to study the effects of surfactant on particle deposition and on the subsequent particle–cell interaction *in vitro*. However, although particle–surfactant interaction is a very important issue which also needs to be considered in nanotoxicology, very few studies addressing this topic have been published yet. Other studies with airway epithelial cells cultured in a biphasic manner also showed well-differentiated cell cultures (Ehrhardt *et al.*, 2002; Mathia *et al.*, 2002). In these studies the air-exposed cultures exhibited a clear epithelial morphology and integrity similar to *in situ* conditions.

That the choice of the exposure method is of great importance was recently shown by Holder and co-workers. Completely different dose responses were found when either air-liquid or submerged cultures of 16HBE14o cells were exposed to aerosolized diesel particles (Holder *et al.*, 2008b). Although exposure by each method caused a slight decrease in cell viability and increased levels of IL-8, the response to air-liquid exposure occurred at doses several orders of magnitude lower than exposure to particles in suspension.

19.4.3.4   3D Models

Studies have shown that when cells are removed from their host tissue and grown as monolayers on impermeable surfaces, they undergo dedifferentiation and lose specialized functions, which is thought to be, in part, due to the disassociation of cells from their native three-dimensional (3D) tissue structure *in vivo* (Freshney, 2000). In order to preserve the properties of *in vivo* tissue, a 3D lung aggregate model has recently been established, whereby A549 cells were cultured on a rotating-wall vessel bioreactor. The 3D aggregates were compared with the conventional A549 monolayers. An increased expression of epithelial cell-specific markers and decreased expression of cancer-specific markers was found in the 3D culture. Therefore, it has been suggested that the 3D models represent a more physiologically relevant model (Carterson *et al.*, 2005) .

Not only is 3D structure important, but co-cultures of different cell types have also been shown to have an influence on the outcome of the results. For instance, in the airway mucosa epithelial cells, macrophages and dendritic cells continuously cross-talk *in vivo* through intercellular signalling to maintain homeostasis and to coordinate immune responses (Roggen *et al.*, 2006). *In vitro* models of mucosal surfaces are now in use, particularly to characterize the mechanism of particle sampling by intraepithelial dendritic cells (Rescigno *et al.*, 2001a,b; Rothen-Rutishauser *et al.*, 2005). Recently we developed a triple cell co-culture *in vitro* model of the human airway wall, to study the cellular interplay and the cellular response of epithelial cells, human blood monocyte-derived macrophages and dendritic cells to particles (Figure 19.3) (Blank *et al.*, 2007; Rothen-Rutishauser *et al.*, 2005, 2007, 2008a). In this model, monolayers of two different epithelial cell lines, A549 (Lieber *et al.*, 1976) and 16HBE14o epithelia (Forbes, 2000), were grown on a microporous membrane in a two-chamber system. After isolation and differentiation of human blood-derived monocytes into macrophages and dendritic cells, they were added at the apical side and at the basal side of the epithelium, respectively. After the triple cell co-culture was established, cell densities of macrophages and dendritic cells within the culture were quantified using the specific surface markers CD14 and CD86 for the labelling of macrophages and dendritic cells, respectively; the quantitative occurrence of macrophages and dendritic cells resembled very closely the *in vivo* situation (Blank *et al.*, 2007). After its thorough evaluation, this model was exposed to particles (either airborne or suspended in medium) of different materials (polystyrene, titanium dioxide) and of different sizes (≤ 1 μm) (Blank *et al.*, 2007; Rothen-Rutishauser *et al.*, 2007, 2008b). Translocation and cellular localization of particles were studied, as well as the effects of particles on cellular interplay and signalling. Recently, we have shown that dendritic cells and macrophages collaborate as sentinels against fine particles by building a transepithelial interdigitating network of cell processes (Blank *et al.*, 2007), whereas the nanosized material has different translocation characteristics (Rothen-Rutishauser *et al.*, 2007). This triple cell co-culture system might be used for other epithelial models, for example, the gastrointestinal tract or the skin, by replacing the lung epithelial cells by any other epithelial cell type.

19.5   Conclusion

There is a need to use cell cultures to assess as a first step the toxicity of combustion-derived ultrafine particles, and of already-existing or newly developed engineered nanoparticles. The discussed cell culture models may help to elucidate mechanisms of particle–cell interactions in the lung, which are assumed to induce toxic reactions. Despite a number of limitations, cell cultures, and in particular cell lines, offer the opportunity for high-throughput

Figure 19.3   *Laser scanning microscopy images of the triple cell co-culture model. Epithelial cells (dark grey, volume rendering), macrophages (white, surface rendering; black arrows), and dendritic cells (dark grey, surface rendering; white arrow) are shown. The same data-set is shown from top (A), from bottom (B), and without epithelial cells from top (C). Reproduced and adapted from Rothen-Rutishauser et al. (2008b), with permission from ALTEX.*

screening of large numbers of newly developed particles, in special nanoparticles, within a short time. An essential disadvantage is that cell culture models often do not exhibit all the differentiated and functional characteristics of the corresponding native epithelium or entire organ. Therefore, a model for a certain developmental problem or scientific question should be selected very carefully, with consideration being given to its limitations, the experimental design and the interpretation of results.

References

Andre E, Stoeger T, Takenaka S, Bahnweg M, Ritter B, Karg E, Lentner B, Reinhard C, Schulz H, Wjst M (2006) Inhalation of ultrafine carbon particles triggers biphasic pro-inflammatory response in the mouse lung. *Eur Respir J* **28**, 275–285.

Araujo JA, Barajas B, Kleinman M, Wang X, Bennett BJ, Gong KW, Navab M, Harkema J, Sioutas C, Lusis AJ, Nel AE (2008) Ambient particulate pollutants in the ultrafine range promote early atherosclerosis and systemic oxidative stress. *Circ Res* **102**, 589–596.

Aufderheide M (2005) Direct exposure methods for testing native atmospheres. *Exp Toxicol Pathol* **57** (Suppl. 1), 213–226.

Aufderheide M, Mohr U (2000) CULTEX – an alternative technique for cultivation and exposure of cells of the respiratory tract to airborne pollutants at the air/liquid interface. *Exp Toxicol Pathol* **52**, 265–270.

Aufderheide M, Knebel JW, Ritter D (2003) Novel approaches for studying pulmonary toxicity in vitro. *Toxicol Lett* **140–141**, 205–211.

Ayres JG, Borm P, Cassee FR, Castranova V, Donaldson K, Ghio A, Harrison RM, Hider R, Kelly F, Kooter IM, Marano F, Maynard RL, Mudway I, Nel A, Sioutas C, Smith S, Baeza-Squiban A, Cho A, Duggan S, Froines J (2008) Evaluating the toxicity of airborne particulate matter and nanoparticles by measuring oxidative stress potential – a workshop report and consensus statement. *Inhal Toxicol* **20**, 75–99.

Bivas-Benita M, Romeijn S, Junginger HE, Borchard G (2004) PLGA-PEI nanoparticles for gene delivery to pulmonary epithelium. *Eur J Pharm Biopharm* **58**, 1–6.

Blank F, Rothen-Rutishauser BM, Schurch S, Gehr P (2006) An optimized in vitro model of the respiratory tract wall to study particle cell interactions. *J Aerosol Med* **19**, 392–405.

Blank F, Rothen-Rutishauser B, Gehr P (2007) Dendritic cells and macrophages form a transepithelial network against foreign particulate antigens. *Am J Respir Cell Mol Biol* **36**, 669–677.

Borm PJA, Kreyling W (2004) Toxicological hazards of inhaled nanoparticles – Potential implications for drug delivery. *J Nanosci Nanotechnol* **4**, 521–531.

Borm P, Klaessig FC, Landry TD, Moudgil B, Pauluhn J, Thomas K, Trottier R, Wood S (2006) Research strategies for safety evaluation of nanomaterials, Part V: Role of dissolution in biological fate and effects of nanoscale particles. *Toxicol Sci* **90**, 23–32.

Brown DM, Wilson MR, MacNee W, Stone V, Donaldson K (2001) Size-dependent proinflammatory effects of ultrafine polystyrene particles: a role for surface area and oxidative stress in the enhanced activity of ultrafines. *Toxicol Appl Pharmacol* **175**, 191–199.

Brzoska M, Langer K, Coester C, Loitsch S, Wagner TO, Mallinckrodt C (2004) Incorporation of biodegradable nanoparticles into human airway epithelium cells – in vitro study of the suitability as a vehicle for drug or gene delivery in pulmonary diseases. *Biochem Biophys Res Commun* **318**, 562–570.

Carterson AJ, Honer zu BK, Ott CM, Clarke MS, Pierson DL, Vanderburg CR, Buchanan KL, Nickerson CA, Schurr MJ (2005) A549 lung epithelial cells grown as three-dimensional aggregates: alternative tissue culture model for Pseudomonas aeruginosa pathogenesis. *Infect Immun* **73**, 1129–1140.

Chen HW, Su SF, Chien CT, Lin WH, Yu SL, Chou CC, Chen JJ, Yang PC (2006) Titanium dioxide nanoparticles induce emphysema-like lung injury in mice. *FASEB J* **20**, 2393–2395.

Cozens AL, Yezzi MJ, Kunzelmann K, Ohrui T, Chin L, Eng K, Finkbeiner WE, Widdicombe JH, Gruenert DC (1994) CFTR expression and chloride secretion in polarized immortal human bronchial epithelial cells. *Am J Respir Cell Mol Biol* **10**, 38–47.

Dick CA, Brown DM, Donaldson K, Stone V (2008) The role of free radicals in the toxic and inflammatory effects of four different ultrafine particle types. *Inhal Toxicol* **15**, 39–52.

Donaldson K, Tran L, Jimenez LA, Duffin R, Newby DE, Mills N, MacNee W, Stone V (2005) Combustion-derived nanoparticles: a review of their toxicology following inhalation exposure. *Part Fibre Toxicol* **2**, 10.

Duffin R, Tran L, Brown D, Stone V, Donaldson K (2007) Proinflammogenic effects of low-toxicity and metal nanoparticles in vivo and in vitro: highlighting the role of particle surface area and surface reactivity. *Inhal Toxicol* **19**, 849–856.

Ece GD, Shah LK, Devalapally H, Amiji MM, Carrier RL (2008) A model predicting delivery of saquinavir in nanoparticles to human monocyte/macrophage (Mo/Mac) cells. *Biotechnol Bioeng* **101**, 1072–1082.

Ehrhardt C, Fiegel J, Fuchs S, Abu-Dahab R, Schaefer UF, Hanes J, Lehr CM (2002) Drug absorption by the respiratory mucosa: cell culture models and particulate drug carriers. *J Aerosol Med* **15**, 131–139.

Forbes I (2000) Human airway epithelial cell lines for in vitro drug transport and metabolism studies. *Pharm Sci Technol Today* **3**, 18–27.

Foster KA, Oster CG, Mayer MM, Avery ML, Audus KL (1998) Characterization of the A549 cell line as a type II pulmonary epithelial cell model for drug metabolism. *Exp Cell Res* **243**, 359–366.

Freshney RI (2000) *Culture of Animal Cells: A Manual of Basic Technique*. Wiley-Liss: New York.

Gardner DR, Kennedy GL (1993) Methodologies and technology for animal inhalation toxicology studies. In *Toxicology of the Lung* (2nd edn), Gardner DR, Crapo JD, McClellan RO (eds). Raven Press: New York.

Gehr P (1994) *Anatomy and Morphology of the Respiratory Tract. Human Respiratory Tract Model for Radiological Protection*. ICRP Publication 66, Smith H (ed.). Annals of the ICRP: Pergamon.

Gehr P, Bachofen M, Weibel ER (1978a) The normal human lung: ultrastructure and morphometric estimation of diffusion capacity. *Respir Physiol* **32**, 121–140.

Gehr P, Hugonnaud C, Burri PH, Bachofen H, Weibel ER (1978b) Adaptation of the growing lung to increased Vo2: III. The effect of exposure to cold environment in rats. *Respir Physiol* **32**, 345–353.

Gehr P, Schürch S, Berthiaume Y, Im Hof V, Geiser M (1990) Particle retention in airways by surfactant. *J Aerosol Med* **3**, 27–43.

Gehr P, Green FH, Geiser M, Im Hof V, Lee MM, Schurch S (1996) Airway surfactant, a primary defense barrier: mechanical and immunological aspects. *J Aerosol Med* **9**, 163–181.

Gonzalez-Flecha B (2004) Oxidant mechanisms in response to ambient air particles. *Mol Aspects Med* **25**, 169–182.

Goulaouic S, Foucaud L, Bennasroune A, Laval-Gilly P, Falla J (2008) Effect of polycyclic aromatic hydrocarbons and carbon black particles on pro-inflammatory cytokine secretion: impact of PAH coating onto particles. *J Immunotoxicol* **5**, 337–345.

Grenha A, Grainger CI, Dailey LA, Seijo B, Martin GP, Remunan-Lopez C, Forbes B (2007) Chitosan nanoparticles are compatible with respiratory epithelial cells in vitro. *Eur J Pharm Sci* **31**, 73–84.

Gruber FP, Hartung T (2004) Alternatives to animal experimentation in basic research. *ALTEX* **21** (Suppl. 1), 3–31.

Gstraunthaler G, Hartung T (2002) Good cell culture practice: good laboratory practice in the cell culture laboratory for the standardization and quality assurance of in vitro studies. In *Cell Culture Models of Biological Barriers. In-vitro Test Systems for Drug Absorption and Delivery*, Lehr C-M (ed.). Taylor and Francis: New York; 112–120.

Handler JS, Green N, Steele RE (1989) Cultures as epithelial models: porous-bottom culture dishes for studying transport and differentiation. *Methods Enzymol* **171**, 736–744.

Herzog E, Casey A, Lyng FM, Chambers G, Byrne HJ, Davoren M (2007) A new approach to the toxicity testing of carbon-based nanomaterials – the clonogenic assay. *Toxicol Lett* **174**, 49–60.

Hohr D, Steinfartz Y, Schins RP, Knaapen AM, Martra G, Fubini B, Borm PJ (2002) The surface area rather than the surface coating determines the acute inflammatory response after instillation of fine and ultrafine TiO_2 in the rat. *Int J Hyg Environ Health* **205**, 239–244.

Holder AL, Lucas D, Goth-Goldstein R, Koshland CP (2008a) Cellular response to diesel exhaust particles strongly depends on the exposure method. *Toxicol Sci* **103**, 108–115.

Holder AL, Lucas D, Goth-Goldstein R, Koshland CP (2008b) Cellular response to diesel exhaust particles strongly depends on the exposure method. *Toxicol Sci* **103**, 108–115.

Iijima S (1991) Helical microtubules of graphitic carbon. *Nature* **354**, 56–58.

Im Hof V, Gehr P, Gerber V, Lee MM, Schurch S (1997) In vivo determination of surface tension in the horse trachea and in vitro model studies. *Respir Physiol* **109**, 81–93.

Jang M, Ghio AJ, Cao G (2006) Exposure of BEAS-2B cells to secondary organic aerosol coated on magnetic nanoparticles. *Chem Res Toxicol* **19**, 1044–1050.

Kaiser JP, Wick P, Manser P, Spohn P, Bruinink A (2008) Single walled carbon nanotubes (SWCNT) affect cell physiology and cell architecture. *J Mater Sci Mater Med* **19**, 1523–1527.

Kemp SJ, Thorley AJ, Gorelik J, Seckl MJ, O'Hare MJ, Arcaro A, Korchev Y, Goldstraw P, Tetley TD (2008) Immortalisation of human alveolar epithelial cells to investigate nanoparticle uptake. *Am J Respir Cell Mol Biol* **39**, 591–597.

Kreyling WG, Semmler M, Erbe F, Mayer P, Takenaka S, Schulz H, Oberdorster G, Ziesenis A (2002) Translocation of ultrafine insoluble iridium particles from lung epithelium to extrapulmonary organs is size dependent but very low. *J Toxicol Environ Health A* **65**, 1513–1530.

Li N, Xia T, Nel AE (2008) The role of oxidative stress in ambient particulate matter-induced lung diseases and its implications in the toxicity of engineered nanoparticles. *Free Radic Biol Med* **44**, 1689–1699.

Lieber M, Smith B, Szakal A, Nelson-Rees W, Todaro G (1976) A continuous tumor-cell line from a human lung carcinoma with properties of type II alveolar epithelial cells. *Int J Cancer* **17**, 62–70.

Limbach LK, Li Y, Grass RN, Brunner TJ, Hintermann MA, Muller M, Gunther D, Stark WJ (2005) Oxide nanoparticle uptake in human lung fibroblasts: effects of particle size, agglomeration, and diffusion at low concentrations. *Environ Sci Technol* **39**, 9370–9376.

Limbach LK, Wick P, Manser P, Grass RN, Bruinink A, Stark WJ (2007) Exposure of engineered nanoparticles to human lung epithelial cells: influence of chemical composition and catalytic activity on oxidative stress. *Environ Sci Technol* **41**, 4158–4163.

Mathia NR, Timoszyk J, Stetsko PI, Megill JR, Smith RL, Wall DA (2002) Permeability characteristics of calu-3 human bronchial epithelial cells: in vitro-in vivo correlation to predict lung absorption in rats. *J Drug Target* **10**, 31–40.

Maynard AD, Aitken RJ (2007) Assessing exposure to airborne nanomaterials: Current abilities and future requirements. *Nanotoxicol* **1**, 26–41.

Mills NL, Amin N, Robinson SD, Anand A, Davies J, Patel D, de la Fuente JM, Cassee FR, Boon NA, MacNee W, Millar AM, Donaldson K, Newby DE (2006) Do inhaled carbon nanoparticles translocate directly into the circulation in humans? *Am J Respir Crit Care Med* **173**, 426–431.

Muhlfeld C, Gehr P, Rothen-Rutishauser B (2008a) Translocation and cellular entering mechanisms of nanoparticles in the respiratory tract. *Swiss Med Wkly* **138**, 387–391.

Muhlfeld C, Rothen-Rutishauser B, Blank F, Vanhecke D, Ochs M, Gehr P (2008b) Interactions of nanoparticles with pulmonary structures and cellular responses. *Am J Physiol Lung Cell Mol Physiol* **294**, L817–L829.

Muller J, Huaux F, Moreau N, Misson P, Heilier JF, Delos M, Arras M, Fonseca A, Nagy JB, Lison D (2005) Respiratory toxicity of multi-wall carbon nanotubes. *Toxicol Appl Pharmacol* **207**, 221–231.

Nel A, Xia T, Madler L, Li N (2006) Toxic potential of materials at the nanolevel. *Science* **311**, 622–627.

Nemmar A, Hoet PH, Vanquickenborne B, Dinsdale D, Thomeer M, Hoylaerts MF, Vanbilloen H, Mortelmans L, Nemery B (2002) Passage of inhaled particles into the blood circulation in humans. *Circ* **105**, 411–414.

Nemmar A, Hamoir J, Nemery B, Gustin P (2005) Evaluation of particle translocation across the alveolo-capillary barrier in isolated perfused rabbit lung model. *Toxicol* **208**, 105–113.

Nicod LP (2005) Lung defenses: an overview. *Eur Respir Rev* **95**, 45–50.

Oberdörster G, Maynard A, Donaldson K, Castranova V, Fitzpatrick J, Ausman K, Carter J, Karn B, Kreyling W, Lai D, Olin S, Monteiro-Riviere N, Warheit D, Yang H (2005a) Principles for characterizing the potential human health effects from exposure to nanomaterials: elements of a screening strategy. *Part Fibre Toxicol* **2**, 8.

Oberdörster G, Oberdörster E, Oberdörster J (2005b) Nanotoxicology: An emerging discipline evolving from studies of ultrafine particles. *Environ Health Perspect* **113**, 823–839.

Oberdörster G, Stone V, Donaldson K (2007) Toxicology of nanoparticles: A historical perspective. *Nanotoxicol* **1**, 2–25.

Ochs M, Weibel E (2008) Functional design of the human lung for gas exchange. In *Fishman's Pulmonary Diseases and Disorders* (4th edn), Fishman AP, Elias JA, Fishman JA, Grippi MA, Senior RM, Pack A (eds). McGrawHill: New York.

Park S, Lee YK, Jung M, Kim KH, Chung N, Ahn EK, Lim Y, Lee KH (2007) Cellular toxicity of various inhalable metal nanoparticles on human alveolar epithelial cells. *Inhal Toxicol* **19** (Suppl. 1), 59–65.

Pauluhn J (2005) Overview of inhalation exposure techniques: strengths and weaknesses. *Exp Toxicol Pathol* **57** (Suppl. 1), 111–128.

Pauluhn J, Mohr U (2000) Inhalation studies in laboratory animals – current concepts and alternatives. *Toxicol Pathol* **28**, 734–753.

Peters A, Wichmann HE, Tuch T, Heinrich J, Heyder J (1997) Respiratory effects are associated with the number of ultrafine particles. *Am J Respir Critical Care Med* **155**, 1376–1383.

Poland CA, Duffin R, Kinloch I, Maynard A, Wallace WA, Seaton A, Stone V, Brown S, MacNee W, Donaldson K (2008) Carbon nanotubes introduced into the abdominal cavity of mice show asbestos-like pathogenicity in a pilot study. *Nat Nanotechnol* **3**, 423–428.

Pope ICA, Dockery DW, Schwartz J (1995) Review of epidemiological evidence of health effects of particulate air pollution. *Inhalation Toxicology* **7**, 1–18.

Porter AE, Gass M, Muller K, Skepper JN, Midgley P, Welland M (2007) Visualizing the uptake of C60 to the cytoplasm and nucleus of human monocyte-derived macrophage cells using energy-filtered transmission electron microscopy and electron tomography. *Environ Sci Technol* **41**, 3012–3017.

Reddel RR, Ke Y, Gerwin BI, McMenamin MG, Lechner JF, Su RT, Brash DE, Park JB, Rhim JS, Harris CC (1988) Transformation of human bronchial epithelial cells by infection with SV40 or adenovirus-12 SV40 hybrid virus, or transfection via strontium phosphate coprecipitation with a plasmid containing SV40 early region genes. *Canc Res* **48**, 1904–1909.

Rescigno M, Rotta G, Valzasina B, Ricciardi-Castagnoli P (2001a) Dendritic cells shuttle microbes across gut epithelial monolayers. *Immunobiol* **204**, 572–581.

Rescigno M, Urbano M, Valzasina B, Francolini M, Rotta G, Bonasio R, Granucci F, Kraehenbuhl JP, Ricciardi-Castagnoli P (2001b) Dendritic cells express tight junction proteins and penetrate gut epithelial monolayers to sample bacteria. *Nat Immunol* **2**, 361–367.

Ritter D, Knebel J, Aufderheide M (2004) Comparative assessment of toxicities of mainstream smoke from commercial cigarettes. *Inhal Toxicol* **16**, 691–700.

Roggen EL, Soni NK, Verheyen GR (2006) Respiratory immunotoxicity: an in vitro assessment. *Toxicol In Vitro* **20**, 1249–1264.

Rothen-Rutishauser BM, Kiama SG, Gehr P (2005) A three-dimensional cellular model of the human respiratory tract to study the interaction with particles. *Am J Respir Cell Mol Biol* **32**, 281–289.

Rothen-Rutishauser BM, Schurch S, Haenni B, Kapp N, Gehr P (2006) Interaction of fine particles and nanoparticles with red blood cells visualized with advanced microscopic techniques. *Environ Sci Technol* **40**, 4353–4359.

Rothen-Rutishauser B, Muhlfeld C, Blank F, Musso C, Gehr P (2007) Translocation of particles and inflammatory responses after exposure to fine particles and nanoparticles in an epithelial airway model. *Part Fibre Toxicol* **4**, 9.

Rothen-Rutishauser B, Blank F, Muhlfeld C, Gehr P (2008a) In vitro models of the human epithelial airway barrier to study the toxic potential of particulate matter. *Expert Opin Drug Metab Toxicol* **4**, 1075–1089.

Rothen-Rutishauser B, Mueller L, Blank F, Brandenberger C, Muehlfeld C, Gehr P (2008b) A newly developed in vitro model of the human epithelial airway barrier to study the toxic potential of nanoparticles. *ALTEX* **25**, 191–196.

Rotoli BM, Bussolati O, Bianchi MG, Barilli A, Balasubramanian C, Bellucci S, Bergamaschi E (2008) Non-functionalized multi-walled carbon nanotubes alter the paracellular permeability of human airway epithelial cells. *Toxicol Lett* **178**, 95–102.

Sakagami M (2006) In vivo, in vitro and ex vivo models to assess pulmonary absorption and disposition of inhaled therapeutics for systemic delivery. *Adv Drug Deliv Rev* **58**, 1030–1060.

Sallusto F, Lanzavecchia A (1994) Efficient presentation of soluble antigen by cultured human dendritic cells is maintained by granulocyte/macrophage colony-stimulating factor plus interleukin 4 and downregulated by tumor necrosis factor alpha. *J Exp Med* **179**, 1109–1118.

Sallusto F, Cella M, Danieli C, Lanzavecchia A (1995) Dendritic cells use macropinocytosis and the mannose receptor to concentrate macromolecules in the major histocompatibility complex class II compartment: downregulation by cytokines and bacterial products. *J Exp Med* **182**, 389–400.

Schins RP, Knaapen AM (2007) Genotoxicity of poorly soluble particles. *Inhal Toxicol* **19** (Suppl. 1), 189–198.

Schulz H, Harder V, Ibald-Mulli A, Khandoga A, Koenig W, Krombach F, Radykewicz R, Stampfl A, Thorand B, Peters A (2005) Cardiovascular effects of fine and ultrafine particles. *J Aerosol Med* **18**, 1–22.

Schürch S, Gehr P, Im Hof V, Geiser M, Green F (1990) Surfactant displaces particles toward the epithelium in airways and alveoli. *Resp Physiol* **80**, 17–32.

Shapiro DL, Nardone LL, Rooney SA, Motoyama EK, Munoz JL (1978) Phospholipid biosynthesis and secretion by a cell line (A549) which resembles type II aleveolar epithelial cells. *Biochim Biophys Acta* **530**, 197–207.

Stearns RC, Paulauskis JD, Godleski JJ (2001) Endocytosis of ultrafine particles by A549 cells. *Am J Respir Cell Mol Biol* **24**, 108–115.

Steimer A, Haltner E, Lehr CM (2005) Cell culture models of the respiratory tract relevant to pulmonary drug delivery. *J Aerosol Med* **18**, 137–182.

Stoeger T, Reinhard C, Takenaka S, Schroeppel A, Karg E, Ritter B, Heyder J, Schulz H (2006) Instillation of six different ultrafine carbon particles indicates a surface area threshold dose for acute lung inflammation in mice. *Environ Health Perspect* **114**, 328–333.

Tippe A, Heinzmann U, Roth C (2002) Deposition of fine and ultrafine aerosol particles during exposure at the air/cell interface. *Aerosol Science* **33**, 207–218.

Unfried K, Albrecht C, Klotz LO, von Mikecz A, Grether-Beck S, Schins RP (2007) Cellular responses to nanoparticles: target structures and mechanisms. *Nanotoxicol* **1**, 1–20.

Veranth JM, Kaser EG, Veranth MM, Koch M, Yost GS (2007) Cytokine responses of human lung cells (BEAS-2B) treated with micron-sized and nanoparticles of metal oxides compared to soil dusts. *Part Fibre Toxicol* **4**, 2.

Vinzents PS, Moller P, Sorensen M, Knudsen LE, Hertel O, Jensen FP, Schibye B, Loft S (2005) Personal exposure to ultrafine particles and oxidative DNA damage. *Environ Health Perspect* **113**, 1485–1490.

Voisin C, Aerts C, Jakubczak E, Houdret JL, Tonnel TB (1977a) Effects of nitrogen dioxide on alveolar macrophages surviving in the gas phase. A new experimental model for the study of in vitro cytotoxicity of toxic gases [author's transl.]. *Bull Eur Physiopathol Respir* **13**, 137–144.

Voisin C, Aerts C, Jakubczk E, Tonnel AB (1977b) La culture cellulaire en phase gazeuse. Un nouveau modele experimental d'etude in vitro des activites des macrophages alveolaires. *Bull Eur Physiopathol Respir* **13**, 69–82.

Waldman WJ, Kristovich R, Knight DA, Dutta PK (2007) Inflammatory properties of iron-containing carbon nanoparticles. *Chem Res Toxicol* **20**, 1149–1154.

Warheit DB, Webb TR, Colvin VL, Reed KL, Sayes CM (2008) Pulmonary bioassay studies with nanoscale and fine-quartz particles in rats: toxicity is not dependent upon particle size but on surface characteristics. *Toxicol Sci* **95**, 270–280.

Wick P, Manser P, Limbach LK, Dettlaff-Weglikowska U, Krumeich F, Roth S, Stark WJ, Bruinink A (2007) The degree and kind of agglomeration affect carbon nanotube cytotoxicity. *Toxicol Lett* **168**, 121–131.

Wiebert P, Sanchez-Crespo A, Falk R, Philipson K, Lundin A, Larsson S, Moller W, Kreyling WG, Svartengren M (2006) No significant translocation of inhaled 35-nm carbon particles to the circulation in humans. *Inhal Toxicol* **18**, 741–747.

Wottrich R, Diabate S, Krug HF (2004) Biological effects of ultrafine model particles in human macrophages and epithelial cells in mono- and co-culture. *Int J Hyg Environ Health* **207**, 353–361.

20

Iron Oxide Magnetic Nanoparticle Nanotoxicity: Incidence and Mechanisms

Thomas R. Pisanic, Sungho Jin and Veronica I. Shubayev

20.1 Introduction

Nanotechnology holds great promise for a wide range of applications and has emerged as one of the largest areas of active study in the applied sciences. Tremendous effort is underway toward the development of new nanomaterials and uses thereof in fields ranging from consumer electronics and alternative energy to biomedicine. Magnetic nanomaterials represent a subclass within the overall category of nanomaterials and are already widely used in many applications, particularly in the biomedical sciences. Of those magnetic materials currently employed in biomedicine, none is more widely used than iron oxide magnetic nanoparticles (IOMNPs). IOMNPs are attractive because of their intrinsic magnetic characteristics coupled with the fact that they have historically been considered biocompatible and nontoxic to humans, due to their demonstrated relatively large LD50 and apparent lack of acute toxicity *in vivo*. Recently, however, multiple studies, particularly *in vitro*, have called these claims into question. The purpose of this summary is to review comprehensively the reported incidences of IOMNP-induced nanotoxicity as well as the potential mechanisms of this toxicity, and determine how these studies should be considered in the larger light of IOMNP use in biomedicine.

Nanotoxicity: From In Vivo *and* In Vitro *Models to Health Risks*   Edited by Saura Sahu and Daniel Casciano
© 2009 John Wiley & Sons, Ltd

20.2   Iron Oxide Magnetic Nanoparticles in Current Biomedical Applications

The last 25 years have seen a tremendous expansion in the use and development of IOMNPs for a large number of biomedical applications. The majority of these applications can be broken down into a few major categories: cell separation, magnetic resonance imaging (MRI) contrast enhancement, cell labelling and tracking via MRI, magneto-fection, drug delivery and magnetic hyperthermia. While numerous other biomedical uses have been reported and are currently in development, these categories have become at least reasonably well-established and account for the major portion of the current literature.

One of the first biomedical applications of magnetic nanoparticles to develop and fully mature was *in vitro* magnetic cell sorting and separation (Radbruch *et al.*, 1994; Thiel *et al.*, 1998). In biomedicine it is often desirable to isolate particular cells of interest, such as stem cells, from a larger, heterogeneous population. By attaching antibodies against specific phenotypic markers to magnetic nanoparticles, the surfaces of particular cell phe-notypes can be tagged and isolated from these populations by separation via magnetic field gradients. The majority of these cell separations are used for quantification and analysis, although it may be desirable in some instances to use the isolated cells of interest for *in vivo* applications.

Another field that has matured and continues to be an area of highly active research is the use of IOMNPs as contrast agents in magnetic resonance imaging (MRI) (Bulte and Kraitchman, 2004; Corot *et al.*, 2006). In these applications, IOMNPs are intravenously injected into patients and preferentially accumulate in certain regions of the body, such as the liver or lymph nodes. IOMNPs act to modify proton relaxation times, thereby acting to modulate local resonance signals and provide increased contrast. In addition to general contrast enhancement, IOMNPs can also be used to monitor cells of interest within the body. By delivering sufficient amounts of IOMNPs into these cells of interest such as stem cells, the cells, which now contain high intracellular iron concentrations, can then be located and tracked throughout the body via high-resolution MRI.

Enhanced gene transfection is another area in which the unique properties of IOMNPs have been utilized to provide advantages over existing methodologies (Dobson, 2006). The efficiency and utility of gene transfection is often undermined by poor transfection kinet-ics, resulting in inadequate gene transfection rates. In order to circumvent poor diffusive kinetics, vector DNA can be noncovalently attached to magnetic nanoparticles, such as IOMNPs, and regiospecifically concentrated to the surface of cells in a method known as magnetofection (MF). MF can readily increase transfection efficiency several thousand fold over nonmagnetic methods.

Similar to the attachment of genes in MF, drugs can also be attached to the surface of (or encapsulated with) IOMNPs for selective and enhanced delivery of therapeutic agents to desired locales within the body (MacBain *et al.*, 2008; Namdeo *et al.*, 2008). The primary impetus of this technique is the reduction in necessary systemic drug doses/concentrations and adverse effects for healthy tissues. The effectiveness of this approach, however, is reliant upon the general assumption of the lack of undesirable toxicity of the delivery vehicle itself: the IOMNPs.

One other growing field of use for IOMNPs is magnetic hyperthermia (MH) (Gazeau *et al.*, 2008). Here, magnetic nanoparticles are concentrated at the site of tumors and exposed to a high-frequency alternating magnetic field, resulting in heat generation by the particles themselves due to magneto-hysteretic losses. Since the nanoparticles are nominally localized to only the tumor region, the thermally induced cellular death of other tissues is limited. This technique, however, also relies upon the assumption that locally high concentrations of MNPs will not appreciably negatively affect those tissues surrounding the tumor.

While many of these applications are showing great promise for their intended purposes, one of their underlying assumptions is the biocompatibility of iron oxide and, in particular, IOMNPs. However, as in all fields of biomedicine, the principle of *primum non nocere*, that is, to first do no harm, requires that all measures be taken to evaluate these presumptions. Likewise, a careful and consistent review of recent literature on the biocompatibility of iron oxide nanoparticles is in order.

20.3   Nanotoxicology for Iron Oxide Magnetic Nanoparticles

With even a perfunctory look at the history of medicine it becomes glaringly obvious that many social and technological advances and even medical treatments have later been shown to be major health hazards. Many foods, pesticides, manufacturing processes and byproducts, and numerous pharmaceuticals have been linked to the major causes of many of the health problems now confronting humanity. This, in spite of the fact that in many of these cases, even after careful evaluation, little to no immediate toxicity to humans was originally observed. But with time and the expansion of understanding of more complex biological mechanisms, the adverse consequences of their (often widespread) use became readily apparent. So that history does not 'repeat itself', this problem must be addressed and not become only an afterthought in the development of nanotechnology.

20.3.1   Iron Oxide Magnetic Nanoparticles for Biomedicine

Prior to 1990, two major applications for engineered IOMNPs developed: *in vitro* cell sorting and *in vivo* MRI contrast enhancement. Chronologically speaking, the former developed first and was first demonstrated in the late 1970s and further developed in the early 1980s by Molday and colleagues (Molday *et al.*, 1977; Molday and Molday, 1984). Iron oxide nanoparticles were synthesized and coated with polymers such as methacrylate and, ultimately, dextran, along with surface antigen specific antibodies for the tagging and magnetic isolation of cell phenotypes. This technique has become commercially well established and is readily used in many biomedical laboratories.

The observation of iron oxide as a potential MRI contrast agent, on the other hand, seemed to arise in the more serendipitous observation of changes in proton relaxation times in the livers of patients with iron overload disease states (Stark *et al.*, 1985). Later it was shown that these 'states' could be artificially induced via administration of iron oxide particles (Renshaw *et al.*, 1986), which also naturally accumulated preferentially in the liver as well as the spleen (Weissleder *et al.*, 1987a,b, 1988). Borrowing from synthesis techniques previously developed for IOMNPs for cell-sorting, efforts in the mid to late 1980s saw the development of (dextran-coated) magnetic nanoparticles specifically engineered for improved biocompatibility and blood circulation lifetimes in MRI (Weissleder *et al.*, 1990a,b).

20.4   Toxicity Studies

20.4.1   A Brief History

Following the development of IOMNPs as MRI contrast agents, the ostensible premise of iron oxide nanoparticle biocompatibility emerged from the marrying of two major observations: firstly, the high LD50 and initial iron metabolism/tissue distribution studies for the seminally developed IOMNPs for MRI (Weissleder *et al.*, 1989); and, secondly, the previously established lack of toxicity characteristics of molecular iron (except at exceedingly high concentrations) (Jacobs and Worwood, 1980). What followed was a *non sequitur* assumption that if IOMNPs were seemingly sufficiently acutely nontoxic and rapidly cleared from the body, and, in addition, systemic iron levels remained significantly lower than the LD50 values established for aqueous iron, then the IOMNPs should therefore be considered 'biocompatible'. What failed to be taken into account, however, were the unique characteristics and surface chemical effects of nanomaterials and IOMNPs that only recently are beginning to be studied and observed.

As will be shown in the remainder of this chapter, recent studies are beginning to call into question the biocompatibility of at least certain incarnations of IOMNPs. Both *in vitro* and *in vivo* studies from various investigators have demonstrated that the presence of IOMNPs can exert toxic effects at levels far lower than those established for molecular iron.

The original impetus for studies of IOMNP biocompatibility came from their development as MRI contrast agents. Yet although IOMNPs have been in development for *in vivo* biomedical use for over 20 years, until the last five years or so the literature has been mostly devoid of independent studies on the toxic effects of these particle formulations. In general the literature has been bimodal in its evaluation of IOMNPs, with those groups developing the particles for specific applications such as MRI contrast enhancement and cell labelling purporting the safety of their formulations, as opposed to the slowly increasing number of independent studies that report specific toxicities in their evaluations. Here, we are primarily concerned with evaluating the literature that has definitively observed toxic or negative effects of IOMNPs upon biological systems in an effort to determine how these reports can, if possible, be linked together to form an overall picture of the biocompatibility, or lack thereof, of IOMNPs.

20.4.1.1   The Emergence of IOMNP Toxicity Studies

Initial studies of IOMNP toxicity often assumed *a priori* that iron oxide was nontoxic and biocompatible. Many were performed as complementary studies by the developers of the MRI contrast agents in question, and generally concluded that IOMNPs were completely safe for cells up to any reasonable concentration that might occur in the body.

Of the relatively few IOMNP studies published prior to 2000, the majority were performed using dextran-coated monocrystalline iron oxide nanoparticles (MION), specifically those developed by Weissleder and colleagues. These initial studies were conducted with the intent of investigating the mechanism of IOMNP cellular uptake that had been incidentally observed in various cell types (Moore *et al.*, 1997) or for showing the utility of these particles and their potential use for nonspecifically labelling particular cells of interest such as macrophages (Weissleder *et al.*, 1997) or neurons (Neuwelt *et al.*, 1994). Initially, while lysosomal trafficking of MION was noted via fluorescent microscopy, no attempts were made at determining potential cellular toxicities. These interesting labelling and uptake studies further engendered research aimed toward the development of IOMNPs specifically tailored to deliver sufficient amounts of IOMNPs into cells to allow for their detection and tracking, ultimately *in vivo,* via high-resolution MRI. Acknowledging at least the possibility of a negative effect of such high intracellular IOMNP concentrations, Weissleder and colleagues noted that, in their experiments, no particle-mediated cytotoxicity could be observed (Moore *et al.*, 1997). These studies continued on and culminated in the successful detection of TAT-mediated magnetically labelled individual stem cells *in vivo* (Lewin *et al.*, 2000; Bulte *et al.*, 2001).

Many types of IOMNP coatings were developed as a primary consequence of the relative plethora of studies aimed at loading and targeting cells with IOMNPs. Most IOMNP surface coatings that have been used in biomedical applications fall into (at least) one of five categories: natural or synthetic polymers; proteins and antibodies; silica; electrostatic stabilizers; and cell-penetrating peptides (CPPs). Figure 20.1 illustrates the general structure of and differences between these coatings, as well as references to studies (covered herein) that have utilized them. As will be shown, it is these coatings that are instrumental in determining the type and extent of interactions with biological structures. Likewise, IOMNPs can be coated with hybrids of these categories (e.g. polymer-CPPs and polymer-proteins) as a means of combining the particular physico-chemical properties of the coatings, such as colloidal stability, protein function and cellular targeting, to form novel nanostructures tailored precisely to interact with these structures in specific biomedical applications (Gupta *et al.*, 2007; McCarthy *et al.*, 2007).

20.4.2   Studies Reporting *In Vitro* Toxicity

By as early as 1996, at least one group reported that certain types of coated IOMNPs showed significant toxicities, measuring the detrimental effects that the nanoparticles had towards human granulocytes *in vitro* (Mueller *et al.*, 1996). Yet it was not until several years later that interest slowly began to develop in the broader scientific community. The seminal papers on the magnetic labelling of cells brought more interest into the development of various coatings that might enhance or modify the intracellular uptake of IOMNPs. Additionally, as a means of distinguishing the properties of various IOMNPs, the nature and consequence of intracellular delivery slowly began to be more closely observed. In 2003, Hilger *et al.* observed significant drops in cell survival rates for human adenocarcinoma cells exposed to various IOMNP preparations, particularly those coated with cationic surfactants (Hilger *et al.*, 2003). Based upon the assumption of the safety of iron oxides, the authors attributed the observed cytotoxicities to cellular membrane binding, chemical formulation components other than iron oxide, or the generation of a non-physiological pH. In the same year came other independent reports of IOMNP toxicity. In a pair of reports, Berry *et al.* (2003, 2004a) showed that both uncoated and dextran-coated IOMNPs caused

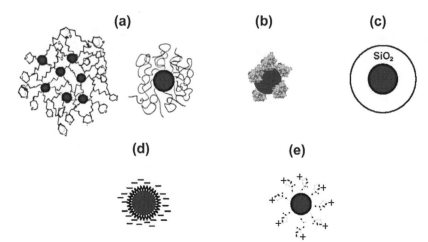

Figure 20.1 *Major types of IOMNP coatings used for biomedical applications. IOMNPs must be coated in order to remain colloidally stable in physiological solutions, as well as to endow them with properties suitable for particular biomedical applications. These coatings include (select examples utilizing each type are cited): (a) natural or synthetic polymers, such as dextran or PEG (Molday and Molday, 1984; Weissleder et al., 1990b; Mueller et al., 1996; Bulte et al., 2001; Huth et al., 2004; Gupta and Curtis, 2004a; Bourrinet et al., 2006; Hu et al., 2006; Muller et al., 2007); (b) proteins, such as albumin or antibodies (Berry et al., 2003, 2004bb; Gupta et al., 2003; Gupta and Curtis, 2004b); (c) silica (SiO_2) (Kim et al., 2006; Lu et al., 2007); (d) electrostatic stabilizers, which are typically negatively charged, such as DMSA or citric acid (Lacava et al., 1999a,1999bb; Stroh et al., 2004; Pisanic et al., 2007; de Freitas et al., 2008; Wilhelm and Gazeau, 2008); and (e) positively-charged cell-penetrating peptides, such as the TAT peptide (Lewin et al., 2000; Zhao et al., 2002; Gupta et al., 2007).*

varying degrees of cell death and induced vacuole formation and clear disruptions in the cytoskeleton of dermal fibroblasts. Once again, given the previously purported safety and biocompatibility of dextran coated IOMNPs, these authors were unclear of the cause of these observations and suggested that further studies into the stability of dextran coatings be implemented. These observations were reconfirmed in other reports by the Curtis group that clearly demonstrated cytotoxicity and cytoskeletal disruption by uncoated IOMNPs with fibroblasts (Gupta and Curtis, 2004a,b; Gupta and Gupta, 2005a). In one study, the group modulated the uptake mechanism and subsequent cytoskeletal disruption of the IOMNPs by coating the IOMNPs with a hydrophilic polysaccharide (Gupta and Gupta, 2005a). Similarly, in other studies they specifically determined to prevent the apparent endocytosis-mediated cytotoxic effects by coating the IOMNPs with different proteins, and showed that the cell response, including gene expression, could be directly modulated by the choice of coating (Gupta *et al.*, 2003; Berry *et al.*, 2004b).

Due to the increase in studies on IOMNP and other nanoparticle cytotoxicities, general scientific scrutiny on the emerging field of nanotoxicology has intensified in the past few years (Donaldson *et al.*, 2004; Service, 2005). As other studies began to show that cells loaded with large amounts of IOMNPs could be tracked via MRI, interest in the potential effects that this loading might have upon cellular health increased. In one study, van den Bos *et al.* (2003) showed that the clinically-used IOMNP formulation, Feridex, demonstrated significant toxic effects upon macrophages, including decreased proliferation and cell death. Further investigation revealed that the cause of toxicity was directly attributable to oxidative stress and the generation of free radicals. Similarly, in a study specifically evaluating the effect that magnetic labelling might have upon cells, Stroh *et al.* (2004) confirmed that the delivery of large amounts of citrate-coated IOMNPs into cells resulted in a significant increase in protein oxidation and oxidative stress. The group positively confirmed that iron was the source of reactive oxygen species (ROS) by showing a dramatic reduction in these levels via coadministration of an iron chelator.

Broader evaluations on the cytotoxicity of metallic and non-metallic oxides appeared in 2005. Indicative of the changing scientific climate, multiple studies of several nanoparticle formulations, including IOMNPs, which were previously considered innocuous to humans, were compared with nanoparticles of the well-known carcinogen asbestos (Soto *et al.*, 2005, 2007; Brunner *et al.*, 2006). In a pair of transmission electron microscopic (TEM) and MTT assay evaluations of the toxicity of various nanoparticles, including bare IOMNPs, Soto and colleagues noted that murine macrophage cells exposed to bare IOMNPs showed cytoxicities nearing 90% of the asbestos controls (Soto *et al.*, 2005, 2007). Similarly, in a study by Brunner *et al.* (2006), IOMNPs showed a cell-specific response, clearly showing toxicity equal to that of asbestos toward human mesothelioma cells (yet little toward rat fibroblasts). The authors postulated that the EC50 value in this study ($\sim$100 μM), roughly 40 times lower than those published for iron ions, was due to massive Haber-Weiss reactions resulting from the rapid uptake and intracellular transportation of nanoparticles that is probably vastly different from aqueous iron ions. In another broad-spectrum toxicity test of metal oxide nanoparticles by Jeng and Swanson (2006), bare IOMNPs showed significant morphological effects, but only moderate toxicity upon a neuroblastoma cell line. The authors measured mitochondrial function via MTT assay, and at the highest concentrations tested ([Fe] $\sim$2.5 mM) showed that IOMNPs had a statistically significant effect upon mitochondrial function. Au *et al.* (2007) studied the effects of a proprietary chemical formulation of IOMNPs upon astrocytes *in vitro* at a specified concentration ($\sim$175 μM) and also found significant effects upon mitochondrial function as well as decreased cell viability. Similarly, our group also developed a quantifiable model cell system and tested the effect of a well-published anionic IOMNP formulation upon the various cell functions of a pheochromocytoma neuronal-type cell line (Pisanic *et al.*, 2007). We found that the particles elicited a dose-dependent ([Fe] = 0.15–15 mM) diminishing ability of the cells to either survive or demonstrate normal biological responses and morphologies.

Table 20.1 lists those studies that have conclusively observed IOMNP-mediated nanotoxicity *in vitro*. It should be noted that there exist multiple other *in vitro* studies that have demonstrated little, no, or more moderate toxicities of IOMNPs as well as various other nanoparticle formulations (Sun *et al.*, 2005; Auffan *et al.*, 2006; Hu *et al.*, 2006; de Freitas *et al.*, 2008; Petri-Fink *et al.*, 2008), many of which are summarized in more general reviews of nanotoxicity (Lewinski *et al.*, 2008). While not to discount the validity of such studies, these ambiguous cases are beyond the current scope. An interesting caveat exists, however, in that there are often important differences between studies that appear to have drastic consequences on the outcome of the determination of IOMNP nanotoxicity. The first is the obvious fact that many studies are conducted on vastly different cell types, from liver

Table 20.1   Summary of reports demonstrating in vitro IOMNP toxicity.

Particle coating	Particle size (nm)	Cell type	Observed toxic effects	Reference
Polylactide, glycolide, lipids	400–800	Human granulocytoma	Dose-dependent reduced cell viability.	Mueller et al. (1996)
Cationic, anionic, starch	8, 10, 220	Human adeno-carcinoma	Strong decrease in cell survival rates, abnormal subcellular structures.	Hilger et al. (2003)
Bare, dextran	8–10	Human fibroblast line	Apoptosis, altered morphology and behavior.	Berry et al. (2004a)
Bare, dextran, albumin	8–10	Human fibroblast line	Alterations in proliferation, behavior and morphology.	Berry et al. (2003)
Bare, PEG	10, 40–50	Human fibroblast line	Reduced cell adherence, altered behavior and morphology.	Gupta and Curtis (2004a)
Bare, lactoferrin, ceruloplasmin	14	Human fibroblast line	Coating-dependent reduction in adherence and viability. Alterations in morphology.	Gupta and Curtis (2004b)
Bare, pullalan	14, 42	Human fibroblast line	Coating-dependent reduction in adherence and viability. Alterations in morphology.	Gupta and Gupta (2005a)
Dextran, liposomes	80–120	Rabbit skeletal myoblasts	Dextran IOMNPs induced generation of free radicals, reduced proliferation and death.	van den Bos et al. (2003)
Citrate	9	Rat macrophage line	Induction of oxidative stress.	Stroh et al. (2004)
Bare	50	Human mesothelioma, Rat fibroblast line	Cell-specific toxicity, reduction of mesothelioma viability.	Brunner et al. (2006)
Bare	50	Murine alveolar macrophages line	Reduction in cell viability similar to asbestos controls.	Soto et al. (2005)
Bare	5–140	Murine alveolar macrophage line, Human macrophage line and Human epithelial line	Reduction in cell viability comparable to asbestos for all three lines tested.	Soto et al. (2007)
Bare	25	Mouse neuroblastoma	Alterations in mitochondrial function (LDH leakage) at higher concentrations.	Jeng and Swanson (2006)

(Continued)

Table 20.1   *(Continued)*

Particle coating	Particle size (nm)	Cell type	Observed toxic effects	Reference
Proprietary	Proprietary	Rat astrocytes	Cell detachment, reduction in viability and evidence of significant mitochondrial uncoupling via MTS assay.	Au *et al.* (2007)
DMSA	8–15	Murine pheochromo-cytoma	Dose-dependent reductions in adhesion, viability and biological response to cue. Alterations in morphology.	Pisanic *et al.* (2007)
DMSA	6	Human fibroblasts	Decrease in viability and mitochondrial activity at higher concentrations.	Auffan *et al.* (2006)
DMSA, citric acid, lauric acid	??	Human melanoma	Citric and lauric acid coated IOMNPs showed dose-dependent toxicity (via MTT assay) and alterations in morphology. Lauric acid particles induced apoptotic features.	de Freitas *et al.* (2008)
Glucose, maltose lactose	1.6–2.1	Human fibroblast line	Cell morphology affected differentially based upon saccharide coating. Maltose coating caused significant decrease in viability.	de le Fuente *et al.* (2007)

PEG, polyethylene glycol; DMSA, dimercaptosuccinic acid; PVA, polyvinyl alcohol; A-PVA, vinyl alcohol/vinyl amine copolymer; PEI, polyethyleneimine.

and neuronal cell lines to various macrophage type cells. Secondly, and perhaps equally as important, is the fact that many authors may fail to report the precise chemical composition or coatings of the IOMNPs used in their experiments. As mentioned previously, this chemical composition and coating can have drastic consequences for nanoparticle stability, aggregate size, and the level and type of cellular interaction, significantly affecting the fate and extent of IOMNP internalization (Raynal *et al.*, 2004). These differential considerations were recently highlighted in a study of the effects of IOMNPs and other nanoparticle preparations upon multiple cell lines by Díaz *et al.* (2008). While the authors reported significant toxicity and ROS production in response to IOMNPs, similar to Jeng and Swanson (2006) the results were cell-specific, with the responses of the four tested cell lines differing drastically. It was additionally shown that the number of nanoparticles per cell (independent of concentration) as well as the number of cells tested might also affect the results of cytotoxicological evaluations. The authors concluded that not only was it not possible to find a direct correlation between ROS production and cell toxicity, but also that strict standardization is necessary for useful nanotoxicological evaluation. Similarly, de la Fuente *et al.* (2007), in a study of IOMNPs coated in various similar simple saccharides, showed that even the most seemingly minute changes in IOMNP coating can drastically affect cell responses and viability. It is these and similar issues that represent some of the challenges facing the field of nanotoxicology, as will be further discussed later on.

20.4.3   Studies Reporting *In Vivo* Toxicity

The first published methodical study of IOMNP *in vivo* toxicity was performed by Weissleder *et al.* (1989) on his dextran-coated AMI-25 formulation. Following intravenous administration in rats and beagle dogs, within 4 hours roughly 90 % of the IOMNPs were found sequestered in the liver and spleen of the animals and did not appear to cause any acute or subacute toxic effects, even at high concentrations. The results of this study seemed quite conclusive as to the apparent complete biocompatibility of (at least these) IOMNPs for *in vivo* use at the tested concentrations. For the next ten years, based upon this assumption, the literature was, similar to the *in vitro* case, almost completely silent on the possibility of *in vivo* IOMNP toxicity.

While the vast majority of studies into the *in vivo* toxic effects of IOMNPs have either assumed or confirmed these initial conclusions (Chung, 2002), there have been a few notable exceptions, which are detailed here and listed in Table 20.2. By as early as 1999, preliminary studies began to report on the toxic effects of IOMNP formulations when administered *in vivo*. In a pair of papers, Lacava *et al.* (1999a,b) reported that ionically-stabilized IOMNPs injected intraperitoneally into mice caused severe inflammatory reactions and ultimately resulted, in one case, in the death of a third of the test animals. Several years later, similar

Table 20.2   *Summary of reports demonstrating in vivo IOMNP toxicity.*

Particle coating	Particle size (nm)	Animal; administration	Observed toxic effects	Reference
Bare, tartrate	10	Mouse; intraperitoneal	Cell death, apoptosis, mutagenicity, severe inflammatory reactions	Lacava *et al.* (1999a)
Bare, citric acid	10	Mouse; intraperitoneal	Mutagenicity, apoptosis and severe inflammatory reactions. Citric IOMNPs caused mortality in 2/6 animals.	Lacava *et al.* (1999b)
DMSA	9.4	Mouse; lung	IOMNP aggregation in blood vessels, organ parenchyma and cells. Inflammation, leukocyte infiltration.	Garcia *et al.* (2005)
Silica	50	Mouse; intraperitoneal	IOMNPs crossed blood brain and blood testes barriers. Particles remained in various organs >4 weeks.	Kim *et al.* (2006)
Oleic and pluronic acid	193	Rat; intravenous	Transient increase in oxidative stress (several days). Significant increase in iron levels.	Jain *et al.* (2008)
Dextran	30	Mouse, rat, rabbit, dog, monkey; intravenous	Long-term increase in iron levels, maternotoxicity, teratogenicity and moderate systemic toxicity symptoms; neurovegetation, alterations in neurobehavior	Bourrinet *et al.* (2006)

results were obtained by the same group in a study of the effects of injection of anionic IOMNPs into the lungs of mice, resulting in readily observable inflammation (Garcia *et al.*, 2005). In a separate but similar study, Kim *et al.* (2006) also investigated the effects of intraperitoneal injection of silica-coated IOMNPs into mice. While no apparent toxic effects were observed in this study, it was incidentally observed that by four weeks post-injection, the IOMNPs remained localized in many organs including the liver, lungs, kidneys, spleen, heart, testes, uterus and, quite notably, the brains of the mice. This observation has been confirmed in other IOMNP studies (Jain *et al.*, 2008) and may have particular consequences for the development of neurodegenerative disease that will be expounded upon later.

Currently, the only IOMNPs used clinically as MRI contrast agents are the SPIO (superparamagnetic iron oxide) and USPIO (ultrasmall superparamagnetic iron oxide) type dextran-coated iron oxides. Likewise there exist many reports in the literature of their use in various MRI applications. While very few of these studies note any adverse side-effects (Corot *et al.*, 2006), a notable exception is the report of preclinical safety studies of a USPIO formulation that showed low to moderate toxicity at lower concentrations, but significant toxic effects at high doses (Bourrinet *et al.*, 2006). The authors reported induction of neurobehavioral and neurovegetative effects at moderate doses, as well as reproductive toxicity including fetal malformations and teratogenicity in rats and rabbits, yet concluded that these findings should be considered acceptable within the overall biocompatible profile.

Despite the somewhat alarming observations described above, the number of studies further investigating potential IOMNP-mediated nanotoxicity, particularly *in vivo,* has been markedly deficient. Yet in spite of the lack of studies, there still remains a significant body of work from various studies ranging from particle–cell interactions to particle toxicology that can be pieced together to form a better understanding of both the demonstrated and potential nanotoxicity of IOMNPs.

20.5   Mechanisms of IOMNP Nanotoxicity

20.5.1   Initial Interactions and Entry into Cells

Within the body or *in vitro* culture, nanoparticles inevitably interact with numerous biomolecules and cells. Interactions with the surfaces of these cells provide opportunity for passage across the plasma membrane and entry into the cytoplasm via various mechanisms. These interactions and mechanisms act as a 'gateway' into the cellular milieu and a host of potentially toxic reactions within the cell. There are, in short, six major pathways of entry into the cell interior: phagocytosis, clathrin-mediated endocytosis, caveolae-mediated endocytosis, non-specific endocytosis, macropinocytosis, and passive diffusion across the membrane (Unfried *et al.*, 2007). Studies have repeatedly revealed that the mechanism of IOMNP or other nanoparticle entry into cells is highly reliant upon the physico-chemical characteristics, such as size and charge, of the nanoparticle as well as its surface characteristics (Storm *et al.*, 1995; Thorek and Tsourkas, 2008). In fact, depending on their coatings, all of these pathways have been observed as routes of entry for various different IOMNPs. Figure 20.2 contains a schematic and general overview of these routes of entry, as well as citations of examples within the IOMNP literature.

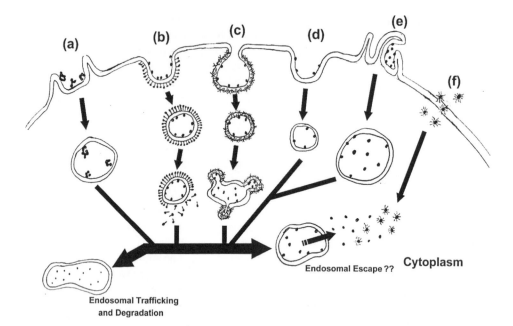

Figure 20.2 *Primary pathways of IOMNP entry into the cell (select relevant examples of IOMNP studies are cited). In (a) phagocytosis, larger or aggregates of IOMNPs are engulfed by actin-dependent extrusions of the plasma membrane, leading to the formation of phagosomes, early endosomes and degradation via endosomal processing (Schulze et al., 1995; Yamada et al., 1998; Nakayama et al., 2000; Zhang et al., 2002; Park, 2003; Raynal et al., 2004; Rogers and Basu, 2005; von zur Muhlen et al., 2007). The process of (b) clathrin-mediated endocytosis begins with the interaction of the IOMNPs with surface receptors leading to the recruitment of clathrin and the formation of invaginations of the plasma membrane (Berry et al., 2003; Wilhelm et al., 2003; Lanone and Boczkowski, 2006; Harush-Frenkel et al., 2007; Lu et al., 2007). These invaginations are internalized before shedding their clathrin coating and progressing toward early endosomes. Similarly, (c) caveolae-mediated endocytosis begins by the formation of caveolin-coated, flask-shaped invaginations in the plasma membrane. IOMNPs can interact with these structures and enter via endocytosis of caveolin-coated vesicles (Harush-Frenkel et al., 2007). The vesicles typically proceed to fuse together to form caveosomes before later endosomal trafficking. (d) Non-specific endocytosis includes pathways not involving clathrin or caveolae, but ultimately proceeds to undergo typical endosomal trafficking (Lanone and Boczkowski, 2006; Lu et al., 2007). The process of (e) macropinocytosis is typically induced by CPP-coated IOMNPs (Wilhelm et al., 2002; Gupta et al., 2007; McCarthy et al., 2007). Actin-dependent protrusions engulf associated and nearby IOMNPs, as well as the surrounding fluids, leading to the formation of special endosomes called macropinosomes and later endosomal trafficking. Finally, certain (membrane soluble) IOMNPs may directly pass through the membrane into the cytoplasm via (f) passive diffusion across the membrane (Zhang et al., 2002; Gupta and Curtis, 2004a). With the exception of the last pathway, each of these processes is thought normally to result in degradation of the IOMNPs (or other cargo) in late endosomes and lysosomes. However, multiple studies have shown that nanoparticles and PM do reach the cytoplasm and the cellular constituents through mechanisms such as diffusion across the membrane or leakage from endosomes.*

When considering entry into cells, it may be useful to consider entry into so-called 'professional' phagocytic cells such as macrophages independently from other types of non-phagocytic cells. It is well established that macrophages, the professional phagocytic cells of the immune system, are a primary response to the presence of foreign particles/matter *in vivo,* and the case of IOMNPs is no exception. Several studies *in vivo* have specifically observed massive macrophage infiltration and uptake of IOMNPs (Moore *et al.*, 1997; Lacava *et al.*, 1999b; Ruehm *et al.*, 2001; Kooi *et al.*, 2003; Raynal *et al.*, 2004; Garcia *et al.*, 2005; Muldoon *et al.*, 2005; Bourrinet *et al.*, 2006). Consequently, interest in the mechanism of uptake by macrophages developed prior interest in other cells. Early investigations into the mechanisms of IOMNP macrophage interactions by Weissleder and colleagues (Schulze *et al.*, 1995; Moore *et al.*, 1997) revealed that modifications to the surface of SPIOs, such as opsonization, may not only result in different amounts of IOMNPs being delivered into the cell, but also in different mechanisms of delivery. In the latter study, entry into macrophages was increased six-fold simply through passively coating the particles with protein (opsonization). The authors concluded that the observed increase in opsonized particle uptake was presumably receptor-based, due to the prominence of scavenger and complement receptors on the surface of the cells (Park, 2003). Likewise, Zhang *et al.* (2002) and Hu *et al.* (2006) observed that IOMNPs coated with polyethylene glycol (PEG) could reduce protein adsorption compared with bare particles and resulted in significantly less uptake by macrophages *in vitro*.

Several investigators have sought to investigate further the mechanism(s) of entry of SPIOs and other various IOMNPs into macrophages as a means of potentially manipulating these interactions in order to control the fate of IOMNPs *in vivo*. In a study seeking to elucidate the exact mechanism of uptake by these cells, Raynal *et al.* (2004) observed the differences in uptake between ferumoxide (superparamagnetic iron oxide; SPIOs) and ferumoxtran-10 (ultrasmall superparamagnetic iron oxide; USPIOs). Ostensibly, the only difference between these IOMNPs is their size, with the former having an average *hydrodynamic* diameter ranging between 120 and 180 nm and the latter between 15 and 30 nm. This difference, however, resulted in drastically different uptake rates in this study, which seems to at least partially explain the discrepancy in their half-lives *in vivo* (<10 minutes for ferumoxide (Nakayama *et al.*, 2000) vs. 24 to 36 hours for ferumoxtran-10). The authors also studied the mechanism of ferumoxide uptake and, using inhibition studies, revealed that the ferumoxide particles seem to undergo scavenger receptor A (SR-A) mediated endocytosis by macrophages *in vitro*. This class of receptors is primarily involved in the removal of cellular and other debris *in vivo,* and also serves a role in host defense. Significantly, the defense response by SR-As has been implicated in the development of neurological diseases such as Alzheimer's disease (Yamada *et al.*, 1998). While the authors did not report the mechanism of Ferumoxtran-10 uptake, other recent studies have indicated that these particles undergo endocytosis via mediation through a different cell surface receptor, namely Mac-1 (CD11b/CD18), and may indicate that particle size determines with which receptors the IOMNPs interact (von zur Muhlen *et al.*, 2007). The authors additionally showed that USPIOs are endocytosed more readily by activated macrophages than non-activated ones, indicating that not only cell type, but also cell phenotype and changes in membrane receptor expression can significantly influence the uptake of IOMNPs. This additionally corroborates the findings in a study by Rogers and

Basu (2005) where activated macrophages exhibited different uptake profiles depending upon their exposure to various cytokines.

Barring a few select studies, the precise uptake mechanisms of IOMNPs into non-phagocytic cells have not been systematically investigated. What has been gathered, however, is that there can be dramatic differences in the rate of entry into the cells simply based upon the choice of the many available IOMNP coatings (Gupta and Gupta, 2005b). In general, the lowest levels of intracellular delivery have been reported in studies of IOMNPs coated with polymers, particularly natural polysaccharides such as dextran (including SPIO and USPIO) and pullalan (Gupta and Gupta, 2005a). The polymer coatings are typically neutrally charged in order to stabilize the ferrofluid from aggregation by sterically repelling the other IOMNPs. This neutral steric repulsion is also effective at reducing particle–cell interactions and, thus, effectively reduces entry into both phagocytic and non-phagocytic cells alike when compared with bare and charged particles (Chouly *et al.*, 1996). Similar to macrophages, even the modest uptake of dextran-coated particles in non-phagocytic cells has recently been shown to be dependent upon the size of the nanoparticles themselves (Thorek and Tsourkas, 2008). In other studies involving the use of synthetic polymers such as polyethylene glycol (PEG), researchers have observed reduced uptake in some cells and *increased* uptake in others via non-endocytotic mechanisms, including passive diffusion through the plasma membrane (Zhang *et al.*, 2002; Gupta and Curtis, 2004a). The increased uptake in these cells has not been well-characterized, and has been ostensibly attributed to either the membrane-soluble nature of PEG (Zhang *et al.*, 2002) or simply increased fluid-phase endocytosis (Gupta and Curtis, 2004a). In certain other applications (e.g. magnetofection) the polymeric coating may actually be endowed with charge in order to *facilitate* IOMNP–cell interactions and promote entry into the cell (Huth *et al.*, 2004; Pan *et al.*, 2008). Huth *et al.* (2004) used polethylenimine (PEI) coated IOMNPs as a vehicle for significantly enhancing gene transfection via both clathrin and caveolae-mediated endocytosis. Likewise, Harush-Frenkel *et al.* (2007) also investigated the use of charged polymeric coatings, but blocked the clathrin pathway and showed that IOMNPs could still undergo endocytosis via an independent (nonspecific) pathway.

Studies of cell interactions and labelling involving the use of proteins and antibodies as IOMNP coatings have shown that they typically enter through classical endocytotic mechanisms, which can, in some cases, be modulated or blocked by the choice of protein coating. For example, Berry *et al.* (2003) observed reduced levels of clathrin-mediated endocytosis in albumin-coated IOMNPs vs. bare and dextran-coated particles. Similarly, specific receptors, such as folate receptors, have been targeted by coating IOMNPs with complementary ligands, resulting in the induction of receptor-mediated endocytosis (Zhang *et al.*, 2002; Kohler *et al.*, 2005). In other studies, Curtis and colleagues demonstrated that targeting of certain cell surface receptors, such as the insulin receptor, or coating with iron binding proteins, such as transferrin, could act to *prevent* endocytosis (Gupta *et al.*, 2003; Berry *et al.*, 2004b). Other types of IOMNP coatings such as silica have been shown to be innately internalized by many cell types, including multiple organ systems *in vivo* (Kim *et al.*, 2006) and stem cells *in vitro* (Lu *et al.*, 2007). Using clathrin inhibitors, further investigation of silica IOMNP–cell interactions by Lu *et al.* (2007) revealed clathrin-mediated endocytosis of these nanoparticles as the likely route of entry into mesenchymal stem cells.

Over the last several years there have been many reports of experiments aimed at delivering very high levels of IOMNPs into cells via various surface coatings in order to track cells of interest via high-resolution MRI. In general, the most successful delivery methods have utilized coatings of highly charged molecules such as CPPs, the TAT peptide in particular (Lewin *et al.*, 2000; Zhao *et al.*, 2002), as well as strongly anionic coatings such as dimercaptosuccinic acid (DMSA) (Wilhelm *et al.*, 2002, 2003) or citric acid (Stroh *et al.*, 2004; de Freitas *et al.*, 2008). While TAT-coated IOMNPs would be expected to be electrostatically attracted to the dominant negative charge of the cell surface (due to sialic acid residues) and enter the cell via macropinocytosis (Gump and Dowdy, 2007), it is interesting that anionic coatings also seem to be effective at coordinating the particles to isolated positively charged regions on the cell surface. Further studies by Wilhelm and Gazeau (2008) have revealed that the anionic magnetic nanoparticles (AMNPs) enter the cell through both clathrin-mediated endocytosis as well as nonspecific endocytosis. The observed efficiency of these and similar particles, however, may be somewhat of a double-edged sword, as highly electrostatic coatings have shown a disproportionately high incidence rate of toxicity compared with other coatings, demonstrating toxicity in multiple *in vitro* and *in vivo* studies (see Tables 20.1 and 20.2).

20.5.2   Working Models of Nanotoxicity

The majority of potentially toxic interactions (i.e. the mechanisms of nanotoxicity) are thought to occur once the iron oxide nanoparticles reach the interior of the cell. Still in its infancy, the field of nanotoxicology owes most of its beginnings and working models to the field of modern particle toxicology. Particle toxicology arose from studies of the effects of inhaled 'ultrafine particles' such as coal dust and asbestos, as well as ambient air particulate matter (PM). The field of PM toxicology and its intimate relation to nanotoxicology has recently been the subject of several reviews, ranging from their historic development (Stone and Donaldson, 2007) to overviews of emerging toxicological concepts (Lanone and Boczkowski, 2006; Unfried *et al.*, 2007). Interestingly, some of the major constituents of PM are transition metals and it is the oxides of these transition metals, including IOMNPs, that are thought to be primarily responsible for the oxidative stress caused by inhalation of PM. Much work has gone into the study of the mechanisms of PM-induced health effects and the role that transition metals may play in the generation of free radicals. There is little reason to doubt that similar mechanisms of toxicity might be involved in the case of engineered IOMNPs.

The intrinsic geometric characteristics of IOMNPs and other nanoparticles endow them with their potent chemical reactivity. As the physical size of particles decreases, the surface area to volume ratio exponentially increases, as do the surface molecules that are free to react with the cellular milieu. It is the sheer number of IOMNPs coupled with the reactivity of the transition metal surface molecules that give rise to the massive oxidizing capabilities of the nanoparticles. While the mammalian cellular machinery includes defenses against oxidative stress such as glutathione (GSH) and antioxidant enzymes, these defenses can be overmatched by the formidable oxidizing capacity of the nanoparticles.

There are thought to be at least four primary sources of oxidative stress in response to PM (or IOMNPs): direct generation of reactive oxygen species (ROS) from the surface

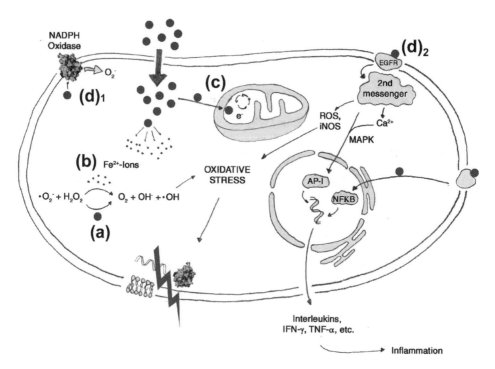

Figure 20.3 *Potential mechanisms of IOMNP toxicity and induction of oxidative stress; examples of IOMNP or PM studies reporting these mechanisms are cited. Upon passage through the plasma membrane via the mechanisms described in Figure 20.2, IOMNPs can interact with cellular constituents through a number of ways. The IOMNPs can (a) directly generate ROS from their surfaces via Harber-Weiss reactions or, alternatively, through (b) iron that has leached from the particle (van den Bos et al., 2003; Cadenas, 2004; Stroh et al., 2004; Jain et al., 2008). The IOMNP can also directly interact with and (c) alter mitochondrial or other organelle functions (Arimoto et al., 2005; Au et al., 2007; Soto et al., 2007). Lastly, the particles may (d)$_1$ directly interact with redox active proteins such as NADPH oxidase (Arimoto et al., 2005; Amara et al., 2007) and (d)$_2$ induce cell signalling pathways that lead to activation of inflammatory cells, generation of ROS and reactive nitrogen species such as nitric oxide (NO) (Lacava et al., 1999a,b; Jimenez et al., 2000; Schins and Donaldson, 2000; Zhou et al., 2003; Chaves et al., 2005; Garcia et al., 2005; Naveau et al., 2006; Sigliente et al., 2006; Zhai et al., 2008; Zhu et al., 2008). Reactive species such as ROS and NO cause oxidative damage to cellular components such as proteins, lipids and DNA that lead to numerous pathologies.*

of the IOMNPs; generation of ROS via leaching of iron molecules from the iron oxide particles; altered mitochondrial and other organelle function; and induction of cell signalling pathways and activation of inflammatory cells resulting in the generation of ROS and reactive nitrogen species such as nitric oxide (NO) (Risom *et al.*, 2005). Figure 20.3 shows a schematic of these interactions, as well as references to the IOMNP-related studies that have demonstrated them. Based upon the results of the various *in vitro* studies covered herein, it appears that engineered IOMNPs have exhibited the potential to induce oxidative stress via all four of these mechanisms. Thus it would be useful to review these mechanisms and how studies may indirectly or directly support their existence as potential sources of observed IOMNP nanotoxicity.

20.5.2.1 Generation of ROS from IOMNPs

It is well-documented that transition metals such as iron can generate ROS through Harber-Weiss type reactions:

$$\bullet O2- + H2O2Fe \rightarrow \bullet OH + OH- + O2 \tag{1}$$

In this reaction, superoxide, produced via normal metabolic processes and redox cycling within the cell, reacts with hydrogen peroxide to form the hydroxyl free radical. The catalyst at the center of this reaction, however, is the iron that has been brought into the cell by the IOMNP nanoparticle, on the surface of the particles themselves or molecular iron that has leached from the nanoparticles. It is the reactions that these molecules undergo that can ultimately be to blame for the excessive production of the hydroxyl free radicals responsible for oxidative stress and damage to the cell. These reactions are known as Fenton chemistry and can be described by the equations:

$$Fe^{3+} + \bullet O_2^- \longrightarrow Fe^{2+} + O_2| \tag{2}$$
$$Fe^{2+} + H_2O_2 \longrightarrow Fe^{3+} + \bullet OH + OH^- \tag{3}$$

The potent hydroxyl free radical generated by these reactions is then free to attack any biomolecule (e.g. lipids, proteins, DNA, etc.) within diffusive distance.

Several studies already mentioned have shown direct evidence of ROS damage by IOM-NPs. In particular, the study by van den Bos *et al.* (2003) directly showed an IOMNP dose-dependent increase in lipid peroxidation, while Stroh *et al.* (2004) measured significant increases in both lipid and protein oxidation. Likewise, Jain *et al.* (2008) noted considerable increases in lipid peroxidation in several tissues *in vivo* following administration of IOMNPs to Sprague-Dawley rats. As an additional corollary, Alekseenko *et al.* (2008) also studied the effects of IOMNPs upon neuronal cells. In contrast to others, however, they studied the effects of ferritin, the natural iron storage protein that exists in cells of all types and contains a 7 nm iron oxide core, surrounded by a protein coat (Theil *et al.*, 2006). In these studies ferritin was found to generate ROS directly in rat synaptosomes, which the authors concluded could ultimately lead to neurodegeneration *in vivo*.

20.5.2.2 Alteration of Organelle and Mitochondrial Function

Mitochondria produce energy via the citric acid cycle and are critically dependent upon redox reactions from the respiration chain. Electrons that leak from this process have been shown to be a constant source of superoxide and hydrogen peroxide in eukaryotes (Cadenas, 2004) and are free to react with iron via Fenton chemistry. Perhaps more importantly, however, are the potential interactions of the IOMNPs with the mitochondria themselves. Nanoparticles have been implicated to be in direct contact with, and to produce damage within, mitochondria (Sioutas *et al.*, 2003). Given this proximity to the mitochondria, it is highly likely that the redox active surface of IOMNPs could significantly affect electron flow and act to alter mitochondrial functionality.

A common assay employed for testing cell viability is the MTT (and MTS) assay. The assay indirectly measures viability by testing for the presence of active reductase enzymes within the mitochondria of living cells, ultimately causing a color change in the MTT reagent upon its reduction. Although commonly used as a viability assay, the MTT assay more specifically represents a measure of mitochondrial function as opposed to cell viability. Since changes in mitochondrial reductive capacity do not always covary with viability, the MTT assay can in principle be used to measure alteration of mitochondrial function in *living* cells. Likewise, given the likely negative influence that nanoparticles such as IOMNPs have upon mitochondrial redox reactions, there is reason to believe that the utility of MTT in evaluating IOMNP cytotoxicity may be suboptimal. Several studies described above utilized the MTT assay to assess mitochondrial function; however, the authors typically assumed nonviability in those cells exhibiting reduced reductive activity (Gupta and Curtis, 2004b; Hussain *et al.*, 2005; Au *et al.*, 2007; Soto *et al.*, 2007).

Another potential intracellular target for IOMNP-associated toxicity is the plasma membrane and proteins. In addition to induction of cell signalling pathways, IOMNP-induced redox reactions can activate and upregulate plasma membrane proteins such as nicotinamide adenine dinucleotide phosphate (NADPH) oxidase (Amara *et al.*, 2007) and its analogs (Arimoto *et al.*, 2005), thereby inducing generation of the oxidase product, superoxide anion, O_2^-. Importantly, this activation is particularly known to occur within phagocytic cells such as the previously mentioned macrophages (Park, 2003), which are known to take up IOMNPs reliably *in vitro* and within the body (Storm *et al.*, 1995; Moore *et al.*, 1997; Weissleder *et al.*, 1997; Ruehm *et al.*, 2001; Zhang *et al.*, 2002; Kooi *et al.*, 2003; Raynal *et al.*, 2004; Soto *et al.*, 2005, 2007; Muller *et al.*, 2007).

20.5.2.3   Induction of Cell Signalling and Inflammatory Pathways

While activation of macrophage transmembrane proteins such as NADPH oxidase can cause direct production of ROS and oxidative stress within individual cells, it is the activations of proteins involved in signalling pathways, such as epidermal growth factor receptor (EGFR) and those associated with the nuclear factor-kappa B (NFκB) pathway, that are thought to engender the development of the potentially larger detrimental effects *in vivo* (Donaldson *et al.*, 2003). Stimulation of these enzymes can result in the induction of signalling cascades and, ultimately, a much broader inflammatory response. For example, PM is known to directly or indirectly (through oxidative stress) activate EGFR (Zanella *et al.*, 1999; Sydlik *et al.*, 2006), inducing generation of IP3 leading towards increased ROS production and calcium production (Robison *et al.*, 1995; Barthel and Klotz, 2005), as well as activation of inducible nitric oxide synthase (iNOS) (Korhonen *et al.*, 2001). Increases in cytosolic calcium coupled with EGFR tyrosine kinase activity lead to phosphorylation of a family of mitogen-activated protein kinases (MAPK) including c-Jun N-terminal kinase (JNK1/2), p38 kinase and extracellular signal regulated kinase (ERK1/2) resulting in the activation of the transcription factor, activating protein-1 (AP-1) (Albrecht *et al.*, 2004). It is activation of AP-1 that ultimately leads to the production of inflammatory mediators and inflammation. Likewise, NFκB, the oxidative stress response transcription factor found ubiquitously within eukaryotes (Schreck *et al.*, 1992), has also been shown to be activated by numerous PM, including iron oxide (Jimenez *et al.*, 2000; Zhu *et al.*, 2008), and also results in induction of inflammatory pathways including production of various cytokines such as various interleukins, interferon gamma (IFN-γ), tumor growth factor beta (TGF-β) and tumor necrosis factor alpha (TNF-α) (Schins and Donaldson, 2000; Albrecht *et al.*, 2004).

There has been little direct evidence of the induction of inflammatory pathways and cytokines by *engineered* IOMNPs *in vitro* or *in vivo;* however, there are a few notable exceptions. Two recent *in vitro* studies of the effect of IOMNP loading upon macrophage function have revealed modification of cellular behaviors as well as modulated cytokine expression. Siglienti *et al.* (2006) observed that loading macrophages with either SPIO or USPIO resulted in enhanced IL-10 production and inhibition of TNF-α, indicating potential immunomodulatory capabilities. Hsiao *et al.* also studied the response of SPIO loading upon macrophages and found that high doses/levels of SPIO induced the secretion of TNF-α and resulted in production of nitric oxide. Similarly, in another study in cooperation with Gazeau, Naveau *et al.* (2006) showed that IOMNP (AMNP) labelling of human gingival fibroblasts resulted in significant signs of inflammation, including increased expression of interleukins 1 and 4, as well as secretion of matrix metalloproteinases (MMPs).

Only a select number of *in vivo* studies have demonstrated inflammatory responses to IOMNPs. In particular, the previously mentioned studies by Lacava *et al.* of intraperitoneal and lung injection of DMSA and citric acid coated IOMNPs into mice resulted in massive macrophage infiltration, severe inflammatory reactions, IL-1 and 10 production, and apoptosis (Lacava *et al.*, 1999a,b; Garcia *et al.*, 2005; Chaves *et al.*, 2005). Incidentally, comparable studies of the effect of iron PM inhalation have produced similar inductions of cytokine expression (Zhou *et al.*, 2003), as well as severe inflammation and associated immune responses (Zhu *et al.*, 2008). Lastly, Zhai *et al.* (2008) studied the effects of subcutaneous injection of dextran-coated IOMNPs into mice and, after histological evaluation, found moderate inflammation, including local infiltration by leukocytes and macrophages.

20.6   Long-Term Nanotoxicity of IOMNPs – General and Neurological Toxicity

Due to the relatively recent advent of IOMNPs engineered for *in vivo* use, there are currently no studies investigating the long-term effects that exposure to them might have. There may, however, be knowledge that can be extrapolated from analogous studies in PM toxicology that might indicate potential long-term implications. For example, the well-studied oxidative stress paradigm is known to be involved in the development of numerous disorders such as: pulmonary injury and morbidity (Li *et al.*, 2008), atherosclerosis (Bonomini *et al.*, 2008), cardiovascular disease (Dhalla *et al.*, 2000), diabetes mellitus (Maiese *et al.*, 2007), reproductive dysfunction (Turner and Lysiak, 2008), cancer (Gago-Dominguez and Castelao, 2008), and others (Cachofeiro *et al.*, 2008; Chrissobolis and Faraci, 2008).

One system that has been shown to be exceptionally susceptible to transition metals such as iron and the effects of oxidative stress is the neurological system (Galazka-Friedman, 2008). In particular, it has become well-established that iron and iron accumulation is associated with multiple neurodegenerative diseases including Parkinson's and Alzheimer's disease (Barnham and Bush, 2008). Since the discovery of iron oxide particles in the brains of humans over 15 years ago by Kirschvink *et al.* (1992), investigators have repeatedly reported the presence of high levels of accumulated iron in the brains of patients suffering from neurodegenerative disorders. Currently it is unclear as to where this excess iron originates; however, several theories have been postulated including derivation from exogenous sources such as inhaled PM (Calderon-Garciduenas *et al.*, 2002; Mohan-Kumar *et al.*, 2008) which may also have implications for the use of engineered IOMNPs (Alekseenko *et al.*, 2008). Iron has traditionally been thought to cross the blood–brain barrier primarily through a transferrin-mediated route through the capillary endothelium; however, there are multiple other routes including passage through olfactory nerves or circumventricular organs (Begley, 2004). Additionally, iron has also been shown to be transported via ferritin (Fisher *et al.*, 2007) and may represent a method of transport from dissolved iron derived from IOMNPs to the brain. This fact was highlighted in the study by Alekseenko *et al.* (2008), who demonstrated that the iron nanoparticles within ferritin proteins could generate ROS within brain synapses, as well as interfere with the transmission of neuronal signals. Furthermore, several studies on the biodistribution of IOMNPs and other nanoparticles have shown particle migration to the brains of exposed subjects (Kim *et al.*, 2006; Jain *et al.*, 2008; Kwon *et al.*, 2008). This, coupled with the fact that nanoparticles have also been shown to be selectively toxic to neuronal type cells (Block *et al.*, 2004; Pisanic *et al.*, 2007; Ali and Hussain, 2008), gives rise to concern regarding the potential effects that exposure to IOMNPs, whether deliberate, as in the case of MRI applications, or incidental, such as environmental PM, might have.

The brain is especially susceptible to the effects of oxidative stress, as it is highly reliant upon iron for its high respiratory activity, as well as myelinogenesis and the production of numerous neurotransmitters. This high oxidative metabolism and resulting generation of ROS make the brain particularly susceptible to imbalances in the oxidative equilibrium that are thought to be related to, or cause, neurodegeneration and associated disorders. While excess iron can directly produce toxic effects on neurons according to the classical oxidative stress paradigm, it is thought that one of the prime mediators and culprits of ROS-mediated neurotoxicity are microglia (Block *et al.*, 2007). Microglia are the professional phagocytic cells of the central nervous system, and serve many of the same functions as macrophages do for the rest of the body, including phagocytosis, antigen presentation and cytokine release. Exposure to metal oxides has been shown to specifically activate these cells *in vitro* and result in the generation of ROS that can damage and kill dopaminergic neurons (Block *et al.*, 2004; Long *et al.*, 2006). But what is particularly troublesome is that damaged or dead neurons, particularly those from an instigating stimulus, such as might be derived from IOMNP exposure, can also further activate microglia, inducing a condition known as reactive microgliosis that results in a vicious and destructive cycle progressing to neurodegenerative disease (Block *et al.*, 2007). Given simply the preliminary evidence derived from fields such as particle toxicology, brain neurotoxicity is just one of many potential long-term effects of IOMNP exposure that can only be speculated upon. While the proposition of these potential long-term effects may be somewhat a matter of conjecture, it is nonetheless important to bear them in mind when designing methods and appropriate model systems for IOMNP and other nanoparticle nanotoxicological evaluations. Likewise, it is imperative that specific and deliberate guidelines for the evaluation of nanomaterials be put in place in order to guard against unforeseen consequences of their use.

20.7   Considerations for the Future of IOMNP Nanotoxicological Testing

One thing that can certainly be surmised from a review of the literature regarding IOMNP nanotoxicology is that there exists a wide range of observations and conclusions regarding their toxicity. While many of these reports may appear contradictory, there may be a number of reasons for these discrepancies. One important distinction is that each IOMNP formulation is unique and may have slightly or even vastly different physico-chemical characteristics. Not only that, but even subtle differences in the synthesis of nominally identical formulations may exert effects on the properties such as the uniformity and coating stability between manufactured lots. Likewise, it is imperative that, prior to testing of any nanoparticle formulation, these attributes be analyzed and recorded. While tedious, this precaution may allow for future, time-saving guidelines for IOMNP production and safety. In order to adequately assess and compare the nanotoxicity of a given IOMNP formulation, the starting and final materials must reach a degree of uniformity. The second and previously stated difference in published studies, particularly those *in vitro*, is the heterogeneity of the cell types tested. This is due to the relative immaturity of and lack of standardized practice within the field of nanotoxicology. Once again, in order to quantify or even compare toxicological profiles of different IOMNP formulations, a standard set of well-studied cell lines should be established for nanotoxicological testing. Likewise, within these cell lines, *quantifiable* parameters and endpoints also need to be established for each implemented cell line. As an example, in the previously mentioned study by Hussain *et al.* (2005), the effects of various nanoparticle formulations upon *physiologically relevant* parameters of liver cells such as mitochondrial function, GSH levels, and the generation of ROS were determined. Likewise, in our evaluation of the effects of AMNPs upon the pheochromocytoma line, PC12, a cell line that has been widely studied and characterized (Tischler and Greene, 1978), we used readily quantifiable criteria such as viability, neurite length and neurite frequency to determine the effects of IOMNP exposure (Pisanic *et al.*, 2007). Other potentially relevant criteria that have been used in others' toxicity studies of this cell line include measures of protein expression levels such as growth-associated protein (GAP)-43 or cellular products, such as dopamine (Hussain *et al.*, 2006). In order to bring a higher level of consistency, it is likewise recommended that these criteria be developed and established by panels of experts within the fields of both nanotechnology and toxicology.

A third issue that should be addressed is simply the overall deficiency of thorough independent investigations of the effects of IOMNP exposure, particularly upon systems that are known to be susceptible to the effects of oxidative stress, such as the central nervous system. If the disproportion between the number of independent *in vitro* IOMNP nanotoxicity studies and the numerous proposed applications can be considered poor at best, the number of independent *in vivo* nanotoxicity studies can be considered abysmal. Given this discrepancy, a concerted effort between academia, industry and governmental institutions to evaluate IOMNPs and other nanomaterials must begin, and has likewise been recommended by numerous experts (Donaldson *et al.*, 2004; Service, 2005; Lanone and Boczkowski, 2006; Nel *et al.*, 2006; Medina *et al.*, 2007).

20.8   Conclusions

IOMNPs exhibit many unique properties such as superparamagnetism, remote manipulation, magnetic detection and the ability to be used within the innate magnetic field transparency of biological tissues. These attributes offer numerous advantages and opportunities for nanobiotechnology, and likewise have engendered innumerable demonstrations and propositions for their use in biomedical applications. Initial studies into the nanotoxicity of IOMNPs resulted in a preliminary endorsement of biocompatibility that has been all but assumed since. However, a growing list of literature in the field of nanotoxicology is now beginning to evaluate this tenet. Multiple studies have demonstrated that, in many instances, IOMNPs can and do induce nanotoxicological effects. While overall the exact mechanisms of demonstrated and potential IOMNP toxicities have yet to be adequately elucidated, the extension of valuable principles from more mature fields such as particle toxicology will likely shed light on them in the near future. It is imperative that proper precautions and tests be put into place and implemented in order to protect not only human subjects, but also the future of the nanotechnology industry. Failure to do so may ultimately result in widespread public distrust of and opposition to the future development of potentially ground-breaking medical advancements.

References

Albrecht C, Borm PJA, Unfried K (2004) Signal transduction pathways relevant for neoplastic effects of fibrous and non-fibrous particles. *Mutat Res-Fundam Mol Mech Mutagen* **553**, 23–35.

Alekseenko AV, Waseem TV, Fedorovich SV (2008) Ferritin, a protein containing iron nanoparticles, induces reactive oxygen species formation and inhibits glutamate uptake in rat brain synaptasomes. *Brain Res* **1241**, 193–200.

Ali S, Hussain S (2008) Nanoparticles and nanomaterials: Friend or foe. *Toxicology Letters* **180**, S21–S21.

Amara N, Bachoual R, Desmard M, Golda S, Guichard C, Lanone S, Aubier M, Ogier-Denis E, Boczkowski J (2007) Diesel exhaust particles induce matrix metalloprotease-1 in human lung epithelial cells via a nadp(h) oxidase/nox4 redox-dependent mechanism. *Am J Physiol-Lung Cell Mol Physiol* **293**, L170–L181.

Arimoto T, Kadiiska MB, Sato K, Corbett J, Mason RP (2005) Synergistic production of lung free radicals by diesel exhaust particles and endotoxin. *Am J Respir Crit Care Med* **171**, 379–387.

Au C, Mutkus L, Dobson A, Riffle J, Lalli J, Aschner M (2007) Effects of nanoparticles on the adhesion and cell viability on astrocytes. *Biol Trace Elem Res* **120**, 248–256.

Auffan M, Decome L, Rose J, Orsiere T, De Meo M, Briois V, Chaneac C, Olivi L, Berge-Lefranc JL, Botta A, Wiesner MR, Bottero JY (2006) In vitro interactions between DMSA-coated maghemite nanoparticles and human fibroblasts: A physicochemical and cyto-genotoxical study. *Environ Sci Technol* **40**, 4367–4373.

Barnham KJ, Bush AI (2008) Metals in Alzheimer's and Parkinson's diseases. *Current Opinion in Chemical Biology* **12**, 222–228.

Barthel A, Klotz LO (2005) Phosphoinositide 3-kinase signaling in the cellular response to oxidative stress. *Biol Chem* **386**, 207–216.

Begley DJ (2004) Delivery of therapeutic agents to the central nervous system: The problems and the possibilities. *Pharmacol Ther* **104**, 29–45.

Berry CC, Wells S, Charles S, Curtis ASG (2003) Dextran and albumin derivatised iron oxide nanoparticles: Influence on fibroblasts in vitro. *Biomaterials* **24**, 4551–4557.

Berry CC, Wells S, Charles S, Aitchison G, Curtis ASG (2004a) Cell response to dextran-derivatised iron oxide nanoparticles post internalisation. *Biomaterials* **25**, 5405–5413.

Berry CC, Charles S, Wells S, Dalby MJ, Curtis ASG (2004b) The influence of transferrin stabilised magnetic nanoparticles on human dermal fibroblasts in culture. *Int J Pharmaceutics* **269**, 211–225.

Block ML, Wu X, Pei Z, Li G, Wang T, Qin L, Wilson B, Yang J, Hong JS, Veronesi B (2004) Nanometer size diesel exhaust particles are selectively toxic to dopaminergic neurons: The role of microglia, phagocytosis, and nadph oxidase. *Faseb J* **18**, 1618–1620.

Block ML, Zecca L, Hong JS (2007) Microglia-mediated neurotoxicity: Uncovering the molecular mechanisms. *Nat Rev Neurosci* **8**, 57–69.

Bonomini F, Tengattini S, Fabiano A, Bianchi R, Rezzani R (2008) Atherosclerosis and oxidative stress. *Histology and Histopathology* **23**, 381–390.

Bourrinet P, Bengele HH, Bonnemain B, Dencausse A, Idee JM, Jacobs PM, Lewis JM (2006) Preclinical safety and pharmacokinetic profile of ferumoxtran-10, an ultrasmall superparamagnetic iron oxide magnetic resonance contrast agent. *Invest Radiol* **41**, 313–324.

Brunner TJ, Wick P, Manser P, Spohn P, Grass RN, Limbach LK, Bruinink A, Stark WJ (2006) In vitro cytotoxicity of oxide nanoparticles: Comparison to asbestos, silica, and the effect of particle solubility. *Environ Sci Technol* **40**, 4374–4381.

Bulte JW, Douglas T, Witwer B, Zhang SC, Strable E, Lewis BK, Zywicke H, Miller B, van Gelderen P, Moskowitz BM, Duncan ID, Frank JA (2001) Magnetodendrimers allow endosomal magnetic labeling and in vivo tracking of stem cells. *Nat Biotechnol* **19**, 1141–1147.

Bulte JWM, Kraitchman DL (2004) Iron oxide mr contrast agents for molecular and cellular imaging. *NMR Biomed* **17**, 484–499.

Cachofeiro V, Goicochea M, de Vinuesa SG, Oubina P, Lahera V, Luno J (2008) Oxidative stress and inflammation, a link between chronic kidney disease and cardiovascular disease. *Kidney International* **74**, S4–S9.

Cadenas E (2004) Mitochondrial free radical production and cell signaling. *Mol Aspects Medicine* **25**, 17–26.

Calderon-Garciduenas L, Azzarelli B, Acuna H, Garcia R, Gambling TM, Osnaya N, Monroy S, Tizapantzi MD, Carson JL, Villarreal-Calderon A, Rewcastle B (2002) Air pollution and brain damage. *Toxicol Pathol* **30**, 373–389.

Chaves SB, Silva LP, Lacava ZCM, Morais PC, Azevedo RB (2005) Interleukin-1 and interleukin-6 production in mice's lungs induced by 2,3meso-dimercaptosuccinic-coated magnetic nanoparticles. *J Applied Physics* **97**, 10Q915.1–10Q915.3

Chouly C, Pouliquen D, Lucet I, Jeune J, Jallet P (1996) Development of superparamagnetic nanoparticles for MRI: Effect of particle size, charge and surface nature on biodistribution. *Journal of Microencapsulation* **13**, 245–255.

Chrissobolis S, Faraci FM (2008) The role of oxidative stress and NADPH oxidase in cerebrovascular disease. *Trends Mol Med* **14**, 495–502.

Chung SM (2002) Safety issues in magnetic resonance imaging. *J Neuro-Ophthal* **22**, 35–39.

Corot C, Robert P, Idee JM, Port M (2006) Recent advances in iron oxide nanocrystal technology for medical imaging. *Advanced Drug Delivery Reviews* **58**, 1471–1504.

de Freitas ERL, Soares PRO, Santos RD, dos Santos DL, da Silva JR, Porfirio EP, Bao SN, Lima ECD, Morais PC, Guillo LA (2008) In vitro biological activities of anionic gamma-Fe_2O_3 nanoparticles on human melanoma cells. *J Nanosci Nanotechnol* **8**, 2385–2391.

de la Fuente JM, Alcantara D, Penades S (2007) Cell response to magnetic glyconanoparticles: Does the carbohydrate matter? *IEEE Transactions on Nanobioscience* **6**, 275–281.

Dhalla NS, Temsah RM, Netticadan T (2000) Role of oxidative stress in cardiovascular diseases. *J Hypertension* **18**, 655–673.

Diaz B, Sanchez-Espinel C, Arruebo M, Faro J, de Miguel E, Magadan S, Yague C, Fernandez-Pacheco R, Ibarra MR, Santamaria J, Gonzalez-Fernandez A (2008) Assessing methods for blood cell cytotoxic responses to inorganic nanoparticles and nanoparticle aggregates. *Small* **4**, 2025–2034.

Dobson J (2006) Gene therapy progress and prospects: Magnetic nanoparticle-based gene delivery. *Gene Ther* **13**, 283–287.

Donaldson K, Stone V, Borm PJA, Jimenez LA, Gilmour PS, Schins RPF, Knaapen AM, Rahman I, Faux SP, Brown DM, MacNee W (2003) Oxidative stress and calcium signaling in the adverse effects of environmental particles (pm10). *Free Radic Biol Med* **34**, 1369–1382.

Donaldson K, Stone V, Tran CL, Kreyling W, Borm PJA (2004) *Nanotoxicology.* Occup Environ Med **61**, 727–728.

Fisher J, Devraj K, Ingram J, Slagle-Webb B, Madhankumar AB, Liu X, Klinger M, Simpson IA, Connor JR (2007) Ferritin: A novel mechanism for delivery of iron to the brain and other organs. *Am J Physiol-Cell Physiol* **293**, C641–C649.

Gago-Dominguez M, Castelao JE (2008) Role of lipid peroxidation and oxidative stress in the association between thyroid diseases and breast cancer. *Crit Reviews Oncology Hematology* **68**, 107–114.

Galazka-Friedman J (2008) Iron as a risk factor in neurological diseases. *Hyperfine Interactions* **182**, 31–44.

Garcia MP, Parca RM, Chaves SB, Silva LP, Santos AD, Lacava ZGM, Morais PC, Azevedo RB (2005) Morphological analysis of mouse lungs after treatment with magnetite-based magnetic fluid stabilized with DMSA. *J Magnetism and Magnetic Mat* **293**, 277–282.

Gazeau F, Levy M, Wilhelm C (2008) Optimizing magnetic nanoparticle design for nanothermotherapy. *Nanomedicine* **3**, 831–844.

Gump JM, Dowdy SF (2007) TAT transduction: The molecular mechanism and therapeutic prospects. *Trends Mol Med* **13**, 443–448.

Gupta AK, Berry C, Gupta M, Curtis A (2003) Receptor-mediated targeting of magnetic nanoparticles using insulin as a surface ligand to prevent endocytosis. *IEEE Transactions on Nanobioscience* **2**, 255–261.

Gupta AK, Curtis ASG (2004a) Surface modified superparamagnetic nanoparticles for drug delivery: Interaction studies with human fibroblasts in culture. *J Mater Sci-Mater Med* **15**, 493–496.

Gupta AK, Curtis ASG (2004b) Lactoferrin and ceruloplasmin derivatized superparamagnetic iron oxide nanoparticles for targeting cell surface receptors. *Biomaterials* **25**, 3029–3040.

Gupta AK, Gupta M (2005a) Cytotoxicity suppression and cellular uptake enhancement of surface modified magnetic nanoparticles. *Biomaterials* **26**, 1565–1573.

Gupta AK, Gupta M (2005b) Synthesis and surface engineering of iron oxide nanoparticles for biomedical applications. *Biomaterials* **26**, 3995–4021.

Gupta AK, Naregalkar RR, Vaidya VD, Gupta M (2007) Recent advances on surface engineering of magnetic iron oxide nanoparticles and their biomedical applications. *Nanomedicine* **2**, 23–39.

Harush-Frenkel O, Debotton N, Benita S, Altschuler Y (2007) Targeting of nanoparticles to the clathrin-mediated endocytic pathway. *Biochem Biophys Res Commun* **353**, 26–32.

Hilger I, Fruhauf S, Linss W, Hiergeist R, Andra W, Hergt R, Kaiser WA (2003) Cytotoxicity of selected magnetic fluids on human adenocarcinoma cells. *Journal of Magnetism and Magnetic Materials* **261**, 7–12.

Hsiao JK, Chu HH, Wang YH, Lai CW, Choii PT, Hsieh ST, Wang JL, Liu HM (2008) Macrophage physiological function after superparamagnetic iron oxide labeling. *NMR Biomed* **21**, 820–829.

Hu FX, Neoh KG, Cen L, Kang ET (2006) Cellular response to magnetic nanoparticles "PEGylated" via surface-initiated atom transfer radical polymerization. *Biomacromolecules* **7**, 809–816.

Hussain SM, Hess KL, Gearhart JM, Geiss KT, Schlager J (2005) In vitro toxicity of nanoparticles in brl 3a rat liver cells. *Toxicol Vitro* **19**, 975–983.

Hussain SM, Javorina AK, Schrand AM, Duhart HM, Ali SF, Schlager J (2006) The interaction of manganese nanoparticles with pc-12 cells induces dopamine depletion. *Toxicol Sci* **92**, 456–463.

Huth S, Lausier J, Gersting SW, Rudolph C, Plank C, Welsch U, Rosenecker J (2004) Insights into the mechanism of magnetofection using pei-based magnetofectins for gene transfer. *J Gene Medicine* **6**, 923–936.

Jacobs A, Worwood M (1980) *Iron in Biochemistry and Medicine.* Academic Press: London.

Jain TK, Reddy MK, Morales MA, Leslie-Pelecky DL, Labhasetwar V (2008) Biodistribution, clearance, and biocompatibility of iron oxide magnetic nanoparticles in rats. *Molecular Pharmaceutics* **5**, 316–327.

Jeng HA, Swanson J (2006) Toxicity of metal oxide nanoparticles in mammalian cells. *J Environ Science Health Part A - Toxic/Hazardous Substances & Environ Engineering* **41**, 2699–2711.

Jimenez LA, Thompson J, Brown DA, Rahman I, Antonicelli F, Duffin R, Drost EM, Hay RT, Donaldson K, MacNee W (2000) Activation of nf-kappa b by pm10 occurs via an iron-mediated mechanism in the absence of i kappa b degradation. *Toxicol Appl Pharmacol* **166**, 101–110.

Kim JS, Yoon TJ, Kim BG, Park SJ, Kim HW, Lee KH, Park SB, Lee JK, Cho MH (2006) Toxicity and tissue distribution of magnetic nanoparticles in mice. *Toxicol Sci* **89**, 338–347.

Kirschvink JL, Kobayashikirschvink A, Woodford BJ (1992) Magnetite biomineralization in the human brain. *Proc Nat Ac Sci USA* **89**, 7683–7687.

Kohler N, Sun C, Wang J, Zhang M (2005) Methotrexate-modified superparamagnetic nanoparticles and their intracellular uptake into human cancer cells. *Langmuir* **21**, 8858–8864.

Kooi ME, Cappendijk VC, Cleutjens K, Kessels AGH, Kitslaar P, Borgers M, Frederik PM, Daemen M, van Engelshoven JMA (2003) Accumulation of ultrasmall superparamagnetic particles of iron oxide in human atherosclerotic plaques can be detected by in vivo magnetic resonance imaging. *Circulation* **107**, 2453–2458.

Korhonen R, Kankaanranta H, Lahti A, Lahde M, Knowles RG, Moilanen E (2001) Bi-directional effects of the elevation of intracellular calcium on the expression of inducible nitric oxide synthase in j774 macrophages exposed to low and to high concentrations of endotoxin. *Biochemical Journal* **354**, 351–358.

Kwon JT, Hwang SK, Jin H, Kim DS, Mina-Tehrani A, Yoon HJ, Chop M, Yoon TJ, Han DY, Kang YW, Yoon BI, Lee JK, Cho MH (2008) Body distribution of inhaled fluorescent magnetic nanoparticles in the mice. *J Occup Health* **50**, 1–6.

Lacava ZGM, Azevedo RB, Lacava LM, Martins EV, Garcia VAP, Rebula CA, Lemos APC, Sousa MH, Tourinho FA, Morais PC, Da Silva MF (1999a) Toxic effects of ionic magnetic fluids in mice. *J Magnetism Magnetic Mat* **194**, 90–95.

Lacava ZGM, Azevedo RB, Martins EV, Lacava LM, Freitas MLL, Garcia VAP, Rebula CA, Lemos APC, Sousa MH, Tourinho FA, Da Silva MF, Morais PC (1999b) Biological effects of magnetic fluids: Toxicity studies. *J Magnetism Magnetic Mat* **201**, 431–434.

Lanone S, Boczkowski J (2006) Biomedical applications and potential health risks of nanomaterials: Molecular mechanisms. *Curr Mol Med* **6**, 651–663.

Lewin M, Carlesso N, Tung CH, Tang XW, Cory D, Scadden DT, Weissleder R (2000) TAT peptide-derivatized magnetic nanoparticles allow in vivo tracking and recovery of progenitor cells. *Nature Biotechnology* **18**, 410–414.

Lewinski N, Colvin V, Drezek R (2008) Cytotoxicity of nanoparticles. *Small* **4**, 26–49.

Li N, Sioutas C, Cho A, Schmitz D, Misra C, Sempf J, Wang MY, Oberley T, Froines J, Nel A (2003) Ultrafine particulate pollutants induce oxidative stress and mitochondrial damage. *Environ Health Perspect* **111**, 455–460.

Li N, Xia T, Nel AE (2008) The role of oxidative stress in ambient particulate matter-induced lung diseases and its implications in the toxicity of engineered nanoparticles. *Free Radic Biol Med* **44**, 1689–1699.

Long TC, Saleh N, Tilton RD, Lowry GV, Veronesi B (2006) Titanium dioxide (p25) produces reactive oxygen species in immortalized brain microglia (bv2): Implications for nanoparticle neurotoxicity. *Environ Sci Technol* **40**, 4346–4352.

Lu CW, Hung Y, Hsiao JK, Yao M, Chung TH, Lin YS, Wu SH, Hsu SC, Liu HM, Mou CY, Yang CS, Huang DM, Chen YC (2007) Bifunctional magnetic silica nanoparticles for highly efficient human stem cell labeling. *Nano Lett* **7**, 149–154.

Maiese K, Chong ZZ, Shang YC (2007) Mechanistic insights into diabetes mellitus and oxidative stress. *Current Medicinal Chemistry* **14**, 1729–1738.

McBain SC, Yiu HHP, Dobson J (2008) Magnetic nanoparticles for gene and drug delivery. *Int J Nanomed* **3**, 169–180.

McCarthy JR, Kelly KA, Sun EY, Weissleder R (2007) Targeted delivery of multifunctional magnetic nanoparticles. *Nanomedicine* **2**, 153–167.

Medina C, Santos-Martinez MJ, Radomski A, Corrigan OI, Radomski MW (2007) Nanoparticles: Pharmacological and toxicological significance. *Br J Pharmacol* **150**, 552–558.

MohanKumar SMJ, Campbell A, Block M, Veronesi B (2008) Particulate matter, oxidative stress and neurotoxicity. *Neurotoxicology* **29**, 479–488.

Molday RS, Yen SPS, Rembaum A (1977) Application of magnetic microspheres in labeling and separation of cells. *Nature* **268**, 437–438.

Molday RS, Molday LL (1984) Separation of cells labeled with immunospecific iron dextran microspheres using high-gradient magnetic chromatography. *FEBS Lett* **170**, 232–238.

Moore M, Weissleder R, Bogdanov A (1997) Uptake of dextran-coated monocrystalline iron oxides in tumor cells and macrophages. *JMRI-J Magn Reson Imaging* **7**, 1140–1145.

Mueller RH, Maassen S, Weyhers H, Specht F, Lucks JS (1996) Cytotoxicity of magnetite-loaded polylactide, polylactide/glycolide particles and solid lipid nanoparticles. *Int J Pharmaceutics (Amsterdam)* **138**, 85–94.

Muldoon LL, Sandor M, Pinkston KE, Neuwelt EA (2005) Imaging, distribution, and toxicity of superparamagnetic iron oxide magnetic resonance nanoparticles in the rat brain and intracerebral tumor. *Neurosurgery* **57**, 785–796.

Muller K, Skepper JN, Posfai M, Trivedi R, Howarth S, Corot C, Lancelot E, Thompson PW, Brown AP, Gillard JH (2007) Effect of ultrasmall superparamagnetic iron oxide nanoparticles (ferumoxtran-10) on human monocyie-macrophages in vitro. *Biomaterials* **28**, 1629–1642.

Nakayama M, Yamashita Y, Mitsuzaki K, Yi T, Arakawa A, Katahira K, Nakayama Y, Takahashi M (2000) Improved tissue characterization of focal liver lesions with ferumoxide-enhanced t1 and t2-weighted MR imaging. *J Magnetic Res Imaging* **11**, 647–654.

Namdeo M, Saxena S, Tankhiwale R, Bajpai M, Mohan YM, Bajpai SK (2008) Magnetic nanoparticles for drug delivery applications. *J Nanosci Nanotechnol* **8**, 3247–3271.

Naveau A, Smirnov P, Menager C, Gazeau F, Clement O, Lafont A, Gogly B (2006) Phenotypic study of human gingival fibroblasts labeled with superparamagnetic anionic nanoparticles. *Journal of Periodontology* **77**, 238–247.

Nel A, Xia T, Madler L, Li N (2006) Toxic potential of materials at the nanolevel. *Science* **311**, 622–627.

Neuwelt EA, Weissleder R, Nilaver G, Kroll RA, Romangoldstein S, Szumowski J, Pagel MA, Jones RS, Remsen LG, McCormick CI, Shannon EM, Muldoon LL (1994) Delivery of virus-sized iron-oxide particles to rodent CNS neurons. *Neurosurgery* **34**, 777–784.

Pan XG, Guan J, Yoo JW, Epstein AJ, Lee LJ, Lee RJ (2008) Cationic lipid-coated magnetic nanoparticles associated with transferrin for gene delivery. *Int J Pharmaceutics* **358**, 263–270.

Park JB (2003) Phagocytosis induces superoxide formation and apoptosis in macrophages. *Exp Mol Med* **35**, 325–335.

Petri-Fink A, Steitz B, Finka A, Salaklang J, Hofmann H (2008) Effect of cell media on polymer coated superparamagnetic iron oxide nanoparticles (spions): colloidal stability, cytotoxicity, and cellular uptake studies. *Eur J Pharm Biopharm* **68**, 129–137.

Pisanic TR, Blackwell JD, Shubayev VI, Finones RR, Jin S (2007) Nanotoxicity of iron oxide nanoparticle internalization in growing neurons. *Biomaterials* **28**, 2572–2581.

Radbruch A, Mechtold B, Thiel A, Miltenyi S, Pfluger E (1994) High-gradient magnetic cell sorting. In *Methods in Cell Biology, Vol. 42: Flow Cytometry*, Darzynkiewicz Z, Robinson JP, Crissman HA (eds). Academic Press: SanDiego; 388–405.

Raynal I, Prigent P, Peyramaure S, Najid A, Rebuzzi C, Corot C (2004) Macrophage endocytosis of superparamagnetic iron oxide nanoparticles – mechanisms and comparison of ferumoxides and ferumoxtran-10. *Invest Radiol* **39**, 56–63.

Renshaw PF, Owen CS, McLaughlin AC, Frey TG, Leigh JS (1986) Ferromagnetic contrast agents – a new approach. *Magn Reson Med* **3**, 217–225.

Risom L, Moller P, Loft S (2005) Oxidative stress-induced DNA damage by particulate air pollution. *Mutat Res-Fundam Mol Mech Mutagen* **592**, 119–137.

Robison TW, Zhou HF, Forman HJ (1995) Modulation of adp-stimulated inositol phosphate-metabolism in rat alveolar macrophages by oxidative stress. *Arch Biochem Biophys* **318**, 215–220.

Rogers WJ, Basu P (2005) Factors regulating macrophage endocytosis of nanoparticles: Implications for targeted magnetic resonance plaque imaging. *Atherosclerosis* **178**, 67–73.

Ruehm SG, Corot C, Vogt P, Kolb S, Debatin JF (2001) Magnetic resonance imaging of atherosclerotic plaque with ultrasmall superparamagnetic particles of iron oxide in hyperlipidemic rabbits. *Circulation* **103**, 415–422.

Schins RPF, Donaldson K (2000) Nuclear factor kappa-b activation by particles and fibers. *Inhal Toxicol* **12** (Suppl. 3), 317–326.

Schreck R, Albermann K, Baeuerle PA (1992) Nuclear factor kappa-b – an oxidative stress-responsive transcription factor of eukaryotic cells (a review). *Free Radic Res Commun* **17**, 221–237.

Schulze E, Ferrucci JT, Poss K, Lapointe L, Bogdanova A, Weissleder R (1995) Cellular uptake and trafficking of a prototypical magnetic iron-oxide label in-vitro. *Invest Radiol* **30**, 604–610.

Service RF (2005) Nanotechnology – calls rise for more research on toxicology of nanomaterials. *Science* **310**, 1609–1609.

Siglienti I, Bendszus M, Kleinschnitz C, Stoll G (2006) Cytokine profile of iron-laden macrophages: Implications for cellular magnetic resonance imaging. *J Neuroimmunol* **173**, 166–173.

Soto KF, Carrasco A, Powell TG, Garza KM, Murr LE (2005) Comparative in vitro cytotoxicity assessment of some manufactured nanoparticulate materials characterized by transmission electron microscopy. *J Nanopart Res* **7**, 145–169.

Soto K, Garza KM, Murr LE (2007) Cytotoxic effects of aggregated nanomaterials. *Acta Biomater* **3**, 351–358.

Stark DD, Moseley ME, Bacon BR, Moss AA, Goldberg HI, Bass NM, James TL (1985) Magnetic-resonance imaging and spectroscopy of hepatic iron overload. *Radiology* **154**, 137–142.

Stone V, Donaldson K (2007) Toxicology of nanoparticles: A historical perspective. *Nanotoxicology* **1**, 2–25.

Storm G, Belliot SO, Daemen T, Lasic DD (1995) Surface modification of nanoparticles to oppose uptake by the mononuclear phagocyte system. *Advanced Drug Delivery Reviews* **17**, 31–48.

Stroh A, Zimmer C, Gutzeit C, Jakstadt M, Marschinke F, Jung T, Pilgrimm H, Grune T (2004) Iron oxide particles for molecular magnetic resonance imaging cause transient oxidative stress in rat macrophages. *Free Radic Biol Med* **36**, 976–984.

Sun R, Dittrich J, Le-Huu M, Mueller MM, Bedke J, Kartenbeck J, Lehmann WD, Krueger R, Bock M, Huss R, Seliger C, Grone HJ, Misselwitz B, Semmler W, Kiessling F (2005) Physical and biological characterization of superparamagnetic iron oxide- and ultrasmall superparamagnetic iron oxide-labeled cells – a comparison. *Invest Radiol* **40**, 504–513.

Sydlik U, Bierhals K, Soufi M, Abel J, Schins RPF, Unfried K (2006) Ultrafine carbon particles induce apoptosis and proliferation in rat lung epithelial cells via specific signaling pathways both using egf-r. *Am J Physiol-Lung Cell Mol Physiol* **291**, L725–L733.

Theil EC, Matzapetakis M, Liu XF (2006) Ferritins: Iron/oxygen biominerals in protein nanocages. *J Biol Inorg Chem* **11**, 803–810.

Thiel A, Scheffold A, Radbruch A (1998) Immunomagnetic cell sorting – pushing the limits. *Immunotechnology* **4**, 89–96.

Thorek DLJ, Tsourkas A (2008) Size, charge and concentration dependent uptake of iron oxide particles by non-phagocytic cells. *Biomaterials* **29**, 3583–3590.

Tischler AS, Greene LA (1978) Morphologic and cytochemical properties of a clonal line of rat adrenal pheochromocytoma cells which respond to nerve growth factor. *Lab Invest* **39**, 77–89.

Turner TT, Lysiak J (2008) Oxidative stress: A common factor in testicular dysfunction. *J Andrology* **29**, 488–498.

Unfried K, Albrecht C, Klotz L-O, Von Mikecz A, Grether-Beck S, Schins RPF (2007) Cellular responses to nanoparticles: Target structures and mechanisms. *Nanotoxicology* **1**, 52–71.

van den Bos EJ, Wagner A, Mahrholdt H, Thompson RB, Morimoto Y, Sutton BS, Judd RM, Taylor DA (2003) Improved efficacy of stem cell labeling for magnetic resonance imaging studies by the use of cationic liposomes. *Cell Transplant* **12**, 743–756.

von zur Muhlen C, von Elverfeldt D, Bassler N, Neudorfer I, Steitz B, Petri-Fink A, Hofmann H, Bode C, Peter K (2007) Superparamagnetic iron oxide binding and uptake as imaged by magnetic resonance is mediated by the integrin receptor mac-1 (cd11b/cd18): Implications on imaging of atherosclerotic plaques. *Atherosclerosis* **193**, 102–111.

Weissleder R, Hahn PF, Stark DD, Rummeny E, Saini S, Wittenberg J, Ferrucci JT (1987a) MR imaging of splenic metastases – ferrite-enhanced detection in rats. *Am J Roentgenol* **149**, 723–726.

Weissleder R, Stark DD, Compton CC, Wittenberg J, Ferrucci JT (1987b) Ferrite-enhanced MR imaging of hepatic lymphoma – an experimental-study in rats. *Am J Roentgenol* **149**, 1161–1165.

Weissleder R, Stark DD, Rummeny EJ, Compton CC, Ferrucci JT (1988) Splenic lymphoma – ferrite-enhanced MR imaging in rats. *Radiology* **166**, 423–430.

Weissleder R, Stark DD, Engelstad BL, Bacon BR, Compton CC, White DL, Jacobs P, Lewis J (1989) Superparamagnetic iron-oxide – pharmacokinetics and toxicity. *Am J Roentgenol* **152**, 167–173.

Weissleder R, Elizondo G, Wittenberg J, Lee AS, Josephson L, Brady TJ (1990a) Ultrasmall superparamagnetic iron-oxide – an intravenous contrast agent for assessing lymph-nodes with MR imaging. *Radiology* **175**, 494–498.

Weissleder R, Elizondo G, Wittenberg J, Rabito CA, Bengele HH, Josephson L (1990b) Ultrasmall superparamagnetic iron-oxide – characterization of a new class of contrast agents for MR imaging. *Radiology* **175**, 489–493.

Weissleder R, Cheng HC, Bogdanova A, Bogdanov A (1997) Magnetically labeled cells can be detected by MR imaging. *JMRI-J Magn Reson Imaging* **7**, 258–263.

Wilhelm C, Gazeau F, Roger J, Pons JN, Bacri JC (2002) Interaction of anionic superparamagnetic nanoparticles with cells: Kinetic analyses of membrane adsorption and subsequent internalization. *Langmuir* **18**, 8148–8155.

Wilhelm C, Billotey C, Roger J, Pons JN, Bacri JC, Gazeau F (2003) Intracellular uptake of anionic superparamagnetic nanoparticles as a function of their surface coating. *Biomaterials* **24**, 1001–1011.

Wilhelm C, Gazeau F (2008) Universal cell labelling with anionic magnetic nanoparticles. *Biomaterials* **29**, 3161–3174.

Yamada Y, Doi T, Hamakubo T, Kodama T (1998) Scavenger receptor family proteins: Roles for atherosclerosis, host defence and disorders of the central nervous system. *Cell Mol Life Sci* **54**, 628–640.

Zanella CL, Timblin CR, Cummins A, Jung M, Goldberg J, Raabe R, Tritton TR, Mossman BT (1999) Asbestos-induced phosphorylation of epidermal growth factor receptor is linked to c-fos and apoptosis. *Am J Physiol-Lung Cell Mol Physiol* **277**, L684–L693.

Zhai Y, Wang XL, Wang XM, Xie H, Gu HC (2008) Acute toxicity and irritation magnetic of water-based dextran-coated magnetic fluid injected in mice. *J Biomed Mater Res Part A* **85A**, 582–587.

Zhang Y, Kohler N, Zhang MQ (2002) Surface modification of superparamagnetic magnetite nanoparticles and their intracellular uptake. *Biomaterials* **23**, 1553–1561.

Zhao M, Kircher MF, Josephson L, Weissleder R (2002) Differential conjugation of TAT peptide to superparamagnetic nanoparticles and its effect on cellular uptake. *Bioconjugate Chemistry* **13**, 840–844.

Zhou YM, Zhong CY, Kennedy IM, Pinkerton KE (2003) Pulmonary responses of acute exposure to ultrafine iron particles in healthy adult rats. *Environ Toxicol* **18**, 227–235.

Zhu MT, Feng WY, Wang B, Wang TC, Gu YQ, Wang M, Wang Y, Ouyang H, Zhao YL, Chai ZF (2008) Comparative study of pulmonary responses to nano- and submicron-sized ferric oxide in rats. *Toxicology* **247**, 102–111.

21

Toxicity Testing and Evaluation of Nanoparticles: Challenges in Risk Assessment

David Y. Lai and Philip G. Sayre

21.1 Introduction

A nanoparticle/nanomaterial is generally defined as a particle/material having a physico-chemical structure greater than typical atomic/molecular dimensions but at least one dimension smaller than 100 nm. It includes particles/materials engineered or manufactured by humans on the nanoscale with specific physico-chemical composition and structure to exploit properties and functions associated with its dimensions. Some of the common nanoparticle types are: (1) carbon-based materials (e.g. nanotubes, fullerenes), (2) metal-based materials (e.g. nanogold, nanosilver, quantum dots, metal oxides), and (3) dendrimers (e.g. dendritic forms of ceramics).

Because of their unique physico-chemical, electrical, mechanical and thermal properties, nanotechnology and nanomaterials are being incorporated into all spheres of our life through their applications in electronics, pesticides, consumer products, and the chemical and pharmaceutical industries. The toxic effects of many of the engineered nanoparticles have not been characterized. However, studies of asbestos and other particles for the past several decades have revealed important mechanisms of their carcinogenic and toxic actions. Based on analogies drawn to fibers and particles and what we know about their toxicity and mechanisms, it seems possible that some nanomaterials may act similarly to those carcinogenic fibers and particles. This concern is enhanced by the unusual physico-chemical properties of engineered nanomaterials potentially associated with unique adverse effects on biological systems. As the particle size is decreased within the nanoscale range,

Nanotoxicity: From In Vivo *and* In Vitro *Models to Health Risks*   Edited by Saura Sahu and Daniel Casciano

fundamental physical and chemical properties appear to change – often displaying different chemical and electronic properties/reactivity, and toxicological properties that differ from their bulk materials. Nanoparticles may translocate through membranes and have the potential to reach inside biomolecules and other organs in addition to those which are the portals of entry, a situation not possible for larger particles. Therefore, there are concerns about the health effects of nanoparticles, and experts are in agreement that the adverse effects of nanoparticles cannot be predicted from the known toxicity of material of macroscopic size.

A challenge facing the regulatory agencies, academia and industries is the development of testing strategies to characterize the hazard potential of the increasing numbers of nanomaterials. In 2005, a screening strategy was developed by an International Life Sciences Institute Research Foundation/Risk Science Institute (ILSI) workgroup for a hazard identification process of nanomaterial risk assessment. The workgroup proposed a comprehensive array of *in vitro* and *in vivo* assays to investigate the toxicity of diverse nanomaterials in routes of entry and target tissues (Oberdörster *et al.*, 2005a). More recently, a number of base set hazard tests have also been developed as a component of a nanoparticle risk management framework to meet regulatory requirements or voluntary program needs. Short-term *in vitro* tests of toxicity can provide a rapid and relatively inexpensive way to assess the potential toxicity of large numbers of untested nanoparticles. However, there are a number of inherent issues of *in vitro* test systems that result in false positives and false negatives, and recent studies have shown little correlation between *in vitro* and *in vivo* toxicity of some nanomaterials. Whereas inhalation is the preferred method of pulmonary exposure for hazard identification, it is not always feasible to test nanoparticles by inhalation, and alternative exposure methods (e.g. intratracheal instillation, intratracheal inhalation, pharyngeal/laryngeal aspiration) have been used by various investigators. However, many of these alternative exposure methods have shortcomings. For example, they deliver nanoparticles as a bolus to the lung airways and bypass the upper respiratory tract, which may lead to differences in the doses, rate, and/or distribution of materials in the lungs. The pulmonary response could potentially be caused by the high single dose of nanomaterials, leading to artificial effects due to large mass and agglomeration formed during exposure procedures which may not be relevant to the chronic lower exposure levels in occupational and environmental settings. The issues of *in vitro* and *in vivo* testing are discussed. Special attention is paid to the complexities encountered with the evaluation of the test data and assessment of the potential hazard/risk of nanoparticles.

21.2   Possible Mechanisms of Action and Basis of Human Health Concern

A number of mineral fibers and particles including asbestos and crystalline silica have been shown to be carcinogenic or possibly carcinogenic to humans (International Agency for Research on Cancer (IARC), 1997, 2002). Studies of asbestos and other fibers have shown that both physical and chemical parameters of fibers are related to their biological activity: fiber geometry and dimensions, biopersistence in the lungs, chemical composition, and surface reactivity (Kane *et al.*, 1996). In general, fibers with a smaller diameter will penetrate deeper into the lungs, while long fibers (longer than the diameter of alveolar macrophages) will only be cleared slowly. In addition to fiber length, chemical factors play an important role in fiber durability and biopersistence; fibers with high alkali or alkali earth oxide contents and low contents of Al_2O_3, Fe_2O_3 and TiO_2 tend to have low durability and hence low biopersistence (Searl, 1994). For fibers with the same dimensions, those with a higher surface charge density have been shown to be more carcinogenic, suggesting that the carcinogenicity of fibers may also be related to surface properties such as surface chemistry, reactivity, and surface area of the fiber (Bonneau et al., 1986a,b). It has been suggested that the carcinogenicity of some fibers may be a function of the aspect ratio, the dose, and other chemical properties (e.g. surface charge density) of the fiber/particle; a sufficient quantity of short, thin fibers/particles may also be carcinogenic.

For particles, surface activity is the fundamental aspect of their toxicity. Studies of mineral particles have demonstrated that the toxic and carcinogenic effects are related to the surface area and surface activity of inhaled particles such as crystalline silica and TiO_2 (Oberdörster et al., 1994; Fubini, 1997; Clouter et al., 2001; Duffin et al., 2002; Warheit et al., 2006, 2007a). Surface modification of silica affects its cytotoxicity, inflammation responses and fibrogenicity (Fubini, 1997; Duffin et al., 2002). Particle surface characteristics, therefore, are considered to be key factors in free radicals and reactive oxygen species (ROS) formation and in the development of fibrosis and cancer by quartz (crystalline silica) (Fubini, 1997).

The mechanisms leading to the development of lung cancers and/or malignant mesothelioma by exposure to fibers and particles are not clearly understood. Although direct genotoxicity and other hypothesized mechanisms may play a role and cannot be ruled out in fiber/particle carcinogenesis, more recent experimental evidence has shown an association of chronic inflammation and fibrosis with cancer and suggested an indirect mechanism involving persistent inflammation with the release of cytokines, growth factors and ROS to be more important in fiber and particle carcinogenesis (Woo et al., 1988; IARC, 1997). A sequential pattern of cellular responses to inhaled fibers/particles is induced in the lung: (a) aggregation of alveolar macrophages; (b) phagocytosis of the particles by macrophages; (c) necrosis/lysis of macrophages with the release of cellular content, including the ingested particles; (d) accumulation of other macrophages and fibroblasts; (e) persistent inflammation of epithelial cells and production of collagen; (f) fibrosis; and (g) tumors. The recruitment and activation of alveolar macrophages and inflammatory cells in response to persistent fibers in the lungs is accompanied by release of cytokines and ROS that could induce oxidative stress and damage DNA. It is likely that fibers/particles are genotoxic via the production of these reactive oxygen species (IARC, 1997; Topinka et al., 2006). Cytokines and growth factors released from alveolar macrophages and inflammatory cells may contribute to proliferation of preneoplastic cells which lead to tumor development.

Based on analogies with fibers and particles and what we know about their toxicity and mechanisms, there are concerns about the health effects of nanoparticles. Because of their nanoscale and unique physico-chemical properties, there are additional concerns about nanoparticles. Many of the special properties of nanoparticles are due to the nano-size and an extremely large surface-to-volume ratio relative to bulk materials. As a particle decreases in size, the surface area increases and a greater proportion of atoms/molecules are found at the surface compared with those inside. Thus nanoparticles have a much larger surface area per unit mass and a higher potential for biological interaction compared with larger particles. The increase in the surface-to-volume ratio results in the increase of the particle surface energy which may become reactive. Therefore, as materials reach the nano-scale, they often display different chemical and electronic properties/reactivity and can have toxicological properties that differ from their bulk materials. Data from limited pulmonary studies in rats have demonstrated that exposures to some metal/metal oxide nanoparticles produced enhanced toxicity responses when compared with larger-sized particles of similar chemical composition. For example, even a typically inert bulk compound, such as gold, can elicit a biological response when it is introduced as a nanomaterial (Goodman *et al.*, 2004). With the particle size reduced from 17 µm (surface area: 3.99×10^2 cm^2/g) to 23.5 nm (surface area: 2.95×10^5 cm^2/g), the toxicity (LD$_{50}$) of copper particles sharply increased from >5000 mg/kg to 413 mg/kg in mice (Chen *et al.*, 2006). In a subchronic inhalation study in rats, ultrafine TiO$_2$ particles (20 nm) have been shown to elicit a persistently higher inflammatory reaction in the lungs compared with larger sizes (250 nm) of TiO$_2$ (Oberdörster *et al.*, 1994; Oberdörster, 2000). Shvedova *et al.* (2005) reported unusual inflammatory and fibrogenic pulmonary responses to specific nanomaterials, suggesting that they may injure the lung by new mechanisms. Similarly, studies conducted by Lam *et al.* (2004) and Warheit *et al.* (2004), examining the pulmonary toxicity of carbon nanotubes, have provided evidence that manufactured nanomaterials can display unique toxicity. As particle size decreases, the toxicity of particles generally increases.

Carbon nanotubes have features of both fibers and nanoparticles; they may therefore exhibit some of their effects through oxidative stress and inflammation (Donaldson *et al.*, 2006). Nanoparticles of various chemical compositions have been shown to generate reactive oxygen species (ROS) in both *in vivo* and *in vitro* studies (Brown *et al.*, 2001; Wilson *et al.*, 2002). Li *et al.* (2003) demonstrated that nanoparticles preferentially mobilize to mitochondria. Since mitochondria are redox active organelles, nanoparticles may alter ROS production and interfere with antioxidant defenses. Nanoparticles have also been implicated in interfering with cell signalling via ROS-mediated activation of cytokine gene expression (Brown *et al.*, 2004; Pacurari *et al.*, 2008). Therefore, like fibers and particles, ROS release following exposure to nanoparticles may contribute to oxidative stress, DNA damage, and proliferation of preneoplastic cells leading to cancer. Indeed, a two-year inhalation study has shown a statistically significant increase in lung cancer in rats exposed to nanosized (15–40 nm) TiO$_2$ at an average concentration of 10 mg/m^3 (Heinrich *et al.*, 1995). Exposing the mesothelial lining of the body cavity of mice (as a surrogate for the mesothelial lining of the chest cavity) to long multiwalled carbon nanotubes (MWCNTs) has resulted in asbestos-like, length-dependent pathogenic behavior (Poland *et al.*, 2008).

Recent studies have shown that nanoparticle injury can also proceed by nonoxidant paradigms that are specific to their novel properties. One example is the release of toxic ions when the thermodynamic properties of a material (including surface free energy) favor particle dissolution to a suspending medium or biological environment. ZnO nanoparticle dissolution can induce cytotoxicity and apoptosis in mammalian cells due to the release of Zn^{2+} ions under aqueous conditions (Jeng and Swanson, 2006; Xia *et al.*, 2008). Another paradigm of nanoparticle toxicity is the ability of some nanoparticles to form a corona with proteins, which leads to adverse biological effects through protein unfolding, fibrillation, thiol cross-linking and loss of enzyme activity (Vertegel *et al.*, 2004; Chen and von Mikecz, 2005; Cedervall *et al.*, 2007).

Inhaled particles in the nanosize range can certainly deposit in all parts of the respiratory tract including the alveolar region of the lungs. Dependent upon the specific application, oral, dermal and other routes of exposure are also possible for nanoparticles. Because of their small size, they may pass into cells directly through the cell membrane or penetrate the skin and distribute throughout the body once translocated to the blood circulation. There is evidence that nanoparticles can translocate from the portal of entry, the respiratory tract, via different pathways to other organs/tissues, making them uniquely different from larger-sized particles in that they may induce direct adverse responses in remote organs. For example, there are data from animal studies showing possible translocation of inhaled nanoparticles to the nervous system and other organs/tissues (Oberdörster *et al.*, 2002, 2004). Therefore, there is also concern for the toxic/carcinogenic effects of nanoparticles on extra-pulmonary organs/tissues.

21.3   Testing Strategies for Human Health Effects

The toxic effects of nanoparticles have not been characterized, but it is generally believed that nanoparticles can have toxicological properties that differ from their bulk materials. A number of testing strategies have been developed for the evaluation of the potentially toxic effects of nanoparticles.

21.3.1   International Life Sciences Institute (ILSI)

In 2005, the International Life Sciences Institute Research Foundation/Risk Science Institute (ILSI) convened a group of international experts in fiber and particle toxicology to develop a broad data-gathering strategy for the initial hazard identification process of nanomaterial risk assessment. The panel proposed a comprehensive array of *in vitro* and *in vivo* assays to investigate the toxicity of diverse nanoparticles in routes of entry and target tissues (Oberdörster *et al.*, 2005a). This is the first proposed toxicity testing strategy for nanoparticles. Specific testing protocols were not detailed in this report, but are assumed to follow or be modified from existing protocols established for testing fibers and particles.

21.3.1.1   *In Vitro* Assays

Exposure to nanoparticles via the respiratory tract is a major concern, so lung epithelial cells, macrophages, immune cells and fibroblasts are key targets for nanoparticle effects. Because of possible penetration of the skin and cell membrane, nanoparticles may get into the circulatory system and translocate to other organs/tissue, so *in vitro* tests are also suggested for portal-of-entry toxicity for skin and the mucosal membranes, and target organ toxicity for endothelium, blood, spleen, liver, nervous system, heart and kidney. The *in vitro* systems, effects and endpoints for nanomaterial testing are summarized in Table 21.1.

In addition, a number of non-cellular assays for assessment of durability, protein interactions, complement activation, and free radical production were recommended. Recently, Warheit *et al.* (2007a) have investigated pulmonary toxicity in rats with three forms of ultrafine-TiO_2 particles and found differential responses related to their surface properties as determined by the vitamin C yellowing assay (Rajih *et al.*, 1999), which measures the chemical reactivity of the sample toward an antioxidant, specifically a vitamin C derivative. With greater chemical reactivity, the yellowing of the test sample will increase providing a higher Δb. The Δb value was determined by comparing the color of the test and blank films. The Δb value of P25 ultrafine-TiO_2 (80/20 anatase/rutile) has been shown to be 23.8, whereas that of ultrafine rutile is 1.2 and that of fine rutile is 0.4. The findings are

Table 21.1   In vitro *systems for nanomaterial testing (modified from Oberdörster et al.,*
2005a), Particle and Fibre Toxicology, copyright 2005 Oberdörster
http://www.particleandfibretoxicology.com/content/2/1/8

Portals of entry	Cell/tissue type	Effect	Endpoint
Lung	Epithelium	Toxicity	Trypan blue, LDH, apoptosis
		Inflammation	Gene expression, oxidative stress, signal transduction pathways
		Translocation	Transfer of nanoparticles across membranes
		Carcinogenesis	Genotoxicity, comet assay, 8-OHdG, *hprt* assay, proliferation assay
	Macrophages	Toxicity	Trypan blue, LDH, apoptosis
		Chemotaxis	Chemotaxis assay
		Phagocytosis	Particle uptake into cells, cytoskeletal staining
		Inflammation	Gene expression, oxidative stress, signal transduction pathways
	Immune cells	Immune response	Cytokine profile, adjuvant effects
	Endothelium	Inflammation	Adhesion molecules, oxidative stress
		Coagulation	Von Willebrand factor, tissue factor
	Fibroblasts	Inflammation	Oxidative stress, cytokine profile, gene expression profile
		Fibrosis	Collagen synthesis, cell proliferation
	Lung slices	Inflammation	Oxidative stress, signal transduction pathway, immunohistopathology
		Translocation	Particles across membranes
		Fibrosis	Collagen synthesis
Skin	Cell systems (e.g. HEK)	Cytotoxicity	Cell viability – MTT, neutral red,
		Inflammation	Cytokine profile
	Flow-through diffusion systems	Absorption	
	Isolated skin flap model	Absorption, cytotoxicity, inflammation	Glucose utilization, any other markers depending on endpoints (cytokine profiles, histopath, etc.)
Mucosa	Intestinal epithelium (GI tract)	Cytotoxicity	Cell viability – MTT, neutral red, trypan blue
			Apoptosis
		Inflammation	Cytokine profile, oxidative stress, signal transduction pathway
		Translocation	Permeability assays
	GALT	Inflammation	Cytokine profile, oxidative stress
		Immune response	Adjuvant effects
	Buccal epithelium (oral cavitiy)	Cytotoxicity	Cell viability – MTT, neutral red, trypan blue
			Apoptosis
		Inflammation	Cytokine profile, oxidative stress, signal transduction pathway
		Translocation	Permeability assays

(Continued)

Table 21.1 *(Continued)*

Portals of entry	Cell/tissue type	Effect	Endpoint
	Vaginal epithelium (reproductive system)	Cytotoxicity	Cell viability – MTT, neutral red, trypan blue Apoptosis
		Inflammation	Cytokine profile, oxidative stress, signal transduction pathway
		Translocation	Permeability assays
Target organ toxicity Endothelium	Endothelial cells (e.g. HUV-EC-C)	Cytotoxicity	Cell viability – MTT, neutral red, trypan blue Apoptosis
		Homeostasis	Oxidative stress, gene expression profile
		Translocation	Permeability assays
Blood	Red blood cells, platelets, bone marrow (megakaryocytes)	Inflammation/ immune response	Platelet activation Cytokine/chemokine release from leukocytes Oxidative stress Complement activation
		RBC/particle interactions	
Liver	Hepatocytes	Toxicity	Cell viability – MTT, neutral red, trypan blue Apoptosis
	Kupffer cells	Inflammation	Cytokine profile, oxidative stress, signal transduction pathway, gene expression
	Isolated perfused liver slices	Coagulation Translocation, distribution	Von Willebrand factor, tissue factor Histopathology
	Liver slices	Toxicity studies	Cytotoxicity, P450 assay, ATP assays, GSH content
	Collagen sandwich cultures	Toxicity studies	Cytotoxicity, P450 assay, ATP assays, GSH content
Spleen	Lymphocytes	Immune response	Cytokine profile
Central and autonomic nervous system	Neuronal cells	Toxicity	Cytotoxicity – Trypan blue, LDH Apoptosis
		Inflammation	Cytokine profile, oxidative stress, signal transduction pathway, gene expression
		Translocation	Gene expression, microscopic examination
	Astroglial, microglial cells	Inflammation	Cytokine profile, oxidative stress, signal transduction pathway, gene expression

(Continued)

Table 21.1 *(Continued)*

Portals of entry	Cell/tissue type	Effect	Endpoint
Heart	Cardiomyocytes	Toxicity	Cytotoxicity – Trypan blue, LDH Apoptosis
		Inflammation	Cytokine profile, oxidative stress, signal transduction pathway, gene expression
		Function	Beat – rhythm testing
Kidney	Cell (e.g. HK-2, MDCK, LCC-PK1)	Toxicity	Cytotoxicity – Trypan blue, LDH Apoptosis
		Inflammation	Cytokine profile, oxidative stress, signal transduction pathway, gene expression

consistent with others (e.g. Fubini, 1997; Clouter *et al.*, 2001) that surface reactivity is most important for the toxicity/carcinogenicity of some particles such as crystalline silica and TiO_2. The vitamin C yellowing assay using P25 ultrafine-TiO_2 as a positive control and ultrafine rutile TiO_2 as a negative control may be a useful non-cellular assay for testing the relative surface reactivity of other metal oxide nanoparticles.

21.3.1.2 *In Vivo* Assays

Currently, little information is available regarding airborne levels of nanomaterials generated during production and processing, or quantities which may be aerosolized into the environment. Nonetheless, due to their small size, aerosolization of respirable nanomaterials is likely, either as singlet or as aggregated particles, and the inhalation route of human exposures is a concern. Skin is also a target for nanoparticles, especially from nanoparticles in cosmetics and sunscreen (Davis, 1994). The use of nanotechnology in food products is expected to grow (Food Safety Authority of Ireland (FSAI), 2008). Furthermore, it is possible that during the life of a nanomaterial (production, application, disposal, etc.) it may appear in the water supply or be inadvertently ingested. Therefore, oral exposure to the nanomaterial is also a concern. Depending on nanoparticle size and surface chemistry, exposure to nanoparticles via the portal-of-entry also includes a probability of translocation to other organs and tissues. In this test strategy, a two-tier approach was recommended for *in vivo* assays. Tier 1 *in vivo* assays are proposed for pulmonary, oral and skin exposures. Tier 1 evaluations include markers of inflammation, oxidant stress, and cell proliferation in portal-of-entry and selected remote organs and tissues. The Tier 1 *in vivo* assay systems, study designs and endpoints for evaluating the hazard potential of nanoparticles are summarized in Table 21.2.

Tier 2 evaluations are proposed only for pulmonary exposures. A 28-day inhalation study by nose-only or whole-body exposure in male or female rats with up to 3 months observation is recommended for evaluating the toxicity and carcinogenic potential of nanomaterials. Evaluation includes markers of damage, oxidant stress, cell proliferation, the degree/intensity and duration of pulmonary inflammation, and cytotoxic effects and histopathology of pulmonary and ex-pulmonary organs/tissues.

Table 21.2 Tier 1 in vivo assay systems for nanoparticles (Summarized from Oberdörster et al., 2005a), Particle and Fibre Toxicology, copyright 2005 Oberdörster http://www.particleandfibretoxicology.com/content/2/1/8

Exposure	Method	Design	Endpoint
Pulmonary	Inhalation	Species: rat or mouse Sex: M or F Concentrations: 3 Duration: 14 days with responses monitored at 24 hr, 7 days and 28 days postexposure	1. Bronchoalveolar lavage (BAL) analysis for inflammatory and cytotoxic effects 2. Oxidative stress markers 3. Histopathology of pulmonary and ex-pulmonary organs/tissues 4. Cell proliferation 5. Other organ-specific endpoints
	Intratracheal instillation; pharyngeal/laryngeal aspiration	Single exposure in M/F rat/mouse at three concentration levels with responses monitored at 24 hr, 7 days, and 28 days postexposure	1. Bronchoalveolar lavage (BAL) analysis for inflammatory and cytotoxic effects 2. Oxidative stress markers 3. Histopathology of pulmonary and ex-pulmonary organs/tissues 4. Cell proliferation 5. Other organ-specific endpoints
Oral	Gavage	A single dose in M/F rat/mouse	1. Feces should be collected for 4 days postexposure to determine amount of material eliminated 2. GALT, mesenteric lymph nodes and liver should be analyzed for presence of nanoparticles 3. Evaluate systemic toxicity if significant absorption is evident
Skin	Topical	Three doses applied with an occlusion device on normal and abraded skin of 4–6 rats or pigs. Dose for 5 or 7 days and up to 28 days	1. Draize test score for erthema and edema at 0.5, 1, 2,4, 8 and 24 hr 2. Identify cellular changes and localize the particles in slin biopsies by TEM 3. For repeated exposure, hematology, clinical chemistry, local lymph nodes and immunotox. should be performed at termination

When inhalation exposure is not feasible, the panel concluded that other pulmonary exposure methods may be acceptable. Some of these methods include intratracheal instillation (Warheit *et al.*, 2004; Muller *et al.*, 2005), intratracheal inhalation (Osier and Oberdörster, 1997; Elder *et al.*, 2004), and pharyngeal/laryngeal aspiration (Rao *et al.*, 2003). As in inhalation studies, evaluation of both pulmonary and ex-pulmonary toxic endpoints/markers and histopathology are recommended. If the results of the subchronic inhalation study indicate that the nanoparticles have carcinogenic potential, a two-year inhalation study bioassay in rats may be warranted.

The Tier 2 studies are aimed at obtaining additional information with respect to the deposition, translocation and biopersistence of the test material to aid in interpreting the results for risk assessment. Tier 2 studies will also provide additional information to either characterize further effects seen in Tier 1 studies, or to obtain new data using specific models of susceptibility. Respective animal models include exposures of senescent, transgenic and knockout animals, and animals with compromised organ systems (e.g. hypertension; diabetes models; immuno-compromised, infectivity models). In addition to obtaining information for identifying nanoparticle hazards, Tier 2 studies will also provide data on underlying mechanisms which can be used in concert with the mechanistic *in vitro* studies. Following protocols similar to the OECD Guideline 422 for Testing of Chemicals, Tier 2 evaluations for pulmonary exposures could include potential effects on the reproductive tract, placenta and fetus.

21.3.2   DuPont's Environmental, Health and Safety Framework

An environmental, human health and safety (EHS) framework, which includes a base set of toxicity screening studies, was developed for assessing the potential hazard of nano-based products by the DuPont Haskell Laboratory (Warheit *et al.*, 2007b). These minimum base set toxicity tests include a single-dose instillation, or 28-day pulmonary/oral study in rats with postexposure observation periods for up to 3 months, a dermal and an eye irritation test in rabbits, a skin sensitization study in mice, an acute oral toxicity study in rats, and two *in vitro* genotoxicity studies (Table 21.3).

Table 21.3   *Minimum base set mammalian toxicity tests (Summarized from Warheit* et al., *2007b), with permission from Elsevier*

Toxicity test	Design	Endpoint
Pulmonary bioassay	Intratracheal instilled to male rats at doses of 1 or 5 mg/kg	Evaluated for BALF inflammation markers, cell proliferation, and histopathology at post-instillation of 24 hr, 1 week, 1 month and 3 months
Skin irritation	Applied as a single 0.5 g dermal dose to the shaved skin of three male rabbits for 4 hr	Evaluated by Draize score for sign of dermal irritation about 60 min., and 24, 48 and 72 hr after removal of test substance
Skin sensitization	Local lymphnode assay (LLNA) in mice	Evaluated for cell proliferation in the draining auricular lymph nodes of the ears and compared with the vehicle control group
Acute oral toxicity	Up and down procedure: rats were dosed by gavage one at a time at a minimum of 48 hr intervals	Observed for mortality, body weight effects, and clinical signs for 14 days after dosing
Eye irritation	Administered to one eye of each of 3 rabbits	The conjunctiva, iris and cornea were evaluated and scored about 1, 24, 48 and 72 hr after administration
Genotoxicity	1. Ames strains and E. coli were tested with or without metabolic activation	1. Gene mutation
	2. Chinese hamster ovary (CHO) cells with or without metabolic activation	2. Chromosomal aberrations

If there is evidence of toxicity from the short-term toxicity studies in the base set, more extensive studies may be conducted. These include: chronic (>1 year) inhalation/oral toxicity studies, chronic dermal irritation/sensitization studies, developmental and reproductive toxicity studies, and immunotoxicity. The test guidelines developed by the US Environmental Protection Agency (EPA, 2007a) and Organization for Economic Co-operation and Development (OECD, undated) for testing chemicals are used for these tests.

21.3.3 Nanotechnology Characterization Laboratory (NCL)

In addition to physico-chemical characterizations, a standardized assay cascade for safety and efficacy testing of nanoparticles intended for cancer therapeutics and diagnostics has been developed by the Nanotechnology Characterization Laboratory (NCL) of the National Cancer Institute (NCI). The safety assay cascade includes various *in vitro* and *in vivo* toxicity testing for use in support of submission of investigational new drug (IND) applications with the Food and Drug Administration (FDA). The toxicity tests are intended to investigate the pharmacology, tissue distribution and clearance, blood contact properties, immunotoxicity, cytotoxicity, mechanistic toxicology, and sterility of the nanomaterials (NCL, undated). Some of the *in vitro* and *in vivo* toxicity tests are listed in Table 21.4.

Table 21.4 Toxicity assays of nanomaterials by the Nanotechnology Characterization Laboratory (NCL).

In vitro	In vivo
Phase I/II enzyme induction/suppression	Initial disposition studies Tissue distribution Clearance Half-life Plasma AUC
Oxidative stress: GSH hemeostasis Lipid peroxidation ROS	Toxicity studies Blood chemistry Hematology Gross pathology Histopathology
Cytotoxicity (necrosis) MTT and LDH release	Immunotoxicity 28-day screen Repeated-dose immunogenicity
Cytotoxicity (apoptosis) Caspase 3 activation	GLP studies PK/ADME Expanded single dose acute toxicity
Immunotoxicity Plasma protein binding Hemolysis Platelet aggregation Coagulation Complement activation Leukocyte proliferation Macrophage/neutrophil function Cytotoxicity of NK cells	

Understanding that standard methods are important for the regulatory agencies to eval-uate nanotechnology drugs, NCL has been actively developing standard protocols, not only to characterize nanomaterials, but also to assess safety and toxicity. So far, NCL has developed more than 30 protocols in its assay cascade (NCL, undated). All toxicity assay protocols developed are subject to extensive in-house and external validation, and to regular revision to ensure applicability to various types of nanomaterials. Three of the protocols are becoming American Society for Testing and Materials (ASTM) standards: (1) proto-col for examining the destruction of red blood cells (hemolytic potential); (2) method for evaluating a common side-effect of anticancer drug, that is, inhibition (or stimulation) of the maturation of certain bone marrow cells (macrophages); and (3) method for evaluating toxicity on liver carcinomas and kidney cells (NCL, undated).

21.4   Issues of Short-term *In Vitro* Assays

Short-term *in vitro* tests of toxicity can provide a rapid and relatively inexpensive way to assess the potential toxicity of large numbers of untested nanoparticles. *In vitro* studies also allow specific biological and mechanistic pathways to be isolated and tested under controlled conditions to provide initial data on the comparative toxicity of nanomaterials of diverse sizes and shapes. With appropriate matched controls, *in vitro* assays are useful for screening and prioritizing substances for *in vivo* assays. Based on the understanding of the mechanisms of toxicity caused by fibers and particles, it may be possible to identify shorter, simpler *in vitro* tests (or battery of tests) that are adequately predictive of a chronic inhalation bioassay; successful development of such an *in vitro* system will reduce animal use for toxicity testing of nanoparticles.

A review of the toxicity studies of fibers by an expert panel concluded that no single short-term *in vitro* test (or battery of tests) can be used to predict the carcinogenicity po-tential of fibrous particles (Bernstein *et al.*, 2005). At present, *in vitro* test systems also appear to have limited usefulness for hazard identification of nanoparticles due to a num-ber of inherent issues resulting in false positives and false negatives. Some generic issues associated with the *in vitro* approach include: (i) high-dose effects – effects observed at high-dose levels used in *in vitro* assays may not extrapolate to low-dose effects *in vivo*; (ii) time course effects – short-term *in vitro* endpoints (e.g. release of inflammatory me-diates, cell proliferation) may not be predictive of long-term physiological effects; and (iii) cell line effects – toxic responses may differ using different cell lines. Specific issues related to *in vitro* toxicity testing of nanaoparticles are:

- A number of endpoints employ the measurement of a cellular product, such as release of a protein. Recent data from several research groups (e.g. Brown *et al.*, 2000; Kim *et al.*, 2003; Dutta *et al.*, 2007) have demonstrated that various types of nanoparticles can adsorb key proteins such as albumin, LDH, fibronectin, and TGF-β, leading to confounding endpoint measurements.
- Some nanomaterials such as carbon nanotubes have been shown to interfere with the MTT (mitochondrial reduction of tetrazolium) cytotoxicity assay by absorbing the reduced formazan dye, resulting in an underestimation of cytotoxic potency (Worle-Knirsh *et al.*, 2006).

- Many assays employ the measurement of a colored or fluorescent product. For instance, fluorescent nanoparticles such as quantum dots may interfere with the product used to quantity specific cellular responses (Monteiro-Riviere and Inman, 2006).
- Under *in vitro* cell culture conditions ('wet phase'), physico-chemical characterization of particles including particle size are likely to change from the powder form ('dry phase'). The type and composition of culture medium (e.g. addition of serum) can affect toxicity measurements – probably due to influences affecting agglomeration and/or surface chemistry of nanoparticles (Murdock *et al.*, 2008).
- Use of organic solvents for creating suspensions or dispersive agents/surfactants to maintain the nanoparticles from forming aggregates may not be relevant to normal exposure conditions, and these agents may have biological activity that can confound the findings. Incomplete removal of the organic solvent tetrahydrofuran used to create water-soluble suspensions of C_{60} is believed to contribute to the cytotoxicity of C_{60} in human cells (Andrievsky *et al.*, 2005).
- New mechanisms may be missed leading to false negatives. For instance, while inflammation and oxidative stress have been identified as possible mechanisms underlying the etiology of nanoparticles, the toxicity of cationic dendrimers appears to be related not to oxidative stress generation, but to disruption of cell membrane integrity through interaction of the positive charge terminal group with the anionic lipids of the cell membrane (Mecke *et al.*, 2006).

Therefore, depending on the type of cells, the duration of exposure, the concentration of nanoparticles and the composition of the culture media, testing of the same nanomaterial can have different outcomes. Recent studies have shown little correlation between *in vitro* and *in vivo* toxicity of some nanomaterials. For instance, Sayes *et al.* (2007a) assessed the capacity of *in vitro* screening studies to predict *in vivo* pulmonary toxicity of several fine or nanosized particles in rats, including carbonyl iron, crystalline and amorphous silica, and zinc oxide. For the *in vitro* component of the study, different culture conditions were utilized. In the *in vivo* component of the study, rats were exposed by intratracheal instillation to each of the materials. Following exposures, the lungs of exposed rats were lavaged and endpoints were measured at numerous time points post-exposure. When considering the range of toxicity endpoints, the comparisons of *in vivo* and *in vitro* measurements demonstrated little correlation. Similarly, whereas nano-C_{60} and $C_{60}(OH)_{24}$ were reported to be toxic to a number of cell types *in vitro* (Sayes *et al.*, 2004), there was no evidence of adverse effects in lung tissues at three months post-instillation exposure to doses up to 3 mg/kg of the two types of fullerenes in rats (Sayes *et al.*, 2007b). *In vitro* assays of oxidant stress also failed to predict the progressive interstitial fibrotic response to inhalation exposure to single-walled carbon nanotubes (SWCNTs) (Kagan *et al.*, 2006; Shvedova *et al.*, 2008).

Therefore, mechanistic pathways elucidated and positive results obtained from *in vitro* test systems need further validation in animal models. With appropriate matched controls, *in vitro* assays may be useful for supporting health hazard concerns and prioritizing further animal studies if the results are positive. If the results from *in vitro* assays are negative, further testing in appropriate *in vivo* systems may be warranted for further evaluation.

21.5   Issues of *In Vivo* Assays

Inhalation is the 'gold standard' with regard to method of exposure of the respiratory tract for hazard identification of fibers and particles and to obtain dose-response data in quantitative risk assessment, since inhalation is the normal physiological route for delivery of nanoparticles into the lungs. A number of barriers, however, exist for conducting inhalation studies. Inhalation studies are costly, use large numbers of animals, take a long time to complete, and require sophisticated exposure facilities. Technically, because of the high propensity of hydrophobic nanomaterials to agglomerate, preparation of adequate aerosol dispersions suitable for inhalation studies is immensely difficult. As a result, alternative exposure methods (e.g. intratracheal instillation, intratracheal inhalation, pharyngeal/laryngeal aspiration) have been used by various investigators for delivery of particles to the lungs.

Intratracheal instillation is a method whereby particles can be administered directly into the lower respiratory tract of rats (Driscoll *et al.*, 2000; Warheit *et al.*, 2004; Muller *et al.*, 2005). However, this non-inhalation method bypasses the upper respiratory tract which can be a potentially important target site for inhaled nanoparticles. Furthermore, this exposure method delivers nanoparticles as a bolus to the lung airways; the pulmonary response could potentially be caused by the high single dose of nanomaterials – artificial effects of large mass and agglomeration formed during exposure procedures may not be relevant to chronic lower exposure in occupational and environmental settings. Asphyxia has been reported in rats following intratracheal instillation of single-walled carbon nanotubes (SWCNTs) as a result of formation of large agglomerations in suspension (Warheit *et al.*, 2004).

Studies have been conducted to compare responses of the rat lung to inhaled and intractracheally instilled quartz and titanium dioxide. In an earlier study (Driscoll *et al.*, 1991), intratracheal instillation resulted in greater response than inhalation at the same lung burden of quartz or titanium dioxide. The greater responses after instillation were believed to be due to differences in the dose rate and/or distribution of materials in the lungs. A comparison of the inflammation responses of the lungs following instillation and inhalation exposures to quartz also showed different patterns of injury in rats, but with interstitial granulomas only observed in animals exposed by inhalation (Henderson *et al.*, 1995). A more recent study (Warheit *et al.*, 2005), however, found that the results generated from the instilled TiO_2 particles (about 300–400 nm) were consistent with findings from an inhalation assay of virtually identical particles.

The technique of pharyngeal/laryngeal aspiration involves placing a suspension of the nanoparticles on the back of the tongue and the pulling of the tongue which results in a reflex gasp and aspiration of the droplets (Rao *et al.*, 2003). This technique is reputed to give better dispersion throughout the lung than instillation. Comparision of the inhalation protocols with results on pharyngeal aspiration in a recent study showed that inhalation exposure to SWCNTs resulted in pulmonary responses very similar to those following pharyngeal aspiration; all pulmonary responses were retained or enhanced. Overall, SWCNT inhalation exposure was much more potent than aspiration of a bolus dose of SWCNT (Shvedova *et al.*, 2008).

During intratracheal inhalation exposure, rats are ventilated at a constant rate with an aerosol of the test particle. To avoid the large-scale deposition that occurs in the rodent nose, animals breathe the aerosolized particles but the nasal pathway is circumvented. This method has the advantages of using small amounts of particles (Osier and Oberdorster, 1997; Elder *et al.*, 2004). However, like the intratracheal instillation method, it deposits particles in the lower respiratory tract and bypasses the upper respiratory tract which can be a potentially important target site for inhaled nanoparticles. The response of rats exposed to 'fine' (about 250 nm) and 'ultrafine' (about 21 nm) TiO_2 particles by intratracheal inhalation and intratracheal instillation has been compared. The results showed a decreased pulmonary response in rats receiving particles through inhalation when compared with those receiving particles through instillation (Osier and Oberdorster, 1997).

With appropriate dosing schemes and match controls, it appears that some of these methods are reasonable alternatives to more costly and time-consuming inhalation studies and may be useful for hazard identification or establishing the relative toxicity ranking among different nanomaterials for further testing. If the studies show positive responses in the pulmonary tract and/or ex-pulmonary organs/tissues, the data may be considered supportive for a health hazard concern. If no toxic responses are noted in both the pulmonary tract and the ex-pulmonary organs/tissues, further testing may not be precluded since these exposure methods are not physiological routes for human exposure, and the exposure periods in these method may be too short. Currently, the toxicity databases on nanomaterials are limited; more data on nanomaterials may be needed to validate the negative results of studies by these exposure methods.

21.6   Challenges in Risk Assessment

When considering risks of individual materials, possibly the first challenge lies in defining these materials as a group in order to apply nanomaterial-specific considerations. Nomenclature conventions are important to eliminate ambiguity when communicating differences between nanomaterials and bulk materials, and in reporting for regulatory purposes. Consistent nomenclature allows associations to be drawn between the potential hazards of particles as reported in public literature or other sources. At present, there is no harmonized terminology/nomenclature for clearly defining nanomaterials or standardized methods for accurately characterizing many of their physical and general properties.

One of the first definitions of nanotechnology was established by the National Nanotechnology Initiative (NNI; see www.nano.gov) for US government research on the applications and implications of nanomaterials. It defined nanotechnology as the understanding and control of matter at dimensions between approximately 1 and 100 nanometers, where unique phenomena enable 'novel applications'. Similarly, the International Organization for Standardization (ISO) defined 'nanoscale' as a size range from approximately 1 nm to 100 nm, but noted that while the size could vary beyond these limits, 'special properties', which include those that are not extrapolations from a larger size, are key. Thus, it also defined a 'nano-object' as a material with one, two or three external dimensions in the nanoscale, and went on to define several other specific nanoparticles (nanofiber, nanotube, nanoplate, nanowire and quantum dot; ISO, 2008). On the other hand, the definition of a nanoparticle provided by the American Standards of Testing Materials (ASTM) focuses more on particle dimensions and less on unique properties: a 'nanoparticle' is defined as a subclassification of ultrafine particle with lengths in two or three dimensions greater than 0.001 micrometer (1 nm) and smaller than about 0.1 micrometer (100 nm) and which may or may not exhibit a

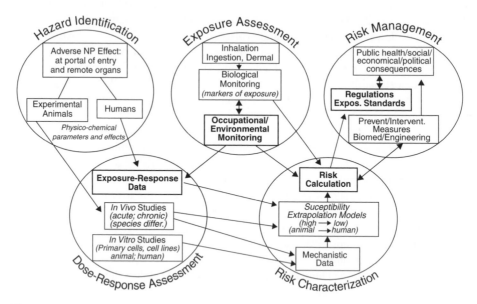

Figure 21.1   *Components of the risk assessment process (Oberdörster et al., 2005b). Reproduced with permission from Environmental Health Perspectives*

size-related intensive property. It notes that the length scale may be a hydrodynamic diameter or a geometric length appropriate to the intended use of the nanoparticle (ASTM, 2008).

Many US regulatory definitions for oversight of nanomaterials have adopted definitions similar to those of the NNI and ISO. In addition to defining nano by its scale, the 'novel applications'/'special properties' component of the materials was also considered. However, the current understanding of the 'novel applications'/'special properties' is incomplete; it is often unclear what exactly these 'novel applications'/'special properties' are, how they are measured and they compare from one type of nanomaterial to another. Without further definition, the concept presents materials classification and regulatory problems.

The risk assessments paradigm described by the National Research Council (NRC, 1983) for commercial chemical assessments is anticipated to be appropriate for the assessment of nanomaterials, although nanomaterials may have unique aspects to their estimations of hazard, dose-responses and exposures due to their differing physico-chemical properties. A detailed examination of the components of the risk assessment process (Figure 21.1) reveals large data gaps required for determining the risk of most nanomaterials (Oberdörster *et al.*, 2005b). Unlike bulk chemicals and pesticides, the data needed for specific nanomaterials such as toxicity/carcinogenicity data, adsorption/distribution/metabolism/excretion, and occupational/environmental monitoring information are unavailable at this time. Furthermore, generalized exposure scenarios for nanomaterials have also not been developed for the risk assessments of nanomaterials from a life-cycle perspective – from manufacture, to storage, shipping, incorporation into a commercial product such as an electronic device or paint component, use by consumers, and disposal (USEPA, 2007b). Without reliable data on effects and exposure, in-depth risk assessments of nanomaterials for regulation cannot be conducted at this time. Some of the reasons for the lack of adequate data to assess potential risks of nanoparticle are discussed below.

21.6.1 Characterization of Physico-chemical Properties

Because of their nanoscale and unique physico-chemical properties, it is believed that some nanoparticles can have toxicological properties that differ from their bulk materials. A number of studies have demonstrated that nanoparticle toxicity is extremely complex, and there is a strong likelihood that biological activity of nanoparticles will depend on a variety of physico-chemical properties such as particle size, shape, agglomeration state, crystal structure, chemical composition, surface area and surface properties (chemistry, surface charge, etc.) (Nemmar *et al.*, 2003; Sayes *et al.*, 2006a,b; Isakovic *et al.*, 2006; Warheit *et al.*, 2006, 2007a).

Based on the possible mechanisms of action and limited toxicity database of some nanomaterials, it appears that surface reactivity may be one of the critical factors in the toxicity of nanomaterials. While the nanosize (and thus increased surface area) alone may contribute to increased surface reactivity for some nanomaterials, other physico-chemical properties (e.g. dimension and shape, agglomeration state, chemical composition, crystal structure, surface coating) may be modifiers and important determinants of their surface reactivity and thus their cytotoxic/carcinogenic potential. A number of studies have shown that both physical and chemical factors are important determinants of nanoparticle toxicity, and that changes in the size, shape and other physico-chemical properties of a nanoparticle could result in changes in their adverse effects (Figure 21.2).

Chemical composition is one of the important parameters for the characterization of nanomaterials, which comprise nearly all substance classes, such as metal/metal oxides, organic/inorganic compounds and polymers, as well as biomolecules. Dependent on the particle surface chemistry, reactive groups on a particle surface will certainly modify the biological effects. For example, pulmonary exposure to nanoscale TiO_2 particles did not produce more cytotoxic or inflammatory effects to the lungs of rats compared with crystalline silica (which despite being larger in size, has much higher surface reactivity), indicating that toxicity is not always dependent upon particle size and surface area (Warheit *et al.*, 2006). Differing degrees of inflammation and lung injury have been induced upon instillation of nanoparticles of different compositions (e.g. NiO_2, Co_3O_4, TiO_2 and carbon black) in rat lungs (Dick *et al.*, 2003). Exposure to nanoparticles of the same composition but different crystal structures can also produce differential pulmonary effects. The cytotoxicity, cell proliferation, inflammation and histopathological responses in the lungs were compared with rats instilled intratracheally with P25 ultrafine TiO_2 nanoparticles (80/20 anatase/rutile) and two ultrafine rutile TiO_2 nanoparticle types of similar sizes and surface areas. Exposure to P25 ultrafine TiO_2 nanoparticles (80/20 anatase/rutile) produced marked pulmonary inflammation, cytotoxicity and adverse lung tissue effects. In contrast, only transient inflammation was produced following exposure to the two ultrafine rutile TiO_2 nanoparticles (Warheit *et al.*, 2007a). In an *in vitro* study using human lung epithelial cells or human dermal fibroblasts, anatase TiO_2 nanoparticles have been shown to be more chemically reactive and 100 times more cytotoxic than rutile TiO_2 nanoparticles of similar particle sizes (Sayes *et al.*, 2006a). The greater cytotoxic responses of anatase TiO_2 nanoparticles were attributed to a higher production of reactive oxygen species due to a

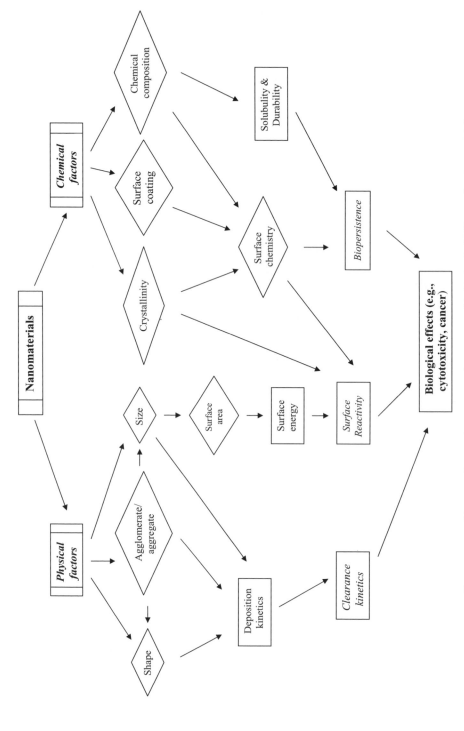

Figure 21.2 *Physico-chemical properties that can modify biological effects of nanomaterials.*

superior photocatalyst of the anatase crystal phase and differences inherent in the crystal structures of the two crystal phases.

Under ambient conditions, some nanoparticles can form aggregates or agglomerates. These aggregates/agglomerates have various forms, from dendritic structures to chain or spherical structures. Different aggregate/agglomerate structures/states of SWCNTs have been associated with distinct regional responses of mice lungs because of their differing size, shape and deposition/clearance kinetics (Shvedova *et al.*, 2005). To maintain the characteristics of nanoparticles, they are often stabilized with coatings or molecular adducts to prevent aggregation/agglomeration. The properties of nanoparticles can be significantly altered by surface modification, and the distribution of nanoparticles in the body strongly depends upon the surface characteristics. For instance, hydroxylated fullerene (C_{60}) has been demonstrated to be much less toxic than unsubstituted fullerene and induce distinct types of cell death by different mechanisms (Isakovic *et al.*, 2006). On the other hand, as the degree of sidewall functionalization increases, the SWCNT cytotoxic response of cells in culture decreases (Sayes *et al.*, 2006b). Changes of surface properties by coating of various nanoparticles (e.g. nano-CdSe/ZnS) with different types and concentrations of surfactants have also been shown to change their body distribution and their effects on biological systems significantly (Araujo *et al.*, 1999; Kirchner *et al.*, 2004).

Nanoscale materials are known to have various shapes such as spherical, needle-like, tubes, platelets, and so on. The shape of nanomaterials may have effects on the kinetics of deposition and absorption in the body. The importance of biopersistence to fiber carcinogenesis is well-documented (e.g. Bernstein *et al.*, 2001). The biopersistence of fibers is determined by their solubility/durability and clearance kinetics. Most nanoparticles are poorly soluble. However, dissolution of some nanoparticles occurs in culture medium or biological fluids, and cellular uptake, subcellular localization and toxic effects can be affected by the solubility/durability of nanoparticles. For instance, dissolved and non-dissolved ZnO have been shown to have different cellular uptake pathways and cytotoxicity due to their differing ability in releasing the toxic Zn^{2+} ions and production of ROS and oxidative stress (Xia *et al.*, 2008).

A challenge facing hazard identification/risk assessment of engineered nanoparticles is the wide diversity and complexity of the types of materials with varying physico-chemical properties. Even for the same type of nanomaterial, different methods of synthesizing, processing and functionalizing will generate products of different composition, size, shape and surface chemistry, with different types and levels of residual impurities, all of which can potentially affect their toxicity. Therefore, the toxicity evaluation of nanomaterials requires comprehensive physico-chemical characterization of each of them, and the test outcomes of one type of nanomaterial will not be appropriate to assess the toxicity of the same type/class of nanomaterial produced or processed by a different method. The specific material's properties are needed to characterize the particles in relevant toxicity tests, and complete and accurate characterization of the test material is essential to provide a basis for understanding the key properties that determine their biological effects. Characterization of the physico-chemical properties is a key element in all three test strategies for toxicity testing of nanoparticles (Oberdörster *et al.*, 2005a; Warheit, *et al.*, 2007b; NCL, undated). However, physico-chemical characterization of particles at the nanoscale presents unique challenges as properties of many nanomaterials can change through their life cycle and in various environmental and biological media.

The possibility of physico-chemical changes in the material before and after administration in a test system presents a challenge in identifying the key characteristics for nanomaterial toxicity. As-produced SWCNT material, for instance, has been shown to be a complex matrix of carbon nanotubes, nanoropes, non-tubular carbon and metal catalyst nanoparticles of different sizes. The pulmonary toxicity of as-produced SWCNTs released during manufacture and handling will depend on the partitioning and arrangement of these components within airborne particles and is expected to vary significantly by aerosol particle size and production batch (Maynard et al., 2007). The importance of characterizing the as-produced nanomaterial accurately is illustrated by the recent findings of Mitchell et al. (2007). In their publication they found that MWCNTs did not elicit significant lung toxicity in mice after 14 days of exposure to a maximum concentration of 5 mg/m^3. These results were in marked contrast to the findings of other investigators testing carbon nanotubes under similar conditions where significant lung responses were noted. The lower potency of the 'MWCNTs' seen in Mitchell et al. (2007) was later found to be due to the testing of carbon nanofibers instead of MWCNTs.

Under ambient conditions, some nanoparticles can form aggregates or agglomerates, and this aggregation/agglomeration raises concerns when considering size-dependent toxicity, specific surface area toxicity, and dose-dependency for in vitro and in vivo studies. In liquid media or biological fluids/tissues, as in air, nanoparticles can also undergo physico-chemical changes such as changes in size, composition, aggregation and agglomeration state. An example of physico-chemical changes to nanoparticles in liquid media is provided by Sayes et al. (2007a). In this study, carbonyl iron, crystalline silica, amorphous silica, nanosized zinc oxide, and fine-sized zinc oxide particles were measured in the dry state, in water, and in PBS and F-12K cell culture media. The mean particle size and size distribution generally increased in water relative to the dry state. The same trend was seen when the mean size and size distribution for particles in water were compared with those in the cell culture media. Aggregation was most pronounced in the F-12K medium. For all particles, surface charges decreased in the cell culture media, except for the crystalline silica particles. Murdock et al. (2008) also noted that stock solutions of various nanoparticles used for toxicological studies experienced significant changes in particle agglomeration and surface charge over time.

The association of nanoparticles with proteins in vivo has been known for some time, and adsorbed proteins can play an important role in directing the disposition and toxicity of nanomaterials such as SWCNTs and nanosized silica (Dutta et al., 2007). There is also evidence that some ultrafine particles can adsorb components of pulmonary surfactant onto the particle surface, conditioning the particle surfaces and affecting their distribution and biological effects (Wallace et al., 2007).

Realizing that the determination of physico-chemical characteristics of nanoparticles at several different time-points is critical to the outcomes of the risk assessment process, the ILSI panel (Oberdörster et al., 2005a) recommended the following characterization framework:

(1) characterization of as-produced or supplied materials;
(2) characterization of administered material;
(3) characterization of material following administration;
(4) human exposure characterization.

Some key characteristics were identified and some measurement methods and tools were recommended (Oberdörster *et al.*, 2005a). However, so far only a few reference materials with standard characteristics have been developed and used by the Nanotechnology Characterization Laboratory for toxicity testing (NCL, undated). Characterizing nanomaterials following administration in a test system provides the best data on dose and properties related to the responses observed. However, detection of nanomaterials, and measurements of changes to their physico-chemical properties are more challenging when the materials are in a tissue or biological fluid. Currently, many characterization methods exist for measuring particles in powder form, but their reliability, precision and accuracy on nanoscale materials are often called into question, and there are no well-defined techniques for characterization of nanomaterials in aqueous or biological solutions.

Whether a material can form aggregates or agglomerates upon exposure to environmental media or body fluids has a bearing on risk assessment due to differences in size, surface area and shape. An 'aggregate' is defined as a discrete group of particles in which the various individual components are not easily broken apart, such as in the case of primary particles that are strongly bonded together (for example, fused, sintered or metallically bonded particles). In contrast, the term 'agglomerate' refers to a group of particles held together by relatively weak forces (for example, Van der Waal's or capillary forces) that may break apart into smaller particles upon processing. Due to incomplete characterization, the two terms have often been used interchangeably and not clearly defined in the literature, which precludes accurate risk assessment.

Further research is needed to identify the key element(s) of engineered nanoparticles that may contribute to nanoparticle toxicity, and better understand the mechanisms of nanoparticle toxicity for risk assessment and structure–activity relationships (SAR) models development. Research is also needed to develop analytical methods for identifying, characterizing and measuring nanoparticles in biological systems, the environment and the workplace.

21.6.2   Testing Methodologies

Although a large number of peer-reviewed articles documenting toxicity test results of various nanoparticles have been published in recent years, the scientific community has yet to determine which nanomaterials are, and which are not, hazardous to humans or the environment. Conflicting results are often reported for nanomaterials of the same class/subclass or presumably identical materials. This is not too surprising since for materials even with the same chemical composition, they can have a difference in size/size distribution, surface areas and agglomeration states, which can influence their toxicological effects (Figure 21.2). Another reason for the conflicting results may be due to the lack of standardized testing methodologies/protocols for assessing the potential hazards of nanoparticles. Although the ILSI panel (Oberdörster *et al.*, 2005a) developed a screening strategy with a comprehensive array of *in vitro* and *in vivo* assays to investigate the toxicity of nanoparticles, no standard test protocols were recommended because standard test protocols for nanoparticles were not developed. Data from *in vitro* tests have limited usefulness for hazard identification of nanoparticles because of various issues including differing designs and conduct of the tests by different laboratories. Depending on the type of cells, the duration of exposure, the concentration of nanoparticles and the composition of the culture media, testing of the same nanomaterial can have different outcomes. The DuPone EHS framework (Warheit *et al.*, 2007b) used test guidelines developed by the US Environmental Protection Agency (EPA, 2007a) and Organization for Economic Cooperation and Development (OECD, undated). However, these guidelines are not specific for testing nanoparticles and require modifications.

Many methodologies that set inhalation exposure levels for workers and the general population, such as EPA's Reference Concentration (RfC) process (EPA, 1994), require the results of an inhalation study conducted in accordance with established test guidelines (e.g. OECD or EPA) for dose-response analysis. Using data from intratracheal instillation or pharyngeal aspiration studies for setting regulatory standards may not be appropriate because of the issues involved in these nonphysiological exposure methods (see discussion below). However, conducting chronic inhalation studies in rodents for every new nanomaterial introduced is impractical. These studies are very costly, use hundreds of animals, often take more than three years to complete, require sophisticated exposure facilities, and are technically difficult. Therefore, very few studies are available for industrial nanomaterials that report subchronic or chronic inhalation toxicity data. The only animal test data available from chronic inhalation testing of a nanomaterial are from Heinrich *et al.* (1995); these data were used in setting a draft NIOSH Recommended Exposure Level or REL (NIOSH, 2005) for nanosized titania.

21.6.3   Toxicokinetics

There is an urgent need for toxicokinetic data for nanoparticles: the detection and measurement of nanoparticles at their deposition sites, the dose-response, fate and persistence of nanoparticles in humans. As shown in Figure 21.2, a number of physico-chemical properties can influence the deposition and clearance kinetics of nanoparticles. However, it is unclear to what extent the different nanoparticle characteristics affect their absorption, distribution, metabolism and excretion (ADME). It has been found that for certain nanoparticles the clearance mechanism may be less effective than for larger particles after deposition in the respiratory tract. Their small size helps them to enter the cells by endocytosis and reach the circulating system, eventually reaching various potential target sites (Limbach *et al.*, 2004; Oberdörster *et al.*, 2005b). Dissolution appears to be one of the key elements for determining the biological fate and effects of some nanoscale materials (Borm *et al.*, 2006). All nanoparticles, upon exposure to tissues and fluids of the body, will immediately adsorb onto the surface of some of the macromolecules that they encounter. The specific features of this adsorption process will depend upon the surface characteristics of the particles, including surface chemistry and surface energy (Sousa *et al.*, 2004). The existence of both passive surface layers and surface active agents may compromise the risk evaluation of nanoparticles. At present, interactions between nanoparticles and target cells on the cellular and molecular levels are poorly understood. Important research needs for risk assessment include the transport of nanoparticles in the human body and the mechanisms of interaction at the subcellular level.

The nanoparticle dose at a target site (internal dose) in the body is critical for quantitative risk assessment of nanomaterials. Some methods using electron microscopy analysis are available to detect nanoparticles in urine, blood and other organs. However, these methods are not quantitative, and can only provide information on whether specific nanoparticles are present or absent in the biological matrix. New detection methods to determine the ADME of nanoparticles need to be developed and validated. The limited data available and the data gaps in toxicokinetics of nanoparticles have been reviewed (Hagens *et al.*, 2007).

21.6.4   Dose-response metrics

The dosimetry of nanoparticles has important impacts on both exposure assessment and study design, and defining the most appropriate dosimetry is a challenge. It is generally believed that for particles/nanoparticles, particle surface area is a more appropriate dosemetric than the traditional mass-based measure of dose (mg/kg). For instance, in a subchronic inhalation study in rats, ultrafine TiO_2 particles (20 nm) have been shown to elicit a persistently higher inflammatory reaction in the lungs compared with larger-sized (250 nm) TiO_2 when both types of particles were instilled at the same mass dose. As with some other larger toxic particles, a correlation between particle surface area and toxic effects was observed, suggesting that the particle surface area rather than the mass of the retained TiO_2 particles is the most relevant dose metric (Oberdörster *et al.*, 1994; Oberdörster, 2000). This is consistent with the mechanistic consideration that cellular responses may be related to surface area of a nanoparticle due to effects arising from particle surface chemistry/reactivity. Alternatively, the observed toxic responses may be a result of obstruction of cellular processes by the physical presence of poorly-soluble particles in sufficient numbers to cover the cellular surface of alveolar macrophages. Studies by Moss (2006) confirmed the occurrence of this alternative mechanism, and re-analysis of data by Oberdörster *et al.* (1994) suggested that both nanoparticle number and surface area are relevant dose metrics for ultrafine TiO_2 particles.

A recent study exploring the applicability of different physical exposure metrics to a range of nanoparticle classes and relating them to health impact-relevant attributes showed that no single method for monitoring nanoaerosol exposure will suit all nanomaterials (Maynard and Aitken, 2007). Similarly, Wittmaack (2007) noted that all three dose metrics (mass, number and surface area) worked well to describe the nanoparticle dose for quantifying lung inflammation response to nanoparticle exposure. As it is still unclear which chemical and/or physical characteristics contribute to the observed toxic responses, the relevant inhalation dosimetry in risk assessment of a nanoparticle may be surface area, particle number or even mass per volume; the complexity of particle properties and mechanisms of action preclude generalization to all nanoparticles, and make it a scientific challenge to select the best dose metric for toxicity testing and quantitative risk assessment of specific nanoparticles.

21.6.5   Exposure

Human exposure to nanoparticles is expected to accelerate as the development of new nanomaterials for consumer and food products, drug-delivery systems and other applications continues to grow. Inhalation represents an important route of human exposure to any airborne particles, including nanoparticles. The gastrointestinal tract is a possible portal of entry for nanoparticles since there are increasing uses of nanotechnology in food products (FSAI, 2008). The skin of workers is a potential route for exposure to nanoparticles during their manufacture. The general public may also be exposed to nanoparticles through the dermal route, as many textiles and cosmetics such as sunscreens contain nanoparticles (Davis, 1994).

Exposure information for recently-manufactured airborne nanoparticles is currently represented by very few studies due to the lack of inexpensive real-time monitoring instruments and methods. Maynard *et al.* (2004) found that airborne concentrations of SWCNTs during handling in a production facility were lower than 53 μg/m^3, and glove deposits were estimated at between 0.2 and 6 mg per hand. Other estimates of carbon nanotubes in the workplace have been published since (Han *et al.*, 2008), but these estimates are confounded to some degree due to background levels of particulates. NIOSH (2006) conducted monitoring in a university laboratory which was producing carbon nanofibers. In this study they found that most handling processes did not release carbon nanofibers. However, wet sawing of composite material containing nanofibers and transferring carbon nanofibers to a mixing vessel did raise the airborne particle concentrations. Also, carbon nanofiber materials were tracked out of the laboratory and into office space at the facility. While data on exposures in the workplace that are useful in estimating worker exposures are limited at best, there are no field monitoring-derived data for manufactured nanomaterials in the aquatic or terrestrial environment (Klaine *et al.*, 2008), making it even more difficult to estimate exposures to the general population.

21.7   Conclusions and Perspectives

Based on analogies to asbestos and crystallized silica and what we know about their toxicity and mechanisms, there are concerns about the health effects of nanoparticles. Because of their nanoscale and unique physico-chemical properties, there are additional concerns regarding nanoparticles for human health effects, and experts are in agreement that the adverse effects of nanoparticles cannot be predicted from the known toxicity of material of macroscopic size. Thus, it has been a challenge facing the regulatory agencies, academia and industries to assess the potential health risks of increasing numbers of nanoparticles in existence.

A number of studies have demonstrated that nanoparticle toxicity is extremely complex, and there is a strong likelihood that the biological activity of nanoparticles will depend on a variety of physico-chemical properties such as particle size, shape, agglomeration state, crystal structure, chemical composition, surface area and surface properties. A challenge facing hazard identification/risk assessment of engineered nanoparticles is the wide diversity and complexity of the types of materials with varying physico-chemical properties. As toxicity evaluation of nanomaterials requires comprehensive physico-chemical characterization of each of them at the nanoscale, this characterization presents unique challenges as properties of many nanomaterials can change through their life cycle and in various environmental and biological media. At present, there is no harmonized terminology/nomenclature for clearly defining nanomaterials or standardized methods for accurately characterizing many of their properties in biological matrices, the environment and the workplace.

Several screening strategies consisting of *in vitro* and *in vivo* assays to investigate the toxicity of diverse nanomaterials have been developed for the hazard identification process of nanomaterial risk assessment. However, *in vitro* test systems have limited usefulness in this regard due to a number of inherent issues. Meanwhile, conducting chronic inhalation studies in rodents for every new nanomaterial is impractical. Furthermore, there are currently no standardized testing methodologies/protocols for assessing the potential hazards of nanoparticles. Although a large number of peer-reviewed articles documenting toxicity test results of various nanoparticles have been published in recent years, the scientific community has yet to determine which nanomaterials are, and which are not, hazardous to humans or the environment. Conflicting results are often reported for nanomaterials of the same class/subclass or seemingly identical materials.

The complexity of particle properties and mechanisms of action also preclude generalization to all nanoparticles regarding the best dose metric for toxicity testing and quantitative risk assessment of specific nanoparticles. Currently, there is an urgent need for toxicokinetic data for nanoparticles; yet, new detection methods to determine the ADME of nanoparticles await development and validation. The impacts of varying physical and chemical properties of nanomaterials on the exposure outcomes are poorly understood. Generalized exposure scenarios and adequate exposure assessment techniques and tools for measuring nanoparticles in indoor and ambient environments have not been developed for the risk assessment of nanomaterials from a life-cycle perspective. As a result, reliable information on all components of the risk assessment process – hazard identification, dose-response assessment, risk characterization and exposure assessment – are not available for nanomaterials.

For the last few years, research efforts have been initiated by governments of several countries and international organizations (including the US, European Union and Japan) to develop priorities for risk-based safety evaluations for nanomaterials (e.g. Thomas and Sayre, 2005; Thomas et al., 2006; EPA, 2007b). As there is an urgent need to develop standardized methods for toxicity testing of nanoparticles, efforts are being made to review and modify the current OECD and EPA test guidelines (OECD, undated; EPA, 2007a). The availability of reference materials would be important to benchmark the adverse effects once standardized methods for toxicity testing of nanoparticles are developed. In 2006, a project was proposed to develop reference materials for engineered nanoparticle toxicology and metrology by the Nanotechnology Research Coordination Group (NRCG) of the UK (Aitken et al., 2007). A similar project was initiated in the US by the National Institute of Science and Technology (NIST) and the NCL. In December 2007, a set of reference standards for nanoparticles aimed at the biomedical research community was released. The new reference materials consist of colloidal gold nanoparticles with nominal diameters of 10, 30 and 60 nm in suspension (NCL, undated).

Recently, an international group of nanotoxicologists called the International Alliance for Nano Environment, Human Health and Safety Harmonization (IANH) has assembled to establish testing protocols that will enable reproducible toxicological testing of nanomaterials at the cell and animal levels (Maynard, 2008). A set of round robin experiments will be conducted by researchers who share identical nanomaterials, cells and biological systems and use a common protocol. The results of independently conducted tests will be compared, and any deviations resulting from errors in applying the protocol will be tracked and eliminated. Therefore, the outcome will be a core set of rationally designed protocols that enable scientists from different laboratories to obtain identical and reproducible results. With the efforts of this and other international research programs, it is hoped that reliable risk and safety evaluations for nanomaterials will be developed in the near future.

In light of the advances in molecular biology, system biology, genomics, proteomics, bioinformatics and high-throughput techniques, the National Research Council (NRC, 2007) has proposed a new strategy for toxicity testing of chemicals in the twenty-first century. New toxicity testing systems will mainly rely on predictive, high-throughput assays using primary cells or cell lines to evaluate relevant perturbations in key toxicity pathways. Recently, the EPA has launched a large-scale ToxCast project (Dix *et al.*, 2007) to develop predictive high-throughput screening (HTS) and genomic bioactivity signatures useful for screening toxicants, characterizing toxicity pathways and prioritizing further testing. There is promise that high throughput screening may also open new insights into ways to classify large numbers of nanomaterials according to their relative toxicity and possible modes of action. Indeed, a generalizable systematic approach for assessing the biological activity of nanomaterials by multiple physiological cell-based assays, in multiple cell types, and at multiple doses has been proposed. Based on similarities in their toxicity profiles, clustering methods can then be used to classify nanomaterials into groups, and nanomaterials that cause similar biological effects *in vitro* can be identified for further studies *in vivo*. Shaw *et al.* (2008) have shown that metal and metal oxide nanomaterials can be clustered into meaningful structure activity groupings by using such high-throughput techniques that represent a broad array of cellular physiological responses. As low-throughput *in vitro* screening methods continue to improve (e.g. Herzog *et al.*, 2007), use of genomics and proteomics analyses may assist in the interpretation of responses elicited by various types of nanomaterials (Oberdörster *et al.*, 2005a). All these new approaches and strategies in the twenty-first century should help to address some of the challenges facing risk assessment of nanoparticles.

Disclaimer: The scientific views expressed in this chapter are solely those of the authors and do not necessarily reflect the views and policies of the US Environmental Protection Agency.

References

Aitken RJ, Hankin SM, Tran CL, Donaldson K, Stone V, Cumpson P, Johnstone J, Chaudhry Q, Cash S (2007) *REFNANO: Reference materials for engineered nanoparticle toxicology and metrology*, IOM, UK. Final report on project CB01099, 21 August 2007.

Andrievsky G, Klochkov V, Derevyanchenko L (2005) Is the C60 fullerene molecule toxic? *Fuller Nanotub Carbon Nanostruct* **13**, 363–376.

Araujo L, Lobenberg R, Kreuter J (1999) Influence of the surfactant concentration on the body distribution of nanoparticles. *J Drug Target* **6**, 373–385.

ASTM (2008) *Standard Terminology Relating to Nanotechnology*. ASTM Committee E 2456-06.

Bernstein DM, Riego-Sintes JM, Ersboell BK, Kunert J (2001) Biopersistence of synthetic mineral fibers as a predictor of chronic inhalation toxicity in rats. *Inhal Toxicol* **13**, 823–849.

Bernstein D, Castranova V, Donaldson K, Fubini B, Hesterberg T, Kane A, Lai D, McConnell EE, Muhle H, Oberdörster G, Olin S, Warheit D (2005) Testing of fibrous particles: Short-term assays and strategies. Report of an ILSI Risk Science Institute Working Group. *Inhalation Toxicol* **17**, 1–41.

Bonneau L, Suquet H, Malard C, Pezerat H (1986a) Studies on surface properties of asbestos, I: Acute sites on surface of chrysotile and amphiboles. *Environ Res* **41**, 251–267.

Bonneau L, Malard C, Pezerat H (1986b) Studies on surface properties of asbestos, II: Role of dimensional characteristics and surface properties of mineral fibers in the induction of pleural tumors. *Environ Res* **41**, 268–275.

Borm P, Klaessig FC, Landry TD, Moudgil B, Pauluhn J, Thomas K, Trottier R, Wood S (2006) Research strategies for safety evaluation of nanomaterials, Part V: Role of dissolution in biological fate and effects of nanoscale particles. *Toxicol Sci* **90**, 23–32.

Brown DM, Stone V, Findlay P, MacNee W, Donaldson K (2000) Increased inflammation and intracellular calcium caused by ultrafine carbon black is independent of transition metals or other soluble components. *Occup Environ Med* **57**, 685–691.

Brown DM, Wilson MR, MacNee W, Stone V, Donaldson K (2001) Size-dependent proinflammatory effects of ultrafine polystyrene particles: a role for surface area and oxidative stress in the enhanced activity of ultrafines. *Toxicol Appl Pharmacol* **175**, 191–199.

Brown DM, Donaldson K, Borm PJ, Schins RP, Dehnhardt M, Gilmour P *et al.* (2004) Calcium and ROS-mediated activation of transcription factors and TNF-alpha cytokine gene expression in macrophages exposed to ultrafine particles. *Am J Physiol Lung Cell Mol Physiol* **286**, L344–L353.

Cedervall T, Lynch I, Lindman S, Berggard T, Thukin E, Nilsson H, Dawson KA, Linse S (2007) Understanding the nanoparticle-proterin corona using methods to quantify exchange rates and affinity to proteins for nanoparticles. *Proc Natl Acad Sci USA* **104**, 2050–2055.

Chen M, von Mikecz A (2005) Formation of nucleoplasmic protein aggregates impairs nuclear function in response to SiO_2 nanopartiocles. *Exp Cell Res* **305**, 51–62.

Chen Z, Meng H, Xing GM, Chen CY, Zhao YI, Jia G, Wang TC, *et al.* (2006) Acute toxicological effects of copper nanoparticles *in vivo. Toxicol Lett* **163**, 109–120.

Clouter C, Brown D, Höhr D, Borm P, Donaldson K (2001) Inflammatory effects of respirable quartz collected in workplaces versus standard DQ12 quartz: particle surface correlates. *Toxicol Sci* **63**, 90–98.

Davis DA (1994) Sunscreen oddities. *Drug Cosmetics Ind* **155**, 20.

Dick CAJ, Brown DM, Donaldson K, Stone V (2003) The role of free radicals in the toxic and inflammatory effects of four different ultrafine particle types. *Inhal Toxicol* **15**, 39–52.

Dix DJ, Houck KA, Martin MT, Richard AM, Setzer RW, Kavlock RJ (2007) The ToxCast program for prioriting toxicity testing of environmental chemicals. *Toxicol Sci* **95**, 5–12. Available at http://www.epa.gov/ncct/toxcast/

Donaldson K, Aitken R, Tran L, Stone V, Duffin R, Forrest G, Alexander A (2006) Carbon nanotubes: A review of their properties in relation to pulmonary toxicology and workplace safety. *Toxicol Sci* **92**, 5–22.

Driscoll KE, Lindenschmidt RC, Murer JK, Perkins L, Perkins M, Higgins I (1991) Pulmonary response to inhaled silica or titanium dioxide. *Toxicol Appl Pharmacol* **111**, 201–210.

Driscoll KE, Costa DL, Hatch G, Henderson R, Oberdorster G, Salem H, Schesinger RB (2000) Intratracheal instillation as an exposure technique for the evaluation of respiratory tract toxicity: uses and limiotations. *Toxicol Sci* **55**, 24–35.

Duffin R, Tran CL, Clouter A, MacNee W, Stone V, Donaldson K (2002) The importance of surface area and specific reactivity in the acute pulmonary inflammatory response to particles. *Ann Occup Hyg* **46**, 242–245.

Dutta D, Sunddaram SK, Teeguarden JG, Riley BJ, Fifiels LS, Jacobs JM, Addleman SR, Kaysen GA, Moudgil BM, Weber TJ (2007) Adsorbed proteins influence the biological activity and molecular targeting of nanomaterials. *Toxiocol Sci* **100**, 303–315.

Elder AC, Gelein R, Oberdorster G, Finketstein J, Motter R, Wang Z (2004) Efficient depletion of alveolar macrophages using intratracheally inhaled aerosols of liposome-encapsulated clodronate. *Exp Lung Res* **30**, 105–112.

Environmental Protection Agency (EPA) (1994) *Methods for derivation of inhalation reference concentrations and application of inhalation dosimetry*. EPA/600/8-90/066F. Environmental Protection Agency: Washington, DC.

Environmental Protection Agency (EPA) (2007a) *Health effects test guidelines*. OPPTS 870 series. Available at http://www.epa.gov/opptsfrs/publications/OPPTS_Harmonized/870_Health_Effects_Test_Guidelines/Drafts/

Environmental Protection Agency (EPA) (2007b) *Nanotechnology White Paper, EPA 100/B-07/001*. Office of the Science Advisor, Science Policy Council. Environmental Protection Agency: Washington, DC. Available at www.epa.gov/osa

Food Safety Authority of Ireland (2008) *The relevance for food safety of applications of nanotechnology in the food and feed industries*. Food Safety Authority of Ireland: Dublin. Available at http://www.fsai.ie/resources_and_publications/scientific_reports.html

Fubini B (1997) Surface reactivity in the pathogenic response to particulates. *Environ Health Persp* **105S**, 1013–1020.

Goodman CM, McCusker CD, Yilmaz T, Rotello VM (2004) Toxicity of gold nanoparticles functionalized with cationic and anionic side chains. *Bioconjug Chem* **15**, 897–900.

Hagens WI, Oomen AG, de Jong WH, Cassee FR, Sips AJM (2007) What do we (need to) know about the kinetic properties of nanoparticles in the body? *Regul Toxicol Pharmacol* **49**, 217–229.

Han JH, Lee E, Lee J, So K, Lee Y, Bae G, Lee S-B, Ji J, Cho M, Yu I (2008) Monitoring multiwalled carbon nanotube exposure in carbon nanotube research facility. *Inhalation Toxicology* **20**, 741–749.

Heinrich U, Fuhst R, Rittinghausen S, Creutzenberg O, Bellmann B, Koch W, Levsen K (1995) Chronic inhalation exposure of Wistar rats and two different strains of mice to diesel engine exhaust, carbon black, and titanium dioxide. *Inhal Toxicol* **7**, 533–556.

Henderson R, Driscoll K, Harkema J, Lindenschmidt RC, Chang I, Maples K, Barr E (1995) A comparison of the inflammation response of the lungs to inhaled vesus instilled particles in F344 rats. *Fund Appl Toxicol* **24**, 183–197.

Herzog E, Casey A, Lyng FM, Chambers G, Byrne HJ, Davoren M (2007) A new approach to the toxicity testing of carbon-based nanomaterials – the clonogenic assay. *Toxicol Lett* **174**, 49–60.

International Agency for Research on Cancer (IARC) (1997) Silica, some silicates, coal dust and para-aramid fibrils. In *IARC Monographs on the Evaluation of Carcinogenic Risks to Humans*. International Agency for Research on Cancer, Vol. **68**: Lyon, France.

International Agency for Research on Cancer (IARC) (2002) Man-made vitreous fibers. In *IARC Monographs on the Evaluation of Carcinogenic Risks to Humans*. International Agency for Research on Cancer, Vol. **81**: Lyon, France.

ISO (2008) *Nanotechnologies – Terminology and definitions for nano-objects – nanoparticle, nanofiber, and nanoplate*. ISO Technical Committee 229, International Organization for Standardization: Geneva.

Isakovic A, Markovic Z, Todorovic-Markovic B, *et al.* (2006) Distinct cytotoxic mechanisms of pristine versus hydroxylated fullerene. *Toxicol Sci* **91**, 173–183.

Jeng HA, Swanson J (2006) Toxicity of metal oxide nanoparticlesin mammalian cells. *J Environ Sci Pt A* **41**, 2699–2771.

Kagan VE, Tyurina YY, Tyurin VA, Kondura NV, Potapovich AI, Osipov AN, Kisin ER, Schwegler-Berry D, Mercer R, Castranova V, Shvedova AA (2006) Direct and indirect effect of single walled carbon nanotubes on RAW 264.7 macrophages: Role of iron. *Toxicol Lett* **165**, 88–100.

Kane AB, Boffetta P, Saracci R, Wilbourn JD (eds) (1996) *Mechanism of Fibre Carcinogenesis*. IARC Scientific Publications No. 140. International Agency for Research on Cancer: Lyon, France.

Kim H, Liu X, Kobayashi T, Kohyama T, Wen FQ, Romberger DJ, *et al.* (2003) Ultrafine carbon black particles inhibit human lung fibroblast-mediated collagen gel contraction. *Am J Respir Cell Mol Biol* **28**, 111–121.

Kirchner C, Liedl T, Kudera S, Pellegrino T, Javier AM, Gaub HE, Stolzle S, Fertig N, Parak WJ (2004) Cytotoxicity of colloidal CdSe and CdSe/ZnS nanoparticles. *Nano Lett* **5**, 331–338.

Klaine S, Alvarez P, Batley G, Fernandez T, Handy R, Lyon D, Mahendra S, Mclaughlin M, Lead J (2008) Nanomaterials in the environment: behavior, fate, bioavailability, and effects. *Environmental Toxicology and Chemistry* **27**, 1825–1851.

Lam CW, James JT, McCluskey R, Hunter RL (2004) Pulmonary toxicity of single-walled carbon nanotubes in mice 7 and 90 days after intratracheal instillation. *Toxicol Sci* **77**, 126–134.

Li N, Sioutas C, Cho A, Schmitz D, Misra C, Sempf J, *et al.* (2003) Ultrafine particulates pollutants induce oxidative stress and mitochondria damage. *Environ Health Persp* **111**, 455–460.

Limbach LK, Li Y, Grass RN, Brunner TI, Hintermann MA, Muller M, Gunter D, Starr WI (2004) Oxide nanoparticle uptake 9n human lung fibroblasts: effects of particle size, agglomeration, and diffusion at low concentration. *Environ Sci Technol* **38**, 5208–5216.

Maynard A (2008) Nanotoxicologists self-assemble. *Nanotechnology* Sept. 9, 2008. Available at http://2020science.org/2008/09/09/nanotoxicogists-self-assemble/

Maynard AD, Baron P, Foley M, Shvedova A, Kisin E, Castranova V (2004) Exposure to carbon nanotube material: aerosol release during the handling of unrefined singlewalled carbon nanotube material. *J Toxicology Environ Health Part A* **67**, 87–107.

Maynard AD, Aitken RJ (2007) Assessing exposure to airborne nanomaterials. Current ability and future requirements. *Nanotoxicology* **1**, 26–41.

Maynard AD, Ku BK, Emery M, Stolzenburg M, McMurry PH (2007) Measuring particle size-dependent physicochemical structure in airborne single walled carbon nanotube agglomerates. *J Nanoparticle Res* **9**, 85–92.

Mecke A, Orr BG, Banaszak Holl MM, Baker JR (2006) Lipid bilayer disruption by polyamidoamine dendrimers: the role of generation and capping group. *Langmuir* **21**, 10348–10354.

Mitchell L, Gao J, Vander Wal R, Gigliotti A, Burchiel S, McDonald J (2007) Pulmonary and systemic immune response to inhaled multiwalled carbon nanotubes. *Toxicol Sci* **100**, 203–214.

Monteiro-Riviere NA, Inman AO (2006) Challenges for assessing carbon nanomaterial toxicity to the skin. *Carbon* **44**, 1070–1078.

Moss OR (2006) Insights into the health effects of nanoparticles: why numbers matter. *CIIT Activities* **26**(2), May–Sept.

Muller J, Huaux F, Moreau N, Misson P, Heilier J-F, Delos M, Arras M, Fonseca A, Nagy JB, Lison D (2005) Respiratory toxicity of multi-wall carbon nanotubes. *Toxicol Appl Pharmacol* **207**, 221–231.

Murdock RC, Braydich-Stolle L, Schlager AM, Hussain SM (2008) Characterization of nanomaterial dispersion in solution prior to *in vitro* exposure using dynamic light scattering technique. *Toxicol Sci* **101**, 239–253.

Nanotechnology Characterization Laboratory (NCL) (undated) *Assay Cascade Protocols.* National Cancer Institute, Nanotechnology Characterization Laboratory. Available at http://ncl.cancer.gov/working_assay-cascade.asp.

National Institute of Occupational Safety and Health (NIOSH) (2005) *Draft NIOSH Current Intelligence Bulletin, Evaluation of health hazard and recommendations for occupational exposure to titanium dioxide.* National Institute of Occupational Safety and Health: Cincinnati, Ohio. Available at http://www.cdc.gov/niosh/review/public/TIo2/default.html

National Institute of Occupational Safety and Health (NIOSH) (2006) *Health Hazard Evaluation Report HETA #2005-0291-3025.* University of Dayton Research Institute. National Institute of Occupational Safety and Health: Cincinnati, Ohio. Available at http://www.cdc.gov/niosh/hhe/reports

National Research Council (NRC) (1983) *Risk Assessment in the Federal Government: Managing the Process.* National Research Council, National Academy Press: Washington, DC.

National Research Council (NRC) (2007) *Toxicity Testing in the 21st Century: A Vision and A Strategy.* National Research Council, National Academy Press: Washington, DC.

Nemmar A, Hoylaerts MF, Hoet PHM, Vermylen J, Nemery B (2003) Size effect of intratracheally instilled particles on pulmonary inflammation and vascular thrombosis. *Toxicol Appl Pharmacol* **186**, 38–45.

Oberdörster G (2000) Toxicology of ultrafine particles: in vivo studies. *Phil Trans R Soc Lond A* **358**, 2712–2740.

Oberdörster G, Ferin J, Lehnert BE (1994) Correlation between particle size, in vivo particle persistence, and lung injury. *Environ Health Persp* **102**(Suppl. 5), 173–179.

Oberdörster G, Sharp Z, Atudorei V, Elder A, Gelein R, Lunts A, *et al.* (2002) Extrapulmonary translocation of ultrafine carbon particle following whole-body inhalation exposure of rats. *J Toxicol Environ Health A* **65**, 1531–1543.

Oberdörster G, Sharp Z, Atudorei V, Elder A, Gelein R, Kreyling W, *et al.* (2004) Translocation of inhaled ultrafined particles to the brain. *Inhal Toxicol* **16**, 437–445.

Oberdörster G, Maynard A, Donaldson K, Castranova V, Fitzpatrick JW, Ausman K, Carter J, Karn B, Kreyling W, Lai D, Olin SS, Monteiro-Riviere NA, Warheit DB, Yang H (2005a) Principles for characterizing the potential human health effects from exposure to nanomaterials: Elements of a screening strategy. *Particle and Fibre Toxicol* **2**. Available at http://www.particleand fibretoxicology.com/content/2/1/8

Oberdörster G, Oberdörster E, Oberdörster J (2005b) Nanotoxicology: An emerging discipline evolving from studies of ultrafine particle. *Environ Health Persp* **113**, 823–839.

OECD (undated) *OECD Guidelines for the Testing of Chemicals, Section 4, Health Effects.* Available at http://puck.sourceoecd.org/vl=5507610/nw=1/rpsv/cw/vhosts/oecdjpurnals/1607310x/vln4/contp1-1.htm.

Osier M, Oberdörster G (1997) Intratracheal inhalation vs intratracheal instillation: differences in particle effects. *Fund Appl Toxicol* **40**, 220–227.

Pacurari M, Yin XJ, Zhao J, Ding M, Leonard SS, Schwegier-Berry D, Ducatman BS, Sbarra S, Hoover MD, Castranova V, Vallyathan V (2008) Raw single-wall carbon nanotubes induce oxidative stress and activate MAPKs, AP-1, NF-κB, and Akt in normal and malignant human mesotheliol cells. *Environ Health Persp* **116**, 1211–1217.

Poland CA, Duffin R, Kinloch I, Maynard A, Wallace WAH, Seaton A, Stone V, Brown S, MacNee W, Donaldson K (2008) Carbon nanotubes introduced into the abdominal cavity of mice show asbestos-like pathogenicity in a pilot study. *Nature Nanotechnology* **3**, 423–428.

Rajih T, Nedeljkovic M, Chen LX, Poluektov O, Thurnauer MC (1999) Improving optical and charge separation properties of nanocrystalline TiO_2 by surface modification with Vitamin C. *J Phys Chem B* **103**, 3515.

Rao GVS, Tinkle S, Weissman DN, Antonini JM, Kashon ML, Salmen R, Battelli LA, Willard PA, Hubbs AF, Hoover MD (2003) Efficacy of a technique for exposing the mouse lung to particles aspirated from the pharynx. *J Toxicol Environ Health* **A66**, 1441–1452.

Sayes C, Fortner J, Guo W, Lyon D, Boyd A, Ausman K, *et al.* (2004) The differential cytotoxicity of water-soluble fullerenes. *Amer Chem Soc* **4**, 1881–1887.

Sayes CM, Wahi R, Kurian PA, Liu Y, West JL, Ausman KD, Warheit DB, Colvin VL (2006a) Correlating nanoscale titania structure with toxicity: A cytotoxicity and inflammatory response study with human dermal fibroblasts and human lung epithelial cells. *Tox Sci* **92**, 174–185.

Sayes CM, Liang F, Hudson JL, Mendez J, Guo W, Beach JM, Moore VC, Doyle CD, West JL, Billups WE, Ausman KD, Warheit DB, Colvin VL (2006b) Functionalization density dependence of single-walled carbon nanotubes cytotoxicity in vitro. *Toxicol Lett* **162**, 135–142.

Sayes CM, Reed KL, Warheit DB (2007a) Assessing toxicity of fine and nanoparticles: comparing *in vitro* measurements to *in vivo* pulmonary toxicity profiles. *Toxicol Sci* **97**, 163–180.

Sayes CM, Marchione AA, Reed KL, Warheit DB (2007b) Comparative pulmonary toxicity assessment of C60 water suspensions in rats: few differences in fullerene toxicity in vivo in contrast to in vitro profiles. *Nano Lett* **7**, 2399–2406.

Searl A (1994) A review of the durability of inhaled fibers and options for the design of safer fibers. *Ann Occup Hyg* **38**, 839–855.

Shaw SY, Westly EC, Pittet MJ, Subramanian A, Schreiber SL, Weissleder R (2008) Perturbational profiling of nanomaterial biological activity. *Proceed Nat Acad Sci* **105**, 7387–7392.

Shvedova AA, Kisin ER, Mercer R, *et al.* (2005) Unusual inflammatory and fibrogenic pulmonary responses to single-walled carbon nanotubes in mice. *Am J Physiol-Lung Cell Mol Physiol* **289**, 698–708.

Shvedova AA, Kisin ER, Murray AR, Johnson VJ, Gorelik O, *et al.* (2008) Inhalation vs. aspiration of single-walled carbon nanotubes in C57BL/6 mice: inflammation, fibrosis, oxidative stress, and mutagenesis. *Am J Physiol Lung Cell Mol Physiol* **295**, L552–L565.

Sousa SR, Ferreira PM, Saramago B, Melo LV, Barbosa MA (2004) Human serum adsorption on TiO_2 from single protein solutions and from plasma. *Langmuir* **20**, 9745–9754.

Thomas K, Sayre P (2005) Research strategies for safety evaluation of nanomaterials, Part I: Evaluating the human health implications of ecposure to nanoscale materials. *Toxicol Sci* **87**, 316–321.

Thomas K, Aguar P, Kawasaki H, Morris J, Nakanishi J, Savage N (2006) Research strategies for safety evaluation of nanomaterials, Part VIII: International efforts to develop risk-based safety evaluations for nanomaterials. *Toxicol Sci* **92**, 23–32.

Topinka J, Loli P, Dusinska M, Hurbánková M, Kováčiková Z, Volkovová K, Kažimírová A, Barančoková M, Tatrai E, Wolff T, Oesterle D, Kyrtopoulos SA, Georgiadis P (2006) Mutagenesis by man-made mineral fibers in the lung of rats. *Mutat Res* **595**, 174–183.

Vertegel AA, Siegel RW, Dordick JS (2004) Silica nanoparticle size influences the structure and enzymatic activity of adsorbed lysozyme. *Langmuir* **20**, 6800–6807.

Wallace WE, Keane MJ, Murray DK, Chisholm WP, Maynard AD, Ong T-M (2007) Phospholipid lung surfactant and nanoparticles surface toxicity: Lessons from diesel soots and silicate dusts. *J Nanoparticle Res* **9**, 23–38.

Warheit DB, Laurence BR, Reed KL, Roach DH, Reynolds GA, Webb TR (2004) Comparative pulmonary toxicity assessment of single-wall carbon nanotubes in rats. *Toxicol Sci* **77**, 117–125.

Warheit DB, Brook WJ, Lee KP, Webb TR, Reed KL (2005) Comparative pulmonary toxicity inhalation and instillation studies with different TiO_2 particle formulations: impact of surface treatments on particle toxicity. *Toxicol Sci* **88**, 514–524.

Warheit D, Webb TR, Sayes CM, Colvin VL, Reed KL (2006) Pulmonary instillation studies with nanoscale TiO_2 rods and dots in rats. Toxicity is not dependent upon particle size and surface area. *Toxicol Sci* **91**, 227–236.

Warheit D, Webb TR, Reed KL, Frerichs S, Sayes CM (2007a) Pulmonary toxicity study with three forms of ultrafine-TiO_2 particles: differential responses related to surface properties. *Toxicology* **230**, 90–104.

Warheit D, Hoke RA, Finlay C, Donner EM, Reed KL, Sayes CM (2007b) Development of a base set of toxicity tests using ultrafine TiO_2 particles as a component of nanoparticle risk management. *Toxicol Lett* **171**, 99–110.

Wilson M, Lightbody JH, Donaldson K, Stone V (2002) Oxidative interactions between ultrafine particles and metals in vitro and in vivo. *Toxicol Appl Pharmacol* **184**, 172–179.

Wittmaack K (2007) Search of the most relevant parameter for quantifying lung inflammatory response to nanoparticle exposure: particle number, surface area, or what? *Environ Health Persp* **115**, 187–194.

Woo Y-T, Lai DY, Arcos JC, Argus MF (1988) Foreign-body carcinogens: fibers, silica, and implants. In *Chemical Induction of Cancer*, Vol. **IIIC**. Academic Press: San Diego; 508–559.

Worle-Knirsh JM, Pulskamp K, Krug HF (2006) Oops they did it again! Carbon nanotubes hoax scientists in viability assays. *Nano Lett* **6**, 1261–1268.

Xia T, Kovochich M, Liong M, Mädler L, Gilbert B, Shi H, Yeh JI, Zink JI, Nel AE (2008) Comparison of the mechanism of toxicity of zinc oxide and cerium oxide nanoparticles based on dissolution and oxidative stress properties. *ACS Nano* **2**, 2121–2134.

22

Evaluating Strategies For Risk Assessment of Nanomaterials

Nastassja Lewinski, Huiguang Zhu and Rebekah Drezek

22.1 Introduction

Nanotechnology is a field in constant motion, with new articles on novel particles, syntheses, applications and safety data being published continuously. The establishment of the
National Nanotechnology Initiative (NNI) in 2001 provided significant federal support for
the development of new nanotechnology applications, and now as these applications mature, understanding the environmental, health and safety implications of nanotechnology
has also risen as a priority (NNI, 2008). With previous technologies, less attention was
given to preventative actions to reduce risk; however, in recent years, society has become
more technologically conscious, resulting in an increase in public demand for extensive
safety evidence on the products and environments to which people are exposed (Hansen
et al., 2008). As a result, with emerging technologies such as nanotechnology, there has
been a push towards characterizing risks to human health, or conducting risk assessments,
in tandem with technology research and development (Wilsdon, 2004; Morgan, 2005; Renn
and Roco, 2006; Davis, 2007; Bauer *et al.*, 2008).

With the increasing number of consumer goods and developing medical interventions
containing nanomaterials, more questions are being raised as to the safety of these products
and the potential impact they may have on human health (Maynard, 2006). Evaluation of
the potential hazards of nanotechnology and its products is an emerging area in toxicology,
with the majority of nanotechnology safety data encompassing nanomaterials. Nanomaterials encompass a multitude of classes, which contain different subclasses and countless
modified versions. Currently, there is a limited range of nanomaterial toxicity data available, with only a few nanomaterial types already tested (Lewinski *et al.*, 2008). Given

Nanotoxicity: From In Vivo *and* In Vitro *Models to Health Risks*   Edited by Saura Sahu and Daniel Casciano
© 2009 John Wiley & Sons, Ltd

the large number of compounds to evaluate, conducting routine toxicology testing on all nanomaterials would be an overwhelming endeavor, as it would entail *in vitro* and *in vivo* testing covering many endpoints and mechanisms of action. As many commercial chemicals have yet to be thoroughly assessed, it may also be impractical. Therefore, as more risk and safety data on nanotechnology is collected and discussed, it becomes increasingly important that scientists identify the particular types of nanomaterials that require more extensive investigation through comprehensive risk assessment (Davis, 2007).

As the field of nanotechnology continues to grow, risk assessors face several challenges. The development of new and existing nanomaterials continues to outpace environmental, health and safety (EHS) data collection. In addition, people question whether standard testing protocols are adequate to determine health hazards and toxicity levels for nanomaterials, and whether the tests being conducted are representative of actual exposure conditions. The goal of this chapter is to provide an overview of the traditional risk assessment process, the status of nanotechnology risk assessment, and the research needs to make significant progress towards understanding the risks that nanomaterial exposure presents.

22.2   What is Risk Assessment?

Although researchers commonly use the term 'risk assessment', its definition is often misconstrued. In general, the term is broad and applies not only to human health but also to several other areas such as economics, environment, security and transportation. For the purposes of this chapter, only health risk assessment in relation to nanotechnology will be discussed. The fundamental concept behind health risk assessments is that without exposure, there is no risk. As Paracelsus recognized 500 years ago, 'All things are poison and nothing is without poison, only the dose permits something not to be poisonous' (Hunter, 2008). The structured process of human health risk assessment traditionally involves four parts as proposed by the National Research Council (NRC) in 1983 (National Academy of Sciences, 1983). These include (1) identifying and defining the hazard, (2) quantifying the relationship between dose and toxic response, (3) determining the routes and conditions of exposure, and (4) characterizing and analyzing the risk of exposure (Morris and Willis 2007; National Academy of Sciences, 1994).

Before describing the aforementioned risk assessment steps, the terminology that will be used throughout must be defined. The term 'risk' is defined as the probability of adverse effects due to exposure to a 'hazard', a hazardous chemical, biological or physical agent (Christensen *et al.*, 2003). The term 'adverse effect' refers to 'a biochemical change, functional impairment, or pathologic lesion that affects the performance of the whole organism, or reduces an organism's ability to respond to an additional environmental challenge' (Environmental Protection Agency, 2008d). A 'risk assessment' therefore is the methodology or process used to generate a quantitative estimate of risk (National Academy of Sciences, 1983, 1994). When conducting a risk assessment, both the potential short- and long-term consequences of exposure for human health as well as the likelihood of environmental release and the resulting ecological impact must be considered. A risk assessment therefore provides information concerning the safety of commercial substances, ensuring that their manufacture is conducted in a manner that reduces risk. 'Risk management' is the process of selecting control measures to reduce a hazard, chance of exposure or adverse effects,

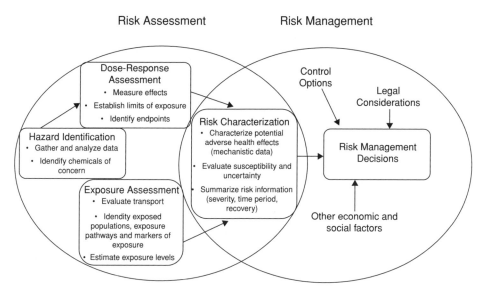

Risk Assessment   Risk Management

Figure 22.1   *Risk analysis paradigm (modified from National Academy of Sciences, 1983). Reprinted with permission from the National Academies Press, Copyright 1983, National Academy of Sciences*

and evaluating the results of its implementation. Finally, a more global term that includes evaluation of all relevant attributes of hazards, risks, adverse effects, events and conditions that lead to adverse effects, and populations or environments that influence or experience adverse effects, is 'risk analysis' (see Figure 22.1) (Schierow, 1996).

22.2.1   Hazard Identification

Generally, the first step of a risk assessment is hazard identification. According to the NRC, hazard identification 'entails identification of the contaminants that are suspected to pose health hazards, quantification of the concentrations at which they are present in the environment, a description of the specific forms of toxicity (neurotoxicity, carcinogenicity, etc.) that can be caused by the contaminants of concern, and an evaluation of the conditions under which these forms of toxicity might be expressed in exposed humans...' (National Academy of Sciences, 1994). Hazard identification of a substance uses available data from *in vitro* and *in vivo* studies as well as models, such as quantitative structure–activity relationships (QSAR). Not only the hazard of the substance is considered, but its reaction intermediates and metabolic products are also included. Characteristics of hazardous materials include toxicity, bioavailability, potential for bioaccumulation and persistence (van Leeuwen and Vermeire, 1995).

Identifying the types and extent of adverse effects associated with exposure is an important aspect of hazard assessment. Traditional hazard and risk assessments rely on data from standard toxicological studies whose primary goal is to determine levels of exposure that result in certain defined endpoints, usually lethality or different disease outcomes. More rigorous tests also seek data on precursor biological events that may play a role in disease progression, such as altered metabolism, cell signalling, hormone trafficking and gene expression. In the case of nanomaterials, some are hazardous due to their chemical composition. For example, several quantum dots are synthesized from known hazardous chemicals, such as cadmium, selenium or lead, which makes them potentially hazardous as well. Others are made of what are considered non-hazardous chemicals, such as fullerenes which are carbon-based, and therefore would be expected to be less hazardous. However, studies have indicated that these carbon nanomaterials can induce cytotoxic responses from cells *in vitro* (Jia *et al.*, 2005a; Magrez *et al.*, 2006; Panessa-Warren *et al.*, 2006). This then raises the question as to what property inherent in the nanoscale material makes it different from its bulk counterpart to elicit different toxic responses. Further discussion on this topic can be found in section 22.4.

22.2.2   Dose-Response Assessment

Following hazard assessment, dose-response assessment:

> 'entails a further evaluation of the condition under which the toxic properties of a chemical might be manifested in exposed people, with particular emphasis on the quantitative relation between the dose and the toxic response. The development of this relationship may involve the use of mathematical models. This step may include an assessment of variations in response, for example, differences in susceptibility between young and old people.' (National Academy of Sciences, 1994)

This assessment examines the levels at which certain toxicity endpoints occur, including acute toxicity, chronic toxicity, immunotoxicity, neurotoxicity, mutagenicity, reproductive/developmental toxicity and carcinogenicity. In addition, dose-response testing looks for threshold levels such as the 'no observable adverse effect' level (NOAEL), the 'lowest observable adverse effect' level (LOAEL), and/or develops an alternative 'benchmark dose' (BMD). While the NOAEL and LOAEL values are based on discrete doses from a study, the BMD is an estimated dose corresponding to a particular response level, such as the effective dose at the 10% level, derived from the dose-response curve model (Setzer and Kimmel, 2003). The advantages to using BMD are that it is not affected by sample size as with NOAEL and it takes into account the entire range of dose-response data collected (Castorina and Woodruff, 2003). Ultimately, these values are used to determine reference levels of exposure, such as oral reference dose or inhalation reference concentration.

22.2.3   Exposure Assessment

There is no risk without exposure, and exposure assessment 'involves specifying the population that might be exposed to the agent of concern, identifying the routes through which exposures can occur, and estimating the magnitude, duration, and timing or the doses that people might receive as a result of their exposure' (National Academy of Sciences, 1994). Possible exposure pathways include inhalation, oral and dermal, with the addition of direct exposure in the case of medical administrations such intradermal, intraperitoneal and intravenous injections. For indirect, environmental exposure, fate and transport of the substance through the environment as well as the points of entry into the environment may also be considered in exposure assessment. In addition to routes of exposure, varying the duration of exposure can result in different outcomes; therefore, acute and chronic exposure effects must also be evaluated to resolve any influence on the endpoints of interest. For example, some effects are reversible once exposure to the chemical is stopped. In this case, constant chronic exposure testing would be important to better assess the potential for an adverse response to exposure.

22.2.4   Risk Characterization

The final step of risk assessment, which synthesizes data from the hazard, dose-response and exposure assessments, is risk characterization which 'develop[s] a qualitative or quantitative estimate of the likelihood that any of the hazards associated with the agent of concern will be realized in exposed people...[and] also include[s] a full discussion of the uncertainties associated with the estimates of risk' (National Academy of Sciences, 1994). By identifying the potential source, population and pathways of exposure, this makes it possible to distinguish between substances that present little to no concern from those that are likely to cause adverse effects. However, the quality of this analysis depends upon the availability and quality of the data on the material in question. Poor information can be worse than no information at all. Despite the importance of risk assessment, thorough evaluation of many engineered nanomaterials is still lacking, and due to the lack of standard material characterization criteria, the available data are not necessarily suitable for risk assessment.

Under ideal conditions, a risk assessment gathers, organizes and summarizes all of the important information relevant to the potential hazard (Schierow 1996). It includes qualitative and quantitative data on the characteristics of the hazard, exposure conditions and potential effects, as well as, in some cases, discussion on any scientific uncertainties. However, as risk assessment is a means of scientific inquiry, the quality of information provided can vary from comprehensive to basic, objective to biased. When there is no past experience with a hazard, there is no basis for any quantitative estimate, and if there is experience but no record, risk estimates are likely to be unreliable (Schierow 1996). Finally, risk assessment ultimately is used to guide risk analysis and risk management decisions about acceptable risk (Figure 22.1). Often this overlaps with risk–benefit analysis, which takes into account several overarching factors including technological feasibility, cost, societal impact, legislative robustness, research uncertainties and assumptions of risk management.

22.3   Current Status of Nanomaterial Environmental, Health and Safety Data

As the field of nanotechnology continues to grow, more emphasis is being placed on learning from past mistakes of other emerging technologies in order to 'get nanotechnology right the first time' (Balbus et al., 2006). Therefore health and environmental implications, specifically toxicity, has become a major area of nanotechnology research to understand the potential impact before widespread application and use. Unfortunately, despite the early start towards hazard identification, thorough characterization of the nanomaterials tested was not always carried out, and therefore more rigorous studies on well-characterized nanomaterials are still needed in order to understand the mechanism behind any biological effects observed, and to build quantitative structure–activity relationships. Despite these shortcomings, a vast amount of data has been collected over the past few years, with twice as many studies published in 2007 compared with 2005 (Buzea et al., 2007).

Table 22.1   OECD list of 14 priority manufactured nanomaterials (OECD, 2008a).

Carbon-based	Metal-based	Other
Carbon black	Aluminium oxide	Silicon dioxide
Fullerenes (C60)	Cerium oxide	Polystyrene
Single-walled carbon nanotubes (SWCNTs)	Iron nanomaterials	Dendrimers
Multi-walled carbon nanotubes (MWCNTs)	Silver nanomaterials	Nanoclays
	Titanium dioxide	
	Zinc oxide	

Nanomaterials are generally classified based on their chemical composition, shape and dispersion, and it has been suggested that composition, dissolution, surface area and surface characteristics, size, size distribution and shape are the major factors influencing their toxicity (Warheit *et al.*, 2007). However, composition and surface characteristics are not very explicit terms and could include properties such as molecular structure, purity, surface chemistry, zeta potential (surface charge), catalytic activity (e.g. redox potential) and solubility. Taking into account these variations, researchers are presented with a daunting number of samples to test for each nanomaterial. Fourteen representative engineered nanomaterials have been identified by the Organization for Economic Cooperation and Development (OECD) as priority nanomaterials for evaluation and are listed in Table 22.1 (OECD, 2008a). Yet many other nanomaterials also exist, including other fullerenes (C70, C78, C84), metal nanomaterials (gold, ruthenium, germanium oxide, nickel carbide) and quantum dots (CdSe, CdTe, ZnSe, InGaAs, PbS, PbSe) to name but a few, and they may also require focused EHS testing if widely used in the future (Singh and Nalwa, 2007).

Specific details about the physical and chemical properties of nanomaterials and how they may change over time should be addressed before any risk assessment, which can later guide development of more detailed profiles of nanomaterial properties and their hazard and exposure potential. Identifying trends relating chemical structure and function to biological activity is of particular interest in order to facilitate predictive modelling, and therefore reduce the need for case-by-case testing. Some examples of structure-related properties include the increased particle surface reactivity due to the high surface area to volume ratios of nanomaterials. This increased reactivity, one of the advantages of using nanoscale materials for certain applications, may also result in toxicity. Some nanomaterials have been approved by the US Food and Drug Administration (FDA) and are currently being used for medical applications, such as iron oxide nanomaterials for magnetic resonance imaging (MRI) of the liver (Bulte and Kraitchman, 2004; de Vries *et al.*, 2005). However, there still remain questions as to how well-suited existing tests are for nanomaterials, and various toxicity testing continues to increase. The following sections explore the current status of nanomaterial EHS research addressing both external and internal routes of exposure, as depicted in Figure 22.2.

22.3.1   External Exposure

Human lungs, skin and gastrointestinal (GI) tract are in constant contact with the environment, making them the first ports of entry for environmental contaminants, including nanomaterials. External exposure therefore constitutes the amount of a substance that

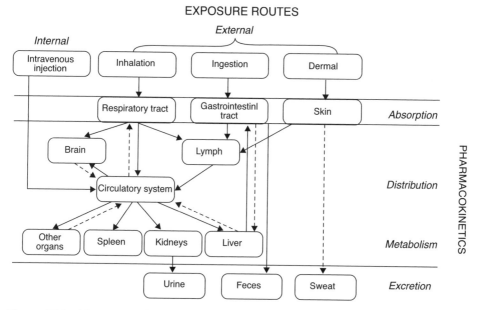

EXPOSURE ROUTES

Figure 22.2 *Nanomaterial exposure and transport pathways. Black lines indicate routes reported in published in vivo studies; dashed lines indicate additional possible routes (modified from Hagens* et al. *(2007), with permission from Environmental Health Perspectives and Oberdörster* et al. *(2005b) with permission from Elsevier)*

reaches the epithelium of these organs. Although toxicity to these points of entry is of concern, systemic toxic effects are greater in severity. Taking this into account, not only should ports of entry – lungs, skin, gut – be considered in testing, but also potential target tissues – endothelium, blood cells and vessels, liver, spleen, brain and fetus – due to nanomaterial translocation. Since the lungs, skin and GI tract naturally serve as outer physical barriers, the likelihood of internal exposure depends on the ability of the substance to be absorbed or cross barrier cell layers at the site of exposure (Wood, 2003). This section focuses on the studies to date that investigate the effects of nanomaterials after inhalation, dermal contact and oral intake.

22.3.1.1   Inhalation

Inhalation is viewed as the route of nanomaterial exposure of significant concern, since inhaled particulates are known to induce various respiratory conditions, and compared with dermal and oral exposure, inhalation is more likely to result in a significant systemic dose (Hoet *et al.*, 2004; Oberdörster *et al.*, 2005b; Hoyt and Mason, 2008). As seen in Table 22.2, only a few types of nanomaterials have been tested for pulmonary effects. This may be due to the higher chance of aerosolization of these nanomaterials compared with others. For example, many engineered nanomaterials are produced in solution and therefore would need to be dried in order to pose an inhalation risk. However, carbon nanomaterials can be produced in the solid phase and could, upon agitation, form aerosols and be released into the air. Raw HiPco and laser-synthesized carbon nanotube (CNT) samples have been

Table 22.2 Summary of selected studies on nanomaterial exposure via intratracheal instillation or aerosol inhalation.

Model system	NP	Average NP size	Concentration	Duration	Results	Reference
Young Wistar rats	C60	N/A	10 mg/kg	3 days	Modified C60 attenuates ischemia-reperfusion induced lung injury	Lai et al. (2003)
Kunming mice	MWCNTs	50 nm (od), 10 μm (l)	0.5 mg/ml, 13–80 mg/m³	8, 16, 24 days	Inflammation of bronchi lining and mechanical lesions of surrounding alveolar netting structures, aerosol inhalation had less effect	Li et al. (2007a)
C57Bl/6 mice	MWCNTs	10–20 nm (d), 5–15 μm (l)	0.3–5 mg/m³	6 hr/day, 7 or 14 days	Alveolar macrophages contained particles, no inflammation or tissue damage observed	Mitchell et al. (2007)
Sprague-Dawley rats	MWCNTs (15 layer avg.)	~5 nm (id), ~10 nm (od)	0.5, 2, 5 mg/animal	3, 15, 60 days	MWCNTs are biopersistent, stimulate TNF-a production, inflammation and fibrosis	Muller et al. (2005)
Crl:CD (SD)IGS BR rats	Nano-C60	160 nm	0.2, 0.4, 1.5, 3 mg/kg	24 hr, 1 wk, 1 mo, 3 mo	Transient inflammation, no adverse lung effects	Sayes et al. (2007)
B6C3F mice	SWCNTs	N/A	2, 10 mg/ml	7, 90 days	Unpurified SWCNTs used, dose-dependent epitheloid granuloma formation, interstitial and peribrochial inflammation	Lam et al. (2004)
Crl:CD(SD)IGS BR rats	SWCNTs	1.4 nm (d), >1 μm (l)	1, 5 mg/kg	1 day, 7 day, 1 mo, 3 mo	Unpurified SWCNTs used, non-dose-dependent multifocal granuloas, possible regression of lesions from 1–3 months	Warheit et al. (2004)
C57BL mice	SWCNTs	1–4 nm, 14.3 nm or 2.14 μm (d)	10, 20, 40 μg/animal	1, 3, 7, 28, 60 days	Purified SWCNTs used, inflammation occurred with increases in BAL levels, dose-dependent interstitial fibrosis	Shvedova et al. (2005)

Animal	Material	Size	Dose	Duration	Findings	Reference
Sprague-Dawley rats	TiO_2	290 nm (d)	4 mg/kg in 200 µl	4, 24, 48, 74 hr	GCH and Muc5ac gene expression induced within 24 hr, associated with elevated IL-13	Ahn et al. (2005)
C57Bl/6 mice	TiO_2	2–5 nm (d)	8.88 mg/m^3	4 hr/day for 1 or 10 days	120–130 nm aggregates form, acute exposure had no adverse effects, subacute exposure had significant inflammatory response	Grassian et al. (2007)
Wistar rats	TiO_2, Ni, Co NPs, quartz	62.3, 45.3, 46.1 cm^2 (s.a.)	250 µg/ml	18–24 hr	Size dependence observed for inflammation and neutrophil influx, quartz highly reactive – more than other NPs	Duffin et al. (2007b)
Fischer 344 rats	C60	55 nm, 930 nm	2.2 mg/m^3 (nano), 2.35 mg/m^3 (micro) 0.5 mg/kg	3 hr/day for 10 days	Minimal toxicity, undetected in blood	Baker et al. (2008)
ICR mice	SWCNTs	N/A		3, 14 days	Increased macrophage infiltration during short term, granuloma formation after 1 wk	Chou et al. (2008)
C57BL/6 mice	SWCNTs	0.69 µm	10 µg/lung	1, 7 days, 1 month	Lung deposits generally <1 µm, little macrophage uptake observed	Mercer et al. (2008)
Wistar rats	NiO	20 nm (reported), 139 nm (measured in chamber)	10^5 particles/cm^3	4 wk, examined 4 days, 1 mo, 3 mo after	Slight inflammation observed, clearance observed, though some persistence	Oyabu et al. (2007)

found to produce large airborne particulates after gentle agitation, with the particles of respirable size, less than 10 μm, produced with more vigorous agitation (Baron *et al.*, 2003; Maynard and Kuempel, 2005). Furthur detailed discussion on the generation of airborne nanomaterials is beyond the scope of this chapter.

When evaluating the risk due to nanomaterial inhalation, it is important to recognize that exposure of the lower regions of the lungs is the ultimate concern, as it is in this region that cytotoxicity would have greater detrimental effects and translocation is more likely to occur. Inhaled air passes through various respiratory structures (e.g. the nasal cavity, pharynx, larynx, trachea) until it reaches the lungs, by which time it has lost most of the large particles it carries. The uptake of inhaled nanomaterials depends upon the distribution in the lung. Larger or heavier particles are more likely to be deposited in the airways, or conducting zone, with mucus in the tracheobronchial region facilitating clearance of particles in this area (Hayes, 2001). Smaller particles, generally less than 5 microns in diameter, can achieve deposition in the lower respiratory zone of the lungs containing the alveolar beds, with clearance mediated by alveolar macrophages (Hoet *et al.*, 2004). Although absorption can occur throughout the respiratory tract, the epithelial barrier layer thickness differs in the airways and the alveoli, decreasing from the bronchia to the alveoli. For example, the bronchial epithelium has ∼60 μm thickness while the alveoli consist of a single cell layer with thickness of 0.1–0.2 μm (Patton, 1996). For a more extensive description of the deposition of nanomaterials in the lungs, the reader is referred to a recent review by Yang *et al.* (2008).

The potential hazards associated with inhalation of particulates, including nanoparticles, have been discussed extensively in the literature, and therefore the remainder of this section will overview two major outcomes of concern: the onset of lung disease and particle translocation. The effects from particulate inhalation are dependent upon the particle dose, deposition, dimension and durability (Borm and Kreyling, 2004). Toxicity is determined by the deposited dose, which is influenced by the size of the particles and the external exposure concentration. Therefore, exposure to high concentrations of nanoparticles could lead to a toxic effect as their small size makes them more likely to be deposited in the lower gas exchange region of the lungs. Once deposited, the durability of the particles dictates the accumulation potential and persistence of the particles in the lungs. In addition to these factors, particle charge has also been found to influence deposition, with charged nanomaterials having higher deposition efficiencies compared with nanomaterials with no charge. For example, nanomaterials with polar surfaces were found to have different translocation rates across lung epithelium in a hamster model (Nemmar *et al.*, 2001).

Pulmonary inflammation and fibrosis are well-known consequences of particulate exposure; however, studies comparing submicron and nanoparticles of the same chemical have observed that this occurs to a greater extent with nanoparticles (Donaldson *et al.*, 2001; Oberdörster *et al.*, 2005b). The onset of inflammation is grossly distinguished by irritation and swelling of lung tissue, while fibrosis constitutes increased collagen content and/or structural alterations to the lung (Card *et al.*, 2008). On a cellular level, lung inflammation is often the result of oxidative stress. Reactive oxygen species generation at the surface of nanomaterials leads to the depletion of antioxidants and initiation of lipid peroxidation which triggers a series of cellular events, involving biomarkers such as nuclear factor kB (NF-kB) and activator protein-1 (AP-1), which ultimately result in inflammation (Duffin *et al.*, 2007a). The consequences of chronic inflammation include lung disease, cardiovascular disease and cancer (Buzea *et al.*, 2007). Studies on inflammation and fibrosis resulting from nanomaterial exposure have focused upon the effects of carbon nanotubes, carbon black, fullerenes, silica, and metal-based nanoparticles such as titanium dioxide and silver, as seen in Table 22.2.

Nanomaterials can also escape macrophage phagocytosis and endocytosis, resulting in inflammation and fibrosis or problems in other organs due to translocation. Size is an influential factor regulating the uptake of deposited particles by alveolar macrophages. Unlike micron-sized particles, nanoparticles with diameters less than 70 nm are observed to elude alveolar macrophages and therefore are able to gain access to the pulmonary interstitium, and potentially enter capillary blood flow (Moghimi and Hunter, 2001). Conversely, long nanofibers measuring more than 20 μm in one dimension are too long to be phagocytosed by macrophages, which are smaller in diameter. Therefore, these particles clear much more slowly, persist in the lungs and induce non-specific responses, such as chemokine and cytokines release, which can result in inflammation and fibrosis (Borm and Kreyling, 2004; Hoet *et al.*, 2004). Several reviews on carbon nanotube toxicity to the lungs can be found; however, recently highlighted in the literature were two published studies presenting evidence that multiwalled carbon nanotubes (MWCNTs) can promote mesothelioma (Borm and Kreyling, 2004; Maynard and Kuempel, 2005; Oberdörster *et al.*, 2005b; Donaldson *et al.*, 2006; Bergamaschi *et al.*, 2006; Lam *et al.*, 2006; Duffin *et al.*, 2007a; Poland *et al.*, 2008; Takagi *et al.*, 2008). In both studies, intraperitoneal injection was used to expose the mesothelium directly to MWCNTa in solution. For a positive control, different types of asbestos were used. Takagi *et al.* (2008) tested MWCNTs, crocidolite and C60. At ten days, fibrinous deposits and excess abdominal fluid were observed, and by week 24, 100 % mortality was reached for the MWCNT-treated group (Takagi *et al.*, 2008). The fullerene-exposed mice did not produce the same symptoms. From these observations, the biopersistance of MWCNTs due to their size was suggested as a possible predictive factor for mesotheliomagenic potential. Poland *et al.* (2008) similarly tested five suspensions: long, straight and short tangled MWCNTs, long and short amosite, and carbon black nanomaterial. The presence of inflammation and granulomas was only found in mice exposed to the long straight fibers, for both MWCNTs and asbestos after seven days. From these observations, the mechanism of lung injury was suggested to be hindered phagocytosis.

The systemic effects of nanomaterial exposure will be discussed in further detail in section 22.3.2; however, a brief discussion on the concerns of potential cardiovascular damage due to nanoparticle translocation in the lungs is presented here. If nanoparticles enter the blood circulation, there is potential for them to cause injury to blood vessels or promote blood clot formation; however, this may be particle specific, as seen in the studies using comparably sized polystyrene and diesel exhaust particles (Nemmar *et al.*, 2004). For single-walled carbon nanotubes (SWCNTs), increased artherosclerosis was found in mice after intrapharyngeal instillation (Li *et al.*, 2007b). However, these effects may be due to the method of exposure, as different effects are observed after intratracheal instillation of nanomaterials compared with aerosol exposure (Li *et al.*, 2007a).

22.3.1.2   Dermal

Direct dermal contact with nanomaterials is likely to occur through usage of clothing, cosmetics, wound dressings or dermal drug delivery treatments containing nanomaterials (Muller *et al.*, 2002; Vanrolleghem and Lee, 2003; Emerich and Thanos, 2007; Qi *et al.*,

2007; Maneerung *et al.*, 2008). In the case of dermal drug delivery, nanomaterials are designed for enhanced skin penetration; however, for other cases such as occupational or environmental exposure, this is an undesired effect. Although research in both areas significantly contributes towards understanding nanomaterial interactions with skin, this section focuses on unassisted nanomaterial penetration through skin. The skin, when intact, is generally an effective barrier to foreign substances compared with the lungs and gastrointestinal tract, which are considered more vulnerable. Skin consists of several layers; starting from the outer layer these are the stratum corneum, epidermis, dermis, and subcutaneous layers. The stratum corneum consists of a thick, 15–100 µm layer of keratinized, dead cells that are closely packed together (Rushmer *et al.*, 1966). This outermost layer serves as a protective covering of the body and restricts the penetration of particles greater than 1 µm (Gwinn and Vallyathan, 2006).

Due to controversies arising from their use in sunscreens, much data has been collected on the absorption and penetration of titanium dioxide and zinc oxide nanoparticles through skin (Nohynek *et al.*, 2007). In a pilot study, increased titanium levels in the epidermis and dermis was found comparing unexposed and sunscreen-treated skin (Tan *et al.*, 1996). Subsequent studies have shown little or no penetration of both micron- and nanosized TiO_2 and ZnO beyond the stratum corneum layer (Schulz *et al.*, 2002; Gamer *et al.*, 2006; Cross *et al.*, 2007). Nanomaterial penetration into the skin has been suggested to be size-dependent – smaller nanomaterials are more likely to penetrate into the skin deeper than larger ones (Hoet *et al.*, 2004). Quantum dots with hydrodynamic diameters up to 45 nm have been reported to penetrate through intact pig skin after 24-hr exposure (Ryman-Rasmussen *et al.*, 2006). Charge may also play a role, as 50 and 500 nm negatively-charged fluorescent latex particles were found to have similar permeation to the epidermis (Kohli and Alpar, 2004). However, other reports on metal oxide and other metal-based nanoparticles indicate the presence of nanoparticles in the hair follicle shaft and skin furrows, which are several microns deep and naturally accessible to particles, but no epidermal penetration (Baroli *et al.*, 2007; Lademann *et al.*, 2007; Nohynek *et al.*, 2007; Zhang *et al.*, 2008).

The previously mentioned experiments did not take into account skin movement, particle charge or skin integrity. Submicron particles with sizes between 500–1000 nm were shown to penetrate through the epidermis after application of a flexing motion (Tinkle *et al.*, 2003). Another study similarly found skin penetration to the stratum granulosum layer of the epidermis for peptide-modified C60 fullerene after mechanical flexing (Rouse *et al.*, 2007). These two studies suggest that movement may enhance nanomaterial absorption through the skin, compared with transport observed after nanomaterial application to flat skin. Dermal penetration of nanomaterials past the epidermis suggests that nanomaterials could enter systemic circulation from this point of entry. This could result in translocation of nanomaterials to internal organs of the body, such as the liver, lymph nodes and kidneys, as observed in mice intradermally injected with quantum dots (Gopee *et al.*, 2007).

In addition to testing the potential hazard of nanomaterial translocation after dermal absorption, cutaneous toxicity should also be evaluated. Free radical formation has been observed with metal and carbon-based nanomaterials, and could promote skin aging or apoptosis (Wakefield *et al.*, 2004; Sayes *et al.*, 2005; Tian *et al.*, 2006). *In vitro* photo-toxicity, genotoxicity and photo-genotoxicity tests on pristine and coated titanium dioxide nanomaterials used in sunscreens have resulted in negative findings (Nohynek *et al.*, 2007). However, the majority of these tests were conducted on rutile TiO_2 nanoparticles with only a few anatase forms tested. The crystalline form of TiO_2 has been correlated to the photoactivity, with the anatase form exhibiting higher reactive oxygen species generation compared with the rutile form (Sayes *et al.*, 2006). This consideration should be taken into account when interpreting the findings. Reactive oxygen species (ROS) generation is not exclusive to TiO_2 nanoparticles and has also been suggested to be the mechanism behind toxicity of C60 and MWCNT nanoparticles on human fibroblasts and keratinocytes (Fumelli *et al.*, 2000; Shvedova *et al.*, 2003; Sayes *et al.*, 2006). However, compared with *in vitro*, little *in vivo* work has been done in examining the toxicity of nanomaterials to skin. Table 22.3 provides data from *in vivo* dermal studies for several types of nanomaterials. Overall, these limited *in vivo* studies have identified only mild irritation as an adverse response to topical nanomaterial application.

22.3.1.3   Gastrointestinal

Oral exposure tends to be overlooked, as ingestion of substances is thought to occur only by intentional means and can therefore be controlled. In addition, many materials have low bioavailability, making them poorly absorbed through the intestinal lumen and therefore unlikely to reach target organs to produce toxic effects. However, ingestion of food products or drinking water purposely containing or contaminated with nanomaterials could occur, as well as inadvertent occupational ingestion (Cherrie *et al.*, 2006; Zhu *et al.*, 2008). Although local nanomaterial interactions throughout the entire alimentary canal could result after ingestion, of more concern is delivery into the circulation, with absorption most likely to take place in the intestines.

Limited work has been conducted on nanomaterial uptake in the gut; however, data on polystyrene latex microparticle uptake after oral administration indicate that nanoparticles could be absorbed across the gastrointestinal tract (Delie, 1998). GI uptake of nanomaterials requires transport through mucus and cellular barriers, and the extent of uptake is dependent upon particle size (Jani *et al.*, 1990; Desai *et al.*, 1996). Nanoparticles can cross the small intestine by persorption, or translocate from the lumen of the intestinal tract through intestinal lymphatic tissue, Peyer's patches and enterocytes, and further distribute inside the body through the lymphatics to the liver and spleen (Hillery *et al.*, 1994; Desai *et al.*, 1996; Hillyer and Albrecht, 2001; Hussain *et al.*, 2001). For example, studies on 60 nm polystyrene spheres have found 10 % of the administered dose to be absorbed or adsorbed to the stomach and intestines, with 60 % of this absorbed in the Peyer's patches (Hillery *et al.*, 1994; Florence *et al.*, 1995). However, no absorption is also likely, and excretion exclusively through the feces has been observed after oral administration of radioactively labelled C60 and radioactive iridium nanoparticles (Yamago *et al.*, 1995; Kreyling *et al.*, 2002).

Although some absorption data on naturally occurring and engineered particles have been collected, the mechanism of translocation remains unclear (Hussain *et al.*, 2001). Generalizing polystyrene particle data, the factors identified to affect uptake and translocation of nanoparticles after oral administration include size, surface chemistry, shape, stability, dose, length of administration and methods used to quantify absorption (Hussain *et al.*, 2001; Florence, 2005). Specifically, increased uptake was found for particles with diameters less than 1 μm and it has been suggested that this becomes less pronounced as the diameter decreases below 50 nm (Florence *et al.*, 2000). Positively charged particles were

Table 22.3 Summary of selected studies on nanomaterial exposure via topical or intradermal administration.

Model system	NP	Surface coating	Average NP size	Concentration	Duration	Results	Reference
CD-1 mice	C60 (in benzene)	N/A	N/A	200 µg	72 hr, 24 wks	In benzene, no acute toxic effects	Nelson et al. (1993)
Hairless mice	C60 (in toluene)	N/A	N/A	1.25 mM	20 wks	Erythema observed but no tumor growth	Moriguchi et al. (1999)
Rabbit eye and human	C60	N/A	N/A	N/A	72 hr, 96 hr	No signs of irritation or acute response	Huczko et al. (1999)
Rabbit eye and human	CNTs	N/A	N/A	N/A	72 hr, 96 hr	No signs of irritation or acute response	Huczko and Lange (2001)
Male Wistar rats	MWCNTs	N/A	20–40 nm (d), 0.5–5 µm (l)	0.1 mg	1, 4 wk	Higher inflammatory response for longer tubes, suggested to be due to macrophage inability to envelope larger particles	Sato et al. (2005)
Female SKH-1 hairless mice	CdSe/CdS QD	PMAO-PEG	TEM: 8.4 nm (l), 5.8 nm (w); DLS: 39 nm (hd); SEC: 37 nm (hd)	48 pmol	0, 4, 8, 12, 24 hr	~40% lost from injection site, 7.5% found in liver, lymph nodes and kidneys	Gopee et al. (2007)
Yorkshire pig skin	CdSe/CdS QD	PMAO, PMAO-PEG	39–40 nm	6.67, 3.33 or 1.67 nM	4 hr	QD-COOH were present in greater quantity and accumulated in the capillaries compared with QD-PEG	Lee et al. (2007)
Yorkshire pig skin	CdSe/CdS QD	PMAO-PEG	8.4 nm (l), 5.78 nm (w); 39–40 nm (hd)	1, 2, 10 mM; 1.25–10 nM	1–48 hr	Localization isolated to stratum corneum and hair follicles, cytotoxicity observed at 1.25 nM, signs of inflammatory effects observed	Zhang et al. (2008)

Skin	Nanoparticle	Coating	Size	Dose	Time	Observations	Reference
Yorkshire pig skin	CdSe/ZnS QD	PEG, carboxylic acid, PEG-amine	4.6 nm, 12 nm (l) 6 nm (w); 14–45 nm (hd)	62.5 pmol/cm^2	8, 24 hr	Some epidermal and dermal penetration observed after 24 hr	Ryman-Rasmussen et al. (2006)
SKH-1 wild-type mice skin	CdSe/ZnS QD	Carboxyl	20–33 nm	3 pmol/cm^2	8, 24 hr	QD found in skin defects, folds, and hair follicles; QD penetrate as deep as dermis in UVR exposed skin but at low levels	Mortensen et al. (2008)
Full thickness human abdominal skin	g-maghemit, iron	Sodium bis(2-ethylhexyl) sulfosuccinate, tetramethylammonium hydroxide	TEM: 5.9 nm; DLS: 48.6, 1238.2, 13326.8 nm; TEM: 4.9, 12.8, 23.3 nm DLS: 4.8, 9.6, 82.6 nm	200 µl	3, 6, 12 and 24 hr	NPs <10 nm penetrate skin through stratum corneum lipidic matrix and hair follicle orifices, reaching the stratum granulosum and hair follicles	Baroli et al. (2007)
Pig ear skin, human calf skin	PLGA NPs	Fluorescein	320 nm	2 µg/cm^2	1, 24, 72, 120 hr	Tape stripping, hair follicle penetration 1 hr, release out few days	Lademann et al. (2007)
Human skin	TiO$_2$	N/A	10–50 nm (microfine)	8%	2-6 wk	sunscreen treated skin had higher Ti concentrations than control	Tan et al. (1996)
Human forearm skin	TiO$_2$	Trimethyloctylsilane, Al$_2$O$_3$/SiO$_2$	20 nm, 110-15 nm, 100 nm	160 µg/cm^2	36 hr	TiO$_2$ only found in stratum corneum	Schulz et al. (2002)
Human abdominal skin	ZnO	Siliconate	TEM: 15–40 nm; BET: 30 nm, XRD: 26 nm; PCS: 30 nm	10 µl/cm^2	12, 24 hr	No significant difference in Zn penetrating epidermal membrane for two sunscreens and untreated membranes over 24 hr, isolated mainly to upper stratum corneum	Cross et al. (2007)

found to be absorbed more effectively through the GI tract relative to neutral or negatively charged particles (Hussain *et al.*, 2001; Florence, 2005).

Beyond latex microspheres, studies investigating the differences between uptake of micro- and nanoscale metal particles have also been published. For copper nanoparticles with a diameter of 23.5 nm, the LD50 was found to be 413 mg/kg which was closer to that of ionic copper (110 mg/kg) compared with micron-sized copper particles (>5000 mg/kg) (Chen *et al.*, 2006). Injuries to the kidneys, liver and spleen, such as color change and necropsy, were observed in nano-copper exposed mice, but not in mice exposed to 17 μm copper particles (Chen *et al.*, 2006). Zinc nanoparticles were found to induce lethargy, nausea, vomiting, diarrhea and growth retardation in exposed mice, with less severe symptoms observed in the mice exposed to zinc microparticles (Wang *et al.*, 2006). Acute toxicity of 25, 80 and 155 nm TiO_2 nanoparticles was investigated, with no toxicity observed for all three samples after oral exposure to a single 5 g/kg TiO_2 dose. Particle accumulation in the liver was found for the 80 nm TiO_2 *in vivo*; however, this was not observed for the 25 nm TiO_2, which instead accumulated in the spleen, kidneys, and lung tissues (Wang *et al.*, 2007). Gold nanoparticles have also been tested with persorption through the Peyer's patch regions and uptake through enterocytes observed (Hillyer and Albrecht, 2001). Low acute toxicity after oral feeding of selenium, zinc, iron and silicon nanoparticles has also been found (Jia *et al.*, 2005b; Cha and Myung 2007; Rohner *et al.*, 2007). These and other studies involving oral administration of nanomaterials are summarized in Table 22.4.

22.3.2   Internal Exposure

Although internal exposure can occur as a consequence of external exposure, significant internal concentrations of nanomaterials are more likely to result from use of medical interventions. Numerous biomedical applications utilizing nanomaterials are being commercialized or developed in order to revolutionize the current state of healthcare (Yang *et al.*, 2007; Xia *et al.*, 2008). Several avenues currently under investigation include improved medical treatment and diagnostics, such as targeted and controlled-release drug delivery, as well as novel biosensors and optical diagnostic techniques (Koo *et al.*, 2005). For biomedical purposes, especially *in vivo* applications such as those mentioned, toxicity is a critical factor to consider when evaluating their potential. These medical applications of nanomaterials involve deliberate, direct administration of these materials to people, and while the administered dose, or exposure, can be controlled, it is important to determine what dose levels are safe. Few studies have examined the *in vivo* pharmacokinetics of nanomaterials using animal models to identify their organ localization, metabolism pathways and excretion routes. However, due to incomplete characterization in some cases, the properties attributed to the observed effects and possible toxicity cannot be resolved. Despite these limitations, the preliminary data from these investigations do reveal some general trends, and this section focuses on the effects resulting from internal nanomaterial exposure.

22.3.2.1   Distribution

Once absorbed or injected, nanomaterials present in the blood stream can distribute to the tissues, often initially to organs with rich blood supplies, such as the spleen, kidneys, liver, heart, lungs and brain (Hoet *et al.*, 2004; Kim *et al.*, 2008). The rate of nanomaterial

Table 22.4 Summary of selected studies on nanomaterial exposure via oral feeding or gavage.

Model system	NP	Average NP size	Concentration	Duration	Results	Reference
ddY mice	C60	N/A	18 kBq (0.18 MBq/ml)	48 hr	97% excreted in feces after 48 hr	Yamago et al. (1995)
Sprague-Dawley rats	C60	N/A	2 g/kg	14 days	No toxicity observed	Mori et al. (2006)
Sprague-Dawley rats	Dendrimer	2.5 nm	14, 28 mg/kg	3, 6, 24 hr	After 6 hr highest in small intestine (15%), large intestine (5%) and blood (3%), ~2% reached liver, spleen and kidneys, absorbed through Peyer's patches > enterocytes	Florence et al. (2000)
BALB/c mice	Au	4, 10, 28, 58 nm diameter	2×10^5 parts per billion ng/gwater	7 days	Presence of particles in extruding enterocytes dependent on size; persorption in ileum and Peyer's patches	Hillyer and Albrecht (2001)
WKY/NCrl BR rats	Ir192	15, 80 nm	5 kBq (5 ml)	3 days	After 6 hr no radioactivity detected in urine or internal organs, entire dose found in feces within 2–3 days	Kreyling et al. (2002)
ICR mice	Cu	23.5 nm, 17 μm	108–1080 mg/kg	N/A	LD50 nano-copper: 413 mg/kg; ion-copper: 110 mg/kg; micro-copper: >5000 mg/kg	Chen et al. (2006)
CD-ICR mice	Zn	58 nm, 1.08 μm	5 g/kg	2 wks	Possible intestinal obstruction and behavioral changes for nano-zinc exposed	Wang et al. (2006)
BALB/c mice	Zn, Fe, Si	300 nm (Zn), 100 nm (Fe), 10–110 nm (Si)	2.5 g	3 days	Low toxicity observed; liver, heart and spleen showed some nonspecific hemorrhage, lymphocytic infiltration, and medullary congestion	Cha and Myung (2007)
CD-1 ICR mice	TiO$_2$	20, 80, 155 nm	5 g/kg	2 wks	No acute toxicity observed; retained in liver, kidneys, spleen and lungs; some hepatic damage, nephrotoxicity and myocardial damage suggested	Wang et al. (2007)
Sprague-Dawley rats	Ag	60 nm	30 mg/kg, 300 mg/kg, 1000 mg/kg	28 days	Exposure to >300 mg Ag NP may result in slight liver damage; no genetic toxicity in vivo; dose-dependent accumulation of Ag in all the tissues examined	Kim et al. (2008)

clearance from circulation is dependent upon the size and surface characteristics of the nanomaterials. Vascular fenestrations differ for different organs, ranging from less than 6 nm to 400 nm in different animal models, and can change with the presence of diseased tissue (Gaumet *et al.*, 2008). In addition, depending upon the surface chemistry, plasma proteins can adsorb to the nanoparticle surface promoting opsonization, or phagocytosis by immune cells, which results in clearance through the reticuloendothelial system (RES), including organs such as the liver and spleen (Owens and Peppas, 2006). Table 22.5 summarizes the findings of several *in vivo* biodistribution studies to date, indicating the blood circulation half-life values and the organ localization for various nanomaterials.

In vivo testing of several nanomaterials has indicated that nanomaterials administered via intravenous or intraperitoneal injection localize in several of the organs rich in blood supply or associated with the RES. Specifically the liver, lungs, kidneys and spleen contain significant levels of nanomaterials, with the highest doses found in the liver. Evidence of accumulation is reported for several of the nanomaterials tested, although in the majority of the studies cited the time endpoint was short, only a few hours to a day. In order to confirm any potential bioaccumulation, more chronic or long-term studies are needed. Interestingly, the few studies that tracked nanomaterial distribution for as long as one month continued to detect levels of nanomaterials with little to no observed toxic effects. Examining the dose levels tested, the values are in the high mg/kg–g/kg range. Although LD50 values could not be estimated due to lack of lethality, these doses correspond to moderate toxicity as classified using the Hodge and Sterner scale (see Table 22.6).

Since delivery to target sites is hindered by rapid clearance from the blood stream and RES uptake, nanomaterials developed for medical applications are often surface coated to increase the circulation half-life. A common coating, also used to improve the pharmacokinetics of pharmaceuticals such as peptide drugs, involves binding poly(ethylene glycol) (PEG) chains to the nanomaterial surface (Harris and Chess, 2003). This increases the size to prevent filtration by excretory organs and suppress opsonization, to enhance retention of the nanomaterial in the circulation, thus facilitating enhanced delivery to the desired target sites. Prolonged circulation in the bloodstream increases the time for nanomaterials to contact blood components; therefore, hematocompatibility of nanomaterials is critical, since adverse effects due to poor blood compatibility include hemolysis, thrombogenicity and complement activation (Dobrovolskaia *et al.*, 2008).

With reduced opsonization, nanomaterial biodistribution may also shift away from RES organs or accumulation may be decreased. Studies using quantum dots coated with PEG have shown reduced accumulation in the liver and spleen, and their presence in the intestines suggests excretion from the liver (Akerman *et al.*, 2002; Ballou *et al.*, 2004). For quantum dots coated with non-PEG coatings, mercaptoundecanoic acid and bovine serum albumin, sequestration in RES organs was observed (Fischer *et al.*, 2006). However, despite reduced accumulation, PEG-coated particles are still sequestered. In addition to PEG-coated quantum dots, gold nanoparticles coated with PEG also accumulate in RES organs; however, much smaller percentages of the injected dose ($\sim$1 % total) were found in these organs after short-term exposure compared with quantum dots, which exhibited no evidence of clearance (Sonavane *et al.*, 2008). Nanomaterial localization in the kidney is expected, as the kidneys serve as a clearance route from the body. Further discussion on kidney clearance of nanomaterials is presented in the next section.

Table 22.5 Summary of selected studies on nanomaterial exposure via intravenous injection.

Model system	NP	Surface coating	Average NP size	Concentration	Duration	Average blood $t_{1/2}$	Results	Reference
ddY mice	Au	N/A	10, 50, 100, 200 nm	1 g/kg	24 hr	N/A	Accumulation in liver highest, followed by lungs, spleen and kidneys, however at low (<1% administered dose) levels	Sonavane et al. (2008)
ddY mice	Au nanorods	PEG	65 nm (l) 11 nm (w)	0.5–0.9 mM Au	72 hr	1 hr	After 72 hr, no Au found in blood, 35% of dose found in liver, small amounts found in lungs, kidneys, spleen	Niidome et al. (2006)
Female albino mice	Au nanoshells	PEG	120 nm	2.4×10^{11} particles/ml	4 hr to 28 days	3.7 hr	Limit in mice muscle tissue about 70 pg, accumulating in the RES organs, 1–10 ppm levels found in bone, muscle, kidney, lung	James et al. (2007)
BALB/c mice	Au3Cu nanoshells	PEI/PAA	55 nm	2, 20, 40 mg/kg	30 days	N/A	100, 83 and 67% survival for the 3 doses from lowest to highest, Au and Cu present in urine 3 hr postinjection	Su et al. (2007)
C57BL/6J mice	Au-dendrimer	Amine terminated (+), sodium carboxylated (0), carboxylated (+)	5, 11, 22 nm	16 mg/kg	7 days	N/A	Au in kidneys after 4 days suggest some sequestered, excretion of (+) > (−) & (0) charged NPs; > concentration of (+) NPs in kidneys versus liver and spleen	Balogh et al. (2007)
Sprague-Dawley rats	C60	14C, ammonium salt derivative	300 nm	0.2–1.6 mCi	120 hr	<1 min	>95% (>50% ammonium salt functionalized) found in liver; no radioactivity found in urine or feces	Bullard-Dillard et al. (1996)

(Continued)

Table 22.5 *(Continued)*

Model system	NP	Surface coating	Average NP size	Concentration	Duration	Average blood $t_{1/2}$	Results	Reference
Sprague-Dawley rats	C60	MSAD	N/A	15 mg/kg	24 hr	6.8 hr	15 mg/kg well tolerated, urine did not contain detectable levels, high concentrations found in liver, binding studies indicate ~99% sample bound to plasma proteins	Rajagopalan et al. (1996)
Nude mice	CdSe	PEG, 64Cu	12, 21 nm (d)	5.55 MBq/ 25 pmol	10 min, 30 min, 4.5 h, 12 h, 36 h	N/A	Both methods show rapid uptake by the liver (27.4–38.9 %ID/g) (%ID/g is percentage injected dose per gram tissue) and spleen (8.0–12.4 %ID/g). Size has no influence on biodistribution within the range tested here. Pegylated QD have slightly slower uptake into liver and spleen (6 vs 2 min) and show additional low-level bone uptake (6.5–6.9 %ID/g). No evidence of clearance from these organs was observed.	Schipper et al. (2007)
BALB/c nu/nu mice	CdSe/ZnS	MAA, GFE, F3, LyP-1, PEG	3.5 nm; 5.5 nm (uncoated)	100–200 mg QDs in 0.1–0.2 μl solution	5 or 20 min	N/A	Specific tissue targeting is achievable; accumulation in liver and spleen; estimate PEG and peptide surface reduce liver and spleen uptake by ~95%	Akerman et al. (2002)

Animal model	Material	Coating	Size	Dose	Time	Half-life	Observations	Reference
BALB/c mice	CdSe/ZnS	PAA, mPEG-750, mPEG-5000, COOH-PEG-3400	N/A	50–500 pmol QDs in 50–200 μL saline	1–3 hr at 1 min interval	3.2 min (750); 12.6 min (3400); 71 min (5000)	QDs in endosomes in liver, spleen and bone; fluorescence at 1 mo. similar to 24 h signal, mPEG-5000 QDs have lower accumulation	Ballou et al. (2004)
Sprague–Dawley rats	CdSe/ZnS	MUA, BSA	25, 80 nm	5 nmol in 0.2 ml solution	90 min	38.7 min (BSA); 58.5 min (MUA)	QDs uptaken in Kupffer cells, majority uptake in liver with some in spleen, lungs, kidneys, lymph nodes, bone marrow	Fischer et al. (2006)
Sprague–Dawley rats	CdSe/ZnS	DHLA, DHLA-PEG, Cysteine, Cysteamine	TEM: 2.85–4.31 nm, DLS: 4.64–7.22 nm, GFC: 4.36–8.65 nm	10 pmol/g	4 hr	48 min (4.36 nm) to 20 hr (8.65 nm)	Renal clearance observed for particles with HD <5.5 nm as well as lower liver and spleen uptake	Choi et al. (2007)
Sprague–Dawley rats	CdTe	N/A	6 nm	2 mM 1 ml/kg	24 hr	N/A	Few signs of toxicity, but changes in locomotor activity were observed	Zhang et al. (2007)
ICR mice	CdTe/ZnS	Methoxy-PEG-5000	13 nm	40 pmol	28 days	18.5 hr	100 % dose recovered in carcass after 1 and 28 days (44–50 % muscle, skin, bone; 29–40 % liver; 4.8–5.2 % spleen; 1.5–9.1 % kidneys)	Yang et al. (2007)
Mice	Iron NP	MION-47, amino-CLIO, Tat-CLIO	>3 nm	8.80–6.11 %	24 hr		Nanoparticles removed by phagocytes of reticuloendothelial, MION long blood half life	Wunderbaldinger et al. (2002)

(Continued)

Table 22.5 (*Continued*)

Model system	NP	Surface coating	Average NP size	Concentration	Duration	Average blood $t_{1/2}$	Results	Reference
New Zealand rabbits	SWCNTs	N/A	~1 nm (d), ~300 nm (l)	7.5 ml bolus containing 75 μg, ~20 μg/kg	24 hr	1 hr	Significant fluorescence of nanotubes found only in the liver; absence of acute toxicity	Cherukuri *et al.* (2006)
Nude mice	SWCNTs	PL-PEG 2000 & 5400, 64Cu	1–5 nm (d), 100–300 nm (l)	200–300 μCi	24 hr	0.5 hr (PEG-2000); 2 hr (PEG-5400)	No obvious toxicity observed; retained radioactivity suggests slow excretion; ~20 % ID/g (PEG-5400) & ~35 % ID/g (PEG-2000) in liver	Liu *et al.* (2007)
Male Wistar rats (strain Crl : WI (Han))	TiO$_2$	N/A	20–30 nm	5 mg/kg	28 days	N/A	TiO$_2$ retained in liver over 28 days; no observed toxic effects; initial blood conc. ~5000 mg/g; ~120 mg/g liver (day 1, 14, 28), ~0.65 mg/g kidney (day 1)	Fabian *et al.* (2008)

Table 22.6   *Hodge and Sterner combined relative toxicity classes (Hodge and Sterner, 1949). Reprinted by permission of Taylor & Francis, http://www.informaworld.com*

| Category | Routes of administration | | | |
	Oral LD50 (single dose to rats) mg/kg	Inhalation LC50 (4 hour exposure to rats) ppm	Dermal LD50 (single application to rabbits) mg/kg	Amount for average adult
Extremely toxic	<1	<10	<5	A grain, a taste
Highly toxic	1 to 50	10–100	1 to 43	1 tsp, 4 cc
Moderately toxic	50–500	100–1000	44–340	1 oz., 30 g
Slightly toxic	500–5000	1000–10,000	350–2810	1 cup, 250 g
Practically nontoxic	5000–15,000	10,000–100,000	2820–22,590	1 quart, 1000 g
Relatively harmless	>15,000	>100,000	>22,600	>1 quart

22.3.2.2 Metabolism and Excretion

An important determinant of toxicity for pharmaceuticals, and analogously nanomaterials for medical applications, is dose. Issues related to what constitutes or best describes nano-material dose will be discussed in section 22.4. However, other factors that also determine what dose needs to be administered include the pharmacokinetics and bioavailability of the substance. For example, protein adsorption to nanomaterials was previously mentioned to influence their uptake. Another consequence of protein adsorption could be an alteration in their bioactivity; however, the effects of this are still relatively unknown (Lundqvist *et al.*, 2004). Also previously mentioned, nanomaterials are often surface-modified (PEGylation) to alter their pharmacokinetic profile, or bioconjugated to elicit bioactivity. Metabolism of these functional groups on surface-modified nanomaterials by enzymes in the human body could occur, such as for example, the protein cap of a functionalized quantum dot could be cleaved by proteases (Hardman, 2006). Non-specific cellular internalization of nanomate-rials has also been reported, and this could result in their breakdown inside lysosomes due to lysosomal metabolism via the acidic pH or lysosomal enzymes (Chang *et al.*, 2006). In the case of metal nanomaterials, free metal ions would be released and enzymes, such as metallothionein, would contribute towards metabolism (Nordberg, 1998). However, these possible metabolic outcomes are merely speculative and have yet to be experimentally confirmed.

As with other chemicals, it is expected that nanomaterials that enter systemic circula-tion are excreted through the kidneys and liver, unless they are cleared by other defense mechanisms from the port of entry, and this has been confirmed in some preliminary biodis-tribution studies. Evidence of renal clearance from the body has been found for fullerenes, SWCNTs, gold nanoparticles and nanoshells, CdSe/ZnS quantum dots and dendrimers (Rajagopalan *et al.*, 1996; Wang *et al.*, 2004; Lee *et al.*, 2005; Singh *et al.*, 2006). Interest-ingly, for 5 nm gold nanoparticles, sequestration in the kidney tissue has been suggested, as elevated Au levels continued to be detected 4 days post-injection (Balogh *et al.*, 2007). However, for cysteine-coated quantum dots with hydrodynamic diameters below 5.5 nm, up to ~80 % of the injected dose could be found present in the urine. While presence of nanomaterials in the urine have been reported, few of these studies collected quantitative data on the amount excreted in the urine, leaving the need for further quantitative excretion studies.

Although presence in the liver is found for all nanomaterials, it is not clear whether nano-materials are eliminated via hepatic clearance. Fluorescence from PEG-coated quantum dots in the intestines of exposed mice suggests excretion from the liver (Akerman *et al.*, 2002; Ballou *et al.*, 2004). However, other studies show no appreciable change in the levels of nanomaterials found in the liver throughout the course of both short- and long-term studies (Liu *et al.*, 2007; Fabian *et al.*, 2008). These levels range from around 30 % of the injected dose to as much as 95 % of the injected dose (Bullard-Dillard *et al.*, 1996; Niidome *et al.*, 2006). This suggests that while excretion may occur, it is slow due to liver sequestration after RES uptake. Specifically, liver macrophages such as Kupffer cells have been identified in some studies as the primary cellular location in the liver (Sadauskas *et al.*, 2007). However, the mechanism of liver macrophage-mediated excretion of nanomaterials has yet to be investigated.

Despite observations of renal and hepatic clearance, complete clearance has not been reported for intravenous or intraperitoneal administered nanomaterials. Rather, after the final time-point, at most one month post-injection, lower levels are observed but similar distributions of the nanomaterials throughout the animal still remain. Some studies that quantify the percentage of the administered dose in the feces found no detectable levels during the test period and reported virtually no excretion (Bullard-Dillard *et al.*, 1996; Fischer *et al.*, 2006; Yang *et al.*, 2007). The mechanism governing nanomaterial excretion remains largely unknown, and it is possible that not all nanomaterials will be eliminated from the body. Accumulation could take place as suggested by the lack of complete clear-ance in some cases. At low concentrations or with single exposure, toxicity may be absent or minimal; however, following a high dose or chronic exposure, the accumulation dose could lead to significant toxicity. As mentioned above, bioaccumulation and persistence are two characteristics of hazardous substances; however, data are not yet available regarding the extent and consequences of nanomaterial accumulation *in vivo*.

22.4 Uncertainties in Nanomaterial Risk Assessment and Current Efforts to Address these Challenges

In order to maximize the utility of any risk assessment, the sources of uncertainty need to be identified and their impact evaluated to incorporate this factor into the assessment. No decision lacks uncertainty, and since risk assessment involves estimating the probability and consequences of a certain outcome, any uncertainty must be clearly understood. Some uncertainties can be resolved by collecting additional data; however, questions on the accuracy of the measurements, the conditions in which the data were collected, and even the appropriateness of the model used are often harder to resolve. Thus, when conducting a risk assessment these limitations should be acknowledged.

Before discussing the limitations in current nanomaterial risk assessment, it is important to recognize the inherent restrictions of the general risk-assessment process (Sexton *et al.*,

1995). It can be argued that the methodologies used in risk assessment are not rigorous or suitable enough to assess beyond immediate or short-term risks. In addition, risk assessment is viewed by some as an oversimplification of the problem, since it generally focuses on one hazard and one effect at a time, and therefore does not take into account the toxicity of complex mixtures or chemical interactions. Since most, if not all, risk-assessment data are derived from laboratory experiments, extrapolation of the data to represent relevant conditions also adds to the uncertainty. Extrapolating *in vitro* to *in vivo* levels, animal to human responses, or high/threshold to low/non-threshold dose exposures, are issues with all risk assessments, not just nanomaterials.

Unfortunately, nanomaterial risk assessment suffers from insufficient amounts of data. Because of this, there remains uncertainty regarding the influence of particle characteristics on toxicity, especially the impact of size; the mechanisms behind toxicity; the properties influencing fate and transport in the body, particularly those that affect translocation potential; and the metrics to use for exposure measurement. Although additional studies continue to be conducted, with the ongoing introduction of new nanomaterials and coating methods, the current case-by-case testing strategy is inefficient for the number of nanomaterials to be tested. This section discusses these challenges to current nanomaterial research approaches, as well as current and potential strategies aimed at addressing these challenges.

22.4.1 Lack of Rigorous Nanomaterial Characterization

The utility of data in published nanomaterial studies can be limited due to poor characterization of the nanomaterials used. In order to elucidate the physico-chemical properties that correlate with certain biological effects, nanomaterial characterization is essential; however, what constitutes full characterization has not been standardized. Although a vast array of characterization data can be collected, this rigorous approach would not only be time-intensive but would also require a large amount of resources. Hypothesis-driven material characterization would reduce the number of measurements, but the question again becomes how much data is sufficient to make a valid correlation? Due to these considerations, standardization of nanomaterial nomenclature, characterization and measurement has become a major research priority (Environmental Protection Agency, 2008c; NIOSH, 2008; National Nanotechnology Initiative, 2008).

22.4.1.1 Standard Nomenclature and Characterization

Organizations such as ASTM International, the American National Standards Institute (ANSI) and the International Organization for Standardization (ISO) are actively involved in nanomaterial standards development. ASTM International developed a set of standardized nanotechnology terminology or nomenclature which was published in 2007 (ASTM, 2007). Similarly, ISO standards on terminology and definitions for nano-objects as well as health and safety practices in occupational settings were published in 2008 (ISO, 2008a,b). However, a list of necessary information on a nanomaterial for adequate identification or labelling has not been developed. Physico-chemical property characterization is critical to distinguish between different nanomaterials, to determine which properties contribute to any observed biological effects, as well as to enable future reproduction and inter-laboratory validation. Suggested properties to incorporate in nanomaterial identification and characterization include: technical/commercial name, composition (e.g. chemical formula, purity), molecular/crystalline structure, size (or surface area) distribution, agglomeration/aggregation state, surface chemistry (e.g. coating or modification), zeta potential (surface charge), solubility, porosity, and catalytic activity (e.g. redox and radical formation potential) (Warheit *et al.*, 2007; OECD, 2008a) .

In addition to establishing a standard set of properties to measure, standardization of the measurement techniques used to quantify these properties is also critical to achieving consistency and facilitating inter-laboratory comparisons. For example, it is without question that particle size should be reported, but this measurement can be determined using several techniques. These include estimation from sizing curves derived from absorbance spectra, direct visualization using atomic force or electron microscopy, analytical ultracentrifugation, differential mobility analysis, size exclusion chromatography, dynamic light scattering for spherical particles, and small-angle X-ray scattering (Oberdörster *et al.*, 2005a). Therefore, quantifying nanoparticle size using all of these techniques would not only be unnecessary but also wasteful of resources. However, no one technique is capable of fully characterizing a sample. Consequently, the extent of nanomaterial characterization reported in published studies tends to reflect the instrumentation capabilities of the corresponding research facility. Without the establishment of standard measurement techniques, nanomaterial toxicology literature will continue to include studies with poor or insufficient characterization that add little to understanding the mechanisms of nanomaterial toxicity.

22.4.1.2   Development of Certified Reference Nanomaterials

At this time, no standardized guidelines for the synthesis and production of nanomaterials are available, and instead various nanomaterial synthesis protocols with multiple approaches to synthesizing the same nanomaterial are described in a vast array of publications. However, successful reproduction of these methods is highly dependent upon the experience level of the chemist, and even experienced chemists can lack consistency and precision between multiple syntheses. Therefore, there is a lack of quality control with the nanomaterials being used in risk-assessment related studies, and consequently there is a push towards the development of nanomaterial reference standards. The National Institute of Standards and Technology (NIST) with the National Cancer Institute's Nanotechnology Characterization Laboratory (NCL) have developed their first set of citrate-stabilized gold nanoparticle reference standards to be used for instrument calibration, measurement qualification and *in vitro* characterization experiments.

This development of reference nanomaterials could also address the need for adequate experimental controls. A challenge with testing nanomaterials is that their physical properties such as size and shape, in addition to their chemical composition, can influence their toxicity profile. Therefore, in order to resolve the source of toxicity, control particles are needed. Early studies only compared unexposed and nanomaterial-exposed samples to assess toxicity. While this can provide information regarding concentration limits such as the median and maximum lethal doses, little information as to what property of the nanomaterial leads to the observed toxicity can be ascertained. More recent studies include comparison with the bulk chemical counterpart or larger particles of the same chemical makeup (Chen *et al.*, 2006; Wang *et al.*, 2006, 2007). In addition, silica and quartz nanomaterials of similar size and shape to the test nanomaterial sample have been used as positive controls (Warheit *et al.*, 2005). Ultimately, a set of reference nanomaterials for research laboratories to use either as the test sample or as controls in their studies would eliminate uncertainty due to inconsistencies between synthesis and characterization protocols between research groups.

22.4.2 Relevance to Actual Exposure Conditions

It is generally recognized that human exposure to nanomaterials could be incidental, occupational or intentional during their synthesis, production, use and disposal (Stern and McNeil, 2008). As discussed in section 22.3, potential portals of nanomaterial entry include the lungs, skin and GI tract, as well as various medical administration routes. However, current methods of estimating human nanomaterial exposure effects focus on testing individual nanomaterials on particular cell lines, as only a limited amount of *in vivo* data has been collected. In addition, a majority of these studies analyze the health effects after acute, high-dose exposures, as these test methodologies are easier to control and are well established. Consequently, nanomaterial exposure studies often inadequately account for common, real-life conditions, such as variable durations and multiple incidents of exposure. In order to better understand the interactions of nanomaterials in the body, it is essential that exposure tests not only determine the boundaries or limits of exposure, but also reflect actual exposure conditions.

22.4.2.1 Characterization and Detection in Biological Samples

Although several techniques exist for nanomaterial characterization, challenges still remain for characterization of nanomaterials in biological settings, both *in vitro* and *in vivo*. The influence of biological media on nanomaterials is also not well understood. Proteins, such as albumin, in human serum could alter the observed effects of nanomaterials in biological samples. It is known that protein adsorption occurs with administration of pharmaceuticals or implantation of medical devices, and this adsorbed layer may affect the response profile. Protein adsorption to the nanomaterial surface may influence the transport properties as well as the toxicity, as the fraction of nanomaterials with an adsorbed protein layer, or 'protein corona', would be larger and could exhibit less reactivity due to protein interaction with the surface coating ligands (Lundqvist *et al.*, 2008). It is not known whether the interaction between the proteins and nanomaterials is reversible, promotes nanomaterial aggregation, influences nanomaterial metabolism or affects protein structure (Lynch and Dawson, 2008). Since the nanomaterial surface is what the cell initially encounters upon nanomaterial exposure, more research is required to examine the interaction of proteins and other biomolecules with nanomaterials.

Despite the small size of nanomaterials directly after synthesis, interactions of nanomaterials with biological media can alter the surface characteristics, often resulting in aggregate formation which increases the effective size of the exposed material. Aggregation of nanomaterials may also affect their distribution and toxicity profiles. Therefore even with rigorous characterization of the particles before administration, it remains questionable as to whether this form is representative of what would be encountered in an exposure condition. In order to reflect actual exposure conditions, characterization measurements could be taken at several points in the life cycle of nanomaterials such as the dry, wet, *in vitro, in vivo* and *ex vivo* phases. However, this would take up significant time and resources; therefore, a focus on adapting and utilizing high throughput screening methods could make this endeavor more efficient (Xia *et al.*, 2008).

Current measurement techniques present additional challenges when determining nano-material exposure in biological contexts. For example, the size of nanomaterials is often measured using electron microscopy; however, preparation for electron microscopy requires dehydration of the sample, which can alter the size of nanomaterials that are surface coated for water solubilization or bioactivity. Dynamic light scattering measures the size of particles in solution; however, this technique is limited to spherical nanoparticles with sizes generally greater than a few nanometers, depending on the scattering properties of the particles concerned (Xu, 2008). Other techniques such as size exclusion chromatography are also used; however, all of these sizing methods require an isolated nanomaterial solution. While nanomaterials as prepared need no further processing, nanomaterials in biological samples would need to be isolated from the biological matrix. Not only would these procedures vary depending on the media, but they could also alter the nanomaterial surface properties. In addition, there still remains the issue of differentiating between naturally occurring and engineered nanomaterials within a sample. While some nanomaterials possess distinguishing properties such as magnetism or unique optical signatures, these may not facilitate isolation from naturally occurring nanomaterials for characterization. Many if not all of these issues would need to be resolved in order to achieve accurate sizing of nanomaterials in biological media. Since size is recognized to influence the bioactivity of a material, developing methods to characterize nanomaterials accurately in biological samples is a current research focus (Environmental Protection Agency, 2008c; National Nanotechnology Initiative, 2008).

Particularly for medical applications of nanomaterials, it is important to monitor dose. This means quantitative measurement of nanomaterials in media such as saliva, blood, urine, feces and tissue. Most methods to detect nanomaterials in biological media only determine whether or not nanomaterials are present in the sample, and are not sensitive enough to provide quantitative information such as concentration. Although inductively coupled plasma mass spectroscopy has been used to quantify metal nanomaterial concentrations in animal exposure studies, this method is limited to metal-containing nanomaterials. In addition, there is currently no method to determine the amount of internalized nanomaterials *in vivo*. As it has been suggested that the nanomaterial concentration inside versus outside the cell better predicts nanomaterial toxicity, it may be useful if this could be monitored *in vivo* (Chang *et al.*, 2006). Therefore, improved *in vivo* nanomaterial detection methods yielding quantitative information are essential to understand better the potential dose effects.

22.4.2.2   Identify Appropriate Endpoints

For a majority of the nanomaterials being used today, acute high-dose exposure data have been collected. However, it is unlikely that the majority of human exposures to nanomaterials will involve high concentrations occurring over a short period of time. Therefore, chronic low-dose exposure studies may be more appropriate for determining nanomaterial health effects. For this case, time serves as an influential endpoint. Not only can the dose duration influence the observed effects, but also the frequency of doses during a period of time can contribute when testing repeated dosing. With low-dose exposures, conventional endpoints such as inflammation, mutagenicity and mortality may not occur during the testing period. Therefore, alternative endpoints such as transcriptional or proteomic changes may be more suitable. Toxicogenomic profiles for many chemicals are being developed to identify the intracellular events that lead to toxic response and disease (Boverhof and Zacharewski, 2006; National Academy of Sciences, 2007). A new direction for nanomaterial risk assessment research could be to establish toxicogenomic fingerprints for nanomaterials, to use to identify potential disease pathways associated with nanomaterial exposure.

22.4.3   Poor Dissemination and Evaluation of Existing Data

Understanding the human health effects resulting from nanomaterial exposure is important regardless of the type or extent of use. Since public perception often dictates the success of any new technology, the full potential of nanomaterial-based technologies can only be achieved by maintaining positive perception. Despite the increased amount of nanomaterial exposure data published over the past three years, there still remains uncertainty regarding nanomaterial safety. This could be attributed to the lack of widespread availability or quality of existing data. Strategies to facilitate dissemination and evaluation of existing data could include comprehensive data collection, targeted research, increased peer review, and standardized reporting.

22.4.3.1   Information Sharing and Standardization of Reporting

An effective approach to improving risk assessment is to improve the quality and comprehensiveness of available data. Critical reviews, while still necessary, only capture a snapshot of the state of emerging technologies such as nanotechnology, since these technologies are continually in flux. Instead, to accelerate closing the current knowledge gaps, a comprehensive database needs to be created, made publicly available and actively populated, or provide quasi-real time updates by the research community to address the rapid advancement of the fiel,d and facilitate timely development of standards and policies.

However, currently available databases such as those provided by the International Council on Nanotechnology (ICON) at Rice University, the Nanoscale Science and Engineering Center (NSEC) at the University of Wisconsin-Madison, the Project on Emerging Nanotechnologies (PEN) at the Woodrow Wilson Center for Scholars, and the Science to Achieve Results (STAR) grant listings through the EPA, all only serve as repositories of publication abstracts related to nanomaterial environmental, health and safety research (Environmental Protection Agency, 2008a). There still lacks the establishment of an open access database of actual experimental methods and data extracted from these publications. A similar database to NIOSH's Hazardous Substances Data Bank or EPA's Integrated Risk Information System could be developed, and NIOSH has spearheaded the start of a Nanoparticle Information Library although the format of this database is currently unknown (Environmental Protection Agency, 2008b).

While databases are often suggested to consolidate large volumes of data for cross-comparative analysis, standardization in reporting is critical for any database to be useful. For example, some nanomaterial exposure studies do not report parameters such as size, or values are reported without acknowledging the method used (Zhang *et al.*, 2007; Chou *et al.*, 2008). Although this need relates to the standardization of nanomaterial characterization addressed above, one approach to encouraging this change would be to engage major scientific journals and government research agencies in requiring such particle characterization for publication and grant reporting.

22.4.3.2   Development of Predictive Models

The weakest link in risk assessment is often considered the exposure assessment step (Wakefield, 2000). Common approaches to modelling exposure from a variety of sources are needed, as comprehensive testing is costly, often involving large-scale *in vivo* studies that would need to be conducted on every nanomaterial of concern. Predictive models that correlate existing data with similar variants could also help to identify nanomaterials that should be given priority in safety evaluations. Whether this can be achieved by extrapolating from existing data and identifying appropriate parameters to modify existing models is still questionable. The use of physico-chemical data to classify nanomaterials and their group toxicity effects is a critical next stage of nanomaterial toxicology development. As more data is collected in these coming years, a move towards developing a similar research program to ToxCast, which forecasts toxicity based on bioactivity profiles, for nanomaterials may be beneficial (Dix *et al.*, 2007). It still remains a major challenge to the science and engineering research communities to develop predictive tools based on structure/function relationships in which nanomaterial toxicity can be modelled, and the current research direction aims to address this need (National Nanotechnology Initiative, 2008).

22.4.3.3   Coordinated Research Direction

Ultimately, in order to effectively address the research demands outlined throughout this section, collaboration is needed, especially given the amount of data to collect and process related to nanotechnology EHS research. To organize and direct current research efforts, the establishment of a federal EHS research plan has been recommended in several publications and conference reports (Oberdörster *et al.*, 2005a; Maynard *et al.*, 2006; Balbus *et al.*, 2007). Despite federal funding of nanotechnology EHS research since the 2005 fiscal year, the objectives of this funding have only been defined in the 2008 NNI report *Strategy for Nanotechnology-Related Environmental, Health, and Safety Research* (National Nanotechnology Initiative, 2008). Five key research categories were identified to be coordinated by the agencies in parentheses: (1) instrumentation, metrology and analytical methods (NIST); (2) nanomaterials and human health (NIH); (3) nanomaterials and the environment (EPA); (4) human and environmental exposure assessment; (5) risk management methods (FDA, EPA) (National Nanotechnology Initiative, 2008). In addition, the OECD is spearheading efforts for international EHS research collaboration with the aim of addressing similar issues as those outlined throughout this section (OECD, 2008b). Although it is premature to evaluate these new initiatives, increased oversight of nanomaterial EHS research may promote more efficient use of available resources as well as timely generation of safety data.

22.5   Conclusion

Despite the heightened concern and current uncertainty behind nanomaterial safety, there are many beneficial uses of nanomaterials, adding new or improved technologies to resolve today's challenges. Given their increasing presence in the consumer market, nanomaterial risk communication to the public will impact upon the future direction of nanomaterial research. Therefore while the accuracy of currently available nanomaterial toxicity data may be questionable, the interpretation of these data is more influential. Early risk assessment evaluates hazard potential by establishing exposure limits to help define the system of exposure; however, more important is to determine how these effect levels compare relative to known toxic substances. In addition, risk assessment not only involves identifying hazards, but also exposures to those hazards. This raises the question as to whether there is a real need to capture everything in nanomaterial risk assessments. If exposure is highly unlikely for a certain nanomaterial, then rigorous testing of that nanomaterial may not be necessary. As a result the trend in nanomaterial safety research has been to test each nanomaterial on a case-by-case basis; however, this has also limited our progress towards development of predictive modelling. Therefore, it is also important to continue to review present efforts in order to ensure research is addressing the questions, and the right questions are being answered.

This is by no means an exhaustive review of the entire literature concerning nanomaterial risk assessment. Rather the aim was to provide a primer on the issues faced when applying the current risk assessment paradigm to nanomaterials. It is important to keep in mind that decision-making requires timely answers to questions about risks and hazards to public health in order to mitigate future or current potential harm. As described in the previous section, there is a need for the careful utilization of limited resources, as well as proactive information-sharing and continual reevaluation in nanomaterial risk assessments. From a researcher's perspective, it is important that sound scientific evidence be the basis of the evaluation and prioritization of the policies guiding the development of emerging technologies; however, in order to achieve this end, the research community must organize and coordinate their efforts.

Acknowledgements

The authors would like to thank Clare Ouyang for her assistance. This work was supported by the Center for Biological and Environmental Nanotechnology (NSF Award Number EEC-0118007).

References

Ahn MH, Kang CM, Park CS, Park SJ, Rhim T, Yoon PO, Chang HS, Kim SH, Kyono H, Kim KC (2005) Titanium dioxide particle-induced goblet cell hyperplasia: association with mast cells and IL-13. *Respiratory Research* **6**, 34.

Akerman ME, Chan WCW, Laakkonen P, Bhatia SN, Ruoslahti E (2002) Nanocrystal targeting in vivo. *P Natl Acad Sci USA* **99**, 12617–12621.

ASTM (2007) *Standard Terminology Relating to Nanotechnology*. ASTM Standard E2456-06. ASTM International: Philadelphia. Available at http://www.astm.org/Standards/E2456.htm

Baker GL, Gupta A, Clark ML, Valenzuela BR, Staska LM, Harbo SJ, Pierce JT, Dill JA (2008) Inhalation toxicity and lung toxicokinetics of C60 fullerene nanoparticles and microparticles. *Toxicol Sci* **101**, 122–131.

Balbus JM, Florini K, Denison RA, Walsh SA (2006) Getting it right the first time – Developing nanotechnology while protecting workers, public health, and the environment. *Living in a Chemical World: Framing the Future in Light of the Past* **1076**, 331–342.

Balbus JM, Maynard AD, Colvin VL, Castranova V, Daston GP, Denison RA, Dreher KL, Goering PL, Goldberg AM, Kulinowski KM *et al.* (2007) Meeting report: hazard assessment for nanoparticles – report from an interdisciplinary workshop. *Environmental Health Perspectives* **115**, 1654–1659.

Ballou B, Lagerholm BC, Ernst LA, Bruchez MP, Waggoner AS (2004) Noninvasive imaging of quantum dots in mice. *Bioconjugate Chem* **15**, 79–86.

Balogh L, Nigavekar SS, Nair BM, Lesniak W, Zhang C, Sung LY, Kariapper MS, El-Jawahri A, Llanes M, Bolton B *et al.* (2007) Significant effect of size on the in vivo biodistribution of gold composite nanodevices in mouse tumor models. *Nanomedicine* **3**, 281–296.

Baroli B, Ennas MG, Loffredo F, Isola M, Pinna R, Lopez-Quintela MA (2007) Penetration of metallic nanoparticles in human full-thickness skin. *J Investigative Dermatology* **127**, 1701–1712.

Baron P, Maynard A, Foley M (2003) *Evaluation of Aerosol Release During the Handling of Unrefined Single Walled Carbon Nanotube Material.* NIOSH: Cincinnati, OH.

Bauer C, Buchgeister J, Hischier R, Poganietz WR, Schebek L, Warsen J (2008) Towards a framework for life cycle thinking in the assessment of nanotechnology. *Journal of Cleaner Production* **16**, 910–926.

Bergamaschi E, Bussolati O, Magrini A, Bottini M, Migliore L, Bellucci S, Iavicoli I, Bergamaschi A (2006) Nanomaterials and lung toxicity: interactions with airways cells and relevance for occupational health risk assessment. *Int J Immunopathology Pharmacology* **19**(suppl.), 3–10.

Borm PJA, Kreyling W (2004) Toxicological hazards of inhaled nanoparticles – Potential implications for drug delivery. *J Nanoscience Nanotechnology* **4**, 521–531.

Boverhof DR, Zacharewski TR (2006) Toxicogenomics in risk assessment: applications and needs. *Toxicol Sci* **89**, 352–360.

Bullard-Dillard R, Creek KE, Scrivens WA, Tour JM (1996) Tissue sites of uptake of 14C-labeled C60. *Bioorganic Chemistry* **24**, 376–385.

Bulte JW, Kraitchman DL (2004) Iron oxide MR contrast agents for molecular and cellular imaging. *NMR in Biomedicine* **17**, 484–499.

Buzea C, Pacheco I, Robbie K (2007) Nanomaterials and nanoparticles: sources and toxicity. *Biointerphases* **2**, MR17–MR71.

Card JW, Zeldin DC, Bonner JC, Nestmann ER (2008) Pulmonary applications and toxicity of engineered nanoparticles. *Am J Physio-Lung Cellular Molecular Physio* **295**, L400–L411.

Castorina R, Woodruff T (2003) Assessment of potential risk levels associated with US Environmental Protection Agency Reference Values. *Environmental Health Perspectives* **111**, 1318–1325.

Cha KE, Myung H (2007) Cytotoxic effects of nanoparticles assessed in vitro and in vivo. *J Microbiology Biotechnology* **17**, 1573–1578.

Chang E, Thekkek N, Yu WW, Colvin VL, Drezek R (2006) Evaluation of quantum dot cytotoxicity based on intracellular uptake. *Small* **2**, 1412–1417.

Chen Z, Meng H, Xing G, Chen C, Zhao Y, Jia G, Wang T, Yuan H, Ye C, Zhao F *et al.* (2006) Acute toxicological effects of copper nanoparticles *in vivo*. *Toxicology Letters* **163**, 109–120.

Cherrie JW, Semple S, Christopher Y, Saleem A, Hughson GW, Philips A (2006) How important is inadvertent ingestion of hazardous substances at work? *Ann Occup Hyg* **50**, 693–704.

Cherukuri P, Gannon CJ, Leeuw TK, Schmidt HK, Smalley RE, Curley SA, Weisman RB (2006) Mammalian pharmacokinetics of carbon nanotubes using intrinsic near-infrared fluorescence. *P Natl Acad Sci USA* **103**, 18882–18886.

Choi HS, Liu W, Misra P, Tanaka E, Zimmer JP, Ipe BI, Bawendi MG, Frangioni JV (2007) Renal clearance of quantum dots. *Nature Biotechnology* **25**, 1165–1170.

Chou CC, Hsiao HY, Hong QS, Chen CH, Peng YW, Chen HW, Yang PC (2008) Single-walled carbon nanotubes can induce pulmonary injury in mouse model. *Nano Lett* **8**, 437–445.

Christensen FM, Andersen O, Duijm NJ, Harremoes P (2003) Risk terminology – a platform for common understanding and better communication. *J Hazardous Materials* **103**, 181–203.

Cross SE, Innes B, Roberts MS, Tsuzuki T, Robertson TA, McCormick P (2007) Human skin penetration of sunscreen nanoparticles: in-vitro assessment of a novel micronized zinc oxide formulation. *Skin Pharmacology and Physiology* **20**, 148–154.

Davis JM (2007) How to assess the risks of nanotechnology: learning from past experience. *J Nanoscience Nanotechnology* **7**, 402–409.

Delie F. 1998. Evaluation of nano- and microparticle uptake by the gastrointestinal tract. *Advanced Drug Delivery Reviews* **34**(2–3): 221–233.

Desai MP, Labhasetwar V, Amidon GL, Levy RJ. 1996. Gastrointestinal uptake of biodegradable microparticles: Effect of particle size. *Pharmaceutical research* **13**(12): 1838–1845.

de Vries IJM, Lesterhuis WJ, Barentsz JO, Verdijk P, van Krieken JH, Boerman OC, Oyen WJG, Bonenkamp JJ, Boezeman JB, Adema GJ *et al.* (2005) Magnetic resonance tracking of dendritic cells in melanoma patients for monitoring of cellular therapy. *Nature Biotechnology* **23**, 1407–1413.

Dix DJ, Houck KA, Martin MT, Richard AM, Setzer RW, Kavlock RJ (2007) The ToxCast program for prioritizing toxicity testing of environmental chemicals. *Toxicological Sciences* **95**, 5–12.

Dobrovolskaia MA, Aggarwal P, Hall JB, McNeil SE (2008) Preclinical studies to understand nanoparticle interaction with the immune system and its potential effects on nanoparticle biodistribution. *Molecular Pharmaceutics* **5**, 487–495.

Donaldson K, Stone V, Duffin R, Clouter A, Schins R, Borm P (2001) The quartz hazard: effects of surface and matrix on inflammogenic activity. *J Environ Pathol Toxicol Oncol* **20**(Suppl. 1), 109–118.

Donaldson K, Aitken R, Tran L, Stone V, Duffin R, Forrest G, Alexander A (2006) Carbon nanotubes: a review of their properties in relation to pulmonary toxicology and workplace safety. *Toxicological Sciences* **92**, 5–22.

Duffin R, Mills NL, Donaldson K (2007a) Nanoparticles – a thoracic toxicology perspective. *Yonsei Medical Journal* **48**, 561–572.

Duffin R, Tran L, Brown D, Stone V, Donaldson K (2007b) Proinflammogenic effects of low-toxicity and metal nanoparticles in vivo and in vitro: highlighting the role of particle surface area and surface reactivity. *Inhal Toxicol* **19**, 849–856.

Emerich DF, Thanos CG (2007) Targeted nanoparticle-based drug delivery and diagnosis. *J Drug Targeting* **15**, 163–183.

Environmental Protection Agency (2008a) Science to Achieve Results Nanotechnology Research Grants. Available at http://www.epa.gov/ncer/rfa/

Environmental Protection Agency (2008b) Integrated Risk Information System. Available at http://cfpub.epa.gov/ncea/iris/index.cfm

Environmental Protection Agency (2008c) *Draft Nanomaterial Research Strategy (NRS)*. Available at http://epa.gov/ncer/nano/publications/nano_strategy_012408.pdf

Environmental Protection Agency (2008d) Glossary of IRIS Terms. Available at http://www.epa.gov/IRIS/help_gloss.htm

Fabian E, Landsiedel R, Ma-Hock L, Wiench K, Wohlleben W, van Ravenzwaay B (2008) Tissue distribution and toxicity of intravenously administered titanium dioxide nanoparticles in rats. *Archives of Toxicology* **82**, 151–157.

Fischer HC, Liu LC, Pang KS, Chan WCW (2006) Pharmacokinetics of nanoscale quantum dots: In vivo distribution, sequestration, and clearance in the rat. *Adv Funct Mater* **16**, 1299–1305.

Florence AT (2005) Nanoparticle uptake by the oral route: Fulfilling its potential? *Drug Discovery Today: Technologies* **2**, 75–81.

Florence AT, Hillery AM, Hussain N, Jani PU (1995) Factors affecting the oral uptake and translocation of polystyrene nanoparticles – histological and analytical evidence. *J Drug Targeting* **3**, 65–70.

Florence AT, Sakthivel T, Toth I (2000) Oral uptake and translocation of a polylysine dendrimer with a lipid surface. *J Control Release* **65**, 253–259.

Fumelli C, Marconi A, Salvioli S, Straface E, Malorni W, Offidani AM, Pellicciari R, Schettini G, Giannetti A, Monti D *et al.* (2000) Carboxyfullerenes protect human keratinocytes from ultraviolet-B-induced apoptosis. *J Investigative Dermatology* **115**, 835–841.

Gamer AO, Leibold E, van Ravenzwaay B (2006) The in vitro absorption of microfine zinc oxide and titanium dioxide through porcine skin. *Toxicology in Vitro* **20**, 301–307.

Gaumet M, Vargas A, Gurny R, Delie F (2008) Nanoparticles for drug delivery: The need for precision in reporting particle size parameters. *Eur J Pharm Biopharm* **69**, 1–9.

Gopee NV, Roberts DW, Webb P, Cozart CR, Siitonen PH, Warbritton AR, Yu WW, Colvin VL, Walker NJ, Howard PC (2007) Migration of intradermally injected quantum dots to sentinel organs in mice. *Toxicol Sci* **98**, 249–257.

Grassian VH, O'Shaughnessy PT, Adamcakova-Dodd A, Pettibone JM, Thorne PS (2007) Inhalation exposure study of titanium dioxide nanoparticles with a primary particle size of 2 to 5 nm. *Environ Health Perspect* **115**, 397–402.

Gwinn MR, Vallyathan V (2006) Nanoparticles: health effects – Pros and cons. *Environ Health Perspect* **114**, 1818–1825.

Hagens WI, Oomen AG, de Jong WH, Cassee FR, Sips AJAM (2007) What do we (need to) know about the kinetic properties of nanoparticles in the body? *Regulatory Toxicology and Pharmacology* **49**, 217–229.

Hansen SF, Maynard A, Baun A, Tickner JA (2008) Late lessons from early warnings for nanotechnology. *Nature Nanotechnology* **3**, 444–447.

Hardman R (2006) A toxicologic review of quantum dots: toxicity depends on physicochemical and environmental factors. *Environ Health Perspect* **114**, 165–172.

Harris JM, Chess RB (2003) Effect of pegylation on pharmaceuticals. *Nature Reviews Drug Discovery* **2**, 214–221.

Hayes A (2001) *Principles and Methods of Toxicology*. Taylor & Francis: Philadelphia.

Hillery AM, Jani PU, Florence AT (1994) Comparative, quantitative study of lymphoid and nonlymphoid uptake of 60 nm polystyrene particles. *J Drug Targeting* **2**, 151–156.

Hillyer JF, Albrecht RM (2001) Gastrointestinal persorption and tissue distribution of differently sized colloidal gold nanoparticles. *J Pharmaceutical Sciences* **90**, 1927–1936.

Hodge H, Sterner J (1949) Tabulation of toxicity classes. *American Industrial Hygiene Association Quarterly* **10**, 93–96.

Hoet PH, Bruske-Hohlfeld I, Salata OV (2004) Nanoparticles – known and unknown health risks. *J Nanobiotechnology* **2**, 12.

Hoyt VW, Mason E (2008) Nanotechnology: Emerging health issues. *J Chem Health Safety* **15**, 10–15.

Huczko A, Lange H, Calko E (1999) Fullerenes: Experimental evidence for a null risk of skin irritation and allergy. *Fullerene Science and Technology* **7**, 935–939.

Huczko A, Lange H (2001) Carbon nanotubes: Experimental evidence for a null risk of skin irritation and allergy. *Fullerene Science and Technology* **9**, 247–250.

Hunter P (2008) A toxic brew we cannot live without – Micronutrients give insights into the interplay between geochemistry and evolutionary biology. *EMBO Reports* **9**, 15–U15.

Hussain N, Jaitley V, Florence AT (2001) Recent advances in the understanding of uptake of microparticulates across the gastrointestinal lymphatics. *Advanced Drug Delivery Reviews* **50**, 107–142.

ISO (2008a) *Nanotechnologies – Terminology and definitions for nano-objects – Nanoparticle, nanofibre and nanoplate*. ISO/TS 27687:2008. International Organization for Standardization: Geneva. Available at http://www.iso.org/iso/iso_catalogue/catalogue_tc/

ISO (2008b) *Nanotechnologies – Health and safety practices in occupational settings relevant to nanotechnologies*. ISO/TR 12885:2008. International Organization for Standardization: Geneva. Available at http://www.iso.org/iso/iso_catalogue/catalogue_tc/

James W, Hirsch L, West J, O'Neal P, Payne J (2007) Application of INAA to the build-up and clearance of gold nanoshells in clinical studies in mice. *J Radioanalytical Nuclear Chem* **271**, 455–459.

Jani P, Halbert GW, Langridge J, Florence AT (1990) Nanoparticle uptake by the rat gastrointestinal mucosa - quantitation and particle-size dependency. *J Pharmacy Pharmacology* **42**, 821–826.

Jia G, Wang HF, Yan L, Wang X, Pei RJ, Yan T, Zhao YL, Guo XB (2005a) Cytotoxicity of carbon nanomaterials: Single-wall nanotube, multi-wall nanotube, and fullerene. *Environmental Science Technology* **39**, 1378–1383.

Jia X, Li N, Chen J (2005b) *A subchronic toxicity study of elemental Nano-Se in Sprague-Dawley rats.* Life Sciences **76**, 1989–2003.

Kim YS, Kim JS, Cho HS, Rha DS, Kim JM, Park JD, Choi BS, Lim R, Chang HK, Chung YH *et al.* (2008) Twenty-eight-day oral toxicity, genotoxicity, and gender-related tissue distribution of silver nanoparticles in Sprague-Dawley rats. *Inhal Toxicol* **20**, 575–583.

Kohli AK, Alpar HO (2004) Potential use of nanoparticles for transcutaneous vaccine delivery: effect of particle size and charge. *Int J Pharmaceutics* **275**, 13–17.

Koo OM, Rubinstein I, Onyuksel H (2005) Role of nanotechnology in targeted drug delivery and imaging: a concise review. *Nanomedicine* **1**, 193–212.

Kreyling WG, Semmler M, Erbe F, Mayer P, Takenaka S, Schulz H, Oberdorster G, Ziesenis A (2002) Translocation of ultrafine insoluble iridium particles from lung epithelium to extrapulmonary organs is size dependent but very low. *J Toxicology Environ Health-Part A* **65**, 1513–1530.

Lademann J, Richter H, Teichmann A, Otberg N, Blume-Peytavi U, Luengo J, Weiss B, Schaefer UF, Lehr CM, Wepf R *et al.* (2007) Nanoparticles – an efficient carrier for drug delivery into the hair follicles. *Eur J Pharm Biopharm* **66**, 159–164.

Lai YL, Murugan P, Hwang KC (2003) Fullerene derivative attenuates ischemia-reperfusion-induced lung injury. *Life Sci* **72**, 1271–1278.

Lam CW, James JT, McCluskey R, Hunter RL (2004) Pulmonary toxicity of single-wall carbon nanotubes in mice 7 and 90 days after intratracheal instillation. *Toxicol Sci* **77**, 126–134.

Lam CW, James JT, McCluskey R, Arepalli S, Hunter RL (2006) A review of carbon nanotube toxicity and assessment of potential occupational and environmental health risks. *Critical Reviews in Toxicology* **36**, 189–217.

Lee C, MacKay J, Frechet J, Szoka F (2005) Designing dendrimers for biological applications. *Nature Biotechnology* **23**, 1517–1526.

Lee HA, Imran M, Monteiro-Riviere NA, Colvin VL, Yu WW, Riviere JE (2007) Biodistribution of quantum dot nanoparticles in perfused skin: evidence of coating dependency and periodicity in arterial extraction. *Nano Lett* **7**, 2865–2870.

Lewinski N, Colvin V, Drezek R (2008) Cytotoxicity of nanoparticles. *Small* **4**, 26–49.

Li JG, Li WX, Xu JY, Cai XQ, Liu RL, Li YJ, Zhao QF, Li QN (2007a) Comparative study of pathological lesions induced by multiwalled carbon nanotubes in lungs of mice by intratracheal instillation and inhalation. *Environmental Toxicology* **22**, 415–421.

Li Z, Hulderman T, Salmen R, Chapman R, Leonard SS, Young SH, Shvedova A, Luster MI, Simeonova PP (2007b) Cardiovascular effects of pulmonary exposure to single-wall carbon nanotubes. *Environ Health Persp* **115**, 377–382.

Liu Z, Cai WB, He LN, Nakayama N, Chen K, Sun XM, Chen XY, Dai HJ (2007) In vivo biodistribution and highly efficient tumour targeting of carbon nanotubes in mice. *Nature Nanotechnology* **2**, 47–52.

Lundqvist M, Sethson I, Jonsson BH (2004) Protein adsorption onto silica nanoparticles: conformational changes depend on the particles' curvature and the protein stability. *Langmuir* **20**, 10639–10647.

Lundqvist M, Stigler J, Elia G, Lynch I, Cedervall T, Dawson KA (2008) Nanoparticle size and surface properties determine the protein corona with possible implications for biological impacts. *Proc Natl Acad Sci USA* **105**, 14265–14270.

Lynch I, Dawson KA (2008) Protein-nanoparticle interactions. *Nano Today* **3**, 40–47.

Magrez A, Kasas S, Salicio V, Pasquier N, Seo JW, Celio M, Catsicas S, Schwaller B, Forro L (2006) Cellular toxicity of carbon-based nanomaterials. *Nano Lett* **6**, 1121–1125.

Maneerung T, Tokura S, Rujiravanit R (2008) Impregnation of silver nanoparticles into bacterial cellulose for antimicrobial wound dressing. *Carbohydrate Polymers* **72**, 43–51.

Maynard A (2006) *Nanotechnology: A Research Strategy for Addressing Risk*. Woodrow Wilson International Center for Scholars: Washington, DC.

Maynard AD, Kuempel ED (2005) Airborne nanostructured particles and occupational health. *J Nanoparticle Research* **7**, 587–614.

Maynard AD, Aitken RJ, Butz T, Colvin V, Donaldson K, Oberdorster G, Philbert MA, Ryan J, Seaton A, Stone V et al. (2006) Safe handling of nanotechnology. *Nature* **444**, 267–269.

Mercer RR, Scabilloni J, Wang L, Kisin E, Murray AR, Schwegler-Berry D, Shvedova AA, Castranova V (2008) Alteration of deposition pattern and pulmonary response as a result of improved dispersion of aspirated single-walled carbon nanotubes in a mouse model. *Am J Physiology* **294**, L87–97.

Mitchell LA, Gao J, Wal RV, Gigliotti A, Burchiel SW, McDonald JD (2007) Pulmonary and systemic immune response to inhaled multiwalled carbon nanotubes. *Toxicol Sci* **100**, 203–214.

Moghimi SM, Hunter AC (2001) Capture of stealth nanoparticles by the body's defences. *Crit Rev Ther Drug Carrier Syst* **18**, 527–550.

Morgan K. 2005. Development of a preliminary framework for informing the risk analysis and risk management of nanoparticles. *Risk Analysis* **25**, 1621–1635.

Mori T, Takada H, Ito S, Matsubayashi K, Miwa N, Sawaguchi T (2006) Preclinical studies on safety of fullerene upon acute oral administration and evaluation for no mutagenesis. *Toxicology* **225**, 48–54.

Moriguchi T, Yano K, Hokari S, Sonoda M (1999) Effect of repeated application of C-60 combined with UVA radiation onto hairless mouse back skin. *Fullerene Science and Technology* **7**, 195–209.

Morris J, Willis J (2007) *Nanotechnology White Paper*. Environmental Protection Agency: Washington, DC.

Mortensen L, Oberdorster G, DeLouise LA (2008) Nanoparticle skin penetration – effect of UVR. *J Investigative Dermatology* **128**, S208–S208.

Muller J, Huaux F, Moreau N, Misson P, Heilier JF, Delos M, Arras M, Fonseca A, Nagy JB, Lison D (2005) Respiratory toxicity of multi-wall carbon nanotubes. *Toxicol Appl Pharmacol* **207**, 221–231.

Muller RH, Radtke M, Wissing SA (2002) Solid lipid nanoparticles (SLN) and nanostructured lipid carriers (NLC) in cosmetic and dermatological preparations. *Advanced Drug Delivery Reviews* **54**, S131–S155.

National Academy of Sciences (1983) *Risk Assessment in the Federal Government: Managing the Process*. National Research Council, National Academy of Sciences: Washington, DC.

National Academy of Sciences (1994) *Science and Judgment in Risk Assessment*. National Research Council, National Academy of Sciences: Washington, DC.

National Academy of Sciences (2007) *Applications of Toxicogenomic Technologies to Predictive Toxicology and Risk Assessment*. National Research Council, National Academy of Sciences: Washington, DC.

National Nanotechnology Initiative (2008) *The National Nanotechnology Initiative: Strategy for Nanotechnology-related Environmental, Health, and Safety Research*. National Science and Technology Council. Available at http://www.nano.gov/NNI_EHS_Research_Strategy.pdf

Nelson MA, Domann FE, Bowden GT, Hooser SB, Fernando Q, Carter DE (1993) Effects of acute and subchronic exposure of topically applied fullerene extracts on the mouse skin. *Toxicology and Industrial Health* **9**, 623–630.

Nemmar A, Vanbilloen H, Hoylaerts MF, Hoet PH, Verbruggen A, Nemery B (2001) Passage of intratracheally instilled ultrafine particles from the lung into the systemic circulation in hamster. *Am J Respiratory Crit Care Medicine* **164**, 1665–1668.

Nemmar A, Hoylaerts MF, Hoet PHM, Nemery B (2004) Possible mechanisms of the cardiovascular effects of inhaled particles: systemic translocation and prothrombotic effects. *Toxicology Letters* **149**, 243–253.

Niidome T, Yamagata M, Okamoto Y, Akiyama Y, Takahashi H, Kawano T, Katayama Y, Niidome Y (2006) PEG-modified gold nanorods with a stealth character for in vivo applications. *J Control Release* **114**, 343–347.

NIOSH (2008) *Strategic Plan for NIOSH Nanotechnology Research: Filling the Knowledge Gaps.* Available at http://www.cdc.gov/niosh/topics/nanotech/strat_planINTRO.html

Nohynek GJ, Lademann J, Ribaud C, Roberts MS (2007) Grey goo on the skin? Nanotechnology, cosmetic and sunscreen safety. *Crit Rev Toxicol* **37**, 251–277.

Nordberg M (1998) Metallothioneins: historical review and state of knowledge. *Talanta* **46**, 243–254.

Oberdörster G, Maynard A, Donaldson K, Castranova V, Fitzpatrick J, Ausman K, Carter J, Karn B, Kreyling W, Lai D *et al.* (2005a) Principles for characterizing the potential human health effects from exposure to nanomaterials: elements of a screening strategy. *Part Fibre Toxicol* **2**, 8.

Oberdörster G, Oberdörster E, Oberdörster J (2005b) Nanotoxicology: an emerging discipline evolving from studies of ultrafine particles. *Environ Health Persp* **113**, 823–839.

OECD (2008a) *List of Manufactured Nanomaterials and List of Endpoints for Phase One of the OECD Testing Programme.* Organisation for Economic Co-operation and Development. Available at http://www.oecd.org/document/53/0,3343,en_2649_37015404_37760309_1_1_1_1,00.html

OECD (2008b) *OECD Working Party on Nanotechnology.* Organisation for Economic Co-operation and Development. Available at http://www.oecd.org/document/35/0,3343,en_21571361_41212117_42378531_1_1_1_1,00.html

Owens DE, Peppas NA (2006) Opsonization, biodistribution, and pharmacokinetics of polymeric nanoparticles. *Int J Pharm* **307**, 93–102.

Oyabu T, Ogami A, Morimoto Y, Shimada M, Lenggoro W, Okuyama K, Tanaka I (2007) Biopersistence of inhaled nickel oxide nanoparticles in rat lung. *Inhal Toxicol* **19**(Suppl. 1), 55–58.

Panessa-Warren BJ, Warren JB, Wong SS, Misewich JA (2006) Biological cellular response to carbon nanoparticle toxicity. *J Phys-Condens Mat* **18**, S2185–S2201.

Patton JS (1996) Mechanisms of macromolecule absorption by the lungs. *Advanced Drug Delivery Reviews* **19**, 3–36.

Poland CA, Duffin R, Kinloch I, Maynard A, Wallace WA, Seaton A, Stone V, Brown S, Macnee W, Donaldson K (2008) Carbon nanotubes introduced into the abdominal cavity of mice show asbestos-like pathogenicity in a pilot study. *Nature Nanotechnology* **3**, 423–428.

Qi KH, Chen XQ, Liu YY, Xin JH, Mak CL, Daoud WA (2007) Facile preparation of anatase/SiO_2 spherical nanocomposites and their application in self-cleaning textiles. *Journal of Materials Chemistry* **17**, 3504–3508.

Rajagopalan P, Wudl F, Schinazi RF, Boudinot FD (1996) Pharmacokinetics of a water-soluble fullerene in rats. *Antimicrobial Agents and Chemotherapy* **40**, 2262–2265.

Renn O, Roco MC (2006) Nanotechnology and the need for risk governance. *J Nanoparticle Research* **8**, 153–191.

Rohner F, Ernst FO, Arnold M, Hilbe M, Biebinger R, Ehrensperger F, Pratsinis SE, Langhans W, Hurrell RF, Zimmermann MB (2007) Synthesis, characterization, and bioavailability in rats of ferric phosphate nanoparticles. *J Nutr* **137**, 614–619.

Rouse JG, Yang JZ, Ryman-Rasmussen JP, Barron AR, Monteiro-Riviere NA (2007) Effects of mechanical flexion on the penetration of fullerene amino acid-derivatized peptide nanoparticles through skin. *Nano Letters* **7**, 155–160.

Rushmer RF, Buettner KJ, Short JM, Odland GF (1966) The skin. *Science* **154**, 343.

Ryman-Rasmussen JP, Riviere JE, Monteiro-Riviere NA (2006) Penetration of intact skin by quantum dots with diverse physicochemical properties. *Toxicol Sci* **91**, 159–165.

Sadauskas E, Wallin H, Stoltenberg M, Vogel U, Doering P, Larsen A, Danscher G (2007) Kupffer cells are central in the removal of nanoparticles from the organism. *Particle Fibre Toxicol* **4**, 10.

Sato Y, Yokoyama A, Shibata K, Akimoto Y, Ogino S, Nodasaka Y, Kohgo T, Tamura K, Akasaka T, Uo M *et al.* (2005) Influence of length on cytotoxicity of multi-walled carbon nanotubes against human acute monocytic leukemia cell line THP-I in vitro and subcutaneous tissue of rats in vivo. *Molecular Biosystems* **1**, 176–182.

Sayes CM, Gobin AM, Ausman KD, Mendez J, West JL, Colvin VL (2005) Nano-C-60 cytotoxicity is due to lipid peroxidation. *Biomaterials* **26**, 7587–7595.

Sayes CM, Wahi R, Kurian PA, Liu YP, West JL, Ausman KD, Warheit DB, Colvin VL (2006) Correlating nanoscale titania structure with toxicity: A cytotoxicity and inflammatory response study with human dermal fibroblasts and human lung epithelial cells. *Toxicol Sciences* **92**, 174–185.

Sayes CM, Marchione AA, Reed KL, Warheit DB (2007) Comparative pulmonary toxicity assessments of C60 water suspensions in rats: few differences in fullerene toxicity in vivo in contrast to in vitro profiles. *Nano Lett* **7**, 2399–2406.

Schierow L (1996) *The Role of Risk Analysis and Risk Management in Environmental Protection.* Congressional Research Service. Available at http://www.fas.org/spp/civil/crs/94-036.htm

Schipper ML, Cheng Z, Lee S-W, Bentolila LA, Iyer G, Rao J, Chen X, Wu AM, Weiss S, Gambhir SS (2007) MicroPET-based biodistribution of quantum dots in living mice. *J Nucl Med* **48**, 1511–1518.

Schulz J, Hohenberg H, Pflucker F, Gartner E, Will T, Pfeiffer S, Wepf R, Wendel V, Gers-Barlag H, Wittern KP (2002) Distribution of sunscreens on skin. *Adv Drug Deliver Rev* **54**, S157–S163.

Setzer R, Kimmel C (2003) Use of NOAEL, benchmark dose and other models for human risk assessment of hormonally active substances. *Pure Applied Chemistry* **75**, 2151–2158.

Sexton K, Reiter LW, Zenick H (1995) Research to strengthen the scientific basis for health risk assessment: a survey of the context and rationale for mechanistically based methods and models. *Toxicology* **102**, 3–20.

Shvedova AA, Castranova V, Kisin ER, Schwegler-Berry D, Murray AR, Gandelsman VZ, Maynard A, Baron P (2003) Exposure to carbon nanotube material: assessment of nanotube cytotoxicity using human keratinocyte cells. *J Toxicol Environ Health* **66**, 1909–1926.

Shvedova AA, Kisin ER, Mercer R, Murray AR, Johnson VJ, Potapovich AI, Tyurina YY, Gorelik O, Arepalli S, Schwegler-Berry D *et al.* (2005) Unusual inflammatory and fibrogenic pulmonary responses to single-walled carbon nanotubes in mice. *Am J Physiology* **289**, L698–708.

Singh R, Pantarotto D, Lacerda L, Pastorin G, Klumpp C, Prato M, Bianco A, Kostarelos K (2006) Tissue biodistribution and blood clearance rates of intravenoulsy administered carbon nanotube radiotracers. *PNAS* **103**, 3357–3362.

Singh S, Nalwa HS (2007) Nanotechnology and health safety – toxicity and risk assessments of nanostructured materials on human health. *Journal of Nanoscience and Nanotechnology* **7**, 3048–3070.

Sonavane G, Tomoda K, Sano A, Ohshima H, Terada H, Makino K. 2008. In vitro permeation of gold nanoparticles through rat skin and rat intestine: effect of particle size. *Colloids and Surfaces B-Biointerfaces* **65**, 1–10.

Stern ST, McNeil SE (2008) Nanotechnology safety concerns revisited. *Toxicological Sciences* **101**, 4–21.

Su CH, Sheu HS, Lin CY, Huang CC, Lo YW, Pu YC, Weng JC, Shieh DB, Chen JH, Yeh CS (2007) Nanoshell magnetic resonance imaging contrast agents. *J Am Chem Soc* **129**, 2139–2146.

Takagi A, Hirose A, Nishimura T, Fukumori N, Ogata A, Ohashi N, Kitajima S, Kanno J (2008) Induction of mesothelioma in p53+/− mouse by intraperitoneal application of multi-wall carbon nanotube. *J Toxicological Sciences* **33**, 105–116.

Tan MH, Commens CA, Burnett L, Snitch PJ (1996) A pilot study on the percutaneous absorption of microfine titanium dioxide from sunscreens. *Australasian J Dermatology* **37**, 185–187.

Tian FR, Cui DX, Schwarz H, Estrada GG, Kobayashi H (2006) Cytotoxicity of single-wall carbon nanotubes on human fibroblasts. *Toxicol in Vitro* **20**, 1202–1212.

Tinkle SS, Antonini JM, Rich BA, Roberts JR, Salmen R, DePree K, Adkins EJ (2003) Skin as a route of exposure and sensitization in chronic beryllium disease. *Environ Health Persp* **111**, 1202–1208.

van Leeuwen K, Vermeire T (1995) *Risk Assessment of Chemicals: An Introduction.* Kluwer Academic Publishers: Boston.

Vanrolleghem PA, Lee DS (2003) On-line monitoring equipment for wastewater treatment processes: state of the art. *Water Science and Technology* **47**, 1–34.

Wakefield G, Green M, Lipscomb S, Flutter B (2004) Modified titania nanomaterials for sunscreen applications – reducing free radical generation and DNA damage. *Materials Science and Technology* **20**, 985–988.

Wakefield J (2000) Human exposure assessment: Finding out what's getting in. *Environ Health Persp* **108**, A24–A26.

Wang B, Feng W-Y, Wang T-C, Jia G, Wang M, Shi J-W, Zhang F, Zhao Y-L, Chai Z-F (2006) Acute toxicity of nano- and micro-scale zinc powder in healthy adult mice. *Toxicology Letters* **161**, 115–123.

Wang H, Wang J, Deng X, Sun H, Shi Z, Gu Z, Liu Y, Zhao Y (2004) Biodistribution of carbon single-wall carbon nanotubes in mice. *J Nanoscience Nanotechnology* **4**, 1019–1024.

Wang J, Zhou G, Chen C, Yu H, Wang T, Ma Y, Jia G, Gao Y, Li B, Sun J *et al.* (2007) Acute toxicity and biodistribution of different sized titanium dioxide particles in mice after oral administration. *Toxicology Letters* **168**, 176–185.

Warheit DB, Laurence BR, Reed KL, Roach DH, Reynolds GAM, Webb TR (2004) Comparative pulmonary toxicity assessment of single-wall carbon nanotubes in rats. *Toxicological Sciences* **77**, 117–125.

Warheit DB, Brock WJ, Lee KP, Webb TR, Reed KL (2005) Comparative pulmonary toxicity inhalation and instillation studies with different TiO₂ particle formulations: impact of surface treatments on particle toxicity. *Toxicol Sci* **88**, 514–524.

Warheit DB, Borm PJA, Hennes C, Lademann J (2007) Testing strategies to establish the safety of nanomaterials: Conclusions of an ECETOC workshop. *Inhal Toxicol* **19**, 631–643.

Wilsdon J (2004) The politics of small things: nanotechnology, risk, and uncertainty. *Technology and Society Magazine, IEEE* **23**, 16–21.

Wood PJ (2003) Protective mechanisms of the body. *Anaesthesia & Intensive Care Medicine* **4**, 339–341.

Wunderbaldinger P, Josephson L, Weissleder R (2002) TAT peptide directs enhanced clearance and hepatic permeability of magnetic nanoparticles. *Bioconjugate Chem* **13**, 264–268.

Xia M, Huang R, Witt KL, Southall N, Fostel J, Cho MH, Jadhav A, Smith CS, Inglese J, Portier CJ *et al.* (2008) Compound cytotoxicity profiling using quantitative high-throughput screening. *Environ Health Persp* **116**, 284–291.

Xu R (2008) Progress in nanoparticles characterization: sizing and zeta potential measurement. *Particuology* **6**, 112–115.

Yamago S, Tokuyama H, Nakamura E, Kikuchi K, Kananishi S, Sueki K, Nakahara H, Enomoto S, Ambe F (1995) In vivo biological behavior of a water-miscible fullerene: 14C labeling, absorption, distribution, excretion and acute toxicity. *Chemistry & Biology* **2**, 385–389.

Yang RH, Chang LW, Wu JP, Tsai MH, Wang HJ, Kuo YC, Yeh TK, Yang CS, Lin P (2007) Persistent tissue kinetics and redistribution of nanoparticles, quantum dot 705, in mice: ICP-MS quantitative assessment. *Environ Health Persp* **115**, 1339–1343.

Yang W, Peters JI, Williams RO (2008) Inhaled nanoparticles – a current review. *Int J Pharmaceutics* **356**, 239–247.

Zhang LW, Yu WW, Colvin VL, Monteiro-Riviere NA (2008) Biological interactions of quantum dot nanoparticles in skin and in human epidermal keratinocytes. *Toxicol Applied Pharmacol* **228**, 200–211.

Zhang Y, Chen W, Zhang J, Liu J, Chen G, Pope C (2007) In vitro and in vivo toxicity of CdTe nanoparticles. *J Nanoscience Nanotechnology* **7**, 497–503.

Zhu H, Han J, Xiao JQ, Jin Y (2008) Uptake, translocation, and accumulation of manufactured iron oxide nanoparticles by pumpkin plants. *J Environmental Monitoring* **10**, 713–717.

23

Strategies for Risk Assessment of Nanomaterials

Hae-Seong Yoon, Hyun-Kyung Kim, Dong Deuk Jang and Myung-Haing Cho

23.1 Introduction

'There is plenty of room at the bottom' is an often-quoted prophetic remark of physicist Richard Feynman (1960), who speculated on the potential of nanotechnology. The manipulation of individual atoms can create very small structures with properties significantly different from the bulk material with the same composition. Forty years after this statement was made, nanotechnology has emerged and progressed rapidly. More than 600 manufacturer-identified nanotechnology-based consumer products are currently available on the market (Woodrow Wilson International Center for Scholars, 2008). Nanotechnology is considered one of the most significant innovations and milestones to enhance our quality of life since the Industrial Revolution (Gwinn and Vallyathan, 2006). However, nanotechnology presents new challenges in terms of understanding and evaluating its potential adverse health effects. Indeed, widespread applications of a new class of high-tech materials have raised concerns regarding the occupational, consumer, environmental safety and human health impacts (Hood, 2004).

The term 'nanomaterial' refers to materials with one or more components with at least one dimension <100 nm. Nanomaterials include nanoparticles as a subset of materials that are defined as a single particle with a diameter <100 nm. Manufactured nanoparticles are particles that are engineered/manufactured intentionally with specific physico-chemical compositions to achieve specific mechanical, optical, electrical and magnetic properties (Borm *et al.*, 2006).

Most nanomaterials currently used can be organized into four types: carbon-based, metal-based, dendrimers and composites (Environmental Protection Agency, 2007).

Nanotoxicity: From In Vivo *and* In Vitro *Models to Health Risks*   Edited by Saura Sahu and Daniel Casciano
© 2009 John Wiley & Sons, Ltd

Carbon-based materials are found in the form of hollow spheres, ellipsoids or cylinders. Spherical and ellipsoidal forms are called fullerenes, while the cylindrical forms of carbon nanomaterials are referred to as nanotubes. Metal-based nanomaterials can be categorized into quantum dots, nanogold, nanosilver and metal oxides, such as titanium dioxide. A quantum dot is a packed spherical nanocrystal composed of hundreds to thousands of atoms with a size varying from a few nanometers to a few hundred nanometers. Quantum dots have optical properties and the capacity to form covalent bonds with low-weight molecules such as peptides, antibodies and nucleic acids, providing advantages for probes in diagnostic medical imaging as well as in a variety of treatments. Dendrimers are nano-sized polymers based on branched units of unspecified chemistry. These materials have been utilized in catalysis on account of their numerous chain ends with specific chemical functions. In addition, in three-dimensional structures, dendrimers contain interior cavities into which certain molecules could be fitted, which can provide a useful characteristic for drug delivery. Nanocomposites combine nanoparticles with other nanoparticles or with larger, bulk-type materials to produce new composites with interesting properties (Environmental Protection Agency, 2007).

23.2   Risk Assessment of Nanomaterials

Health risk assessments are essential for the introduction of new technologies and the application of new classes of materials. Regarding nanotechnology, workers, including researchers and manufacturers, handling nanomaterials are at greatest risk of the adverse effects of nanomaterials. However, as the applications of nanomaterials grow, the risk of exposure to the general public is also on the rise (Borm *et al.*, 2006). Fundamentally, risk assessment of nanomaterials follows the paradigm of traditional risk assessment for conventional chemicals (National Research Council, 1983, 1994; Environmental Protection Agency, 2007). However, the experimental data and research on nanomaterials over a decade have supported the idea that traditional methodologies for risk assessment may be either insufficient or unsuitable for assessing nanomaterials, mainly due to their novel chemical and physical properties. Accordingly, new strategies and methodologies for assessing the risk of nanomaterials are needed. The development of the methodologies for nanomaterials should consider several factors, including:

(1) surface area and number of particles per unit volume;
(2) novel physico-chemical properties;
(3) physico-chemical parameters for dose metrics to determine the dose-response relationship for toxicity;
(4) agglomeration and disagglomeration states;
(5) coating and impurities on or within the surface of nanoparticles;
(6) biological processes, such as translocation, cellular uptake and toxicological mechanisms;
(7) reference materials for evaluating nanoparticles (Scientific Committee on Emerging and Newly Identified Health Risks, 2007).

23.2.1   Physico-chemical Properties Considered for Risk Assessment

The physico-chemical properties of nanomaterials are unique and significantly different from micrometer-sized materials with an identical chemical composition. Characterization of the physico-chemical properties is an important step in assessing the risks of nanomaterials. A wide range of physical and chemical properties should be considered when characterizing the hazards of nanomaterials. These include the chemical composition, surface chemistry, particle size, size distribution, shape, surface area, surface charge, surface morphology, surface coating, rheology, porosity, crystallinity and amorphocity, stoichiometry, dissolution kinetics and solubility, hydrophilicity or hydrophobicity, primary nanoparticles, aggregates and impurities (Oberdörster *et al.*, 2005b; Scientific Committee on Consumer Products, 2007; Scientific Committee on Emerging and Newly Identified Health Risks, 2007).

There is a critical need to understand which physico-chemical properties of nanomaterials are important for determining their toxicity or impact on human health. The particle size and surface area have been well characterized and are known to have an effect on the pulmonary toxicity of nanoparticles (Borm *et al.*, 2006). Three main factors, chemical composition, potential adhesion to cells or cellular incorporation, and forms of nanoparticles, contribute to the toxicity of nanoparticles absorbed into the body (Kirchner *et al.*, 2005). Based on the literature, a number of factors are likely to influence the toxicity of nanoparticles: particle number and size distribution, dose of particles to the target tissue, surface treatment, aggregate/agglomerate state, surface charge, particle shape and/or electrostatic attraction potential, and method of particle synthesis.

23.2.2   Toxicity

Two lines of research have been proposed for risk assessment of nanomaterials (Balbus *et al.*, 2007): (1) to generate data on nanomaterials for workers or the general public being significantly exposed; and (2) to develop more appropriate methods for accurate and efficient testing of the hazards and exposure. A tiered approach should help to achieve more efficient hazard screening, but there are some challenges that exist in toxicity studies, such as extrapolating between species, including animals to humans, using in *vitro* assays to predict the *in vivo* toxicity, and assessing the inter-individual variability (Environmental Protection Agency, 2008).

Standardization should be achieved in several areas related to toxicity studies, including the dose metrics, methods of toxicity tests, and reference materials. Standardized toxicity tests are required to assess the risk of nanomaterials. Although there is some controversy regarding the most relevant toxicity tests, there appears to be consensus in that a battery of short-term *in vitro* and *in vivo* tests will be necessary. Reference materials must be developed and used to facilitate the appropriate interpretation of the test results and compare the data from individual tests. There are six major characteristics that need to be determined for reference/test materials: the aerodynamic equivalent diameter, absolute length, specific surface area, number of particles per unit mass, concentration of bulk and/or surface contaminants, and polymorphic composition (Environmental Protection Agency, 2007). In addition, a reference bank of the materials is needed for the development of standardized well-characterized reference nanomaterials. Three main criteria have been suggested for the selection of nanomaterials contained in the bank: (1) a high production volume of industrial nanomaterials based on the scale of production and likelihood of exposure; (2) hypothesis-driven, based on how their physico-chemical properties are expected to interact with a biological system, to provide useful answers on particular toxicology questions; and (3) a selection based on standardized comparative studies with distributed analysis (Scientific Committee on Emerging and Newly Identified Health Risks, 2007).

23.2.2.1   A Multi-tiered Strategy for Assessing Health Effects of Nanomaterials

A multi-tiered approach is recommended for evaluating the health effects of nanomaterials, and is driven by several critical factors: the diversity of nanomaterials; the cost and availability of nanomaterials; and the need to develop alternative approaches, assays or methods that predict the health effects resulting from exposure to nanomaterials (Environmental Protection Agency, 2008).

Tier 1: In vitro *studies.* In vitro studies form an essential part of risk assessment-directed research paradigms and are also vital components of all tiered approaches for assessing the toxicity of nanomaterials (Holsapple *et al.*, 2005; Oberdörster *et al.*, 2005b; Nel *et al.*, 2006). *In vitro* studies are used for the following: to screen and rank the relative toxicities of a variety of nanomaterials rapidly; to conduct rapid comparative toxicity studies between nanomaterials and the corresponding bulk size materials; and conduct rapid assessment of the changes in nanomaterial toxicity following their interactions with environmental media. In addition, *in vitro* tests can perform ADME (absorption, distribution, metabolism and excretion) studies at the cellular and intracellular levels, and evaluate the carcinogenic, pulmonary, immunological, neurological, reproductive, cardiovascular and developmental effects of nanomaterials *in vitro* (Environmental Protection Agency, 2008).

These studies help us to understand the modes of action and mechanisms of toxicity at the molecular level, and are useful for obtaining mechanistic information on the nanomaterials of interest. For these purposes, studies using the cells/tissues of initial contact (respiratory tract, gastrointestinal tract, skin) and cells/tissues from the potential target organs are appropriate (Environmental Protection Agency, 2008). The study designs should reflect the *in vivo* circumstances as closely as possible.

Tier 2: In vivo *studies.* Tier 2 studies can examine the animal/*in vivo* toxicity and biokinetics of nanomaterials. These *in vivo* studies are generally guided by the results from *in vitro* studies concerning the prioritizing or ranking of nanomaterials and designing studies, such as appropriate nanomaterial exposure concentrations and the health endpoints to monitor. *In vivo* studies can provide information on the cancer, pulmonary, dermal and gastrointestinal toxicities related to the initial deposition of nanomaterials by various exposure routes. In addition, these studies can evaluate the immunological, neurological, reproductive, cardiovascular and developmental toxicity to determine the systemic toxicity of nanomaterials. The results from *in vivo* studies can be compared with those from *in vitro* studies to identify those *in vitro* assays that best correlate with the *in vivo* toxicity or health effects of nanomaterials (Environmental Protection Agency, 2008).

Tier 3: Physico-chemical characterization studies. Tier 3 studies examine the physical and chemical properties of nanomaterials that are associated with their *in vitro* and *in vivo* toxicity. These studies use non-cellular methods to evaluate the surface reactivity of nano-materials, as well as understand their interactions with biological molecules and/or fluids in order to determine the surface properties and interactions that most affect their *in vitro* and *in vivo* biokinetics (Environmental Protection Agency, 2008). The surface of a single nanoparticle can be analyzed using the following techniques: zeta potentials, secondary ion mass spectroscopy, X-ray photoelectron spectroscopy, thermogravimetry, atomic force microscopy and scanning tunnelling microscopy, and particle surface reactivity (Borm et al., 2006).

23.2.2.2   Translocation (ADME)

Nanomaterials can translocate from the portal of entry to the blood and other organs (Scientific Committee on Emerging and Newly Identified Health Risks, 2007). Such translocation is essential for determining the ultimate toxicity and target organs of nanomaterials. In the case of systemically absorbed materials, the liver might be the primary translocation site, with spleen, bone marrow, heart, kidney, bladder and brain as the secondary sites of organs (Nemmar et al., 2001, 2002; Takenaka et al., 2004). Although nanomaterials generally have three possible portals of entry – the lungs, skin and gut – they can also enter the body through the nasal epithelium and be transported by the olfactory nerve. Translocation can be affected by various physico-chemical properties, including size, surface charge and shape.

The major portal of entry for airborne particles is the respiratory tract. Insoluble particles deposit at sites throughout the respiratory tract, which depends on their aerodynamic behavior and size, and then may be cleared by macrophages and the mucociliary system (Scientific Committee on Emerging and Newly Identified Health Risks, 2007). However, certain particles can enter the interstitium of the lungs and cause an interstitial inflammatory response, subsequently transferring to the draining lymph nodes (Elder et al., 2006). The soluble components of particles may gain access to the blood. The organic components released from particles may be metabolized in the liver or lungs. Nanoparticles can access cellular tissue locations that are not accessible to larger particles, and have been found in the mitochondria (Li et al., 2003) and nucleus (Chen and von Mickecz, 2005). Nanoparticles may enter via the olfactory and trigeminal nerve endings found in the nose and naso-pharynx and also enter the blood stream, either directly through the lung or possibly the lymphatic system. The fate of nanoparticles in the blood is not completely understood. However, uptake by the reticulo-endothelial system of the spleen, bone marrow and liver sinusoids has been demonstrated.

The exposure (dose) and toxicokinetics of a substance, which is expressed as absorption, distribution, metabolism and excretion (ADME), are very important for assessing the effects. The general framework of ADME studies of nanomaterials should be similar (Balbus et al., 2007). However, analytical techniques for detecting nanomaterials and their byproducts when metabolized in biological systems are quite challenging. Therefore, toxicokinetic studies are hampered by the difficulty in tracing the nanomaterials in the body using the current detection techniques (Scientific Committee on Emerging and Newly Identified Health Risks, 2007).

Imaging techniques, such as real-time *in vivo* imaging, can be used for certain nanoma-terials with special properties, such as quantum dots and magnetic nanocrystals. However, such imaging techniques are not generally quantitative and should be confirmed by other analyses. Currently, nanoparticles that lack intrinsic fluorescence or other properties that aid visualization pose an enormous challenge for assessing the ADME characteristics (Scientific Committee on Emerging and Newly Identified Health Risks, 2007).

23.2.2.3   Hazard Identification

The toxicological tests required for hazard identification of nanomaterials are not well established and await validation. Nanomaterials can elicit a variety of tissue responses such as cell activation, generation of reactive oxygen species, inflammation and cell death (Ahn *et al.*, 2005; Chen *et al.*, 2006; Xia *et al.*, 2006). The primary mechanism of action by inhalation or dermal routes of exposure appears to be free radical generation and oxidative stress associated with surface reactivity (Oberdörster *et al.*, 2005a). For example, the oxidative stress associated with TiO_2 nanoparticles results in early inflammatory responses, such as an increase in polymorphonuclear cells, impaired macrophage phagocytosis, and/or fibroproliferative changes in rodents (Bermudez *et al.*, 2004).

With respect to the mutagenicity, genotoxicity and carcinogenicity of nanoparticles, it is important to be cautious when interpreting and extrapolating the experimental data of nanomaterials, especially with an *in vitro* investigation. For blood, the markers of thrombosis and atherogenesis need to be considered. In addition, the potential degenerative effects and oxidative stress on the brain should be assessed for hazard identification (Scientific Committee on Emerging and Newly Identified Health Risks, 2007).

23.2.2.4   Dose-Response Relationship (Dose Metrics)

Three dose metrics – particle number, surface area, and particle mass – have been proposed as appropriate metrics for assessing the toxicity or health impact (Environmental Protection Agency, 2007). The relevant inhalation dosimetry in risk assessment of nanomaterials might be the surface area or particle number rather than the mass per volume or per body weight, even though there is a great deal of complexity in the other properties that make generalization to all nanomaterials difficult (Yamamoto *et al.*, 2003; Tsuji *et al.*, 2006). Although there is some debate, it is believed that the particle surface area is a more appropriate dose metric for evaluating the dose-response relationships than the particle mass or particle number.

23.2.3   Exposure Assessment

Exposure assessment is the process of quantitatively and/or qualitatively evaluating intake and/or uptake of potential hazardous materials in the population of interest from food, water, air, soil and other sources (World Health Organization, 2005). In order to conduct exposure assessment, it is necessary to have sufficient information on the sources, amounts and routes of exposure of the chemical of interest to the population of interest. Exposure assessment of nanomaterials adopts the same scheme used to assess conventional chemicals. However, the physico-chemical properties of nanomaterials determine their own unique behavior, which can affect the pattern of exposure and toxicity in both living organisms and the environment. Accordingly, individual characteristics including size, surface area, charge and solubility also need to be considered when evaluating exposure to nanomaterials more accurately.

23.2.3.1 Nanomaterial Exposure Sources

Most nanomaterials consist of metals, metal oxides, silicon and carbon, which form different types of structures, as described below.

Quantum dots. Quantum dots are considered artificial atoms, which form nanometer-sized structures made from materials such as silicon (Royal Society, 2004). They behave like atoms and exhibit quantum effects when their size is small enough (Alivisatos, 1996). Quantum dots can emit light at different wavelengths depending on their size (Gao *et al.*, 2005). Therefore, quantum dots are used not only in optical devices but also in composites, solar cells and biological labels (Fu *et al.*, 2005; Weng and Ren, 2006; Li *et al.*, 2007).

Nanofilms. Nanofilms are one-dimensional thin layers, which are applied in the manufacture of electronic devices and in chemistry (Royal Society, 2004). Because nanofilms possess a large surface area and/or specific reactivity, their enhanced activity and selectivity can result in economic and resource savings. Therefore they are applied routinely to the production of fuel cells and catalysts. Nanofilms have potential uses in medical fields, such as coatings for medical implant devices, scaffolds for tissue engineering, coatings for targeted drug delivery, and artificial cells for oxygen therapeutics (Srivastava and McShane, 2005; Haynie *et al.*, 2006; Nakahara *et al.*, 2007).

Nanotubes. Carbon nanotubes are long, thin cylindrical atomic layers of graphite. Two types of carbon nanotubes are available, single-walled and multi-walled. Single-walled carbon nanotubes (SWCNTs) have a single cylindrical wall, whereas multi-walled nanotubes (MWCNTs) consist of cylinders within cylinders. Carbon nanotubes have extremely high mechanical strength and electric conductivity. Therefore, they are mainly used in electronics and optoelectronics (Royal Society, 2004; Bandaru, 2007). Inorganic nanotubes and inorganic fullerene-like materials were discovered soon after the development of carbon nanotubes, and exhibit excellent lubricating properties and catalytic reactivity (Tenne and Rao, 2004). In addition, inorganic nanotubes are attractive vehicles for drug/gene delivery because of their hollow structures with pores and ease of surface functionalization (Son *et al.*, 2007).

Nanowires. Nanowires are self-assembled ultrafine wires composed of quantum dots. Nanowires of silicon, gallium nitride and indium phosphide have excellent optical, electronic and magnetic properties and are used in semiconductors (Fan *et al.*, 2006; Lew Yan Voon *et al.*, 2008). Nanowires have promising applications in high-density data storage media as well as in electronic/optoelectronic devices (Varfolomeev *et al.*, 2005; Tang *et al.*, 2008; Wang, 2008).

Nanoparticles. Nanoparticles are tiny particles <100 nm in diameter. Manufactured nanoparticles usually have different characteristics from larger particles of the same material. For example, unlike larger particles, nanoparticles of titanium dioxide and zinc oxide are transparent in the visible spectrum but absorb UV light, making them suitable as sunscreen ingredients (Nohynek *et al.*, 2007; Peng *et al.*, 2008). Nanoparticles also have a large surface area, and can enhance the reactivity as an efficient catalyst (Wang, 2006; Kim et al., 2007; Liu et al., 2008).

Fullerenes. Fullerenes have the structure of an elongated sphere of carbon atoms formed by interconnecting six- and five-member rings. The first isolated fullerene, C_{60}, is composed of 60 carbon atoms, containing 20 hexagons and 12 pentagons with the configuration of a closed cage. Fullerenes have magnetic properties that have great potential for high-density magnetic storage media and electronic circuits (Tang et al., 2007). Fullerenes are also used as ball bearings to lubricate surfaces and vehicles for drug delivery (Bakry et al., 2007; Dutta, 2007).

Dendrimers. Dendrimers are spherical polymeric molecules that can carry out specific chemical functions on account of the numerous chain ends on their surface. Moreover, dendrimers contain interior cavities that are capable of holding and carrying other molecules. The main applications of dendrimers are coatings and inks. Dendrimers can also play a role of nanoscale carrier, such as in drug delivery systems (Agarwal et al., 2008; Qian and Yang, 2008; Bonacucina et al., 2009).

23.2.3.2   Commercial Products using Nanotechnology

Sunscreens and cosmetics. Sunscreens and cosmetics currently comprise one of the largest inventories of nanomaterial-containing products. Nanoparticles of titanium dioxide or zinc oxide are transparent in the visible spectrum, but absorb and reflect ultraviolet rays. Therefore, these nanoparticles are attractive ingredients for sunscreen products (Nohynek et al., 2007; Peng et al., 2008). Nanosized iron oxide is used as a coloring agent in some lipsticks, and nanosized chitosan is added to hair conditioners and skin creams to enhance absorption. Many cosmetics contain nanoparticles of active substances, such as coenzyme 10, oils and vitamins, which are believed to penetrate skin more readily.

Textiles. Nanoscience leads to the production of stain- and wrinkle-resistant clothing, as well as wind-, water- and shrink-proof fabrics and hydrophilic wear. A variety of products have been manufactured from fabrics possessing these characteristics, such as pants, jackets, shirts, coats, neckties and gloves. Socks, sleepers, shoe pads and underwear containing silver nanoparticles have been released into the market on account of the antibacterial and antifungal properties of silver.

Personal care. Hair dryers and irons employ the benefits of nanotechnology, in that nano-sized zirconia enhances the ion conductivity and heat resilience. and nano-silver prevents growth of bacteria and fungi. Soap is made with nano-germanium or nano-silver to improve the cleansing efficiency. Nano-silver and nano-gold are used in the production of toothpaste and toothbrushes on account of their ability to disinfect bacteria in the mouth. Hair-care products, such as shampoo and conditioner, also employ nanotechnology by using silver or active ingredients.

Sporting goods. A high-performance ski wax containing nanoparticles is used to generate a hard and fast-gliding surface. Tennis rackets with carbon nanotubes enhance the torsion and flex resistance, giving more power through rigidity. Long-lasting tennis balls, which are made by coating the inner core with clay polymer nanocomposites, have twice the lifetime of conventional balls.

Health. Wound dressings, which maintain safe bactericidal concentrations of silver with nanocrystalline technology, can yield antimicrobial barrier protection. Mosquito repellent sprays containing a microcapsulated fragrance can maintain the effect of the repellent for a long time. Nanotechnology has been applied to filtration systems to increase the intrinsic filtration efficiency, resulting in the improved elimination of potentially harmful airborne contaminants.

Food and beverages. There are very few food products using nanotechnology on the market. Nanodrops of canola oil have been developed as a liquid carrier that allows healthy components, such as vitamins, minerals and phytochemicals, to pass through the digestive system effectively. The nano-grade tea powder can release all the essences effectively, thereby boosting adsorption and the annihilation of viruses, free radicals, cholesterol and blood fat. In addition, selenium in nano-tea can be supplied ten times more efficiently to the body than conventional infusion methods. A chocolate drink powder consisting of nano-sized clusters can contain nutritional supplements and reduce surface tension, yielding increased wetness and nutrient absorption. Nano-silver coatings are used in the manufacture of antibacterial kitchen and tableware, as well as non-stick coated pans.

Electronics and computers. Computer hardware, such as flash memory and hard disk drive, is built using nano-sized silicon and copper wiring technology. Nano-silver mobile phones, which remain free of bacteria and odor, are available on the market. Nanomaterials are also used in batteries to hold high energy or increase battery life. Nano-silver technology is used in large kitchen appliances, laundry care products, air conditioners, and air purifiers to prevent bacterial infections.

Glasses. Windows coated with titanium dioxide are highly water-repellent and antibacterial, and coatings based on nanoparticulate oxides can catalytically destroy chemical agents. Nano-scale layers on glass surfaces are wear- and scratch-resistant, and improve adhesion.

Cutting tools. Cutting tools made from nanomaterials of tungsten carbide, tantalum carbide or titanium carbide have superior wear- and erosion-resistance. An example is a drill to bore holes in circuit boards.

23.2.3.3 Environmental Sources of Nanomaterials

Nanoparticles also exist in nature, such as the products from volcanic eruptions and forest fires. However, the intentional application of nanomaterials is increasing in a variety of fields, and there is more potential for the environmental release of nanomaterials. The potential sources of release include every procedure in the manufacture, storage, transport, consumption, disposal and recycling of products containing nanomaterials. Nanosized particles can deposit and accumulate in the air, soil and water, and can be transported to different media.

Depending on the chemical and physical characteristics, the behavior of nanomaterials in the environment is different. For example, airborne particles with a small size (diameters <80 nm) agglomerate rapidly to form larger particles, whereas large particles (>2000 nm) tend to settle (Aitken *et al*, 2004). Intermediate-sized particles (80 nm < diameter < 2000 nm) remain suspended in the air for the longest time, which means that they may have a greater chance of being inhaled (Bidleman, 1988; Preining, 1998). The size distribution of nanomaterials in the air may depend upon the time elapsed after their release, as well as on the atmospheric conditions. Therefore, all these characteristics need to be evaluated in order to determine the biologically relevant exposure.

Nanomaterials can be adsorbed in the soil more strongly than larger particles on account of the large surface area (Environmental Protection Agency, 2007). On the other hand, nanosized particles can move further before being adsorbed on the soil because they are small enough to pass through the spaces between soil particles. The soil properties can also affect the behavior of nanomaterials.

Nanomaterials in water may have a different fate depending on their properties, such as solubility, dispersability, and interactions with the aquatic environment. Nanoparticles may settle on the bottom of a water environment more slowly than larger particles, thereby increasing the possibility of exposure to the underwater biota (Oberdörster *et al.*, 2005a). Nanomaterials in water may undergo hydrolysis and/or photolysis. However, some nanoparticles, such as fullerenes, can be stabilized in water because they build up aqueous colloids containing nanocrystalline aggregates (Fortner *et al.*, 2005). This reaction may delay the removal of nanoparticles from aquatic environments, which can increase the likelihood of exposure.

The mechanism for abiotic and biotic degradation of nanomaterials is not completely understood. Hydrolysis or photolysis can degrade nanoparticles in aqueous conditions. Biodegradation may be affected by the chemical and physical properties of nanoparticles as well as by the environmental conditions. Nanomaterials composed of metals or ceramics are unlikely to be biodegraded. On the other hand, fullerenes have been reported to be degraded by wood decay fungi (Filley *et al.*, 2005). The degraded products may have different toxicity from the original compounds. For example, some types of quantum dots, which were degraded under photolytic and oxidative conditions, were transformed into a metalloid core with greater toxicity (Hardman, 2006). Therefore, further research will be needed in this area.

The uptake of nanomaterials by bacteria or other microorganisms indicates the potential for bioaccumulation through the food chain (Biswas and Wu, 2005). However, not enough is known to draw meaningful conclusions on the bioaccumulation of nanomaterials.

23.2.3.4   Routes of Exposure

Inhalation. Nanomaterials in the air are inhaled and deposited in the respiratory tract. Depending on their properties, nanomaterials inhaled from the air can reach the deep lung. Usually, inhaled molecules can be taken up by cells, such as macrophages and leukocytes, which is a process the human body uses to remove extraordinary invaders. However, molecules small enough to pass through the cell membrane can enter the cell and interfere with the cellular functions (Renwick *et al.*, 2001). A rat lung exposure model showed that titanium dioxide particles with a diameter <100 nm crossed the cellular membranes and did not elicit phagocytic mechanisms (Geiser *et al.*, 2005).

The amount of particles likely to be deposited in the lungs increases with decreasing particle size (Royal Society, 2004). However, the overall number, surface area and specific properties of individual nanoparticles can be other critical points that affect exposure and toxicity. For example, nanotubes have a fibrous shape and nanometer dimensions similar to asbestos. Therefore, they are expected to reach the deep lung without dissolution when inhaled. However, single-walled nanotubes tend to clump into large masses and are unlikely to be dispersed in the air (Maynard et al., 2004; Warheit et al., 2004). Despite the similar structure and size, the exposure and toxicity pattern of nanotubes may be different from those of asbestos because of their unique properties.

Many studies have shown that nanoparticles can enter the bloodstream from the lungs and be translocated to other organs (Takenaka et al., 2001; Nemmar et al., 2002; Oberdörster et al., 2002; Kwon et al., 2008). A fraction of inhaled nanosized titanium dioxide was transported from the airway lumen to the connective tissues, and subsequently released into the systemic circulation of rats (Muhlfeld et al., 2007). Another animal study showed that nanomaterials can be translocated to secondary organs, such as liver, spleen and heart, even though the inhaled particles were mainly cleared through feces (Kreyling et al., 2002). Nanosized particles inhaled in humans were also shown to be translocated into the bloodstream and liver (Brown et al., 2002; Nemmar et al., 2002). Moreover, the nanoparticles deposited in the nasal region may enter the brain by transport through the olfactory nerves (Hunter and Dey, 1998; Kreyling et al., 2002; Oberdörster et al., 2004, 2005a; Elder et al., 2006). Kleinman et al. (2008) reported that inhaled ultrafine particles affected the inflammatory processes of the central nervous system through the MAP kinase signalling pathways. Peters et al. (2006) suggested that inhaled ultrafine particles might have neurodegenerative consequences.

Ingestion. Nanoparticles of silicates, titanium dioxide or silver are ingested through foods and toothpaste, at the level of 10^{12}–10^{14} particles (Lomer et al., 2004). Depending on the size, surface charge, shape and other properties, the ingested particles may be translocated from the gastrointestinal (GI) tract to the blood. Eventually, the blood-circulated nanomaterials are likely to be retained in other organs, such as the liver, spleen and kidney (Jani et al., 1994; Wang et al., 2007). An animal study demonstrated that the transported nanomaterials from the GI tract might induce hepatic- and nephro-toxicity in a rat (Wang et al., 2007). However, Semmler et al. (2004) reported that ultrafine metal particles administered esophageally were fully excreted in the feces. Therefore, not all nanomaterials ingested appear to accumulate in the body.

Dermal exposure. Despite the increasing market for cosmetics containing nanomaterials, the potential for dermal penetration of nanoparticles is still controversial. Size and charge appear to affect the transport of nanoparticles through the skin (Kohli and Alpar, 2004; Ryman-Rasmussen et al., 2006). The penetration of small particles, such as quantum dots with a 7 nm diameter, into the dermis has been demonstrated (Ryman-Rasmussen et al., 2006). The transported nanomaterials can encounter dendritic cells and enter the blood circulation through the lymph nodes (Kohli and Alpar, 2004). However, nanosized titanium dioxide in sunscreens was deposited on the outermost surface of the stratum corneum but was not detected in the deeper stratum corneum layers, which suggests that it does not penetrate the dermis (Pflucker et al., 1999, 2001; Nohynek et al., 2007).

Other routes. Parenteral exposure can occur from the use of medical products containing nanoparticles, and ocular exposure is possible when nanoparticles in the air enter the eyes.

23.2.3.5   Exposure Scenarios

Workplace. Airborne and dermal exposure to nanoparticles is likely to occur during the generation of the products containing nanomaterials. In general, the production process for nanoparticle-containing goods is well established. Therefore, it is relatively safe from the risk of nanomaterial exposure, except in the case of an accident, such as leakage from a reactor. The subsequent steps, such as filling, pouring, bagging, transfer and clean-up, are more likely to cause contamination and expose workers to nanomaterials than production itself. Exposure can also occur during machine repair, destruction and recycling (National Institute for Occupational Safety and Health, 2004, 2005a,b). Therefore, the exposure scenario needs to include each potential step and combine the exposure measures of all steps.

Ingestion and ocular exposure are additional exposure routes in the workplace. Hand-to-mouth contact of workers may result in the ingestion of nanoparticles. Moreover, workers may take in food or mucus contaminated with nanomaterials in the workplace. Airborne nanomaterials, such as powders and mists, can come in contact with the eyes and cause ocular exposure, particularly in workplace settings, such as the manual application of spray coatings.

The workplace is expected to be fully furnished with the appropriate protective equipment and tools, such as air filtration systems, protective gloves and masks. The efficiency of protective equipment can affect the degree of exposure. The behavioral patterns of workers may also influence the level of occupational exposure, for example, the wearing of protective equipment, eating at the workplace, or the frequency of hand-washing.

General population. The general population may be exposed to nanomaterials via environmental release and the use of products containing nanomaterials. As described above, the sources of environmental releases include every procedure from the manufacture to recycling of products containing nanomaterials. Nanosized particles can be discharged into the environment and may come in contact with people through inhalation, ingestion and dermal exposure. Inhalation is likely to be the main exposure route for environmental nanomaterials (Royal Society, 2004).

Many consumer products containing nanoparticles are currently on the market, and can lead to direct or indirect contact with consumers during their use. Although there are a few products that cause olfactory exposure of nanomaterials, such as aerosols or sprays containing nanosized compounds, the main exposure routes of nanomaterials from consumer products appear to be through the skin and ingestion. Dermal exposure can occur through the use of cosmetics, textiles, medical patches and sporting goods containing nanomaterials, even though nanoparticles bound in a matrix or fixed in the products are not likely to be released. Food and beverages, personal care products, as well as kitchen and tableware, can result in the ingestion of nanomaterials. Nanoparticles can be the main functional ingredients of the product, or be an additive that is unlikely to give rise to exposure. The exposure pattern may be different for each individual product. Therefore, the specific characteristics and consumer usage pattern of each product need to be determined before drafting an exposure scenario.

23.2.3.6   Conducting Exposure Assessments

Exposure assessment measures the intake and/or uptake of potentially hazardous materials in the exposure media over a designated period of time. In order to assess the level of exposure, all the data need to be obtained, such as the concentration of the chemical of interest in all potential exposure media, the activity patterns of the population of interest, and other exposure factors, such as body weight, inhalation rate and dermal absorption rate. All the values can be measured or surveyed directly using multiple samples or populations, but models are sometimes used to estimate the exposure.

The most probable scenario that reflects the particular situation of interest needs to be selected for a more accurate exposure assessment. In this scenario, exposed populations and exposure routes are defined, and the exposure is calculated from the appropriate equations using the collected data. For example, the following equations can be used to determine the ingestion, inhalation and dermal absorption of nanoparticles (quoted from example exposure scenarios, Environmental Protection Agency, 2004a):

Ingestion:

$$ADD = \frac{C * IR * EF * ED}{AT}$$

ADD = potential average daily dose from the ingestion of the contaminated material of interest (mg/kg/day)
C   = concentration of contaminant in the material of interest (mg/g)
IR   = intake rate of the material of interest (g/kg/day)
EF   = exposure frequency (days/year)
ED   = exposure duration (years)
AT   = averaging time (days)

Inhalation:

$$C_{adjusted} = \frac{C * ET * EF * ED}{AT}$$

$C_{adjusted}$ = concentration of the contaminants in air adjusted (mg/m^3)
C   = concentration of contaminants in air (mg/m^3)
ET   = exposure time (hr/day)
EF   = exposure frequency (days/year)
ED   = exposure duration (years)
AT   = averaging time (hours)

Dermal absorption:

$$ADD = \frac{C * SA/BW * AF * EF * ED * ABS}{AT}$$

ADD   = potential average daily dose from dermal contact with the contaminated material of interest (mg/kg/day)
C   = concentration of contaminant in the material of interest (mg/kg)
SA/BW = surface area of the skin that contacts the material of interest (cm^2/event) divided by body weight (kg)

AF   = adherence factor (mg/cm^2)
EF   = exposure frequency (events/yr)
ED   = exposure duration (years)
ABS   = absorption fraction; this value is chemical-specific
AT   = averaging time (days)

23.2.4 Risk Characterization

Based on previous studies, the risk from bulk materials should not be extrapolated directly to nanomaterials. The toxic effects of manufactured nanomaterials may be reinforced by the characteristics introduced for the intended functions of nanomaterials. In most cases, there is insufficient information available on the toxic effects and exposure levels to make an appropriate risk characterization of nanomaterials. Since the underlying mechanisms for the exposure processes and toxic effects of nanomaterials in humans are not completely understood, these inherent uncertainties would be most important for estimating the risk. These uncertainties are related with the following: the persistence of nanomaterials in the environment, relevance of the exposure routes to individual circumstances, metrics for the exposure measurements, the mechanisms for translocation and degradation of nanoparticles in the body, and toxic mechanisms of nanoparticles (Scientific Committee on Emerging and Newly Identified Health Risks, 2007).

   Not all individuals in the population respond to nanomaterial exposure in the same way or to the same degree (Environmental Protection Agency, 2004b). Although susceptibility factors are largely unknown, certain individuals are likely to be susceptible to the adverse effects of nanomaterials based on their predisposition factors (Pope, 2000; Samet *et al.*, 2000). For example, exposure to nanoparticles may exacerbate the disease of individuals with asthma or chronic obstructive pulmonary disease (COPD). These effects may be associated with the oxidative and pro-inflammatory effects of nanoparticles, and inflamed lungs are more permeable to nanoparticles. The deposition of nanoparticles is usually enhanced in patients with COPD, presumably due to their abnormal airway. Diabetic patients are also a potential risk group, probably due to an endothelial dysfunction. In addition, smokers may be considered a risk group because of their enhanced uptake of nanoparticles (Scientific Committee on Emerging and Newly Identified Health Risks, 2007).

23.3 Conclusions

Nanotechnology represents a revolutionary fast-growing field, and there has been a significant increase in the number of consumer products based on nanotechnologies on the market each year. There is no agreement on the risks and benefits of nanomaterials. However, there are significant concerns about the potential adverse effects of nanomaterials on human health, with very limited information on their toxicity.

   Considerable work remains to be done in order to evaluate properly the risk of nanomaterials, and such work will be a very challenging and daunting job. Several papers have reviewed the characterization, fate and toxicity of nanomaterials, and proposed research needs or gaps for risk assessment (Holsapple *et al.*, 2005; Morgan, 2005; Thomas and Sayre, 2005; Borm *et al.*, 2006; Powers *et al.*, 2006; Tsuji *et al.*, 2006). In order to facilitate risk assessment-oriented studies, there is a fundamental requirement to measure and characterize nanomaterials and their byproducts in a range of media, including biological media. In particular, standardized protocols for determining the hazard and exposure as well as characterization of the nanomaterials will be necessary for assuring the quality of the data and its consistency. Test methods for determining the toxicity and exposure scenario should also be developed in a suitable and effective manner. Such research should be carried out within a risk assessment paradigm for nanomaterials, in order to address the key questions and to fill the gaps in knowledge.

Acknowledgements

This research was supported by grant 08181KFDA481 from the Korea Food and Drug Administration.

References

Agarwal A, Asthana A, Gupta U, Jain NK (2008) Tumour and dendrimers: a review on drug delivery aspects. *J Pharm Pharmacol* **60**, 671–688.

Ahn MH, Kang CM, Park CS, Park SJ, Rhim T, Yoon PO, Chang HS, Kim SH, Kyono H, Kim KC (2005) Titatium dioxide particle-induced goblet cell hyperplasia: association with mast cells and IL-3. *Respir Res* **6**, 34.

Aitken RJ, Creely KS, Tran CL (2004) *Nanoparticle: An Occupational Hygiene Review*. Research Report 274. Institute of Occupational Medicine for the Health and Safety Executive: London.

Alivisatos AP (1996) Perspectives on the physical chemistry of semiconductor nanocrystals. *J Phys Chem* **100**, 13226–13239.

Bakry R, Vallant RM, Najam-ul-Hag M, Rainer M, Szabo Z, Huck CW, Bonn GK (2007) Medicinal applications of fullerenes. *Int J Nanomed* **2**, 639–649.

Balbus JM, Maynard AD, Colvin VL, Castranova V, Daston GP, Denison RA, Dreher KL, Goering PL, Goldberg AM, Kulinowski KM, Monteiro-Riviere NA, Oberdörster G, Omenn GS, Pinkerton KE, Ramos KS, Rest KM, Sass JB, Silbergeld EK, Wong BA (2007) Meeting report: Hazard assessment for nanoparticles – report from an interdisciplinary workshop. *Environ Health Perspect* **115**, 1654–1659.

Bandaru PR (2007) Electrical properties and applications of carbon nanotube structures. *J Nanosci Nanotechnol* **7**, 1239–1267.

Bermudez E, Mangum JB, Wong BA, Asgharian B, Hext PM, Warheit DB, Everitt JI, Moss OR (2004) Pulmonary responses of mice, rats and hamsters to subchronic inhalation of ultrafine titanium dioxide particles. *Toxicol Sci* **77**, 347–357.

Bidleman TF (1988) Atmospheric processes, wet and dry deposition of organic compounds are controlled by their vapor-particle partitioning. *Environ Sci Technol* **22**, 361–367.

Biswas P, Wu C-Y (2005) Nanoparticles and the environment. *J Air Waste Manage Assoc* **55**, 708–746.

Bonacucina G, Cespi M, Misici-Falzi M, Palmieri GF (2009) Colloidal soft matter as drug delivery system. *J Pharm Sci* **98**, 1–42.

Borm PJA, Robbins D, Haubold S, Kuhlbusch T, Fissan H, Donaldson K, Schins R, Stone V, Kreyling W, Lademann J, Krutmann J, Warheit D, Oberdörster E (2006) The potential risks of nanomaterials: a review carried out for ECETOC. *Part Fibre Toxicol* **3**, 11.

Brown JS, Zeman KL, Bennett WD (2002) Ultrafine particle deposition and clearance in the healthy and obstructed lung. *Am J Respir Crit Care Med* **166**, 1240–1247.

Chen HW, Su SF, Chien CT, Lin WH, Yu SL, Chou CC, Chen JJ, Yang PC (2006) Titanium dioxide nanoparticles induce emphysema-like lung injury in mice. *FASEB J* **20**, 2392–2395.

Chen M, von Mickecz A (2005) Formation of nucleoplasmic protein aggregates impairs nuclear function in response to SiO_2 nanoparticles. *Exp Cell Res* **305**, 51–62.

Dutta RC (2007) Drug carriers in pharmaceutical design: promises and progress. *Curr Pharm Des* **13**, 761–769.

Elder A, Gelein R, Silva V, Feikert T, Opanashuk L, Carter J, Potter R, Maynard A, Ito Y, Finkelstein J, Oberdörster G (2006) Translocation of inhaled ultrafine manganese oxide particles to the central nervous system. *Environ Health Perspect* **114**, 1172–1178.

Environmental Protection Agency (2004a) *Example Exposure Scenarios*. National Center for Environmental Assessment: Washington, DC.

Environmental Protection Agency (2004b) *Office of Research and Development, Air Quality Criteria for Particulate Matter*. Report number EPA/600/P-99/002a,bF.

Environmental Protection Agency (2007) *Nanotechnology White Paper*. Science Policy Council, EPA: Washington, DC. Available at http://es.epa.gov/ncer/nano/publications/whitepaper 12022005.pdf

Environmental Protection Agency (2008) *Nanomaterial Research Strategy*. Office of Research and Development: Washington, DC.

Fan HJ, Werner P, Zacharias M (2006) Semiconductor nanowires: from self-organization to patterned growth. *Small* **2**, 700–717.

Feynman RP (1960) There's plenty of room at the bottom. *Engineering and Science* Feb. 1960, 22–36.

Filley TR, Ahn M, Held BW, Blanchette RA (2005) Investigations of fungal mediated (C60-C70) fullerene decomposition. Preprints of extended abstracts presented at the ACS National Meeting, American Chemical Society, Division of Environmental Chemistry: **45**, 446–450.

Fortner JD, Lyon DY, Sayes CM, Boyd AM, Falkner JC, Hotze EM, Alemany LB, Tao YJ, Guo W, Ausman, KD, Colvin VL, Hughes JB (2005) C_{60} in water: nanocrystal formation and microbial response. *Environ Sci Technol* **39**, 4307–4316.

Fu A, Gu W, Larabell C, Alivisatos AP (2005) Semiconductor nanocrystals for biological imaging. *Curr Opin Neurobiol* **15**, 568–575.

Gao X, Yang L, Petros JA, Marshall FF, Simons JW, Nie S (2005) In vivo molecular and cellular imaging with quantum dots. *Curr Opin Biotechnol* **16**, 63–72.

Geiser M, Rothen-Rutishauser B, Kapp N, Schurch S, Kreyling W, Schulz H, Semmler M, Im Hof V, Heyder J, Gehr P (2005) Ultrafine particles cross cellular membranes by nonphagocytic mechanisms in lungs and in cultured cells. *Environ Health Perspect* **113**, 1555–1560.

Gwinn MR, Vallyathan V (2006) Nanoparticles: Health effects – pros and cons. *Environ Health Perspect* **114**, 1818–1825.

Hardman R (2006) A toxicological review of quantum dots: toxicity depends on physicochemical and environmental factors. *Environ Health Perspect* **114**, 165–172.

Haynie DT, Zhang L, Zhao W, Rudra JS (2006) Protein-inspired multilayer nanofilms: science, technology and medicine. *Nanomedicine* **2**, 150–157.

Holsapple MP, Farland WH, Landry TD, Monteiro-Riviere NA, Carter JM, Walker NJ, Thomas KV (2005) Research strategies for safety evaluation of nanomaterials, part II: Toxicological and safety evaluation of nanomaterials, current challenges and data needs. *Toxicol Sci* **88**, 12–17.

Hood E (2004) Nanotechnology: Looking as we leap. *Environ Health Perspect* **112**, A741–A749.

Hunter DD, Dey RD (1998) Identification and neuropeptide content of trigeminal neurons innervating the rat nasal epithelium. *Neuroscience* **83**, 591–599.

Jani PU, McCarthy DE, Florence AT (1994) Titanium dioxide (rutile) particles uptake from the rat GI tract and translocation to systemic organs after oral administration. *Int J Pharm* **105**, 157–168.

Kim KJ, Kim YH, Ahn HG (2007) Catalytic performance of nanosized Pt-Au alloy catalyst in oxidation of methanol and toluene. *J Nanosci Nanotechnol* **7**, 3795–3799.

Kirchner C, Liedl T, Kudera S, Pellegrino T, Munoz Javier A, Gaub HE, Stolzle S, Fertig N, Parak WJ (2005) Cytototoxicity of colloidal CdSe and CdSe/ZnS nanoparticles. *Nano Lett* **5**, 331–338.

Kleinman MT, Araujo JA, Nel A, Sioutas C, Campbell A, Cong PQ, Li H, Bondy SC (2008) Inhaled ultrafine particulate matter affects CNS inflammatory processes and may act via MAP kinase signaling pathways. *Toxicol Lett* **178**, 127–130.

Kohli AK, Alpar HO (2004) Potential use of nanoparticles for transcutaneous vaccine delivery: effect of particle size and charge. *Int J Pharm* **275**, 13–17.

Kreyling WG, Semmler M, Erbe F, Mayer P, Takenaka S, Schulz H, Oberdorster G, Ziesenis A (2002) Translocation of ultrafine insoluble iridium particles from lung epithelium to extrapulmonary organs is size dependent but very low. *J Toxicol Environ Health A* **65**, 1513–1530.

Kwon JT, Hwang SK, Jin H, Kim DS, Minai-Tehrani A, Yoon HJ, Choi M, Yoon TJ, Han DY, Kang YW, Yoon BI, Lee JK, Cho MH (2008) Body distribution of inhaled fluorescent magnetic nanoparticles. *J Occup Health* **50**, 1–6.

Lew Yan Voon LC, Zhang Y, Lassen B, Willatzen M, Xiong O, Eklund PC (2008) Electronic properties of semiconductor nanowires. *J Nanosci Nanotechnol* **8**, 1–26.

Li N, Sioutas C, Cho A, Schmitz D, Mistra C, Sempf J, Wang M, Oberley T, Froines J, Nel A (2003) Ultrafine particulate pollutants induce oxidative stress and mitochondrial damage. *Environ Health Perspect* **111**, 455–460.

Li ZB, Cai W, Chen X (2007) Semiconductor quantum dots for *in vivo* imaging. *J Nanosci Nanotechnol* **7**, 2567–2581.

Liu J, He F, Durham E, Zhao D, Roberts CB (2008) Polysugar-stabilized Pd nanoparticles exhibiting high catalytic activities for hydrodechlorination of environmentally deleterious trichloroethylene. *Langmuir* **24**, 328–336.

Lomer MC, Hutchinson C, Volkert S, Greenfield SM, Catterall A, Thompson RP, Powell JJ (2004) Dietary sources of inorganic microparticles and their intake in healthy subjects and patients with Crohn's disease. *Br J Nutr* **92**, 947–955.

Maynard AD, Baron PA, Foley M, Shvedova AA, Kisin ER, Castranova V (2004) Exposure to carbon nanotube material: aerosol release during the handling of unrefined single-walled material. *J Toxicol Environ Health A* **67**, 87–107.

Morgan K (2005) Development of a preliminary framework for informing the risk analysis and risk management of nanoparticles. *Risk Anal* **25**, 1621–1635.

Muhlfeld C, Geiser M, Kapp N, Gehr P, Rothen-Rutishauser B (2007) Re-evaluation of pulmonary titanium dioxide nanoparticle distribution using the "relative deposition index": evidence for clearance through microvasculature. *Part Fibre Toxicol* **4**, 7.

Nakahara Y, Matsusaki M, Akashi M (2007) Fabrication and enzymatic degradation of fibronectin-based ultrathin films. *J Biomater Sci Polym Ed* **18**, 1565–1573.

National Institute for Occupational Health and Safety (2004) *Nanotechnology and Workplace Safety and Health*. Available at http://www.cdc.gov/niosh/docs/2004-175/

National Institute for Occupational Health and Safety (2005a) *Approaches to Safe Nanotechnology: an Information Exchange with NIOSH*. Available at http://www.cdc.gov/niosh/topics/nanotech/safenano/pdfs/approaches_to_safe_nanotechnology_28november2006_updated.pdf

National Institute for Occupational Health and Safety (2005b) *Strategic Plan for NIOSH Nanotechnology Research Program: Filling the Knowledge Gaps*. Available at http://www.cdc.gov/niosh/topics/nanotech/strat_plan.html

National Research Council (NRC) (1983) *Risk Assessment in the Federal Government: Managing the Process*. National Academy Press: Washington, DC.

National Research Council (NRC) (1994) *Science and Judgment in Risk Assessment*. National Academy of Sciences: Washington, DC.

Nel A, Xia T, Madler L, Ning L (2006) Toxic potential of materials at the nanolevel. *Science* **311**, 622–627.

Nemmar A, Vanbilloen H, Hoylaerts MF, Hoet PHM, Verbruggen A, Nemery B (2001) Passage of intratracheally instilled ultrafine particles from the lung into the systemic circulation in hamster. *Am J Respir Crit Care Med* **164**, 1665–1668.

Nemmar A, Hoet PH, Vanquickenborne B, Dinsdale D, Thomeer M, Hoylaerts MF, Vanbilloen H, Mortelmans L, Nemerv B (2002) Passage of inhaled particles into the blood circulation in humans. *Circulation* **105**, 411–414.

Nohynek GJ, Lademann J, Ribaud C, Roberts MS (2007) Grey goo on the skin? Nanotechnology, cosmetics and sunscreen safety. *Crit Rev Toxicol* **37**, 251–277.

Oberdörster G, Sharp Z, Atudorei V, Elder A, Gelein R, Lunts A, Kreyling W, Cox C (2002) Extrapulmonary translocation of ultrafine carbon particles following whole-body inhalation exposure of rats. *J Toxicol Environ Health A* **65**, 1531–1543.

Oberdörster G, Shapr Z, Atudorei V, Elder A, Gelein R, Kreyling W, Cox C (2004) Translocation of inhaled ultrafine particles to the brain. *Inhal Toxicol* **16**, 437–445.

Oberdörster G, Oberdörster E, Oberdörster J (2005a) Nanotoxicology: an emerging discipline evolving from studies of ultrafine particles. *Environ Health Perspect* **113**, 823–839.

Oberdörster G, Maynard A, Donaldson K, Castranova V, Fitzpatrick J, Ausman K, Carter J, Karn B, Kreyling W, Lai D, Olin S, Monteiro-Riviere N, Warheit D, Yang H; ILSI Research Foundation/Risk Science Institute Nanomaterial Toxicity Screening Working Group (2005b) Principles for characterizing the potential human health effects from exposure to nanomaterials: elements of a screening strategy. *Part Fibre Toxicol* **2**, 8–43.

Peng CC, Yang MH, Chiu WT, Chiu CH, Yang CS, Chen KC, Peng RY (2008) Composite nano-titanium oxide-chitosan artificial skin exhibits strong wound-healing effect-an approach with anti-inflammatory and bacterial kinetics. *Macromol Biosci* **8**, 316–327.

Peters A, Veronesi B, Calderon-Garciduenas L, Gehr P, Chen LC, Geiser M, Reed W, Rothen-Rutishauser B, Schurch S, Schulz H (2006) Translocation and potential neurological effects of fine and ultrafine particles a critical update. *Part Fibre Toxicol* **3**, 13.

Pflucker F, Hohenberg H, Holzle E, Will T, Pfeiffer S, Wepf R, Diembeck W, Wenck H, Gers-Barlag H (1999) The outermost stratum corneum layer is an effective barrier against dermal uptake of topically applied micronized titanium dioxide. *Int J Cosmet Sci* **21**, 399–411.

Pflucker F, Wendel V, Hohenberg H, Gartner E, Will T, Pfeiffer S, Wepf R, Gers-Barlag H (2001) The human stratum corneum layer: an effective barrier against dermal uptake of different forms of topically applied micronised titanium dioxide. *Skin Pharmacol Appl Skin Physiol* **14**(Suppl. 1), 92–97.

Pope CA III (2000) Epidemiology of fine particulate air pollution and human health: Biologic mechanisms and who's at risk? *Environ Health Perspect* **108**(Suppl. 4), 713–723.

Powers KW, Brown SC, Krishna VB, Wasdo SC, Moudgil BM, Robert SM (2006) Research strategies for safety evaluation of nanomaterials, part VI: characterization of nanoscale particles for toxicological evaluation. *Toxicol Sci* **90**, 296–303.

Preining O (1998) The physical nature of very, very small particles and its impact on their behavior. *J Aerosol Sci* **29**, 481–495.

Qian L, Yang X (2008) Dendrimer films as matrices for electrochemical fabrication of novel gold/palladium bimetallic nanostructures. *Talanta* **74**, 1649–1653.

Renwick LC, Donaldson K, Clouter A (2001) Impairment of alveolar macrophage phagocytosis by ultrafine particles. *Toxicol App Pharmacol* **172**, 119–127.

Royal Society (2004) *Nanoscience and Nanotechnologies: Opportunities and Uncertainties*. Policy Document 19/04. Royal Society: London. Available at http://royalsociety.org/document.asp?id=2023

Ryman-Rasmussen JP, Riviere JE, Monteiro-Riviere NA (2006) Penetration of intact skin by quantum dots with diverse physicochemical properties. *Toxicol Sci* **91**, 159–165.

Samet JM, Zeger SL, Dominici F, Curriero F, Coursac I, Dockery DW, Schwartz J, Zanobetti A. 2000. The national morbidity, mortality, and air pollution study. Part II: Morbidity and mortality from air pollution in the United States. *Res Rep Health Eff Inst* **94**(Pt 2): 5–70.

Scientific Committee on Consumer Products (2007) *Preliminary Opinion on Safety of Nanomaterials in Cosmetic Products*. European Commission. Available at http://ec.europa.eu/health/ph_risk/committees/04_sccp/docs/sccp_o_099.pdf

Scientific Committee on Emerging and Newly-Identified Health Risks (2007) *The appropriateness of the risk assessment methodology in accordance with the Technical Guidance Documents for new and existing substances for assessing the risks of nanomaterials*. European Commission: Brussels.

Semmler M, Seitz J, Erbe F, Mayer P, Heyder J, Oberdörster G, Kreyling WG (2004) Long-term clearance kinetics of inhaled ultrafine insoluble iridium particles from the rat lung, including transient translocation into secondary organs. *Inhal Toxicol* **16**, 453–459.

Son SJ, Bai X, Lee SB (2007) Inorganic hollow nanoparticles and nanotubes in nanomedicine Part 1. Drug/gene delivery applications. *Drug Discov Today* **12**, 650–656.

Srivastava R, McShane MJ (2005) Application of self-assembled ultra-thin film coatings to stabilize macromolecule encapsulation in alginate microspheres. *J Microencapsul* **22**, 397–411.

Takenaka S, Karg E, Roth C, Schulz H, Ziesenis A, Heinzmann U, Schramel P, Hevder J (2001) Pulmonary and systemic distribution of inhaled ultrafine silver particles in rats. *Environ Health Perspect* **109**(Suppl. 4), 547–551.

Takenaka S, Karg E, Kreyling WG, Lentner B, Schulz H, Ziesenis A, Schramel P, Heyder J (2004) Fate and toxic effects of inhaled ultrafine cadmium oxide particles in the rat lung. *Inhal Toxicol* **16**(Suppl. 1), 83–92.

Tang CF, Deng H, Tang B, Cheng H, Wang JC, Chen JJ (2008) Non-linear optical properties of zinc oxide nanowires. *J Nanosci Nanotechnol* **8**, 1150–1154.

Tang J, Xing G, Zhao F, Yuan H, Zhao Y (2007) Modulation of structural and electronic properties of fullerene and metallofullerenes by surface chemical modification. *J Nanosci Nanotechnol* **7**, 1085–1101.

Tenne R, Rao CNR (2004) Inorganic nanotubes. *Phil Trans R Soc Lond A* **362**, 2099–2125.

Thomas K, Sayre P (2005) Research strategies for safety evaluation of nanomaterials, Part I: Evaluating the human health implication of exposure to nanoscale materials. *Toxicol Sci* **87**, 316–321.

Tsuji JS, Maynard AD, Howard PC, James JT, Lam C, Warheit DB, Santamaria AB (2006) Research strategies for safety evaluation of nanomaterials, Part IV: risk assessment of nanoparicles. *Toxicol Sci* **89**, 42–50.

Varfolomeev A, Pokalyakin V, Tereshin S, Zaretsky D, Bandvopadhvav S (2005) Switching time of nanowire memory. *J Nanosci Nanotechnol* **5**, 753–758.

Wang J, Zhou G, Chen C, Yu H, Wang T, Ma Y, Jia G, Gao Y, Li B, Sun J, Li Y, Jiao F, Zhao Y, Chai Z (2007) Acute toxicity and biodistribution of different sized titanium dioxide particles in mice after oral administration. *Toxicol Lett* **168**, 176–185.

Wang P (2006) Nanoscale biocatalyst systems. *Curr Opin Biotechnol* **17**, 574–579.

Wang ZL (2008) Oxide nanobelts and nanowires-growth, properties and applications. *J Nanosci Nanotechnol* **8**, 27–55.

Warheit DB, Laurence BR, Reed KL, Roach DH, Reynolds GAM, Webb TR (2004) Comparative toxicity assessment of single-wall carbon nanotubes in rats. *Toxicol Sci* **76**, 117–125.

Weng J, Ren J (2006) Luminescent quantum dots: a very attractive and promising tool in biomedicine. *Curr Med Chem* **13**, 897–909.

World Health Organization (2005) Food safety risk analysis. Part 1: an overview and framework manual. Provisional edition. FAO and WHO: Rome. Available at http://www.fsc.go.jp/sonota/foodsafety_riskanalysis.pdf

Woodrow Wilson International Center for Scholars (2008) *Nanotechnology Consumer Products Inventory*. Available at http://www.nanotechproject.org/inventories/consumer

Xia T, Kovochich M, Brant J, Hotze M, Sempf J, Oberley T, Sioutas C, Yeh JI, Wiesner MR, Nel AE (2006) Comparison of the abilities of ambient and manufactured nanoparticles to induce cellular toxicity according to an oxidative stress paradigm. *Nano Lett* **6**, 1794–1807.

Yamamoto A, Honma R, Sumita M, Hamawa T (2003) Cytoxicity evaluation of ceramic particles of different sizes and shapes. *J Biomed Mater Res A* **68**, 244–256.

24

Metal Nanoparticle Health Risk Assessment

Mario Di Gioacchino, Nicola Verna, Rosalba Gornati, Enrico Sabbioni and Giovanni Bernardini

24.1   Introduction

Nanomaterials are a diverse class of small-scale (<100 nm) substances formed by molecular-level engineering to achieve unique mechanical, optical, electrical and magnetic properties. Nanomaterials are expected to improve virtually all types of products (Royal Society, 2004), and commercialization of products that exploit these unique properties is increasing. However, these same properties present new challenges to understanding, predicting and managing potential adverse health effects following exposure (Hood, 2004).

Widespread application of nanomaterials confers enormous potential for human exposure and environmental release. Like genetically modified organisms, the future of nanotechnology will depend upon public acceptance of the risks versus benefits. Even beyond public acceptance, experience with past 'miracle' materials (e.g. asbestos) advises caution in using novel substances without fully evaluating potential health risks.

The first issue to consider to assess the risks from nanoparticles (NPs) to man is the lack of knowledge of their toxicity, the first difficulty being the lack of standards in nanotechnology to support legislation and regulation, risk assessment and communication.

Priorities for nanotechnology standardization are: (a) systematic terminology for materials composition and features; (b) metrology/methods of analysis/standard test methods (particle size and shape, particle number and distribution, particle mass); and (c) toxicity effects/risk assessment (health and safety, reference standards for testing and controls, and testing methods for toxicity).

Nanotoxicity: From In Vivo *and* In Vitro *Models to Health Risks*   Edited by Saura Sahu and Daniel Casciano
© 2009 John Wiley & Sons, Ltd

The greatest current risk is to the occupational health of workers involved in research and manufacture of NPs and nanofibres. However, as applications of nanomaterials increase, the risk of exposure to the general public will grow. It will be necessary to monitor products that incorporate NPs and nanofibres throughout their life, from manufacture to disposal and waste processes. Some products will involve direct delivery of NPs to humans, for therapy and diagnostics or application of cosmetics to the skin. In some cases there could be unintentional uptake, for example, ingestion of NPs used in food packaging technology.

The wide variety of routes by which NPs could be taken up by the body complicates the definition of NPs to be used in risk assessment. It is probably necessary to consider multi-component and multi-phase particles of any size and composition that can be absorbed by the body. The overall risk of exposure to nanofibres may be lower than for NPs because it appears to be more difficult to generate aerosols of nanofibres.

Epidemiological studies consistently show that an increase in atmospheric particulate concentrations leads to a short-term increase in morbidity and mortality. Inhalation is the most significant exposure route for the unintentionally-generated particles. Regulation is aimed mainly at particulate matter <2.5 µm (PM2.5) in the environment, and more recently attention has focused on ultrafine particles (UFPs), whose diameters of < 0.1 µm (PM0.1) are consistent with definitions of nanoparticles.

For most manufactured NPs no toxicity data are available. Experimental and toxicological works have been performed with a small set of NPs, mainly carbon black (CB), titanium dioxide (TiO_2), iron oxides and amorphous silica. These particles were considered to be so-called 'nuisance dusts' until it was observed that upon prolonged exposure in rats, inflammation and lung tumours can occur, and further, that the overall mortality increases by 0.9 % in relation to a 10 µg/m^3 increase in the concentration of environmental particles (Pope et al., 2003).

Experimental toxicological studies (with CB, TiO_2) have indicated that NPs cause such adverse effects at lower dose levels than their fine counterparts, but so far few human studies have been able to investigate this (Kuempel et al., 2006; Warheit et al., 2007). The key question is whether and how the different pieces of toxicological and epidemiological evidence on different NPs can be mutually used or whether a more targeted and systematic approach is necessary.

24.2   Risk Assessment

Depending on the conditions of manufacture, formulation, use and final disposal, a risk assessment of NPs may need to address:

(1) Worker safety. Typically workers are exposed to higher levels of chemicals and for more prolonged periods of time compared with the general population, and this will probably be the case for NP production.
(2) Safety of consumers using products that contain NPs.
(3) Safety of local human populations due to chronic or acute release of NPs from industry.
(4) The potential for human re-exposure through the environment. Particular attention should be focused on products that are deliberately used in nanoparticle form in the environment, such as biocides, or environment-improving agents.

(5) The environmental and human health risks involved in the disposal or recycling of nanoparticle-dependent products.

One or more of these risk assessments may be omitted, if there are valid reasons to conclude that no exposure will occur. In principle, the traditional risk assessment procedure is an appropriate tool for assessing the risks from exposure to NPs under specified exposure conditions.

The traditional risk assessment methodology comprises the following stages: (a) exposure assessment; (b) hazard identification; (c) hazard characterization; (d) risk characterization. This framework has not yet been applied to NPs, in terms of either their potential human or environmental impacts, for a number of related reasons. There is an unclear situation with regard to regulatory requirements for risk assessment. As a consequence, there are no official guidelines on what constitutes an appropriate testing regimen.

24.3 Exposure Assessment

To date there is no good evidence of a specific particle size, shape and surface charge at which altered penetration of cell membranes occurs. It is biologically plausible that, as the particle size decreases, a sudden increase in absorption and/or toxicity arises. It is important to establish whether this is typical for all NPs or specific for selected ones. It has also been found that, for certain NPs, the clearance mechanisms may be less effective than for larger particles. For example, impaired phagocytosis has been observed in a macrophage cell line containing NPs (Renwick *et al.*, 2001). If this finding reflects a more general phenomenon, NPs would need to be considered to have the potential for bioaccumulation in humans and possibly in other species and in the environment.

What is clear from the published literature on human toxicology is that the expression of the exposure dose in terms of unit weight, which is the established practice in toxicology, is often not appropriate when studying the toxicity of NPs. Instead, either total surface area, or number of particles, or a combination of surface area and number of particles, should be used.

24.3.1 Exposure Scenarios

Most human individuals are routinely exposed to particles in the ambient atmosphere, primarily from diesel fumes and other combustion processes. Initially only about 10 nm in diameter, these rapidly coalesce to produce larger aggregates of up to about 100 nm, which may remain in the air for days or weeks. In a normal room, the air can contain 10,000 to 20,000 NPs/cm^3, whilst these figures can reach 50,000 NPs/cm^3 in a wood and 100,000 NPs/cm^3 in urban streets. These concentrations imply that every hour, individuals breath millions of NPs, and it is estimated that at least half of these reach the alveoli. While ingestion and skin penetration are potential exposure routes for engineered nanomaterials (Oberdörster *et al.*, 2005), the inhalation route for airborne nanomaterials has perhaps received the most attention (Ayres *et al.*, 2008; Schulte *et al.*, 2008a,b). For some materials, studies have shown that the toxicity of inhaled particles increases as particle size becomes smaller and as the overall surface area of inhaled material becomes larger (Oberdörster *et al.*,

2005; Ayres *et al.,* 2008; Schulte *et al.,* 2008a,b). The possible routes of exposure can be through ingestion, skin contact and inhalation.

24.3.1.1   Gastrointestinal Exposure

It was recognized by Kumagai in 1926 (cited in Salata, 2004) that particles could translocate from the lumen of the intestinal tract via aggregations of intestinal lymphatic tissue (Peyer's patches), containing M cells. It is now known that uptake of inert particles can occur not only through immune cells present in Peyer's patches, but also through enterocytes, and to a lesser extent across para-cellular pathways (Aprahamian *et al.,* 1987). However, once again data in the literature on potential exposure through the gastro-intestinal tract are very scarce.

NPs can enter the digestive tract by direct ingestion, for example, as food constituents (such as colorants – titanium oxide), pharmaceuticals, in water or cosmetics (toothpaste, lipstick) (Lomer *et al.,* 2002), and dental prosthesis debris (Ballestri *et al.,* 2001). In addition, NPs can be cleared from the respiratory tract via the mucociliary escalator and ingested. Moreover, in the intestinal lumen, endogenous NPs can be present due to calcium and phosphate secretion (Lomer *et al.,* 2004). On the basis of a market basket study in the UK, it has been estimated that TiO_2 with a mean diameter of 200 nm (about 200 mg/day per person) would account for about one thousand particles per day (Lomer, 2000). The use of specific alimentary products, such as salad dressing containing a nanoparticle TiO_2 whitening agent, can increase the daily average intake by more than 40-fold. A database of foods and pharmaceuticals containing NPs has been reported by Lomer *et al.* (2004).

It is important to consider what happens with food containing NPs. Will NPs remain in the intestinal tract or will they move on into the body? The intestine should take up nutrition while protecting the body from unwanted substances in the food. It is not known whether NPs are regarded as 'unwanted substances' and excreted or not.

The rapid transit of material through the intestinal tract (on the order of hours), together with the continuous renewal of epithelium, led to the hypothesis that nanomaterials will not remain there for indefinite periods. The extent of particle absorption in the gastrointestinal (GI) tract is affected by size, surface chemistry and charge, length of administration, and dose (Hoet *et al.,* 2004).

When ingested, NPs enter the stomach and are submitted to usual digestive processes that may attack them. Meng *et al.* (2007) have shown that 23.5 nm Cu NPs consume the hydrogen ions in the stomach more quickly than microparticles, converting the Cu NPs into cupric ions whose toxicity is very high. NPs surviving to the gastric digestion can be absorbed in the enteric tract.

Jani *et al.* (1990) have proved the translocation of NPs in rats orally administered with polystyrene microspheres in the size range 50 nm to 3 µm. A consistent absorption was observed, with a systemic distribution to liver, spleen, blood and bone marrow. Particles larger than 100 nm did not reach the bone marrow, and those larger than 300 nm were absent from blood. No particles were detected in heart or lung tissue. The uptake was mainly via M-cells (specialized phagocytic enterocytes) of the Peyer's patches, with translocation into the mesenteric lymph and then to systemic organs.

Rats, fed with a suspension of 6–9 µm down to 5–30 nm Fe particles, revealed Fe NPs within the tissues of the duodenum (brush border, lateral intercellular spaces of the mucosal cells, mitochondrial cristae and cytoplasm of both mucosal and stromal cells).

The observations indicated that nano-sized Fe particles may be taken up by the mucosa, and that the passage of such particles across the epithelial barrier may take place through both a paracellular as well as a transcytotic process (McCullough *et al.*, 1995).

Another possibility for intestinal uptake of NPs is via enterocytes (Hoet *et al.*, 2004). In a study performed in mice orally administered with 4 and 58 nm gold NPs, Hillyer and Albrecht (2001) showed the capture of gold NPs by the intestine, their passage through the blood and their translocation to the brain, lungs, heart, kidneys, intestine, stomach, liver and spleen. The uptake occurred by persorption through holes created by extruding enterocytes. This effect was inversely proportional to the size of the NPs: the smaller the particle, the greater was the passage.

Papers dealing with the fate of ingested NPs can often be criticized for the imprecision with which the particles are defined, because many of the reported NPs are actually over 100 nm in size. Jani *et al.* (1994) and Böckmann *et al.* (2000) orally administered larger particles (500 nm and 160–380 nm) of TiO_2 to rats and volunteers, respectively, and titanium was found in the blood and in the liver.

The translocation of NPs from the GI tract to other organs through the blood is not so clear. Other researchers have failed to show it. After oesophageal administration of 30 nm ^{192}Ir particles by gavage, virtually all the isotope was excreted by faeces within 2–3 days (Semmler *et al.*, 2004). No ^{192}Ir was detected in urine during the observation period, and no radioactivity was detected in any organ or tissue, suggesting a very poor absorption of such particles by the intestinal lumen with no translocation to blood and other tissues. Similar findings have been reported by Kreyling *et al.* (2002).

At present, the contrasting results of the literature concerning the uptake of NPs via the GI tract is likely to depend not only on particle size (Hillyer and Albrecht, 2001), but also on other modifying characteristics such as the surface charge (Jani, 1989), the hydrophilicity, biological coatings with attachment of ligands (Hussain and Florence, 1998; Shakweh *et al.*, 2005) or chemical coating with surfactants (Hillery *et al.*, 1994), which offers the possibility for site-specific targeting of different regions of the GI tract.

24.3.1.2   Dermal Exposure

There is very little data in the literature on potential exposure through the skin, even though nanomaterials have been used in cosmetics and pharmaceuticals for many years. Currently, most of the dermal exposure concerns skin preparations that use NPs. In theory, harmful effects arising from skin exposure may either occur locally within the skin, or alternatively the substance may be absorbed through the skin and disseminate via the bloodstream, possibly causing systemic effects, although there is no evidence of this as yet.

The outer portion of the epidermis is a variably thick keratinized layer of dead cells and it is difficult for ionic compounds and water soluble molecules to pass it (Hoet *et al.*, 2004). In theory, NPs might penetrate the skin by entering between or through epithelial cells (inter- and intra-cellular routes) or via the skin appendages (hair follicles, sebaceous and sweat glands) (Borm *et al.*, 2006). Three pathways of penetration across the skin have been identified: intercellular, transfollicular and transcellular (Scientific Committee on Consumer Products, 2007).

The passive transport of NPs through the intact stratum corneum is considered highly unlikely because of the matrix of corneocytes, lipid bilayers within the intercellular spaces, and the physiological environment below the stratum corneum containing high levels of proteins. If the skin is damaged, and the normal barrier disrupted, then the probability of entry of particles may be substantially increased. Thus, mechanical deformation is capable of transporting particles through the stratum corneum and into the epidermis and dermis (Oberdörster *et al.*, 2005).

The question of the transport of NPs through skin is still controversial, and published results are not in complete agreement, due also to a confusion concerning particle size. The studies conducted to date diverge and allow no sound scientific conclusions on the cutaneous absorptive potential of NPs. Despite this, some products containing NPs are commercially available, such as certain lipsticks containing iron oxides, and even sun creams which contain the so-called 'micronised particles' of TiO_2 and ZnO (Scientific Committee on Consumer Products, 2007).

Some studies report that NPs are able to penetrate the stratum corneum (Borm *et al.*, 2006; Oberdörster *et al.*, 2005). A number of laboratory studies have tested nanometal penetration through intact human, pig or mouse skin. Such studies have shown that very few particles, if any, reach living skin cells (Menzel *et al.*, 2004; Gamer *et al.*, 2006; Baroli *et al.*, 2007; Cross *et al.*, 2007; Mavon *et al.*, 2007). In this context, the European Union-funded NanoDerm project conducted a series of experimental studies over three years and found no evidence of dermal penetration in intact human and pig skin using a variety of analytical techniques, TiO_2 formulations and test conditions (Nanoderm, 2007). In healthy skin, TiO_2 was generally detected in the topmost layers of the stratum corneum disjunctum. Penetration occurs via mechanical action and no diffusive transport takes place.

Recently, Kiss *et al.* (2008) have given evidence that TiO_2 NPs *in vivo* do not penetrate through the intact epidermal barrier. However, while the application of a sunscreen containing 8 % 10–15 nm TiO_2 NPs onto the skin of humans showed no penetration, such NPs in oil-in-water emulsions showed penetration, particularly on hairy skin at the hair follicle sites or pores (Tsuji *et al.*, 2006). The penetration of TiO_2 NPs from sunscreen through the stratum corneum and at hair follicles has been reported by Gwinn and Vallyathan (2006), Schulz *et al.* (2002) and Tsuji *et al.* (2006). A study has shown that 7 nm quantum dots (CdSe) can cross the epidermis, directly translocating to the derma (Ryman-Rasmussen *et al.*, 2006).

Baroli *et al.* (2007) studied the human skin penetration and permeation of Fe NPs. They have shown that NPs were able to penetrate the hair follicle and stratum corneum, occasionally reaching the viable epidermis. The small dimensions of their particles might have been responsible for the exceptionally deep penetration to the viable epidermis.

NPs are commonly used in sunscreens and other cosmetics, and since sun creams are often applied to sun-damaged skin, the effect of ultraviolet radiation on penetrated NPs is a concern. In this context, Mortensen *et al.* (2008) investigated the penetration of quantum dots into skin of mice with and without UV exposure. Although low levels of penetration were seen in both the UV-exposed and non-exposed mice, qualitatively higher levels of penetration were observed in the UV-exposed mice.

Hair follicles make up 0.1 % of the skin surface and can be potential openings for deeper movement of NPs into the skin. Many efforts have been directed to understanding particle penetration in the hair follicle, which have shown that particles with diameters ranging between 7 µm and 20 nm are almost exclusively found in the hair follicle infundibulum and below (Meidan *et al.*, 2005; Vogt *et al.*, 2006; Lekki *et al.*, 2007). However, penetration studies for zinc and titanium NPs show follicular accumulation, but not penetration into deeper tissues (Lademann *et al.*, 2006, 2007).

Although cosmetics are meant to be used on normal skin, it is known that they are commonly also applied to non-healthy and/or physiologically compromised skin. A large portion of the European population suffers from atopic syndrome and 2 % has psoriasis. In such groups, the barrier properties of the skin may be impaired. Within the EC-funded NanoDerm-project (NanoDerm, 2007), the potential penetration of formulations containing TiO_2 NPs in some individuals with psoriatic skin was investigated. The psoriatic skin tested by NanoDerm does not completely address the potential for penetration in damaged skin. Sunburned skin might be more permeable, as might be the skin of children or the elderly, or the skin in regions of the body where it is naturally thinner. The NanoDerm assessment concluded that sunscreen containing titanium NPs should not be applied to open wounds, and called for more study of psoriatic skin, which has no outer protective barrier.

Possible transdermic transport in the case of broken skin that facilitates the entry of a wide range of larger particles (500 nm to 7 μm) was reported by Oberdörster *et al.* (2005). The breaching by dextran particles (500 nm) (Tinkle *et al.*, 2003) and primary and aggregated functionalised buckyballs (3.5 nm) following flexing the skin (Rouse *et al.*, 2007) has been described. In both studies, NPs were found in deeper dermal layers when the skin was flexed. For more detailed information on the safety of NPs in cosmetic products, the reader may refer to the *Opinion on Safety of Nanomaterials in Cosmetics Products* (Scientific Committee on Consumer Products, 2007).

NPs are also of growing interest for topical treatment of skin diseases to increase skin penetration of drugs and to reduce side-effects. A recent study (Küchler *et al.*, 2009) has investigated the effect of particle size and structure on skin penetration of particles by loading nile red to dendritic core-multishell nanotransporters (CMS; 20–30 nm) and solid lipid NPs (SLNPs; 150–170 nm). The results show that CMS nanotransporters can favour the penetration of a model dye into the skin even more than SLNPs, which may reflect size effects.

Conclusions. A comparison of previous and more recent literature data concerning NPs and skin should be made very carefully, taking into account experimental settings and investigation techniques, as well as type of skin, donor species, evaluation of skin integrity by means other than ocular observation, NP type and dimensions: parameters which, if not appropriately considered, may lead to erroneous evaluations (Baroli, 2008). In addition, many researchers think that NPs cannot penetrate skin based on the idea that its function is to protect us from the external environment, and because of the contradictory or negative results on sunscreens supplemented with TiO_2 and ZnO NPs (Nohynek *et al.*, 2007). However, at present, this conclusion is questionable. In fact, NPs that are used in sunscreens have been specifically formulated to stick to the skin (Nohynek *et al.*, 2008) and penetration experiments have been performed using healthy skin and lab conditions; thus, the results have been obtained under experimental conditions far from reality (i.e. damaged skin, seaside conditions comprising high hydration of skin, higher skin temperature, different skin vascularization, and sunburn; Baroli, 2008). Such NPs are dispersed in any possible medium, and it is unrealistic to base their safe evaluation on the extrapolation of scientifically weak and contradictory results available for TiO2 and ZnO NPs.

Also, current investigations of NP penetration into the skin use static imaging technology, which does not detect small quantities of NPs that may reach the vascular bed of the dermis. If the skin is exposed to large amounts of NPs, even small fractions may become an important source for secondary target organs, for which no information about the associated risk is available. Taking all this information together, we think that, currently, there are large data gaps on skin penetration of NPs, and this precludes a correct risk assessment for the nanomaterials in cosmetic products.

24.3.1.3   Lung Exposure

The respiratory tract is an important route of unintended exposure of NPs. Their deposition in the alveolar region offers the possibility of their absorption in the lung. Although NPs tend to agglomerate, potentially making their aerodynamic characteristics similar to those of larger particles (Jefferson, 2000), size remains a key characteristic which correlates with toxic responses. In this section attention is focused on the deposition and distribution of NPs in the lung, and their translocation and distribution.

Epithelial translocation and distribution.   Some studies have shown that NPs deposited in the respiratory tract readily translocate from the alveolar region to epithelial and interstitial sites. Mühlfeld *et al.* (2008a) have pointed out that when NPs effectively reach the alveolar region of the lungs, the size of the epithelial surface that may simultaneously come into contact with the particles amounts to approximately $140 \, m^2$. Once in the alveoli, NPs have a high probability of encountering the alveolar epithelium, also because the uptake of NPs by alveolar macrophages seems to play a more minor role than for larger-sized particles (Takenaka *et al.,* 2006; Geiser *et al.,* 2008). The next step would be translocation of NPs over the alveolar epithelium, deposition in the interstitium, or translocation over the capillary endothelium, entering the circulation.

A pioneering investigation on the effect of the size of inhaled NPs on their fate in lungs concerns the exposure of rats via the nose to $^{239}PuO_2$ NPs (10–30 nm) or microparticles, showing that the deposition in the deep compartments of the lungs was greater for the NPs than for the microparticles (Kanapilly and Diel, 1980). Inhalation by rats of TiO_2 particles in the range 12–250 nm led to a presence of these particles in the pulmonary interstitium (Ferin *et al.,* 1992). Another milestone study demonstrated that the retention kinetics of rats exposed by inhalation to 20 nm and 250 nm TiO_2 particles were different for the two differently-sized compounds, the clearance of the 20 nm NPs being significantly slower than the 250 nm particles (Oberdörster *et al.,* 1994). In a recent study, Nurkiewicz *et al.* (2008) at equivalent pulmonary loads showed that 21 nm TiO_2 NPs produce greater remote microvascular dysfunction as compared with 1 μm TiO_2 particles.

There is evidence of the greater ability of NPs to enter interstitial spaces after alveolar deposition, compared with bulk particles. Mühlfeld *et al.* (2007) verified previous suggestions on the free movement of TiO_2 NPs among four lung compartments (air-filled spaces, epithelium/endothelium, connective tissue, capillary lumen) in correlation with compartment size. He concluded that 22 nm TiO_2 NPs do not move freely between pulmonary tissue compartments, although they can pass from one compartment to another with relative ease. This finding contradicts the former concept of unrestricted NP movement (Geiser *et al.,* 2005), rather suggesting a controlled translocation process. For a more detailed review on interactions of NPs with pulmonary structures, the reader may refer to Mühlfeld *et al.* (2008b).

Translocation into circulatory system and distribution.   Despite a growing body of literature, there is still debate over whether NPs deposited and absorbed in the lung can cross the air–blood barrier, and translocate into the blood circulation in significant amounts.

Differently from microparticles, inhaled NPs can diffuse more like gas molecules and deposit anywhere in the respiratory tract. Like gases, simply because of their size, NPs can pass through the lungs into the bloodstream, reaching sensitive sites such as bone marrow, liver, kidneys, spleen and heart (Mühlfeld *et al.*, 2008a).

Exposure via inhalation by rats of 40 and 51 nm CdO NPs resulted in an efficient deposition in the lungs (Takenaka *et al.*, 2004). Elevated levels of Cd in blood, liver and kidneys were observed, indicating systemic translocation of a fraction of deposited Cd. However, apparent systemic translocation of Cd took place only in animals exposed to a high concentration that induced lung injury.

A significant uptake of 200 nm solid lipid NPs (SLNPs) radiolabelled with ^{95}Tc into the lymphatics after inhalation, and a high rate of distribution in periaortic, axillar and inguinal lymph nodes, was observed by Videira *et al.* (2002).

Studies on rats exposed to 15 nm Ag NPs by inhalation, or to agglomerated Ag NPs or Ag ions intratracheally, showed how the particle size and the tendency of particles to form agglomerates affect the distribution pathway in the lungs (Takenaka *et al.*, 2001). In such compartments, the amounts of silver from NPs or ions decreased rapidly with time after the end of exposure, while most instilled agglomerated NPs were phagocytized by alveolar macrophages and retained in the lung. Two mechanisms were suggested. Whereas agglomerated NPs remain partly undissolved in alveolar macrophages, single NPs are dissolved rapidly in the lung and silver enters the blood capillaries by diffusion. Alternatively, Ag NPs cross the alveolar wall, gaining entrance to the blood capillaries. Interestingly, in the blood, the significant amounts of silver detected initially decreased rapidly, which shows that systemic distribution occurred. Trace of silver were found in the liver, kidney, spleen, brain and heart, while the nasal cavities, especially the posterior portion, and lung-associated lymph nodes showed relatively high concentrations of silver.

Exposure via inhalation by rats of 40 and 51 nm CdO resulted in efficient deposition in the lung (Takenaka *et al.*, 2004). Elevated levels of Cd in the blood, liver and kidneys was observed, indicating systemic translocation of a fraction of deposited Cd from the lung. However, apparent systemic translocation of Cd took place only in animals exposed to a high concentration that induced lung injury.

Geiser *et al.* (2005) and Mühlfeld *et al.* (2007) provided evidence that only a small fraction of 22 nm TiO$_2$ NPs inhaled by rats translocate to the circulation, reaching extrapulmonary organs via the bloodstream. In contrast, Nemmar *et al.* (2002), after inhalation exposure of five volunteers to ^{99m}Technetium-radiolabelled carbon particles (Technegas, <100 nm), showed substantial radioactivity over the liver and other areas of the body. However, in this study, leaching of the radio-label from the particles was not considered. Brown *et al.* (2002) were not able to find any detectable particulates (<2 % of inhaled ^{99m}Technetium-radiolabelled carbon aerosol, limit of detection) when the data were corrected for leaching of the radio-label off the particles. These findings were confirmed by Mills *et al.* (2006) who carried out an experiment by inhalation of ten volunteers with ^{99m}Technetium-radiolabelled carbon particles using a design very similar to that of Nemmar *et al.* (2002). In contrast to Nemmar's study, Mills' results did not support the hypothesis that inhaled Technegas carbon NPs pass directly from the lungs into the systemic circulation with no accumulation of radioactivity detected over the liver or spleen. Other studies failed to confirm Nemmar's results, and detected only a low degree of translocation of inhaled NPs to the bloodstream and other organs.

Kreyling *et al.* (2002) showed that, in rats, the GI tract and feces were the predominant ways of clearance of inhaled 15 and 80 nm ^{192}Ir NPs, and that less than 1 % of the deposited particles were recovered into secondary organs such as liver, spleen, heart and brain. They also showed that the translocated fraction of the 80 nm NPs was about an order of magnitude less than that of 15 nm NPs.

Inhalation exposure of rats to nanoscale [13]C (20–29 nm) showed a translocation of such material to the liver, with no significant increase of NPs detected in other organs, suggesting a direct input into the blood compartment from the respiratory tract (Oberdörster *et al.*, 2002). A study of the role of alveolar macrophages in the fate of NPs in the lung using 16 nm gold NPs (Takenaka *et al.*, 2006) confirmed a significant but low translocation of NPs from lung to blood entering systemic pathways.

Kwon *et al.* (2008) examined the body distribution of inhaled 50 nm fluorescent magnetic NPs in mice. Magnetic resonance imaging and confocal laser scanning microscope analysis showed that NPs were distributed in various organs, including the liver, testis, spleen, lung and brain, indicating that these NPs could penetrate the blood–brain barrier.

Translocation into the central nervous system. The translocation of NPs from the olfactory epithelium to the olfactory bulb and the brain seems a peculiar property of inhaled NPs. Translocation along olfactory nerves was previously described for nasally deposited colloidal 50 nm gold NPs in squirrel monkeys (de Lorenzo and Darin, 1970). Oberdörster *et al.* (2002) showed that after inhalation of (20–29 nm) ^{13}C NPs, a translocation from the olfactory mucosa of the rat to the olfactory bulb occurred, providing a portal of entry into the central nervous system (CNS) for solid NPs. The same group (Oberdörster *et al.*, 2004) performed experiments in rats inhaled with 36 nm ^{13}C NPs. During inhalation about 20 % of the particles deposited into the nasopharyngeal region reached the olfactory mucosa and further translocated to the olfactory bulb where they persisted. The NPs were also able to cross the blood–brain barrier in some regions, targeting the CNS.

The translocation of NPs to the brain was also found in mice exposed by inhalation to 20–200 nm TiO^2 NPs. Such translocation along the nervous system to the brain was size-dependent (Wang *et al.*, 2005). To determine whether olfactory translocation also occurs with agglomerated NPs, Elder *et al.* (2006) exposed rats to agglomerates of 30 nm MnO_2 NPs via inhalation. The study showed that the olfactory neuronal pathway is efficient for translocating manganese to the CNS, and that this can result in inflammatory changes. However, whether entire particles were translocated or solubilised manganese was transported to the CNS must be a matter for future research.

A study by Yu *et al.* (2007) further confirmed the ability of inhaled gold NPs to accumulate in the olfactory bulb and to translocate into systemic organs of rats, indicating a cumulative effect of gold exposure and stressing the importance of kinetic factors in the systemic distribution of NPs. More recently, Tang *et al.* (2008) investigated the accumulation of silver NPs in the brain of rats, and their effect on the blood-brain barrier. The results showed that Ag NPs could traverse the blood-brain barrier and move into the brain in the form of particles, and that such NPs can induce neuronal degeneration and necrosis by accumulating in the brain over a long period of time.

Conclusions. We can argue that it is currently accepted that the translocation of NPs from lung to other tissues can occur. However, only a small fraction of inhaled NPs translocate across the air–blood barrier, entering the circulation and reaching other organs. Further studies are needed to establish the significance of this translocation. In particular, research priorities should concern:

(1) study of the metabolic fate of the translocated NPs;
(2) an assessment of nanocardiovascular toxicity, because the cardiovascular system is currently considered an important potential target for NPs;
(3) study of the toxic effects of the translocated NPs with the help of *in vitro* systems such as perfused organs and cell cultures; effects investigated would involve systems which are particularly sensitive to long-term low-dose exposure, such as the CNS and reproductive system;
(4) study of the mechanisms of NP penetration into cells and their intracellular distribution in relation to the immunotoxic responses and other effects, particularly the production of reactive oxygen species (ROS) and inflammation proteins.

24.3.2   Sampling of NPs

The exposure assessment should answer six basic questions:

- How, when and where does exposure occur?
- What or who is exposed?
- How much exposure occurs?
- How does exposure vary?
- How uncertain are exposure estimates?
- What is the likelihood that exposure will occur?

The greatest exposure to NPs is mainly related to workers during production and transfer of the intermediate or final product to other handling steps. Exposure of the public to NPs can only occur through the product or release of NPs to the environment. However, no systematic approach related to the control of NP production and products exists to our knowledge. During a 2004 workshop in Brussels, experts of the European Commission recommended the development of a nomenclature for intermediate and finished engineered nanomaterials, assigning a universally recognized Chemical Abstract Service (CAS) Number to engineered NPs (Nanotechnologies, 2004).

Separate factors have to be investigated and discussed to assess the exposure potential. Relevant factors are the probability of exposure, the extent of exposure (time and concentration) and the uptake route (inhalation, transdermal, ingestion). No systematic approach is currently available for assessing the probability of exposure related to NP production and handling processes.

For nanometre-sized aerosols, measurement of mass is not sufficient: number or area will increase as size decreases (Kittelson, 1998). Oberdörster *et al.* (2005) showed the tremendous differences in mass concentrations and surface areas for particles. In general, the use of mass concentration data alone is insufficient, and the number concentration and/or surface area need to be included. Serita *et al.* (1999) exposed rats to metallic nickel NPs at around the Japanese occupational exposure level (OEL). This OEL was based on particles larger than nanoscale, but exposure to concentrations around the OEL ($1.4 \, \text{mg/m}^3$) in the form of NPs caused severe lung injury after a single exposure.

A systematic review of applied production methods, products and their handling must be compiled before any specific recommendations can be given related to the probability of exposure. Unfortunately, no personal sampler exists to specifically measure the concentration (either mass or number concentration) of particles below 100 nm diameter. The exposure can currently only be deduced/calculated based on the limited measurements conducted so far (ISSA Chemistry Section, 2001; Kuhlbusch et al., 2004; Schulte et al., 2008b; Fujitani et al., 2008).

The ideal sampler to measure biologically relevant exposure to NP aerosols would be a personal sampling device which collects the relevant size fraction and provides an instantaneous measure of a sample surface area, or which facilitates the off-line analysis of the sample to provide a measure of surface area. Continued research is needed, and the collection of such information for exposure registries may be useful for future epidemiological studies.

One of the key questions related to nanomaterial exposure is the particle parameter to be measured. Possible parameters could be number concentration, surface area, mass concentrations, weighted size distribution, state of agglomeration, surface reactivity (e.g. ability to produce radicals, zeta potential), chemical composition and morphology. It has to be noted that mass concentrations of ultrafine particles are generally extremely low. Number concentrations, in contrast, are mainly dominated by particles in the ultrafine size range. This is the main parameter currently used in exposure measurements.

Sampling of NPs is a great challenge. The sampling strategy should ensure that the particle collection methods, including location, represent as accurately as possible the real exposure at the site in question, and methods should be developed and chosen according to the size and nature of the particles under investigation. The separation of NPs from larger particles by inertial impaction can only be achieved at a relatively high pressure drop. Considering that typical ambient atmospheric NP concentrations are less than 1 $\mu g/m^3$, collection of filter samples for gravimetric analysis and chemical characterization is only feasible with certain high-volume sampling techniques (Sarnat et al., 2003). In addition, the discrimination between existing ambient particles and engineered NPs is an important factor in the sampling strategy.

Technologies to measure some of these metrics for NPs in situ have been identified, for example to determine number concentration. However, these are not readily available, particularly in a form which may be used to measure personal exposure on a routine basis. Most instruments capable of detecting one or various characteristics are large and cumbersome devices that are not very suitable even as static monitors.

Instruments to determine number concentration are usually based on the principle of growing the particles in a saturated vapor atmosphere so that they become large enough to be detected by optical counters (Dahmann, 2001). Condensation nuclei or particle counters (CNC or CPC) can detect particles in situ in the size range 3–3000 nm.

Number-weighted size distributions of ultrafine particles can be measured in situ using a scanning mobility particle sizer (SMPS) or an electrical low-pressure impactor (ELPI). A differential mobility analyser (DMA) selects airborne particles of uniform sizes by selective stripping according to electrical mobility. A 13-stage low-pressure impactor (LPI) classifies the particles into different aerodynamic diameter sizes (range 6.8 nm to 10 μm). Low-pressure cascade impactors (e.g. Berner LPI) or micro-orifice cascade impactors (e.g.

MOUDI) enable the capture of aerosols on the impactor stages (Hewitt, 1995; Gijsbers *et al.,* 2002).

For the measurement of ultrafine particle concentration in terms of the surface area metric, only two instrument types have been identified: the epiphaniometer and the aerosol diffusion charger/electrometer (Maynard, 2003). A method to estimate the surface area of airborne particles from the results of direct reading instruments is based on the fractal dimensions of the particles. The fractal dimension is related to the space-filling properties of particles (Rogak *et al.*, 1993).

As stated, particle identification (e.g. morphology and geometry), identification of single particles and agglomerates, and determination of optical diameters can be derived from electron microscope analysis.

Particle size-selective measurement of particle concentration seems to be limited in accuracy for both particle size distribution (i.e. SMPS results) and for particle number concentration (i.e. ELPI results). The results of the measurements can confirm the suspected temporal and spatial variation in (number) concentration and aerosol size distribution. From the results it can be determined that both location and time greatly affect both parameters (Brouwer *et al.,* 2004). Collection of (ultrafine) particles from the workplace air and further analysis generated useful additional information. It is important that the sampling of aerosols should be either (ultrafine) size-selective (e.g. ELPI samples) or from the breathing zone (e.g. personal air samples). Conventional static size-selective sampling methods, where the samples are analysed gravimetrically, are very much hindered by detection limits. Moreover, only a few stages of LPIs cover the whole NP range: in the case of the Berner LPI, only the last stage and the back-up filter.

Use of low pressure stages in cascade impactors allows the collection of particles as small as 50 nm in devices such as the electrical low-pressure impactor. The hypersonic impactors are capable of collecting particles down to 50 nm, and focusing impactors are capable in principle of operating below 10 nm. However, deposition forces are necessarily high, leading to the possibility of particle damage.

Each of the measurement methods has its drawbacks, but when used in combination they may give full insight into the presence of ultrafine particle aerosols in the workplace.

Both the workplace study and the experimental study showed that there are spatial variations in both particle number concentration and size distribution. Therefore, the use of static samplers at fixed locations hampers the interpretation of the results for personal exposure of ambulatory workers. Even for workers who are positioned at fixed workstations, the interpretation will be very inaccurate. Further research is needed to enable accurate quantification of personal exposure of workers, and to identify determinants of exposure (Brouwer *et al.,* 2004)

To ameliorate the accuracy of risk assessment we need the development of a model describing the dispersion and transformation of NPs and their agglomerates in the working environment. Further, it is crucial to determine standard measurement methods and strategies to harmonize exposure data for risk assessment and to enable the development of safety standards; further, it is essential to standardize the number concentration measurements including definition of the lower and upper particle size range determined at a defined relative humidity.

24.4   Hazard Identification and Characterization

Health effects data on workers exposed to NPs are limited because of the incipient nature of the field, the relatively small number of workers potentially exposed to date, and the lack of time for chronic disease to develop and be detected. Human data derive from exposures to ultrafine and fine particles, which have been assessed in epidemiological air pollution studies and in studies of occupational cohorts exposed to mineral dusts, fibers, welding fumes, combustion products, and poorly soluble, low-toxicity particulates such as titanium dioxide and carbon black (Maynard and Kuempel, 2005; Nel *et al.*, 2006). Many data, essentially related to exposure to engineered NPs, also derive from animal studies (Donaldson *et al.*, 2004, 2006; Lam *et al.*, 2004, 2006; Warheit *et al.*, 2004; Oberdörster *et al.*, 2005; Shvedova *et al.*, 2005; Elder *et al.*, 2006). A strong positive correlation exists between the surface area, oxidative stress, and proinflammatory effects of NPs in the lung (Oberdörster *et al.*, 2005; Nel *et al.*, 2006); however, the extrapolation of animal studies to humans needs a prudent evaluation.

Although the findings are not conclusive, various studies of engineered NPs in animals raise concerns about the existence and severity of hazards posed to exposed workers (Kipen and Laskin, 2005). Possible adverse effects include the development of fibrosis and other pulmonary effects after short-term exposure to carbon nanotubes (Oberdörster *et al.*, 2005; Shvedova *et al.*, 2005; Lam *et al.*, 2006), the translocation of NPs to the brain via the olfactory nerve, the ability of NPs to translocate into the circulation, and the potential for NPs to activate platelets and enhance vascular thrombosis (Radomski *et al.*, 2005). For poorly soluble, low-toxicity dusts such as titanium dioxide, smaller particles in the nanometer size range appear to cause an increase in risk for lung cancer in animals on the basis of particle size and surface area (Heinrich *et al.*, 1995; Tran *et al.*, 2000; Oberdörster *et al.*, 2005).

None of these findings are conclusive about the nature and extent of the hazards, but they may be sufficient to support precautionary action. Ultimately, the significance of hazard information depends on the extent to which workers are exposed to the hazard. A need has been identified for NP-specific risk assessments that will be unique to nanotechnology (Scientific Committee on Emerging and Newly Identified Health Risks, 2005).

It is evident that there are sufficient data in the literature to conclude that, from a risk assessment point of view and for some types of NP at least, it is not valid to rely entirely on toxicological findings from testing the component of a NP of interest in another physical form.

An approach to hazard identification and characterization for a chemical of interest could be the following:

(1) If there are considerable available data in the literature, the hazardous properties of its nanoforms should be evaluated in a test battery. In this case the question is what further information is required to supplement this in order to provide the necessary confidence in the safety of the NP product. It must be reiterated that it is not scientifically valid to rely exclusively on the properties of the chemical in other physical forms for risk assessment purposes.
(2) If NPs have very similar hazard properties to other physical forms, further work on hazard assessment on the NPs may not be necessary.

(3) If the NP form has substantially different properties and no information is available on its biological properties, suitable exposure methods should be used to evaluate their toxicity.

The question that needs to be addressed in this case is: what is the full package of tests that needs to be conducted? The selection of this test battery should be informed by knowledge of the chemical, physical and biological properties, along with data on the same chemical in other physical forms. *In vitro* tests play an important role in this screening process; in principle, combined with information on the surface chemistry, these tests could provide an important early indicator of the differences or similarities in potential hazard between the NP form of a substance and other physico-chemical forms. However, characterization of the uptake, distribution, deposition and retention of NPs and the comparison with their larger-sized counterparts may require an *in vivo* approach.

Screening assessments of exposures to the more studied NPs could be conducted by developing toxicity benchmarks using the weight of evidence from studies of: (a) nanoscale forms in the toxicological and pharmacological literature; (b) fine-scale forms corrected for the proportionally greater surface area of nanoscale particles; (c) more toxic particles such as UFP; and (d) the toxicology and epidemiology of metal fumes. Uncertainties in such assessments will have to be considered given data limitations; however, collectively, the available studies are beginning to reveal important features necessary for initial risk assessments of specific NPs.

24.5   Risk Characterization

Chemicals in NP form may result in changes in both exposure (including environmental fate and persistence, uptake, metabolism, clearance and bioaccumulation) and the nature and magnitude of the adverse effects. Due to the lack of available data on the risk characterization of different NPs, no generic conclusions are possible at this stage. Consequently, each product and process that involves NPs must be considered separately in terms of:

- worker safety during the manufacture of NPs;
- safety of consumers using products that contain NPs;
- safety of local populations due to chronic or acute release of NPs from manufacturing and /or processing facilities;
- potential human health risk for re-exposure through the environment due to disposal or recycling of NP-dependent products.

In the absence of suitable hazard data, a precautionary approach should be adopted. In fact, NPs are likely to be highly biopersistent in humans and/or in environmental species. It should also be noted that there is no reliable information on the effect of the simultaneous exposure to multiple forms of NPs. It would be appropriate to assume that the effects are additive, or there could be interactions between NPs and other stressors (either physical, chemical or biological) which should be considered on a case-by-case basis.

In view of the potential for NPs to penetrate proteins, nucleic acids and other biological molecules, it is possible that unique adverse effects never previously observed for chemicals in other physical forms could occur.

The main source of information on the potential for adverse human health effects with NPs are the epidemiological studies of airborne particles in ambient air. These have shown that smaller particles of low solubility (less than 1 μm) are substantially more toxic than larger particles. Furthermore, it has been found that as far as ambient air pollution with fine particles is concerned, there is a population subgroup (including individuals with severe chronic respiratory and heart disease) that is much more sensitive to the adverse effects than the public as a whole (Mark, 2004; Environmental Protection Agency, 2004; Peters, 2005). Whether the same population would be much more sensitive to other forms of airborne NPs is uncertain, but this must be considered to be a real possibility (Aitken *et al.*, 2004).

There is some evidence from published studies that NPs can have a different (greater?) toxicity compared with larger particles of the same substance. There is evidence of a different modulation of cytokine production by mononuclear cells exposed to $CoCl_2$, microparticles and NPs (Petrarca *et al.*, 2006). Co microparticles showed a greater inhibitory effect compared with other Co forms. Its inhibitory activity was detected at all concentrations and towards all cytokines, whereas Co solutions selectively inhibited IL-2, IL-10 and TNF-alpha at maximal concentration. Co nanoparticles induced an increase in TNF-alpha and IFN-gamma release and an inhibition of IL-10 and IL-2: a cytokine pattern similar to that detected in the experimental and clinical autoimmunity. Nanotubes administered intratracheally were found to produce dose-dependent lung lesions differently from carbon black (Lam *et al.*, 2004), and to induce multifocal pulmonary granuloma, effect different from those of quartz, carbon black and graphite (Warheit *et al.*, 2004).

The conclusion from these studies is that NPs have a specific activity/toxicity and supports the case for a separate/additional risk assessment of substances that are in NP form. These differences may be attributable to the fact that they have a much greater surface area to weight ratio than larger particles and, as a consequence, they tend to be more chemically reactive and bind other substances to their surface more effectively. Because of the inverse relationship between particle size and surface area, it is imperative that dose–effect (or concentration–effect) relationships are established as a function of total surface area and/or number of particles, rather than mass units. Furthermore, a comparison should be made between the effects of the conventional and the NP form(s) of the substance.

24.6   Risk Management

In order to evaluate the risk from NPs, characterisation and evaluation of nanotechnology knowledge in relation to the two frames of reference is needed. Risk knowledge has been categorized into simple, complex, uncertain or ambiguous, based upon whether the method of evaluation was scientific (evidence-based) or societal (value-based). *Simple risks* have clear cause–effect relationships for materials and their impact. *Complex risk* refers to the difficulty in identifying the causal links and their effects. There is insufficient knowledge about the cause and effect relationship. *Uncertain risk knowledge* refers to the incompleteness of knowledge, with the available knowledge relying on uncertain assumptions, assertions and predictions. *Ambiguous risk knowledge* has variable interpretations, although it largely denotes a lack of proper understanding of the phenomena and their effects.

The risk management strategies presented in the final phase aim to tackle the hazards to society by setting out measures for avoiding, preventing, reducing, transferring or self-retaining risks. This will require an evolutionary approach, given that nanotechnology is interdisciplinary, its applications span over different sectors, and its development is taking place across the world.

Nanotechnology products will require pre-market testing for health and environmental impact, life-cycle assessment and consideration of secondary risks. In order to deal with exposure risks, nanomaterial monitoring methodologies need to be developed along with methods for reducing exposure (for example, through the use of protective equipment). The institutional risk management strategy emphasises the need for systematic liaison between industry and government, and the need for transparency in decision-making in research, development and investment.

The risk communication is the final stage of risk assessment. All good-practice approaches to communication between all relevant stakeholders would involve objectively stating information about benefits and the unintended side-effects of nanotechnology. The international disclosure of risk information by large multinational companies, and an integrated risk communication program for scientists, regulators and industrial developers, would facilitate the development of new products and their acceptance by society at large. The global nature of technology development would require involvement of all nations, encouraging public–private partnerships, sharing of standards and best practices.

Risk communication includes publication of scientific results. To make scientific data available and comparable, it is important that the specification of the studied NP form is thorough and comprehensive. The description should include the chemical composition of the NPs, including formulation components and impurities, surface chemistry, acidity/basicity, redox potential, reactivity (redox, photoreactivity, etc.) and the nature of any surface coating or adsorbed species. Furthermore, it should be supplied with detailed data on particle size range, along with information on other physical characteristics such as shape, density, surface area and charge, solubility, porosity, roughness morphology, crystallinity and magnetic properties.

At the present time the extent to which the toxicokinetics, environmental distribution and fate of NPs can be predicted from knowledge of their physico-chemical properties is unclear. In view of the limited range of substances as yet produced in NP form, and the potential for most chemical substances or mixtures to be produced in this form, caution needs to be used in extrapolation from published data.

References

Aitken RJ, Creely KS, Tran CL (2004) *Nanoparticles: An Occupational Hygiene Review*. Health and Safety Executive, Research Report 274. HSE Books: London.

Aprahamian M, Michel C, Humbert W, Devissaguet JP, Damge C (1987) Transmucosal passage of polyalkylcyanoacrylate nanocapsules as a new drug carried in the small intestine. *Biol Cell* **61**, 69–76.

Ayres JG, Borm P, Cassee FR, Castranova V, Donaldson K, Ghio A, Harrison RM, Hider R, Kelly F, Kooter IM, Marano F, Maynard RL, Mudway I, Nel A, Sioutas C, Smith S, Baeza-Squiban A, Cho A, Duggan S, Froines J (2008) Evaluating the toxicity of airborne particulate matter and nanoparticles by measuring oxidative stress potential – a workshop report and consensus statement. *Inhal Toxicol* **20**, 75–99.

Ballestri M, Baraldi A, Gatti AM, Furci L, Bagni A, Loria P, Rapaa M, Carulli N, Albertazzi A (2001) Liver and kidney foreign bodies granulomatosis in a patient with malocclusion, bruxism, and worn dentalprostheses. *Gastroenterology* **121**, 1234–1238.

Baroli B (2008) Nanoparticles and skin penetration. Are there any potential toxicological risks? *J Verbr Lebensm* **3**, 330–331.

Baroli B, Ennas MG, Loffredo F, Isola M, Pinna R, López-Quintela AM (2007) Penetration of metallic nanoparticles in human full-thickness skin. *J Investig Dermatol* **127**, 1701–1712.

Böckmann J, Lahl H, Eckert T, Unterhalt B (2000) Titan-Blutspiegel vor und nach Belastungsversuchen mit Titandioxid. *Pharmazie* **55**, 140–143.

Borm PJ, Robbins D, Haubold S, Kuhlbusch T, Fissan H, Donaldson K, Schins R, Stone V, Kreyling W, Lademann J, Krutmann J, Warheit D, Oberdorster E (2006) The potential risks of nanomaterials: a review carried out for ECETOC. *Part Fibre Toxicol* **14**, 3–11.

Brouwer DH, Gijsbers JHJ, Lurvink MWM (2004) Personal exposure to ultrafine particles in the workplace: Exploring sampling techniques and strategies. *Ann Occup Hyg* **48**, 439–453.

Brown JS, Zeman KL, Bennet WD (2002) Ultrafine particle deposition and clearance in the healthy and obstructed lung. *Am J Respir Crit Care Med* **166**, 1240–1247.

Cross SE, Innes B, Roberts MS, Tsuzuki T, Robertson TA, McCormick P (2007) Human skin penetration of sunscreen nanoparticles: in-vitro assessment of a novel micronized zinc oxide formulation. *Skin Pharmacol Physiol* **20**, 148–154.

Dahmann D (2001) Inter-comparison of mobility particle sizers (MPS). *Gefarhrstoff Reinhaltung Luft* **61**, 423–428.

de Lorenzo AJD, Darin J (1970) The olfactory neuron and the blood–brain barrier. In *Taste and Smell in Vertebrates*, Wolstenholme GEW, Knight J (eds). Churchill: London; 151–176.

Donaldson K, Stone V, Tran CL, Kreyling W, Borm PJA (2004) Nanotoxicology. *Occup Environ Med* **61**, 727–278.

Donaldson K, Aitken R, Tran L, Stone V, Duffin R, Forrest G (2006) Carbon nanotubes: a review of their properties in relation to pulmonary toxicology and workplace safety. *Toxicol Sci* **92**, 5–22.

Elder A, Gelein R, Silva V, Feikert T, Opanashuk L, Carter J (2006) Translocation of inhaled ultrafine manganese oxide particles to the central nervous system. *Environ Health Perspect* **114**, 1172–1178.

Environmental Protection Agency (2004) *Terms of the Environment*. US Environmental Protection Agency. Available online at http://www.epa.gov/OCEPAterms/

Ferin J, Oberdörster G, Penney DP (1992) Pulmonary retention of ultrafine and fine particles in rats. *Am J Respir Cell Mol Biol* **6**, 535–542.

Fujitani Y, Kobayashi T, Arashidani K, Kunugita N, Suemura K (2008) Measurement of the physical properties of aerosols in a fullerene factory for inhalation exposure assessment. *J Occup Environ Hyg* **5**, 380–389.

Gamer AO, Leibold E, van Ravenzwaay B (2006) The in vitro absorption of microfine zinc oxide and titanium dioxide through porcine skin. *Toxicol In Vitro* **20**, 301–307.

Geiser M, Rothen-Rutishauser B, Kapp N, Schürch S, Kreyling W, Schulz H, Semmler M, Im Hof V, Heyder J, Gehr P (2005) Ultrafine particles cross cellular membranes by nonphagocytic mechanisms in lungs and in cultured cells. *Environ Health Perspect* **113**, 1555–1560.

Geiser M, Casaulta M, Kupferschmid B, Schulz H, Semmler-Behnke M, Kreyling W (2008) The role of macrophages in the clearance of inhaled ultrafine titanium dioxide particles. *Am J Respir Cell Mol Biol* **38**, 371–376.

Gijsbers JHJ, Lurvink MWM, de Pater AJ, Brouwer DH (2002) *Considerations for the selection of sampling methods and strategy for ultrafine particle aerosols. Part 1: Review of methods*. TNO report V 3915/01. TNO: Zeist, The Netherlands.

Gwinn MR, Vallyathan V (2006) Nanoparticles: health effects – pros and cons. *Environ Health Perspect* **114**, 1818–1825.

Heinrich U, Fuhst R, Rittinghauseen S, Creutzenberg O, Bellmann B, Koch W (1995) Chronic inhalation exposure of Wistar rats and 2 different strains of mice to diesel-engine exhaust, carbon black, and titanium dioxide. Inhal Toxicol **7**, 533–556.

Hewitt P (1995) The particle size distribution, density, and specific surface area of welding fumes from SMAW and GMAW mild and stainless steel consumables. *Am Ind Hyg Assoc J* **56**, 128–135.

Hillery AM, Jani PU, Florence AT (1994) Comparative, quantitative study of lymphoid and non-lymphoid uptake of 60 nm polystyrene particles. *J Drug Target* **2**, 151–156.

Hillyer JF, Albrecht RM (2001) Gastrointestinal persorption and tissue distribution of differently sized colloidal gold nanoparticles. *J Pharm Sci* **90**, 1927–1936.

Hoet PHM, Bruske-Hohlfeld I, Salata OV (2004) Nanoparticles – known and unknown health risks. *J Nanobiotechnol* **2**, 12–27.

Hood E (2004) Nanotechnology: Looking as we leap. *Environ Health Perspect* **112**, A741–A749.

Hussain N, Florence AT (1998) Utilizing bacterial mechanisms of epithelial cell entry: invasin-induced oral uptake of latex nanoparticles. *Pharm Res* **15**, 153–156.

ISSA Chemistry Section (2001) Dusts, fumes and mists in the workplace. *Proceedings of the International Symposium, Toulouse, France*. ISSA: 289.

Jani P, Halbert GW, Langridge J, Florence AT (1989) The uptake and translocation of latex nanospheres and microspheres after oral administration to rats. *J Pharm Pharmacol* **41**, 809–812.

Jani P, Halbert GW, Langridge J, Florence AT (1990) The uptake and translocation of latex nanospheres and microspheres after oral administration to rats. *J Pharm Pharmacol* **42**, 821–826.

Jani PU, McCarthy DE, Florence AT (1994) Titanium dioxide (rutile) particle uptake from the rat GI tract and translocation to systemic organs after oral administration. *Int J Pharm* **105**, 157–168.

Jefferson DA (2000) The surface activity of ultrafine particles. *Phil Trans R Soc Lond A* **358**, 2683–2692.

Kanapilly GM, Diel JH (1980) Ultrafine ^{239}PuO$_2$ aerosol generation, characterization and short-term inhalation study in the rat. *Health Phys* **39**, 505–519.

Kipen HM, Laskin DL (2005) Smaller is not always better: nanotechnology yields nanotoxicology. *Am J Physiol Lung Cell Mol Physiol* **289**, L696–L697.

Kiss B, Biro T, Czifra G, Toth B, Kertesz Z, Szikszai Z, Kiss A, Juhasz I, Zouboulis C, Hunyadi J (2008) Investigation of micronized titanium dioxide penetration in human skin xenografts and its effect on cellular functions of human skin-derived cells. *Exp Dermatol* **17**, 659–667.

Kittelson DB (1998) Engines and nanoparticles: a review. *J Aerosol Sci* **29**, 575–588.

Kreyling WG, Semmler M, Erbe F, Mayer P, Takenaka S, Schulz H, Oberdörster G, Ziesenis A (2002) Translocation of ultrafine insoluble iridium particles from lung epithelium to extrapulmonary organs is size dependent but very low. *J Toxicol Environ Health A* **65**, 1513–1530.

Küchler S, Radowski MR, Blaschke T, Dathe M, Plendl J, Haag R, Schäfer-Korting M, Kramer KD (2009) Nanoparticles for skin penetration enhancement – A comparison of a dendritic core-multishell-nanotransporter and solid lipid nanoparticles. *Eur J Pharm Biopharm* **71**, 243–250.

Kuempel ED, Tran CL, Castranova V, Bailer AJ (2006) Lung dosimetry and risk assessment of nanoparticles: evaluating and extending current models in rats and humans. *Inhal Toxicol* **18**, 717–724.

Kuhlbusch TAJ, Neumann S, Fissan H (2004) Number size distribution, mass concentration, and particle composition of PM1, PM2.5, and PM10 in bag filling areas of carbon black production. *J Occup Environ Hyg* **1**, 660–671.

Kwon JT, Hwang SK, Jin H, Kim DS, Minai-Tehrani A, Yoon HJ, Choi M, Yoon TJ, Han DY, Kang YW, Yoon BI, Lee JK, Cho MH (2008) Body distribution of inhaled fluorescent magnetic nanoparticles in the mice. *J Occup Health* **50**, 1–6.

Lademann J, Richter H, Schaefer UF, Blume-Peytavi U, Teichmann A, Otberg N, Sterry W (2006) Hair follicles – a long-term reservoir for drug delivery. *Skin Pharmacol Physiol* **19**, 232–236.

Lademann J, Richter H, Teichmann A, Otberg N, Blume-Peytavi U, Luengo J, Weiss B, Schaefer UF, Lehr CM, Wepf R, Sterry W (2007) Nanoparticles – An efficient carrier for drug delivery into the hair follicles. *Eur J Pharm Biopharm* **66**, 159–164.

Lam C-W, James JT, McCluskey R, Hunter RL (2004) Pulmonary toxicity of single-wall carbon nanotubes in mice 7 and 90 days after intratracheal instillation. *Tox Sci* **77**, 126–134.

Lam CW, James JT, McCluskey RL, Arlli S, Hunter RL (2006) A review of carbon nanotube toxicity and assessment of potential occupational and environmental health risks. *Crit Rev Toxicol* **36**, 159–217.

Lekki J, Stachura Z, Dabros W, Stachura J, Menzel F, Reinert T, Butz T, Pallon J, Gontier E, Ynsa MD, Moretto P, Kerstecz Z, Sziksza Z, Kiss AZ (2007) On the follicular pathway of percutaneous uptake of nanoparticles: Ion microscopy and autoradiography studies. *Nuclear Instruments and Methods in Physics Research Section B, Beam Interactions with Materials and Atoms* **260**, 174–177.

Lomer MCE (2000) Determination of titanium dioxide in foods using inductively coupled plasma optical emission spectrometry. *Analyst* **125**, 2339–2343.

Lomer MCE, Thompson RP, Powell JJ (2002) Fine and ultrafine particles in the diet: influence on the mucosal immune response and association with Crohn's disease. *Proceedings of the Nutrition Society* **61**, 123–130.

Lomer MCE, Hutchinson C, Volkert S, Greenfield SM, Catterall A, Thompson RPH, Powell JJ (2004) Dietary sources of inorganic microparticles and their intake in healthy subjects and patients with Crohn's disease. *Br J Nutrition* **92**, 947–955.

Mark D (2004) *Nanomaterials – A Risk to Health at Work?* Health and Safety Laboratory: Buxton, UK.

Mavon A, Miquel C, Lejeune O, Payre B, Moretto P (2007) In vitro percutaneous absorption and in vivo stratum corneum distribution of an organic and a mineral sunscreen. *Skin Pharmacol Physiol* **20**, 10–20.

Maynard AD (2003) Estimating aerosol surface area from number and mass concentration measurements. *Ann Occup Hyg* **47**, 123–144.

Maynard AD, Kuempel ED (2005) Airborne nanostructured particles and occupational health. *J Nanoparticles Res* **7**, 587–614.

McCullough JS, Hodges GM, Dickson GR, Yarwood A, Carr KE (1995) A morphological and microanalytical investigation into the uptake of particulate iron across the gastrointestinal tract of rats. *J Submicrosc Cytol Pathol* **27**, 119–124.

Meidan VM, Bonner MC, Michniak BB (2005) Transfollicular drug delivery – is it a reality? *Int J Pharm* **306**, 1–14.

Meng H, Chen Z, Xing G, Yuan H, Chen C, Zhao F, Zhang C, Wang Y, Zhao Y (2007) Ultrahigh reactivity and grave nanotoxicity of copper nanoparticle. *J Radioanal Nucl Chem* **272**, 595–598.

Menzel F, Reinert T, Vogt J, Butz T (2004) Investigations of percutaneous uptake of ultrafine TiO_2 particles at the high energy ion nanoprobe LIPSION. *Nuclear Instruments and Methods in Physics Research Section B – Beam Interactions with Materials and Atoms* **220**, 82–86.

Mills NL, Amin N, Robinson SD, Anand A, Davies J, Patel D, de la Fuente JM, Cassee FR, Boon NA, Macnee W, Millar AM, Donaldson K, Newby DE (2006) Do inhaled carbon nanoparticles translocate directly into the circulation in humans? *Am J Respir Crit Care Med* **173**, 426–431.

Mortensen LJ, Oberdörster G, Pentland AP, DeLouise LA (2008) In vivo skin penetration of quantum dot nanoparticles in the murine model: the effect of UVR. *Nano Lett* **8**, 2779–2787.

Mühlfeld C, Geiser M, Kapp N, Gehr P, Rothen-Rutishauser B (2007) Re-evaluation of pulmonary titanium dioxide nanoparticle distribution using the "relative deposition index": Evidence for clearance through microvasculature. *Part Fibre Toxicol* **4**, 7.

Mühlfeld C, Gehr P, Rothen-Rutishauser B (2008a) Translocation and cellular entering mechanisms of nanoparticles in the respiratory tract. *Swiss Med Wkly* **138**, 387–391.

Mühlfeld C, Rothen-Rutishauser B, Vanhecke D, Blank F, Ochs M, Gehr P (2008b) Interaction of nanoparticles with pulmonary structures and cellular responses. *Am J Physiol Lung Cell Mol Physiol* **294**, L817–829.

NanoDerm (2007) Quality of Skin as a Barrier to ultra-fine Particles. Final Report QLK4-CT-2002-02678. Available at http://www.uni-leipzig.de/~nanoderm/Downloads/Nanoderm_

Nanotechnologies (2004) A preliminary risk analysis on the basis of a workshop organized in Brussels on 1–2 March 2004 by the Health and Consumer Protection Directorate General of the European Commission. Available at europa.eu.int/comm/health/ph_risk/documents/ev_20040301_en.pdf

Nel A, Xia T, Mädler L, Li N (2006) Toxic potential of materials at the nanolevel. *Science* **311**, 622–627.

Nemmar A, Hoet PH, Vanquickenborne B, Dinsdale D, Thomeer M, Hoylaerts MF, Vanbilloen H, Mortelmans L, Nemery B (2002) Passage of inhaled particles into the blood circulation in humans. *Circulation* **105**, 411–414.

Nohynek GJ, Lademann J, Ribaud C, Roberts MS (2007) Grey goo on the skin? Nanotechnology, cosmetic and sunscreen safety. *Crit Rev Toxicol* **37**, 251.

Nohynek GJ, Dufour EK, Roberts MS (2008) Nanotechnology, cosmetics and the skin: is there a health risk? *Skin Pharmacol Physiol* **21**, 136–149.

Nurkiewicz TR, Porter DW, Hubbs AF, Cumpston JL, Chen BT, Frazer DG, Castranova V (2008) Nanoparticle inhalation augments particle-dependent systemic microvascular dysfunction. *Part Fibre Toxicol* **5**, 1.

Oberdörster G, Ferin J, Lehnert BE (1994) Correlation between particle size, in vivo particle persistence, and lung injury. *Environ Health Perspect* **102**(Suppl. 5), 173–179.

Oberdörster G, Sharp Z, Atudorei V, Elder A, Gelein R, Lunts A, Kreyling W, Cox C (2002) Extra-pulmonary translocation of ultrafine carbon particles following whole-body inhalation exposure of rats. *J Toxicol Environ Health A* **65**, 1531–1543.

Oberdörster G, Sharp Z, Atudorei V, Elder A, Gelein R, Kreyling W (2004) Translocation of inhaled ultrafine particles to the brain. *Inhal Toxicol* **16**, 437–445.

Oberdörster G, Oberdörster E, Oberdörster J (2005) Nanotoxicology: an emerging discipline evolving from studies of ultrafine particles. *Environ Health Perspect* **113**, 823–839.

Peters A (2005) Particulate matter and heart disease: evidence from epidemiological studies. *Toxicol Appl Pharmacol* **207**, S477–S482.

Petrarca C, Perrone A, Verna N, Verginelli F, Ponti J, Sabbioni E, Di Giampaolo L, Dadorante V, Schiavone C, Boscolo P, Mariani Costantini R, Di Gioacchino M (2006) Cobalt nano-particles modulate cytokine in vitro release by human mononuclear cells mimicking autoimmune disease. *Int J Immunopathol Pharmacol* **19**, 11–14.

Pope CA, Burnett RT, Thurston GD, Thun MJ, Calle EE, Krewski D (2003) Cardiovascular mortality and long-term exposure to particulate air pollution: epidemiological evidence of general pathophysiological pathways of disease. *Circulation* **109**, 71–77.

Radomski A, Jurasz P, Alonso-Escolano P, Drew M, Morandi M, Tadeusz M, *et al.* (2005) Nanoparticle-induced platelet aggregation and vascular thrombosis. *Br J Pharmacol* **146**, 882–893.

Renwick LC, Donaldson K, Clouter A (2001) Impairment of alveolar macrophage phagocytosis by ultrafine particles. *Toxicol Appl Pharmacol* **172**, 119–127.

Rogak SN, Flagan RC, Nguyen HV (1993) The mobility and structure of aerosol agglomerates. *Aerosol Sci Technol* **18**, 25–47.

Rouse JG, Yang J, Ryman-Rasmussen JP, Barron AR, Monteiro-Riviere NA (2007) Effects of mechanical flexion on the penetration of fullerene amino acid-derivatized peptide nanoparticles through skin. *Nano Lett* **7**, 155–160.

Royal Society (2004) *Nanoscience and Nanotechnologies: Opportunities and Uncertainties.* Royal Society, Science Policy Section: London.

Ryman-Rasmussen JP, Riviere JE, Monteiro-Riviere NA (2006) Penetration of intact skin by quantum dots with diverse physicochemical properties. *Toxicol Sci* **91**, 159–165.

Salata O (2004) Applications of nanoparticles in biology and medicine. *J Nanobiotechnology* **30**, 2–3.

Sarnat J, Demokritou P, Koutrakis P (2003) *Measurement of Fine, Coarse and Ultrafine Particles.* School of Public Health, Harvard University: Boston, MA.

Schulte P, Geraci C, Zumwalde R, Hoover M, Kuempel E (2008a) Occupational risk management of engineered nanoparticles. *J Occup Environ Hyg* **5**, 239–249.

Schulte PA, Trout D, Zumwalde RD, Kuempel E, Geraci CL, Castranova V, Mundt DJ, Mundt KA, Halperin WE (2008b) Options for occupational health surveillance of workers potentially exposed to engineered nanoparticles: state of the science. *J Occup Environ Med* **50**, 517–526.

Schulz J, Hohenberg H, Pflucker F, Gartner E, Will T, Pfeiffer S, Wepf R, Wendel V, Gers-Barlag H, Wittern K-P (2002) Distribution of sunscreens on skin. *Adv Drug Deliv Rev* **54**(S1), S157–S163.

Scientific Committee on Consumer Products (2007) European Commission, *Opinion on Safety of Nanomaterials in Cosmetic Products.* Available at http://ec.europa.eu/health/ph_risk/risk_en.htm

Scientific Committee on Emerging and Newly Identified Health Risks (2005) *Opinion on the Appropriateness of Existing Methodologies to Assess the Potential Risks Associated with Engineered and Adventitious Products of Nanotechnologies.* Health & Consumer Protection Directorate-General, European Commission: Brussels.

Semmler M, Seitz J, Erbe F, Mayer P, Heyder J, Oberdörster G, Kreyling WG (2004) Long-term clearance kinetics of inhaled ultrafine insoluble iridium particles from the rat lung, including transient translocation into secondary organs. *Inhal Toxicol* **16**, 453–459.

Serita F, Kyono H, Seki Y (1999) Pulmonary clearance and lesions in rats after a single inhalation of ultrafine metallic nickel at dose levels comparable to the threshold limit value. *Ind Health* **37**, 353–363.

Shakweh M, Besnard M, Nicolas V, Fattal E (2005) Poly(lactide-co-glycolide) particles of different physico-chemical properties and their uptake by Peyer's patches in mice. *Eur J Pharma Biopharmac* **61**, 1–13.

Shvedova AA, Kisin EK, Mercer R, Murray AR, Johnson VJ, Potapovich AI (2005) Unusual inflammatory and fibrogenic pulmonary responses to single-walled carbon nanotubes in mice. *Am J Physiol Lung Cell Mol Physiol* **289**, L698–L708.

Takenaka S, Karg E, Roth C, Schulz H, Ziesenis A, Heinzmann U, Schramel P, Heyder J (2001) Pulmonary and systemic distribution of inhaled ultrafine silver particles in rats. *Environ Health Perspect* **4**, 547–551.

Takenaka S, Karg E, Kreyling WG, Lentner B, Schulz H, Ziesenis A, Schramel P, Heyder J (2004) Fate and toxic effects of inhaled ultrafine cadmium oxide particles in the rat lung. *Inhal Toxicol* **16**(Suppl. 1), 83–92.

Takenaka S, Karg E, Kreyling WG, Lentner B, Möller W, Behnke-Semmler M, Jennen L, Walch A, Michalke B, Schramel P, Heyder J, Schulz H (2006) Distribution pattern of inhaled ultrafine gold particles in the rat lung. *Inhal Toxicol* **18**, 733–740.

Tang J, Xiong L, Wang S, Wang J, Liu L, Li J, Wan Z, Xi T (2008) Influence of silver nanoparticles on neurons and blood-brain barrier via subcutaneous injection in rats. *Appl Surface Sci* **255**, 502–504.

Tinkle SS, Antonini JM, Rich BA, Roberts JR, Salmen R, DePree K, Adkins EJ (2003) Skin as a route of exposure and sensitization in chronic beryllium disease. *Environ Health Perspect* **111**, 1202–1208.

Tran CL, Buchanan D, Cullen RT, Searl A, Jones AD, Donaldson K (2000) Inhalation of poorly soluble particles. II. Influence of particle surface area on inflammation and clearance. *Inhal Toxicol* **12**, 1113–1126.

Tsuji JS, Maynard AD, Howard PC, James JT, Lam C, Warheit DB, Santamaria AB (2006) Research strategies for safety evaluation of nanomaterials, Part IV: risk assessment of nanoparticles. *Toxicol Sci* **89**, 42–50.

Videira MA, Botelho MF, Santos AC, Gouveia LF, de Lima JJ, Almeida AJ (2002) Lymphatic uptake of pulmonary delivered radiolabelled solid lipid nanoparticles. *J Drug Target* **10**, 607–613.

Vogt A, Combadiere B, Hadam S, Stieler KM, Lademann J, Schaefer H, Autran B, Sterry W, Blume-Peytavi U (2006) 40 nm, but not 750 or 1,500 nm, nanoparticles enter epidermal CD1a + cells after transcutaneous application on human skin. *J Invest Dermatol* **126**, 1316–1322.

Wang JX, Chen CY, Sun J, Yu HW, Li YF, Li B, Xing L, Huang YY, He W, Gao YX, Chai ZF, Zhao YL (2005) Translocation of inhaled TiO$_2$ nanoparticles along olfactory nervous system to brain studied by synchrotron radiation X-ray fluorescence. *High Energy Physics & Nuclear Physics* **29**, 76–79.

Warheit DB, Laurence BR, Reed KL, Roach DH, Reynolds GAM, Webb TR (2004) Comparative toxicity assessment of single-wall carbon nanotubes in rats. *Toxicol Sci* **76**, 117–125.

Warheit DB, Webb TR, Reed KL, Frerichs S, Sayes CM (2007) Pulmonary toxicity study in rats with three forms of ultrafine-TiO$_2$ particles: differential responses related to surface properties. *Toxicology* **230**, 90–104.

Yu LE, Yung LL, Ong C, Tan Y, Balasubramaniam KS, Hartono D, Shui G, Wenk MR, Ong W (2007) Translocation and effects of gold nanoparticles after inhalation exposure in rats. *Nanotoxicology* **1**, 235–242.

25

Application of Toxicology Studies in Assessing the Health Risks of Nanomaterials in Consumer Products

J.S. Tsuji, F.S. Mowat, S. Donthu and M. Reitman

25.1 Introduction

Nanotechnology applications have the potential to improve the performance and functionality of virtually every material in our lives, and are increasingly used to enhance consumer products. More than 800 products or product lines containing nanotechnology were on the market as of August 2008, representing a 300 % increase since March 2006 (Project on Emerging Technologies, 2008), or the introduction on average of three to four new products per week. Acceleration in numbers of products containing nanomaterials has attracted considerable concern regarding whether sufficient toxicology studies exist, or could even be conducted to keep pace with the multitude of types of nanomaterials and product applications currently in development and on the market. Although federal funding has increased each year for researching health and safety aspects of nanomaterials, research on such implications still remains a fraction (4 %) of the total budget for nanotechnology research (National Science and Technology Council, 2007). Nevertheless, the success of these novel materials will depend on proactively addressing potential health and environmental risks. Recognition of adverse health effects of other 'miracle materials' (e.g. asbestos, lead), long after widespread use in products, highlights the importance of conducting risk assessments to guide the design and manufacture of safer products.

This chapter describes approaches for using toxicology studies on nanomaterials and related small-scale substances to aid in understanding the possible health risks of their use in consumer products. The overall focus is on approaches for addressing product safety

Nanotoxicity: From In Vivo *and* In Vitro *Models to Health Risks*   Edited by Saura Sahu and Daniel Casciano
© 2009 John Wiley & Sons, Ltd

for consumers, although other issues beyond the scope of this chapter are also necessary for a full product life-cycle assessment (e.g. risks to workers manufacturing products, or to ecological receptors from environmental releases). Examples from the literature on the primary nanomaterials currently in products are used to illustrate various applications and interpretations of toxicity studies for risk assessment; however, this chapter does not attempt to review the complete toxicological literature on nanomaterials. Exposure assessment of nanomaterials of relevance for consumer products is also included as an essential part of interpreting and applying toxicology studies in risk assessment.

25.2 Health Risk Assessment

Health risk assessments are critical for evaluating the potential hazards associated with nanomaterials to address regulatory and public concerns. Although much of the health risk guidance was developed to assess radiation or chemical-contaminated sites under specific federal programs (e.g. Environmental Protection Agency, 1989), the basic elements and framework are broadly applicable, including assessing potential health risks from emerging technologies (i.e. nanotechnology) in consumer products. Beyond the initial hazard identification step, assessment of exposure and toxicity comprise the main components for characterizing potential risks associated with specific substances and products, and aid in understanding data gaps and related uncertainties (National Research Council, 1983). The risk assessment process can be adapted to assessment of the safety of consumer products containing nanomaterials, although some of the questions and issues for consideration will change to be specific for nanomaterials (e.g. Hoet *et al.,* 2004; Oberdörster *et al.,* 2005a; Nel *et al.,* 2006; Maynard, 2006a,b; Thomas *et al.,* 2006; Tsuji *et al.,* 2006; Medley *et al.,* 2007).

The use of health risk assessment in the development of consumer products is an iterative process involving various components of exposure and toxicity based on existing literature, new studies and real-time monitoring, and continual refinement of data (see Figure 25.1). The general approach can be conducted in phases, beginning with a screening assessment using the readily available literature and worst-case assumptions to characterize risks. This initial phase also identifies key areas of uncertainty to focus subsequent investigations to fill data gaps. If the screening assessment notes potential risks, more detailed exposure characterization, measurement, or toxicity testing can be conducted to increase the accuracy of risk estimates. These results then inform decisions on whether to produce the product, or reformulate or change the product to reduce potential risks.

This approach is similar to the broader framework described by the joint recommendations of Environmental Defense (ED) and DuPont, whereby an iterative process is used to assess, prioritize and generate data for identifying and reducing potential environmental, health and safety risks associated with nanoparticles over their full life cycle from manufacture to disposal (Medley *et al.,* 2007).

25.2.1 Exposure Assessment

Exposure assessment, as the 'process of estimating concentration or intensity, duration, and frequency of exposure to an agent that can affect health' (Last, 2001), is the first step in assessing the safety of nanotechnology in products. Given current uncertainties in toxicity

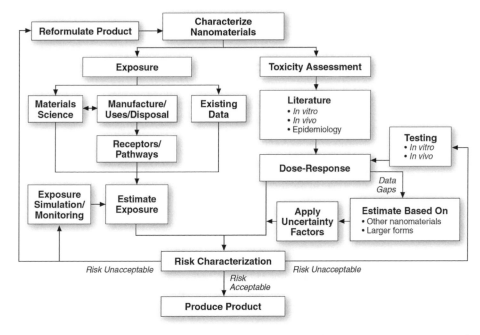

Figure 25.1   *Application of risk assessment to developing products containing nanomaterials.*

for many nanomaterials, particularly those that are engineered, minimizing exposure is of great importance in mitigating health risks (Maynard, 2006a). Although exposure to nano-materials is expected to be higher for workers and researchers who are synthesizing nano-materials from raw materials and mixing or working with nanoscale components, exposure for consumers potentially involves longer duration, varying intensity depending on use patterns and the stage in the product life-cycle, and greater susceptibility for some individuals.

25.2.1.1   Nanotechnology in Products and Current Exposures

The most comprehensive summary of uses of nanotechnology in consumer products is the Nanotechnology Consumer Products Inventory (Inventory), part of the Project on Emerging Nanotechnologies (PEN) maintained online by the Woodrow Wilson International Center for Scholars. As of 21 August 2008, the Inventory reports more than 800 products or product lines containing nanotechnology from 21 countries and produced by 420 different companies (Project on Emerging Nanotechnologies, 2008).[1] The majority of the products on the Inventory (totalling 426) are produced in the US. More than half of the total products listed for all countries are in the category of health and fitness (502 products), with its largest subcategories being personal care (153), cosmetics (126), clothing (115), and sporting goods (82) (Table 25.1).

---

[1] Inclusion of a product in the Inventory is based on three criteria: (a) the product is available and can be readily purchased; (b) nanomaterials in the product are identified either by the product manufacturer or other source; and (c) the description and claims of the nanotechnology used in the product are reasonable.

Table 25.1   Consumer products containing nanomaterials.

Category	Description of products	Example nanomaterials	Number of products
Health and fitness	Cosmetics, clothing, sunscreens, personal care items, filtration equipment, sporting goods	Silver, TiO_2, ZnO, fullerenes, CNTs, silica/silicon, micelles	502
Home and garden	Cleaning products, construction materials, paints, home furnishings	TiO_2, ZnO, silver	91
Food and beverage	Foods, food storage and cooking, supplements	Silver, starch, nanoclays, lipid/carbohydrate micelles	80
Electronics and computers	Computer hardware, cameras and film, displays, communication devices, televisions	Nanowires, coatings, CNTs, nanoceramics, silicon, silver	56
Cross cutting	Coatings or nanofilms on multiple household products, windows, surfaces	Silver, TiO_2, coated nanofibers	52
Automotive	Products for the exterior, maintenance, components, fuel additives, and accessories of vehicles	Nanofibers, cerium oxide, nanoceramics, nanoclays, coatings	43
Appliances	Batteries, heating and cooling devices, large kitchen appliances, power tools, laundry items, clothing care	Silver, nano-phosphate lithium-ion battery	31
Goods for children	Toys, games, sunscreens and personal care for children, and other children's products	Silver, TiO_2, ZnO	18

Note: Categories, product types, and numbers reported as of August 2008 (PEN, 2008). Product numbers within categories reflect some duplication because some products are grouped in more than one category. Nanomaterials are as reported by product information links on PEN (2008).

According to the Inventory, the major types of nanomaterials in products are mainly smaller versions of existing macroscale particles (e.g. metals or metal oxides), but also include engineered carbon-based nanomaterials such as fullerenes and carbon nanotubes (CNTs). Overall, silver (235) is the dominant type of nanomaterial mentioned in product descriptions, followed by carbon (including CNTs and fullerenes; 71), titanium (including titanium dioxide, TiO_2; 38), silica and silicon (31), zinc (including zinc oxide, ZnO; 29), and gold (16). Several other types of nanomaterials/applications (e.g. polymers, clays, quantum dots, organic micelles) are also incorporated in products listed in the Inventory and are noted by other sources.[2] In general, product advertisements do not provide information on

_____

[2] See, for example, www.envirosan.com/, www.voyle.net/Nano%Products%202005-0060.htm, and www.nanogreensciences.com/

product safety, and those that do, claim they are safe but lack detailed information on safety testing. Product information in many cases also lacks specifics on the type of nanoscale substance incorporated in the product. Consumer and regulatory confidence in the safety of products containing nanotechnology will require more information on the assessment of risks.

Nevertheless, exposures to small-scale substances are not entirely novel and numerous examples abound:

- Nanosized particles from natural emissions (e.g. volcanic eruptions, wind-blown dust, forest fires), as well as from and in diesel emissions and air pollution (Oberdörster *et al.,* 2005a).
- Carbonaceous material mixed with silicates and TiO_2 nanocrystals in the lungs of the 5300-year-old mummy discovered in receding ice in the Tyrolean mountain region of Austria (Murr *et al.,* 2004).
- Pigmentary TiO_2 (typically around 200 nm; Morgans, 1990) used in paints and coatings for nearly a century.
- Antimicrobial properties of silver ions and colloids exploited by ancient civilizations for hygienic and medical purposes (Drake and Hazelwood, 2005; Chen and Schluesener, 2008).
- Multiwalled carbon nanotubes (MWCNTs) along with complex nanocrystal aggregates in a 10,000-year-old Greenland ice core (Esquivel and Murr, 2004).
- MWCNTs emitted by natural gas combustion and commonly found in urban air (Lam *et al.,* 2006).

In addition, particles, fibers and small structures have been extensively researched, including nuisance dusts, asbestos (Eastern Research Group, 2003a; Cugell and Kamp, 2004), manmade mineral fibers (Lockey *et al.,* 1996, 2002; International Agency for Research on Cancer, 2002; Eastern Research Group, 2003b), metal fumes (Kuschner *et al.,* 1997), and ultrafine particulates (UFP) (Englert, 2004; Oberdörster *et al.,* 2005a), and a large body of information related to exposure and effects of these substances is available. Moreover, conventional organic molecules have nanoscale dimensions, and colloid and polymer scientists have decades of history with micelles, dendrimers, and other nanoscale materials. The US Environmental Protection Agency (2007) recognizes that the traditional combination of polymeric units into macromolecules (e.g. polyethylene, polystyrene, polyvinyl alcohol, polydimethylsiloxane) does not constitute nanomaterial exposure, but that free nanoscale particles created by conditions of polymerization or post-reaction processing would be considered 'engineered nanoscale nanomaterial'.

25.2.1.2   Exposure Routes of Most Concern

Consumer exposure to nano-containing products could occur through use, misuse or disposal of the product (e.g. Mueller and Nowack, 2008; Benn and Westerhoff, 2008). Potential routes of exposure may include dermal uptake, oral intake via ingestion or mouthing (primarily for children), inhalation, injection (for medical applications), or a combination of pathways. For exposures to occur through these routes the nanomaterial must be released from the product and able to enter the human body. The likelihood of release is specific to both the nanomaterial and the product. Many nanomaterials in products are agglomerated, encapsulated or otherwise physically altered from their original particulate form.

Inhalation and dermal routes have received the greatest attention for nanomaterials in products, given existing knowledge on exposures to small particles. Oral exposures are also a concern for nanomaterials migrating out of food- and food supplement-related applications, as well as children's products. As listed in the Inventory, nano-silver has been included in food containers, food contact articles, and even infant teething rings to impart antimicrobial properties.

25.2.2   Exposure Measurement Techniques

Assessment of consumer exposure depends upon a number of factors including nanomaterial characteristics, type of product matrix or carrier, amount (or dose) of exposure, and the frequency and conditions of use. Several methods are available to quantify exposure to nanomaterials, each with their strengths and limitations (e.g. Aitken *et al.*, 2004; Tsuji *et al.*, 2006; Maynard, 2006b), and research is ongoing in this area, particularly that funded by the National Institute for Occupational Safety and Health (2008). In general, products in which the nanomaterial is encapsulated or inaccessible to the consumer would have low exposure potential under expected uses. Such a distinction is well recognized by various advisory bodies (Scientific Committee on Emerging and Newly Identified Health Risks, 2006; Dekkers *et al.*, 2007), and can be evaluated relatively expediently through various forms of microscopic or spectroscopic analysis either directly, or in combination with materials testing that may include physical and environmental stresses.

25.2.2.1   Characterization of Nanomaterials within Products

Physical, chemical, mechanical and biological properties of materials are sensitive to both external (size, shape and topography) and internal form (crystallinity and microstructure).[3] The structure of substances used in the nanoscale and their interaction with other materials within products determines their innovative properties and performance as well as their potential for exposure.

Although some nanomaterials are used in particulate form, many more are incorporated into binders or matrices that carry or constrain them (e.g. CNTs in tennis rackets; ZnO and TiO_2 in coatings). Once incorporated, they may not retain nano-dimensions or characteristics, such as surface chemistry, and their microstructure may be affected by the formulation, use and aging of the composite system that contains them. Furthermore, the task of characterizing a sufficiently representative portion of the product can be more challenging when a matrix or binder is introduced.

Many of the same methods used to characterize nanomaterials and their interaction with cells or tissues can be applied to formulated products to characterize size, location, distribution and chemistry of nanomaterials in the products, thereby aiding in evaluation of exposure potential. Several of these tools can be used to monitor temporal changes of nanomaterials in the matrix, characteristics of nanomaterials-matrix interfaces, effects of nanomaterials on the matrix stability and other interactions between nanomaterials and matrix (Table 25.2). Scanning electron microscopy (SEM), transmission electron microscopy (TEM), and other microscopy methods as well as selected spectroscopic methods can

---

[3] Some examples include changes in thermal transitions, catalytic activity, optical properties and biomolecular adsorption behavior for nanoscale materials, compared with traditional counterparts.

Table 25.2 Advantages and disadvantages of selected materials characterization techniques for evaluating products containing nanomaterials.

Method	Primary utility	Advantages	Disadvantages
Scanning electron microscopy	Characterize particles within product matrix, elemental composition, topography, encapsulation	Simple, inexpensive, high spatial resolution of surfaces, can handle thicker and larger samples. With variable pressure models, gels and liquids can be imaged	Limited chemical identification, especially of light elements; limited information collected for sample bulk
Transmission electron microscopy	Characterize particles within product matrix, bonding, encapsulation	Very high spatial resolution, can visualize atomic structure and provide chemical information at atomic resolution	Complicated sample preparation; expensive; limited to very thin samples with small sample area; requires high vacuum incompatible with wet samples; possible sample destruction by high energy electron beam
Atomic force microscopy	Image particles on a surface; monitor chemical reactions in liquids	High spatial resolution, provides information on wear/degradation; can examine dry or wet surfaces	Limited scan size, requires small samples with flat surfaces; limited information collected for subsurface voids or particles
Scanning near-field ultrasound holography	Image subsurface features such as voids or near- surface particles; examine particles in cells	Non-destructive subsurface characterization of particles, high spatial resolution	Limited scan size, requires small samples with flat surfaces
X-ray diffraction	Analyze nanomaterial size, crystalline structure, spatial distribution within product matrix	Depth of sample penetration (a few microns); characterization of nanoparticle-matrix interactions including diffusion within matrix over time	Requires high nanoparticle concentration because of low intensity of diffraction peaks, otherwise synchrotron radiation needed; long collection times
X-ray photon spectroscopy	Analyze surface chemistry, coatings	Can determine valency and chemical composition	Must be conducted in ultra-high vacuum (challenging for powders, incompatible with volatile samples); limited spatial resolution (probe size is several millimeters)
Raman spectroscopy	Analyze surface chemistry	Provides information on chemical bonding and local structure; minimal sample preparation, rapid results	Materials must be Raman active; limited spatial resolution

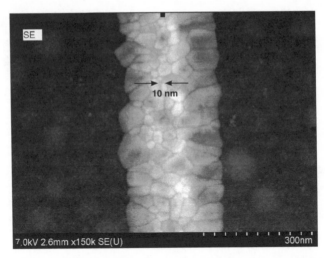

Figure 25.2 *Secondary electron image of polycrystalline ZnO nanostructure taken using Hitachi 4800 model SEM. Structural features as small as 10 nm can be imaged.*

provide a qualitative understanding of the effectiveness of the bonding between nanomaterials and the surrounding material over time, and therefore the effectiveness of encapsulation and potential for particle release. Ultimately, the stability of the nanomaterial–matrix interface controls possible nanomaterial release, leading to human exposure.

Direct imaging in secondary electron mode with SEM can be used to detect nanoparticles near a surface with little sample preparation and spatial resolution as fine as <10 nm (see Figure 25.2), while backscattered electron mode in SEM is particularly useful for examining the top hundreds of nanometers of a surface containing different elements with large atomic number differences, such as clay-filled polymers. Energy dispersive X-ray spectrometry (EDS) maps in SEM can be useful to detect micron-scale non-uniformities in nanomaterial distribution within the product, thereby allowing one to identify 'imperfections' or areas of differing composition. Non-uniformities from clumping may result in less potential for nanomaterials to be released in nano-size, but may have other effects on the desired properties of the formulated product.

TEM allows for higher spatial resolution compared with SEM (approaching sub-angstrom scale imaging), but typically requires more intensive sample preparation and is limited to smaller lateral dimension samples. Two modes of TEM, imaging and diffraction, can be used to collect morphological and crystalline information from the same location of a sample, while the spectroscopic techniques of EDS and electron energy loss spectroscopy (EELS) can provide localized chemical identification (Figure 25.3). Unlike EDS, EELS is sensitive to light elements, including hydrogen and carbon, making it particularly useful for examining organic coatings on nanoparticle surfaces, and understanding nanomaterial–matrix interactions in products. TEM can be used, for example, for visualization of individual nanoparticles, agglomerates, and the distribution and condition of pigments, modifiers and fillers in an organic binder (e.g. paint or coating), thereby providing insight into the effects of mixing conditions and formulation on the form and availability (i.e. exposure potential) of nanoscale components.

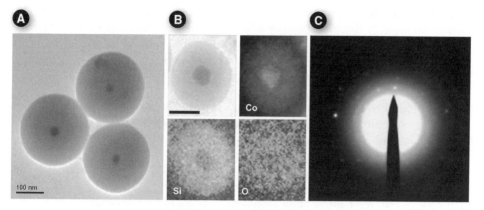

Figure 25.3   *Silica-encapsulated cobalt nanoparticles. (A) TEM image. (B) EELS element maps (scale bar is 100 nm). (C) Diffraction pattern from a single particle confirms the crystalline nature of the cobalt cores (images courtesy of Dr Mohammed Aslam). Reprinted with permission of Dr. Mohammed Aslam, Indian Institute of Technology, Bombay.*

Scanning probe microscopy (SPM) is a class of techniques in which a probe is scanned a few angstroms above the sample surface over relatively small areas ($\leq 100 \ \mu m^2$). The most common technique is atomic force microscopy (AFM). Because AFM is capable of atomic-scale resolution, it is well suited for imaging of nanoparticles on a surface, and can provide information related to erosion or other age-related effects on the surface of an encapsulating matrix. AFM can also be used to monitor morphological changes, and is routinely used in liquids to monitor chemical reactions such as protein crystallization. Scanning near-field ultrasound holography (SNFUH), a relatively new SPM technique based on acoustic interference (Diebold, 2005; Shekhawat and Dravid, 2005), is capable of imaging of features such as voids and nanoparticles that are buried underneath the surface with high spatial resolution (10–100 nm). SNFUH has been used for nondestructive imaging of nanoparticles inside cells (single-walled carbon nanohorns inside mouse erythrocytes; Tetard *et al.*, 2008), an almost impossible task with any other technique. Thus, SNFUH increases the functionality of the regular SPM techniques and may enable direct studies of the fate of nanoscale substances in organisms.

X-rays consist of electromagnetic radiation with sub-angstrom wavelength commonly used to study the crystalline structure of materials. When X-rays impinge on crystalline material, a diffraction pattern consisting of intensity peaks is formed. The diffraction peaks provide a wealth of information related to nanoparticle size, as well as the type of crystal structure and the atomic distances in the material. Thus, X-rays are useful for investigating polymorphs, materials with the same chemical composition but different crystal structures, such as TiO_2. TiO_2 exists as multiple polymorphs: anatase, rutile and brookite. These polymorphs of TiO_2 exhibit distinctly different chemical and physical properties. For example, anatase generally has highly photocatalytic activity compared with the other two forms, an effect that is enhanced by the increased relative surface area of nanoparticles. This property is exploited in self-cleaning films and coatings that degrade adhering organic matter; however, this property can also affect the rate of binder degradation and particle release in coatings.

X-rays can be used to probe the structure of samples in powder or solid form. The penetration depth of X-rays is typically a few microns, such that particles near and below a surface can be systematically analyzed, allowing, for example, the evaluation of the diffusion rates of nanomaterials within a product matrix (Guico *et al.*, 2004), or the study of segregation kinetics associated with *in situ* growth of nanoparticles on a surface (Renaud *et al.*, 2003).

Surface chemistry affects characteristics such as hydrophobicity/hydrophilicity, propensity for agglomeration, surface reactivity, and dispersability and solubility; it is fundamental to interactions with a binder or matrix, and is a factor in toxicological effects. Spectroscopic techniques such as x-ray photon spectroscopy (XPS) and Raman spectroscopy are suited to the study of nanomaterial surface chemistry. For example, XPS can be used to determine chemical composition, oxidation state (e.g. titanium in TiO, Ti_2O_3, Ti_3O_5 and TiO_2; Song *et al.*, 2005), and, more importantly, to quantify relative elemental composition. Because XPS requires little sample preparation and probes the top few nanometers of the sample, it is routinely used to identify the chemical composition of surface layers and to investigate the presence of surface coating layers (e.g. Wu *et al.*, 2004). Raman spectra provide useful insights into chemical bonding at the atomic level, local structure, and morphology of materials that are Raman active (Ferraro and Nakamoto, 1994). For example, although graphite and diamond are chemically just carbon, they are distinguished by unique Raman spectra. Raman spectroscopy and TEM were used to characterize CNTs in homogenated organ samples from mice injected with CNTs (Yang *et al.*, 2008). The presence of CNTs in organs was revealed by a spectral characteristic of graphite carbon. Raman spectroscopy has also been used to study size distributions of semi-conducting nanoparticles in matrices such as glass (Ivanda *et al.*, 2007), and the stability of surfactant-coated, iron-containing nanocolloids in solution (Rabias *et al.*, 2008). The technique requires almost no sample preparation and is relatively quick; spectra can be collected in a matter of seconds to minutes.

25.2.2.2   Measurement of Airborne Exposures

Research is ongoing regarding which dose metric(s) is important for relating airborne nanoparticle exposure to toxicological effects. In addition to the conventional measure of airborne particles based on mass per volume (mg/m^3), measures of surface area and particle number (number concentration in $particles/cm^3$) convey essential data on the nature of exposure. For nanoparticles, surface area is an important exposure metric (Ku and Maynard, 2005; Oberdörster *et al.*, 2005a), because the reactivity of a particle, in part, is related to its surface area, which is high relative to mass for nanoparticles. Other important characteristics such as shape and surface chemistry can be quantified by the characterization techniques as discussed above, or others suited to examining free particles. Several exposure assessment tools are currently available to measure nanoparticles that are free of a retaining matrix, falling loosely into the categories of condensation particle counters (CPC), differential mobility analyzers (DMA), electrical low-pressure impact (ELPI), and aerosol surface area measurement (Rogak *et al.*, 1993; Keller *et al.*, 2001; Ku and Maynard, 2005). In general, measurement is based on nucleation of a particle (CPC), mobility within an electric field (DMA), or impaction by size (ELPI, micro-orifice uniform deposit impactor [MOUDI]). Aerosol surface area measurement is an area of current active research, and some portable, battery-operated instruments have been commercially introduced. These instruments allow for measurement of biologically relevant surface areas of interest for respiratory exposures, providing real-time measurements of respirable particles.

Also of importance in exposure measurement is the ability to distinguish the nanoparticles of interest from 'background' or ambient concentrations of nanoparticles (e.g. from vehicle exhaust or toasting bread). Exposure studies have reported confounding of measurements by nanosized particles emitted by other sources such as building furnaces, rather than from the manufacture of engineered nanomaterials (Maynard and Zimmer, 2002; Kuhlbusch *et al.*, 2004). Ambient particulate matter may also cause nanoparticles to agglomerate (e.g. Seipenbusch *et al.*, 2008), thereby decreasing potential nanoparticle inhalation.

25.2.3   Inhalation Exposure Studies

Few published studies have examined potential exposures from generation of airborne nanoparticles (Table 25.3), and most focus on the workplace, where exposures are expected to be higher than for consumers. The first exposure study from handling engineered nanoparticles characterizes airborne concentrations of 'unprocessed' single-walled CNTs (SWCNTs) in both a laboratory and field setting (Maynard *et al.*, 2004). In the field study, airborne levels were 'very low' in all cases during the removal and handling of SWCNTs from production vessels, with released particles clumping into larger compact masses. The laboratory study likewise indicated low concentrations, and that sufficient mechanical agitation of SWCNTs to produce airborne particles also resulted in rapid agglomeration which reduced airborne levels within minutes (Maynard *et al.*, 2004). Van der Waal's forces among CNTs cause them to agglomerate and form 'ropey' structures in air, thereby limiting inhalation exposure, although concern over dermal exposure was indicated by Maynard *et al.* (2004). Research is continuing at the National Institute for Occupational Safety and Health (NIOSH) to determine the extent that such observations can be generalized to other particles, and to refine methods to characterize nanoscale particles in air (NIOSH, 2007). Other studies have also indicated that agglomeration of inorganic nanoparticles reduces airborne concentrations (Ma-Hock *et al.*, 2007; Seipenbusch *et al.*, 2008). In general, toxicity studies of nanoscale particles (e.g. TiO_2, UFP) report administered particles in air or solution that are larger than the primary particle size.

Another study of airborne particle levels in a fullerene production facility reported short-term increases in $PM_{2.5}$ and the number concentration of submicron-sized particles within the work area (fume hood) during sweeping or vacuuming of fullerenes (Yeganeh *et al.*, 2008). Particle concentrations measured outside the fume hood also showed short-term increases with production activities in the facility; however, photoionization results indicated that particles in the fume hood were largely carbonaceous, and therefore likely to be engineered nanomaterials, whereas those outside were not. $PM_{2.5}$ concentrations in the facility were also influenced by outdoor $PM_{2.5}$ levels. Similar findings were found for PM_{10} in an industrial facility handling powders (Demou *et al.*, 2008). Based on the results, both studies observed that the engineering controls within the facility seemed to be effective in limiting exposure to workers whether via fume hood (Yeganeh *et al.*, 2008) or mask filters (Demou *et al.*, 2008).

Table 25.3 Summary of exposure studies to airborne nanoparticles.

Study	Description	Conclusions
Maynard et al. (2004)	Characterized airborne concentrations of SWCNTs in the laboratory under mechanical agitation and in the field at four production facilities	Exposures were <53 µg/m³; airborne particles rapidly agglomerated into larger 'clumps' (diameter >1 µm)
Hsu and Chein (2007)	Investigated nanoparticle release from a TiO_2-containing coating subjected to mechanical scraping with a fan and UV light to simulate outdoor weathering and erosion	Most particles were 20–150 nm; UV light increased TiO_2 release; substrate type and duration of scraping affected emission rate
Ma-Hock et al. (2007)	Compared aerosols of various inorganic substances generated by different methods	Very few particles in the nanoscale range because of agglomeration
Demou et al. (2008)	Measured exposures to nanoparticles in an industrial facility (including reactor cleaning and handling of powders) using various measurement techniques; evaluated the effectiveness of masks in reducing exposure	Production unit was the major emission source (mean concentration 0.188 mg/m³); masks decreased particle number by >96%
Seipenbusch et al. (2008)	Evaluated simulated platinum nanoparticle aerosols in the workplace over time compared with background aerosols	Released nanoparticles were markedly altered because of coagulation and interaction with background aerosols
Yeganeh et al. (2008)	Evaluated airborne particle levels in a fullerene production facility resulting from production activities as well as from background sources	Short-term increases in $PM_{2.5}$ and number concentration of submicron-sized particles in work areas during cleaning activities

Insight into consumer exposures to nanomaterials from products could be investigated through the use of exposure simulation, as has frequently been used to recreate historical exposures to other airborne substances (e.g. asbestos, solvents, benzene; Esmen and Corn, 1998; Paustenbach et al., 2004; Fedoruk et al., 2005; Mowat et al., 2007; Sheehan et al., 2008). At one of the four facilities tested by Maynard et al. (2004), exposure from removing CNTs from the production vessel was simulated by transferring CNT material between buckets. Ma-Hock et al. (2007) examined the effect of the substances (TiO_2, ZnO, Aerosil fumed silica, silicon dioxide, carbon black, aluminum oxide, copper oxide) and the type of generation methods (dry aerosol generator or liquid nebulizer with drier) on nanoparticle exposure. Overall, agglomeration effects limited nanoparticle exposure. A greater percentage of ZnO was in the nanoscale size range using the liquid nebulizer technique, possibly because of the greater solubility of ZnO (Ma-Hock et al., 2007). Demou et al. (2008) evaluated the effectiveness of mask filters in particle retention under various simulated operating conditions found in the workplace.

An exposure simulation study more specific to consumer exposures attempted to assess whether free nanoparticles could be liberated from a surface coating containing nanoscale TiO_2 as a result of weathering and erosion outdoors (Hsu and Chein, 2007). The test system involved motorized scraping of paint of unreported composition on a single tile subjected to ultraviolet (UV) light and a fan within a small box. Airborne nanoparticles were measured; however, the specific materials system and correlation to likely use conditions are poorly documented (e.g. UV exposure is more likely for an outdoor paint, where the airflow and dissipation volumes would be substantially different than those simulated). Furthermore, because some historically used components of paints have nanoscale dimensions (Morgans, 1990), it is not clear if the particles generated represent a new exposure.

Additional unpublished industry-sponsored research may also exist. Given the need for additional exposure data to aid in shaping future regulations and guidelines to ensure safety and minimize human health risks, several entities (e.g. Royal Society, 2004; Davies, 2006; Maynard 2006c; Medley *et al.,* 2007) have urged that such research be made available, with some, such as the US Environmental Protection Agency's (2007) Voluntary Nanoscale Materials Stewardship Program, offering some protection of confidentiality of trade secrets.

25.2.4   Dermal Exposure Studies

Dermal penetration by nanoscale particles has been researched for over ten years because of the use of 'micronized' (includes particles in the nanoscale range) TiO_2 and ZnO in sunscreen products. These studies using human volunteers or *ex vivo* cadaver or porcine skin preparations have shown little evidence of penetration of the skin by these metal oxide particles to vascularized tissue, although some uncertainties remain because of methodological limitations (see summaries by Gamer *et al.,* 2006; Mavon *et al.,* 2007; Cross *et al.,* 2007; Nohynek *et al.,* 2007, 2008). Such findings would also apply to other consumer uses or personal care products in which relatively inert (i.e. poorly soluble) nanomaterials may contact the skin.

More recent studies have been conducted using quantum dots to more definitively visualize and assess dermal penetration by nanosubstances. A study using a porcine skin preparation indicated that various types of quantum dots penetrated beyond the stratum corneum to the epidermal layer, potentially resulting in systemic exposure (Ryman-Rasmussen *et al.,* 2006), although skin permeability may have been increased by 24-hour exposure to solutions with high alkalinity (Nohynek *et al.,* 2007). A subsequent study by the same research group reported that, under near-neutral pH conditions, a quantum dot of different shape had minimal skin penetration and was limited to the outer layers of the stratum corneum, in the absence of skin damage (Zhang *et al.,* 2008). Studies by the NCTR/FDA (part of the National Technology Program-sponsored nanotechnology toxicity studies) using quantum dots with cadaver skin likewise did not indicate penetration in the absence of dermabrasion and removal of the stratum corneum (e.g. Kraeling *et al.,* 2007).

Some evidence indicates possible penetration to vascularized layers by fullerenes when applied to skin that is repeatedly flexed over 90 minutes, but not in unflexed skin (Rouse *et al.,* 2007). Consumer groups have also raised concerns for those with thinner skin (e.g. children, the elderly) or damaged skin, including from sunburn (e.g. Environmental Working Group, 2008). Thus, the overall evidence indicates negligible dermal penetration resulting from systemic exposure to nanomaterials; however, additional research will need to address the extent of penetration for compromised skin, sensitive individuals, or certain conditions.

25.2.5 Exposure Recommendations

To the extent possible, considerations should be made in the manufacturing phase to impart properties both to the nanomaterial and to the product to minimize exposure. Given uncertainties in quantifying free particle exposure and in assessing toxicity, products that encapsulate or limit liberation of free nanomaterials would greatly reduce exposure. Liquid spray applicators resulting in larger droplets rather than fine aerosols would also decrease inhalation exposure. For consumer product applications in which nanomaterials are encapsulated within structures, such as sports equipment, assessments of exposure would need to evaluate the integrity of encapsulation and the form of nanomaterials during product use, wear, weathering or abuse. Such assessments can be achieved using the well-established discipline of materials science for understanding and characterizing the nature of nanoscale substances within products, as well as the interaction between the product matrix and the nanoscale components.

25.3 Evaluation of Health Effects and Toxicity

Toxicity testing of nanomaterials is generally in the initial state of identifying hazards, in which studies are conducted to produce toxic effects, rather than to define the dose–response relationship or 'no effect' levels. Thus, many studies are *in vitro* (direct application to cell or tissue cultures) or short-term studies in rodents, using extremely high doses and direct administration methods (e.g. intratracheal instillation, injection, oral gavage). Such methods have also been necessary because of the high cost of obtaining uniform sizes of nanomaterials for testing. For the most part, the main types of nanomaterials initially being used in consumer products (e.g. silver, TiO_2, ZnO, or nanomicelles containing existing soap ingredients such as fatty acids or esters) are expected to have relatively low human toxicity. Many of the applications using engineered carbon-based materials that are listed in the PEN Inventory are largely encapsulated or inaccessible to the consumer under normal use (e.g. CNTs in sports equipment). On the other hand, fullerenes appear to be primarily used in certain facial creams and cosmetic products, where they are incorporated within a matrix that may be applied to skin, but would be difficult to inhale as free particles.

Consistent with the exposure routes of concern for nanomaterials, most of the toxicity research has focused on the inhalation pathway. Literature assessing potential health effects via skin contact has focused on dermal penetration, which as reviewed above, indicates little penetration of intact skin by relatively inert nanomaterials, with possible uncertainties regarding certain conditions such as flexion, sunburn, or age-related sensitivities. Fewer studies have investigated the potential health effects associated with oral exposures to nanoparticles, probably because of the lesser concern of this pathway relative to inhalation.

An overview is presented below of the applicability of toxicity studies (with emphasis on exposure routes of primary concern) for assessing health risks for major classes of nanomaterials currently of relevance for consumer products. This discussion also includes applicable knowledge from UFPs, inorganic fume, and asbestos and manmade fibers. This richer literature on small-scale substances serves as a base of knowledge which research can build on to elucidate the key determinants of toxicity and to refine toxicity tests of nanomaterials to assess product safety.

25.3.1 Ultrafine Particulate Literature

Potential concerns regarding nanoparticles are supported by a wealth of literature on health effects of air pollution and occupational exposures to particulate matter, including toxicological studies in animals and epidemiological studies of particulate matter (PM_{10}, $PM_{2.5}$) and UFPs in the environment, and associated trends in morbidity and mortality in urban populations (Oberdörster *et al.*, 1992, 1994, 2000, 2005a; Brown *et al.*, 2001; Witschi and Last, 2001; Yeates and Mauderly, 2001; Donaldson *et al.*, 2002; Donaldson and Stone, 2003; Hofmann *et al.*, 2003; Borm *et al.*, 2004; Englert, 2004). Although UFPs encompass the same-sized particles as nanoparticles (<100 nm), those in urban air largely arise from combustion and thereby contain a variety of chemically reactive substances such as polycyclic aromatic hydrocarbons, metals and inorganic substances. In studying the effect of small-scale particles, experimental studies on UFPs often use carbon and other relatively inert particles such as TiO_2. Studies of UFPs provide considerable insight for evaluating nanoparticles, showing that the site of deposition, clearance ability, mechanisms of action, and toxicological effects depend on characteristics such as particle size, solubility, shape, chemical composition, and surface charge/reactivity.

Studies investigating the effects of UFPs have compared particles of relevance for urban or occupational exposures (e.g. carbon black, platinum) to relatively inert, insoluble particles (e.g. TiO_2, polystyrene, iridium) or to those with known high reactivity (e.g. quartz) (Oberdörster *et al.*, 2000; Brown *et al.*, 2001; Kreyling *et al.*, 2002).[4] Effects noted in inhalation or intratracheal instillation studies include pulmonary inflammation and production of oxidative stress, as well as indications that UFPs would be associated with greater effects in the lung than an equivalent mass of larger-scale particles because of the larger surface area for reactions of smaller particles (Oberdörster *et al.*, 1994).

Studies on UFPs have also helped to outline the main areas of particle deposition and potential uptake in, and translocation from, the respiratory system (Oberdörster *et al.*, 2004). Models of particle deposition in the human respiratory tract (see International Commission on Radiological Protection, 1994) indicate that the region of deposition varies depending upon particle size, with particles larger than 1 μm depositing primarily in the upper respiratory region, but with progressively more deposition in the lower airways and alveolar region with decreased particle size down to about 10 nm (Figure 25.4). Very small particles around 1 nm in size, however, show substantially more deposition by impaction in the upper airways than in the lower alveolar region. Age-related differences in deposition are also predicted by the International Commission on Radiological Protection (1994). For particles <100 nm, deposition decreases with age from infants to adults in the tracheo-broncheal region, is similar among age groups in the nasal pharyngeal region, and is lower for infants and children than adults in the alveolar region. Such age-related differences may have implications for differences in sensitivity to the effects of inhaled nanomaterials.

---

[4] The UFPs used in toxicity studies described below are carbon particles, unless otherwise noted.

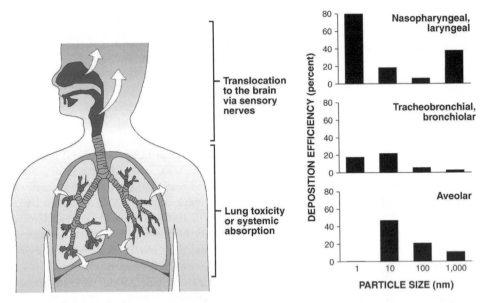

Figure 25.4 *Potential areas of the respiratory tract for nanomaterial translocation and pre-dicted regional deposition by particle size (activity median thermodynamic diameter; Table F1 reference worker in ICRP 1994; separate regional values are combined for nasal pharyngeal and larynx and for tracheobroncheal and bronchiolar).*

Locations of deposition in the respiratory system are of interest for predicting sites for toxicity or for translocation from the respiratory system to the gastrointestinal tract, into systemic circulation, or to the brain via uptake and translocation along sensory nerve axons in the nasal pharyngeal or laryngeal region (Oberdörster *et al.*, 2004). Uptake and translocation along olfactory nerve axons from the nose to the brain (primarily the olfactory region) has been described for UFPs in rats (Oberdörster *et al.*, 2004), gold particles and viruses in non-human primates (summarized by Oberdörster *et al.*, 2004), manganese in rats (Elder *et al.*, 2006), and TiO₂ in mice (Wang *et al.*, 2008), including measures of inflammatory markers in the brain in the latter two studies. The extent to which nanoparticles or solubilized products migrate to other areas of the brain is unclear. Oberdörster *et al.* (2004) and Elder *et al.* (2006) noted that the smaller increases of UFPs or manganese in other regions of the brain following inhalation exposure could also result from translocation to systemic circulation and transfer across the blood–brain barrier. Wang *et al.* (2008), however, noted higher levels of 80 nm TiO₂ particles in the hippocampus than in the olfactory bulb 30 days post-nasal instillation, and stated that this resulted from diffusion from the olfactory bulb.

Although rats have a higher density of olfactory neurons in the nose and a proportionally larger olfactory area in the brain, this pathway of translocation to the brain is reported to be relevant for humans (e.g. Elder *et al.*, 2006). Elevated airborne manganese exposure in workers, for example, affects certain areas of the brain (e.g. parts of the basal ganglia) resulting in a Parkinson-like syndrome termed 'manganism' (e.g. Hua and Huang, 1991; Huang *et al.*, 1998). Thus, research on nanoparticle uptake and translocation may inform our understanding of toxicological mechanisms for previously known effects of airborne agents. Characteristics such as particle size, shape, solubility or reactivity probably affect translocation and subsequent effects, although effects would also be agent-specific.

Translocation of inert nanoparticles from the lung to other parts of the body appears to be lower than initially reported. An often-cited study (Nemmar *et al.*, 2002) showing higher amounts of translocation (e.g. 10 % in the liver; 25 % in the bladder) was reported to have used radiolabels on UFPs that became detached from the particle. Subsequent studies of UFPs with more stable tracers indicated no significant translocation (Wiebert *et al.*, 2006a,b). Other studies reporting translocation of nanoparticles noted relatively small amounts of material in peripheral organs, but a greater degree of translocation with smaller particle size (15 nm vs 80 nm count median diameter; Kreyling *et al.*, 2002). Some differences in extent of translocation are also apparent among types of particles (e.g. 11 % for TiO_2, (Geiser *et al.*, 2005) versus 1−5 % for UFPs (Kreyling *et al.*, 2006)). More soluble nanoscale substances (e.g. manganese, ZnO) would probably translocate more readily to systemic circulation where their toxicity would be related to the substance in dissolved form, rather than as a nanoparticle (Wallenborn *et al.*, 2007).

Systemic toxicity, specifically cardiovascular effects, may also occur without translocation through localized inflammation in the lung and release of cytokines or other biological triggers or messengers. Cardiovascular effects appear to be a general reaction to inhaled nanoparticles as reported for UFPs (Borm, 2002), inert substances such as TiO_2 (Nurkiewicz *et al.*, 2006), and CNTs (Li *et al.*, 2007). Threshold doses for such effects likely exist as indicated by the lack of inflammatory or vascular function changes noted in a study of healthy individuals exposed to specific ambient concentrations of particles in urban air (Bräuner *et al.*, 2008).

25.3.1.1   Inorganic Particulates

In addition to studies on UFPs, which include certain inorganic substances, occupational exposures to fumes in smelters, refineries, pigment manufacturing, or from welding have involved size distributions of particles ranging down into the nanoscale. Based on a wealth of human studies on exposures to metal or metalloid fumes, nanoparticle exposure would probably produce similar effects to those already observed, given sufficient dose (e.g. lung cancer for arsenic, cadmium, hexavalent chromium; central nervous system effects for manganese; and metal fume fever, a transient flu-like condition, for zinc). Exposures to nanoscale TiO_2 and ZnO currently used in consumer products have occurred in pigment manufacturing plants and in zinc refineries for decades. Although some of this exposure also included larger size particles, historical exposures were substantially elevated for all particles across the size distribution. The dose–response relationship for nanoscale particles, however, may be difficult to predict based on historically mixed particle size exposures. Increased translocation out of the lungs by smaller particles may reduce long-term effects in the lungs, for example. Specific toxicity information of relevance for consumer products is listed below for TiO_2, ZnO and silver.

Toxicity of Nanoscale Titanium Dioxide. Among nanomaterials, TiO_2 has the greatest amount of literature for evaluating health effects including epidemiology studies and acute, subacute, subchronic, and chronic toxicity studies, primarily in rats (Heinrich *et al.*, 1995; Bermudez *et al.*, 2004; Hext *et al.*, 2005; National Institute for Occupational Safety and

Health, 2005; Baan *et al.*, 2006; Grassian *et al.*, 2007). These studies collectively indicate that short-term exposures to high amounts of TiO_2 instilled in rat lungs result largely in transient inflammatory effects. Longer-term studies in rats exposed to high airborne TiO_2 concentrations indicate impairment of lung clearance mechanisms, fibrosis, and cancer (Heinrich *et al.*, 1995; Bermudez *et al.*, 2004) by a mechanism that is not considered relevant for humans (i.e. lung overload and greater reactivity in the rat lung; see Fryzek *et al.*, 2003; Borm *et al.*, 2004; Rausch *et al.*, 2004; Hext *et al.*, 2005; Warheit and Frame 2006; Baan *et al.*, 2006).

Warheit *et al.* (2007a) described toxicity information for TiO_2, comprising a minimum base set of tests in accordance with the ED/DuPont framework for responsible development of products containing nanomaterials (Medley *et al.*, 2007). Studies include inhalation toxicity in rats, dermal sensitization and irritation in rabbits and mice, *in vitro* genotoxicity, and aquatic toxicity in three species (Warheit *et al.*, 2007a). The overall results indicate a relatively low level of toxicity to the species tested. Epidemiological studies likewise do not indicate elevated risks of cancer in workers, although some limitations in study design and characterization of particle size have been noted (Fryzek *et al.*, 2003; Boffetta *et al.*, 2004; Hext *et al.*, 2005; National Institute for Occupational Safety and Health, 2005).

The draft risk assessment of TiO_2 by the National Institute for Occupational Safety and Health (NIOSH) (2005) applied the respiratory studies in rats to develop a dose-response assessment for nanoscale and fine-scale (i.e. $>100 \, nm$) TiO_2 to assess health risks of exposure to workers (Kuempel *et al.*, 2006). Specifically, the risk assessment concluded that nanoscale TiO_2 would be associated with a higher risk of lung cancer to workers than the fine-scale TiO_2 at the same mass-based dose, and that toxicity is a function of dose based on surface area. NIOSH (2005), however, also considered the toxicity of TiO_2 as resulting from inert particle surface reactivity rather than from particle chemistry, and that the dose-response was sublinear (Kuempel *et al.*, 2006), such that TiO_2 exposure would have a lower threshold dose below which the cancer risk would be negligible.

NIOSH (2005) based their conclusions on the importance of surface area, and greater toxicity with smaller size, on chronic (Lee *et al.*, 1985; Heinrich *et al.*, 1995) and subchronic (Bermudez *et al.*, 2002, 2004) inhalation studies in rats. Nevertheless, these studies used a more reactive form of TiO_2 (Degussa P25; 80 % anatase, 20 % rutile) for the nanoscale particles (Heinrich *et al.*, 1995; Bermudez *et al.*, 2004) than was used for the fine-scale particles (nearly 100 % rutile) (Lee *et al.*, 1985; Bermudez *et al.*, 2002), thereby confounding this comparison (Warheit *et al.*, 2006a). In addition, surface coatings of TiO_2 particles used in products such as paints have been shown in rat intratracheal instillation studies to reduce toxicity relative to the uncoated type of nanoscale TiO_2 (Degussa P25) commonly used in toxicity studies (Warheit *et al.*, 2005, 2006b, 2007c). The actual toxicity of TiO_2 in products may therefore differ from those reported in toxicity tests, depending on the type of material used.

Anatase TiO_2 nanoparticles (155 nm) were also shown to be slightly more toxic than smaller rutile TiO_2 nanoparticles (80 nm) in producing inflammation and oxidative damage in the olfactory bulb and hippocampus of the brains of mice, following intranasal instillation and subsequent transport to the brain along olfactory neurons (Wang *et al.*, 2008). Whether such exposures would occur in humans exposed to airborne levels of TiO_2 nanoparticles is yet unclear. Studies of workers exposed to aerosols of TiO_2 in the pigment industry have not noted central nervous system effects (NIOSH, 2005), unlike workers exposed to manganese fume. Intranasal instillation of a bolus of particles is also a much higher dose than would occur from inhalation exposure of workers, and far higher than for consumers.

As noted above, dermal penetration of TiO_2 appears to be negligible for intact, uncompromised skin. The literature also generally indicates an absence of adverse skin reactions to micronized or ultrafine TiO_2 used in sunscreens or other skin applications (summarized by Nohynek *et al.,* 2008).

TiO_2 has long been used as an inert ingredient in food products, toothpaste, and other consumer products with potential for oral ingestion. The estimated average daily intake rate of TiO_2 'microparticles' (size not specified) from major food products comprising 93 % of the daily TiO_2 dose is 2.66 mg/person-day, with considerable inter-individual variation (0 to 112 mg/person-day; Lomer *et al.,* 2004). Incidental amounts of TiO_2 may also be ingested inadvertently from use of nanoscale TiO_2 in cleaners or from wear of consumer products containing TiO_2 surface treatments. Some evidence from oral administration of polystyrene microspheres in rats indicates increased gastrointestinal absorption and translocation of smaller particle sizes within the body (Jani *et al.,* 1990).

Among the few oral studies on nanoscale TiO_2, Wang *et al.* (2007) investigated the acute toxicity of various sizes of TiO_2 particles (25, 80 and 155 nm) by gavage administration of a single dose of 5000 mg/kg to mice. No obvious acute toxicity occurred, and no changes in liver, spleen or kidney weights in male mice were observed after 2 weeks. Female mice exposed to 25 or 80 nm particles showed higher liver weight over controls. Certain biochemical parameters in blood serum of primarily female mice also indicated potential effects on the liver, heart and kidney. Histopathology showed effects on the liver and kidneys and slight brain lesion in 80 nm or 155 nm particle-exposed mice, regardless of sex, but not for the 25 nm group. Titanium tissue concentrations were generally highest for the 80 nm group in the liver, followed by lung, spleen, kidney, and to a lesser extent, the brain. Tissue retention for the 25 nm group was generally similar to that of the 155 nm group. Thus, smaller particle size is not necessarily correlated with greater tissue concentrations, retention and toxicity.

Warheit *et al.* (2007a) conducted an acute oral toxicity study of Degussa P25 TiO_2 (median particle size of 140 nm in water) in rats according to Environmental Protection Agency and European Organisation for Economic Co-operation and Development (OECD) guidelines. Particles were 90 % TiO_2 by weight with 7 % alumina and 1 % amorphous silica. A single dose of the TiO_2 particles in water was administered by gavage to one fasted female rat at each dose of 175, 550 or 1750 mg/kg, and three fasted female rats at 5000 mg/kg. No biologically significant changes because of nanoparticle exposure were observed during the 14 days postexposure, nor were gross lesions reported at necropsy.

Toxicity of Nanoscale Zinc Oxide. ZnO generally has low toxicity by all routes of exposure, and has been widely used in consumer products as a pigment, an antimicrobial agent in wound dressings or diaper rash creams, and in sunscreens. Exposure to high airborne levels (typically >15 mg/m^3) of ZnO fume is associated with 'metal fume fever' in workers, typically involving reversible flu-like symptoms (e.g. cough, shortness of breath, fever, chills, nausea and vomiting) (American Conference of Governmental and Industrial Hygienists, 2001). This condition is considered self-limiting, temporary, and non-progressing to pulmonary disease or permanent long-term impairment. Short-term experimental studies in animals and human volunteers report a temporary lung inflammation response, and the combined weight of evidence in animals and humans does not associate zinc or ZnO with cancer or reproductive toxicity, either by inhalation or the oral route of exposure (American Conference of Governmental and Industrial Hygienists, 2001; European Commission, 2004; Agency for Toxic Substances and Disease Registry, 2005). Critical health effects (i.e. those occurring at the lowest doses) associated with absorbed zinc are disruption of the balance of other nutritionally essential elements. Much of this literature is not specific to nanoscale particles of ZnO; however, the particle sizes produced in commercial ZnO production and in products such as sunscreens extend into the nanoscale size range. Short-term inhalation studies in animals or human volunteers of ultrafine or nanoscale ZnO are generally consistent with the body of literature on this compound (e.g. Lam et al., 1985, 1988; Conner et al., 1988; Kuschner et al., 1995, 1997; Wesselkamper et al., 2001, 2005).

Beckett et al. (2005) reported no adverse respiratory, hematological or cardiovascular effects among human subjects (6 men and 6 women) who inhaled 500 μg/m^3 (a tenth of the permissible exposure limit for workers) of fine (291 nm count median diameter) or ultrafine (40 nm count median diameter) ZnO for 2 hours. No significant differences were found in clinical parameters associated with metal fume fever with these exposures. A concentration of 500 μg/m^3 thus appears to be a possible no-observed-adverse-effect level for acute inflammatory responses associated with metal fume fever.

Evidence from multiple published and unpublished studies indicates that ZnO in sunscreens does not cause adverse skin reactions; however, many of these studies used pigment grade ZnO or have incomplete physico-chemical specifications (Scientific Committee on Cosmetic Products and Non-Food Products Intended for Consumers, 2003). The Scientific Committee on Consumer Products (2005) noted that further studies were needed to assess the safety of nanoscale ZnO. In response, ex vivo studies of 80 nm ZnO using porcine skin (Gamer et al., 2006) and 15–40 nm ZnO using a human epidermal membrane (Cross et al., 2007) reported no significant penetration by nanoscale ZnO, thus supporting the conclusions presented by the Scientific Committee on Consumer Products (2005). In addition to sunscreens, ZnO has also been used to promote healing of burns and wounds (Agency for Toxic Substances and Disease Registry, 2005).

Few studies have been conducted on the oral toxicity of zinc or ZnO nanoparticles. In a companion study to Wang et al. (2007), a single acute dose of 5000 mg/kg of zinc nanoparticles (either 58 or 1080 nm) was administered by gavage to mice (Wang et al., 2006). Certain blood enzyme levels were elevated (possibly indicative of effects to the liver and heart) in mice receiving both nano- and microscale zinc powders relative to controls. These elevations were generally higher in mice receiving microscale zinc relative to mice receiving nanoscale zinc. Several other biochemical parameters in blood were significantly elevated in mice receiving the microscale powder, but not in those receiving the nanoscale powder. Kidney parameters measured in blood were elevated for microscale zinc-treated mice relative to nanoscale or control mice; however, severe renal lesions and signs of anemia were reported for the nanoscale zinc-treated mice.

As for the study using a similar high dose of TiO$_2$ in mice, these results are difficult to interpret because of the high dose administered compared to likely exposure to zinc or ZnO in consumer products. For example, if a young child weighing 10 kg were to ingest 5000 mg/kg of zinc in a solution containing 1 % zinc by weight, the child would have to drink 5 kg of solution. Two out of 20 mice receiving the nanozinc powder died within 2 weeks from intestinal obstruction by agglomerated zinc particles (Wang et al., 2006).

More likely effects from overexposure to ingested nanozinc or ZnO would be related to nutritional imbalance from interference with other essential metals (European Commission, 2004; Agency for Toxic Substances and Disease Registry, 2005).

Toxicity of Nanoscale Silver. The antimicrobial effects of silver are related to its ion, which is readily formed by the dissociation of soluble silver salts in solution or from the surface of silver nanoparticles. Silver nitrate solutions (1 %) were used for years in medicine, including as eye drops for all newborns to prevent infection. Silver nanoparticles provide a high surface area for greater release of silver ions per mass of silver. Thus, the existing toxicity literature on soluble silver is of relevance for assessing ions from silver nanoparticles.

Although the antimicrobial properties of silver have been used for centuries for hygienic and medicinal uses with little human toxicity, relatively few studies of relevance for assessing human health effects have been conducted specifically on silver nanoparticles (Chen and Schluesener, 2008; Luoma, 2008). Ancient beliefs in the health benefits of regular ingestion of colloidal silver[5] have persisted to modern times. Other than developing argyria, in which the skin, mucous membranes, and membranes of internal organs retain deposits of silver, relatively few health effects have been reported from such repeated, high-dose exposure. The majority of recent human toxicity literature concerns the medicinal use of silver as an antimicrobial agent in medical devices, bandages, and in creams for wound healing (Atiyeh *et al.,* 2007; Chen and Schluesener, 2008). *In vitro* research suggests that nanosilver has the potential to be toxic to mitochondria and to interact with thiol groups of proteins and enzymes within mammalian cells, thereby depleting antioxidant defenses (Chen and Schluesener, 2008); however, nanosilver would first have to gain access into cells despite various defenses of the body. Unlike other metals, silver in the body appears to be sequestered in lysosomes, benign granules, or within the basement membrane of tissues where they are isolated from cellular and organ functions (Lansdown, 2007; Luoma, 2008). Health effects in humans, such as decreased blood pressure and respiration, and gastrointestinal effects, are typically associated with very high doses of readily absorbed forms of silver such as silver nitrate (Luoma, 2008). Effects on blood pressure and respiration may also be associated more with nitrate than with silver.

Antibiotic resistance to silver is rare, probably because of its broad spectrum of action. Some debate has been ongoing in the medical literature on whether silver controls wound infection and aids healing, or is ineffective or even delays healing, with cytotoxic effects on host cells (Atiyeh *et al.,* 2007; Luoma, 2008; Vermeulen *et al.,* 2008). Antibiotic resistance may be of increased concern with widespread use of nanosilver.

Respiratory effects of silver dusts or fumes have resulted in benign staining of alveoli and bronchial tissues, although metal polishers exposed to metallic silver as well as to other metals have been reported to show respiratory effects similar to those related to particle exposures (e.g. bronchitis, emphysema, and reduction in pulmonary volume; Drake and Hazelwood, 2005). Rats exposed to various concentrations of silver nanoparticles (approximately 49, 133 and 500 μg/m^3) for 6 hours/day, 5 days/week for 90 days showed

---

[5] The size of silver colloids is generally between 1 and 500 nm, although several products including liquid suspensions are advertised with colloid sizes of 0.6 to 25 nm (Luoma, 2008).

dose-dependent increases in inflammatory response and decreases in lung function (i.e. lower tidal and minute volume), although the most obvious changes were in the high dose group, with the low and middle exposure groups being more similar to the unexposed control (Sung *et al.,* 2008). The study system reportedly was able to maintain exposures to silver particles in the intended size range (geometric mean diameter of 18 to 19 nm; range of 6 to 55 nm).

Sustained airborne exposures to unagglomerated particles of silver at the concentrations in this study, or in a historical occupational setting, is an unlikely situation for the current uses of silver in consumer products. For example, the minimum antibacterial effectiveness of a silver solution of about 10 ppm silver (Luoma, 2008) could be considered as a minimum concentration for silver sprays or mists, although ranges as high as 3000 ppm have also been reported (Luoma, 2008). At 1000 ppm silver in an antibacterial spray cleaner, 500 mg/m^3 of product in air is needed to achieve the 500 µg/m^3 administered in the Sung *et al.* (2008) study. Assuming a 10 ppm nanosilver concentration, the aerosolized product concentration would have to be 50,000 mg/m^3. Clearly, inhalation of these air concentrations of a liquid product would be associated with health effects unrelated to the nanoparticles, and may not even be possible.

With the toxicity of silver being less of a concern to humans, much of the focus on the adverse impacts of nanosilver is on environmental effects. For example, widespread use of silver in consumer products, including clothing, cleaning products, washing machines, and even swimming pools (Luoma, 2008), could result in impacts on beneficial bacteria at wastewater treatment plants, in septic tanks, or in discharge water to lakes and rivers. A recent review of the environmental consequences of the use of nanosilver highlighted these concerns (Luoma, 2008). Choi and Hu (2008) also reported that silver nanoparticles in suspension were more toxic to cell cultures of nitrifying bacteria at the same total silver concentration than silver ions, and that toxicity was correlated with amount of silver particles less 5 nm.

Thus, although the active or antimicrobial form of silver nanoparticles may be silver ions, additional research is needed on whether nanoscale silver particles have increased toxicity. Silver nanoparticles may present a higher concentration of silver ions in the 'nano-thin' environment near the particle surface than for the dispersed silver ions in solution. For human exposures, many of the past and current uses of silver, however, involve exposures to nanosilver particles. Environmental concerns rather than toxicity to humans may therefore ultimately influence regulations on the use of silver in consumer products.

25.3.2   Principles from the Mineral Fiber Toxicity Literature

Health effects have been well documented in occupational studies where workers were repeatedly exposed to high concentrations of certain mineral dusts and fibers, such as quartz and asbestos (see Maynard and Kuempel, 2005). For asbestos, the three critical factors related to inhalation risk are fiber length, fiber diameter, and biopersistence. In addition, for asbestos, long, thin fibers are thought to be associated with the highest potency, with some authors assigning zero potency to fibers less than 5 or 10 µm in length (e.g. Berman and Crump, 2003, 2008). Aerodynamic diameter and shape also contribute to toxicity. For asbestos, the softer, curly chrysotile asbestos has far less potency than the stiffer, more needle-like amphibole fibers (Hodgson and Darnton, 2000; Berman and Crump, 2003).

25.3.3   Carbon Nanotubes

Understanding the toxicological potential of CNTs is challenging because of the many possible forms of these engineered materials, with widely varying properties for a variety of purposes. Engineered CNTs can have properties (i.e. high durability with long, thin shape) that are well-known determinants of asbestos toxicity. Although individual CNTs are far smaller than asbestos fibers, CNTs strongly agglomerate into large bundles of fibers of various sizes, including in the asbestos size range of several microns in length (Warheit *et al.*, 2004). In their evaluation of cytotoxicity, Jia *et al.* (2005) showed that carbonaceous nanomaterials with different shapes – SWCNTs, MWCNTs and fullerenes – elicited a range of toxicity, implying that shape is an important consideration in evaluating CNT toxicity. Others have also discussed shape of nanoparticles as a factor influencing health implications (Morgan, 2005).

As with asbestos, the primary mechanism of action for CNTs appears to be related to free radical generation and oxidative stress associated with surface reactivity (e.g. Shvedova *et al.*, 2003; Bottini *et al.*, 2006). Long-term effects of nanomaterials are still relatively unknown, although the long tubular structure of CNTs and asbestos may increase their biopersistence in the human body (Nel *et al.*, 2006). Hesterberg *et al.* (1998) suggested that fibrogenesis and tumorigenesis of asbestos fibers appeared to be more correlated with biodurability in the lungs, rather than the deposition and fiber length.

Concerns related to CNTs thus include (1) their nano-size, and hence high surface area and potentially high reactivity; (2) their fiber-like shape; and (3) their potential for biopersistence. Studies of the effects of CNTs on the respiratory system of rodents indicate the ability to elicit inflammation, granulomas, and formation of fibrous tissue and interstitial fibrosis (Warheit *et al.*, 2004; Muller *et al.*, 2005; Donaldson *et al.*, 2006; Lam *et al.*, 2006; Warheit 2006; Mercer *et al.*, 2008; Shvedova *et al.*, 2009), similar to effects observed following adequate exposure to certain asbestos and fiber types. CNTs in the respiratory tract have also been associated with cardiovascular effects (Li *et al.*, 2007), as noted above for other inert particles in the nanoscale range. The mechanism of action for asbestos is through generation of reactive oxygen species; however, CNTs may be more 'inert' or biocompatible to cellular components. Variation in results of CNT toxicity studies is probably a result of (1) methods of administration resulting in CNTs that are either more agglomerated or more dispersed in the lung; and (2) differences in CNT characteristics. Larger agglomerated forms in the lung tend to be associated more with transient inflammation and foreign body reaction (granuloma formation); whereas more dispersed particles result in CNTs entering the alveolar septum (tissue between adjacent air sacs) where they tend to cause fibrosis (Li *et al.*, 2007; Mercer *et al.*, 2008).

Metal contaminants or impurities resulting from the manufacturing process may contribute to CNT toxicity (Oberdörster *et al.*, 2005a; Nel *et al.*, 2006). Effects of metal content on toxicity have been observed with asbestos, where effects are related to solubility of certain fibers and release of metals. Release of iron from amphiboles is thought to contribute in part to their increased toxicity (Roggli *et al.*, 2004). Indeed, asbestos bodies found in lung tissue are observed to have an iron core, and are considered to result from exposure to amphiboles (e.g. Roggli, 1992; Roggli *et al.*, 2002, 2004; Butnor *et al.*, 2003). It should be noted that the smaller size of nanoparticles may also affect clearance, in that the small size and high surface area may increase dissolution rates.

Surface treatments of engineered particles can also modulate toxicity (Warheit *et al.*, 2003, 2005; Derfus *et al.*, 2004; Morgan, 2005). Such treatments may be exploited in pharmaceutical or medical applications, where functionalized surface groups are attached to CNTs for a specific therapeutic purpose or to modify their water solubility and toxicity (Dumortier *et al.*, 2006). Addition of various functional groups by different reactions appears to have complex effects on surface properties, shape, and toxicity of CNTs as reported by *in vitro* studies of lymphocytes (Dumortier *et al.*, 2006; Bottini *et al.*, 2006; Shvedova *et al.*, 2009).

Two recent studies reported 'asbestos-like' toxicity in rats exposed to various CNTs (Poland *et al.*, 2008; Takagi *et al.*, 2008). Takagi *et al.* (2008) reported that MWCNTs were more potent in causing mesothelioma than asbestos in a study in which male mice of a strain reported to be sensitive to asbestos (as well as to genotoxic carcinogens and reactive oxygen species-related carcinogenesis) received a single intraperitoneal injection of 3 mg/mouse of MWCNTs, fullerenes, or crocidolite asbestos fibers. Approximately 27.5 % of MWCNTs were reported as $\geq 5\,\mu m$ in length with an average width of 100 nm, and, although the MWCNTs were 'rigorously agitated' prior to administration, agglomerates were still found *in vivo*. At the end of the study period (180 days), the group administered MWCNTs had the highest mortality, followed by the crocidolite group. The histopathology results of MWCNT- and crocidolite-treated mice showed 'moderate to severe fibrous peritoneal adhesion' and 'fibrous peritoneal thickening', with a spectrum of peritoneal mesothelial lesions. Histopathology of the fullerene group showed minimal lesions with no indication of mesothelial lesions, and indicated some biodegradation of fullerenes.

In a more comprehensive study at a lower dose, Poland *et al.* (2008) examined the potential for long and short CNTs, tangled CNTs, long and short amosite asbestos fibers, and carbon black to cause pathological responses 'known to be precursors of mesothelioma' after a single injection of 50 µg per mouse into the peritoneal cavity of female mice. They reported that long (>20 µm; 11.5 % in one sample, 77 % in another) MWCNTs exhibit 'asbestos-like, length-dependent, pathogenic behavior' and observed injury to the lining of the abdominal cavity in mice ('mesothelial lining') in the form of inflammation and granulomas at 7 days postinjection. Effects were not observed in mice administered low-aspect-ratio short or tangled MWCNTs.

In both studies, questions regarding the biological plausibility of the mode of administration (e.g. airborne versus direct injection), dose, sources and preparation techniques for CNTs, and ability of CNTs to reach the deep lung and cause potential effects, remain unanswered. These studies, however, indicate parallels between asbestos and CNTs that may begin to establish determinants of toxicity for CNTs and possibly other nanoparticles.

Consistent with the results of Poland *et al.* (2008) regarding MWCNT shape, Tabet *et al.* (2009) reported that agglomerated MWCNT masses showed lower toxicity to human epithelial cells and human mesothelial cells compared with asbestos fibers. MWCNTs dispersed in different solutions and exposed to cells formed tangled masses several microns in diameter, although some primary particles remained. These agglomerates were more numerous and larger in phosphate-buffered saline than in dipalmitoyl lecithin (a component of pulmonary surfactant) or in ethanol. Regardless of dispersant, the MWCNTs decreased cell metabolic activity without affecting cell membrane permeability or apoptosis. Unlike in other studies (summarized by Lam *et al.*, 2006; Helland *et al.*, 2007; Kolosnjaj *et al.*, 2007), no cellular internalization or oxidative stress occurred. By contrast, two asbestos fibers (chrysotile and crocidolite) were internalized in cells, without affecting cell membrane permeability, decreased metabolic activity, and increased apoptosis. Nanoscale carbon black particles were also internalized but without effects. Internalization in cells may depend on the effect of the dispersant media in changing surface properties, whereas lesser toxicity compared with asbestos fibers in this study may be related to differences in shape.

A study by Yang *et al.* (2008) likewise reported that accumulation of SWCNTs in organs of mice (primarily liver, lung and spleen) 3 months post-intravenous administration (40 µg, 200 µg and 1 mg/mouse) resulted in 'low toxicity'. Effects were limited to increases in serum biochemical parameters indicative of possible liver and lung effects, indications of oxidative stress/damage in the liver and lung (chiefly at the highest dose), and slight inflammation and inflammatory cell infiltration in the lung; no changes in serum immunological indicators or apoptosis was reported in the organs examined. The TEM images of the SWCNTs revealed fiber bundles that appear to be more similar to the longer, straighter MWCNTs than the tangled MWCNTs examined by Poland *et al.* (2008), but were considerably shorter in length (10–20 nm in diameter, 2–3 µm in length).

A number of factors (e.g. characteristics of the CNTs, study design, test species) probably influence the observed similarities and differences among studies. Although our understanding of the toxicity of CNTs and, in some cases, of asbestos, is still developing, the emerging trends indicate important characteristics affecting toxicity that should be considered in developing consumer products and for additional toxicology research.

25.3.4   Fullerenes

Along with CNTs, fullerenes (or C60) comprise the other type of commonly used engineered carbon-based nanomaterials in consumer products. As summarized earlier, fullerenes are primarily used in cosmetics because of their antioxidant properties in derivative forms.

Although fullerenes are some of the more cytotoxic and reactive nanomaterials studied, the type used in consumer products is a water-soluble fullerene derivative (e.g. polyhydoxylated fullerene, fullerols, or fullerenols). Pure fullerenes are not water-soluble and are well known for producing reactive oxygen species and hence causing cytotoxicity by oxidative stress. Conversely, fullerols are water-soluble, generally much less cytotoxic *in vitro*, and paradoxically possess antioxidant as well as antimicrobial properties (Isakovic *et al.*, 2006). Fullerol species with more covalently bonded functional groups (e.g. hydroxyls or carboxyls) were reported to be substantially less cytotoxic *in vitro* than less highly derivatized fullerol species (Sayes *et al.*, 2004). Fullerols also are a focus of medical research because of their anti-proliferative effect on certain cancer cell lines (Isakovic *et al.*, 2006). Fullerols are likely to be a continuing focus in toxicity research because of medical and environmental interest. These compounds are formed when fullerenes are released in water and form water-stable aggregates, and thus are the likely compounds in the environment.

25.4   Application of Toxicity Studies in Characterization of Risks

Although sufficient dose-response data are largely lacking for many specific types of nanomaterials, literature is amassing that can be used to begin to assess risk for some uses of nanomaterials in products. Potential health risks of products containing nanosubstances may not differ greatly from existing products not using nanoscale substances. Use of nanotechnology can change the product formulation by decreasing the amount of active ingredient and reducing the use of more toxic solvent carriers.

In assessing risks for consumer products, certain issues must be considered, as discussed below.

25.4.1   Hazard Identification and Characterization of Particle Type

The main concern regarding hazards of nanomaterials is related to their small size, as smaller particles may be, but are not always, more toxic. However, studies of TiO_2 (Yamamoto *et al.*, 2004; Warheit *et al.*, 2006a), quartz (Warheit *et al.*, 2007b) and silver (Hussain *et al.*, 2005) contradict the dogma that smaller size always results in greater toxicity. In summary, size can affect reactivity and transport within the body, but is only one of several determinants of particle toxicity. The other most likely determinants are particle composition, potential for dissolution, surface area and shape (Warheit *et al.*, 2007d). Additional considerations include chemistry, crystal type, and presence of coatings. Chemistry can include substances intentionally added, functional groups, or substances present in nanomaterials that are incidental or impurities. As noted above for TiO_2, the crystal typically used in toxicity testing (uncoated Degussa P25) is more reactive than the type often used in products such as paints. Some self-cleaning films or coatings, however, may incorporate more of the anatase form of TiO_2 with higher photocatalytic activity to degrade organic matter and soil. Coatings such as aluminum silicates used on TiO_2 particles have been shown to reduce toxicity (Warheit *et al.*, 2006a,b); however, the permanence of coatings should be evaluated, including their integrity through use, misuse, wear and weathering.

25.4.2   Exposure Conditions

Similarities and differences in administration and exposure in toxicity studies compared with consumer exposures is of critical importance. Toxicity tests typically employ various means to disperse particles in delivery solutions by adding agents (e.g. tetrahydrofuran) or by sonication and filtration (Lovern and Klaper, 2006; Lovern *et al.*, 2007; Tabet *et al.*, 2009). While liquid products (e.g. sunscreens) may also include dispersants to prevent clumping, intentional dispersion of unagglomerated nanoparticles may not be applicable for most consumer products. Exposure assessment either through modelling releases or exposure simulation provides a means to estimate potential doses to consumers. Material science assessments of the product carrier or other properties of the product matrix can also be useful for evaluating the extent of free nanoparticle release. Even with wear or weathering, the released particles may have attached matrix material that would minimize exposure to nanoparticles. Product design and manufacturing should consider to the extent possible imparting properties both to the nanomaterial and to the product to minimize exposure (e.g. encapsulation, complete mixing, use of liquid sprays rather than aerosols).

25.4.3 Dose

Relevant dose measurements include particle concentration in terms of mass, number, and surface area. Of relevance is not only the dose administered, but also the dose delivered to the target tissue, which depends on bioavailability, transformation and transport. Particle agglomeration may limit the extent of penetration into the body or through other biological barriers. For example, a size limit of 200 nm has been reported for penetration of red blood cell membrane *in vitro* by TiO_2, gold, or fluorescent or polystyrene microspheres (Geiser *et al.,* 2005; Rothen-Rutishauser *et al.,* 2006), and for transport of manganese from nose to brain along sensory axons in rats (Elder *et al.,* 2006). Data generated from pharmaceutical and medical applications of nanotechnology will allow development of physiologically-based pharmacokinetic models to aid understanding of nanomaterials in the body, including effects of particle size and other characteristics. Relatively high doses used in toxicity studies may be placed in perspective by comparing the amount of exposure to a consumer product necessary to achieve the doses used in the toxicity studies, which may be impossible.

25.4.4 Interspecies Differences

Compared with other species, including humans, rats appear to be uniquely sensitive to development of cancer associated with high lung burdens of particles. Even among rodent species, clearance rates and thus susceptibility to lung-overload of particles varies widely (Bermudez *et al.,* 2004). These differences must be considered for each nanoparticle, as well as for each route of exposure (e.g. see discussion of differences related to potential for axonal transport to the brain).

25.4.5 Intraspecies Sensitivity

Sensitive individuals may include those with reactive airway dysfunction syndrome, asthma, cardiovascular disease, and smokers. Additional research is also needed to examine whether the toxicity of specific nanomaterials may be higher for certain life stages (e.g. *in utero*, early childhood, elderly), as well as those with underlying diseases or increased susceptibility. For example, increased sensitivity for subgroups has been reported for UFP exposures in exacerbating asthma and other respiratory conditions in children, and cardiovascular effects in the elderly.

25.4.6 Toxicology Study Design

Many *in vitro* studies of cells and tissues clearly indicate toxicity of nanomaterials at the high concentrations typically used. Although one might therefore conclude that nanotechnology can be toxic to all systems of the body, the pivotal issue is whether such a dose and form of the nanoparticles would actually be experienced by cells and tissues of the body following realistic exposures for consumers. For example, the *in vitro* study by Long *et al.* (2006) provided a hazard assessment of the potential effects of TiO_2 in the brain, which may receive airborne nanoparticles by a direct uptake route through sensory axons in the upper airways as reported in mice (Wang *et al.,* 2008). However, such studies do not indicate whether inhalation of TiO_2 at relevant human exposure levels would cause potential health risks. Positive and negative controls are also needed in studies to assess whether the observed effects of nanomaterials are common to all small particles, are more or less than other particles with known toxicity (e.g. reactive substances such as quartz or asbestos fibers, relatively inert substances such as polystyrene or carbon black; Warheit *et al.*, 2007c; Tabet *et al.*, 2009), or are unique to the particle in question.

Testing strategies for nanomaterials by several scientific forums have recommended initial use of simple *in vitro* testing for screening assessment and to guide decisions on design and conduct of more complex *in vivo* studies as necessary (e.g. Oberdörster et al., 2005b; Warheit *et al.*, 2007d). Rapid screening tests, if predictive of *in vivo* toxicity, could greatly aid assessment of potential health risks of nanomaterials during early product development by decreasing time, expense, and animal use in toxicity testing, and provide the necessary safety information to keep pace with the rate of development of nanotechnology applications. Much of this research focuses on *in vitro* methods to screen for cytotoxicity, cellular interactions, inflammation, or oxidative stress (e.g. Sayes *et al.*, 2007; Rogers *et al.*, 2008; Monteiro-Riviere *et al.*, 2009). Although such tests are showing some utility for screening for toxicity, their results are not necessarily predictive of *in vivo* toxicity and more research is needed to validate tests for specific nanomaterials. For example, essential metals like zinc in cultured cells may indicate a high degree of cytotoxicity, but show few effects *in vivo* (Sayes *et al.*, 2007). Traditional *in vitro* assays using colorimetric or fluorescent dyes as markers of cell viability have also been shown to result in false-positive toxicity in some cases for carbon nanomaterials because of interaction of these materials with the dye (Monteiro-Riviere *et al.*, 2009). Thus, assessments of toxicity will need to rely on several assays with known performance and limitations for nanomaterials, as well as information on the pharmacokinetics of nanomaterials within the body.

In vivo tests provide more realistic data for assessing health risks, although some of the same issues above must be considered in the design of these studies to ensure proper interpretation and characterization of risks.

25.4.7 Dose-Response

Information is accumulating to begin defining health-effect levels for common forms of materials used in consumer products (e.g. TiO_2, ZnO and silver). Development of such levels for the engineered carbonaceous nanomaterials (CNTs, fullerenes) is more difficult because of the multitude of possible characteristics affecting toxicity that will need studies of specific types of materials (or at least groups of materials with similar properties).

25.5 Summary and Conclusions

Potential health risks associated with nanoengineered products (whether actual or not) indicate that researchers should consider product safety early in the process. The available data for risk assessments of products include recent toxicity data on nanomaterials, as well as existing knowledge of exposure and health effects of small-scale materials. For specific product applications, decisions will be required on whether this information is sufficient to be confident of product safety or whether research is needed, and if so, the design of such studies.

Given the current state of the science, with relatively sparse information on engineered nanomaterials, particularly for long-term effects or for sensitive populations, applications with probable low consumer risk would involve:

- particles with more accumulated knowledge indicating generally low human toxicity (e.g. TiO_2, ZnO, silver, or nanomicelles containing existing soap ingredients);
- particles with shapes or biopersistence that are dissimilar to those with known toxicity (e.g. asbestos);
- applications with low potential exposures to free particles because of encapsulation or delivery mechanism;
- products evaluated through exposure modelling or simulation to assess realistic doses to the consumer in comparison to the toxicity literature.

Taken together, the development of consumer products should involve an iterative process for product safety assessment, including phased studies from screening evaluations of exposure and toxicity to gathering more data to assess consumer safety, and redesign of products as necessary.

References

Agency for Toxic Substances and Disease Registry (2005) *Toxicological Profile for Zinc*. Agency for Toxic Substances and Disease Registry: Atlanta. Available at www.atsdr.cdc.gov/toxprofiles/tp60.html [accessed 13 November 2005].

Aitken RJ, Creely KS, Tran CL (2004) *Nanoparticles: An Occupational Hygiene Review*. Research Report 274, Institute of Occupational Medicine for the Health and Safety Executive.

American Conference of Governmental and Industrial Hygienists (2001) *Zinc oxide, fume and dust*. Documentation on TLVs. American Conference of Governmental and Industrial Hygienists.

Atiyeh BS, Costagliola M, Hayek SN, Dibo SA (2007) Effect of silver on burn wound infection control and healing: review of the literature. *Burns* **33**, 139–148.

Baan R, Straif K, Grosse Y, Secretan B, El Ghissassi F, Cogliano V (2006) Carcinogenicity of carbon black, titanium dioxide, and talc. *Lancet Oncol* **7**, 295–296.

Beckett WS, Chalupa DF, Pauly-Brown A, Speers DM, Stewart JC, Frampton MW, Utell MJ, Huang LS, Cox C, Zareba W, Oberdörster G (2005) Comparing inhaled ultrafine versus fine zinc oxide particles in healthy adults: A human inhalation study. *Am J Respir Crit Care Med* **171**, 1129–1135.

Benn TM, Westerhoff P (2008) Nanoparticle silver released into water from commercially available sock fabrics. *Environ Sci Technol* **42**, 4133–4139.

Berman DW, Crump KS (2003) *Final draft: Technical support document for a protocol to assess asbestos-related risk*. EPA# 9345.4-06. US Environmental Protection Agency, Office of Solid Waste and Emergency Response: Washington, DC.

Berman DW, Crump KS (2008) A meta-analysis of asbestos-related cancer risk that addresses fiber size and mineral type. *Crit Rev Toxicol* **38**, 49–73.

Bermudez E, Mangum JB, Asgharian B, Wong BA, Reverdy EE, Janszen DB, Hext PM, Warheit DB, Everitt JI (2002) Long-term pulmonary responses of three laboratory rodent species to subchronic inhalation of pigmentary titanium dioxide particles. *Toxicol Sci* **70**, 86–97.

Bermudez E, Mangum JB, Wong BA, Asgharian B, Hext PM, Warheit DB, Everitt JI, Moss OR (2004) Pulmonary responses of mice, rats, and hamsters to subchronic inhalation of ultrafine titanium dioxide particles. *Toxicol Sci* **77**, 347–357.

Boffetta P, Soutar A, Cherrie JW, Granath F, Andersen A, Anttila A, Blettner M, Gaborieau V, Klug SJ, Langard S, Luce D, Merletti F, Miller B, Mirabelli D, Pukkala E, Adami HO, Weiderpass E (2004) Mortality among workers employed in the titanium dioxide production industry in Europe. *Cancer Causes Control* **15**, 697–706.

Borm PJA (2002) Particle toxicology: from coal mining to nanotechnology. *Inhal Toxicol* **14**, 311–324.

Borm PJ, Schins RP, Albrecht C (2004) Inhaled particles and lung cancer, part B: Paradigms and risk assessment. *Int J Cancer* **110**, 3–14.

Bottini M, Bruckner S, Nika K, Bottini N, Bellucci S, Magrini A, Bergamaschi A, Mustelin T (2006) Multi-walled carbon nanotubes induce T-lymphocyte apoptosis. *Toxicol Lett* **160**, 121–126.

Bräuner EV, Møller P, Barregard L, Dragsted LO, Glasius M, Wåhlin P, Vinzents P, Raaschou-Nielsen O, Loft S (2008) Exposure to ambient concentrations of particulate air pollution does not influence vascular function or inflammatory pathways in young healthy individuals, *Part Fibre Toxicol* **5**, 13.

Brown DM, Wilson MR, MacNee W, Stone V, Donaldson K (2001) Size-dependent proinflammatory effects of ultrafine polystyrene particles: a role for surface area and oxidative stress in the enhanced activity of ultrafines. *Toxicol Appl Pharmacol* **175**, 191–199.

Butnor KJ, Sporn TA, Roggli VL (2003) Exposure to brake dust and malignant mesothelioma: A study of 10 cases with mineral fiber analyses. *Ann Occup Hyg* **47**, 325–330.

Chen X, Schluesener HJ (2008) Nanosilver: A nanoproduct in medical application. *Toxicol Lett* **176**, 1–12.

Choi O, Hu Z (2008) Size dependent and reactive oxygen species related nanosilver toxicity to nitrifying bacteria. *Environ Sci Technol* **42**, 4583–4588.

Conner MW, Flood WH, Rogers AE, Amdur MO (1988) Lung injury in guinea pigs caused by multiple exposures to ultrafine zinc oxide: changes in pulmonary lavage fluid. *J Toxicol Environ Health* **25**, 57–69.

Cross SE, Innes B, Roberts MS, Tsuzuki T, Robertson TA, McCormick P (2007) Human skin penetration of sunscreen nanoparticles: *In-vitro* assessment of a novel micronized zinc oxide formulation. *Skin Pharmacol Physiol* **20**, 148–154.

Cugell DW, Kamp DW (2004) Asbestos and the pleura: a review. *Chest* **125**, 1103–1117.

Davies JC (2006) *Managing the Effects of Nanotechnology*. Woodrow Wilson International Center for Scholars: Washington, DC.

Dekkers S, de Heer C, de Jong WH, Sips AJAM, van Engelen JGM, Kampers FWH (2007) *Nanomaterials in consumer products: Availability on the European market and adequacy of the regulatory framework*. IP/A/ENV/IC/2006-193. European Parliament, Policy Department, Economic and Scientific Policy: Brussels.

Demou E, Peter P, Hellwig S (2008) Exposure to manufactured nanostructured particles in an industrial pilot plant. *Ann Occup Hyg* **52**, 695–706.

Derfus AM, Chan WCW, Bhatia SN (2004) Probing the cytotoxicity of semiconductor quantum dots. *Nano Lett* **4**, 11–18.

Diebold AC (2005) Subsurface imaging with scanning ultrasound holography, *Science* **310**, 61–62.

Donaldson K, Tran CL, MacNee W (2002) Deposition and effects of fine and ultrafine particles in the respiratory tract. *Eur Resp Mon* **7**, 77–92.

Donaldson K, Stone V (2003) Current hypotheses on the mechanisms of toxicity of ultrafine particles. *Ann 1st Super Sanita* **39**, 405–410.

Donaldson K, Aitken R, Tran L, Stone V, Duffin R, Forrest G, Alexander A (2006) Carbon nanotubes: a review of their properties in relation to pulmonary toxicity and workplace safety. *Toxicol Sci* **92**, 5–22.

Drake PL, Hazelwood KJ (2005) Exposure-related health effects of silver and silver compounds: a review. *Ann Occup Hyg* **49**, 575–585.

Dumortier H, Lacotte S, Pastorin G, Marega R, Wu W, Bonifazi D, Briand J-P, Prato M, Muller S, Bianco A (2006) Functionalized carbon nanotubes are non-cytotoxic and preserve the functionality of primary immune cells. *Nano Letters* **6**, 1522–1528.

Eastern Research Group (ERG) (2003a) *Report on the peer consultation workshop to discuss a proposed protocol to assess asbestos-related risk, prepared for the US Environmental Protection Agency.* Office of Solid Waste and Emergency Response, Eastern Research Group.

Eastern Research Group (ERG) (2003b) *Report on the expert panel on health effects of asbestos and synthetic vitreous fibers: The influence of fiber length, prepared for the Agency for Toxic Substances and Disease Registry.* Division of Health Assessment and Consultation: Eastern Research Group.

Elder A, Gelein R, Silva V, Feikert T, Opanashuk L, Carter J, Potter R, Maynard A, Ito Y, Finkelstein J, Oberdörster G (2006) Translocation of inhaled ultrafine manganese oxide particles to the central nervous system. *Environ Health Perspect* **114**, 1172–1178.

Englert N (2004) Fine particles and human health – a review of epidemiological studies. *Toxicol Lett* **149**, 235–242.

Environmental Protection Agency (1989) *Risk assessment guidance for Superfund, Volume 1 – Human health evaluation manual (Part A).* EPA/540/1-89/002. Available at www.epa.gov/oswer/riskassessment/ragsa/pdf/ rags-vol1-pta_complete.pdf.

Environmental Protection Agency (2007) *Concept paper for the nanoscale materials stewardship program under TSCA.* Environmental Protection Agency, Office of Pollution Prevention and Toxics: Washington, DC.

Environmental Working Group (2008) Environmental Working Group's skin deep cosmetic safety database, Sunscreens. Available at www.cosmeticsdatabase.com/special/sunscreens/nanotech.php?nothanks=1 [accessed 3 November 2008].

Esmen NA, Corn M (1998) Airborne fiber concentrations during splitting open and boxing bags of asbestos. *Toxicol Ind Health* **14**, 843–856.

Esquivel EV, Murr LE (2004) A TEM analysis of nanoparticulates in a polar ice core. *Materials Characterization* **52**, 15–25.

European Commission (2004) *European Union Risk Assessment Report – Zinc Oxide, Volume 43*, Munn SJ, Allanou R, Aschberger K, Berthault F, deBruijn J, Musset C, Pakalin S, Paya-Perez A, Pellegrini G, Schwarz-Schulz B, Vegro S (eds). Office for Official Publication of the European Communities: Luxembourg.

Fedoruk MJ, Bronstein R, Kerger BD (2005) Ammonia exposure and hazard assessment for selected household cleaning product uses. *J Exp Anal Environ Epidemiol* **1**, 11.

Ferraro JR, Nakamoto K (1994) *Introductory Raman Spectroscopy.* Academic Press: New York.

Fryzek JP, Chadda B, Marano D, White K, Schweitzer S, McLaughlin JK, Blot WJ (2003) A cohort mortality study among titanium dioxide manufacturing workers in the United States. *J Occup Environ Med* **45**, 400–409.

Gamer AO, Leibold E, van Ravenzwaay B (2006) The *in vitro* absorption of microfine zinc oxide and titanium dioxide through porcine skin. *Toxicol In Vitro* **20**, 301–307.

Geiser M, Rothen-Rutishauser B, Kapp N, Schurch S, Kreyling W, Schulz H, Semmler M, Im Hof V, Heyder J, Gehr P (2005) Ultrafine particles cross cellular membranes by non-phagocytic mechanisms in lungs and in cultured cells. *Environ Health Perspect* **113**, 1555–1560.

Grassian VH, O'Shaughnessy PT, Adamcakova-Dodd A, Pettibone JM, Thorne PS (2007) Inhalation exposure study of titanium dioxide nanoparticles with a primary particle size of 2 to 5 nm. *Environ Health Perspect* **115**, 397–402.

Guico RS, Narayanan S, Wang J, Shull KR (2004) Dynamics of polymer/metal nanocomposite films at short times as studied by x-ray standing waves. *Macromolecules* **37**, 8357–8363.

Heinrich U, Fuhst R, Rittenghausen S, Creutzenberg O, Bellmann B, Koch W, Levsen K (1995) Chronic inhalation exposure of Wistar rats and two different strains of mice to diesel engine exhaust, carbon black, and titanium dioxide. *Inhal Toxicol* **7**, 533–556.

Helland A, Wick P, Koehler A, Schmid K, Som C (2007) Reviewing the environmental and human health knowledge base of carbon nanotubes. *Environ Health Perspect* **115**, 1125–1131.

Hesterberg TW, Hart GA, Chevalier J, Miller WC, Hamilton RD, Bauer J, Thevenaz P (1998) The importance of fiber biopersistence and lung dose in determining the chronic inhalation effects of X607, RCF1, and chrysotile asbestos in rats. *Toxicol Appl Pharmacol* **153**, 68–82.

Hext PM, Tomenson JA, Thompson P (2005) Titanium dioxide: Inhalation toxicology and epidemiology. *Ann Occup Hyg* **49**, 461–472.

Hodgson JT, Darnton A (2000) The quantitative risks of mesothelioma and lung cancer in relation to asbestos exposure. *Ann Occup Hyg* **44**, 565–601.

Hoet PHM, Bruske-Hohlfeld I, Salata OV (2004) Nanoparticles – known and unknown health risks. *J Nanobiotechnol* **2**, 12–26.

Hofmann W, Sturm R, Winkler-Heil R, Pawlak E (2003) Stochastic model of the ultrafine particle deposition and clearance in the human respiratory tract. *Radiat Prot Dosimet* **105**, 77–80.

Hsu LY, Chein HM (2007) Evaluation of nanoparticle emissions of TiO_2 nanopowder coating materials. *J Nanoparticle Res* **9**, 157–163.

Hua MS, Huang CC (1991) Chronic occupational exposure to manganese and neurobehavioral function. *J Clin Exp Neuropsychol* **13**, 495–507.

Huang CC, Chu NS, Lu CS, Chen RS, Calne DB (1998) Long-term progression in chronic manganism: ten years of follow-up. *Neurology* **50**, 698–700.

Hussain SM, Hess KL, Gearhart JM, Geiss KT, Schlager JJ (2005) *In vitro* toxicity of nanoparticles in BRL 3A rat liver cells. *Toxicol In Vitro* **19**, 975–983.

International Agency for Research on Cancer (2002) *IARC monographs on the evaluation of carcinogenic risk of chemicals to humans: Man-made vitreous fibers, Volume 81*. International Agency for Research on Cancer, Lyon, France. Available at www.cie-iarc.fr/htdocs/monographs/vol81/81.html

International Commission on Radiological Protection (1994) *Human Respiratory Tract Model for Radiological Protection*. Elsevier Science: New York.

Isakovic A, Markovic Z, Todorovic-Markovic B, Nikolic N, Sanja Vranjes-Djuric S, Mirkovic M, Dramicanin M, Harhaji L, Raicevic N, Nikolic Z, Trajkovic V (2006) Distinct cytotoxic mechanisms of pristine versus hydroxylated fullerene. *Toxicol Sci* **91**, 173–183.

Ivanda M, Furic K, Music S, Ristic M, Gotic M, Ristic D, Tonejc AM, Djerdj I, Mattarelli M, Montagna M, Rossi F, Ferrari M, Chiasera A, Jestin Y, Righini GC, Kiefer W, Goncalves RR (2007) Low wavenumber Raman scattering of nanoparticles and nanocomposite materials. *J Raman Spectrosc* **38**, 647–659.

Jani P, Halbert GW, Langridge J, Florence AT (1990) Nanoparticle uptake by the rat gastrointestinal mucosa: Quantitation and particle size dependency. *J Pharm Pharmacol* **42**, 821–826.

Jia G, Wang H, Yan L, Pei R, Yan T, Zhao Y, Guo X (2005) Cytotoxicity of carbon nanomaterials: single-wall nanotube, multi-wall nanotube, and fullerene. *Environ Sci Technol* **39**, 1378–1383.

Keller A, Fierz M, Siegmann K, Siegmann HC (2001) Surface science with nanosized particles in a carrier gas. *J Vacuum Sci Technol A* **19**, 1–8.

Kolosnjaj J, Szwarc H, Moussa F (2007) Toxicity studies of carbon nanotubes. In *Bio-Applications of Nanoparticles*, Chan WCW (ed.). Advances in Experimental Medicine and Biology 620. Landes Bioscience and Springer Science-Business Media: New York; 181–204.

Kraeling ME, Gopee HV, Roberts DW, Ogunsola OA, Walker NJ, Yu WW, Colvin VL, Howard PC, Bronaugh RL (2007) Evaluation of *in vitro* penetration of quantum dot nanoparticles into human skin. *Toxicologist* **96**, 289.

Kreyling WG, Semmler M, Erbe F, Mayer P, Takenaka S, Schulz H, Oberdörster G, Ziesenis A (2002) Translocation of ultrafine insoluble iridium particles from lung epithelium to extrapulmonary organs is size dependent but very low. *J Toxicol Environ Health A* **65**, 1513–1530.

Kreyling WG, Semmler-Behnke M, Moller W (2006) Ultrafine particle-lung interactions: Does size matter? *J Aerosol Med* **19**, 74–83.

Ku BK, Maynard AD (2005) Comparing aerosol surface-area measurement of monodisperse ultrafine silver agglomerates using mobility analysis, transmission electron microscopy and diffusion charging. *J Aerosol Sci* **36**, 1108–1124.

Kuempel ED, Tran CL, Castranova V, Bailer AJ (2006) Lung dosimetry and risk assessment of nanoparticles: Evaluating and extending current models in rats and humans. *Inhal Toxicol* **18**, 717–724.

Kuhlbusch TAJ, Neumann S, Fissan H (2004) Number size distribution, mass concentration, and particle composition of PM_1, $PM_{2.5}$, and PM_{10} in bag filling areas of carbon black production. *J Occup Environ Hygiene* **1**, 660–671.

Kuschner WG, D'Alessandro A, Wintermeyer SF, Wong H, Boushey HA, Blanc PD (1995) Pulmonary responses to purified zinc oxide fume. *J Investig Med* **43**, 371–378.

Kuschner WG, Wong H, D'Alessandro A, Quinlan P, Blanc PD (1997) Human pulmonary responses to experimental inhalation of high concentration fine and ultrafine magnesium oxide particles. *Environ Health Perspect* **105**, 1234–1237.

Lam CW, James JT, McCluskey R, Arepalli S, Hunter RL (2006) A review of carbon nanotube toxicity and assessment of potential occupational and environmental health risk. *Crit Rev Toxicol* **36**, 189–217.

Lam HF, Conner MW, Rogers AE, Fitzgerald S, Amdur MO (1985) Functional and morphologic changes in the lungs of guinea pigs exposed to freshly generated ultrafine zinc oxide. *Toxicol Appl Pharmacol* **78**, 29–38.

Lam HF, Chen LC, Ainsworth D, Peoples S, Amdur MO (1988) Pulmonary function of guinea pigs exposed to freshly generated ultrafine zinc oxide with and without spike concentrations. *Am Ind Hyg Assoc J* **49**, 333–341.

Lansdown ABG (2007) Critical observations on the neurotoxicity of silver. *Crit Rev Toxicol* **37**, 237–250.

Last JM (ed.) (2001) *A Dictionary of Epidemiology* (4th edn). Oxford University Press: New York.

Lee KP, Trochimowicz HJ, Reinhardt CF (1985) Pulmonary response of rats exposed to titanium dioxide (TiO_2) by inhalation for two years. *Toxicol Appl Pharmacol* **79**, 179–192.

Li Z, Hulderman T, Salmen R, Chapman R, Leonard SS, Young SH, Shvedova A, Luster MI, Simeonova PP (2007) Cardiovascular effects of pulmonary exposure to single-wall carbon nanotubes. *Environ Health Perspect* **115**, 377–382.

Lockey J, Lemasters G, Rice C, Hansen K, Levin L, Shipley R, Spitz H, Wiot J (1996) Refractory ceramic fiber exposure and pleural plaques. *Am J Respir Crit Care Med* **154**, 1405–1410.

Lockey JE, Lemasters GK, Levin L, Rice C, Yiin J, Reutman S, Papes D (2002) A longitudinal study of chest radiographic changes of workers in the refractory ceramic fiber industry. *Chest* **121**, 2044–2051.

Lomer MC, Hutchinson C, Volkert S, Greenfield SM, Catterall A, Thompson RP, Powell JJ (2004) Dietary sources of inorganic microparticles and their intake in healthy subjects and patients with Crohn's disease. *Br J Nutr* **92**, 947–955.

Long TC, Saleh N, Tilton RD, Lowry GV, Veronesi B (2006) Titanium dioxide (P25) produces reactive oxygen species in immortalized brain microglia (BV2): Implications for nanoparticle neurotoxicity. *Environ Sci Technol* **40**, 4346–4352.

Lovern SB, Klaper R (2006) *Daphnia magna* mortality when exposed to titanium dioxide and fullerene (C60) nanoparticles. *Environ Toxicol Chem* **25**, 1132–1137.

Lovern SB, Strickler JR, Klaper R (2007) Behavioral and physiological changes in *Daphnia magna* when exposed to nanoparticle suspensions (titanium dioxide, nano-C60, and C60HxC70Hx). *Environ Sci Technol* **41**, 4465–4470.

Luoma SN (2008) *Silver Nanotechnologies and the Environment: Old Problems or N Challenges?* Woodrow Wilson International Center for Scholars: Washington, DC.

Ma-Hock L, Gamer AO, Landsiedel R, Leibold E, Frechen T, Sens B, Linsenbuehler M, van Raven-zwaay B (2007) Generation and characterization of test atmospheres with nanomaterials. *Inhal Toxicol* **19**, 833–848.

Mavon A, Miquel C, Lejeune O, Payre B, Moretto P (2007) *In vitro* percutaneous absorption and *in vivo* stratum corneum distribution of an organic and a mineral sunscreen. *Skin Pharmacol Physiol* **20**, 10–20.

Maynard AD (2006a) Nanotechnology: The next big thing, or much ado about nothing? *Ann Occup Hyg* **51**, 1–12.

Maynard AD (2006b) Nanotechnology: Assessing the risks. *Nano Today* **1**(2), 22–33.

Maynard AD (2006c) *Nanotechnology: A Research Strategy for Addressing Risk*. Project on Emerging Nanotechnologies: Washington, DC.

Maynard AD, Zimmer AT (2002) Investigation of the aerosols produced by a high-speed, hand-held grinder using various substrates. *Ann Occup Hyg* **46** (Suppl.1), 320–322.

Maynard AD, Baron PA, Foley M, Shvedova AA, Kisin ER, Castranova V (2004) Exposure to carbon nanotube material: Aerosol release during the handling of unrefined single-walled carbon nanotube material. *J Toxicol Environ Health A* **67**, 87–107.

Maynard AD, Kuempel ED (2005) Airborne nanostructured particles and occupational health. *J Nanoparticle Res* **7**, 587–614.

Medley T, Walsh S, Baier-Anderson C, Balbus J, Carberry J, Denison R, Doraiswamy K, Gannon J, Ruta G, Swain K, Warheit D, Whiting G (2007) *Nano Risk Framework, Environmental Defense—DuPont Nano Partnership*. Environmental Defense: New York and DuPont, Wilmington.

Mercer RR, Scabilloni J, Wang L, Kisin E, Murray AR, Schwegler-Berry D, Shvedova AA, Castra-nova V (2008) Alteration of deposition pattern and pulmonary response as a result of improved dispersion of aspirated single-walled carbon nanotubes in a mouse model. *Am J Physiol Lung Cell Mol Physiol* **294**, L87–L97.

Monteiro-Riviere NA, Inman AO, Zhang LW (2009) Limitations and relative utility of screening assays to assess engineered nanoparticle toxicity in a human cell line. *Toxicol Appl Pharmacol* **234**, 222–235.

Morgan K (2005) Development of a preliminary framework for informing the risk analysis and risk management of nanoparticles. *Risk Anal* **25**, 1621–1635.

Morgans WM (1990) *Outlines of Paint Technology* (3rd edn). Halsted Press, a division of John Wiley & Sons: New York.

Mowat F, Weidling R, Sheehan P (2007) Simulation tests to assess occupational exposure to airborne asbestos from asphalt-based roofing products. *Ann Occup Hyg* **51**, 451–462.

Mueller NC, Nowack B (2008) Exposure modeling of engineered nanoparticles in the environment. *Environ Sci Technol* **42**, 4447–4453.

Muller J, Huaux F, Moreau N, Misson P, Heilier JF, Delos M, Arras M, Fonseca A, Nagy JB, Lison D (2005) Respiratory toxicity of multi-wall carbon nanotubes. *Toxicol Appl Pharmacol* **207**, 221–231.

Murr LE, Esquivel EV, Bang JJ (2004) Characterization of nanostructure phenomena in airborne particulate aggregates and their potential for respiratory health effects. *J Mater Sci Mater Med* **15**, 237–247.

National Institute for Occupational Safety and Health (2005) *NIOSH current intelligence bulletin: Evaluation of health hazard and occupational exposure to titanium dioxide, Draft*. Department of Health and Human Services, Centers for Disease Control and Prevention, National Institute for Occupational Safety and Health. Available at www.cdc.gov/niosh/topics/nanotech [accessed 1 November 2006].

National Institute for Occupational Safety and Health (2007) *Progress toward Safe Nanotechnology in the Workplace*. National Institute for Occupational Safety and Health Nanotechnology Research Center: Cincinnati, OH.

National Institute for Occupational Safety and Health (2008) *Strategic plan for NIOSH Nanotechnology research and guidance: filling the knowledge gap, Draft*. Nanotechnology Research Program, National Institute for Occupational Safety and Health, Centers for Disease Control and Prevention. Available at www.cdc.gov/niosh/topics/nanotech/strat_plan.html [accessed 14 November 2008].

National Research Council (1983) *Risk Assessment in the Federal Government: Managing the Process*. National Research Council, National Academy Press: Washington, DC.

National Science and Technology Council (2007) *Strategic Plan, National Science and Technology Council*. Committee on Technology, Subcommittee on Nanoscale Science, Engineering, and Technology: Washington, DC. Available at http://www.nano.gov/ NNI_Strategic_Plan_2004.pdf.

Nel A, Xia T, Mädler L, Li N (2006) Toxic potential of materials at the nanolevel. *Science* **311**, 622–627.

Nemmar A, Hoet PH, Vanquickenborne B, Dinsdale D, Thomeer M, Hoylaerts MF, Vanbilloen H, Mortelmans L, Nemery B (2002) Passage of inhaled particles into the blood circulation in humans. *Circulation* **105**, 411–414.

Nohynek GJ, Dufour EK, Roberts MS (2008) Nanotechnology, cosmetics and the skin: Is there a health risk? *Skin Pharmacol Physiol* **21**, 136–149.

Nohynek GJ, Lademann J, Ribaud C, Roberts MS (2007) Grey goo on the skin? Nanotechnology, cosmetic and sunscreen safety. *Crit Rev Toxicol* **37**, 251–277.

Nurkiewicz TR, Porter DW, Barger M, Millecchia L, Rao KM, Marvar PJ, Hubbs AF, Castranova V, Boegehold MA (2006) Systemic microvascular dysfunction and inflammation after pulmonary particulate matter exposure. *Environ Health Perspect* **114**, 412–419.

Oberdörster G, Ferin J, Gelein R, Soderholm SC, Finkelstein J (1992) Role of the alveolar macrophage in lung injury: studies with ultrafine particles. *Environ Health Perspect* **97**, 193–197.

Oberdörster G, Ferin J, Lehnert BE (1994) Correlation between particle size, in vivo particle persistence, and lung injury. *Environ Health Perspect* **102** (Suppl. 5), 173–179.

Oberdörster G, Finkelstein JN, Johnston C, Gelein R, Cox C, Baggs R, Elder ACP (2000) *Acute Pulmonary Effects of Ultrafine Particles in Rats and Mice*. Research Report 96. Health Effects Institute: Cambridge.

Oberdörster G, Sharp Z, Atudorei V, Elder A, Gelein R, Kreyling W, Cox C (2004) Translocation of inhaled ultrafine particles to the brain. *Inhal Toxicol* **16**, 437–445.

Oberdörster G, Oberdörster E, Oberdörster J (2005a) Nanotoxicology: An emerging discipline evolving from studies of ultrafine particles. *Environ Health Perspect* **113**, 823–839.

Oberdörster G, Maynard A, Donaldson K, Castranova V, Fitzpatrick J, Ausman K, Carter J, Karn B, Kreyling W, Lai D, Olin S, Monteiro-Riviere N, Warheit D, Yang H (2005b) Principles for characterizing the potential human health effects from exposure to nanomaterials: Elements of a screening strategy. *Part Fibre Toxicol* **2**. DOI: 10.1186/1743-8977-2-8

Paustenbach DJ, Sage A, Bono M, Mowat F (2004) Occupational exposure to airborne asbestos from coatings, mastics, and adhesives. *J Exp Anal Environ Epidemiol* **14**, 234–244.

Poland CA, Duffin R, Kinlock I, Maynard A, Wallace WAH, Seaton A, Stone V, Brown S, MacNee W, Donaldson K (2008) Carbon nanotubes introduced into the abdominal cavity of mice show asbestos-like pathogenicity in a pilot study. *Nature Nanotech* **3**, 423–428.

Project on Emerging Nanotechnologies (2008) *The Nanotechnology Consumer Products Inventory*. Woodrow Wilson International Center for Scholars, Project on Emerging Nanotechnologies:

Washington, DC. Available at: http://www.nanotechproject.org/inventories/consumer/ [accessed 25 December 2008].

Rabias I, Fardis M, Devlin E, Boukos N, Tsitrouli D, Papavassiliou G (2008) No aging phenomena in ferrofluids: The influence of coating on interparticle interactions of maghemite nanoparticles. *ACS Nano* **2**, 977–983.

Rausch LJ, Bisinger EC Jr, Sharma A (2004) Carbon black should not be classified as a human carcinogen based on rodent bioassay data. *Regul Toxicol Pharmacol* **40**, 28–41.

Renaud G, Lazzari R, Revenant C, Barbier A, Noblet M, Ulrich O, Leroy F, Jupille J, Borensztein Y, Henry CR, Deville JP, Scheurer F, Mane-Mane J, Fruchart O (2003) Real-time monitoring of growing nanoparticles. *Science* **300**, 1416–1419.

Rogak SN, Flagan RC, Nguyen HV (1993) The mobility and structure of aerosol agglomerates. *Aerosol Sci Technol* **18**, 25–47.

Rogers EJ, Hsieh SF, Organti N, Schmidt D, Bello D, Rothen-Rutishauser BM (2008) A high through-put *in vitro* analytical approach to screen for oxidative stress potential exerted by nanomaterials using a biologically relevant matrix: Human blood serum. *Toxicol In Vitro* **22**, 1639–1647.

Roggli VL (1992) Asbestos bodies and nonasbestos ferruginous bodies. In *Pathology of Asbestos-Associated Diseases*, Roggli VL, Greenberg SD, Pratt PC (eds). Little, Brown and Company: New York; 39–75

Roggli VL, Sharma A, Butnor KJ, Sporn T, Vollmer RT (2002) Malignant mesothelioma and occupational exposure to asbestos: A clinicopathological correlation of 1445 cases. *Ultrastructural Pathol* **26**, 55–65.

Roggli VL, Oury TD, Sporn TA (eds) (2004) *Pathology of Asbestos-Associated Diseases*. Springer: Berlin.

Rothen-Rutishauser BM, Schurch S, Haenni B, Kapp N, Gehr P (2006) Interaction of fine particles and nanoparticles with red blood cells visualized with advanced microscopic techniques. *Environ Sci Technol* **40**, 4353–4359.

Rouse JG, Yang J, Ryman-Rasmussen JP, Barron AR, Monteiro-Riviere NA (2007) Effects of mechanical flexion on the penetration of fullerene amino acid-derivatized peptide nanoparticles through skin. *Nano Lett* **7**, 155–160.

Royal Society (2004) *Nanoscience and Nanotechnologies: Opportunities and Uncertainties*. The Royal Society and The Royal Academy of Engineering: London.

Ryman-Rasmussen JP, Riviere JE, Monteiro-Riviere NA (2006) Penetration of intact skin by quantum dots with diverse physicochemical properties. *Toxicol Sci* **91**, 159–165.

Sayes CM, Fortner JD, Guo W, Lyon D, Boyd AM, Ausman KD, Tao YJ, Sitharaman B, Wilson LJ, Hughes JB, West JL, Colvin VL (2004) The differential cytotoxicity of water-soluble fullerenes. *Nano Lett* **4**, 1881–1887.

Sayes CM, Reed KL, Warheit DB (2007) Assessing toxicity of fine and nanoparticles: Comparing *in vitro* measurements to *in vivo* pulmonary toxicity profiles. *Toxicol Sci* **97**, 163–180.

SCCNFP (2003) *Opinion concerning zinc oxide, Colipa N°S 76*. European Commission, Scientific Committee on Cosmetic Products and Non-Food Products Intended for Consumers. Available at ec.europa.eu/health/ph_risk/ committees/sccp/documents/out222_en.pdf [accessed 3 January 2009].

SCENIHR (2006) *Modified opinion (after public consultation) on the appropriateness of existing methodologies to assess the potential risks associated with engineered and adventitious products of nanotechnologies, SCENIHR/002/05*. European Commission, Health & Consumer Protection Directorate-General, Directorate C–Public Health and Risk Assessment, Scientific Committee on Emerging and Newly Identified Health Risks. Available at ec.europa.eu/health/ph_risk/committees/04_scenihr/docs/scenihr_o_003b.pdf [accessed 3 January 2009].

Scientific Committee on Consumer Products (2005) *Statement on zinc oxide used in sunscreens, SCCP/0932/05*. European Commission, Scientific Committee on Consumer Products: Brussels.

Seipenbusch M, Binder A, Kasper G (2008) Temporal evolution of nanoparticle aerosols in workplace exposure. *Ann Occup Hyg* **52**, 707–716.

Sheehan P, Malzahn D, Goswami E, Mandel JH (2008) Simulation of benzene exposure during use of a mineral spirit solvent to clean elevator bearing housings. *Human Ecol Risk Assess* **14**, 421–432.

Shekhawat GS, Dravid VP (2005) Nanoscale imaging of buried structures via scanning near-field ultrasound holography. *Science* **310**, 89–92.

Shvedova AA, Castranova V, Kisin ER, Schwegler-Berry D, Murray AR, Gandelsman VZ, Maynard A, Baron P (2003) Exposure to carbon nanotube material: Assessment of nanotube cytotoxicity using human keratinocyte cells. *J Toxicol Environ Health Part A* **66**, 1909–1926.

Shvedova AA, Kisin ER, Porter D, Schulte P, Kagan VE, Fadeel B, Castranova V (2009) Mechanisms of pulmonary toxicity and medical applications of carbon nanotubes: Two faces of Janus? *Pharmacol Thera* **121**, 192–204. DOI: 10.1016/j.pharmthera.2008.10.009.

Song D, Hrbek J, Osgood R (2005) Formation of TiO_2 nanoparticles by reactive-layer-assisted deposition and characterization by XPS and STM. *Nano Lett* **5**, 1327–1332.

Sung JH, Ji JH, Yoon JU, Kim DS, Song MY, Jeong J, Han BS, Han JH, Chung YH, Kim J, Kim TS, Chang HK, Lee EJ, Lee JH, Yu IJ (2008) Lung function changes in Sprague-Dawley rats after prolonged inhalation exposure to silver nanoparticles. *Inhal Toxicol* **20**, 567–574.

Tabet L, Bussy C, Amara N, Setyan A, Grodet A, Rossi MJ, Pairon J-C, Boczkowski J, Lanone S (2009) Adverse effects of industrial multiwalled carbon nanotubes on human pulmonary cells. *J Toxicol Environ Health Part A* **72**, 60–73.

Takagi A, Hirose A, Nishimura T, Fukumori N, Ogata A, Ohashi N, Kitajima S, Kanno J (2008) Induction of mesothelioma in p53+/- mouse by intraperitoneal application of multi-wall carbon nanotube. *J Toxicol Sci* **13**, 105–116.

Tetard L, Passian A, Venmar KT, Lynch RM, Voy BH, Shekhawat G, Dravid VP, Thundat T (2008) Imaging nanoparticles in cells by nanomechanical holography. *Nature Nanotechnol* **3**, 501–505.

Thomas K, Aguar P, Kawasaki H, Morris H, Nakanishi J, Savage N (2006) Research strategies for safety evaluation of nanomaterials, part VIII: International efforts to develop risk-based safety evaluations for nanomaterials. *Toxicol Sci* **92**, 23–32.

Tsuji JS, Maynard AD, Howard PC, James JT, Lam C-W, Warheit DB, Santamaria AB (2006) Research strategies of safety evaluation of nanomaterials, part IV: Risk assessment of nanoparticles. *Toxicol Sci* **89**, 42–50.

Vermeulen H, van Hattem JM, Storm-Versloot MN, Ubbink DT (2007) Topical silver for treating infected wounds. *Cochrane Database of Systematic Reviews* 2007, Issue 1. Art. No.: CD005486. DOI: 10.1002/14651858.CD005486.pub2. John Wiley & Sons, Ltd.

Wallenborn JG, McGee JK, Schladweiler MC, Ledbetter AD, Kodavanti UP (2007) Systemic translocation of particulate matter–associated metals following a single intratracheal instillation in rats. *Toxicol Sci* **98**, 231–239.

Wang B, Feng W-Y, Wang T-C, Jia G, Wang M, Shi J-W, Zhang F, Zhao Y-L, Chai Z-F (2006) Acute toxicity of nano-and micro-scale zinc powder in healthy adult mice. *Toxicol Ltrs* **161**, 115–123.

Wang J, Zhou G, Chen C, Yu H, Wang T, Ma Y, Jia G, Gao Y, Li B, Sun J, Li Y, Jiao F, Zhao Y, Chai Z (2007) Acute toxicity and biodistribution of different sized titanium dioxide particles in mice after oral administration. *Toxicol Lett* **168**, 176–185.

Wang J, Liu Y, Jiao F, Lao F, Li W, Gu Y, Li Y, Ge C, Zhou G, Li B, Zhao Y, Chai Z, Chen C (2008) Time-dependent translocation and potential impairment on central nervous system by intranasally instilled TiO_2 nanoparticles. *Toxicology* **254**, 82–90.

Warheit DB (2006) What is currently known about the health risks related to carbon nanotube exposures? *Carbon* **44**, 1064–1069.

Warheit DB, Reed KL, Webb TR (2003) Pulmonary toxicity studies in rats with triethoxyoctylsilane (OTES)-coated, pigment-grade titanium dioxide particles: Bridging studies to predict inhalation hazard. *Exp Lung Res* **29**, 593–606.

Warheit DB, Laurence BR, Reed KL, Roach DH, Reynolds GAM, Webb TR (2004) Comparative pulmonary toxicity assessment of single-wall carbon nanotubes in rats. *Toxicol Sci* **77**, 117–125.

Warheit DB, Brock WJ, Lee KP, Webb TR, Reed KL (2005) Comparative pulmonary toxicity inhalation and instillation studies with different TiO$_2$ particle formulations: Impact of surface treatments on particle toxicity. *Toxicol Sci* **88**, 514–524.

Warheit DB, Frame SR (2006) Characterization and reclassification of titanium dioxide-related pulmonary lesions. *J Occup Environ Med* **48**, 1308–1313.

Warheit DB, Webb TR, Sayes CM, Colvin VL, Reed KL (2006a) Pulmonary instillation studies with nanoscale TiO$_2$ rods and dots in rats: Toxicity is not dependent upon particle size and surface area. *Toxicol Sci* **91**, 227–236.

Warheit DB, Webb T, Reed K (2006b) Pulmonary toxicity screening studies in male rats with TiO$_2$ particulates substantially encapsulated with pyrogenically deposited, amorphous silica. *Part Fibre Toxicol* **3**, 3.

Warheit DB, Hoke RA, Finlay C, Donner EM, Reed KL, Sayes CM (2007a) Development of a base set of toxicity tests using ultrafine TiO(2) particles as a component of nanoparticle risk management. *Toxicol Lett* **171**, 99–110.

Warheit DB, Webb TR, Colvin VL, Reed KL, Sayes CM (2007b) Pulmonary bioassay studies with nanoscale and fine-quartz particles in rats: Toxicity is not dependent upon particle size but on surface characteristics. *Toxicol Sci* **95**, 270–280.

Warheit DB, Webb TR, Reed KL, Frerichs S, Sayes CM (2007c) Pulmonary toxicity study in rats with three forms of ultrafine-TiO$_2$ particles: Differential responses related to surface properties. *Toxicology* **230**, 90–104.

Warheit DB, Borm PJ, Hennes C, Lademann J (2007d) Testing strategies to establish the safety of nanomaterials: Conclusions of an ECETOC workshop. *Inhal Toxicol* **19**, 631–643.

Wesselkamper SC, Chen LC, Gordon T (2001) Development of pulmonary tolerance in mice exposed to zinc oxide fumes. *Toxicol Sci* **60**, 144–151.

Wesselkamper SC, Chen LC, Gordon T (2005) Quantitative trait analysis of the development of pulmonary tolerance to inhaled zinc oxide in mice. *Respir Res* **6**, 73.

Wiebert P, Sanchez-Crespo A, Seitz J, Falk R, Philipson K, Kreyling WG, Moller W, Sommerer K, Larsson S, Svartengren M (2006a) Negligible clearance of ultrafine particles retained in healthy and affected human lungs. *Eur Respir J* **28**, 286–290.

Wiebert P, Sanchez-Crespo A, Falk R, Philipson K, Lundin A, Larsson S, Moller W, Kreyling WG, Svartengren M (2006b) No significant translocation of inhaled 35-nm carbon particles to the circulation in humans. *Inhal Toxicol* **18**, 741–747.

Witschi HP, Last JO (2001) Toxic responses of the respiratory system. In *Cassarrett and Doull's Toxicology: The Basic Science of Poisons*, Klaassen CD (ed.). McGraw Hill: New York; 515–534.

Wu N, Fu L, Su M, Aslam M, Wong KC, Dravid VP (2004) Interaction of fatty acid monolayers with cobalt nanoparticles. *Nano Lett* **4**, 383–386.

Yamamoto A, Honma R, Sumita M, Hanawa T (2004) Cytotoxicity evaluation of ceramic particles of different sizes and shapes. *J Biomed Mater Res A* **68**, 244–256.

Yang ST, Wang X, Jia G, Gu Y, Wang T, Nie H, Ge C, Wang H, Liu Y (2008) Long-term accumulation and low toxicity of single-walled carbon nanotubes in intravenously exposed mice. *Toxicol Lett* **181**, 182–189.

Yeates DB, Mauderly JL (2001) Inhaled environmental/occupational irritants and allergens: Mechanisms of cardiovascular and systemic responses. *Environ Health Perspect* **109** (Suppl. 4), 479–481.

Yeganeh B, Kull CM, Hull MS, Marr LC (2008) Characterization of airborne particles during production of carbonaceous nanomaterials. *Environ Sci Technol* **42**, 4600–4606.

Zhang LW, Yu WW, Colvin VL, Monteiro-Riviere NA (2008) Biological interactions of quantum dot nanoparticles in skin and in human epidermal keratinocytes. *Toxicol Appl Pharmacol* **228**, 200–211.

26

Safety Assessment of Engineered Nanomaterials in Direct Food Additives and Food Contact Materials

Penelope A. Rice, Kimberly S. Cassidy, Jeremy Mihalov and T. Scott Thurmond

26.1 Introduction

The ability to image, measure, model, and manipulate matter on the nanoscale[1] is leading to new technologies that may affect virtually every sector of our economy and our daily lives. Nanoscale science, engineering and technology are enabling the development of promising new materials and applications across many fields. Nanotechnology applications being developed by the food industry include (Chaudhry *et al.*, 2008):

- nanostructured additives for improved taste, color, flavor, texture, and consistency of food; nanosized or nano-encapsulated nutrients and health supplements for increased absorption and bioavailability;
- nanomaterials incorporated into food packaging for improved mechanical, barrier, and antimicrobial properties;
- nanosensors for tracking and monitoring the condition of food during transport and storage.

---

[1] The NNI defines nanotechnology as the understanding and control of matter at dimensions of roughly 1–100 nm, where unique phenomena enable novel applications. The Food and Drug Administration (FDA) has not adopted a formal, fixed definition of nanotechnology at this time. The FDA Nanotechnology Task Force examined this issue and came to the conclusion that while one definition for 'nanotechnology', 'nanoscale material', or a related term or concept may offer meaningful guidance in one context, that definition may be too narrow or broad to be of use in another. The FDA is continuing to pursue regulatory approaches that take into account the potential importance of material size and the evolving state of the science.

Nanotoxicity: From In Vivo *and* In Vitro *Models to Health Risks*   Edited by Saura Sahu and Daniel Casciano
© 2009 John Wiley & Sons, Ltd

The application of nanotechnology to foods and food packaging has the potential to confer many benefits, including improved sensing of microbial contamination in food, enhanced bioavailability of nutrients, and increased palatability. However, the consumer must be assured that these applications do not create a safety hazard or render the food unsuitable for consumption. Therefore, appropriate safety assurance mechanisms must be in place to ensure that food-related products incorporating nanomaterials meet the FDA criteria of a 'reasonable certainty of no harm'.

The Food and Drug Administration (FDA) of the US regulates a broad range of products including food additives and packaging under the Federal Food, Drug, and Cosmetic Act (FFDCA). The Office of Food Additive Safety (OFAS) in the Center for Food Safety and Nutrition (CFSAN) within the FDA is responsible for ensuring the safety of all substances deliberately added to food (direct food additives), and substances that may become a part of food as a result of migration from food packaging (indirect food additives). The FDA's statutory authority requires premarket authorization for food additives, as stated under Section 409 of the FFDCA, while permitting substances that are 'generally recognized as safe' to be marketed without prior authorization or review by the Agency.

Any substance added to food, directly or indirectly, is a food additive unless the substance is 'generally recognized as safe' (GRAS) for its intended use or is otherwise excluded from the definition of a food additive. An indirect food additive, or food contact substance (FCS), is any substance intended for use as a component of materials used in food packaging or processing. Direct and indirect food additives require premarket approval from the FDA prior to marketing. Direct food and color additives receive premarket approval from FDA via a petition process which results in the publication of regulations authorizing their intended uses. The Food Contact Notification (FCN) process is the primary method of authorizing new uses of a FCS, unless circumstances warrant use of the petition process. GRAS uses of food ingredients do not require premarket authorization by the FDA; however, they may be voluntarily submitted to the Agency for review through a notification procedure.

As is obvious from the examples of nanoscale food applications under development, nanoscale materials could potentially be used in most product areas regulated by the FDA. Although these materials present challenges similar to those posed by products manufactured by conventional technologies, nanoscale materials may present additional challenges due to the fact that properties relevant to product safety and effectiveness may change as size varies within the nanoscale. In order to find ways to address these challenges, the FDA participates in a range of government initiatives, including: the National Science and Technology Council; the Subcommittee on Nanoscale Science, Engineering and Technology (NSET) and its Nanotechnology Environmental and Health Implications (NEHI) working group; and, in particular, the National Nanotechnology Initiative (NNI). The NNI was developed to coordinate multi-agency efforts in the areas of nanoscale science, engineering and technology. Among the goals of this program is to support responsible development of nanotechnology.

The FDA has also formed its own internal working group, the Nanotechnology Executive Task Force (the Task Force), whose primary mission is to address the issue of regulatory assessment of nanomaterials under the Agency's jurisdiction. The established goal of the Task Force is to determine regulatory approaches that would enable the continued development of innovative, safe and effective FDA-regulated products that use nanoscale materials. To this end, in July 2007 it released a report, subsequently published in *Nanotechnology*, which contains a broad overview of the available scientific data for nanomaterials and summarizes the FDA's current thinking on a range of scientific and regulatory issues pertaining to nanoscale materials under the FDA's jurisdiction.[2] In particular, the report addressed the FDA's current thinking on how nanoscale materials fit into the FDA's current safety review paradigms, and made recommendations as to how nanomaterials may be handled within those paradigms. This chapter will explore the Task Force's recommendations, focusing on how those recommendations may be applied to the safety assessment of nanomaterial food additives and food packaging materials. It will also discuss the current state of technology, as it has developed in food-related nanomaterials subsequent to that report, and the current research underway in the Agency regarding human health implications and safety assessment of nanomaterials.

26.2   Experience with Nanotechnology and Food Ingredients/Packaging

The application of nanotechnology to food ingredients and packaging has the possibility to affect a broad range of existing and future products. These applications may influence the manufacture, processing, packaging, intended use, or function of foods and ingredients. Reports indicate that products that claim to utilize nanotechnology are already on the market in various countries, including the US.[3] The products mentioned in this report, which are under the purview of the FDA's food ingredients program, are mainly food contact substances.

A critical responsibility of the FDA is to ensure the safety of food, food ingredients, and food packaging for humans and animals. The Agency performs this function in the US though a number of laws and regulations, which provide premarket approval authority or postmarket oversight for these products. These laws and regulations were constructed without explicit consideration of nanotechnology. However, the language is generalized so as to allow flexibility for the broad range of products reviewed by the Agency and have proven robust over decades of advances in manufacturing technology and review science. In general, the FDA's reviews of food ingredients and packaging components are conducted on a case-by-case basis, where the specifics of each substance can be evaluated in relation to the most relevant science. This type of review is applied to any substance, regardless of its classification, properties, or the type of technology which produced it.

Generally speaking, nanotechnology is a term which describes a broad range of technologies from diverse industries. Simply stated, the scale-size is the inclusive characteristic, with nanotechnology referring to the manipulation of matter on the nanometer scale. Some definitions for nanomaterials broadly include materials which exist at the nanoscale (1 to 100 nm).[4] Some materials which claim to have been produced by nanotechnology may not have novel, size-dependent properties, and therefore do not meet more specific definitions of nanomaterials. As the Agency has not adopted a formal definition for 'nanotechnology',

---

[2] *Nanotechnology: A Report of the US Food and Drug Administration Nanotechnology Task Force*, 25 July 2007, available at http://www.fda.gov/nanotechnology/taskforce/report2007.html

[3] An inventory of nanotechnology-based consumer products currently on the market: Project on Emerging Nanotechnologies, available at http://www.nanotechproject.org/inventories/consumer/

[4] *What is Nanotechnology?* National Nanotechnology Initiative. Available at http://www.nano.gov/html/facts/whatIsNano.html

'nanomaterial' or 'nanoparticle', some ambiguity, therefore, exists in the discussion of what products may currently be classified as nanomaterials or products of nanotechnology.

Many of the traditionally manufactured substances already used in the food industry worldwide are nanoscale. Simple molecules, such as mineral salts, are in the subnanometer range, and simple biomolecules, such as proteins, amino acids, sugars, nucleic acids and polysaccharides, are in the nanometer range. Hence, with respect to size, nanoscale materials are not uncommon and do not necessarily raise unique safety issues when used in food. However, substances with novel properties or functions which are a direct result of their nanoscale size range may present new directions in the food industry. These novel applications could include: detection of contaminants or pathogens; anti-microbial packaging; more effective packaging films or barriers; encapsulation systems for improved bioavailability or sensory properties of nutrients or flavors; and functional ingredients.

To date the FDA has limited experience with food or color additives that fall into the nanosize range where size is critical to the product's technical effect. Titanium dioxide (TiO_2), regulated as a food grade color additive, has a size range that falls into the nanoscale; however, there is no size-related specification in the regulation (21 CFR 73.575). Carbon black (FD&C Black #2) is approved for use in a wide range of cosmetics and does have a nanoscale specification (10–20 nm), although that specification is designed to reduce the possibility of eye irritation from accidental exposure when it is used in eyeliners or brush-on brow or eye shadow (21 CFR 74.2052). Particulate silicon dioxide (SiO_2), approved as an anti-caking agent and for other food-related uses (21 CFR 172.480), also falls into the nanosize range, although size is not a functional property.

Although the FDA's experience with nanotechnology for food additives is limited at present, the Agency continues to study and prepare for submissions of products based on this technology. Perhaps most important among these activities is the Agency's ongoing interactions with product sponsors to develop premarket applications for submission to the Agency. In addition, the Agency is collecting information on the various classes of nanoproducts and their composition in order to be prepared for decision-making and regulation. For example, for some dietary and nutritional supplements, manufacturers market their products in nano-capsules (e.g. nano-emulsions, liposomes, α-cyclodextrin, etc.), which may allow for increased bioavailability for these substances. Manufacturers of flavoring agents may reduce these products to the nano-range to increase the surface area relative to mass, thereby increasing their organoleptic properties and reducing overall cost by using less of the chemicals.

The applications and scientific fundamentals of the wide variety of products that may be included under the umbrella term 'nanotechnology' are generally analogous to those principles which are well understood and currently used in the food industry.

26.3   Exposure Assessment of Nanomaterials

Physico-chemical characteristics and exposure are crucial elements in the hazard identification and risk assessment for any given compound. The chemist on the review team provides crucial information to the toxicologist reviewing the safety of a given substance, by determining the identity and physico-chemical characteristics of the substance in question and by calculating exposures to the substance and any impurities in the substance for which exposure may be expected. Therefore, in order to put the toxicological review process for food ingredients in proper context, the following section will briefly describe the FDA's assessment of the physico-chemical properties and dietary exposure of a given substance, focusing on the application of that review process to nanomaterials. Although the review process herein is described in very general terms, more detailed guidance on this process may be obtained elsewhere (FDA, 1997, 2002, 2006, 2007).[5]

For general classes of direct or indirect food additives, the FDA requires information concerning the identity, physical properties, potential impurities, and potential breakdown products of the additives in question, along with sufficient data to enable calculation of dietary exposures to the additive and impurities in the additive that are expected to be present in the diet. Specifically, the FDA generally requires submission of the following data:

- Information concerning the chemical identity of the additive: the chemical and common name of the additive, the CAS registry number, physical/chemical specifications and stability of the additive, and spectroscopic characterization of the additive.
- Information concerning the composition of the food additive or FCS, including the presence of potential impurities in the finished additive.
- A complete description of the manufacturing process.
- The intended use conditions. For food additives this includes a description of the foods in which the substance is to be used, levels of use in such foods, and the purposes for which the substance is used, and may include a description of the population expected to consume the substance as well as proposed labelling. For FCSs, a description of the intended use conditions includes the types of food expected to be used in contact with the FCS and the maximum temperature and time conditions of the food contact. Any technical effect claims made for the food additive or FCS should be supported by analytical data.

Food additive petitions and GRAS notifications should also include the analytical methods used to detect the additive in food, and substances formed because of its use. From this information, the probable consumption of the substance and the cumulative effect of the substance in the diet are calculated. For FCSs, the migration of the FCS, impurities, and breakdown products from the food contact material to the food itself should be estimated. This may be done by either conducting migration studies or by assuming that all of the FCS, impurity, or degradation product migrates to the food. Migration estimates and intended use conditions are used to identify and estimate consumer exposure to the various substances originating from the FCS. Exposure estimates, expressed as dietary concentration (DC) and estimated daily intake (EDI) values, usually involve combining migrant levels in food with parameters based on the use of articles that might contain the FCS. Cumulative exposure estimates, expressed as the cumulative EDI (CEDI) and derived from the cumulative DC (CDC), are the sum of the exposure estimates from the proposed use and all other permitted uses.

Nanoscale food ingredients and food contact substances should be described in the petition or notification as outlined above. However, nanoscale materials often have chemical, physical, or biological properties that are different from those of their larger counterparts.

---

[5] The petition process is codified in Title 21, Part 171 of the Code of Federal Regulations (21 CFR 171.1 – 171.130).

As these properties may have a significant impact on the interaction of these chemicals with biological systems, the properties of nanoscale food additives and food contact substances should be fully characterized. Important physico-chemical characteristics include:

- particle size and distribution, aggregation and agglomeration characteristics, morphology (e.g. shape, surface area, surface topology, crystallinity, etc.);
- surface chemistry (including zeta potential/surface charge, surface coating and functionalization, catalytic activity, and reactive oxygen species);
- solubility;
- density; and
- porosity.

These characteristics, and the analytical methods available for measuring them, are discussed in the current literature (Oberdörster *et al.*, 2005; Powers *et al.*, 2006; Simon and Joner, 2008). Other special considerations for nanoscale materials include the following:

- Nanoscale materials may agglomerate in food, interact with other components of the food matrix, and interact in the human body following ingestion. Thus, the aggregation and agglomeration characteristics in different media (e.g. aqueous, fatty, acidic, and alcoholic food simulants, and actual food) are also important to understand. The long-term stability of the nanoscale food additive or FCS should also be considered.
- Differences in the manufacturing process for a nanomaterial may also substantially affect the subsequent toxicological properties of the nanomaterial. Thus the manufacture of a nanoscale material should be described in detail. Nanoscale impurities may also arise from the manufacturing process. All potential impurities for which dietary exposure may be expected should be considered in nanoscale food additive petitions and notifications for nanoscale materials.
- If the technical effect of a nanoscale food additive or FCS is dependent upon particle size, data should be presented that demonstrate the specific properties of the nanoscale material that make them useful, as opposed to a material at the bulk or macroscale.
- If migration studies are conducted on a nanoscale FCS, then studies should be validated with a nanomaterial standard or sample. In addition, the agglomeration of the nanoparticles in the food simulant should be considered when evaluating migration.
- We do not expect to change exposure calculation methods when evaluating nanoscale substances. Exposure calculations will continue to be based on mass, as they are now. However, for toxicological considerations it may be necessary to convert this mass-based exposure to other dosimetrics.

The specific data recommendations as outlined above are necessary for proper assessment of the physico-chemical characteristics of the additive and to calculate accurate exposures to the additive and any impurities. This information is critical for use in the risk assessment of these compounds, as detailed in the next section.

26.4   Safety Review of Nanomaterials

The FDA has reviewed an extensive array of toxicity information for various types of compounds as part of its oversight of food ingredients and food contact substances. However, the FDA has had limited experience with reviewing the safety of nanomaterials, as defined by the National Nanotechnology Initiative (NNI). Moreover, while several studies have investigated the toxicity of nanomaterials *in vitro* or *in vivo* via inhalation or parenteral administration, little is currently known about the *in vivo* toxicity of orally-administered nanoparticles. This limited experience in reviewing nanomaterials, coupled with the lack of scientific information concerning oral toxicity, precludes the Agency from issuing specific scientific testing guidance for industry regarding the safety assessment of nanomaterials used in food or food packaging at this time. However, it should be emphasized that the Agency encourages consultation on all matters related to the safety of food ingredients and packaging components, and actively engages with product sponsors to ensure that sufficient data are developed to support safety.

Nanomaterial food ingredients or food contact substances, whether subject to premarket requirements or not, must still meet the same safety standards as other food ingredients. These ingredients must meet the general safety standard of 'reasonable certainty of no harm'. The safety standard is the same for direct food and color additives, food contact substances, and GRAS substances.

The FDA's safety assessment of food ingredients and food contact substances depends on the physico-chemical nature of the compound, the calculated exposures to the compound and constituent impurities, and the available toxicity data for these compounds. Safety evaluations may consider both neoplastic and non-neoplastic endpoints. Neoplastic endpoints may be evaluated directly, via assessment of the ability of the compound to cause cancer in animals, or indirectly, via evaluation of the compound's ability to damage DNA in bacterial or mammalian cells. If appropriate tests show that a food additive is carcinogenic in animals or man, the FDA is prohibited from authorizing its use. In cases where only a component or impurity of the additive may be carcinogenic, the FDA may employ carcinogenic risk assessment to determine safety.

Non-neoplastic endpoints, such as systemic toxicity, reproductive toxicity, or developmental toxicity, are evaluated by using the appropriate *in vivo* studies. Additional information, such as toxicokinetic data, may be evaluated as deemed necessary, based on the context of the safety assessment. The following sections will describe the review process for food contact substances, direct food and color additives and GRAS substances, and explore how these processes might be applied to nanomaterials.

26.4.1   Food Contact Substances

The process by which the FDA evaluates the safety of food packaging constituents, also called FCSs is described in detail elsewhere (Twaroski *et al.*, 2007). This section will briefly summarize the FDA's review process for food contact materials. However, it must be emphasized that each safety assessment is conducted on a case-by-case basis, depending on the physico-chemical nature of the food ingredient and the available data for that compound. Should specific circumstances warrant it, a given substance may require additional information beyond that discussed below in order to establish safety. The reader should keep this fact in mind when reading the summary of OFAS' review process below.

The FDA assesses the safety of all substances expected to be present in the diet as a result of migration from food packaging into food. This includes the FCS itself, breakdown products of the FCS, and impurities (constituents) present in the FCS. These compounds are evaluated using a tiered structure that is dependent upon the calculated exposure to each substance. Based on a general evaluation of unknown chemicals, this structure is designed to address the most relevant toxicological endpoints at a given exposure level. Generally speaking, neoplastic endpoints are evaluated for all compounds for which dietary exposure is expected, even for compounds for which the calculated exposure is very low. Non-neoplastic endpoints are evaluated at exposures greater than 50 ppb. The Agency does not require that specific safety data be developed to support compounds with cumulative exposures less than 0.5 ppb. For cumulative dietary exposures between 0.5 and 50 ppb, the Agency recommends submission of two *in vitro* genotoxicity assays conducted with the compound in question to assess its carcinogenic potential. At cumulative dietary exposures equal to or greater than 50 ppb but less than 1 ppm, submission of subchronic systemic toxicity assays conducted in two species and an *in vivo* micronucleus assay is recommended in addition to the assays recommended for submission for lower exposures. Additional endpoints may be evaluated as warranted using the available data on the compound in question, and exceptions to the tiered structure approach may be necessary on a case-by-case basis. In addition, even if studies are not recommended, all relevant data must be submitted to the FDA and must be considered in the safety assessment. When exposure to the FCS in question is greater than 1 ppm, a full food additive petition could be required. The review paradigm as discussed above may or may not be applicable to nanoscale food contact substances, as discussed in Section 26.3.

26.4.2   Food Ingredients

The FDA is also responsible for overseeing the premarket review of food additives that have a technical effect in food and that are not exempt from premarket review under the FFDCA, and of new color additives to be used in food, cosmetics, drugs and medical devices. Detailed descriptions of the review process for direct food and color additives are available elsewhere (FDA, 1993). Briefly, the safety assessment process for these substances evaluates: (1) the composition and properties of the substance; (2) the amount that would typically be consumed; (3) immediate and long-term health effects; and (4) existing toxicity information. Taking these factors into account, the FDA makes recommendations on the need for additional safety testing for direct food and color additives. The FDA's conclusions are also influenced by how familiar the FDA is with the class of compounds submitted for review. All of these factors are taken into consideration during the FDA's evaluation process.

A number of classes of substances are exempted from the statutory definition of a food additive, and therefore from the premarket approval requirement of food additives. Most such exemptions exclude substances such as pesticides that are regulated under other statutes, but one exemption for 'substances generally recognized as safe (GRAS) for their use in food' is unique. GRAS substances fall under the FDA's regulatory authority for food ingredients, although premarket approval is not required. For GRAS substances, the FDA has established a voluntary notification process whereby sponsors may notify the FDA of independent GRAS determinations for specific uses of food ingredients. Detailed descriptions of the review process for GRAS substances (Gaynor, 2006) are available elsewhere. However, it should be emphasized that, unlike food additive approvals, GRAS determinations can be made by qualified experts outside of government, and the absence of questions from the FDA regarding a GRAS notice does not denote the FDA's agreement with such determinations. For GRAS substances, the FDA may disagree with a notifier's GRAS determination, based on the inadequacy of the toxicological database for the substance or the absence of general recognition of safety among other bases.

Over the years, the FDA has evaluated a wide range of chemicals used as direct additives or food packaging ingredients; however, the use of nanotechnology in food-related products presents a new challenge. Several compounds whose sizes fall within the nano-range have been approved for use by the Agency (e.g. SiO_2, carbon black, titanium dioxide); however, size was generally not considered to be essential for their technical effect. Although the FDA's experience with nanotechnology in food-related applications is very limited at present, the Agency is still working internally and with potential product sponsors to understand and gain answers to safety questions that may arise regarding submissions for products based on this technology. As with other food ingredients and food packaging materials, there needs to be an understanding of the basics of their interactions with biological systems. Also, as with other food ingredients and packaging materials that raise unique safety questions, resolution of those questions may necessitate the submission of additional information beyond what is normally recommended or required for the assessment of a conventional compound at a given exposure. Special considerations for evaluation of the safety of nanomaterials are addressed in detail in the next section.

26.4.3   Special Considerations for Nanomaterials

Developing guidance for industry for submission of nanoproduct-related food ingredients or food packaging materials is a difficult task, given that guidance is usually based upon previous experience with the various classes of substances used as additives or food contact substances. Although the Nanotechnology Task Force's report states that there is no evidence to suggest that nanoscale materials as a group are inherently more hazardous than non-nanoscale materials, the Task Force report also makes clear that the relevant science clearly indicates that a given material's toxicological properties may change when its size is varied from the macroscale to the nanoscale. Although the current state of the science limits the ability of regulators to provide detailed guidance concerning the safety assessment of nanomaterials, the available data on the biological properties of nanomaterials suggests that some nanomaterials may present unique challenges in the assessment of their toxicity. Given this lack of specific safety-related data relevant to the assessment of oral exposure, each food-related product incorporating a nanomaterial will be evaluated on a 'case-by-case' basis. However, consideration of the following factors may be useful when addressing the safety of exposure to a nanoscale compound:

• When assessing the safety of a nanoscale food ingredient, the form of the compound that the consumer will be exposed to in the diet is important. If the consumer will be exposed to the dissociated form of the nanomaterial food ingredient instead of the nanoscale food ingredient itself, it may be appropriately conservative to use available toxicity data for the corresponding dissociated, macroscale compound to support the safety of the nanoscale food ingredient. Should the exposure be expected to be to the nanoscale food ingredient itself, the safety of the nanoscale food ingredient may need to be addressed directly.
• To gain insight into the potential for toxicity of these products, we need to understand what metrics (e.g. size distribution, surface area, zeta potential, porosity, etc.) are important

when assessing their safety Therefore, the nanomaterial should be well-characterized as to its physico-chemical properties. As the FDA generally recommends that the test substance used in toxicity testing be identical to the food ingredient intended for market, the physico-chemical characteristics of the test article should be identical to that of the substance that consumers will be exposed to in the diet (commercial substance), to permit direct extrapolation of the results of toxicity tests to safety assessment of the food ingredient. Close attention should be paid to test article agglomeration/aggregation, zeta potential, functional groups, and other physico-chemical characteristics, both as synthesized and in the testing medium, as these factors may influence the toxicological activity of the test article.

- Although the tested article should, in general cases, be identical to the food ingredient, the notifier may, in certain circumstances, assess the safety of the purified food additive separately from its residual impurities, should the food additive as marketed contain toxicologically-active impurities that would confound interpretation of toxicity studies performed with the commercial product. The notifier may then address the safety of the commercial substance as defined above by lowering or limiting the levels of the problematic residual contaminants. This alternative safety assessment strategy assumes that: (1) the toxicity of the commercial substance is the sum of the individual toxicities of the purified additive and impurities; and (2) a chemically-identical purified food additive may be synthesized and tested such that the test substance has the same toxicological properties as the corresponding component of the mixture. These assumptions may be incorrect for some nanomaterials, however. Studies have demonstrated that purification of some nanomaterials may introduce functional groups that were not present prior to the purification process (Grobert, 2007; Plata et al., 2008). As even slight differences in functional groups can have dramatic effects on the toxicity of highly-similar nanomaterials (Carrero-Sanchez et al., 2006), the purification process may result in a test material that has different toxicological properties from its unpurified counterpart. Therefore, the alternative safety assessment strategy as outlined above may not be applicable to nanomaterials.

- Nanomaterials have been shown to display unique properties in cell-based assays that may affect the ability of the standard array of genotoxicity tests to detect mutagenic activity. For instance, several studies have shown that nanoparticles have the tendency to form aggregates or agglomerates in solution (Sayes et al., 2006, 2007; Jeng and Swanson, 2006), and bind to proteins in various matrices (Borm et al., 2006). These factors may result in formation of macroscale complexes that are inaccessible to the target cells or that are insoluble in the assay medium. This in turn may decrease the sensitivity of the assay by decreasing the dose of the compound delivered to the target cells. Alternatively, the greater surface area to mass ratio of the nanomaterial may result in a greater effective dose being delivered to the target cells, resulting in excess cytotoxicity. Therefore, the nanoscale compound in question should be demonstrated to be compatible with the *in vitro* genotoxicity assays used in the safety assessment. This would entail assessment of precipitation and cytotoxicity of the test substance in the assay, as is recommended for all compounds, as well as assessment of the physico-chemical characteristics of the test material in the test medium and entry of the test article into the target cells. Should the *in vivo* micronucleus assay be the only genotoxicity assay compatible with the nanoscale substance, it is critical to verify that the test substance reached the target site, which may be done via assessment of bone marrow toxicity. However, the one drawback to this approach is that the micronucleus test only measures the endpoint of clastogenicity and is relatively insensitive. Therefore, in consultation with the Agency, the notifier may want to explore other *in vivo* testing paradigms that address the mutagenicity of the test substance.

- In conducting *in vivo* and *in vitro* toxicity studies, careful attention should be paid to the issue of dosimetry. A critical element in determining the toxicity of products in the nano-range is the possible need to shift our traditional reliance on thinking in terms of mass of a substance in assessing toxicity, to using its surface area in our calculations. Some research has shown that surface area can be a better predictor of relative toxicity for materials in the nanoscale range than their mass (Oberdörster *et al.*, 2005). Particular attention should also be paid to the issue of particokinetics for *in vitro* assays (Teeguarden *et al.*, 2007), with careful evaluation of the actual dose that reaches the target cells. The appropriateness of this dose may be evaluated using a mass concentration metric or another metric, such as surface area or particle number. Generally, the greater the applied dose of nanoparticles, the higher the probability that agglomerates will form (Buzea *et al.*, 2007), and precipitation and cytotoxicity will occur. Therefore, the maximum dose limits for the assays currently recommended in Redbook 2000 may not be achievable with certain nanomaterials. In addition, studies have shown that the solvent used in dose formulations of the nanoparticle test substances may have dramatic effects on the nanoparticles' subsequent toxicological properties in a given assay; therefore, appropriate selection of the dosing vehicle for the genotoxicity assays is critical (Baun *et al.*, 2008; Dong *et al.*, 2008). Other concerns also need to be addressed, such as adequate dose-assurance in safety testing (i.e. is the size range consistent for a nanoproduct administered at different dose levels within and across studies?), reproducibility of analytical methodologies for *in vivo* quantification of nanomaterials, and establishment of standards for different classes of nanomaterials.

- Due to the unique physico-chemical properties of nanomaterials, additional toxicological endpoints may need to be considered above those normally evaluated for a given exposure. For instance, to determine whether products in this size range are handled differently by the body than substances in the macro range, *in vivo* data may be generated on their absorption, distribution, metabolism, and excretion (ADME). The inclusion of ADME information in the initial submission will help in understanding how biological systems interact with these materials. In instances where a previously FDA-approved product is enclosed in a nanocapsule to improve its bioavailability, the ADME data could be invaluable for comparing the relative uptake of both forms of the product and determining the need for additional safety data. Many nanomaterials have also been shown to cross the blood–brain (Buzea *et al.*, 2007) and blood–testes barriers (McAuliffe and Perry, 2007), giving these materials access to bodily compartments that are otherwise inaccessible to their macroscale counterparts. Therefore, the safety assessment of a given nanoscale substance may need to address the possibilities of low-dose reproductive toxicity or neurotoxicity.

- Data for appropriate structural analogs may be used in some cases to assess the toxicity of compounds for both neoplastic and non-neoplastic endpoints. This type of analysis is called structure–activity relationship analysis (SAR) and is based on an extensive database of toxicological data for several classes of compounds. Although the Agency has several years of experience using SAR to predict the toxicological properties of molecular compounds, little is currently known regarding the mechanisms of toxicity for nanoscale compounds. Moreover, the available information indicates that small differences in size, shape, impurities, or surface functionalization of nanomaterials can have large impacts on the resultant toxicities of these materials (Sayes *et al.*, 2006; Wilson *et al.*, 2007; Zhu *et al.*, 2006; Lundqvist *et al.*, 2008). Therefore, data generated for similar nanomaterials may not be applicable to the safety assessment of a specific nanomaterial. Until more is known about the interactions of nanomaterials with biological systems, use of SAR analysis in the safety assessment of a nanoscale material may be of limited value. The special considerations in the Chemistry and Toxicology reviews of nanomaterials as discussed above are based on the discussion of these issues in the Nanotechnology Task Force's report. The Task Force's recommendations concerning the assessment of nanomaterials may be summarized as follows:

1. The physico-chemical characteristics of the nanomaterial should be thoroughly assessed.
2. Measurement and reporting methods for nanomaterials should be standardized.
3. The necessity of validation of a given toxicity assay for nanomaterials should be assessed; some nanomaterials may require the development of new toxicity testing paradigms.
4. New dosimetry metrics, such as particle number or surface area, may be necessary.
5. SAR analysis or data generated from 'naturally-occurring' nanomaterials may be useful for assessment of mechanism of action, but not for safety assessment at this time.

26.5   Future Research Needs

To better understand the issues that may arise in evaluating the safety of nanomaterials used as components of food ingredients or packaging, additional data on how these materials interact with biological systems are necessary to determine whether such interactions differ from those of conventional substances, and how such interactions may differ. We also have to determine whether testing methodologies previously used to assess the safety of conventional food ingredients and packaging can be used for comparable nanoscale-based products. Along with validation of safety assessment methodologies, we must develop techniques to generate full nanomaterial characterization profiles and establish which physical/chemical parameters are important for potential toxicity of these materials (e.g. surface characteristics, zeta potential, shape, etc.). It is also important that we understand how these materials behave in both *in vitro* and *in vivo* test systems (e.g. do these materials aggregate or agglomerate when added to test systems, how they interact with gut flora, etc.) before we can start to gain an understanding of how, or whether, they produce toxicity.

In an effort to address some of these needs, the FDA, with the National Toxicology Program, is conducting *in vivo* studies to evaluate the interactions of select nanomaterials with biological systems. These studies using nano-silver and nano-gold are designed to gather data on the ADME characteristics of these materials. In addition, the FDA will be evaluating the utility of a battery of genetic toxicity tests in assessing *in vitro* the mutagenicity and carcinogenicity potentials of nano-silver and nano-TiO$_2$. Nano-silver is of particular interest to OFAS because of its use as an antimicrobial in food contact products.

There are some private sector efforts to deal with the research needs for food-related nanomaterials. The Institute for Food Technologists (IFT), along with the Grocery Manufacturer's Association (GMA), is collaborating with the Nanotechnology Characterization Laboratory (National Cancer Institute) in identifying current and proposed uses of nanomaterials in foods, food and color additives, and packaging, and developing research strategies to address their potential toxicity issues.

Although these studies will contribute much needed information on food-related nanomaterials, much more research may be necessary. Questions on the impact of nanomaterials on the gut immune system and how they are absorbed, distributed, metabolized, and excreted by the body may need to be answered, whether through the development of individual product submissions or through basic research designed to address regulatory issues.

26.6 Conclusions

As other emerging technologies have in the past, nanotechnology poses questions regarding the adequacy and application of regulatory guidance. In recognition of the important role of the science in developing regulatory policies in this area, the rapid growth of the field of nanotechnology, and the evolving state of scientific knowledge relating to this field, a thorough knowledge of the emerging science is needed to enable the Agency to predict and prepare for the types of products the FDA may see in the near future. Although nanoscale materials present regulatory challenges similar to those posed by other products, these challenges may be magnified because the properties of a material relevant to the safety of FDA-regulated products may change in unique ways within the nanoscale range. Based on the conclusions of the FDA's Nanotechnology Task Force report, this chapter has made various recommendations to address potential regulatory issues that may be presented by products that use nanotechnology. A number of recommendations deal with requesting safety data and other information specific to nanoscale materials. Given the rapidly-evolving state of the science, we strongly recommend manufacturers to communicate with the Agency early in the development process for products using nanoscale materials. We hope that the recommendations in this chapter and the Task Force Report, combined with guidance obtained through further consultation with the Agency, will give manufacturers and other interested parties timely information about the FDA's expectations, so as to foster predictability and transparency in the Agency's regulatory processes, while protecting the health of the public.

References

Baun A, Sorensen SN, Rasmussen RF, Hartmann NB, Koch CB (2008) Toxicity and bioaccumulation of xenobiotic organic compounds in the presence of aqueous suspensions of aggregates of nano-C60. *Aquatic Toxicology* **86**, 379–387.
Borm PJA, Robbins D, Haubold S, Kuhlbusch T, Fissan H, Donaldson K, Schins R, Stone V, Kreyling W, Lademann J, Krutmann J, Warheit D, Oberdorster E (2006) The potential risks of nanomaterials: A review carried out for ECETOC. *Particle and Fibre Toxicology* **3**, 11.

Buzea C, Pacheco II, Robbie K (2007) Nanomaterials and nanoparticles: Sources and toxicity. *Biointerphases* **2**: MR17–MR71.

Carrero-Sanchez JC, Elias AL, Mancilla R, Arrellin G, Terrones H, Laclette JP, Terrones M (2006) Biocompatibility and toxicological studies of carbon nanotubes doped with nitrogen. *Nano Letters* **6**, 1609–1616.

Center for Drug Evaluation and Research (2006) *Guidance for Industry and Review Staff: Recommended Approaches to Integration of Genetic Toxicology Study Results.* Available online at http://www.fda.gov/Cder/guidance/6848fnl.htm#A.Weight

Chaudhry Q, Scotter M, Blackburn J, Ross B, Boxall A, Castle L, Aitken R, Watkins R (2008) Applications and implications of nanotechnologies for the food sector. *Food Additives & Contaminants: Part A* **25**, 241–258.

Dong L, Joseph KL, Witkowski CM, Craig MM (2008) Cytotoxicity of single-walled carbon nanotubes suspended in various surfactants. *Nanotechnology* **19**, 1–5.

Food and Drug Administration (1993) " Draft" Redbook: Chapter III: Concern levels and recommended toxicity tests. Available online at http://www.cfsan.fda.gov/~acrobat/red-iii.pdf.

Food and Drug Administration (1997) Guidance for Industry: How to Submit a GRAS Notification, Excerpted from 62 FR 18937- 17 April 1997. Available at http://www.cfsan.fda.gov/~dms/opa-frgr.html.

Food and Drug Administration (2002) Guidance for Industry: Preparation of Food Contact Notifications: Administrative Final Guidance, May 2002. Available at http://www.cfsan.fda.gov/~dms/opa2pmna.html.

Food and Drug Administration (2006) Guidance for Industry: Recommendations for Chemical and Technological Data for Direct Food Additive Petitions, March 2006. Available at http://www.cfsan.fda.gov/~dms/opa2cg4.html.

Food and Drug Administration (2007) Guidance for Industry: Preparation of Premarket Submissions for Food Contact Substances: Chemistry Recommendations, December 2007. Available at http://www.cfsan.fda.gov/~dms/opa3pmnc.html.

Gaynor P (2006) Regulatory Report: FDA's GRAS notification program works. *Food Safety Magazine.* Available at http://www.cfsan.fda.gov/~dms/grasov2.html#authors.

Grobert N (2007) Carbon nanotubes – becoming clean. *Materials Today* **10**, 28–35.

Jeng HA, Swanson J (2006) Toxicity of metal oxide nanoparticles in mammalian cells. *J Environ Sci Health Part A* **41**, 2699–2711.

Lundqvist M, Stigler J, Elia G, Lynch I, Cedervall T, Dawson KA (2008) Nanoparticle size and surface properties determine the protein corona with possible implications for biological impacts. *Proc Natl Acad Sci* **105**, 14265–14270.

McAuliffe ME, Perry MJ (2007) Are nanoparticles potential male reproductive toxicants? A literature review. *Nanotoxicology* **1**, 204–210.

Oberdörster G, Maynard A, Donaldson K, Castranova V, Fitzpatrick J, Ausman K, Carter J, Karn B, Kreyling W, Lai D, Olin S, Monteiro-Riviere N, Warheit D, Yang H (2005) Principles for characterizing the potential human health effects from exposure to nanomaterials: Elements of a screening strategy. *Particle and Fibre Toxicology* **2**, 8.

Plata DL, Gschwend PM, Reddy CM (2008) Industrially-synthesized single-walled carbon nanotubes: compositional data for users, environmental risk assessments, and source apportionment. *Nanotechnology* **19**, 1–14.

Powers KW, Brown SC, Krishna VB, Wasdo SC, Moudgil BM, Roberts SM (2006) Research strategies for safety evaluation of nanomaterials. Part VI. Characterization of nanoscale particles for toxicological evaluation. *Toxicological Sciences* **90**, 296–303.

Sayes CM, Wahi R, Kurian PA, Liu Y, West JL, Ausman KD, Warheit DB, Colvin VL (2006) Correlating nanoscale titania structure with toxicity: A cytotoxicity and inflammatory response study with human dermal fibroblasts and human lung epithelial cells. *Toxicolog Sci* **92**, 174–185.

Sayes CM, Reed KL, Warheit DB (2007) Assessing the toxicity of fine and nanoparticles: Comparing *in vitro* measurements to *in vivo* pulmonary toxicity profiles. *Toxicolog Sci* **97**, 163–180.

Simon P, Joner E (2008) Conceivable interactions of biopersistent nanoparticles with food matrix and living systems following from their physicochemical properties. *J Food Nutr Res* **47**(2), 51–59.

Teeguarden JG, Hinderliter PM, Thrall BD, Pounds JG (2007) Particokinetics *in vitro*: Dosimetry considerations for *in vitro* nanoparticle toxicity assessments. *Toxicolog Sci* **95**, 300–312.

Twaroski ML, Batarseh LI, Bailey AB (2007) Regulation of food contact materials in the USA. In *Chemical Migration and Food Contact Materials*, Barnes KA, Sinclair CR, Watson DH (eds). Woodhead Publishing Ltd.: Cambridge; 17–42.

Wilson MR, Foucaud L, Barlow PG, Hutchinson GR, Sales J, Simpson RJ, Stone V (2007) Nanoparticle interactions with zinc and iron: Implications for toxicology and inflammation. *Toxicol Appl Pharmacol*. DOI: 10.1016/j.taap2007.07.012.

Zhu Y, Ran T, Li Y, Guo J, Li W (2006) Dependence of the cytotoxicity of multi-walled carbon nanotubes on the culture medium. *Nanotechnology* **17**, 4668–4674.

Index

Page numbers in italics refer to tables and figures.

8-hydroxyguanosine (OHdG) 273
16HBE14o cell line 385

α-quartz particles 167–70
α-tocopherol 363

A549 cell line 338, 355, 386, 388, 389
Abraxane 41–2, 43
absorption, nanoparticles 312
absorption spectroscopy 31
absorptive effects 67–9
acetylsalicylic acid (ASA) 215
activated clotting time (ACT) test 307
activated partial thromboplastin time (APTT) test 196, 197, 307
acute high-dose exposure 486
acute respiratory distress syndrome 272
acute toxicity 137, 316–20
adhesion/cohesion profile, nanomedicines 50–1
ADME (absorption, distribution, metabolism, excretion) 503–4
adverse effects 460, 461, 532
aerodynamic equivalent diameter 6
aerosols 8–9, 20, 466–7, 529–30, 531
    exposure assessments 552–3

age-related differences 557
agglomerates/aggregation, nanomaterials 15, 16, 18, 230–2, 446–7, 485
aggregometry impedance 202
air pollution 177, 271–2, 521
air–blood barrier 526
air-borne particles 249, 320, 336, 434, 450, 468, 503, 508, 531
    exposure assessments 552–5
airway epithelial cell lines 385–6, 389
albumin-coated iron oxide magnetic nanoparticles 410
algae 235, 240
alkaline phosphate (ALP) 164, 186
aluminium oxide (Al_2O_3) coating 292
aluminium powder 17
alveolar epithelial cell lines 386–7
alveolar macrophages 173, 251, 273, 290, 291, 294, 429, 449, 526
alveoli 381, 526
Alzheimer's disease 280, 416
ambiguous risk knowledge 534
American Society for Testing and Materials (ASTM) 438, 441–2, 483
AMI-25 406

amphiphilic β-cyclodextrin nanospheres 185
anatase 551
animal models 208–9, 287–8, 289, 290, 382, 436, 448
anionic dendrimers 122
anionic iron oxide magnetic nanoparticles 407
anionic micelles 119
apoptosis 147, 309
aquatic environment 227–41, 508
aquatic organisms 233–40
argyria 184
arsenic 232
asbestos 62–3, 70, 75, 428, 564, 565, 566
asthma 274–5, 512, 569
astrocytes 403
atomic absorption spectroscopy (AAS) 287
atomic force microscopy (AFM) 10, 11, 49, 143, 283–5, 351, 370–1, 549, 551
atopic syndrome 525
Au_{55} clusters 138

B. subtilis 240
background concentrations 553
bacteria 235, 239
batteries 507
BEAS-2B cell line 386
benchmark doses (BMDs) 462
BET specific surface area 8, 9, 19, 32, 283, 352
Bio-Plex suspension array system 146
biodegradation, nanomedicines 48–50, 508
biodistribution (internal exposure) 133–5, 184, 312–15
biomagnification 235
biomarkers 358–65
biomedicine (see medical applications)
biopersistence 64, 445
blood (see also platelets) 192–98, 433, 476
  exposure routes 526–7
blood compatibility 305–8
blood–brain barrier 174–5, 258, 336, 416, 528
blue mussels 233
bovine serum albumin (BSA) 66
bradykinin 197
brain (see central nervous system)
BrdU nucleoside 146
bronchial asthma 274
bronchial epithelium 468
bronchioles 381
bronchoalveolar lavage (BAL) 163, 164, 166, 272, 275, 337–8

Brownian motion 285–6
bulk dry powders 8, 9, 19

C_{60} fullerene 32
  environmental impact 236, 237–8
  exposure routes 173, 256, 470, 471
  immune responses 133
  in vitro experiments 439
  interactions, platelets 210
  risk assessment 471
  toxicity 186, 292
$C_{60}(OH)_{24}$ fullerenol 211–12, 292, 439
Caco-2 cells 129, 130, 343
cadmium 232
cadmium oxide (CdO) nanoparticles 527
cadmium selenide (CdSe) 124
cadmium selenium (Cd-Se) quantum dots 178, 292, 365
Calu-3 cell line 385
canola oil 507
capping agents 31, 35
carbon black 236, 272, 274, 309, 337, 567, 584
  risk assessment 464
carbon nanofibers 450
carbon nanotubes (CNTs) (see also multi-walled carbon nanotubes, single-walled carbon nanotubes) 12, 61, 65–6, 74–5, 76, 430
  absorptive effects 67–9
  biopersistence 64
  characterization 552
  cytotoxicity 124
  dimensions 63
  environmental impact 233, 465
  environmental sources 228, 450
  exposure assessments 554
  history 387
  immune responses 133
  media 136
  natural occurrence 126
  preparation 203
  structure 65, 123–4, 505
  toxicity
    mesothelium 71, 72–4
    pulmonary system 337, 430, 565
  toxicity studies 565–7
  uptake by HeLa cells 138
  uses 280, 336, 506
carbon particles 252, 294, 527
carbon-13 (^{13}C) nanoparticles 174, 528

carbon-based nanomaterials 500
cardiovascular effects 559
carp 232, 236
CD62P expression 213
CD63 surface markers 206
cell adhesion 148
cell counters *201*, 203
cell cultures 380, 383–90, 439
cell cycle 146–7
cell imaging 365–71
cell lines 137–9, 354–8, 380, 383
cell proliferation 164
cell sorting 398, 399, 400
cell stains 140–1, 145
central nervous system (CNS) 172–9
    exposure routes 173–5, 253, 258, 509, 528
    in vitro experiments *433*
    toxicity 175–8, 569
ceramic nanoparticles 172, 249
Ceriodaphnia dubia 235
cerium dioxide (CeO$_2$) nanoparticles 291
cetyl trimethylammonium bromide (CTAB) 125
characterization, nanomaterials 1–24, 29–37,
    351–4, 483–5
    consumer products 548–52
Chemical Abstract Service (CAS) numbers 529
chitosan nanoparticles 47, 123, 130
chronic exposure 54, 56
chronic fatigue immune dysfunction syndrome
    (CFIDS) 310
chronic low-dose exposure 486
chronic obstructive pulmonary disease (COPD)
    512
citrate-coated iron oxide magnetic nanoparticles
    403
citric acid coated iron oxide magnetic
    nanoparticles 415
clathrin-dependent pathways 145
clathrin-mediated endocytosis 410
clay-polymer nanocomposites 507
clearance, nanomaterials 135–6, 250–1, *314*,
    316, 448, 482–3, 528
CMS nanotransporters 525
coagulation, plasma 306–7
coatings 292
cobalt chloride (CoCl$_2$) nanoparticles 534
cobalt (Co) nanoparticles 291, 534
collagen-related peptides (CRPs) 217
colloidal gold 185, 341
colloids 227

colorimetric assays 139–41
complementary activation 133
complex risks 534
computer hardware 507
condensation particle counters (CPCs) 8–9,
    552
conductance 35
conflicting results 447, 451
confocal microscopy 145
consumer products 543–71
contrast agents, MRI 398, 399, 400, 407
copper (Cu) nanoparticles 175–6, 178, 186,
    290, 430, 474
core–shell anatomies 34
corneocytes 523
cosmetics 506, 525
Cremophor EL 41
cryo-transmission electron microscopy
    (cryo-TEM) 34
crystal structures 290
crystalline silica 428
CULTEX system 388
cutaneous toxicity *324–5*
cutting tools 507
cytochrome P450 232, 315
cytokines 415, 429, 534
cytoskeletons 144, 147, 402
cytotoxicity 291, 292, 293–4, 308–11, 353
    in vitro experiments 355–8

daphnia 234
Daphnia magna 240
data analysis 372–3
DCF fluorescence 359
degradation, nanomaterials (*see also*
    biodegradation) 508
delta b* metric 35
dendrimers 130, 131, 135, 213, 439, 500, 506
dendritic cells 304, 385, 389, 509
Derjaguin-Landau-Verwey-Overbeck (DLVO)
    theory 230–1
dermal exposure 128–9, 174, 253, 324, 327,
    338–40, 469–70, 509, 510, 523
    exposure assessments 555–6
    in vitro experiments 339–40
    in vivo experiments 339, 434, *435*
    risk assessment 470–1, *472?–3*
    toxicity studies 561
dermis 338, 509
Desmodesmus subspicatus 240

dextran-coated iron oxide magnetic
    nanoparticles 401–2, 406, 410, 415
diesel exhaust 178, 263, 272, 273
differential display (DD) polymerase chain
    reactions 292–3
differential interference contrast (DIC)
    microscopy 145, 365–6
differential mobility analysers (DMAs) 9, 32,
    530, 552
dipalmitoylphosphatidylcholine (DPPC) 66
dispersal, nanomaterials 14–18, 24
  high aspect ratio nanoparticles 64
dissolution, nanoparticles 287
DNA damage 363–4, 430
DNA microarrays 146, 2932
dose-response assessments 462, 504, 559, 568,
    570
dosimetry 371–2, 381–2, 449, 481, 486, 569
  food additives 591
  in vivo experiments 381–2
dosing intervals, nanomedicines 54–5
dried-TEM 34
drug delivery 45–8, 56, 58, 128, 130, 135, 258,
    398, 470
DuPont 544
DuPont Haskell Laboratory 436, 448
dynamic light scattering (DLS) 33, 207, 285,
    353, 486

electrical applications 507
electrical double layers (EDLs) 231
electron capture 175
electron energy loss spectroscopy (EELS)
    550
electron spin resonance (ESR) 35
elimination (see clearance)
endocytosis 131, 143–5, 234, 236, 408
endothelium 433
energy dispersive X-ray spectrometry (EDS)
    550
enhanced gene transfection 398
enterocytes 522, 523
Environmental Defense (ED) 544
Environmental Protection Agency (EPA) 547,
    555
environmental sources, nanomaterials 507–8,
    510
environmental transformations 232–3
enzyme-linked immunosorbent assay (ELISA)
    146, 202, 207

epidemiological studies 520, 534
epidermal growth factor receptor (EGFR) 414
epidermis 174, 253, 338, 523
epiphaniometers 20
epithelial cells 131, 139, 251, 290, 291, 388,
    523
epithelium 343
equivalent sphere model 6
erythrocytes 305
Escherichia coli 84, 85, 90–2
European Centre of the Validation of
    Alternative Methods (ECVAM) 289
ex vivo experiments 382–3
exposure assessments 462–3, 504–12, 511–12,
    521, 529, 544–8, 568
  food additives 584–6
exposure routes (see also dermal exposure,
    inhalation exposure, oral exposure)
    126–31, 249–58, 449–50, 508–10,
    521–2, 547–8
  risk assessment 464–74, 485
extracellular matrix (ECM) 146, 147, 148
Exubera 44

F-12K medium 446
false negatives 439
Federal Food, Drug, and Cosmetic Act
    (FFDCA) 582, 588
fenestrae 134
Fenton chemistry 413
Feridex 403
ferric oxide (Fe_2O_3) nanoparticles 252, 258
ferritin 413, 416
ferrous oxide (Fe_3O_4) nanoparticles 258
ferumoxide 409
Feynman, Richard 499
fiber physicochemistry 63–4
fibrinogen 195, 197
fibroblast cell lines 387
fibroblasts 139, 339, 402, 429
fibrosis 469
filtration systems 507
flame pyrolysis 11
flow cytometry 202, 205–7
fluorescence microscopy 201, 204, 366–9
fluorescence spectroscopy 31
fluorescent nanoparticles 439
fluorimetry 202
food additives 582–93
  risk assessments 589–92

Food and Drug Administration (FDA) 116–17, 464, 582–3, 585, 587, 588–9, 590, 592
food contact substances (FCSs) 582, 587–8
food products 507, 509
food-borne pathogens 81
freshwater organisms 234, *238*
fullerenes (*see also* C$_{60}$ fullerene) 123, 172, 506
  biodistribution 185, 258
  cytotoxicity 124
  degradation 508
  environmental impact 232, 237–9, 508
  exposure assessments 553, 555
  in vitro experiments 439
  natural occurrence 126
  risk assessment 462, *464*
  toxicity 177, 187, 320
  toxicity studies 567
  uses 556
fullerols 567

G protein-coupled receptors 217
gadolinium-containing nanoparticles (GdNPs) 213
gastrointestinal (GI) tract (*see also* oral exposure)
  toxicity 130, 340–3
  translocation 129–30, 173, 256–7, 290, 471, 509, 522–3
gene expression 292–3
general recognised as safe (GRAS) substances 582, 588–9
genetic toxicity 310–11
genomic analysis 208
germanium nanoparticles 506
gills 233–4, 239
glass surfaces 507
glass wool nanoparticles 233
glomerular capillary 135
glucuronidation 315
glutathione 98, 99–100
glutathione peroxidase 97, 100, 108
glutathione S-transferase 101, 104–5, 107, 109–10
glycol chitosan 185
gold nanocages 280
gold (Au) nanoparticles
  biodistribution 134, 258, 291
  exposure routes 290, 528
  in vitro experiments 358, 368

  interactions
    macrophages 139
    platelets 212
  medical applications 123, 124–5
  toxicity 292
  uses 506
gold nanospheres 304
granulocytes 401
graphene 71, 75
green fluorescent protein (GFP) marker 185, 367
GSH levels 360–2
GSSG levels 361
gut epithelium 234

hair follicles 339, 524
hair-care products 506
Harber-Weiss type reactions 413
hazard identification 461, 504, 532–3, 556, 568, 569
health risk assessments 544
heart *434*
HeLa cells 138
hematocompatibility 476
hemocytes 236
hemolysis 305–6
hemolytic activity 194
hemostasis 196, 198, 199
hepatocytes 135, 315
hepatoxicity (*see* liver)
HepG2 cells 123
high aspect ratio nanoparticles (HARNs) 61–76
  absorptive effects 67–9
  biopersistence 64
  dimensions 63
  dispersal 64
  dosimetry 64
  effects on mesothelium 70–4
  sampling 69–70
  size 64
  surface area 64–7
high-throughput screening 452
HPMA copolymers 119, 134, 135
hydrogen selenide 97
hydrolysis 508
hydroxyl free radicals 413
hydroxylated fullerene 445
hypodermis 338
hypotonic stress response (HSR) 205

immortalized carcinoma cell lines 139
immune responses 132–3, 135
immuno-labelling 366
immunoglobulin 275
immunostaining microscopy *201*
impurities, nanoparticles 353
in situ measurement 530
in vitro experiments 136–48, 248, 287–92,
        343–4, 354–72, 380, 428, 431–4, 438–9,
        447, 450, 502, 570
    biomarkers 358–65
    cell cultures 380, 383–89, 439
    cell imaging 365–71
    cell lines 137–9, 354–8, 380, 383
    characterization experiments 484
    colorimetric assays 139–41
    dosimetry 371–2
    endocytosis 143–5
    hazard identification 533
    plasma membranes 142–3
    standardization 451
in vivo experiments 179, 287–92, 336, 344–5,
        434–6, 439, 440–1, 502, 570
incidental exposure 249
incubation times 137
inducible nitric oxide synthase (iNOS) 414
inductively coupled plasma emission
        spectroscopy (ICP-AES) 34, 145
inductively coupled plasma mass spectrometry
        (ICP-MS) 287, 353
inflammatory responses 146, 192–3, 194–7, 199
ingestion (*see* gastrointestinal tract)
inhalation exposure 131, 173, 184, 249–53,
        336–8, 430–1, 508–9, 526, 527, 556
    exposure assessments 553–5
    in vitro experiments 338
    in vivo experiments 336–8, 434–5, 440
    risk assessment 465–9
inorganic fullerene-like materials 505
inorganic nanotubes 505
insulin 44
internal exposure (*see* biodistribution)
International Alliance for Nano Environment,
        Human Health and Safety
        Harmonization (IANH) 58, 451
International Life Science Institute (ILSI) 428,
        431, 446, 447
International Organization for Standardization
        (ISO) 441, 483
interspecies differences 569

intradermal injection 257
intramuscular injection 257
intraspecies sensitivity 569
intratracheal instillation 163, 337, 435, 440,
        466–7
intravenous injection 258, *477–80*
ionic composition, nanomaterials 231
ionically stabilized iron oxide magnetic
        nanoparticles 406
iridium (Ir) nanoparticles 256, 290, 471, 523
iron (Fe) nanoparticles 229, 324, 522–3, 524
iron oxide magnetic nanoparticles (IOMNPs)
        397–418
    medical applications 398–400
    toxicity 400–418
iron oxide nanoparticles (*see also* ferric oxide
        nanoparticles, ferrous oxide
        nanoparticles)
    biodistribution 258
    environmental sources 248
    medical applications 295
    observation 281
    toxicity
        liver 186
        macrophages 296
    uses 506
isoenzymes 315

kallikrein 197
kallikrein-kinin system 197
keratinocytes 338, 339–40, *342*
kidneys 135, 139, 184, 262, 316, 343, *434*, 476,
        481–2, 562
Kunicki's morphology 204
Kupffer cells 135, 185, 187, 258, 482

labelled nanoparticles 145
lactate dehydrogenase 164, 309
Lactococcus lactis 82
landfill disposal 229
Langerhans cells 339
lanthanide oxide nanoparticles 249
largemouth bass 237
latex nanoparticles
    environmental impact 233, 235
    exposure routes 132, 257, 341
    interactions with platelets 216
LC$_{50}$ test 316
LDH assays 140, 141, *340, 342*, 360, *361*
leukocytes 195, 310

light transmission aggregometry (LTA) 204–5
Limulus Amebocyte Lysate (LAL) assay 304
lipid nanoparticles 307, 527
lipopolysaccharide (LPS) 272–3
liposomes 45–7, 172
    biodegradation 48
    clearance 136
    immune responses 133
    interactions, platelets 213–16
liquid spray applicators 556
liquid-AFM 36
liquid-borne nanoparticles 10
Listeria monocytogenes 84, 85–6, *87*, 90
Live/Dead assay 141
liver
    clearance 135, 482
    exposure routes 184–5, 258, 343, 471, 476
    in vitro experiments *433*
    functions 183, 315
    toxicity 139, 185–7, 320
long fiber amosite (LFA) 72–4
low-pressure impactors (LPIs) 530
luminescent nanoparticles 368
luminometry *202*
lung burden 56–8
lung cancer 430
lung histopathology 165
lung lavage 131, 164, 337
lung retention 43–5, 131
lungs (*see* pulmonary system)
luteinizing hormone-releasing hormone
        (LHRH) 185
lymphatic system 174, 471

macrophage cell lines 387
macrophage overload 57
macrophage phagocytosis 34, 251, 294, 469
macrophages 139, 310, 315, 385, 389, 403, 409,
        414, 429
magnetic hyperthermia (MH) 399
magnetic labelling 403
magnetic nanoparticles 294–5, 397
magnetic resonance imaging (MRI) 294, 398,
        399–400, 401, 407, 464
magnetofection (MF) 398
malondialdehyde (MDA) 362–3
manganese dioxide (MnO_2) nanoparticles 263,
        264
manganese (Mn) nanoparticles 178, 558
manganism 558

MAP kinase 509
mass spectrometry 9
mass/volume diameters 6
medaka 233, 235–6
media 136
medical applications (*see also* nanomedicines)
        115–17, 474, 481, 486
    gold nanomaterials 123, 124–5
    iron oxide magnetic nanoparticles 398–400
    iron oxide nanoparticles 295
    nanofilms 505
    silver nanoparticles 184, 280, *336*
megakaryocytes 198
membrane lysis 143
meningitis 174
mesothelial cell lines 387
mesothelium 70–4, 430
metabolism 315
metal nanomaterials 172, 177, 287
    environmental impact 239
    interactions, platelets 212–13
metal oxide nanomaterials 236, 248–9, 416
metal-based nanomaterials 500
micelles *46*, 48
microbial contamination tests 304–5
microglia 178, 416
micronised nanoparticles 524
microorganisms 235
migration assays 148
mineral fibers 428–9
mitochondria 176, 413?–14, 430
mitogen-activated protein kinase (MAPK) 414
mixed carbon nanomaterials 210–11
monocrystalline iron oxide nanoparticles
        (MIONs) 400
monocytes 385
mononuclear phagocyte system (MPS) 315
morphology, nanoparticles 372
mosquito repellent 507
MRC-9 cell line 387
MSTO-211H cell line 387
MTT assay *127*, 140, 309, *342*, 414, 438
mucoadhesive properties 129
mucosa *432*
mucosal surfaces 389
multi-walled carbon nanotubes (MWCNTs) 67
    environmental impact 239
    interactions, platelets 210–11
    natural occurrence 547
    physico-chemical properties 446

multi-walled carbon nanotubes (*Cont.*)
  risk assessment *464*, 471
  structure 123, 505
  toxicity 74, 75, 124, 339–40, 430, 446, 469
  toxicity studies 566
multiple organ dysfunction (MODS) 192, 194
murine macrophage cells 403
mussels 234, 236

namide adenine dinucleotide phosphate
      (NADPH) oxidase 414
nano-grade tea 507
'nano-objects' (definition) 441
nano-porous alumina membranes 218
nano-Se (*see* selenium nanomaterials)
nano-terms 280
nanoaerosol exposure 449
nanocarriers 43, 44–5, 47, 50–1, 54
nanocrystals *46*, 49
NanoDerm project 524, 525
nanofibers 280
nanofilms 280, 505
nanomedicines (*see also* drug delivery, medical
      applications) 41–58
  adhesion/cohesion profile 50–1
  biodegradability 48–50
  dosing intervals 54–5
  lung burden 56–8
  lung retention 43–5
  purity 51–4
  size 51
  toxicity 58
nanoparticle tracking analysis (NTA) 286
nanoparticles 505–6
Nanotechnology Characterization Laboratory
      (NCL) 437–8, 447, 451, 484
Nanotechnology Consumer Products Inventory
      545, 546, 556
Nanotechnology Research Coordination Group
      (NRCG) 451
nanotubes 505
nanovesicles 47
nanowires 505
National Institute for Occupational Health and
      Safety (NIOSH) 560
National Institute of Standards and Technology
      (NIST) 14, 451, 484
National Nanotechnological Initiative (NNI)
      441, 459, 582
National Research Council (NRC) 442, 461

natural occurrence, nanomaterials 126, 547
NCL preclinical characterization test 309
near-field optical microscopy 281
neurodegenerative diseases 416
neutron capture therapy (NCT) 213
nickel (Ni) nanoparticles 291, 529
nickel oxide (NiO) nanoparticles 263, *264*
nisin 82, 92
NK (natural killer) cells 310
non-uniformities 550
nuclear factor-kappa B (NFκB) pathway 414
nuisance dusts 520
number-weighted size distributions 530

occupational exposure 249, 510, 564–5
ocular exposure 510
olfactory exposure 174, 252–3, 336, 510, 528,
      558
opsonization 131, 133, 195, 409, 476
oral exposure 256, 340–3, 434, 471, 510, 522
  in vitro experiments 343
  in vivo experiments 340–3, 434, *435*
  risk assessment 471–4, *475*
organic dyes 366, 367
organic solvents 439
Organization for Economic Cooperation and
      Development (OECD) 464
overload thresholds 56
oxidative stress 176, 294, 296, 411–12, 416,
      417, 430, 504
ozone 232

P-selectin 206, 216
paclitaxel 41
PAMAM dendrimers 119–21, 130, 217
para-cellular pathways 522
Paracelsus 460
parental exposure 510
Parkinson's disease 416
partial thromboplastin time (PTT) test 196
particle charge 134, 135, 468
particle concentration 34–5, 137, 531
particulate matter (PM) 411, 414–15, 416
PC12 cell line 417
PEGylation 121–2, 124
perfusion chambers 208
peritoneal cavity 70, 71
Peyer's patches 340, 343, 471, 522
PG liposomes 215
phagocytic cells 409, 410, 414

phagocytosis 135, 216, 310, 324, 407
pharmacokinetics (PK) 311–16
pharyngeal/laryngeal aspiration 440–1
phase contrast microscopy *201*, 204, 365
phenanthroline 211
phosphate-buffered saline (PBS) solution
    163–4, 165
photobleaching 368
photolysis 508
photometry *202*
physico-chemical properties, nanomaterials
    443–7, 448, 450, 483, 486, 501, 502–3,
    504
  food additives 591
plasma coagulation system (PCS) 196–7
plasma membranes 142–3
plasmin 197
platelet membrane microparticles 206–7
platelets 198–219
  aggregation 210, 217, 306
  in vitro experiments 200–8
  in vivo experiments 199–200, 208–9, 440–1
  interactions 209–19
    carbon nanoparticles 210–12
    dendrimers 213
    latex nanoparticles 216
    liposomes 213–16
    metal nanoparticles 212–13
  nanomaterial coatings 217
  nanotopography 218
  tissue scaffolds 217–18
PLGA nanoparticles 49, 55, 123
PLT morphology 204, 206
plutonium dioxide ($^{239}PuO_2$) 526
pneumonia 272
polethylenimine-coated iron oxide magnetic
    nanoparticles 410
poly-isobutyl cyanoacrylate (PIBCA) 187
polyalkylcyanoacrylate nanoparticles 133
polyelectrolyte complexation (PEC) fiber 218
polyelectrolyte complexes *46*, 47
polyethylene glycol nanospheres 280
polyglycerol dendrimer (PGLD) 213
polylactic acid (PLA) 82, 92
polymer-coated iron oxide magnetic
    nanoparticles 410
polymerase chain reactions (PCRs) 292–3
polymeric micelles 295
polymeric nanoconstructs 118–19, 315
polymeric nanoparticles *46*, 47–8, 49, 122, 172

polymeric therapeutics 118
polymorphs 551
polysaccharides 233
polystyrene microspheres 522, 561
polystyrene nanoparticles 129, 131, 187, 236,
    257, 341, 383, 471
polyurethane-gold (PU-Ag) nanocomposites
    212
positively charged nanoparticles 134
potentiometric titrations 23
powder sampling 4, 5
prekallikrein 197
preneoplastic cells 430
primary cell cultures 383, 384–5
primary characterization, nanomaterials 30, 32
primum non nocere principle 399
product information 547
professional phagocytic cells 409
Project on Emerging Nanotechnologies (PEN)
    228, 545
protein adsorption 131–2, 134, 481, 485
protein synthesis 145–6
protein–nanoparticle interactions 132
proteins 131–2, 134, 364–5, 446
proteomic analysis 208
prothrombin time (PT) test 196, 197, 307
Pseudokirchneriella subcapitata 240
psoriasis 525
PubMed database 172
pulmonary interstitium 469
pulmonary system (*see also* inhalation
    exposure)
  bioassays 162–5
  clearance 250–1, 291, 381
  ex vivo experiments 382–3
  exposure routes 131, 249–50, 476, 503,
    508–9
  in vitro experiments 380, 383–9, 428, 431,
    432, 439
  in vivo experiments 336, 434–5
  inflammation 468–9
  structure 381
  toxicity 131
    carbon nanotubes 337, 430, 565
    copper nanoparticles 430
    multi-walled carbon nanotubes 446, 469
    quartz 166, 167–70, 290
    silver nanoparticles 563–4
    single-walled carbon nanotubes 337, 338,
    445, 446, 469

pulmonary system (*Cont.*)
    titanium dioxide nanoparticles 131, 165–6,
        168, 259–62, 273, 290, 336–7, 337–8,
        344, 430, 431, 443, 449, 560
    zinc oxide 561
    translocation 131, 173, 251–3, 291, 380, 382,
        503, 509, 526, 527, 558–9
purity, nanomedicines 51–4

quantum dots (QDs) (*see also* zinc oxide
        quantum dots) 124, 129, 172, 500,
        505
    clearance 316
    biodistribution 185
    degradation 508
    environmental impact 235, 237, 240
    exposure routes 327, 470, 524, 555
    risk assessment 462
quartz (*see also* α-quartz particles) 12, 166,
        290, 337, 338, 344, 568
quaternary characterization, nanomaterials 30

rainbow trout 234, 239
Raji cells 343
Raman spectroscopy 31, *549*, 552
reactive oxygen species (ROS) 74, 176, 178,
        187, 294, 296, 358–60, 363, 405, 412,
        413, 414, 416, 430, 471
reactive species (RS) generation 26
red blood cells 194, 305, 569
relative toxicity classes (Hodge and Sterner)
        481
renal excretion, aquatic organisms 237
reptilase 197
reptilase time (RT) test 197
respiratory system (*see* pulmonary system)
respiratory tract (*see also* inhalation exposure)
        381
rethyculo-endothelial system 185
reticuloendothelial system 315, 316, 503
risk analysis 461
risk assessments (*see also* health risk
        assessments) 441–3, 459–89, 500–13,
        520–1
    predictive models 486
    standardization 487
    tiered approaches 501–3
risk characterization 463, 512, 533–4
risk communication 535
risk management 460, 535

Salmonella enterica serovar Enteritidis 84, 85,
        86–90
sampling 4–5, 23, 69–70, 530
scanning electron microscopy (SEM) 10–11, 33,
        145, *201*, 283, *284*, 351, 369, 548–50
scanning mobility particle sizers (SMPSs) 9
scanning near-field optical microscopy (SNOM)
        281
scanning near-field ultrasound holography
        (SNFUH) *549*, 551
scanning probe microscopy (SPM) 551
scintillation *202*
screening assessments 533, 544
SDS-PAGE zymography *202*
Se-methylselenocysteine 105–8, 110
Sears test 19
secondary characterization, nanomaterials 30,
        32, 34, 35
sedimentation 15
selenium 97, 98, 507
    chemopreventive properties 107
    deficiency 98, 99, 100, 108, 109
    supranutritional levels 107, 108, 110
    toxicity 98, 99, 101, 107
selenium nanomaterials (nano-Se) 98–110
    acute toxicity 101, 105–6
    bioavailability 104–5, 107–8
    cytotoxicity 100
    short-term toxicity 101–2, 103, 106
    size effects 108–10
    subchronic toxicity 102
    supranutritional levels 107
selenium oxide nanoparticles 236
selenoenzymes 99, 109
selenomethionine 102–5, 110
serial analysis of gene expression (SAGE) 292,
        293
serum 136–7
serum albumin 115
settling velocities 33
shape, nanomaterials 12–14, 280–5
short fiber amosite (SFA) 72–4
silica coatings 320
silica nanoparticles 125, 126, 132, *340, 342,*
        446
silica nanospheres 338
silica nanotubes 125–6, 144
silica-coated iron oxide magnetic nanoparticles
        407
silicates 509

silicon dioxide (SiO₂) nanoparticles 233, 291, *336*, 338, *340*, 584
silicon (Si) nanoparticles 291
silicosis 125
silver nanocoatings 507
silver (Ag) nanoparticles
    antimicrobial properties 294, 563
    clearance 291
    environmental impact 233, 239, 240, 564
    environmental sources 229
    exposure routes 509, 527, 528
    interactions with alveolar macrophages 291
    natural occurrence 547
    size 568
    toxicity
        alveolar macrophages 291
        central nervous system 178
        liver 320
        pulmonary system 563–4
    toxicity studies 563
    translocation 291
    uses 506, 507
        medical applications 184, 280, *336*
simple risks 534
simultaneous exposure 533
single-walled carbon nanotubes (SWCNTs)
    clearance 316
    environmental impact 232, 234, 237, *238*, 239
    environmental sources 228, 450
    exposure assessments 553
    exposure routes 509
    in vitro experiments 439
    in vivo experiments 440
    physico-chemical properties 445, 446
    risk assessment *464*
    structure 505
    toxicity 124
        blood 210–11
        pulmonary system 337, 338, 445, 446, 469
        relative toxicity 344
        skin 339–40
    toxicity studies 567
size, nanomaterials 5–12, 14, 24, 30, 32–3, 290, 349–50, 441, 446
    high aspect ratio nanoparticles 64
    nanomedicines 51
skin (*see also* dermal exposure)
    in vitro experiments *432*

structure 338, 523
    toxicity 184, 324–7, 339–40
    translocation 525
small interfering RNA (siRNA) 47
soap 506
sodium selenite 98, 99–102
soil contamination 508
solid lipid nanoparticles *46*, 47, 49
solubility 291, 488
sonication 203
specific surface area (SSA) 32, 349, 352
spectroscopy 31
spleen 258, *433*, 471, 476
sports equipment 506–7
stabilizing agents 137
steady-state lung burden 56
sterility, nanoparticles 303–5
Stern layer 286
stratum corneum 253, 338, 339, 470, 509, 523, 524, 555
streptokinase 213
structure–activity relationships (SARs) 69, 591–2
Student's *t* test 373
Stylonychia mytilus 239
subacute toxicity 320, *321–3*
subcutaneous fat 338
submicron particles 470
sunscreens 506, 524, 525, 562
superoxide radical 175
superparamagnetic iron oxide nanoparticles (SPIONs) 185, 315, 407, 409, 415
suppressive subtractive hybridization (SSH) 292, 293
surface area, nanomaterials 19–20, 290
    high aspect ratio nanoparticles 64–7
    nanomedicines 51
surface charge, nanoparticles 20–3
surface chemistry, nanoparticles 31–2, 35, 134, 135, 315, 449, 552
    nanomedicines 44
surface composition, nanoparticles 20
surface reactivity, nanoparticles 23, 434
    nanomedicines 43
surfactants 388
susceptibility factors 512

target sites 448
TAT-coated iron oxide magnetic nanoparticles 411

technetium-radiolabelled carbon nanoparticles 527

tertiary characterization, nanomaterials 30, 33, 35

tetrahydrofuran (THF) 237–8, 439

textiles 506

thioredoxin reductase 97, 104

THP-1 cell line 387

three-dimensional models 388–9

thrombelastography 208

thrombin 196, 197

thrombin time (TT) test 197, 307

thrombocytopenia 215

thrombogenicity 195

thrombosis 306

thromboxan B_2 (TXB$_2$) 207

thyroid deiodinases 97

tissue scaffolds 217–18

titanium dioxide (TiO$_2$) nanoparticles
  clearance 251
  environmental impact 232, 233, 236, *238*, 240
  environmental sources 229, 248, 259
  exposure routes 131, 253, 256, 290, 324, 329, 470, 508–9, 522, 524, 526, 561
  interactions, alveolar macrophages 294
  natural occurrence 547
  physical properties 161, 309, 568
  risk assessment 470–1
  toxicity
    brain 569
    embryos 291
    gastrointestinal tract 343
    kidneys 262
    liver 186, 262
    pulmonary system 131, 165–6, *168*, 259–62, 273, 290, 336–7, 337–8, 344, 430, 431, 443, 449, 560
    skin 339, *342*
  toxicity studies 559–61, 568
  translocation 253
  uses *336*, 339, 506
    food 584

titanium nanoparticles 524

toothpaste 506, 509

ToxCast project, 452

toxicity studies 556–70

toxicogenomic fingerprints 487

toxicokinetic data 448, 451

trachea 381

transcriptomic analysis 208

transition metals 211, 411, 415

translocation, nanomaterials 503–4, 529

transmission electron microscopy (TEM) 10, 11, 33, 145, *201*, 281–3, 351, 548–50, 552

trophic transfer 235

tumor tissue 134

tumors 399

Tyrolean ice mummy 547

ultrafine particles (UFPs) 51, 247–8, 274, 290, 337, 379, 530–1
  toxicity studies 557–64

ultrasmall superparamagnetic iron oxide nanoparticles (USPIONs) 407, 409, 415

uncertain risk knowledge 534

United States Nanotechnology Initiative 171

urine 482

UV irradiation 327

UV light 232

Van der Waal's forces 14, 15, 32

vascular fenestrations 476

viability assays 147

video-enhanced contrast (VEC) microscopy 145

visualization, nanoparticles 145, 550

water contamination 508

water-soluble polymers 119

Western blot test 308

white blood cells (*see* leukocytes)

whole blood aggregometry (WBA) 204–5

wiring technology 507

workplace exposure (*see* occupational exposure)

wound dressings 507

X-ray diffraction (XRD) 33, *549*

X-ray photon spectroscopy (XPS) *549*, 552

X-rays 551–2

zebrafish 233, 237, 240

zeta potentials 20–3, 35, 286, 354

zeta-meters 353–4

zinc nanoparticles 524

zinc oxide (ZnO) 82

zinc oxide (ZnO) fumes 177
zinc oxide (ZnO) nanoparticles
   environmental impact 239–40
   environmental sources 248, 263
   exposure routes 253, 257, 263, 339, 470
   impurities 353
   physico-chemical properties 445

   toxicity 263, 339, 430
   toxicity studies 561–3
   uses *336*, 506
zinc oxide quantum dots (ZnO QDs)
   83–93
   preparation 83–4
zinc (Zn) powder 341